FUNDAMENTALS OF ANATOMY & PHYSIOLOGY

THIRD EDITION

Supplements At-A-Glance

SUPPLEMENT	WHAT IT IS	WHAT'S IN IT
StudyWARE™ CD-ROM StudyWARE	Software program (CD-ROM in the back of the book)	• Quizzes with immediate feedback • Anatomical image labeling • Interactive games • 3-D animations
Study Guide	Print	• Study tips and test-taking strategies • Practice exercises including image labeling, coloring exercises, critical thinking questions, and crossword puzzles • Chapter quizzes • Case Studies
Mobile Downloads DELMAR MOBILE DOWNLOADS	Mobile downloads	• Mobile downloads with key anatomy and physiology terms and art
Anatomy and Physiology Illustrated Flashcards	Print	• Color-coded flashcards correlated to introductory and body system chapters to reinforce body structures
Instructor Resources	CD-ROM	• Electronic Instructor's Manual files • Electronic testbank • Slide presentation created in Power Point® with full-color art and 3-D animations • Conversion grids to help you adjust your lesson plans
WebTutor Advantage	Web access	• On Blackboard and WebCT platforms (other platforms available upon request) • Content and quizzes linked to each chapter • Comprehensive glossary with audio and images • Anatomical image labeling • 3-D animations • StudyWARE™ interactive games • Power Point® slide presentations • Discussion questions • Mid-term and Final exams • Electronic Textbook files
Anatomy and Physiology Image Library, Third Edition	CD-ROM	• Over 1,000 anatomy, physiology and pathology images of different levels of complexity
Medical Terminology Student Theater: An Interactive Video Program	CD-ROM	• Video segments, quizzes and interactive games to reinforce medical terminology

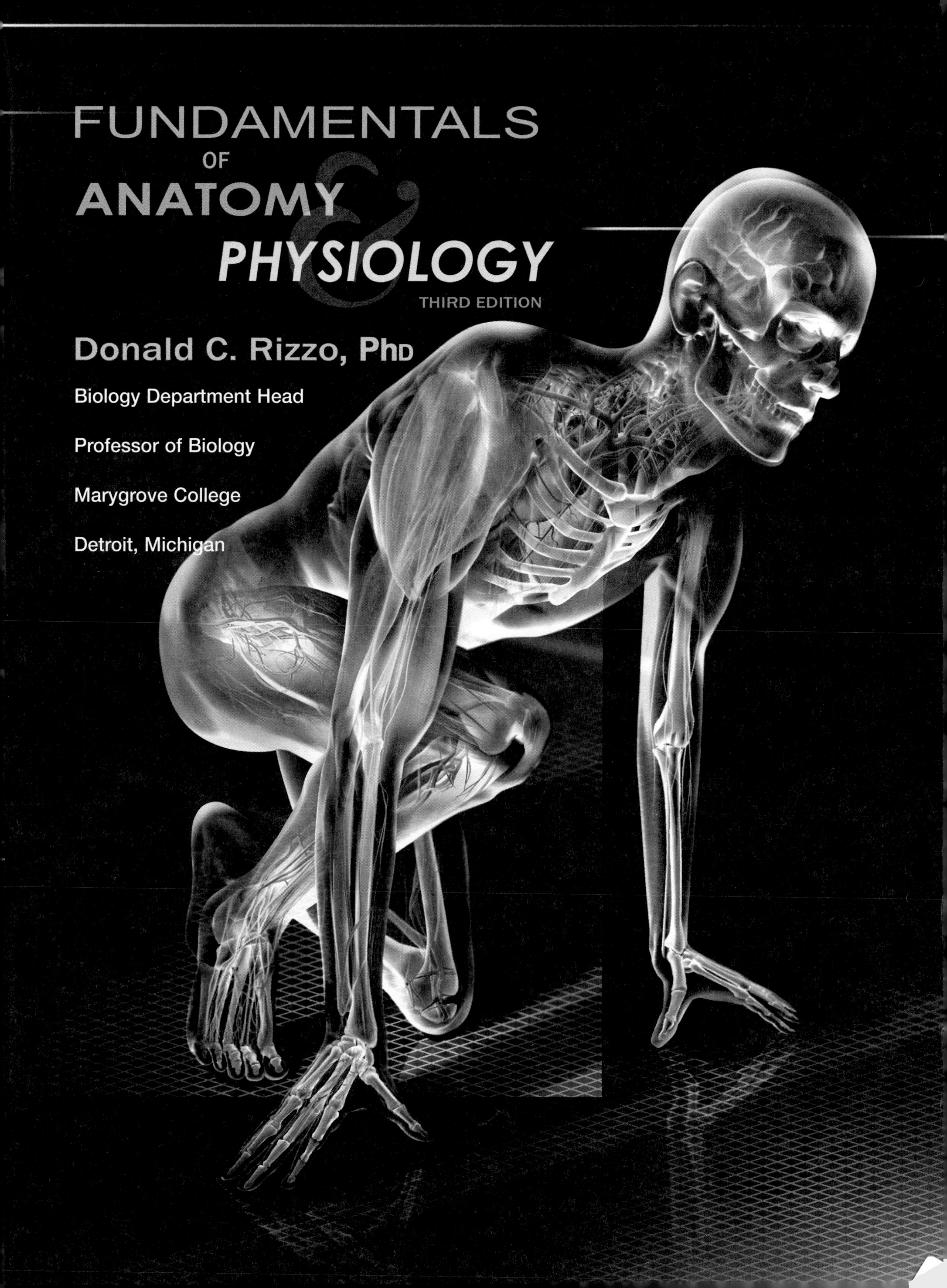

FUNDAMENTALS OF ANATOMY & PHYSIOLOGY

THIRD EDITION

Donald C. Rizzo, PhD

Biology Department Head

Professor of Biology

Marygrove College

Detroit, Michigan

Fundamentals of Anatomy & Physiology, Third Edition

Donald C. Rizzo

Vice President, Career and Professional Editorial: Dave Garza

Director of Learning Solutions: Matthew Kane

Acquisitions Editor: Matthew Seeley

Managing Editor: Marah Bellegarde

Senior Product Manager: Debra Myette-Flis

Editorial Assistant: Samantha Zullo

Vice President, Career and Professional Marketing: Jennifer Baker

Executive Marketing Manager: Wendy Mapstone

Senior Marketing Manager: Kristin McNary

Marketing Coordinator: Scott Chrysler

Production Director: Carolyn S. Miller

Senior Content Project Manager: Kenneth McGrath

Senior Art Director: Jack Pendleton

Technology Project Manager: Benjamin Knapp

Library of Congress Control Number: 2009929876

ISBN-13: 978-1-4390-4458-2
ISBN-10: 1-4390-4458-9

Delmar
5 Maxwell Drive
Clifton Park, NY 12065-2919
USA

Cengage Learning is a leading provider of customized learning solutions with office locations around the globe, including Singapore, the United Kingdom, Australia, Mexico, Brazil, and Japan. Locate your local office at: **international.cengage.com/region**

Cengage Learning products are represented in Canada by Nelson Education, Ltd.

To learn more about Delmar, visit **www.cengage.com/delmar**

Purchase any of our products at your local college store or at our preferred online store **www.ichapters.com**

NOTICE TO THE READER

Printed in the United States of America
1 2 3 4 5 6 7 12 11 10 09

Contents

CHAPTER 3
Cell Structure

CHAPTER 4
Cellular Metabolism and Reproduction: Mitosis and Meiosis

CHAPTER 5
Tissues

CHAPTER 6 The Integumentary System

CHAPTER 7 The Skeletal System

CHAPTER 10

The Nervous System: Introduction, Spinal Cord, and Spinal Nerves

CHAPTER 11

The Nervous System: The Brain, Cranial Nerves, Autonomic Nervous System, and the Special Senses

CHAPTER 12

The Endocrine System

CHAPTER 13
The Blood

CHAPTER 14
The Cardiovascular System

CHAPTER 15
The Lymphatic System

CHAPTER 16
Nutrition and the Digestive System

Chapter 17
The Respiratory System

CHAPTER 18
The Urinary System

CHAPTER 19
The Reproductive System

Preface

TO THE LEARNER

Fundamentals of Anatomy & Physiology, Third Edition, was written and designed for learners pursuing careers in the allied health fields. It is written in clear, concise, and easily understandable scientific language and presupposes no previous biology exposure. This text will guide you along a journey of understanding how the human body operates on a daily basis from birth to death. The writing style and presentation will assist introductory learners with limited backgrounds in the sciences to comprehend the basic concepts of human anatomy and physiology, and the fascinating working mechanisms of our bodies.

Several features are incorporated into each chapter to help you master the content. Review the "How to Use This Book" section on page xxvi for a detailed description and benefit of each feature.

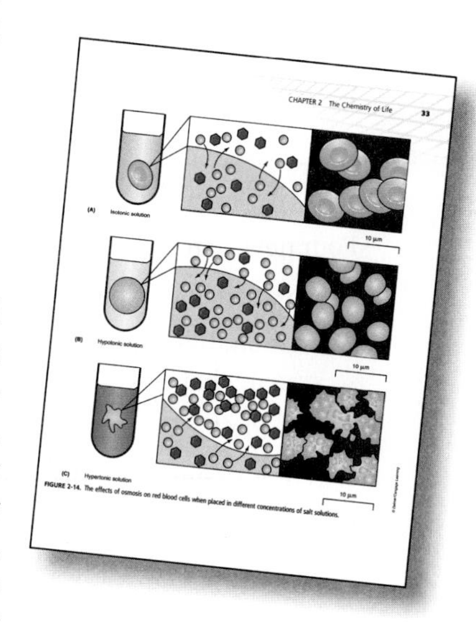

Organization of the Text

Introductory Chapters

The text begins with an introduction to the human body, explaining anatomic terms and the organization of the body from the cellular to the tissue level, how tissues form organs and how organs comprise the various systems of the body. The chemical basis of life is covered in Chapter 2, explaining how elements bond to form molecules like carbohydrates, proteins, fats, and nucleic acids, which are the building blocks of cellular structures. After a discussion of the structure and functions of cells in Chapter 3, Chapter 4 explains how cells convert the foods we eat, via metabolism, into a new form of cellular chemical energy, ATP. This chapter also discusses how cells divide by mitosis, how we pass on our genetic characteristics by meiosis, and the structure of the DNA molecule. After this thorough yet understandable explanation of how cells operate, Chapter 5 describes the anatomy and function of body tissues.

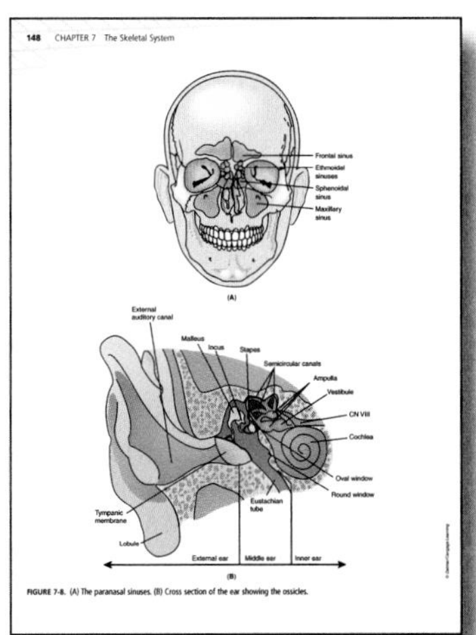

Body System Chapters

Having laid the groundwork for understanding the cellular and tissue levels of organization of the body, the text takes the student on a journey through the various systems of the body. Each system chapter has an introduction to set the stage for explaining in general terms what the system does and the organs it contains. Each organ is discussed in terms of its anatomy and physiology, beginning with the first organ and concluding with the final organ of that system. Beginning on the outside, the first system discussed is the integumentary system in Chapter 6. The skeletal (Chapter 7), articular (Chapter 8), and muscular (Chapter 9) systems are discussed next. These systems operate closely together to allow us to move and respond to changes in our external environment.

The nervous system (Chapters 10 and 11) controls and integrates all other body systems. Chapter 12 discusses the endocrine system, which operates very closely with the nervous system in the chemical control of the body through hormones, helping to maintain the body's internal environment, or homeostasis.

Chapter 13 focuses on the blood and begins the discussion of systems that transport materials through the body. Chapter 14 covers the cardiovascular system, which transports the blood that carries oxygen and nutrients to the body cells, as well as eliminates waste from the body cells. The lymphatic system (Chapter 15) transports fats from the digestive tract to the blood and develops immunities to protect the body from disease. Chapter 16 covers nutrition and the digestive system, which converts the food we eat into a usable form for use by body cells. The respiratory system, which brings in oxygen gas to the body and eliminates carbon dioxide gas, a waste product of cellular metabolism, is discussed in Chapter 17. The urinary system, which filters our blood 60 times a day of the many wastes and excesses that the body does not need, is covered in Chapter 18.

The final chapter of the text is the reproductive system. This system allows us to propagate our species and to pass on our genetic characteristics to our offspring.

Changes to the Third Edition

Several additions and enhancements were made to the textbook and supplements to increase your understanding and broaden your experience of the anatomy and physiology material.

New Features

Four new features were added to enhance your learning and help you put it all together:

- A Search and Explore feature
- Case studies were added to encourage critical thinking of concepts learned in the body system chapters.
- A StudyWARE™ Connection feature
- A Study Guide Practice feature

Mobile Downloads 

Mobile Downloads were added that include audio, illustrations, and animations so you can learn even when you're on the go.

Animations

Several new animations were added to the StudyWARE™, the slides created with PowerPoint, and the Webtutor Advantage.

Chapter-Specific Changes

Following is a list of major chapter-specific changes. Of particular note are the numerous diseases, disorders, and conditions added to the body system chapters.

CHAPTER 1: THE HUMAN BODY

- A new table of the structural levels of organization of the human body was added.
- A discussion of positive and negative feedback was added.

CHAPTER 2: THE CHEMISTRY OF LIFE

- The chemical structure of glycerol and fatty acids was revised.
- Hypotonic solution red blood cells was revised.
- The active transport example was expanded.

CHAPTER 3: CELL STRUCTURE

- Chapter objectives were added.
- The definition of the fluid mosaic pattern of molecules of a cell membrane was expanded.
- New line art showing the structure of a transfer RNA molecule was added.

CHAPTER 4: CELLULAR METABOLISM AND REPRODUCTION: MITOSIS AND MEIOSIS

- New disorders were added including Tay-Sachs disease, Klinefelter's syndrome, and Down syndrome.

CHAPTER 5: TISSUES

- The terms *monocytes, histiocytes*, and *macrophages* were clarified.
- The definition of neuroglia cells was expanded.

CHAPTER 6: THE INTEGUMENTARY SYSTEM

- Images of first-, second-, and third-degree burns were added.
- Images of squamous cell carcinoma, malignant melanoma, and impetigo pustule were added.
- An image of the pustules of chickenpox was added.
- An image of poison oak dermatitis was added.
- Images of paroncyhia, onychomycosis, and onychocryptosis were added.
- New line art showing skin lesions was added.
- New disorders of the integumentary system were added including chickenpox; poison ivy, oak, and sumac dermatitis; paroncyhia, onychomycosis, and onychocryptosis; and skin lesions: papule, macule, wheal, crust, furuncle or boil, bulla or vesicle, pustule, and cyst.

CHAPTER 7: THE SKELETAL SYSTEM

- New line art illustrating the different types of fractures was added.
- New line art of the vertebral column was added.
- New line art showing abnormal curvatures of the spine was added.
- A definition of epidural anesthetics was added.
- New line art of an anterior view of the pelvis was added.
- An image of cleft palate was added.
- New line art of acromegaly was added.
- New disorders of the skeletal system were added including black eye, deviated septum, sinusitis, whiplash, acromegaly, and fractured clavicle.

CHAPTER 8: THE ARTICULAR SYSTEM

- An example of a synchondrosis joint was added.
- New line art showing a bursa of the knee joint was added.
- New disorders of joints were added including gingivitis, hyperextension, and dislocated hip.

CHAPTER 9: THE MUSCULAR SYSTEM

- New line art illustrating the muscles of the abdominal wall was added.
- New disorders of the muscular system were added including rigor mortis, snoring, tetanus, polio, plantar fasciitis, and fibromyalgia.

CHAPTER 10: THE NERVOUS SYSTEM: INTRODUCTION, SPINAL CORD, AND SPINAL NERVES

- The coccygeal spinal nerves are now labeled.

CHAPTER 11: THE NERVOUS SYSTEM: THE BRAIN, CRANIAL NERVES, AUTONOMIC NERVOUS SYSTEM, AND THE SPECIAL SENSES

- New disorders of the nervous system were added including aneurysm, multiple sclerosis, Reye's syndrome, rabies, Tay-Sachs disease, Bell's palsy, concussion, and depression.
- New disorders of the senses were added including motion sickness, cataracts, and glaucoma.

CHAPTER 12: THE ENDOCRINE SYSTEM

- New disorders of the endocrine system were added including diabetes insipidus, seasonal affective disorder, aldosteronism, stress, and adrenogenital syndrome.

CHAPTER 13: THE BLOOD

- Labels were added to the classification of the blood cells figure.
- The life span and functions of blood cells were updated.
- New disorders of blood were added including thrombocytopenia, erythrocytosis, and carbon monoxide poisoning.

CHAPTER 14: THE CARDIOVASCULAR SYSTEM

- Layers of the wall of the heart are now labeled.
- Interior of the heart line art includes additional labels.
- New disorders of the cardiovascular system were added including arrhythmia, heart murmur, stenosed heart valve, incompetent heart valve, angina pectoris, mitral valve prolapse (MVP) syndrome, myocardial infarction, angioplasty, and stents.

CHAPTER 15: THE LYMPHATIC SYSTEM

- Additional labels were added to the lymph node line art.
- New line art was added to illustrate the structure of an antibody.
- New disorders of the lymphatic system were added including bone marrow transplant, cancer and lymph nodes, and systemic lupus erythematosus (SLE).

CHAPTER 16: NUTRITION AND THE DIGESTIVE SYSTEM

- A description of the umami taste was added.
- A radiograph of gallstones in the bile duct was added.
- An image of thrush was added.
- New disorders of the digestive system were added including thrush, tonsillitis, food poisoning, tapeworm infections, pancreatitis, gastritis, gastroesophageal reflux disease (GERD), gastric cancer, and pancreatic cancer.

CHAPTER 17: THE RESPIRATORY SYSTEM

- Line art of anterior and posterior views of the larynx was added.
- New line art showing the position of the vocal cords in the larynx was added.
- A new section on lung capacity was added.

- New disorders of the respiratory system were added including laryngitis, pleurisy, atelectasis, Legionnaires' disease, sudden infant death syndrome (SIDS), influenza (flu), and tuberculosis (TB).

CHAPTER 18: THE URINARY SYSTEM

- New labels were added to the internal anatomy of a kidney line art.
- A discussion of the two types of nephrons—juxtamedullary and cortical—was added.
- Line art was added showing the main functions of the nephrons: filtration, reabsorption, and secretion.
- A discussion of peritoneal dialysis under renal failure was added.
- New disorders of the urinary system were added including hematuria, oliguria, polyuria, pyuria, uremia, polycystic kidney disease, and urinary incontinence.

CHAPTER 19: THE REPRODUCTIVE SYSTEM

- An image of *Trichomonas* was added.
- An image of a penile chancre of syphilis was added.
- New disorders of the reproductive system were added including erectile dysfunction (ED) or impotence, ovarian cancer, menstrual cramps, ectopic pregnancy, and female infertility.

LEARNING SUPPLEMENTS

Fundamentals of Anatomy & Physiology, Third Edition StudyWARE™

The StudyWARE™ CD-ROM offers an exciting way to gain additional practice in learning anatomy and physiology. The quizzes and activities help reinforce even the most difficult concepts. See "How to Use StudyWARE™ to Accompany *Fundamentals of Anatomy & Physiology*, Third Edition" for details.

Study Guide

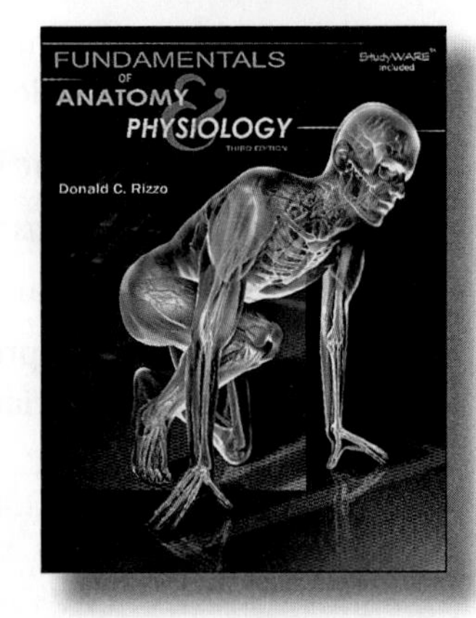

The study guide offers additional practice with exercises corresponding to each chapter in the text, including completion, matching, key terms, art labeling, coloring exercises, critical thinking questions, case studies, crossword puzzles, and chapter quizzes. A section on study tips and test-taking strategies is also included.

ISBN 1-4354-3873-6

Anatomy & Physiology Illustrated Flashcards

Review and learn anatomy and physiology key concepts and terminology with just under 200 full-color flashcards. Anatomy & Physiology Illustrated Flashcards provide mastery of terms and body structures through a series of image labeling and key concept cards. Start by reviewing the anatomy image to study the body structures. Turn the card over to review key concepts or terms related to the body system or individual structures. Next, test yourself with the image labeling cards. Color-coded to keep *like* cards together after separation, the flashcards are organized by introductory and body systems and correlate to chapters in the text where you can access additional information.

ISBN 1-4283-7657-7

TO THE INSTRUCTOR

Rationale and Intended Market

There are many human anatomy and physiology textbooks that instructors can choose for their learners. Most are designed for those with a background in biology and are so extensive in content and coverage that it would take at least a full year to teach all the in-depth subject matter. These texts are designed for biology majors and pre-med learners. There was a need for a textbook that was written for the introductory learner choosing a career in allied health, a book that covers the fundamentals of human anatomy and physiology at a reasonable depth to satisfy the needs of these learners in a one-semester course.

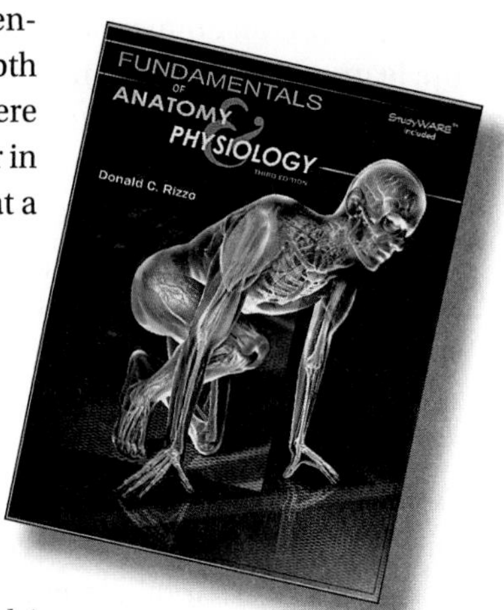

Teaching Support Materials

A number of resource materials are available to accompany this text.

Instructor Resources

The Instructor Resources is a robust computerized tool for your instructional needs! A must-have for all instructors, this comprehensive and convenient CD-ROM contains the following components.

- The Instructor's Manual was created to meet the challenge of teaching. It contains valuable resources for the experienced and novice instructor, including:
  - syllabus for a one-semester course
  - lecture outlines with classroom demonstrations/activities incorporated
  - critical thinking classroom discussion questions
  - answers to review questions in the text
  - answers to exercises and chapter quizzes in the study guide
- ExamView® Computerized Testbank contains over 1,000 questions organized by chapter content, including matching, fill in the blank, multiple choice, and true/false, to assist you in creating chapter, midterm, and final exams.
- Slides created in PowerPoint®, including animations, are designed to aid you in planning your class presentations.

Instructor Resources, ISBN 1-4354-3872-8

Delmar Learning's Anatomy & Physiology Image Library CD-ROM, Third Edition

This CD-ROM includes over 1,050 graphic files. These files can be incorporated into a Power Point®, Microsoft® Word presentation, used directly from the CD-ROM in a classroom presentation, or used to make color transparencies. The Image Library is organized around body systems and medical specialties. The library includes various anatomy, physiology, and pathology graphics of different levels of complexity. Instructors can search and select the graphics that best apply to their teaching situation. This is an ideal resource to enhance your teaching presentation of medical terminology or anatomy and physiology.

ISBN: 1-4180-3928-4

Fundamentals of Anatomy & Physiology Online Course

This fully developed online course introduces learners with little or no prior biology knowledge to the complex and exciting world of anatomy and physiology. The course is a complete interactive online learning solution. Chapter content is organized around body systems and focuses on how each system works together to promote homeostasis. Full-color art, 3-D anatomical animations, audio, and "bite-size" chunks of content fully engage the learner. Interactive games such as image labeling, concentration, and championship reinforce learning. Powerful customization tools allow administrators to individualize the course and assessment tools, while extensive tracking features allow administrators to monitor learner performance and progress.

Anatomy & Physiology Online—Academic Individual Access Code, **ISBN 1-4180-0131-7**
Anatomy & Physiology Online—Academic Institutional Access Code, **ISBN 1-4180-0130-9**

WebTUTOR™ Advantage

Designed to complement the book, WebTUTOR™ is a content rich, Web-based teaching and learning aid that reinforces and clarifies complex concepts. Animations enhance learning and retention of material. The WebCT™ and Blackboard™ platforms also provide rich communication tools to instructors and students, including a course calendar, chat, e-mail, and threaded discussions.

WebTUTOR™ Advantage on WebCT™, **ISBN 1-4354-3875-2**
Text Bundled with WebTUTOR™ Advantage on WebCT™, **ISBN 1-4354-3074-3**
WebTUTOR™ Advantage on Blackboard™, **ISBN 1-4354-3874-4**
Text Bundled with WebTUTOR™ Advantage on Blackboard™, **ISBN 1-4354-3109-X**

ABOUT THE AUTHOR

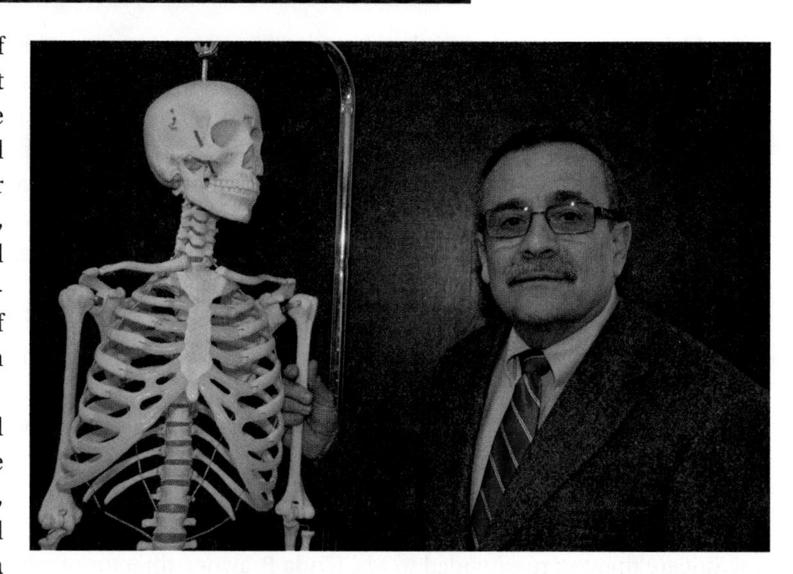

Donald C. Rizzo, Ph.D., is a full professor of biology and head of the biology department at Marygrove College in Detroit, Michigan, where he teaches human anatomy and physiology and medical terminology. He is also responsible for teaching biology II: the unity and diversity of life, principles of biology, parasitology, zoology, and botany. He began his teaching career at Marygrove College in 1974. He was chairperson of the Science and Mathematics Department from 1975–2006 in addition to full-time teaching.

Dr. Rizzo received his B.A. in biology and education in 1968 from Boston State College (now the University of Massachusetts at Boston), M.S. in 1970, and Ph.D. in 1973 from Cornell University in Ithaca, New York. He has been a long term member of the American Association of University Professors and is a member of the American Institute of Biological Sciences and was a past member of the National Association of Science Teachers.

Dr. Rizzo has published in the *Journal of Invertebrate Pathology* and coauthored a computerized test bank for medical terminology. He has developed many teaching aids for his biology classes, including a laboratory manual for parasitology and student study guides for all other classes.

Dr. Rizzo's awards include the Sears Roebuck Foundation Teaching Excellence and Campus Leadership Award in 1990 and the Marygrove College Teacher Scholar Award in 1992. Nominated by his students, he became a member of Who's Who Among American Teachers in 1996, 2000, and 2004. In 2006, he received the Marygrove College Presidential Award for Teaching. In 1990–1996 he was a summer session visiting professor at the University of Michigan Medical School, where he taught the biology component of the post-baccalaureate Pre-Medical Scholarship Program for minority students. He presents at national and international conferences on an inter disciplinary service learning course on "HIV/AIDS: Its Biological and Social Impact" with his two friends and colleagues Professor James Karagon in social work and Dr. Loretta Woodard in literature who, with him, developed and teach this course, every year.

He has conducted biological field work around the globe and participates with students on Study Abroad trips to such places as the Galapagos Islands, South Africa, China, Russia, Europe, and South and Central America. His hobbies include world travel, American art pottery, and American glass. In 2009 he was awarded the Marygrove College Presidential Award for Scholarship.

ACKNOWLEDGMENTS

I would like to acknowledge the dedicated assistance of my friend and colleague, Ms. Teri Miller, our administrative assistant. A very special thank you is also extended to Deb Myette-Flis, my senior product manager, who began working with me on the first edition many years ago and has remained with me as a patient and competent link to Delmar Cengage Learning. She is always available with assistance and answers to my numerous technical questions, providing me with consistent words of encouragement.

I would also like to extend a grateful thank you to my family, friends, administrators, staffs, students, and colleagues at Marygrove College who supported me in this endeavor. Thanks are also extended to my friends Rico and Jess for taking care of the English bulldogs, Bonnie and Clyde, and Cleo the cat when I needed periods of quiet time and solitude to write and edit.

Many thanks are also extended to the instructors from other colleges who reviewed this manuscript. Their constructive suggestions brought new perspectives to topics, and their ideas and comments helped make this third edition the product it is. They each had a new and different perspective that was invaluable to the final editing of this edition.

Sincere thanks are extended to Ms. Linda Brawner, director of educational technology services at Marygrove College, for her assistance with the technical editing of this manuscript.

Reviewers

Anna Marie Avola, PhD
Program Chair
Medical Assisting Program
Hodges University
Naples, Florida

Richard A. Bennett, PhD
Post-Doctoral Fellow
Air Force Research Laboratory—University of Cincinnati
Cincinnati, Ohio

Cherika deJesus, CMA, AAS
Certified Medical Assisting Instructor
Minnesota School of Business
Brooklyn Center Campus
Brooklyn Center, Minnesota

Linda J. Kuhlenbeck, CMA
Medical Assistant Instructor
Brown Mackie College
Hopkinsville, Kentucky

How to Use This Book

Fundamentals of Anatomy & Physiology, Third Edition, helps you understand how the human body is structured, the functions it performs on a daily basis, and how the body systems work together to maintain homeostasis. The following features are integrated throughout the text to assist you in learning and mastering anatomy and physiology core concepts and terms.

The Third edition retained all the successful features of the second edition and added four new features including:

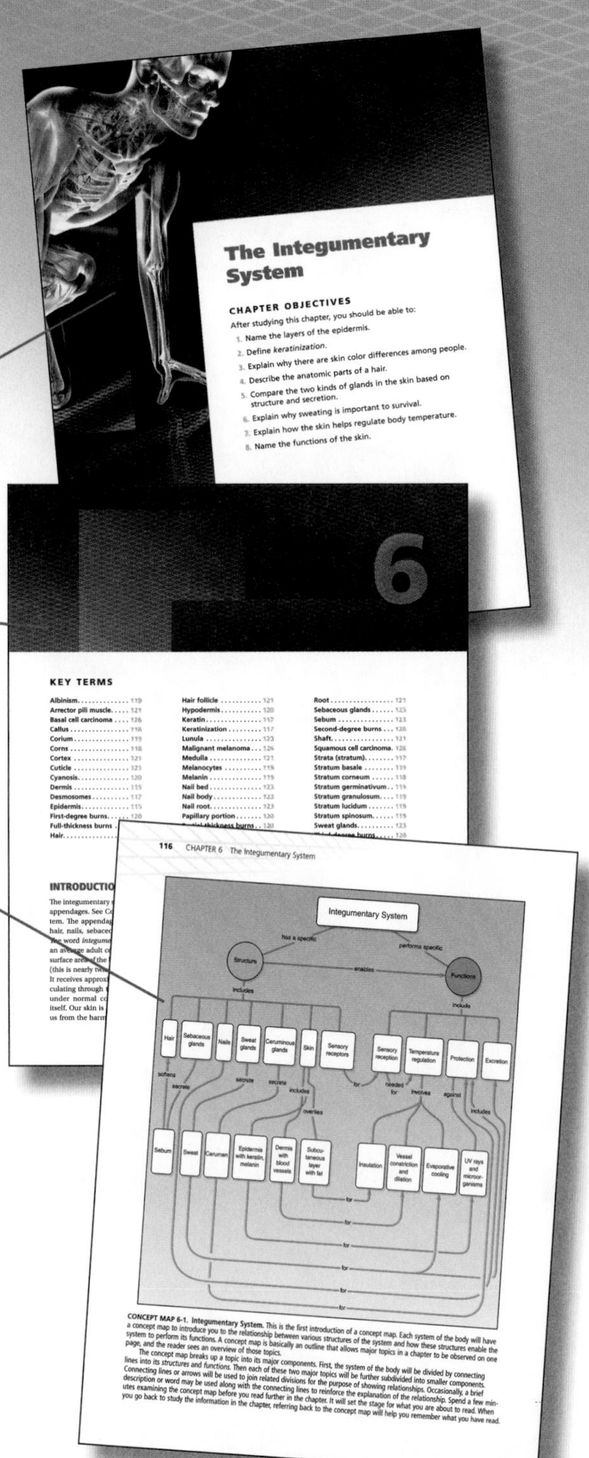

Chapter Objectives

The chapter objectives alert you to concepts you should understand after reading the chapter and completing the review questions.

Key Terms

The list of key terms at the beginning of each chapter references the page number where each term can be found within the text, making locating specific terms to review quick and easy.

Concept Map

Each body system chapter includes a concept map that introduces you to the relationship between various structures of the system and how these structures enable the system to perform its functions. The concept maps help you see the connections between anatomy and physiology of the organs of each body system.

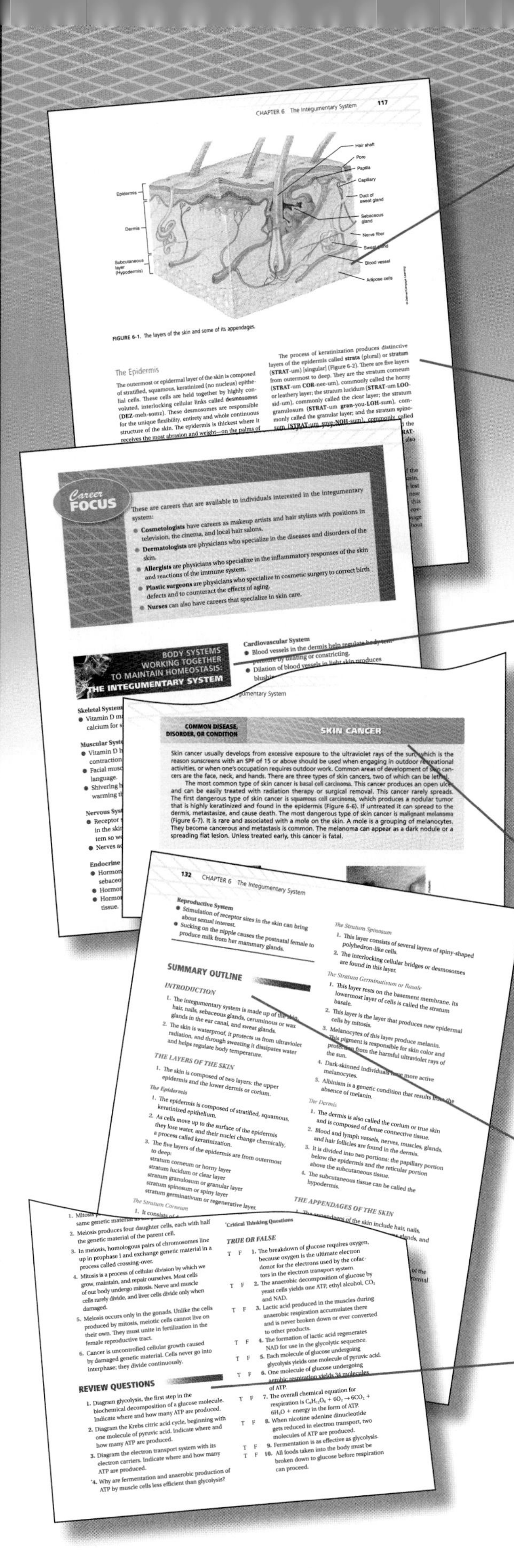

Full-Color Illustrations & Photos

Full-color illustrations and photos provide visual reinforcement of the major concepts covered in each chapter. Color can help you keep a mental picture of the various systems and help reinforce the material you learn in each chapter.

Key Terms

Key terms are highlighted in color, followed by the correct phonetic pronunciation in parentheses and then defined. Proper pronunciation promotes correct spelling of the terms.

Body Systems

This section at the end of each chapter illustrates how each body system works together to maintain the body's internal environment within certain narrow ranges. Seeing each body system's role in maintaining homeostasis helps you see the integration of separate systems into one body.

Special Condition Boxes

Health Alert or Common Disease, Disorder, or Condition boxes provide short descriptions of a significant health alert, common disease, disorder, or condition that can occur in the body system being covered. This information enables you to relate concepts presented in the chapter to real-life situations.

Summary Outline

The summary outline, listing the major topics covered in the chapter, provides a valuable study tool by summarizing the chapter contents.

Review Questions

A variety of exercises provide self-assessment of comprehension of the chapter material. Critical thinking questions allow you to apply concepts learned, and encourage further discussion.

Laboratory Exercises

Essential laboratory exercises at the end of most chapters allow you hands-on experience in the laboratory to observe structures or apply the knowledge learned in the chapter.

As the Body Ages

As the Body Ages feature in all body system chapters discusses physiological changes and effects that aging has on each specific body system.

Career Focus

Career Focus feature introduces learners to health professions related to the chapter content.

Four new features were added to enhance your learning, broaden your experience of the anatomy and physiology material, and help you put it all together.

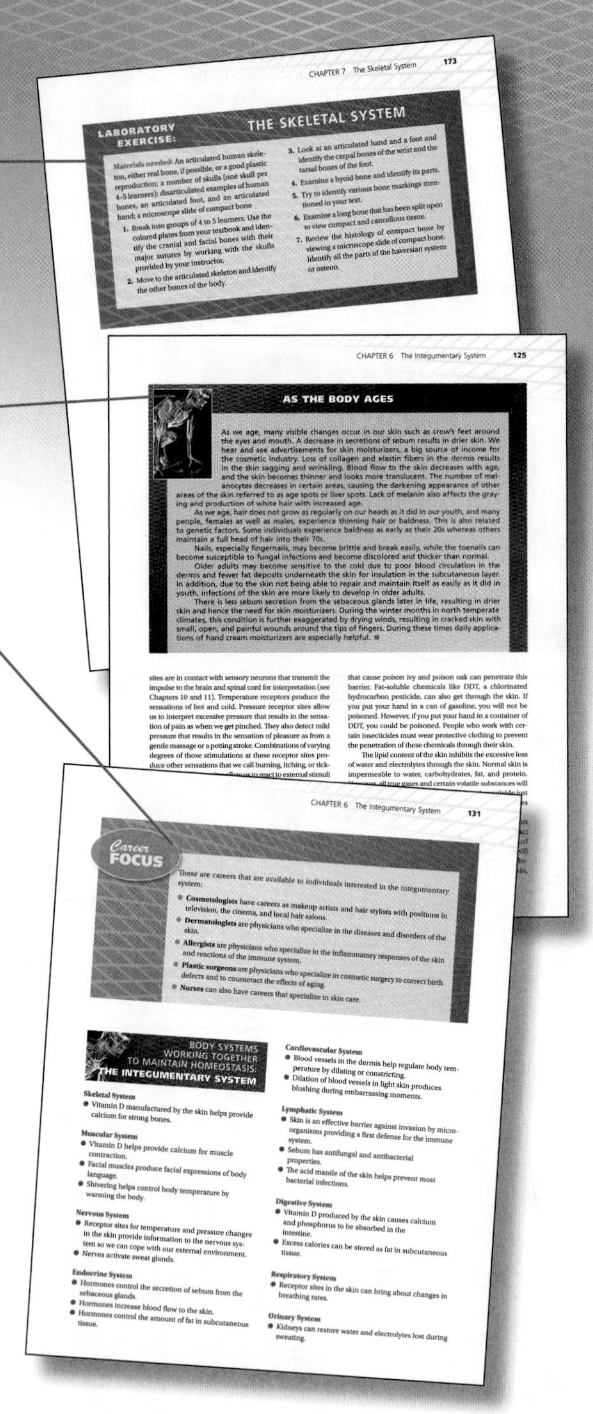

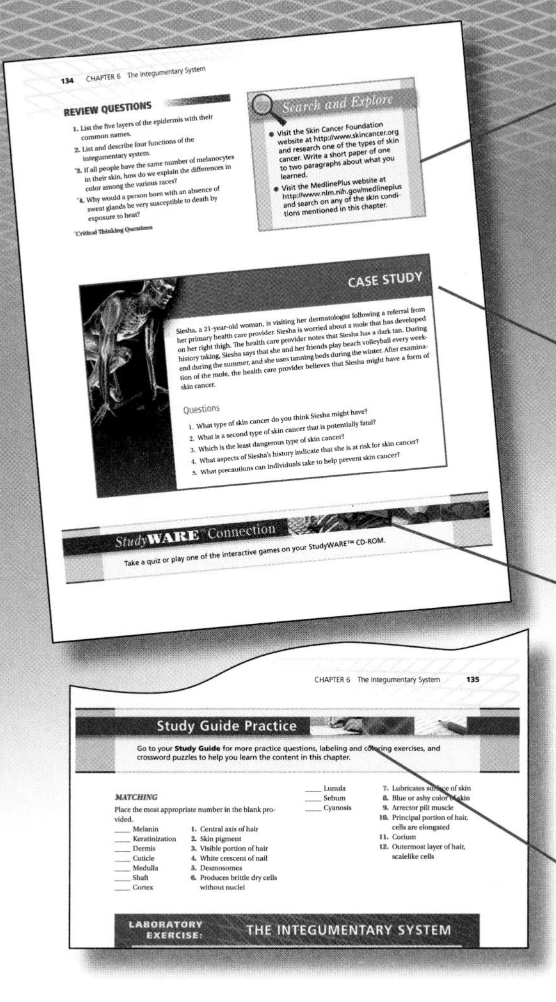

134 CHAPTER 6 The Integumentary System

REVIEW QUESTIONS

1. List the five layers of the epidermis with their common names.
2. List and describe four functions of the integumentary system.
*3. If all people have the same number of melanocytes in their skin, how do we explain the differences in color among the various races?
*4. Why would a person born with an absence of sweat glands be very susceptible to death by exposure to heat?

*Critical Thinking Questions

Search and Explore

- Visit the Skin Cancer Foundation website at http://www.skincancer.org and research one of the types of skin cancer. Write a short paper of one to two paragraphs about what you learned.
- Visit the MedlinePlus website at http://www.nlm.nih.gov/medlineplus and search on any of the skin conditions mentioned in this chapter.

CASE STUDY

Siesha, a 21-year-old woman, is visiting her dermatologist following a referral from her primary health care provider. Siesha is worried about a mole that has developed on her right thigh. The health care provider notes that Siesha has a dark tan. During history taking, Siesha says that she and her friends play beach volleyball every weekend during the summer, and she uses tanning beds during the winter. After examination of the mole, the health care provider believes that Siesha might have a form of skin cancer.

Questions

1. What type of skin cancer do you think Siesha might have?
2. What is a second type of skin cancer that is potentially fatal?
3. Which is the least dangerous type of skin cancer?
4. What aspects of Siesha's history indicate that she is at risk for skin cancer?
5. What precautions can individuals take to help prevent skin cancer?

StudyWARE Connection

Take a quiz or play one of the interactive games on your StudyWARE™ CD-ROM.

CHAPTER 6 The Integumentary System 135

Study Guide Practice

Go to your **Study Guide** for more practice questions, labeling and coloring exercises, and crossword puzzles to help you learn the content in this chapter.

MATCHING

Place the most appropriate number in the blank provided.

____ Melanin
____ Keratinization
____ Dermis
____ Cuticle
____ Medulla
____ Shaft
____ Cortex
____ Lunula
____ Sebum
____ Cyanosis

1. Central axis of hair
2. Skin pigment
3. Visible portion of hair
4. White crescent of nail
5. Desmosomes
6. Produces brittle dry cells without nuclei
7. Lubricates surface of skin
8. Blue or ashy color of skin
9. Arrector pili muscle
10. Principal portion of hair, cells are elongated
11. Corium
12. Outermost layer of hair, scalelike cells

LABORATORY EXERCISE: THE INTEGUMENTARY SYSTEM

Search and Explore

NEW! The **Search and Explore** feature takes you beyond the textbook to expand your learning experience with key word Internet searches, suggested websites to visit with related activities, and brief *human interest* projects designed to add a personal element to your assignments.

Case Study

NEW! A **Case Study** is now included in each body system chapter to encourage you to synthesize material you've learned through critical thinking.

StudyWARE™ Connection

NEW! The **StudyWARE™ Connection** feature directs you to additional learning opportunities such as practice quizzes, animations, image labeling, and other interactive games included on the CD-ROM in the back of the book.

Study Guide Practice

NEW! The **Study Guide Practice** feature reminds you about even more learning tools including practice questions, labeling and coloring exercises, and crossword puzzles in the Study Guide that were created specifically to help you learn anatomy and physiology.

How to Use StudyWARE™ to Accompany Fundamentals of Anatomy & Physiology, Third Edition

Minimum System Requirements

- Operating systems: Microsoft Windows 2000 w/SP 4, Windows XP w/SP 2, Windows Vista w/SP 1
- Processor: Minimum required by Operating System
- Memory: Minimum required by Operating System
- Hard Drive Space:
- Screen resolution: 800 × 600 pixels
- CD-ROM drive
- Sound card and listening device required for audio features
- Flash Player 9. The Adobe Flash Player is free, and can be downloaded from http://www.adobe.com/products/flashplayer/

Setup Instructions

1. Insert disk into CD-ROM drive. The StudyWARE™ installation program should start automatically. If it does not, go to step 2.
2. From My Computer, double-click the icon for the CD drive.
3. Double-click the *setup.exe* file to start the program.

Technical Support

Telephone: 1-800-648-7450

Monday–Friday

8:30 A.M.–6:30 P.M. EST

E-mail: delmar.help@cengage.com

Getting Started

The StudyWARE™ software helps you learn terms and concepts in *Fundamentals of Anatomy & Physiology,* Third Edition. As you study each chapter in the text, be sure to explore the activities in the corresponding chapter in the software. Use StudyWARE™ as your own private tutor to help you learn the material in your *Fundamentals of Anatomy & Physiology,* Third Edition, textbook.

Getting started is easy. Install the software by inserting the CD-ROM into your computer's CD-ROM drive and following the on-screen instructions. When you open the software, enter your first and last name so the software can store your quiz results. Then choose a chapter from the menu to take a quiz or explore one of the activities.

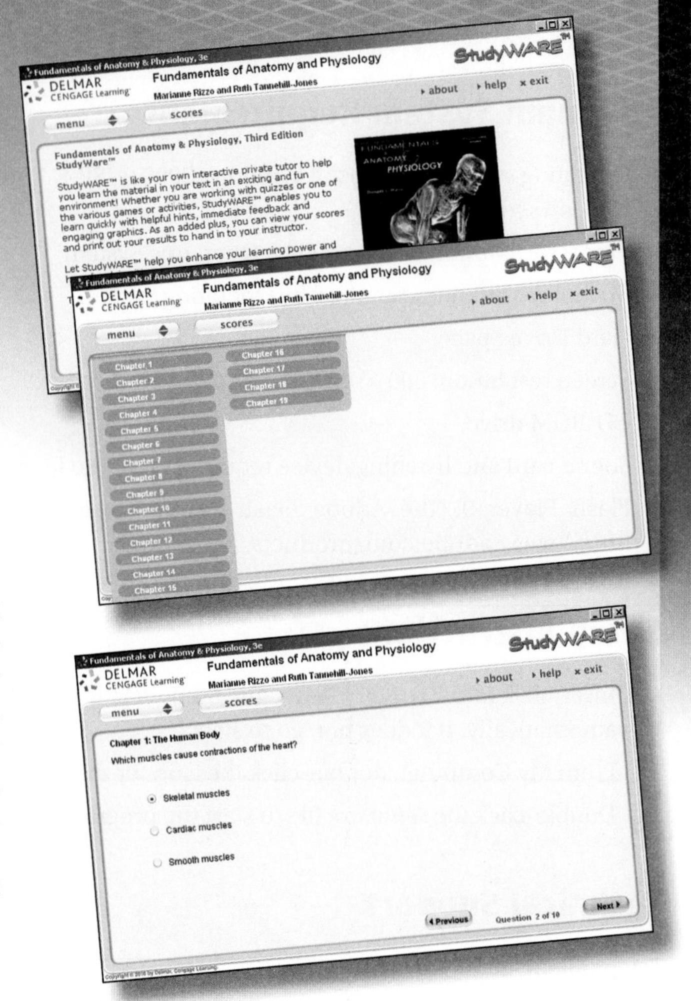

Menus

You can access the menus from wherever you are in the program. The menus include Quizzes and other Activities.

Quizzes

Quizzes include multiple-choice and fill-in-the-blank questions. You can take the quizzes in both practice mode and quiz mode. Use practice mode to improve your mastery of the material. You have multiple tries to get the answers correct. Instant feedback tells you whether you're right or wrong and helps you learn quickly by explaining why an answer was correct or incorrect. Use quiz mode when you are ready to test yourself and keep a record of your scores. In quiz mode, you have one try to get the answers right, but you can take each quiz as many times as you want.

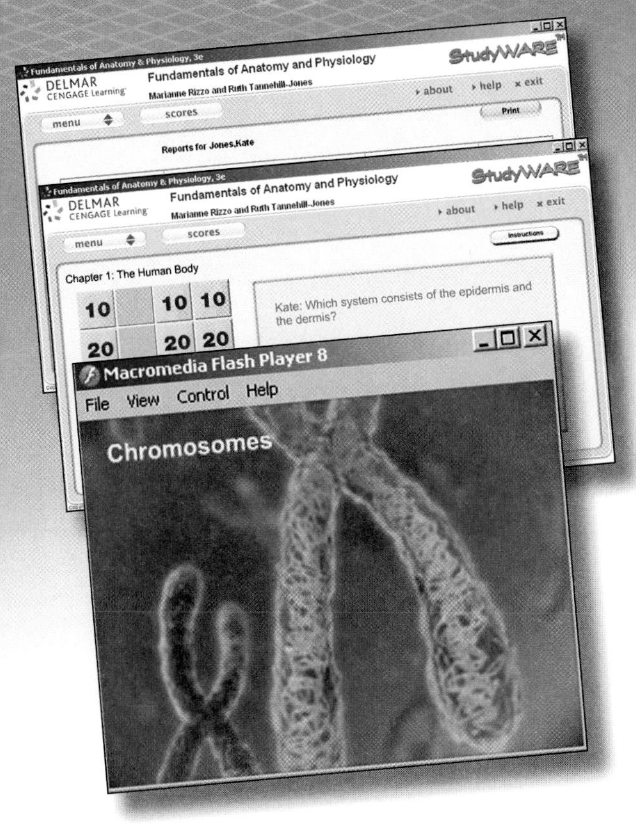

Scores

You can view your last scores for each quiz and print your results to hand in to your instructor.

Activities

Activities include image labeling, hangman, concentration, and championship. Have fun while increasing your knowledge!

Animations

Animations expand your learning by helping you visualize concepts related to anatomy and physiology.

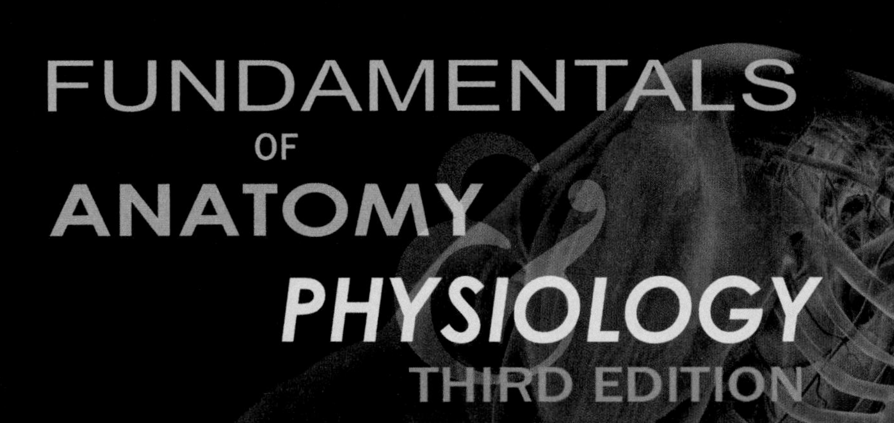
FUNDAMENTALS
OF
ANATOMY &
PHYSIOLOGY
THIRD EDITION

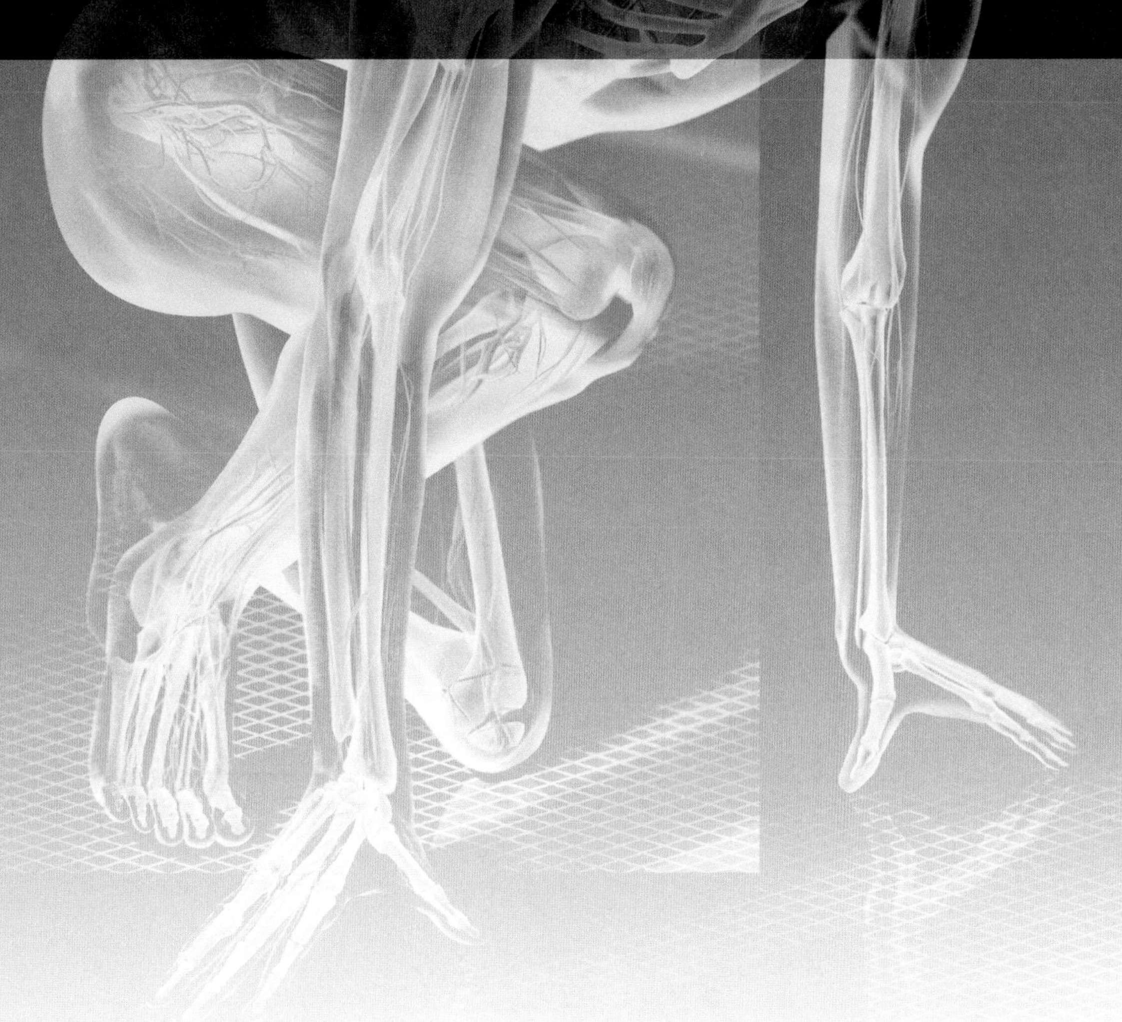

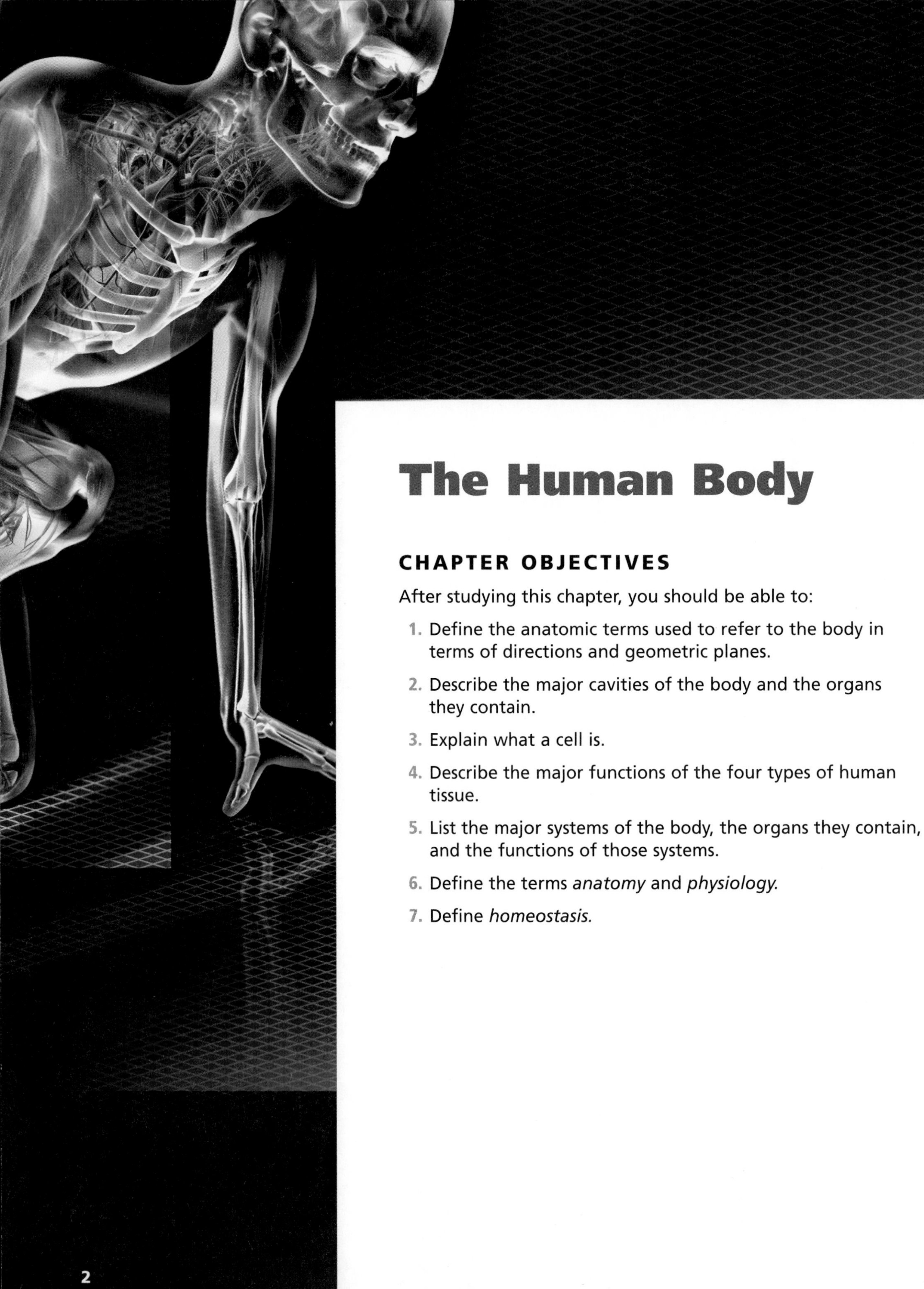

The Human Body

CHAPTER OBJECTIVES

After studying this chapter, you should be able to:

1. Define the anatomic terms used to refer to the body in terms of directions and geometric planes.
2. Describe the major cavities of the body and the organs they contain.
3. Explain what a cell is.
4. Describe the major functions of the four types of human tissue.
5. List the major systems of the body, the organs they contain, and the functions of those systems.
6. Define the terms *anatomy* and *physiology.*
7. Define *homeostasis.*

KEY TERMS

INTRODUCTION

Interest in the human body and how it functions probably developed when our ancestors began to think about the reasons why people became ill and died. All earlier cultures had someone designated as a healer who was responsible for finding plants and herbs that cured body disorders. This healer also was responsible for praying or invoking the assistance of past ancestors to help in the healing process.

As cultures developed and science began to evolve, interest in and knowledge about the human body advanced. Leonardo da Vinci, an Italian (1452–1519), was the first to correctly illustrate the human skeleton with all of its bones. The Flemish anatomist Andreas Vesalius (1514–1564) wrote a book on the human body, and the English anatomist William Harvey (1578–1657) discovered how blood circulates through the body. These are just a few of the many contributors who added to our understanding of the human body and how it functions.

Anatomy is the study of the structure or morphology of the body and how the body parts are organized. **Physiology** is the study of the functions of body parts, what they do and how they do it. These two areas of the organization of the body are so closely associated that it is difficult to separate them. For example, our mouth has teeth to break down food mechanically, a tongue that tastes the food and manipulates it, and salivary glands that produce saliva containing enzymes that break down complex carbohydrates into simple sugars, thus beginning the process of digestion. **Pathology** is the study of the diseases of the body.

We still do not know everything about how the human body functions. Research is still going on today to discover the mysteries of this complex unit we call ourselves.

To facilitate uniformity of terms, scientists have adopted four basic reference systems of bodily organization. These systems are directions, planes, cavities, and structural units. When referring to terms of direction, planes, and cavities, the human body is erect and facing forward. The arms are at the sides and the palms of the hand and feet are positioned toward the front (Figure 1-1). All descriptions of location or position assume the body to be in this posture.

TERMS OF DIRECTION

When an anatomist (one who studies the human body's structures) is describing parts of the body, it is necessary to make reference to their positions in regard to the body as a whole. The following directional terms have been established to facilitate these references. Use Figure 1-2 as your guide as these terms are defined.

Superior means uppermost or above. Example: the head is superior to the neck; the thoracic cavity is superior to the abdominal cavity. **Inferior** means lowermost or below. Example: the foot is inferior to the ankle; the ankle is inferior to the knee. **Anterior** means toward the front. Example: the mammary glands are on the anterior

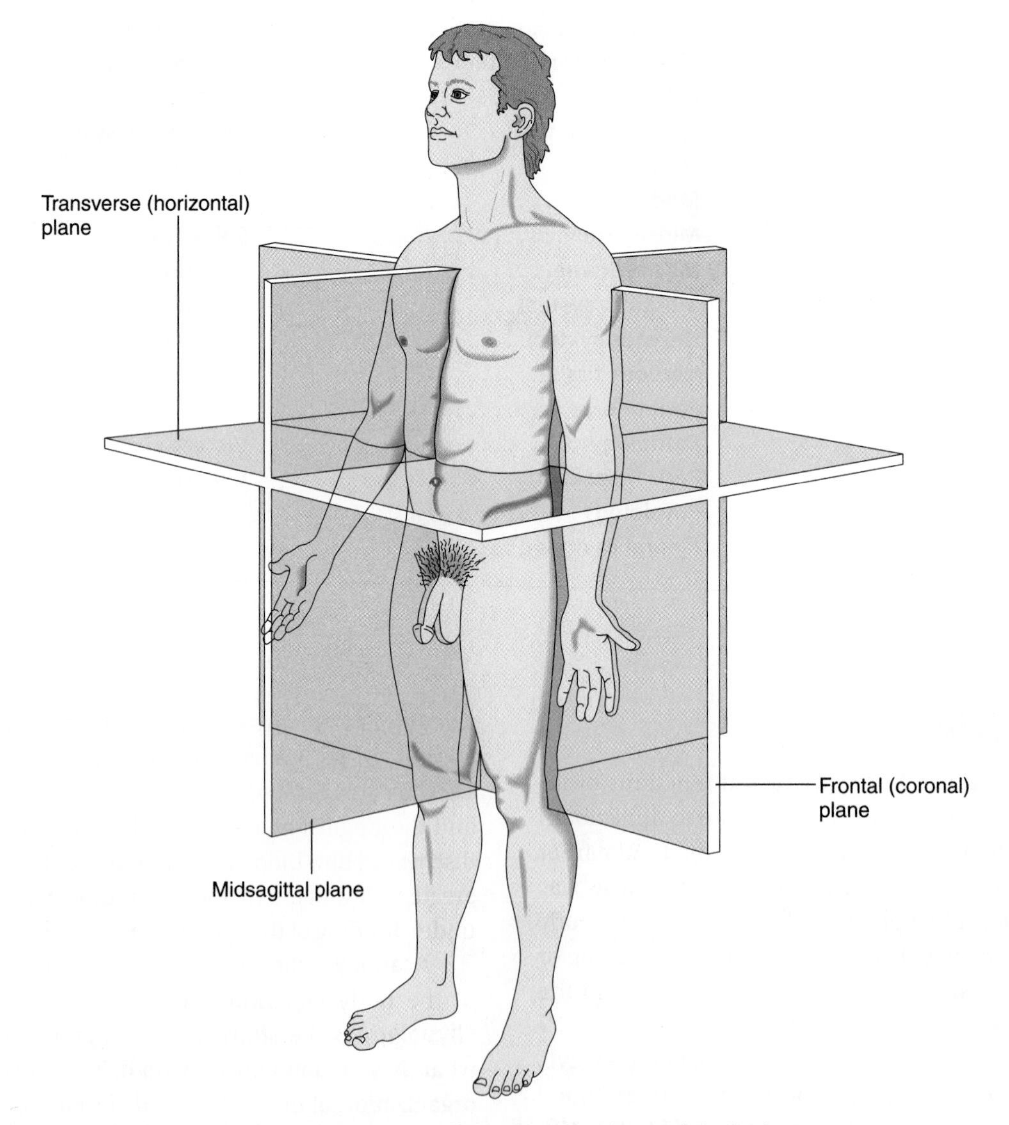

FIGURE 1-1. The human body in correct anatomic position illustrating the planes of the body.

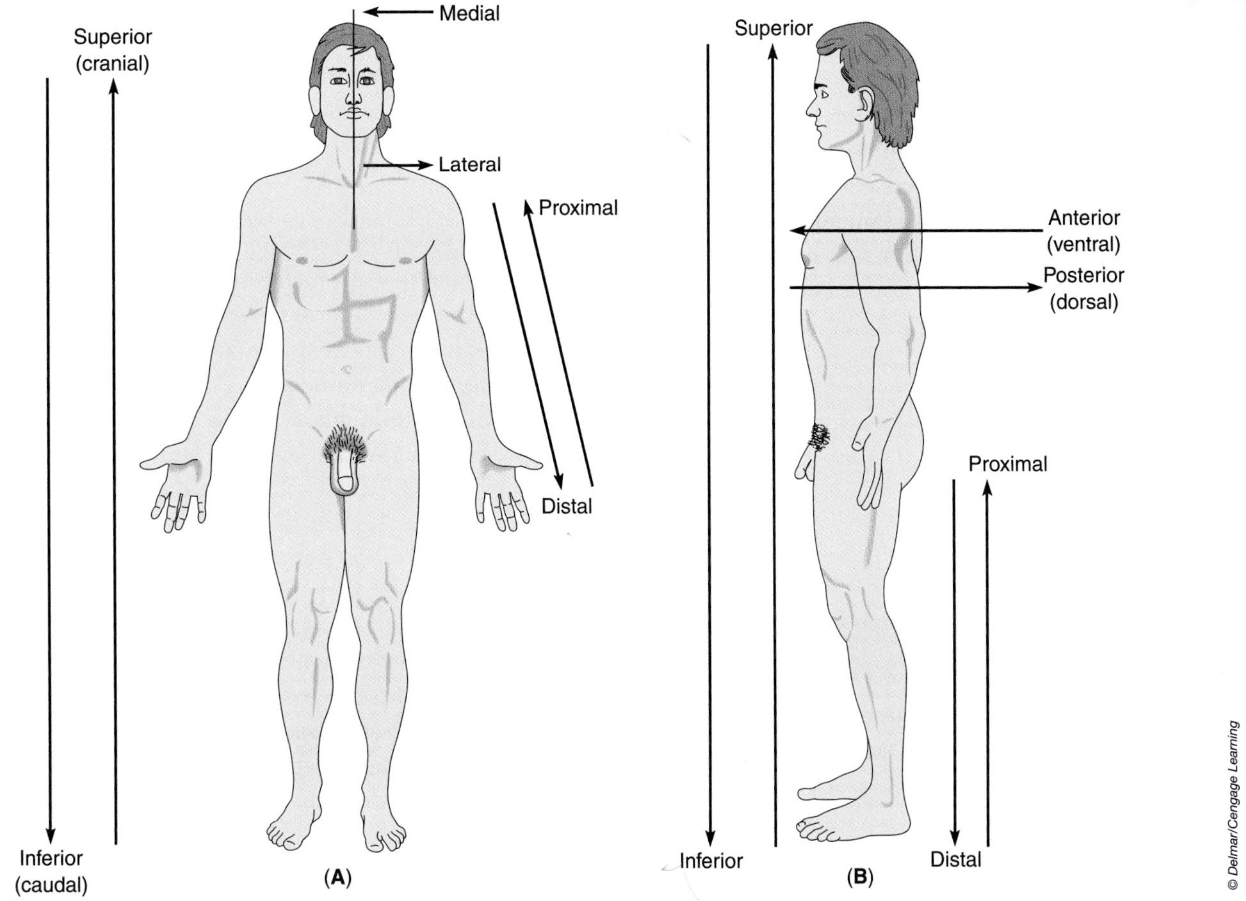

FIGURE 1-2. Directional terms relating to anatomic position. (A) Anterior view of the body. (B) Lateral view of the body.

chest wall. The term **ventral** can also be used for anterior. Ventral means the belly side. **Posterior** means toward the back. Example: the vertebral column is posterior to the digestive tract; the esophagus is posterior to the trachea. The term **dorsal** can also be used for posterior. Dorsal means the back side.

Cephalad (**SEF**-ah-lad) or **cranial** means toward the head. It is synonymous with superior. Example: the thoracic cavity lies cephalad (or superior) to the abdominopelvic cavity. Occasionally, **caudal** (**KAWD**-al) is synonymous with inferior. However, caudal specifically means toward the tail and, as we know, humans do not have tails as adults but we do have tails as developing embryos as do all members of the animal phylum Chordata to which humans belong.

Medial means nearest the midline of the body. Example: the nose is in a medial position on the face, the ulna is on the medial side of the forearm. **Lateral** means toward the side or away from the midline of the body. Example: the ears are in a lateral position on the face; the radius is lateral to the ulna. **Proximal** means nearest the point of attachment or origin. Example: the elbow is proximal to the wrist; the knee is proximal to the ankle. **Distal** means away from the point of attachment or origin. Example: the wrist is distal to the elbow, the ankle is distal to the knee.

*Study***WARE**™ Connection

Play an interactive game labeling directional terms relating to anatomic position on your StudyWARE™ CD-ROM.

PLANES

Occasionally, it is useful to describe the body as having imaginary flat geometric surfaces passing through it called planes (see Figure 1-1). These terms are most useful when describing dissections to look inside an organ or the body as a whole. A **midsagittal** (mid-**SAJ**-ih-tal) plane vertically divides the body through the midline into two equal left and right portions or halves. This is also referred to as a median plane. A **sagittal** plane is any plane parallel to the midsagittal or median plane vertically dividing the body into unequal right and left portions.

A **horizontal** or **transverse** plane is any plane dividing the body into superior and inferior portions. A **frontal** or **coronal** plane is one that divides the anterior (or ventral) and posterior (or dorsal) portions of the body at right angles to the sagittal plane. When organs are sectioned to reveal internal structures, two other terms are often used. A cut through the long axis of an organ is called a longitudinal section, and a cut at right angles to the long axis is referred to as a transverse or cross section.

StudyWARE™ Connection

Watch an animation on body planes on your StudyWARE™ CD-ROM.

CAVITIES

The body has two major cavities: the dorsal cavity and the ventral cavity (Figure 1-3). Each of these is further subdivided into lesser cavities. The organs of any cavity are referred to as the **viscera** (**VISS**-er-ah).

The dorsal cavity contains organs of the nervous system that coordinate the body's functions. It is divided into the **cranial cavity**, which contains the brain, and the **spinal cavity**, which contains the spinal cord.

The ventral cavity contains organs that are involved in maintaining homeostasis or a constant internal environment within small ranges of deviation (Figure 1-4). The first subdivision of the ventral cavity is the **thoracic** (tho-**RASS**-ik) **cavity**. It is surrounded by the rib cage. The thoracic cavity contains the heart in a pericardial sac referred to as the **pericardial cavity**, and the two lungs each covered by the pleural membrane are referred to as the **pleural cavities**. A space called the **mediastinum** (**mee**-dee-ass-**TYE**-num) is found between the two pleural cavities. It contains the heart, thymus gland, lymph and blood vessels, trachea, esophagus, and nerves. The diaphragm muscle separates the thoracic cavity from the abdominopelvic cavity.

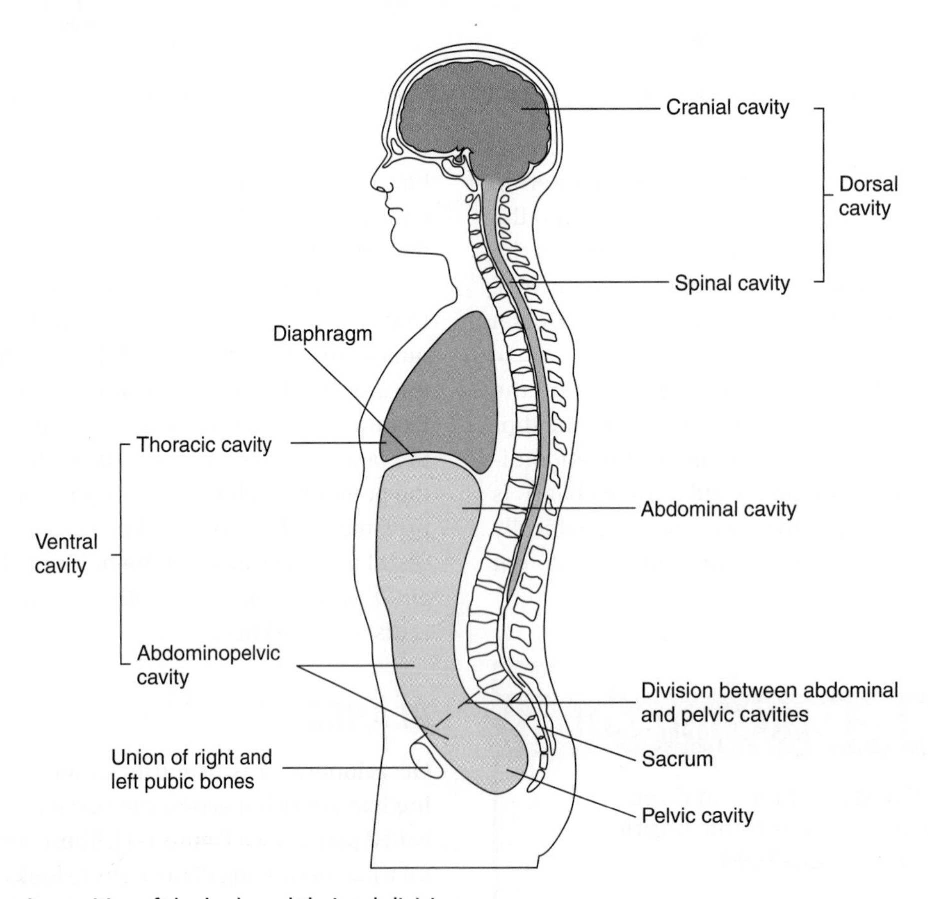

FIGURE 1-3. The major cavities of the body and their subdivisions.

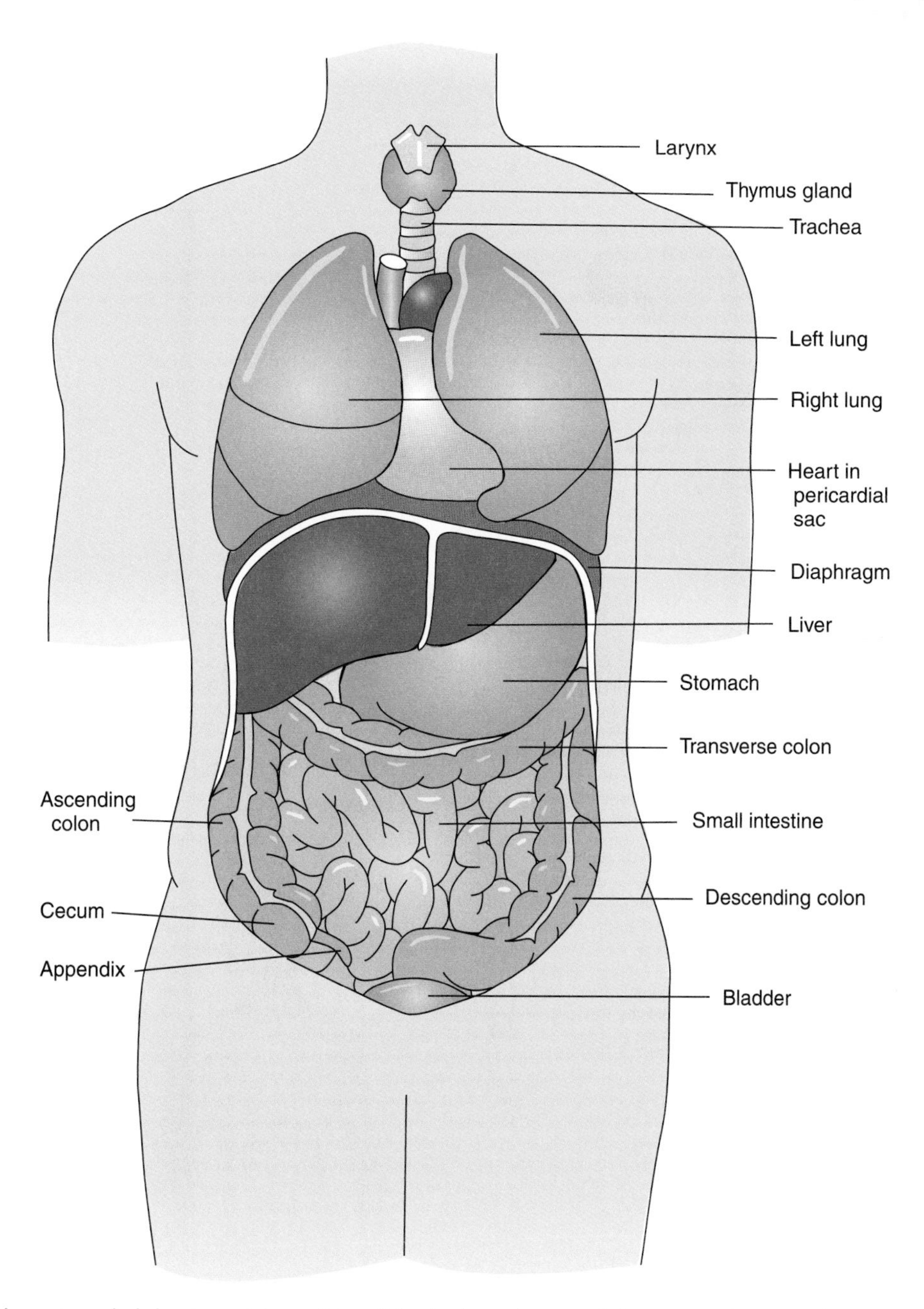

FIGURE 1-4. The thoracic and abdominopelvic cavities of the body and some of the organs they contain.

The **abdominopelvic cavity** is the second subdivision of the ventral cavity. It contains the kidneys, stomach, liver and gallbladder, small and large intestines, spleen, pancreas, and the ovaries and uterus in women.

Two other terms are used when discussing the cavities of the body. The term **parietal** (pah-**RYE**-ehtal) refers to the walls of a cavity. Example: the parietal peritoneum lines the abdominal wall. The term **visceral** refers to the covering on an organ. Example: the visceral peritoneum covers abdominal organs.

STRUCTURAL UNITS

All living material is composed of cells, the smallest units of life. Cells are organized into tissues. Tissues are organized into organs, and organs are part of the major systems of the body (Figure 1-5 and Table 1-1). The cell is the basic unit of biologic organization. The liquid part of a cell is called **protoplasm** (**PRO**-toh-plazm). This protoplasm is surrounded by a limiting membrane, the cell membrane, also called the plasma membrane, which selectively determines what may enter or exit the cell. This proto-

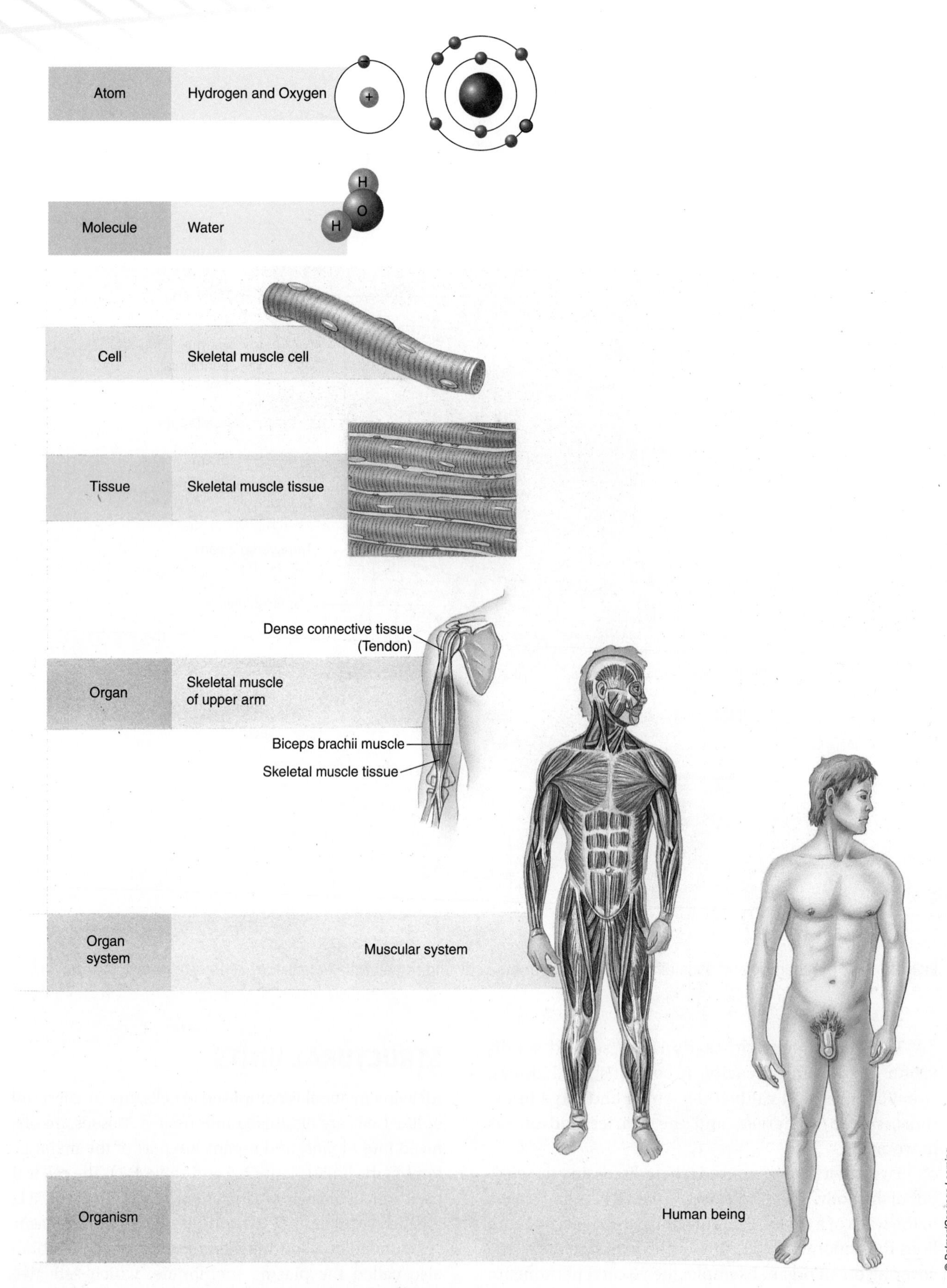

FIGURE 1-5. The structural levels of organization of the body.

Table 1-1 The Structural Levels of Organization of the Human Body

Structural Level	Example
1. Atoms	Atoms are the smallest units of elements, such as carbon, hydrogen, and oxygen.
2. Molecules	Molecules are formed when atoms combine through chemical bonds to form units such as water, sugars, and amino acids.
3. Cells	Cells are the smallest living units of biologic organization made of structures that perform the activities of life, such as the nucleus that controls all the activities of the cell.
4. Tissues	Tissues are made up of similar cells that perform similar functions, such as muscle tissues that cause contraction and movement.
5. Organs	There are four different kinds of tissues (epithelial, connective, muscle, and nervous) that group together in different proportions to make an organ like the stomach, which mixes our food with digestive enzymes.
6. Systems	A group of organs makes up a body system like the nose, pharynx, larynx, trachea, bronchi, and lungs that makes up the respiratory system whose function is to bring in oxygen to the body's cells and take away carbon dioxide gas.
7. Human Organism	All of the organ systems together constitute a functioning human being.

plasm is an aqueous (watery), colloidal (grouping of large molecules) solution of various proteins, lipids, carbohydrates, and inorganic salts that are organized into cellular structures referred to as organelles. These organelles, such as the mitochondria, ribosomes, and lysosomes, among others, are discussed in further detail in Chapter 3.

A cell performs all the activities necessary to maintain life, including metabolism, assimilation, digestion, excretion, and reproduction (see Figure 3-1 in Chapter 3). Different kinds of cells make up a tissue (muscle or bone). Different types of tissues make up an organ (stomach or heart). Finally, organs are grouped into systems (digestive system or nervous system). Each system of the body serves some general function to maintain the body as a whole. All of the diverse tissues of the body can be placed into one of four categories: **epithelial** (**ep**-ih-**THEE**-lee-al), **connective**, **muscle**, or **nervous**. We will study these tissues in greater detail in Chapter 5.

Epithelial tissue covers surfaces and protects (both the outer surface like the skin and inner surfaces of organs like the intestine), forms glands, and lines cavities of the body. It is made up of one or more layers of cells with very little, if any, intercellular material. Connective tissue binds together and supports other tissues and organs. In many instances it is highly specialized (blood, bone, lymphatic tissue). It is made up of different kinds of cells that produce various fibers (elastin and collagen) embedded in a matrix (substance) of nonliving intercellular material. Muscle tissue is characterized by elongated cells (so long in fact they are often referred to as muscle fibers) that generate movement by shortening or contracting in a forcible manner. There are three types of muscle tissue. Skeletal or voluntary muscle pulls on bones and causes body movements. Smooth or involuntary muscle is found in the intestines where it pushes food along the digestive tract. It is also found in arteries and veins where it pushes blood forward. Cardiac muscle is found only in the heart. It is also involuntary and causes contractions of the heart; these contractions pump the blood through thousands of miles of blood vessels. Finally, nervous tissue is composed of nerve cells forming a coordinating system of fibers connecting the numerous sensory (touch, sight) and motor (muscular) structures of the body.

Organs are composed of cells integrated into tissues serving a common function (skin, liver, stomach, heart, lungs). A system is a group of organs.

The **integumentary system** is made up of two layers: the epidermis and the dermis. It includes the skin, hair, nails, sebaceous glands, and sweat glands (see Figure 1-6). Its functions include insulation of the body, protection of the body from environmental hazards such as the ultraviolet radiation of the sun and certain chemicals, and regulation of body temperature and water. It also has receptor sites to detect changes in temperature and pressure.

The **skeletal system** is composed of bones, cartilage, and the membranous structures associated with bones (see Figure 1-6). It protects the soft and vital parts of the body and provides support for body tissues. Its bones act as levers for movement. This system also manufactures blood cells in red bone marrow and stores fat in yellow

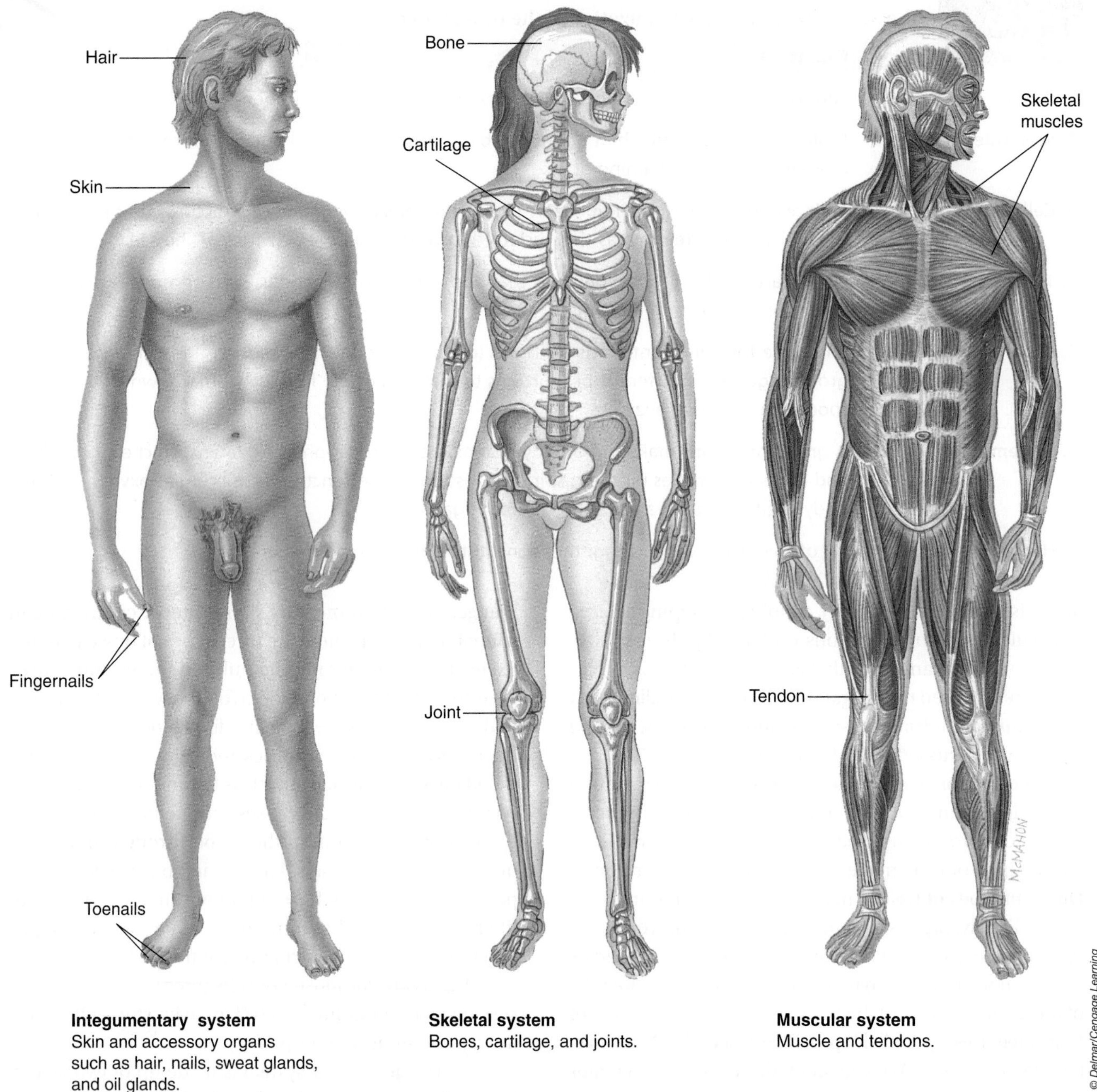

FIGURE 1-6. The integumentary, skeletal, and muscular systems of the body.

bone marrow. Bones store mineral salts like calcium and phosphorous.

The **muscular system** consists of muscles, fasciae (fibrous connective tissues), tendon sheaths, and bursae (fibrous sacs) (see Figure 1-6). Skeletal muscles pull on bones to allow movement; smooth muscle pushes food through the digestive tract and blood through the circulatory system; and cardiac muscle causes contraction of the heart.

The **nervous system** consists of the brain, spinal cord, cranial nerves, peripheral nerves, and the sensory and motor structures of the body (Figure 1-7). Its functions include controlling, correlating and regulating the other systems of the body; interpreting stimuli from the outside world; and controlling the special senses of sight, hearing, taste, and smell.

The **endocrine system** consists of the endocrine (ductless) glands (see Figure 1-7). The master gland, or pituitary, controls the other glands—thyroid, adrenal glands, ovaries, and testes. These glands produce hormones that chemically regulate the body's functions. This system

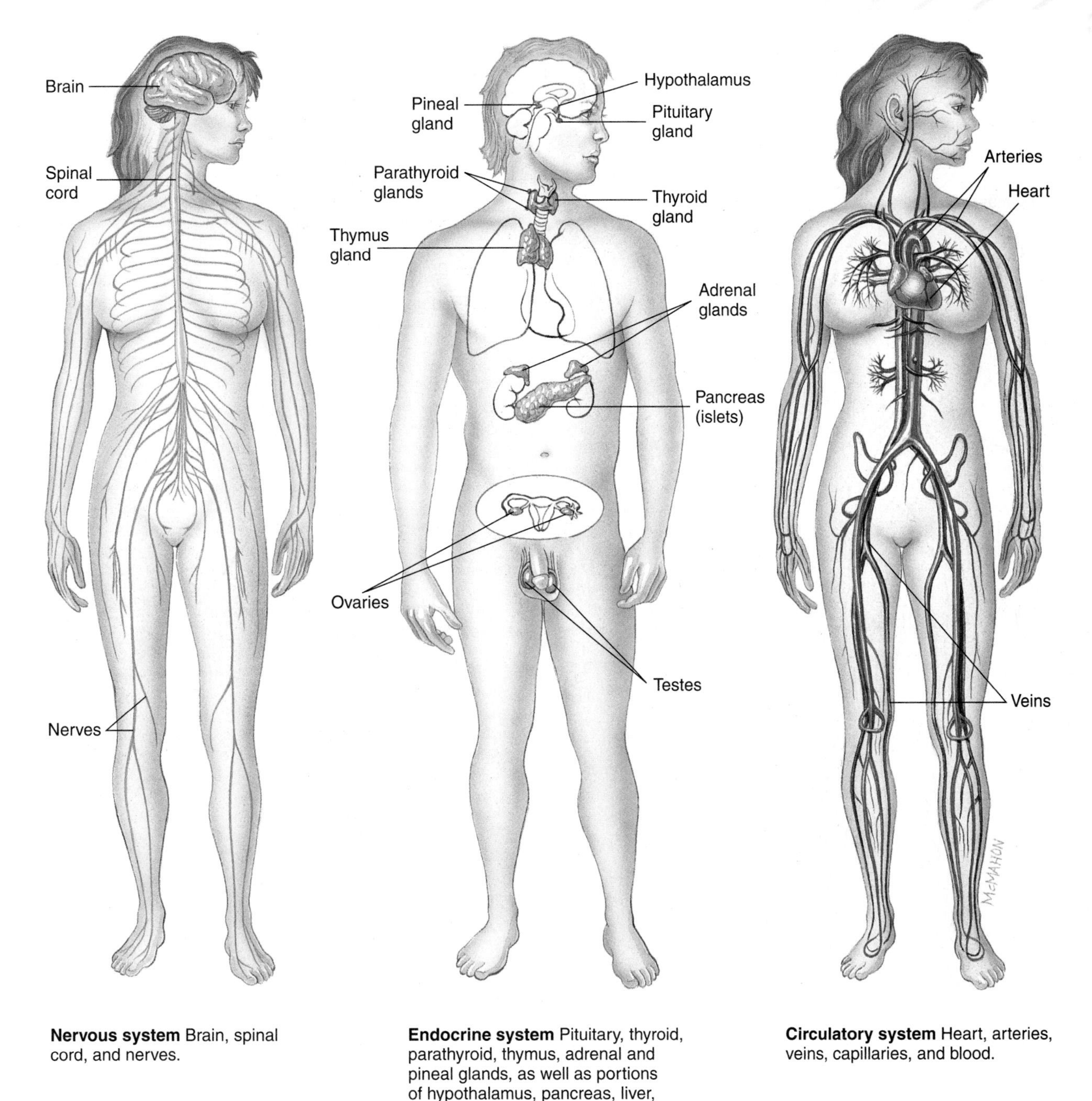

Nervous system Brain, spinal cord, and nerves.

Endocrine system Pituitary, thyroid, parathyroid, thymus, adrenal and pineal glands, as well as portions of hypothalamus, pancreas, liver, kidneys, ovaries, testes, and placenta. Also included are hormonal secretions from each gland.

Circulatory system Heart, arteries, veins, capillaries, and blood.

FIGURE 1-7. The nervous, endocrine, and cardiovascular, or circulatory, systems of the body.

works with the nervous system through the hypothalamus of the brain, which controls the pituitary gland.

The **cardiovascular**, or blood circulatory, **system** consists of the heart, arteries, veins, and capillaries (see Figure 1-7). Its function is to pump and distribute blood, which carries oxygen, nutrients, and wastes to and from the cells of the body.

The **lymphatic**, or immune, **system** is made up of the lymph nodes, the thymus gland, the spleen, and the lymph vessels (see Figure 1-8). Its function is to drain tissue spaces of excess interstitial fluids and absorb fats from the intestine and carry them to the blood. It also protects the body from disease by developing immunities and destroying most invading disease-causing microorganisms.

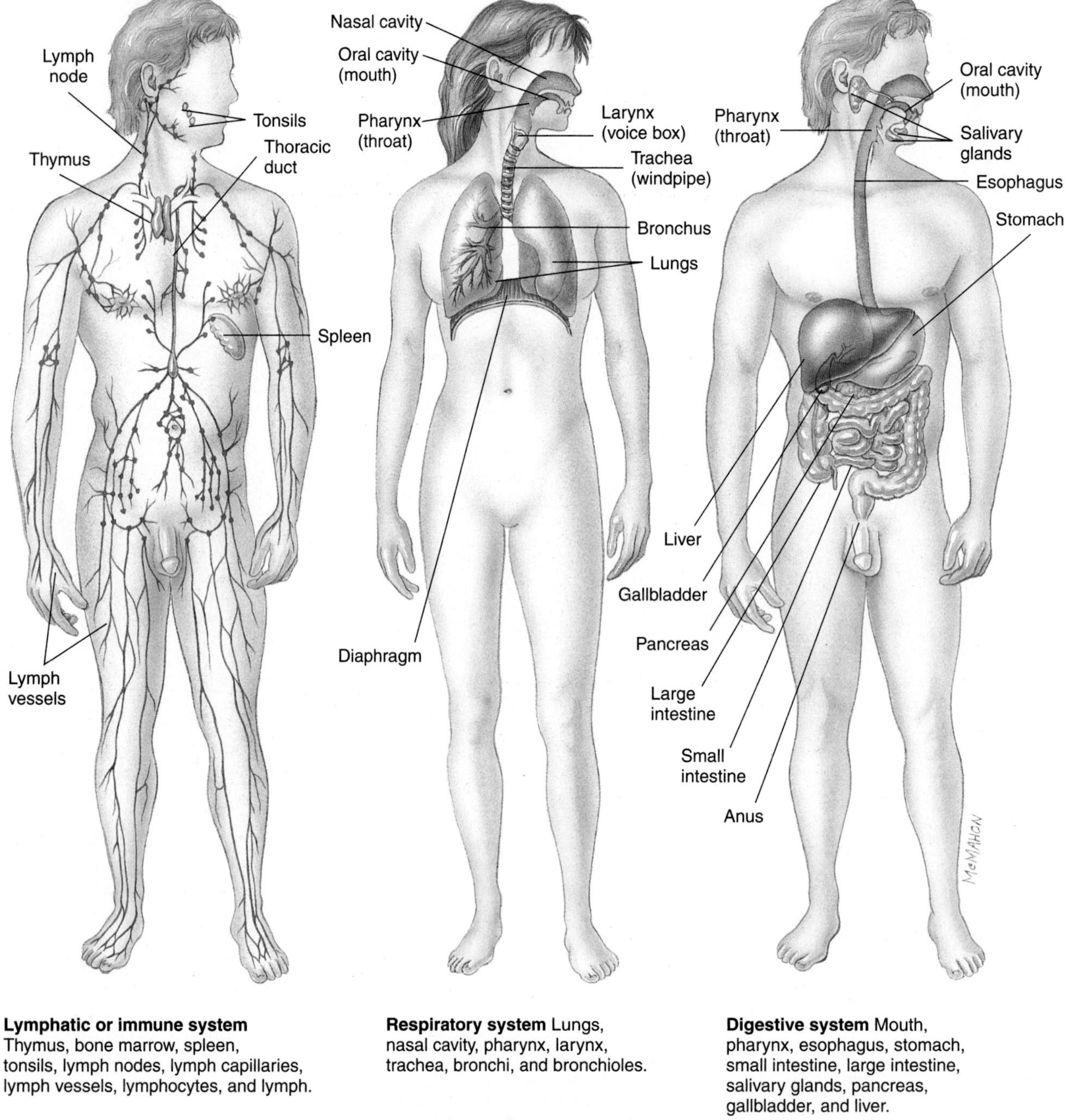

Lymphatic or immune system Thymus, bone marrow, spleen, tonsils, lymph nodes, lymph capillaries, lymph vessels, lymphocytes, and lymph.

Respiratory system Lungs, nasal cavity, pharynx, larynx, trachea, bronchi, and bronchioles.

Digestive system Mouth, pharynx, esophagus, stomach, small intestine, large intestine, salivary glands, pancreas, gallbladder, and liver.

FIGURE 1-8. The lymphatic (or immune), respiratory, and digestive systems of the body.

The **respiratory system** is composed of the nasal cavities, pharynx, larynx, trachea, bronchi, and lungs (see Figure 1-8). It brings oxygen to and eliminates carbon dioxide from the blood.

The **digestive system** includes the alimentary canal (mouth, esophagus, stomach, small and large intestines, rectum, and anus) with its associated glands (salivary, liver, and pancreas) (see Figure 1-8). Its function is to convert food into simpler substances that along with other nutrients can be absorbed by the cells of the body and eliminate indigestible wastes.

The **urinary system** is made up of two kidneys, two ureters, the bladder, and the urethra (Figure 1-9). Its functions include the chemical regulation of the blood, the formation and elimination of urine, and the maintenance of homeostasis.

The **reproductive system** consists of the ovaries, uterine tubes, uterus, and vagina in the female and the

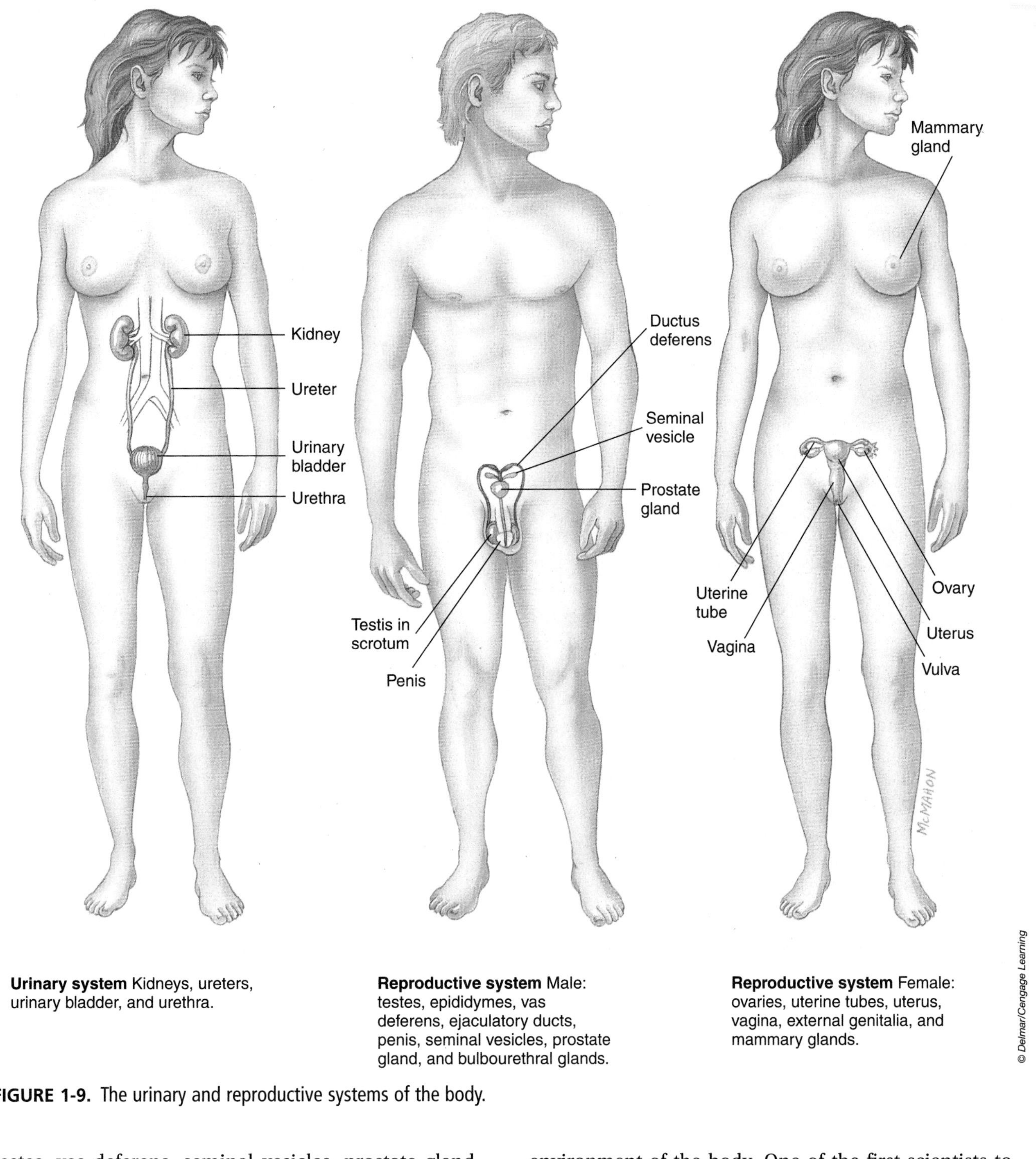

Urinary system Kidneys, ureters, urinary bladder, and urethra.

Reproductive system Male: testes, epididymes, vas deferens, ejaculatory ducts, penis, seminal vesicles, prostate gland, and bulbourethral glands.

Reproductive system Female: ovaries, uterine tubes, uterus, vagina, external genitalia, and mammary glands.

FIGURE 1-9. The urinary and reproductive systems of the body.

testes, vas deferens, seminal vesicles, prostate gland, penis, and the urethra in the male (see Figure 1-9). Its functions include maintenance of sexual characteristics and the perpetuation of our species.

HOMEOSTASIS

Homeostasis (hom-ee-oh-**STAY**-sis) is the maintenance (within varying narrow limits) of the internal environment of the body. One of the first scientists to discuss the significance of homeostasis to the survival of an organism was French scientist Claude Bernard (1813–1878). Homeostasis is essential to survival; hence, many of the body's systems are concerned with maintaining this internal environment. Some examples of homeostasis are blood sugar levels, body temperature, heart rate, and the fluid environment of cells. When homeostasis is maintained, the body is healthy. This is the reason your doctor takes your temperature and blood pressure as part of a routine examination.

We shall examine two examples of maintaining homeostasis. After ingesting a meal, which is predominately carbohydrates (salad, vegetables, bread, and perhaps fruit), the blood glucose level increases dramatically due to the breakdown of the complex carbohydrates by the digestive system into sugars such as glucose. Cells take in the glucose they need carried by the blood, but so much glucose is in the blood that now the pancreas secretes insulin, which moves the excess blood glucose into the liver where it is stored as glycogen, or animal starch. Between meals, when the blood glucose level drops below normal, the pancreas secretes glucagon, which breaks down the glycogen into glucose and returns it to the blood circulatory system for distribution to body cells. Thus, the glucose level in the blood plasma remains at a nearly constant level so that it does not remain elevated after a meal, nor does it drop too low between meals.

Body temperature regulation is another important example of homeostasis. When we go out on a hot summer day and our body temperature rises above 98.6°F, the hypothalamus of the brain detects this change and sends signals to various organs so that we sweat (sweating is a cooling process). As water is excreted by the sweat glands onto the skin, it evaporates in the air (evaporation is a cooling mechanism). In addition, our blood vessels dilate to bring the blood near the skin's surface to dissipate body heat. When our body temperature falls below 98.6°F, such as when we go out on a cold winter day, the hypothalamus sends signals to muscles, causing us to shiver to raise our body temperature; it also causes our blood vessels to constrict to conserve body heat.

Our body must constantly monitor itself to correct any major deviations in homeostasis. It does this by using what is referred to as a *negative feedback loop.* Feedback responses that revise disturbances to our body's condition are examples of negative feedback. A good example of a negative feedback loop is the relationship between your home thermostat and your furnace. You set the thermostat at a temperature of 72°F. When the temperature in your home drops below 72°F, the furnace turns on to raise the house temperature. When the temperature goes above 72°F, the thermostat causes the furnace to turn off. Positive feedback is an increase in function in response to a stimulus. For example, after the first contraction during labor, the uterus continues to contract with more strength and frequency.

Our organ systems help control the internal environment of the body and cells so that it remains fairly constant. Our digestive, urinary, circulatory, and respiratory systems work together so that every cell receives the right amount of oxygen and nutrients, and so waste products are eliminated fairly quickly and do not accumulate to toxic levels. If homeostasis is not maintained, the body will experience disease and, eventually, death.

SUMMARY OUTLINE

INTRODUCTION

1. The four basic reference systems of body organization are directions, planes, cavities, and structural units.

TERMS OF DIRECTION

1. Superior means uppermost or above; inferior means lowermost or below.
2. Anterior means toward the front; ventral is synonymous with anterior. Posterior means toward the back; dorsal is synonymous with posterior.
3. Cephalad or cranial means toward the head; it is synonymous with superior.
4. Medial means nearest the midline; lateral means toward the side.
5. Proximal means nearest the point of attachment; distal means away from the point of attachment.

PLANES

1. A midsagittal or median plane vertically divides the body into equal halves. A sagittal plane is parallel to a median or midsagittal plane.
2. A horizontal or transverse plane divides the body into superior and inferior portions.
3. A frontal or coronal plane divides the anterior or ventral and the posterior or dorsal portions of the body at right angles to the sagittal planes.

CAVITIES

1. The body has two major cavities: the dorsal cavity and the ventral cavity.
2. The dorsal cavity is subdivided into the cranial cavity, which contains the brain, and the spinal cavity, which contains the spinal cord.
3. The ventral cavity is divided into two lesser cavities. The first is the thoracic cavity, which contains the heart in the pericardial cavity and the two lungs each in a pleural cavity. The second is

the abdominopelvic cavity, which contains many of the digestive organs and some urinary and reproductive organs.

4. The term *parietal* refers to the walls of a cavity.
5. The term *visceral* refers to the covering of an organ.

STRUCTURAL UNITS

1. The cell is the basic unit of the body's organization.
2. Different types of cells make up the four tissues of the body: epithelial, connective, muscle, and nervous.
3. Organs are composed of cells integrated into tissues serving a common function.
4. A system is a group of organs that perform a common function.
5. The integumentary system includes skin, hair, nails, and sweat and sebaceous glands. It protects, insulates, and regulates water and temperature.
6. The skeletal system includes bones and cartilage. It allows movement, makes blood cells, stores fat, protects, and supports.
7. The muscular system is made of skeletal, smooth, and cardiac muscle. It causes movement.
8. The nervous system includes the brain, spinal cord, and cranial and spinal nerves. It is the controlling, regulatory, and correlating system of the body.
9. The endocrine system consists of the endocrine glands and their hormones. It regulates chemical aspects of the body in conjunction with the nervous system.
10. The cardiovascular, or blood circulatory, system consists of the heart, arteries, veins, and capillaries. It distributes blood, carrying oxygen, nutrients, and wastes to and from body cells.
11. The lymphatic, or immune, system is made up of the lymph nodes, lymph vessels, the thymus gland, and the spleen. It drains tissues of excess fluids, transports fats, and develops immunities.
12. The respiratory system includes the nose, pharynx, larynx, trachea, bronchi, and lungs. It brings oxygen to and eliminates carbon dioxide from the blood.
13. The digestive system is composed of the organs of the alimentary tract from the lips to the anus and its associated glands. It converts food into simpler substances that can be absorbed along with other nutrients by the body's cells.
14. The urinary system includes the kidneys, ureters, bladder, and urethra. It functions in the chemical regulation of the blood.
15. The reproductive system includes the ovaries, uterine tubes, uterus, and vagina in women and the testes, seminal vesicles, prostate gland, penis, and urethra in men. It maintains sexual characteristics and perpetuates the species.

HOMEOSTASIS

1. Homeostasis is the maintenance of the internal environment of the body within certain narrow ranges.
2. Some examples of homeostasis are blood sugar levels, body temperature, heart rate, and the fluid environment of the cell.

REVIEW QUESTIONS

1. Name the systems of the body and their functions. Indicate what major organs each system contains.
2. The body has two major cavities, each divided into two lesser cavities. List them and explain what each cavity contains.

*3. Discuss how the body maintains homeostasis in terms of the blood glucose level.

*4. Discuss how the body maintains homeostasis in regard to maintaining normal body temperature.

5. Explain what a cell is.
6. List and define the main directional terms of the body.
7. List and define the three planes of division of the body.

*Critical Thinking Questions

MATCHING

Place the most appropriate number in the blank provided.

_____ Superior	1. Toward the back, dorsal
_____ Anterior	2. Uppermost or above
_____ Inferior	3. Toward the side
_____ Posterior	4. Nearest the point of attachment or origin
_____ Medial	5. Any plane dividing the body into superior and inferior
_____ Lateral	
_____ Proximal	

_____ Distal
_____ Horizontal
_____ Midsagittal

6. Away from the point of attachment
7. Toward the front, ventral
8. The plane vertically dividing the body into equal right and left halves
9. Toward the heart
10. Nearest the midline of the body
11. Lowermost or below
12. Frontal

Search the Internet with key words from the chapter to discover additional information and interactive exercises. Key words might include homeostasis, body planes, and anatomic positions.

*Study***WARE**™ Connection

Take a practice quiz and play interactive games that reinforce the content in this chapter on your StudyWARE™ CD-ROM.

Study Guide Practice

Go to your **Study Guide** for more practice questions, labeling and coloring exercises, and crossword puzzles to help you learn the content in this chapter.

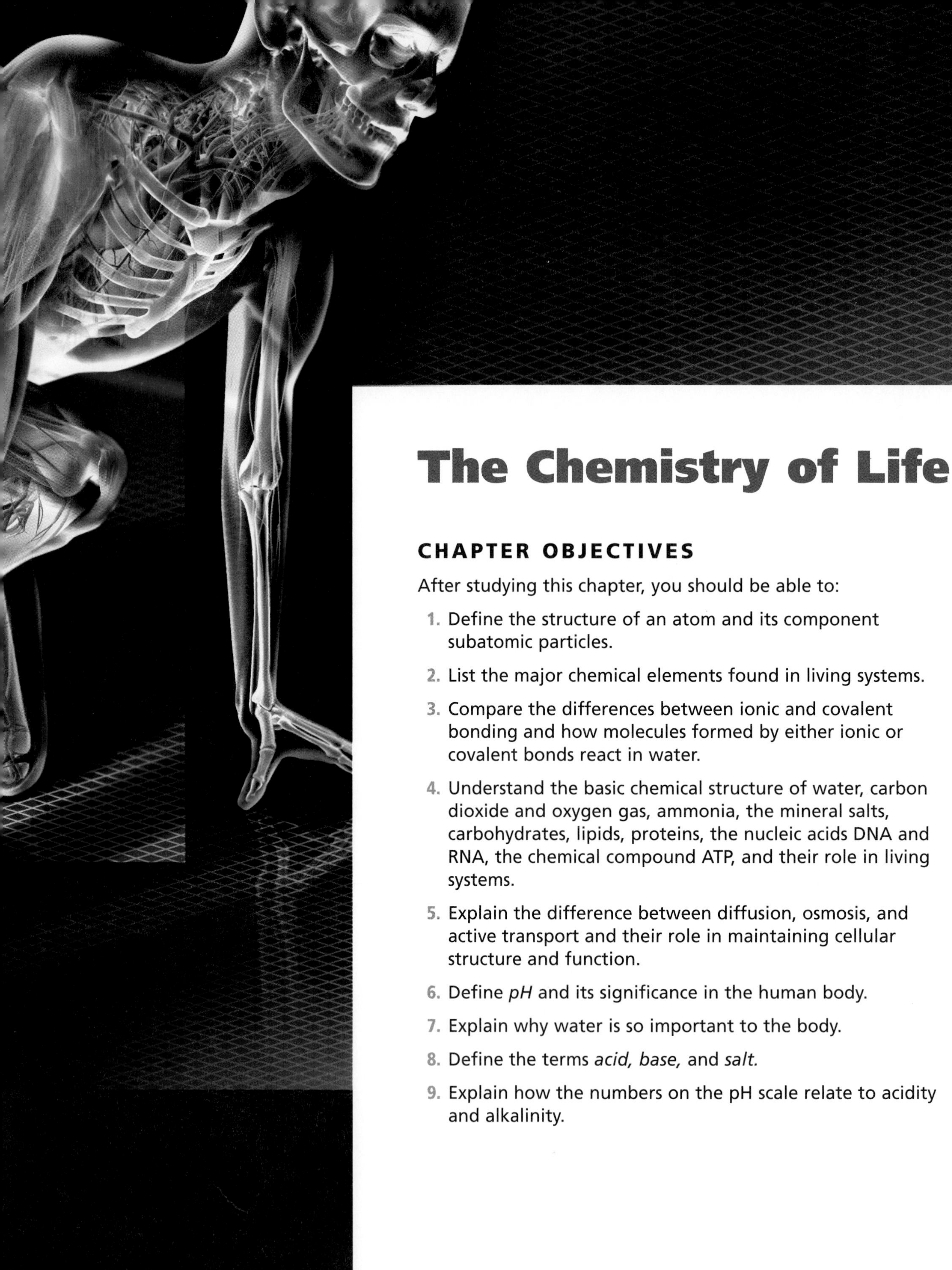

The Chemistry of Life

CHAPTER OBJECTIVES

After studying this chapter, you should be able to:

1. Define the structure of an atom and its component subatomic particles.
2. List the major chemical elements found in living systems.
3. Compare the differences between ionic and covalent bonding and how molecules formed by either ionic or covalent bonds react in water.
4. Understand the basic chemical structure of water, carbon dioxide and oxygen gas, ammonia, the mineral salts, carbohydrates, lipids, proteins, the nucleic acids DNA and RNA, the chemical compound ATP, and their role in living systems.
5. Explain the difference between diffusion, osmosis, and active transport and their role in maintaining cellular structure and function.
6. Define *pH* and its significance in the human body.
7. Explain why water is so important to the body.
8. Define the terms *acid, base,* and *salt.*
9. Explain how the numbers on the pH scale relate to acidity and alkalinity.

2

KEY TERMS

INTRODUCTION

Because all of the structures of the body (cells, tissues, and organs) are composed of chemicals, it is necessary to have a basic understanding of the science of chemistry. In addition, the body functions through chemical reactions. For example, in the digestive process, complex foods are broken down through chemical reactions into simpler substances such as sugars that can be absorbed and used by the body's cells. Later these simple substances are converted into another kind of chemical fuel, **adenosine triphosphate (ATP)** (ah-**DEN**-oh-seen try-**FOS**-fate), which allows the body cells to do work and function. Chemistry is the science that deals with the elements, their compounds, the chemical reactions that occur between elements and compounds, and the molecular structure of all matter. Students of anatomy need to have some basic knowledge of this field of study.

This chapter introduces you to some basic principles of chemistry that will assist in your comprehension of human anatomy and physiology. To understand the human body, it is necessary to understand the chemical basis of life. We will look at the structure of the atom, how atoms interact with one another to form compounds, and how those compounds form the building blocks of life. All nonliving and living things are made of matter. Matter is composed of elements, which are primary substances from which all other things are constructed. Elements cannot be broken down into simpler substances. There are 92 elements that occur naturally. Other elements have been created artificially in the laboratory.

ATOMIC STRUCTURE

Atoms are the smallest particles of an element that maintain all the characteristics of that element and enter into chemical reactions through their electrons. Each atom consists of a relatively heavy, compact central nucleus composed of **protons** and **neutrons**. Lighter particles called **electrons** orbit the nucleus at some distance from its center.

Electrons are practically weightless, and each one carries a negative electrical charge ($^-$). Atomic nuclei are composed of protons and neutrons, except for the hydrogen nucleus, which contains only one proton (Figure 2-1). Each proton and neutron has one unit of atomic weight and is about 1800 times heavier than an electron. Thus, an atom's weight results almost entirely from its protons and neutrons. A proton carries a positive charge ($^+$), whereas a neutron is neutral and has no charge. Like charges repel—they push away from each other. Thus, when you brush your hair on a dry day, like electrical charges build up on the brush and your hair so your hair flies away from the brush. Unlike charges attract. The clinging of clothes taken out of a dryer is due to the attraction of unlike electrical charges.

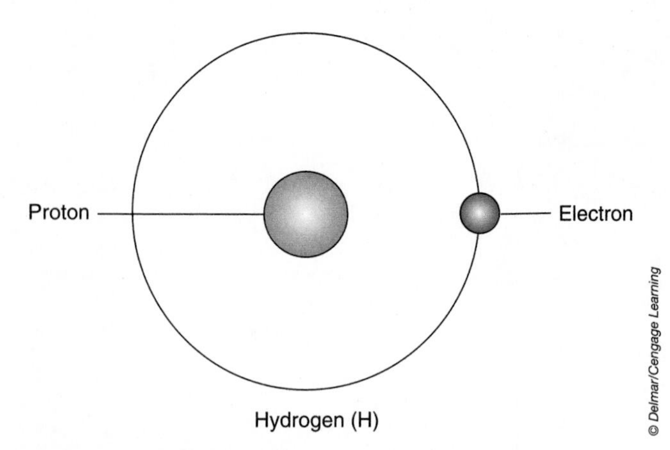

FIGURE 2-1. The hydrogen atom is unique because its nucleus contains only one proton.

ELEMENTS, ISOTOPES, COMPOUNDS

Each element has a distinctive number of protons. An **element** is a substance whose atoms all contain the same number of protons and the same number of electrons. Because the number of protons equals the number of electrons, an atom is electrically neutral. The theory that suggested that all matter consists of atoms was proposed in 1808 by John Dalton (1766–1844). He stated that atoms were responsible for the combinations of elements found in compounds. The atomic theory developed from his proposal. The atomic theory proposed that:

- All matter is made up of tiny particles called atoms.
- All atoms of a given element are similar to one another but different from the atoms of other elements.
- Atoms of two or more elements combine to form compounds.
- A chemical reaction involves the rearrangement, separation, or combination of atoms.
- Atoms are never created or destroyed during a chemical reaction.

In the atoms of some elements, the number of neutrons varies. Carbon is the element found in all living matter. Life on earth is based on the carbon atom. In fact, a whole branch of chemistry called organic chemistry studies the nature of the carbon atom and its chemical reactions. Different atoms of carbon may have different numbers of neutrons. Atoms of carbon may have one of three different atomic weights—12, 13, or 14—depending on the number of neutrons. These different kinds of atoms of the same element are called **isotopes**, and are designated as C^{12}, C^{13} and C^{14}. Each of these isotopes contains six protons and six electrons, but C^{12} has six neutrons, C^{13} has seven neutrons, and C^{14} has eight neutrons. C^{14} is mildly radioactive and is used to estimate the age of fossilized human remains. A radioactive isotope of iodine is used to treat disorders of the thyroid gland. The **atomic number** is the number of protons or the number of electrons. By the late 1800s, scientists discovered similarities in the behavior of the known elements. It was a Russian chemist, Dimitri Mendeleev (1834–1907), who suggested that the elements could be arranged in groups that showed similar physical and chemical properties. From his work, we have the modern **periodic table** of the elements, which arranges the elements by increasing atomic number in such a way that similar properties repeat at periodic intervals (Figure 2-2).

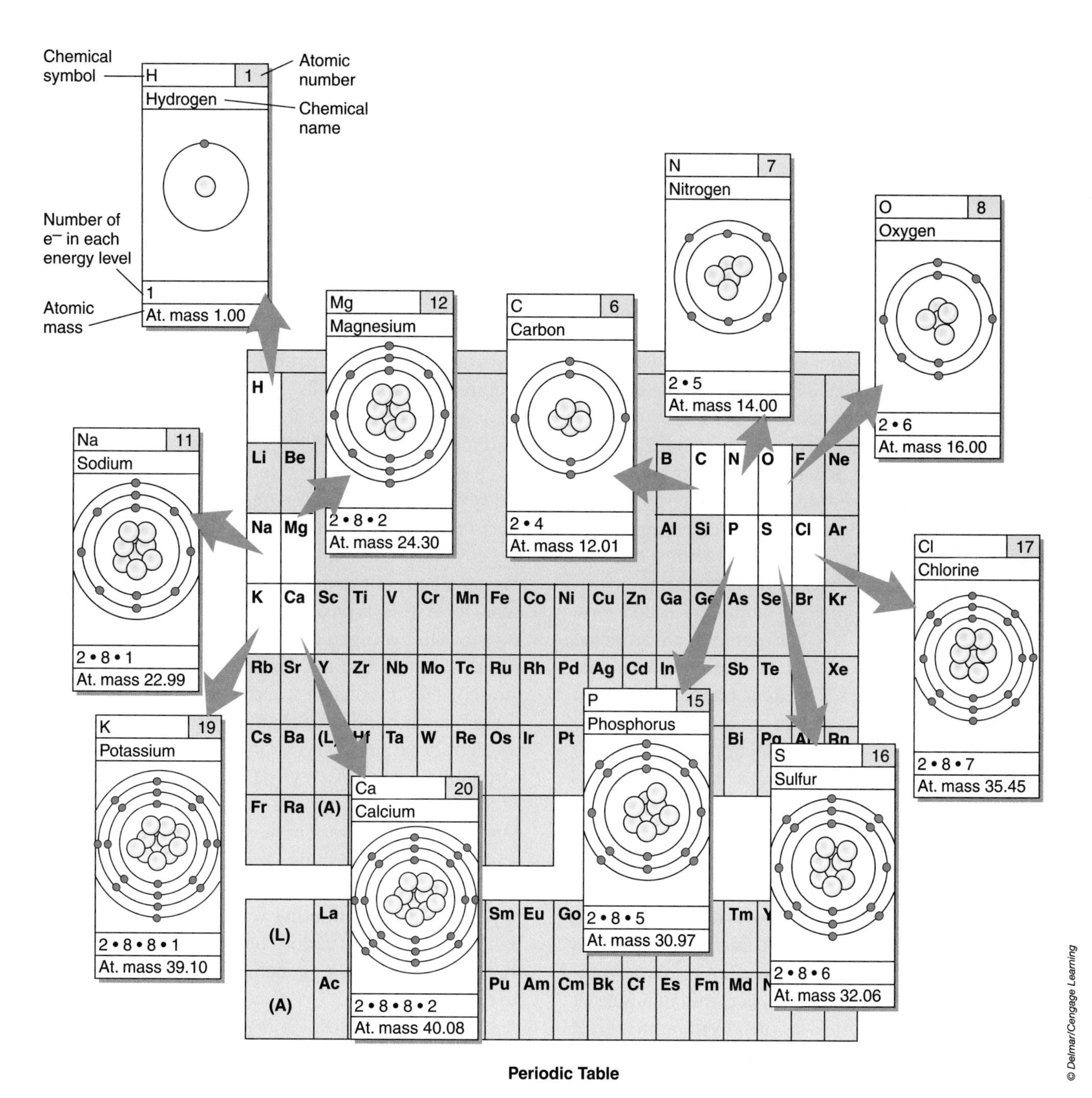

FIGURE 2-2. The periodic table of the elements.

In summary, protons and neutrons make up the nucleus of an atom. Electrons orbit the nucleus. It is impossible to know exactly where any given electron is located at any given moment, but the area where it is found can be referred to as the electron's orbital. **Orbitals** are grouped together to form **energy levels** consisting of electrons. Levels can contain more than one electron. Thus, atoms are represented as a round nucleus (containing protons and neutrons) surrounded by concentric circles representing the energy levels. Carbon has two electrons in the first level and four electrons in the second level. Hydrogen has a single electron in its first level and no other levels. Oxygen has two electrons in the first level and six electrons in the second level.

BONDS AND ENERGY

Atoms combine chemically with one another in one of two ways, that is, they form **bonds**. Chemical bonds are formed when the outermost electrons are transferred (gained or lost) or shared between atoms. When the atoms of two or more different elements combine in this way, a **compound** (such as water, H_2O) is created. This symbol H_2O also represents a **molecule**. A molecule or compound is the smallest combination or particle retaining all the properties of the compound itself.

One type of bond is called an **ionic bond**. This kind of bond is formed when one atom gains electrons while the other atom loses electrons from its outermost level or orbit. Atoms that gain electrons become negatively charged, whereas those that lose electrons become positively charged, each having originally been electrically neutral. The new charged atoms are called **ions**. Negatively charged ions (Cl^-, for example) are attracted to positively charged ions (Na^+). The resulting force that binds these ions together is an ionic bond. Referring to Figure 2-3, notice that the sodium atom has a completely filled innermost level with two electrons, a completely filled second level with eight electrons, but only one electron in its third level. The chlorine atom has a completely filled innermost level with two electrons, a completely filled second level with eight electrons, but only seven in its third level. Because eight electrons fill the outermost level in forming the ionic bond, sodium loses its one electron to the chlorine atom's outermost level, thus filling chlorine's outermost level with eight electrons. The resulting compound, sodium chloride (Na^+Cl^-), is common table salt formed by an ionic bond, held together by the attraction of the opposite electric charges of the ions. When immersed in water, compounds held together by ionic bonds tend to separate or dissociate into their constituent ions because

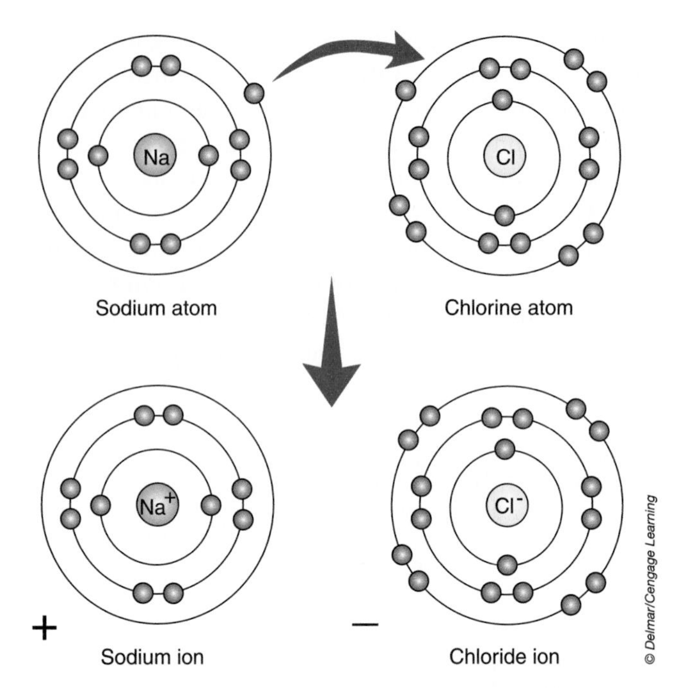

FIGURE 2-3. The formation of an ionic bond between sodium and chloride forms Na^+Cl^- (sodium chloride or table salt).

of the attraction of the water molecule (which we shall discuss later in this chapter). Many of the substances required by human cells exist in nature in ionic form. Some examples are the mineral salts such as sodium, chloride, potassium, calcium, and phosphate.

A second type of bond found in many molecules is the **covalent bond**. In this type of bond, the atoms share electrons to fill their outermost levels. Molecules containing covalent bonds do not dissociate when immersed in water. Four of the most important elements found in cells form this type of bond. They are carbon (C), oxygen (O), hydrogen (H), and nitrogen (N). They constitute about 95% of the materials found in cells. All of the cell's larger molecules, and many of its smaller ones, contain such bonds; for example, the formation of the covalent bond between two hydrogen atoms forms the compound hydrogen gas (H_2) (Figure 2-4).

Another type of bond is the **hydrogen bond**. Hydrogen bonds are very weak bonds and help hold water molecules together by forming a bridge between the negative oxygen atom of one water molecule and the positive hydrogen atoms of another water molecule. Hydrogen bonds also help bind various parts of one molecule into a three-dimensional shape such as a protein molecule like an enzyme.

Elements or molecules furnishing electrons during a reaction are called **electron donors** (e.g., sodium); those that gain electrons during the process are called **electron acceptors** (e.g., chlorine when salt is formed). Some very

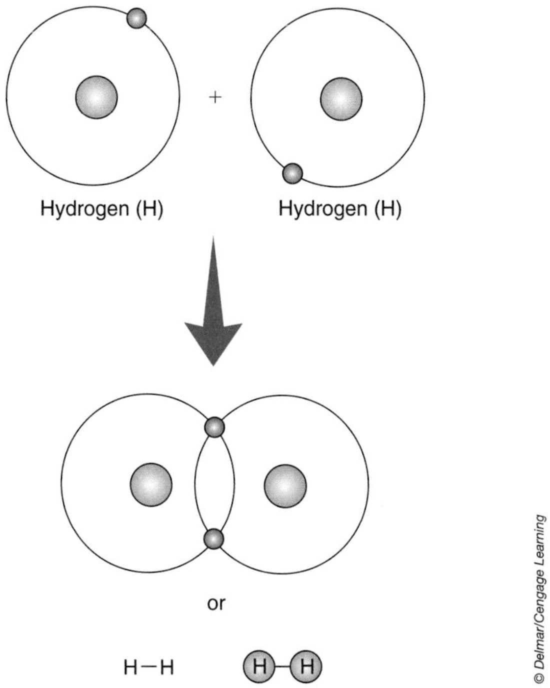

FIGURE 2-4. The formation of compounds through covalent bonding.

special molecules will gain electrons only to lose them to some other molecule in a very short time; these are designated as **electron carriers**. These molecules are discussed in Chapter 4 and are very important in making the cellular energy molecule ATP.

Bonds contain **energy**, the ability to do work. This results from the interaction of the electrons and the nuclei of the bonded atoms. If we measure the amount of energy present between two atoms, we discover that the amount varies as the distance between the atoms changes. When atoms are close to one another, the paths of their electrons overlap. The natural repulsion of these negatively charged electrons tends to drive the two atoms apart. Thus, the amount of energy necessary to keep them together is quite high. This type of bond contains a high degree of energy. If we break these bonds, as in the breakdown of a glucose ($C_6H_{12}O_6$) molecule inside a cell, electron carriers in the cell will use the energy of the released electrons to put together an ATP molecule. ATP is the high-energy fuel molecule that the cell needs to function. This high-energy molecule that is used in the cell is called adenosine triphosphate. This molecule is constantly being created and broken down to release its energy to do the cell's work. It is abbreviated as ATP. It is created by adding a phosphate to adenosine diphosphate. When it is broken down (ATP $\rightarrow$ ADP + PO_4) it releases the energy contained in the phosphate bond. We shall discuss this in further detail in Chapter 4.

COMMON SUBSTANCES IN LIVING SYSTEMS

There are 10 common substances found in living systems. They are water, carbon dioxide gas, molecular oxygen, ammonia, mineral salts, carbohydrates, lipids, proteins, nucleic acids, and adenosine triphosphate.

Water

Water is the most abundant substance in living cells, approximately 60% to 80%; plasma, which is the liquid portion of blood, is 92% water. Water is a small, simple molecule composed of two hydrogen atoms covalently bonded to one oxygen atom. Because the oxygen atom attracts electrons more strongly than do the hydrogen atoms, water molecules are polar with a partial positive charge by the hydrogen atoms and a partial negative charge by the oxygen atom (Figure 2-5). This unique feature of the water molecule determines why ionic bonded molecules dissociate in water. Negatively charged ions (e.g., chloride) are attracted to the positively charged hydrogen atoms, and positively charged ions (e.g., sodium) are attracted to the negatively charged oxygen atoms. Thus, the ionic bonded molecule salt dissociates in water.

Water has a number of roles in cells. It takes part in some reactions, such as photosynthesis in plant cells,

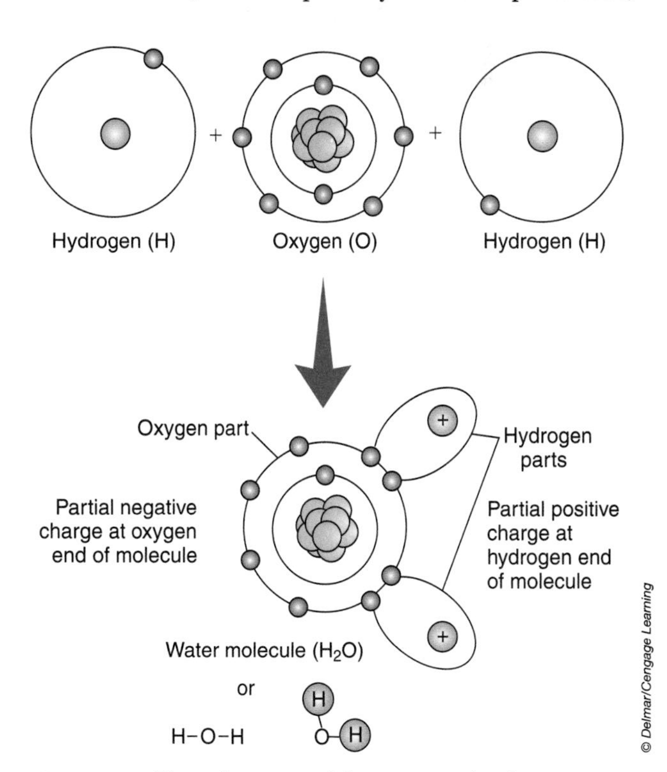

FIGURE 2-5. The uniqueness of the water molecule.

which supplies our earth with molecular oxygen, and respiration in both plant and animal cells, which produces energy.

Photosynthesis:

$$6CO_2 + 12H_2O \rightarrow C_6H_{12}O_6 + 6O_2 + 6H_2O$$

Respiration:

$$C_6H_{12}O_6 + 6O_2 \rightarrow 6CO_2\ 6H_2O + \text{energy in the form of ATP}$$

Digestion of food requires water to break down larger molecules.This is called hydrolysis.Water serves as a medium or **solvent** for other reactions, and water is referred to as the universal solvent. The chemistry of life is dominated by the chemistry of water. Chemical reactions occur in cells between individual atoms, ions, or molecules, not between large aggregations of these particles. It is as these particles move about in the water that they come in contact with other particles and chemical reactions occur. In addition, water is a basis for the transport of materials such as hormones and enzymes in the plasma of blood.

Water also absorbs and releases high levels of heat before its temperature changes, thus helping control normal body temperature. Vigorous exercise liberates heat from contracting muscle cells. This excess heat is absorbed by the water in the cells and then released. Water is part of amniotic fluid and protects the developing fetus. It is also part of the cerebrospinal fluid and protects the brain and spinal cord by functioning as a shock absorber. Finally, water is the base for all body lubricants such as mucus in the digestive tract and synovial fluid in joints.

Carbon Dioxide

The small **carbon dioxide** molecule (CO_2) contains one carbon atom covalently bonded to two oxygen atoms. It is produced as a waste product of cellular respiration and must be eliminated quickly from the body through expiration via the respiratory system and the cardiovascular system. It is also necessary for photosynthesis in plant cells to convert the radiant energy of the sun into usable chemical energy such as glucose for both plant and animal cells. It is also a source of the element carbon, found in all organic compounds of living systems. If carbon dioxide is allowed to accumulate within cells, it becomes toxic by forming carbonic acid as it reacts with water. Hence, we exhale it quickly from the lungs.

Molecular Oxygen

Molecular oxygen (O_2), formed when two oxygen atoms are covalently bonded together, is required by all organisms that breathe air. It is necessary to convert chemical energy (food), such as the energy found in a glucose ($C_6H_{12}O_6$) molecule, into another form of chemical energy, ATP, that can be used by cells to do work. Because O_2 is a product of photosynthesis, it becomes obvious how dependent we animals are on plants for our survival. Without plants there would be no molecular oxygen in our atmosphere, and without O_2 there would be no life on our planet as we know it. The level of O_2 in our atomosphere is maintained at a nearly constant level (about 21% of the gas in the atomosphere is oxygen) by the many different kinds of plants found on our earth.

Ammonia

The **ammonia** molecule (NH_3) comes from the decomposition of proteins via the digestive process and the conversion of amino acids in cellular respiration to ATP molecules. Note that an important element in ammonia is nitrogen. Nitrogen is an essential element in amino acids, which are the building blocks of proteins. Because even a small amount of ammonia is injurious to cells, the human body must quickly dispose of this material. Through enzymes, the liver converts the toxic ammonia to a harmless substance called urea. Because urea is soluble in water, the blood then carries the urea to the kidneys to be filtered and eliminated from the body as urine. Because many plants are able to use NH_3 or the products of bacterial action on NH_3 as a nitrogen source for protein synthesis, ammonia is a common constituent of fertilizers.

Mineral Salts

Mineral salts are composed of small ions. They are essential for the survival and functioning of the body's cells. They function in numerous ways as parts of enzymes or as portions of the cellular environment necessary for enzyme or protein action. Calcium (Ca^+) is necessary for muscle contraction and nervous transmission as well as building strong bones. It is the fifth most abundant element in the body. Phosphate (PO_4^-) is necessary to produce the high-energy molecule ATP. Chloride (Cl^-) is necessary for nervous transmission. Sodium (Na^+) and potassium (K^+) are also necessary for muscle cell contraction and nervous transmission.

Carbohydrates

Carbohydrates (kar-boh-**HIGH**-draytz) are made up of the atoms of carbon, hydrogen and oxygen in a 1:2:1 ratio (e.g., glucose or $C_6H_{12}O_6$). The smallest carbohydrates are the simple sugars that cannot be made to react with water to produce a simpler form. Sugars are generally chains of either five or six carbon atoms. Important five-carbon sugars are **ribose** and **deoxyribose**, which are parts of the RNA and DNA **nucleic acid** molecules. Important six-carbon sugars are **glucose** and **fructose** (the suffix *ose* denotes a sugar) (Figure 2-6). Note the repetition of the H-C-OH unit in the molecule. This is typical of sugars. Starch, **glycogen** (animal starch), cellulose (the material of plant cell walls that forms fiber in our diets), chitin (**KYE**-tin) (the exoskeleton of arthropods such as insects and lobsters), as well as many other complex carbohydrates, are formed by bonding together a number of glucose molecules. Besides glucose there are other six-carbon sugars. Combinations of these with glucose result in another series of sugars such as common table sugar or sucrose, a disaccharide.

Carbohydrates have two important functions: energy storage (sugars, starch, glycogen) and cell strengthening (cellulose of plant cell walls and chitin in the external skeleton of arthropod animals). Energy storage is the more common function of carbohydrates.

Lipids

There are a number of different kinds of **lipids**. Lipids are substances that are insoluble in water. Fats, phospholipids, steroids, and prostaglandins are examples of these different kinds of molecules. We will concentrate on fats, which are a major kind of lipid. Of the fats in the human body, 95% are triglycerides, now called **triacylglycerols** (try-**ass**-il-**GLISS**-er-allz). They consist of two types of building blocks: **glycerol** and **fatty acids.** Glycerol is a simple molecule similar to a sugar except that it has only a three-carbon chain. Each carbon of the chain is bonded to a hydrogen and a **hydroxyl** ($^-$OH) **group** as well as to the carbons of the chain (Figure 2-7). Fatty acids are composed of long chains of carbon atoms of different lengths. All the carbon atoms are bonded to hydrogen atoms

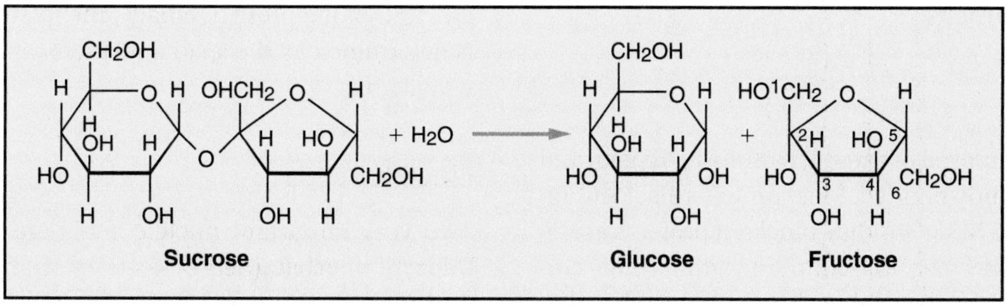

Sucrose + H_2O → Glucose + Fructose

FIGURE 2-6. The chemical structure of the six-carbon sugars, glucose and fructose. When combined, they produce the disaccharide sucrose.

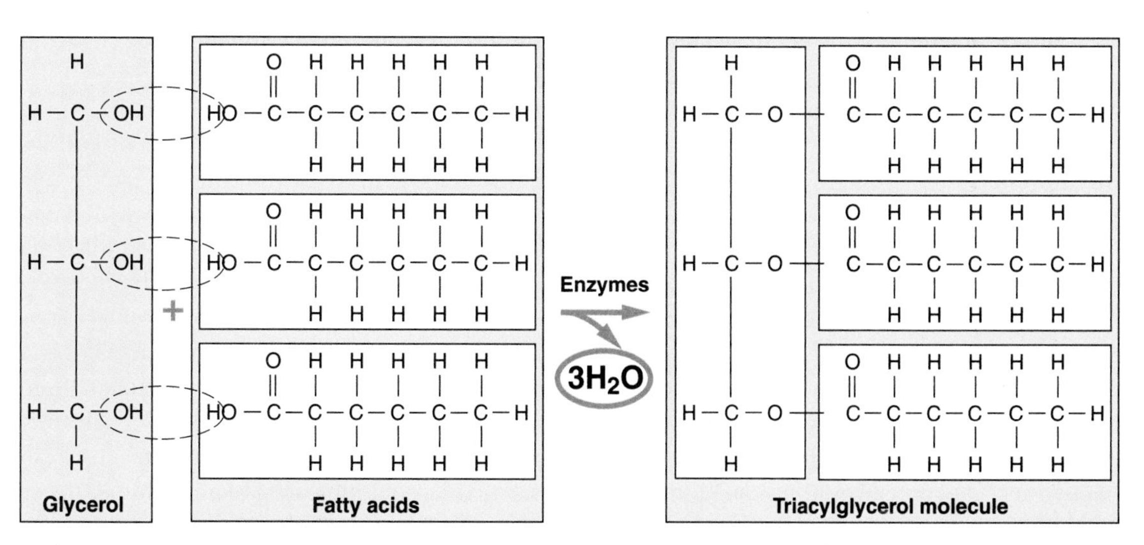

FIGURE 2-7. The structure of a fat like triacylglycerol is composed of a glycerol molecule and fatty acids.

except the carbon at one end of the chain. This carbon atom is bonded to the **carboxyl** (^-COOH) **group**, which makes these molecules slightly acidic. Most naturally occurring fatty acids contain an even number of carbon atoms, 14 to 18. A fatty acid is **saturated** if it contains only single covalent bonds such as those found in whole milk, butter, eggs, beef, pork, and coconut and palm oils. Too much of these fatty acids contributes to cardiovascular disease. Saturated fats tend to be solids at room temperature. However, if the carbon chain has one or more double covalent bonds between the carbon atoms, it is an **unsaturated** fatty acid. These fatty acids are good for you and are found in sunflower, corn, and fish oils. Unsaturated fats tend to be liquids at room temperature. Fats have a number of major roles in the body. Like carbohydrates they contain stored chemical energy. Fat found under the skin acts as an insulator to prevent heat loss. Any animal that lives in the Arctic or Antarctic region (polar bears, seals, whales, or penguins) has a thick layer of insulatory fat. The camel's hump is a thick deposit of fat to protect its internal organs from excessive rises in temperature in the hot desert. Fat also protects organs as a surrounding layer such as the layer around our kidneys to protect them from severe jolts.

Proteins

Proteins are composed of carbon, hydrogen, oxygen, and nitrogen covalently bonded. Most proteins also contain some sulfur. The basic building blocks of proteins are 20 amino acids. They vary in both the length of their carbon chain backbones and the atoms connected to that backbone. However, each amino acid has a carboxyl group (^-COOH), an **amine group** ($^-NH_2$), a hydrogen atom, and the R group. The R group refers to the different types of atoms and length of the chain (Figure 2-8). Covalent bonds form between different amino acids to form proteins. These are referred to as **peptide bonds** (Figure 2-9).

Proteins function in a number of very important ways in the human body. Many are structural proteins. Proteins are part of a cell's membranous structures: plasma membrane, nuclear membrane, endoplasmic reticulum, and mitochondria. In addition, actin and myosin are structural proteins found in a muscle cell. We could not move, talk, breathe, digest, or circulate blood without the proteins actin and myosin. Chemical reactions inside a cell allow a cell to function properly. These chemical reactions would not occur in cells without the assistance of **enzymes.** Enzymes are protein **catalysts,** which increase the rate of a chemical reaction without being affected by the reaction. In addition, our immune system functions because antibodies, which are proteins of a high molecular weight, are formed to combat foreign proteins called antigens that enter the body. Some examples of foreign proteins are bacterial cell membranes, virus protein coats, and bacterial flagella. Finally, proteins are also a source of energy that can be broken down and converted to ATP just like carbohydrates and fats.

Proteins are also discussed in terms of their structure (Figure 2-10). The **primary structure** of a protein is determined by its amino acid sequence. The **secondary structure** is determined by the hydrogen bonds between amino acids that cause the protein to coil into helices or pleated sheets. This shape is crucial to the functioning of proteins. If those hydrogen bonds are destroyed, the protein becomes nonfunctional. Hydrogen bonds can be broken by high temperatures or increased acidity, resulting in changes in pH. The **tertiary structure** is a secondary folding caused by interactions within the peptide bonds and between sulfur atoms of different amino acids. Changes affecting this structure can also affect the function of the protein. Finally, the **quaternary structure** is determined by the spatial relationships between individual units.

Nucleic Acids

Two very important nucleic acids are found in cells. **Deoxyribonucleic** (dee-**ock**-see-rye-boh-noo-**KLEE**-ik) **acid (DNA)** is the genetic material of cells located in the nucleus of the cell. It determines all of the functions and characteristics of the cell. **Ribonucleic** (**rye**-boh-noo-**KLEE**-ik) **acid (RNA)** is structurally related to DNA. Two important types of RNA are **messenger RNA** and **transfer RNA,** which are important molecules necessary for protein synthesis (discussed in Chapter 3).

The nucleic acids are very large molecules made of carbon, oxygen, hydrogen, nitrogen, and phosphorous atoms. The basic structure of a nucleic acid is a chain of **nucleotides.** The DNA molecule is a double helical chain, and the RNA molecules are single chains of nucleotides. A nucleotide is a complex combination of a sugar (deoxyribose in DNA and ribose in RNA), a nitrogen base, and a phosphate group bonded to the sugar. There are two categories of nitrogen bases, which consist of a complex ring structure of carbon and nitrogen atoms. **Purines** consist of a fused double ring of nine atoms. The two purine nitrogen bases are adenine and guanine. **Pyrimidines**

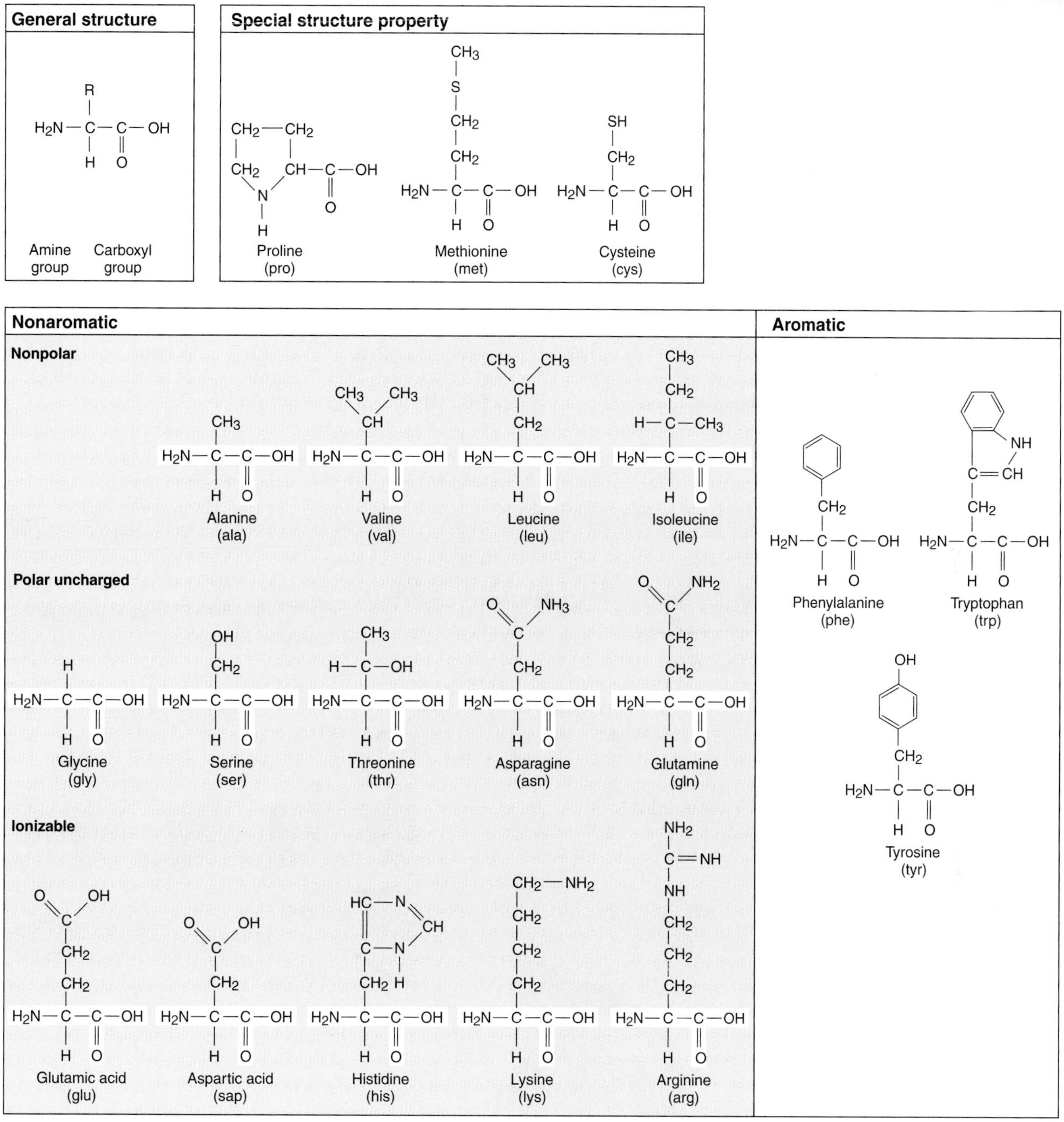

FIGURE 2-8. The general structure of an amino acid and the list of the 20 amino acids found in the human body.

consist of a single ring of six atoms. The three pyrimidine nitrogen bases are thymine, cytosine, and uracil (Figure 2-11). The DNA molecule has adenine, thymine, guanine, and cytosine. The RNA molecule substitutes uracil for thymine and also has adenine, cytosine, and guanine. In the DNA molecule, adenine joins thymine, whereas cytosine always joins guanine in forming the double helical chain. We will discuss this structure in detail in Chapter 4.

Adenosine Triphosphate

Adenosine triphosphate (ATP) is the high-energy molecule or fuel that runs the cell's machinery. All the food we eat (which is a form of chemical energy) must be transformed into another form of chemical energy (ATP) that allows our cells to maintain, repair, and reproduce themselves. The ATP molecule consists of a ribose sugar, the purine adenine, and three phosphate

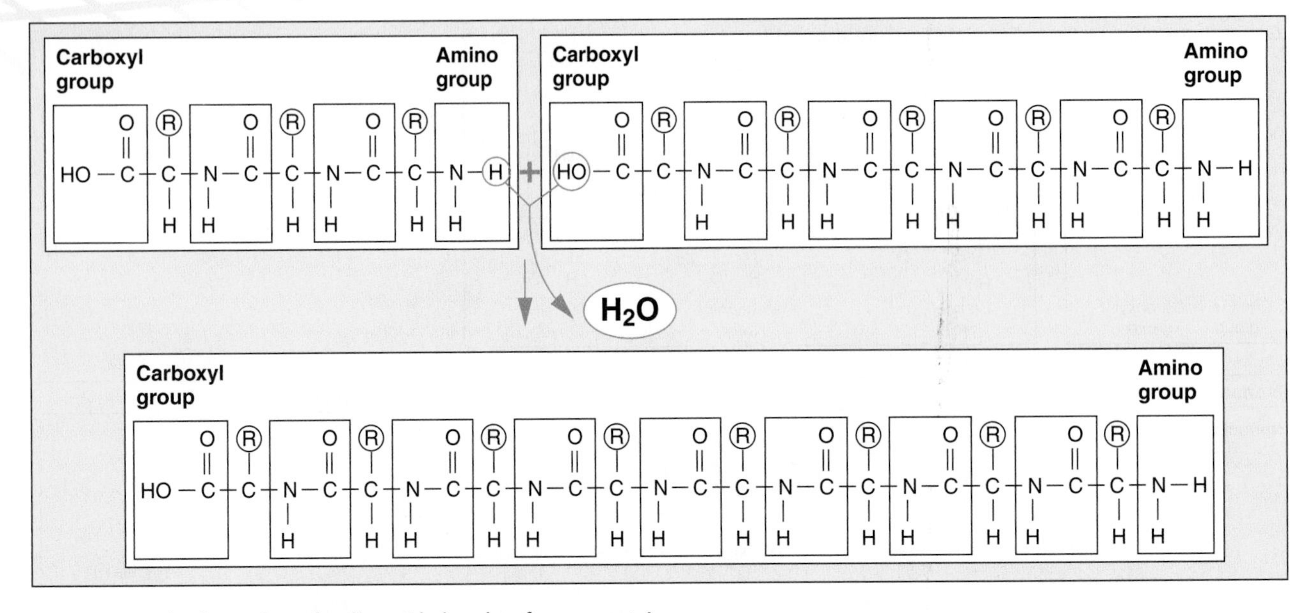

FIGURE 2-9. The formation of a dipeptide bond to form a protein.

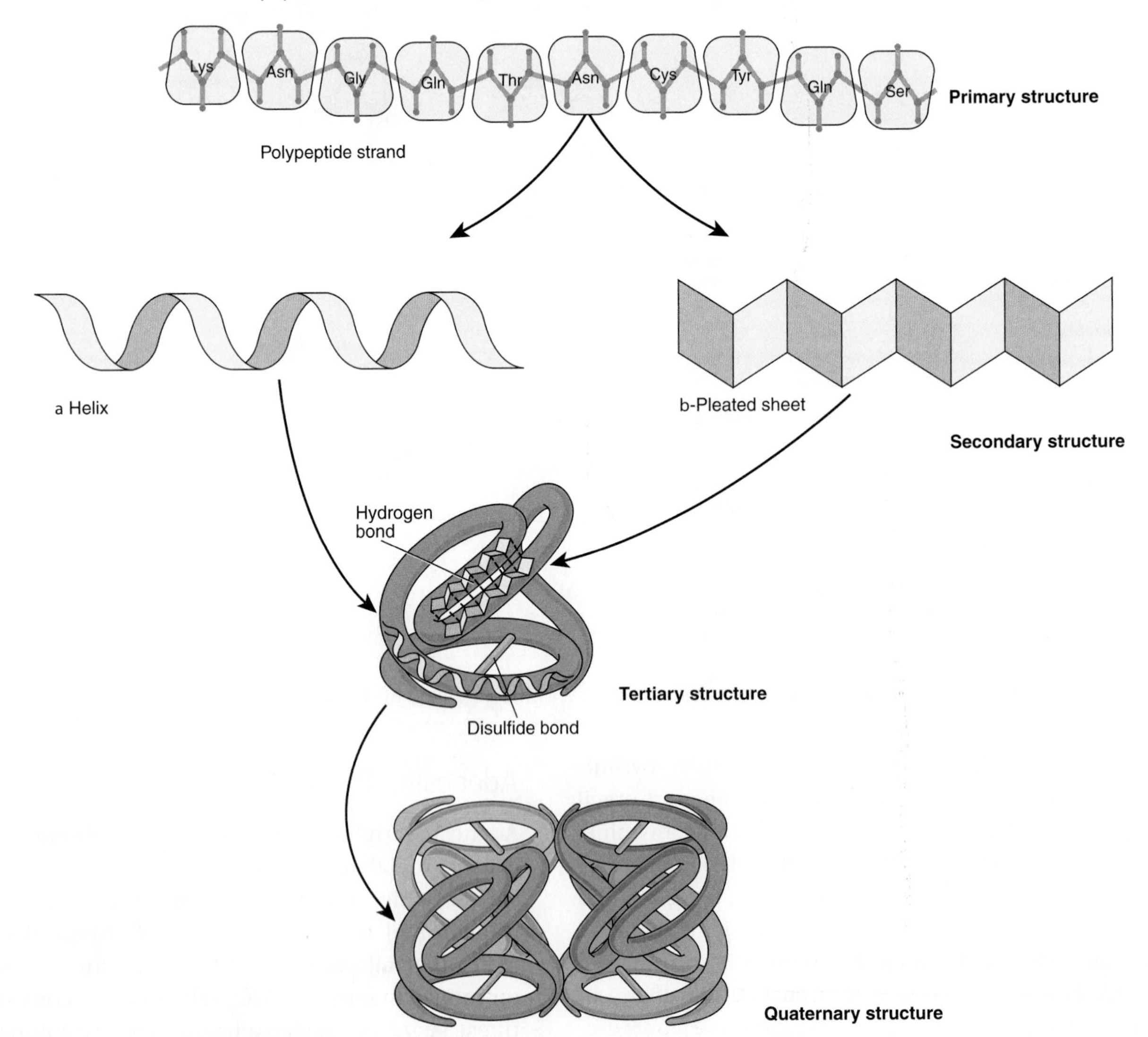

FIGURE 2-10. The four levels of protein structure.

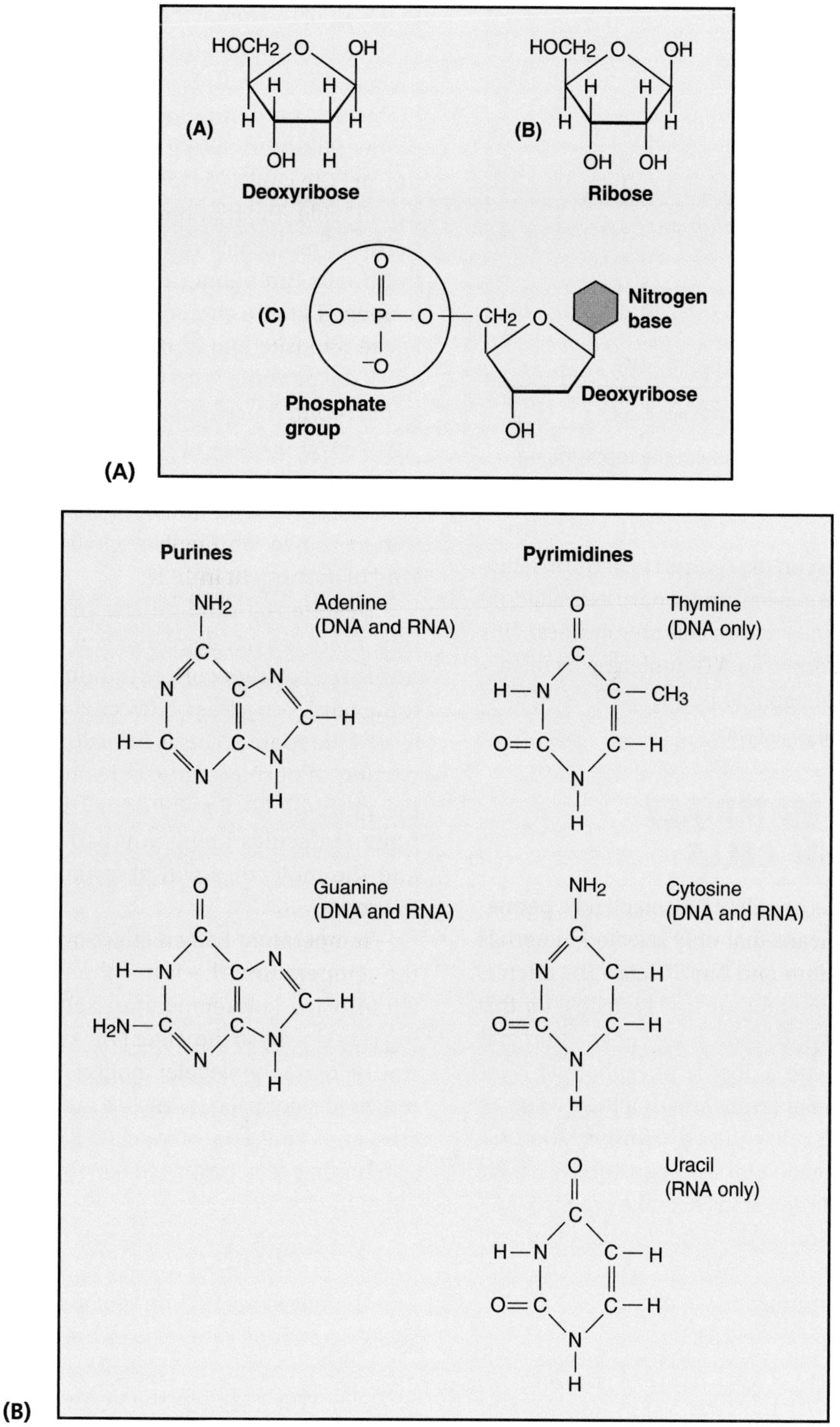

FIGURE 2-11. (A) The structure of a nucleotide and (B) their nitrogen bases.

groups (Figure 2-12). The energy of the molecule is stored in the second and third phosphate groups.

The breakdown of the glucose molecule and other nutrients provides the energy to make ATP molecules (discussed in greater detail in Chapter 4). An ATP molecule is made by putting together an adenosine diphosphate (ADP) with a phosphate group (PO_4): ADP + PO_4 + energy → ATP. The energy stored in the ATP

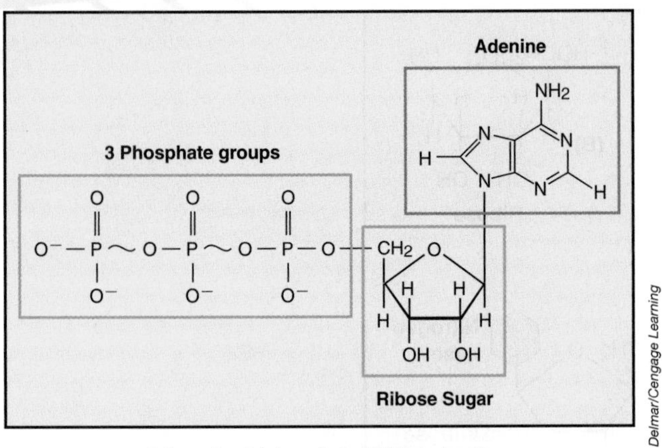

FIGURE 2-12. The structure of the adenosine triphosphate (ATP) molecule.

molecule is then used to run the cell and to perform activities such as structural repair, reproduction, assimilation, and transport of materials across cell membranes. This occurs when we break down an ATP molecule by releasing the energy in the phosphate bonds: ATP $\rightarrow$ ADP + PO_4 + energy (to do cell processes).

MOVEMENT OF MATERIALS INTO AND OUT OF CELLS

The plasma membrane of cells is a **selectively permeable membrane.** This means that only selected materials are capable of getting into and out of cells. The chemical structure of the cell membrane is responsible for this quality. The cell membrane is composed of an outer and inner layer of protein with a double phospholipid layer in between. This chemical arrangement allows water to pass into and out of the cell with ease. However, water is not the only material needed for the cell's survival. Cells need food like sugars, amino acids to make proteins, and nutrients like the mineral salts. Materials pass through the cell's membrane in three different ways: diffusion, osmosis, and active transport.

Diffusion

Diffusion is the movement of molecules through a medium from an area of high concentration of those molecules to an area of low concentration of those molecules. As an example of diffusion, think of a closed perfume bottle in a room. Within the stoppered bottle, perfume molecules are in constant motion; they are in the liquid and the gaseous state. Those in the gaseous state are in faster motion than those in the liquid state. In the air of the room, there are also molecules in motion such as water vapor, oxygen, nitrogen, and carbon dioxide gas. When the perfume bottle is opened, perfume molecules randomly move out of the bottle and randomly bump or collide with those other molecules in the air. The collisions are like bumping billiard balls on a pool table. The random collisions eventually bump the perfume molecules toward the walls of the room and eventually throughout the room. If the perfume bottle is opened at one end of the room and you are standing at the opposite end of the room, you would eventually smell the perfume once the molecules reached your end of the room. A person standing near the perfume bottle when it was opened would smell the perfume molecules before you did. The random collisions of diffusing molecules are referred to as **Brownian movement** after Sir Robert Brown, an English scientist who described this kind of movement in 1827.

Despite the randomness of these collisions, over time there is a net displacement of perfume molecules from areas of high concentration (on and near the perfume bottle) to areas of low concentration (at the other end of the room). This is diffusion. Eventually, the proportion of perfume molecules being bumped back to the perfume bottle will equal the proportion of perfume molecules being bumped away from the bottle and the molecules will be evenly spread throughout the room.

Temperature has an effect on diffusion. The higher the temperature, the faster the movement. Think of a chunk of ice. Low temperature keeps the molecules moving very slowly, so the water is in a solid state. As temperature increases, molecular motion increases and the water moves to a liquid state. The ice melts. Continued heating, such as putting a pot of water on a stove, increases molecular motion even further so that the water becomes water vapor and moves into the gaseous state.

An example of an important diffusion in the human body is the uptake of oxygen by the blood in the lungs and the release of carbon dioxide gas to the lungs from the blood. Blood returning to the lungs is low in oxygen but high in carbon dioxide gas as a result of cellular respiration. When we breathe in air, we take in oxygen gas, so the lungs have lots of oxygen but little carbon dioxide gas. The oxygen moves from an area of high concentration (the lungs) to an area of low concentration (the blood) by diffusion. Similarly, the carbon dioxide gas moves from an area of higher concentration (the blood) to an area of low concentration (the lungs) by diffusion. We exhale to get rid of the carbon dioxide gas now in the lungs.

Osmosis

Osmosis (oz-**MOH**-sis) is a special kind of diffusion. Osmosis pertains only to the movement of water molecules through a selectively permeable membrane (e.g., a plasma membrane) from an area of high concentration of water molecules (e.g., pure water) to an area of low concentration of water molecules (e.g., water to which a **solute** such as salt or sugar has been added).

Osmosis can be demonstrated fairly simply by separating pure distilled water with a selectively permeable membrane (a barrier that will allow only water to pass through it but not solutes such as salt) and adding a 3% salt solution to the water on the other side of the membrane (Figure 2-13). The water level on the solute side will rise, and the water on the pure water side will drop. The rise in water on the open-ended flask tube opposes atmosphere pressure and gravity and will eventually stop rising. At this equilibrium level, the number of water molecules entering the solute area equals the number of water molecules leaving the solute area. The amount of pressure required to stop osmosis is a measure of osmotic pressure. The solution stops rising when the weight of the column equals the osmotic pressure.

The mechanism of osmosis is simple. The salt in the column of water in solution cannot pass through the selectively permeable membrane. Salt is in higher concentration in the solution. Water is in lower concentration in the column because salt has been added to the water. However, the water in the beaker is pure distilled water; there are no solutes in it. The water, which can move through the selectively permeable membrane, causes the observed increased height of the water column in the flask. The water "tries" to equalize its concentration in both the beaker and the flask. Thus, the water moves from an area of high concentration in the beaker through the selectively permeable membrane to an area of low concentration (the salt solution in the flask).

Many biologic membranes are selectively permeable, such as the membranes of cells. The effects of osmosis on red blood cells can easily be demonstrated

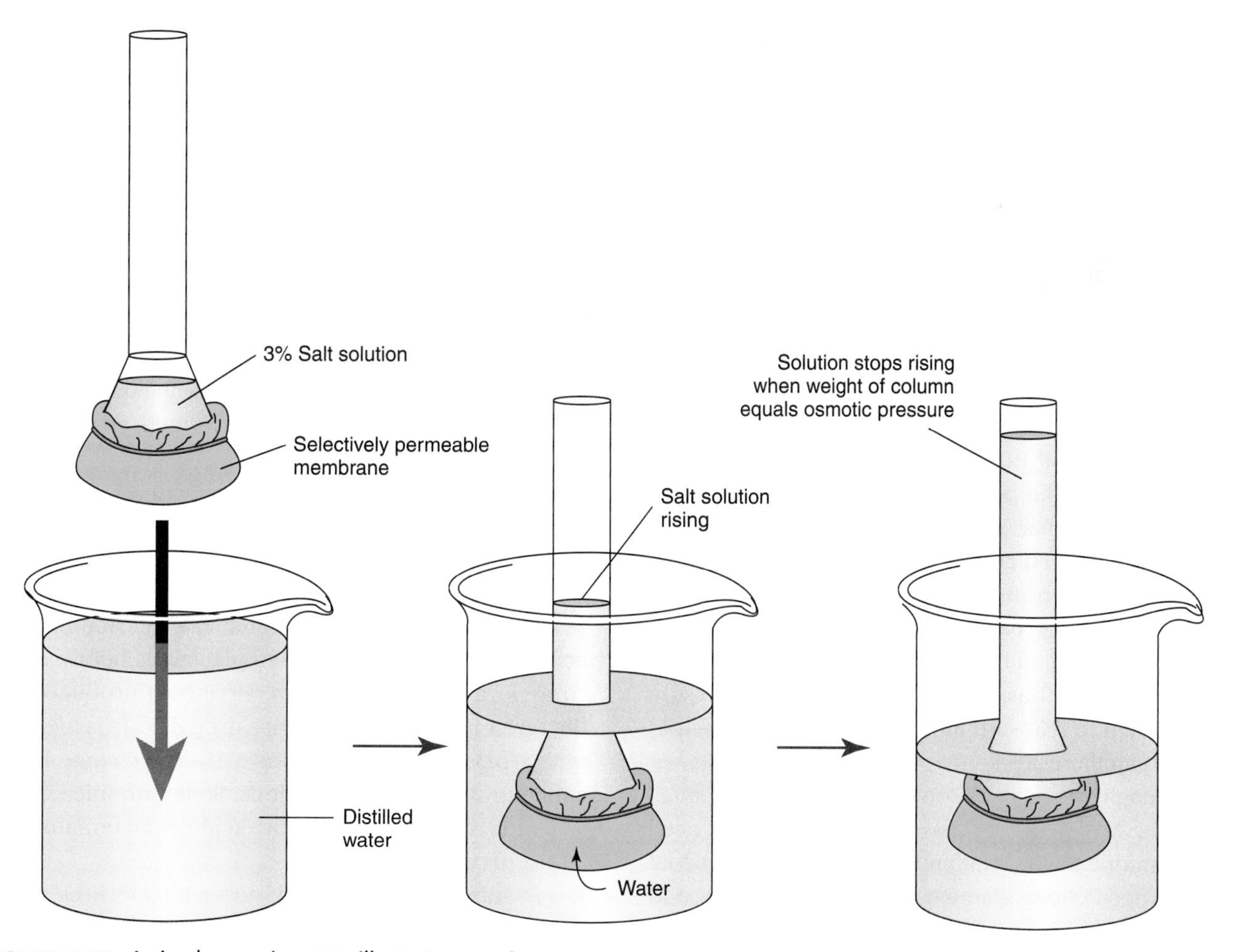

FIGURE 2-13. A simple experiment to illustrate osmosis.

HEALTH ALERT

ACID RAIN

We are all aware of the term **acid rain**. Excess industrial pollutants emitted into the air from coal-fired power plants and automobiles can change the pH of our environment. These pollutants fall back to the earth as acid precipitation (rain, snow, or fog). This *acid rain* can cause respiratory problems when breathed in or gastric problems when reaching the stomach and digestive systems. We are aware of how serious this problem can be when we see lakes whose fish have all been killed or whole forests destroyed near industrial plants with high-pollution rates. Acid precipitation is of global concern.

Buildings and monuments made of limestone (calcium carbonate) are easily eroded by even weak acids. The United States passed and implemented the Clean Air Act of 1990 to help lower the levels of acid precipitation to protect our health and our environment. ■

(see Figure 2-14). If a red blood cell is placed in a normal saline solution (an **isotonic solution**) where the salt concentration outside the red blood cell equals the salt concentration inside the red blood cell, water molecules will pass into and out of the red blood cell at an equal rate, and there will be no observed change in the shape of the red blood cell (see Figure 2-14A). If, however, the red blood cell is placed in pure distilled water (a **hypotonic solution**) where the water molecules are in a higher concentration outside the red blood cell, water will move into the red blood cell, causing it to swell and eventually rupture (see Figure 2-14B). If the red blood cell is placed in a 5% salt solution (a **hypertonic solution**) where there is more water inside the red blood cell than in the solution, the red blood cell will lose water to the solution and will shrivel up or crenulate (see Figure 2-14C).

Because blood in the circulatory system is under pressure due to the beating of the heart, much blood plasma (the fluid part of blood, which is predominantly water with dissolved and colloidal suspended materials in it) is lost into surrounding tissues in the highly permeable one-cell-thick capillaries. Colloidally suspended proteins in the blood cannot pass through the capillary cell membranes; thus, they cause an osmotic pressure large enough to reabsorb most of the fluid that escapes from the capillaries.

Although water and a few other substances with small molecular weights can osmose into the cells that need them, osmotic transportation is insufficient for most of the cell's needs. Sugars, amino acids, larger proteins, and fats are needed by the cell to produce ATP and to maintain and create structure. Cells obtain these nonosmotic or nondiffusable materials by a special mechanism called **active transport**. This mechanism, however, needs energy in the form of ATP to overcome the osmotic/diffusional barriers—another major reason that ATP is so important to a cell's survival. Active transport is the transportation of materials against a concentration gradient or in opposition to other factors that would normally keep the material from entering the cell. Molecules move from an area of low concentration to an area of high concentration (like a food vacuole).

PH

pH is defined as the negative logarithm of the hydrogen ion concentration in a solution: $pH = -\log [H^+]$. Pure water has a pH of 7. Remember that when distilled water (H_2O) dissociates, for every H^+ ion formed, an OH^- ion is also formed. Or, in other words, the dissociation of water produces H^+ and OH^- in equal amounts. Therefore, a pH of 7 indicates neutrality on the pH scale. Figure 2-15 shows the pH of various solutions.

If a substance dissociates and forms an excess of H^+ ions when dissolved in water, it is referred to as an **acid**. All acidic solutions have pH values below 7. The stronger an acid is, the more H^+ ions it produces and the lower its pH value. Because the pH scale is logarithmic, a pH change of 1 means a 10-fold change in the concentration of hydrogen ions. So lemon juice with a pH value of 2 is 100 times more acidic than tomato juice with a pH of 4.

A substance that combines with H^+ ions when dissolved in water is called a **base** or alkali. By combining with H^+ ions, a base therefore lowers the H^+ ion

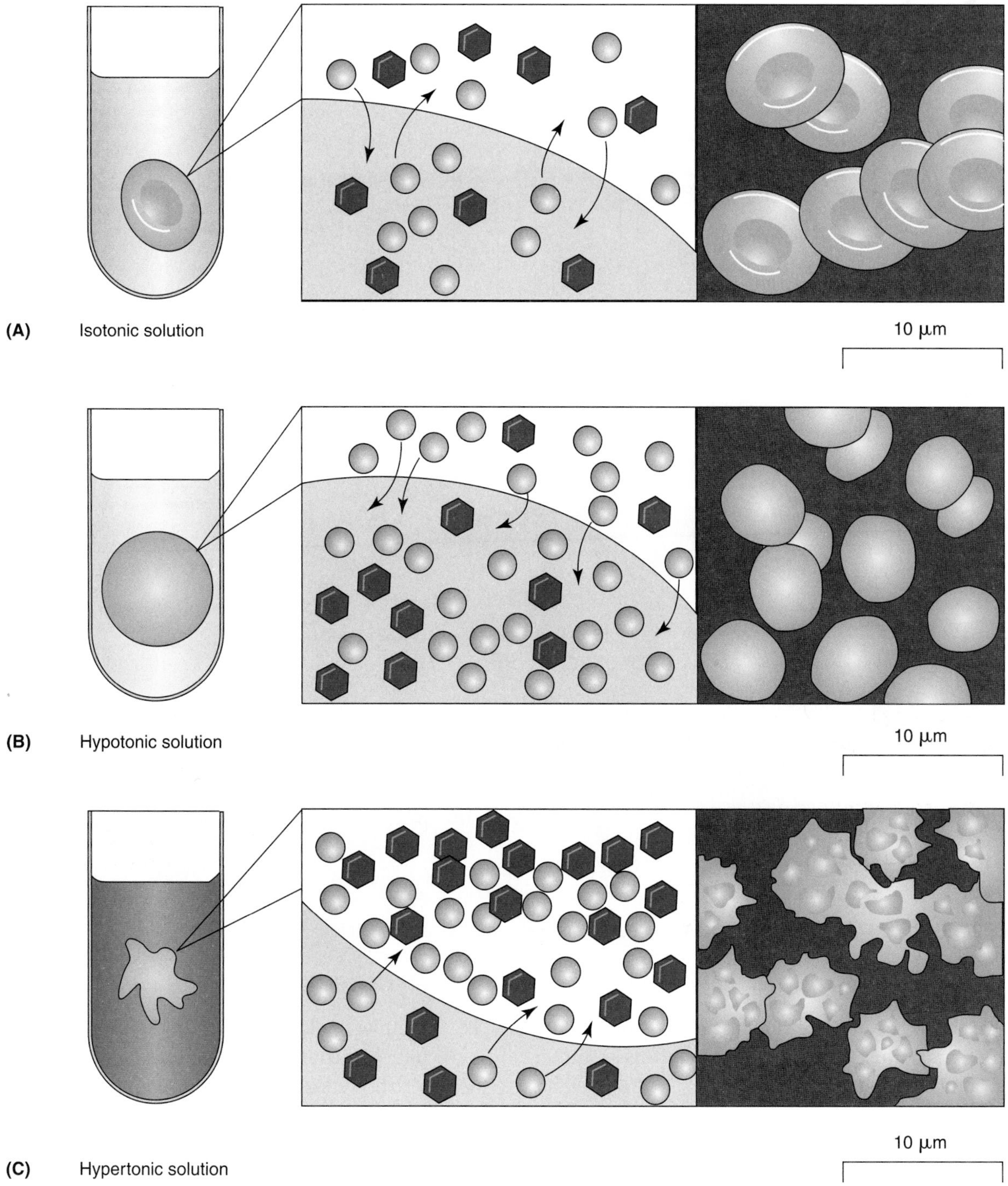

(A) Isotonic solution

10 μm

(B) Hypotonic solution

10 μm

(C) Hypertonic solution

10 μm

FIGURE 2-14. The effects of osmosis on red blood cells when placed in different concentrations of salt solutions.

concentration in that solution. Basic, also called alkaline, solutions have pH values above 7. Seawater with a pH of 8 is 10 times more basic than pure distilled water with a pH of 7. In our bodies, saliva in our mouths has a pH value slightly lower than 7 so it is just slightly acidic, whereas the stomach with its gastric juice and hydrochloric acid is very acidic with a pH value near 1. Our blood on the other hand has a pH value of 7.4, making it just slightly basic. Urine has a pH of 6, which, although acidic, is not as acidic as tomato juice with a pH of 4.

The pH inside most cells and in the fluid surrounding cells is fairly close to 7. Because enzymes are extremely

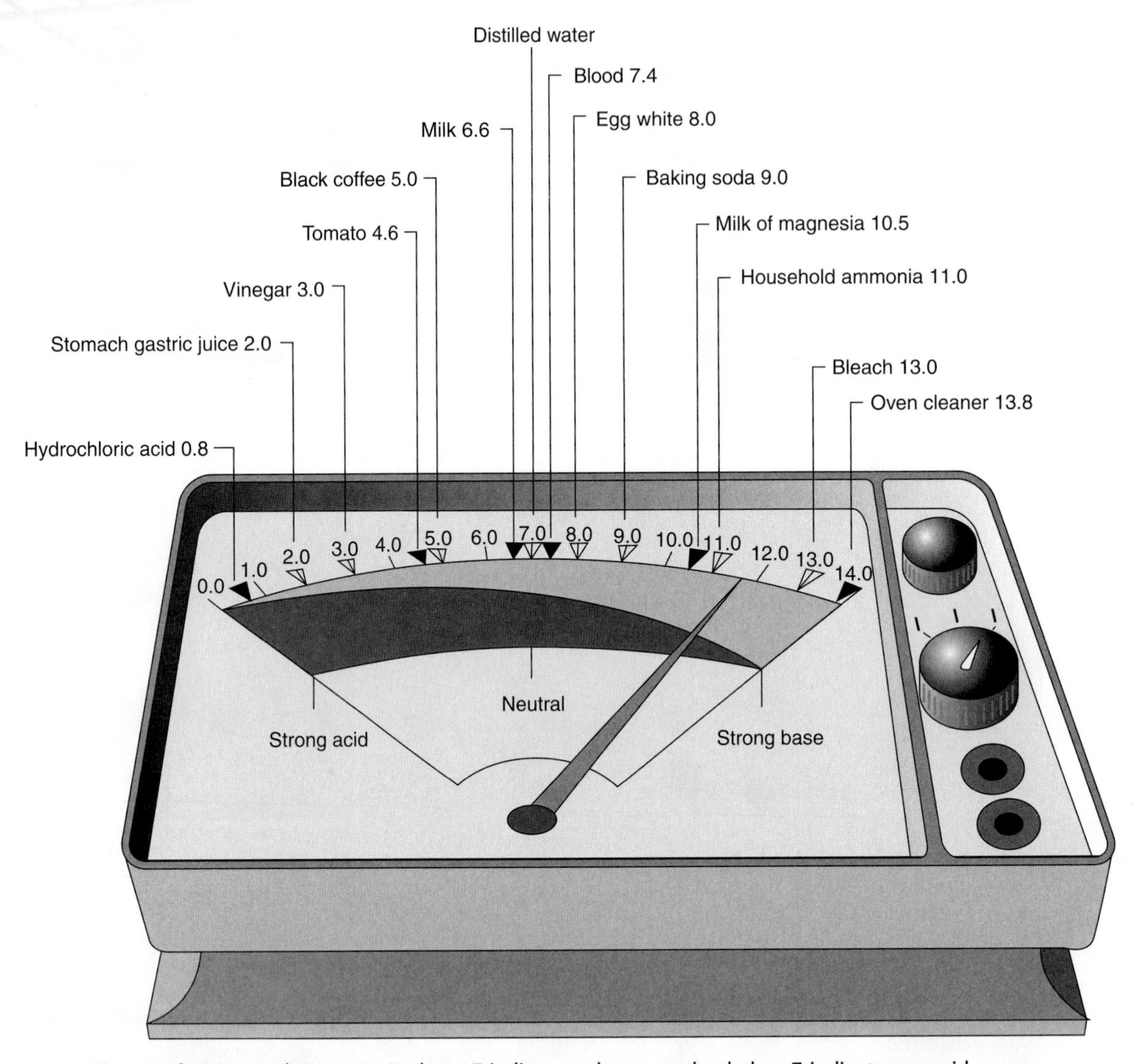

FIGURE 2-15. The pH of various solutions. A pH above 7 indicates a base; a value below 7 indicates an acid.

sensitive to pH, even a small change can render them nonfunctional; thus, our bodies have **buffers**. A buffer is a substance that acts as a reservoir for hydrogen ions, donating them to a solution when their concentration falls, and taking the hydrogen ions from a solution when their concentration rises. Buffers are necessary because the chemical reactions in cells constantly are producing acids and bases. Buffers help maintain homeostasis within cells in regard to pH levels. Most buffers consist of pairs of substances, one an acid and the other a base. For example, the key buffer in human blood is the acid-base pair bicarbonate (a base) and carbonic acid (an acid). Carbon dioxide and water combine chemically to form carbonic acid (H_2CO_3). The carbonic acid then can dissociate in water, freeing H^+ ions and bicarbonate ions HCO_3^-. The blood's pH can be stabilized by the equilibrium between these forward and reverse reactions that interconvert the H_2CO_3 carbonic acid and the HCO_3^- bicarbonate ion (base).

There are many career opportunities for individuals interested in chemistry.

- **Lab Technicians** work as assistant researchers in industry and universities.
- **High School Teachers** specialize in teaching the basic understanding of chemistry to young adolescents.
- **College Professors** conduct research, teach adults upper-level principles of chemistry, and mentor undergraduate and graduate chemical research projects.
- **Organic Chemists** specialize and study the chemistry of the carbon atom and all the compounds that have carbon as part of their molecular structure.
- **Environmental Chemists** study the effects of chemical pollution of the environment.
- **Biochemists** study the chemical basis of life chemistry in genetics, molecular biology, microbiology, or food technology.

SUMMARY OUTLINE

INTRODUCTION

1. Because the body's cells, tissues, and organs are all composed of chemicals and function through chemical reactions, it is necessary to understand some basic chemistry.
2. Chemistry is the science that studies the elements, their compounds, the chemical reactions that occur between elements and compounds, and the molecular structure of all matter.

ATOMIC STRUCTURE

1. Atoms are the smallest particles of elements that maintain all the characteristics of that element and enter into chemical reactions through their electrons.
2. An atom consists of a nucleus containing positively charged protons and neutral neutrons.
3. Electrons have a negative charge and orbit the nucleus of an atom in levels at some distance from the compact heavy nucleus.

ELEMENTS, ISOTOPES, COMPOUNDS

1. An element is a substance whose atoms all contain the same number of protons and the same number of electrons. Atoms are electrically neutral.
2. A compound is a combination of the atoms of two or more elements.
3. An isotope is a different kind of atom of the same element where the number of neutrons in the nucleus varies.
4. Carbon is the element found in all living matter.
5. The periodic table of the elements arranges elements in categories with similar properties.

BONDS AND ENERGY

1. Atoms combine chemically with one another to form bonds by gaining, losing, or sharing electrons.
2. An ionic bond is formed when one atom gains electrons while the other atom in the bond loses electrons. Ionically bonded molecules disassociate when immersed in water. The mineral salts form ionic bonds.
3. A covalent bond is formed when atoms share electrons. Carbon, oxygen, hydrogen, and nitrogen form covalent bonds. These bonds do not dissociate when placed in water.
4. Hydrogen bonds are weak bonds. They help hold water molecules together and bind other molecules into three-dimensional shapes.
5. Molecules furnishing electrons during a chemical reaction are called electron donors; those that gain electrons are called electron acceptors.
6. Bonds contain energy. It is the electrons that contain the energy of a chemical bond.

7. Special molecules called electron carriers accept electrons for a short period of time and use the energy of the electrons to make ATP molecules.

COMMON SUBSTANCES IN LIVING SYSTEMS

The common substances found in living systems are water, carbon dioxide, oxygen, ammonia, mineral salts, carbohydrates, lipids, proteins, nucleic acids, and adenosine triphosphate.

Water

1. Sixty to 80% of a cell is water (H_2O). Water is a slightly polar molecule: the two hydrogen atoms have a partial positive charge and the oxygen atom a partial negative charge. This explains why ionically bonded molecules dissociate when placed in water.
2. Water has many important roles in cells: it takes part in some reactions; it serves as a medium or solvent for other reactions to occur in; it serves as a basis for the transportation of materials; it absorbs and releases heat, maintaining body temperature; it protects; and it is the base for all body lubricants.

Carbon Dioxide

1. Carbon dioxide gas (CO_2) is produced as a waste product of cellular respiration.
2. It is necessary for plants to produce oxygen gas in the photosynthetic reaction, which converts the sun's radiant energy into usable chemical energy like glucose for plant and animal survival.
3. All of the carbon in the carbon-containing molecules of life comes either directly or indirectly from carbon dioxide gas.

Oxygen

1. Molecular oxygen (O_2) is required by all organisms that breathe air.
2. It is necessary for cellular respiration to occur, converting glucose into ATP molecules $C_6H_{12}O_6 + 6O_2 \rightarrow$ ATP (energy) $+ 6CO_2 + 6H_2O$.
3. Oxygen comes from plants in the photosynthesis process $6CO_2 + 12H_2O \rightarrow C_6H_{12}O_6$ (glucose) $+ 6O_2 + 6H_2O$.

Ammonia

1. Ammonia (NH_3) is produced as a by-product of the breakdown of amino acids.
2. Amino acids contain nitrogen and are the building blocks of proteins.
3. Ammonia, which is toxic, is converted to harmless urea by enzymes in our liver.

Mineral Salts

1. The mineral salts are calcium (Ca^+), phosphate ($PO4^-$), chloride (Cl^-), sodium (Na^+), and potassium (K^+).
2. Calcium is needed for muscle contraction and strong bones.
3. Phosphate is needed to make ATP.
4. Sodium, potassium, and chloride are necessary for muscle contraction and nervous transmission.

Carbohydrates

1. Carbohydrates are composed of carbon, hydrogen, and oxygen in a 1:2:1 ratio. The simplest carbohydrates are five and six-carbon sugars.
2. Important five-carbon sugars are deoxyribose and ribose; important six-carbon sugars are glucose and fructose.
3. Carbohydrates have two important functions: energy storage and structural strengthening of the cell.

Lipids

1. Fat is a major type of lipid; 95% of fats in the human body are triacylglycerols, which are composed of glycerol and fatty acids.
2. A fat is called saturated if the fatty acids contain single covalent bonds. These can contribute to cardiovascular disease. A fat is called unsaturated if the fatty acids have one or more double covalent bonds. These are good for you.
3. Fats are a source of energy, act as insulators for the body, and protect organs.

Proteins

1. Proteins contain carbon, oxygen, hydrogen, nitrogen, and sulfur.
2. Amino acids are the building blocks of protein.
3. Proteins are a source of energy. There are structural proteins like actin and myosin in muscle cells, and proteins are an essential part of a cell's membranous structures.
4. Enzymes are protein catalysts that make chemical reactions occur in cells. The functioning of our immune system is based on proteins.
5. Proteins have four types of structure based on bonding: primary (amino acid sequences), secondary (based on hydrogen bonds between amino acids, causing coiling), tertiary (secondary folding based on sulfur atoms), and quaternary (based on spatial relationships between units).

Nucleic Acids

1. Deoxyribonucleic acid (DNA) is the genetic material of the cell found in the nucleus that determines all of the characteristics and functions of the cell.
2. Ribonucleic acid (RNA) exists in two forms necessary for protein synthesis: messenger RNA and transfer RNA.
3. Nucleic acids are composed of chains of nucleotides.
4. A nucleotide is a complex combination of a nitrogen base (purine or pyrimidine), a sugar (deoxyribose), and a phosphate group.
5. The two purine bases are adenine and guanine. The three pyrimidine bases are thymine, cytosine, and uracil (uracil is found in RNA only and is substituted for thymine).

Adenosine Triphosphate

1. ATP is a high-energy molecule that is the fuel that allows cells to function and maintain themselves.
2. The ATP molecule consists of a ribose sugar, adenine, and three phosphate groups. The energy of the molecule is stored in the second and third phosphate groups.
3. The breakdown of the glucose molecule and other nutrients provides the energy to make ATP molecules.

MOVEMENT OF MATERIALS INTO AND OUT OF CELLS

Materials move through plasma membranes in three different ways: diffusion, osmosis, and active transport.

Diffusion

1. Diffusion is the movement of molecules through a medium from an area of high concentration of those molecules to an area of low concentration of those molecules.
2. The random collision of diffusing molecules is called Brownian movement.
3. Increased temperature accelerates the rate of diffusing molecules.
4. An example of diffusion in the human body is the uptake of oxygen by the blood in the lungs and the release of carbon dioxide gas to the lungs from the blood.

Osmosis

1. Osmosis is a special kind of diffusion.
2. Osmosis is the movement of water molecules through a selectively permeable membrane, such as a plasma membrane, from an area of higher concentration of water molecules (e.g., pure water) to an area of low concentration of water molecules (e.g., water to which a solute like salt or sugar has been added).
3. An isotonic solution (e.g., normal saline) is a solution in which the salt concentration outside a cell is the same as that inside a cell. The cell would neither gain nor lose appreciable amounts of water.
4. A hypotonic solution (e.g., pure distilled water) is a solution in which the salt concentration inside the cell is higher than it is outside the cell. The cell would absorb water in such a solution.
5. A hypertonic solution (e.g., a 5% salt solution) is one in which the salt concentration is greater outside the cell than it is inside the cell. The cell would lose water in such a solution.
6. Active transport is the transportation of materials against a concentration gradient in opposition to other factors that would normally keep the material from entering the cell. This mechanism requires energy in the form of ATP and is the main mechanism by which most cells obtain the materials they need for normal functioning.

PH

1. pH is the negative logarithm of the hydrogen ion concentration in a solution: $pH = -\log[H^+]$.
2. If a substance dissociates and forms an excess of H^+ ions when dissolved in water, it is referred to as an acid. Acids have pH values below 7.
3. A substance that combines with H^+ ions when dissolved in water is called a base. Basic solutions have a pH value above 7.
4. Distilled pure water has a pH value of 7 and is neutral.
5. Buffers are special substances that act as reservoirs for hydrogen ions, donating them to a solution when their concentration falls and taking them from a solution when their concentration rises. Buffers help maintain homeostasis within cells in regard to pH levels, keeping them fairly close to 7.

REVIEW QUESTIONS

1. Describe the nature and structure of an atom.
2. List the major chemical elements found in living systems.
*3. Compare ionic and covalent bonding, and indicate which major four elements found in cells bond covalently.
*4. Why is it necessary for a cell to have nucleic acids in its nucleus?
*5. Explain the roles that water plays in living systems.
6. What two major roles do carbohydrates play in living cells?
7. List three functions of fats in the human body.
*8. List four functions of proteins necessary for the function and survival of the human body.
*9. Compare the differences and similarities between osmosis and diffusion and how they function in the body.
*10. What is the significance of active transport to the survival of a cell?
*11. Why is pH important to the maintenance of homeostasis in the body?
12. Discuss the pH of an acid, base, and salt.

***Critical Thinking Questions**

FILL IN THE BLANK

Fill in the blank with the most appropriate term.

1. Molecules that contain carbon, hydrogen, and oxygen are known as ____________. Glucose (a sugar) is one of these molecules.
2. Molecules known as ____________ are the building blocks of protein.
3. All the carbon in the larger organic compounds found in living systems comes directly or indirectly from ____________.
4. The smallest particles of elements that enter into chemical reactions are ____________.
5. An ____________ is a substance whose atoms all contain the same number of protons and the same number of electrons.
6. In the atoms of some elements, the number of neutrons varies. These different kinds of atoms are called ____________.
7. Two kinds of chemical bonds found in living matter are ____________ and ____________ bonds.
8. The most abundant molecule found in living cells is ____________.
9. Carbohydrates have two basic functions: ____________ and ____________.
10. Nucleotides bonded together between the phosphate group of one and the sugar of another form long chain molecules called ____________.

MATCHING

Place the most appropriate number in the blank provided.

_____ NH_3	1. Plant carbohydrate
_____ $C_6H_{12}O_6$	2. Hydroxyl group
_____ Glycogen	3. Amino group
_____ Starch	4. Ammonia
_____ CO_2	5. Glycerol
_____ OH^-	6. Carboxyl group
_____ COOH	7. Inorganic phosphate group
_____ NH_2	8. Animal carbohydrate
_____ PO_4^-	9. Fatty acid
_____ C_5	10. Ribose
	11. Glucose
	12. Carbon dioxide

Search and Explore

- Search the Internet with key words from the chapter to discover additional information and interactive exercises. Key words might include DNA, diffusion, and osmosis.
- Visit http://www.acid-base.com for an acid-base tutorial.

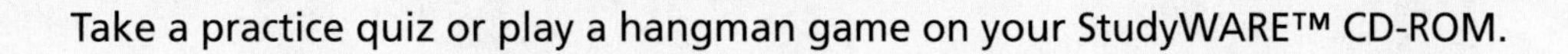

*Study*WARE™ Connection

Take a practice quiz or play a hangman game on your StudyWARE™ CD-ROM.

Study Guide Practice

Go to your **Study Guide** for more practice questions, a labeling exercise, and a crossword puzzle to help you learn the content in this chapter.

LABORATORY EXERCISE: THE CHEMISTRY OF LIFE

Materials needed: A pH meter, osmosis kit, models of chemical molecules

1. pH measurements: With the assistance of a pH meter provided by your instructor, measure the pH of tap water, distilled water, tomato juice, orange juice, apple juice, your saliva, a baking soda solution, ammonia, and household bleach. Prepare a chart to place these items under the basic or acidic category.
2. Demonstration of osmosis: Your instructor will demonstrate the effects of osmosis with the assistance of an osmosis kit from a biologic supply company. The experiment shown in Figure 2-13 can be demonstrated using first a 3% salt solution and then a 3% sugar solution.
3. Examine the chemical models, provided by your instructor, illustrating the molecular structures and bonding of some common substances found in living systems.

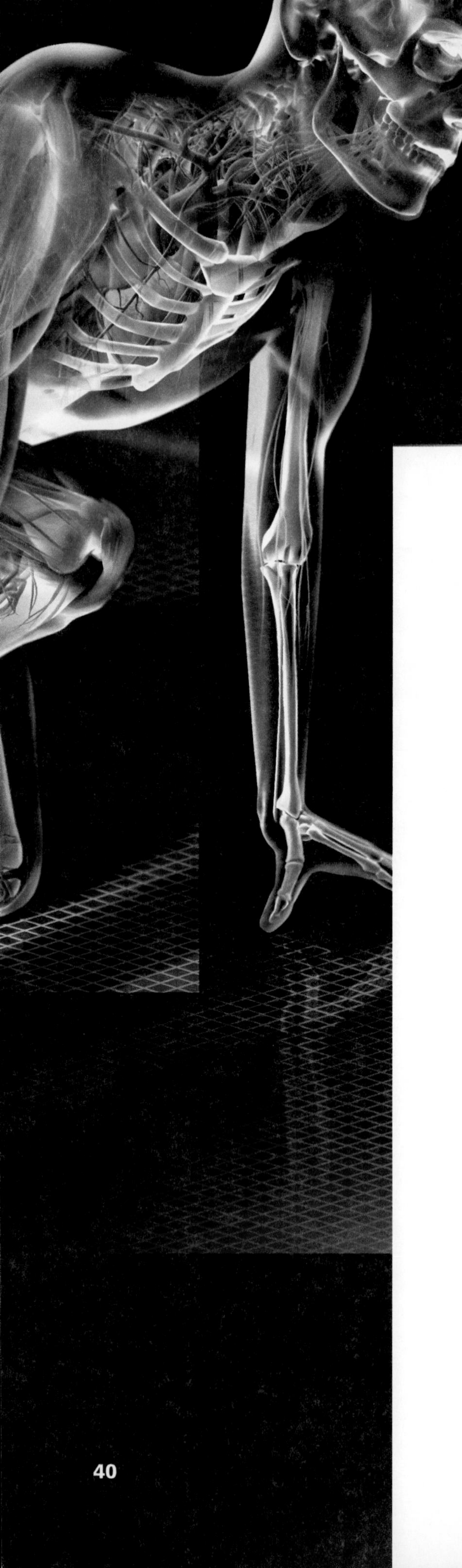

Cell Structure

CHAPTER OBJECTIVES

After studying this chapter, you should be able to:

1. Name the major contributors to the cell theory.
2. State the principles of the modern cell theory.
3. Explain the molecular structure of a cell membrane.
4. Describe the structure and function of the following cellular organelles: nucleus, endoplasmic reticulum, Golgi body, mitochondria, lysosomes, ribosomes, and centrioles.
5. Explain the significance and process of protein synthesis.

3

KEY TERMS

INTRODUCTION

The cell is the basic unit of biologic organization of the human body. During our lifetime our bodies are made up of trillions of cells. Although cells have different functions in the body, they all have certain common structural properties. All cells are composed of **protoplasm,** which is an aqueous colloidal solution of carbohydrates, proteins, lipids, nucleic acids, and inorganic salts surrounded by a limiting cell membrane. This protoplasm (*proto* meaning "first" and *plasm* meaning "formed") is predominantly water with organic compounds in a colloidal suspension and inorganic compounds in solution. These compounds are the building blocks of structures within the protoplasm called **organelles.** Some organelles are common to most cells. Higher cells like those of the human body are called **eukaryotic** cells (eu = true); cells that do not have membrane-bound organelles (e.g., bacteria) are called **prokaryotic** cells.

Organelles that are common to all eukaryotic cells are the nucleus, the mitochondria, the endoplasmic reticulum, ribosomes, the Golgi apparatus, and lysosomes. If a cell has a specialized function that other cells do not have, for example, movement, the cell will have specialized organelles. Cells in our bodies that move materials across their exposed or free surface will be covered with row on row of hundreds of cilia. (For instance, cells in our respiratory tract produce mucus to trap dust and microorganisms that get past the hairs in our nose, then move the material to our throat to be swallowed and passed out through the digestive system.) The human sperm cell, which must travel up the uterus of the female to the upper one-third of the fallopian or uterine tube to fertilize an egg, has a flagellum to propel it along its journey. Plant cells that do photosynthesis (the conversion of light energy into chemical energy, i.e., foods like sugars) have special organelles called chloroplasts.

When one observes a cell under the microscope in a laboratory, the most prominent structure in the cell is the nucleus, which is the control center of the cell. For this reason, the protoplasm of the cell is subdivided into two sections: the protoplasm inside the nucleus is called **nucleoplasm**, and the protoplasm outside the nucleus is called the **cytoplasm**. Cells vary in size and most cells are too small to be seen with an unaided eye. Cells are measured in terms of **microns** (**MY**-kronz), more commonly called **micrometers** (my-**KROM**-ee-terz). One micrometer (μm) equals one-thousandth (10^{-3}) of a millimeter. Most eukaryotic cells range in size from 10 to 100 micrometers in diameter (10 to 100 millionths of a meter). Light microscopes allow us to see general features of cells with magnifications from 10× to 1000×. These are the microscopes we use in the laboratory. However to "see" or study the details of cells, an electron microscope must be used. These microscopes use a beam of electrons to visualize structures and are quite complex. A person must be specially trained to use one of these sophisticated instruments. Our current knowledge of cellular structure comes from research done on cell structure using electron microscopes. The diagrams of a typical animal cell (Figure 3-1) and a typical plant cell (Figure 3-2) illustrate a three-dimensional view of ultrastructure.

*Study*WARE™ Connection

Play an interactive game labeling a typical plant cell on your StudyWARE™ CD-ROM.

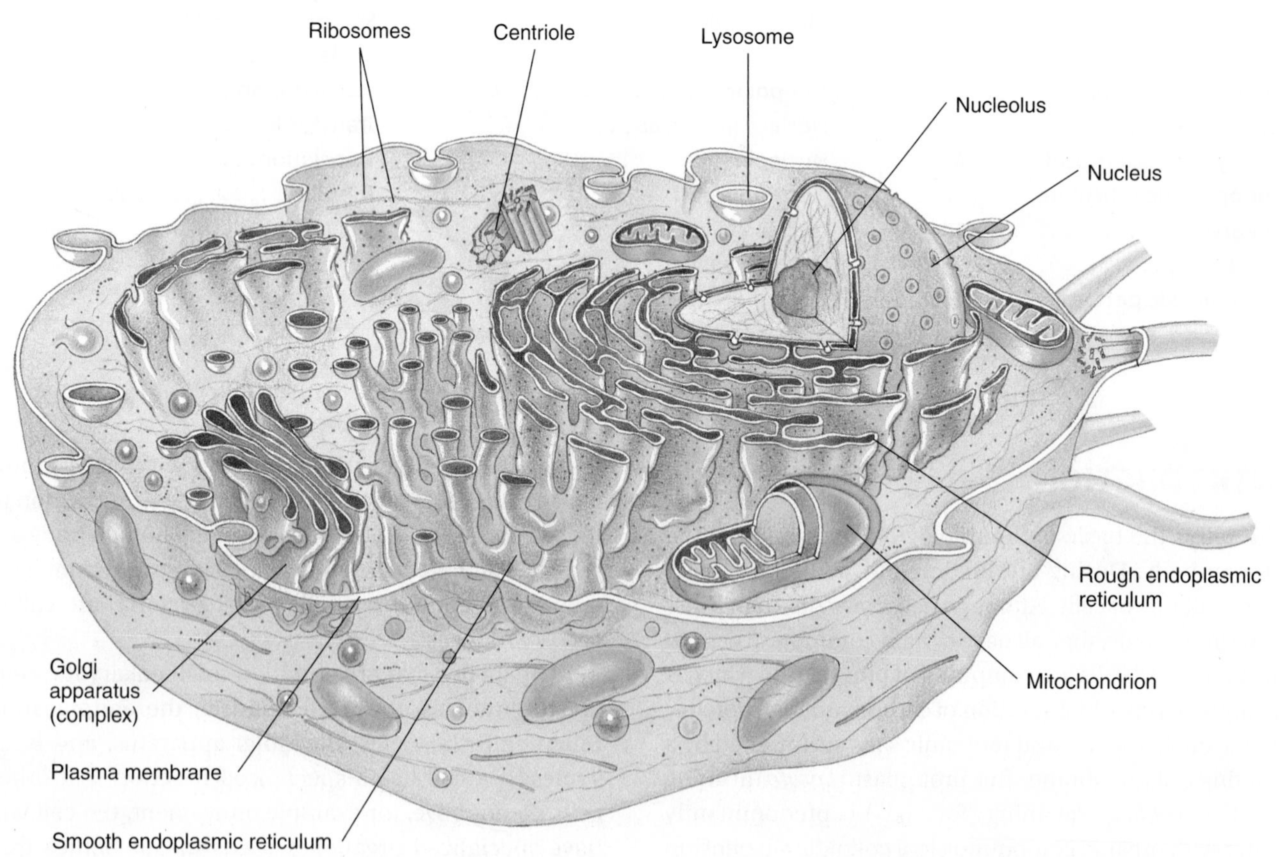

FIGURE 3-1. A diagram of a typical animal cell illustrating a three-dimensional view of cell ultrastructure.

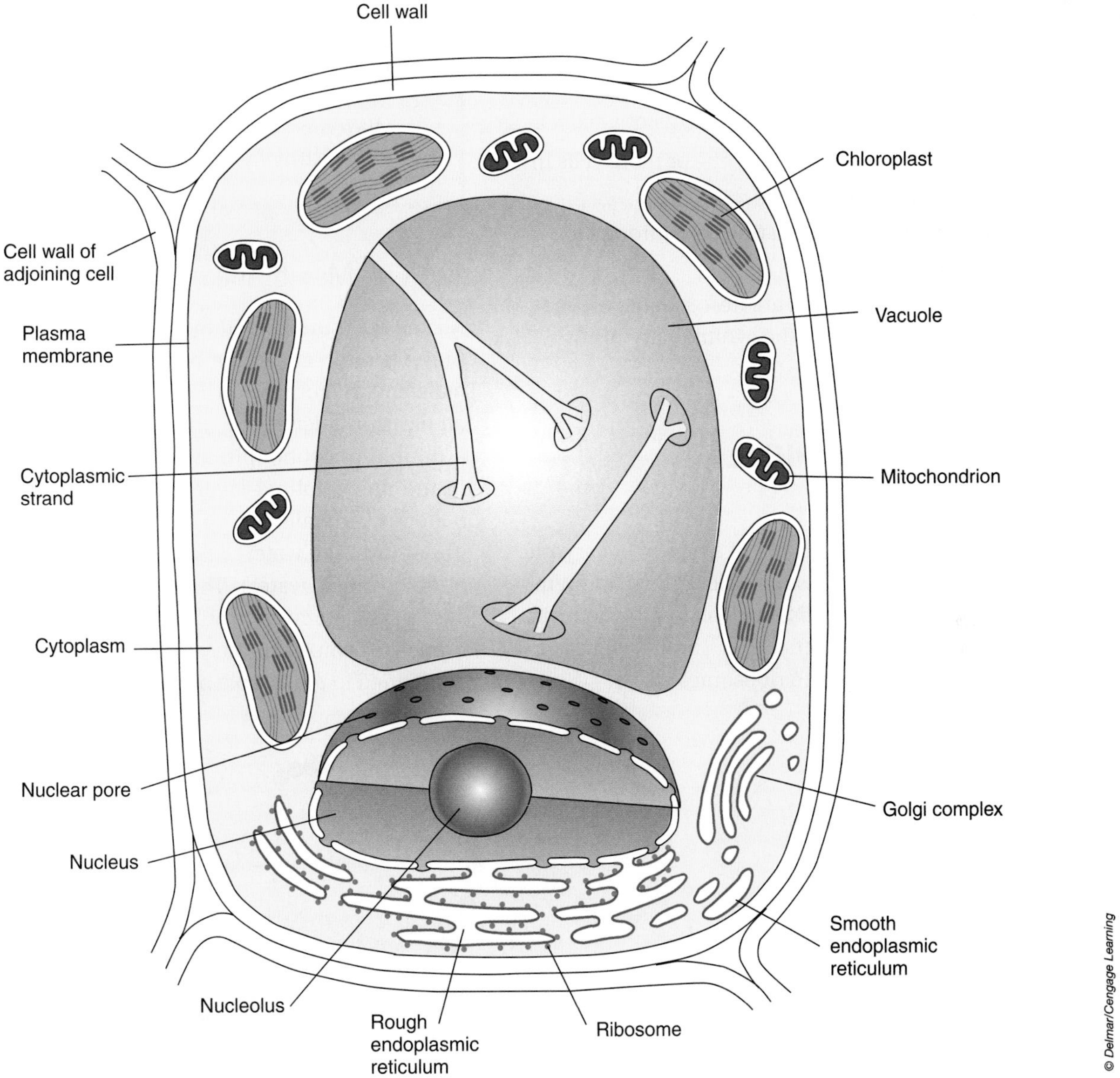

FIGURE 3-2. A diagram of a typical plant cell illustrating a three-dimensional view of cell ultrastructure.

HISTORY OF THE CELL THEORY

Because cells are too small to be seen with the naked eye, they were not observed until the invention of the first microscope in the mid-17th century. Robert Hooke was an English scientist who first described cells in 1665. He built one of the first primitive microscopes to look at a thin slice of cork. Cork is nonliving plant tissue that comes from the bark of trees. Because this was dead plant tissue, he observed the cell walls of dead cells. They resembled tiny rooms, so he called them cellulae (small rooms) from the Latin. Thus, the term *cells* has been used ever since.

Living cells were observed a few years later by the Dutch naturalist Anton von Leeuvenhoek. He observed pond water under his microscope and was amazed at what he saw in what he believed was pure water. He called the tiny organisms in the water animalcules (meaning little animals). It took, however, almost another 150 years before the significance of cells as the building blocks of biologic organization was to take hold.

Two German scientists laid the foundation of what we call today the cell theory. In 1838, Matthias Schleiden, a botanist, after careful study of plant tissues, stated that all plants are composed of individual units called cells. In 1839, Theodor Schwann, a zoologist, stated that all animals are also composed of individual units called cells. Thus the foundation of our modern cell theory was formed.

The modern cell theory consists of the following principles:

1. Cells are the smallest complete living things—they are the basic units of organization of all organisms.
2. All organisms are composed of one or more cells in which all life processes occur.
3. Cells arise only from preexisting cells through the process of cell division.
4. All of today's existing cells are descendants of the first cells formed early in the evolutionary history of life on earth.

ANATOMY OF A TYPICAL EUKARYOTIC CELL

The following structures are parts of a typical eukaryotic cell: cell membrane, cytoplasm, nucleus, nuclear membrane, nucleoplasm, chromatin, nucleolus, mitochondria, lysosomes, endoplasmic reticulum (both rough and smooth), Golgi apparatus, and ribosomes.

StudyWARE™ Connection

Watch an animation of a typical cell on your StudyWARE™ CD-ROM.

The Cell Membrane

All cells are surrounded by a cell membrane. This membrane is often called the **plasma membrane** or the **plasmalemma** (Figure 3-3). Under the high magnification of an electron microscope, this membrane is composed of a double phospholipid layer with proteins embedded in the phospholipid layer. The phospholipids look like balloons with tails. The round balloon-like part is hydrophilic (attracts water) and the double tails are hydrophobic (repels water). This arrangement allows for the easy passage of water molecules through the cell membrane via osmosis (discussed in Chapter 2). The proteins embedded in the double phospholipid layer allow for the

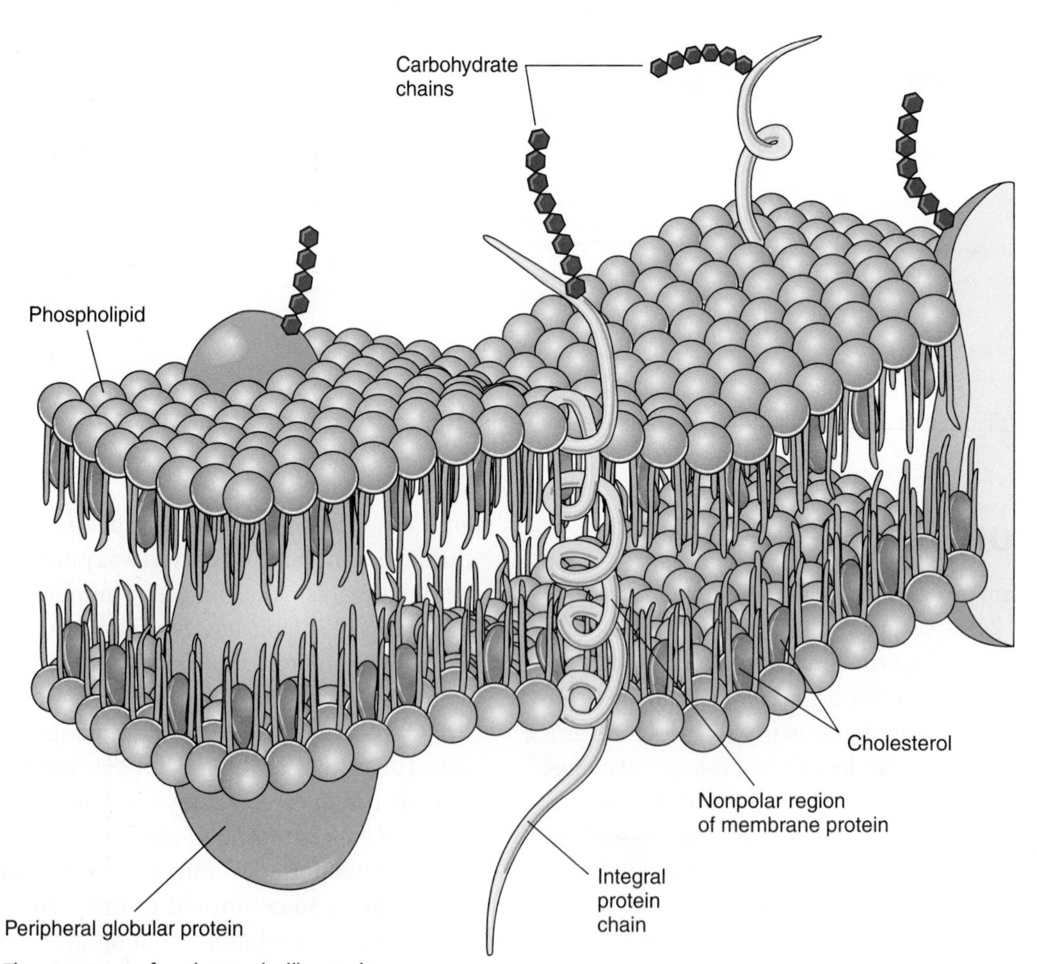

FIGURE 3-3. The structure of a plasma (cell) membrane.

passage of molecules and ions across the cell membrane (Figure 3-4). Some proteins make transport channels for small dissolved ions, others act as enzymes for the active transport of materials into the cell against a concentration gradient and need adenosine triphosphate (ATP) to function, other proteins act as receptor sites for hormones to gain entrance into the cell, and still other proteins act as cell identity markers. In addition, some proteins in the phospholipid layer act as cementing materials for cell adhesion on the outside of cells to hold cells together; others act as structural supports inside the cell attaching to cytoskeleton structures, which hold organelles in place in the cytoplasm. Proteins also make up the structure of the sodium-potassium pump, a unique feature of certain cell membranes like muscle cell membranes and nerve cell membranes (Figure 3-5).

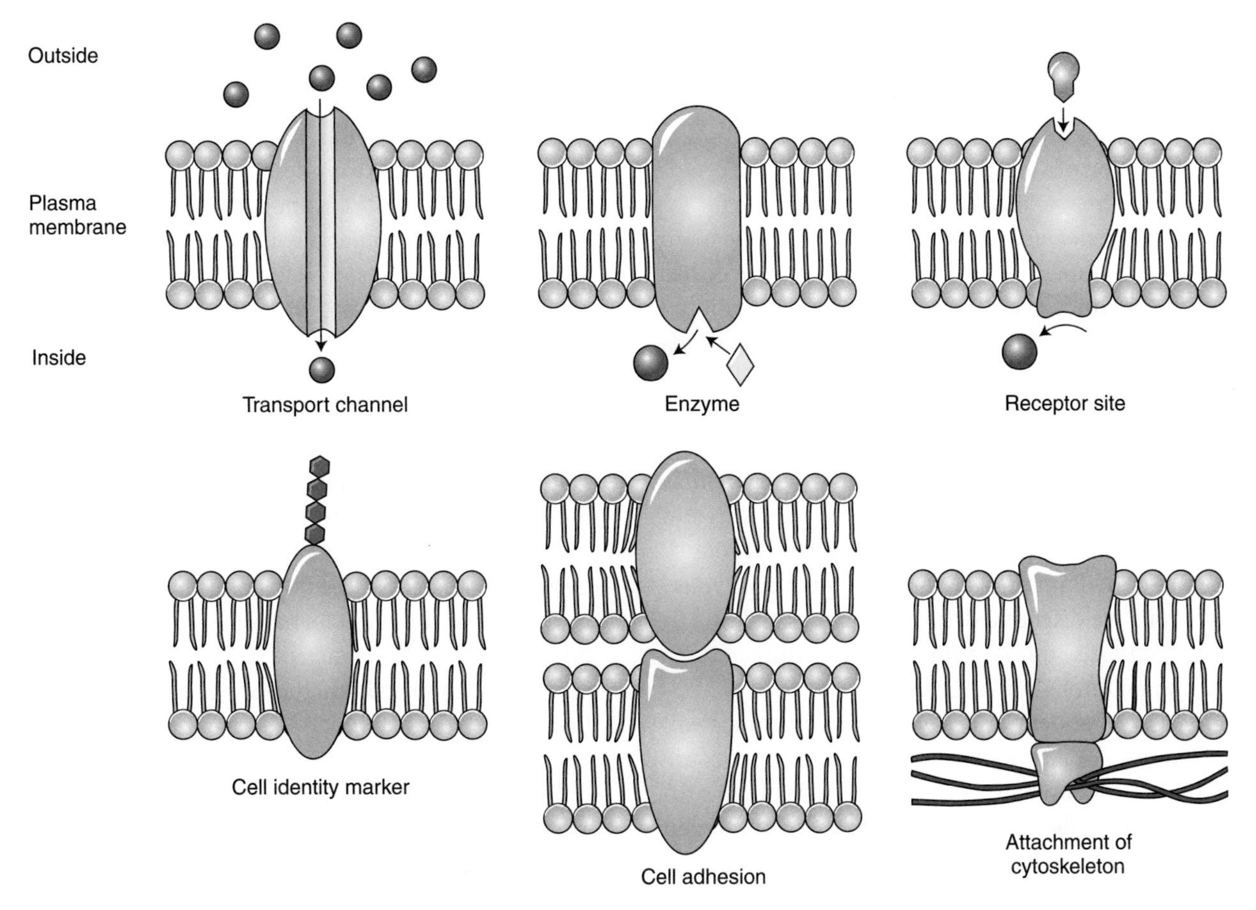

FIGURE 3-4. The functions of proteins embedded in the double phospholipid layer of the cell membrane.

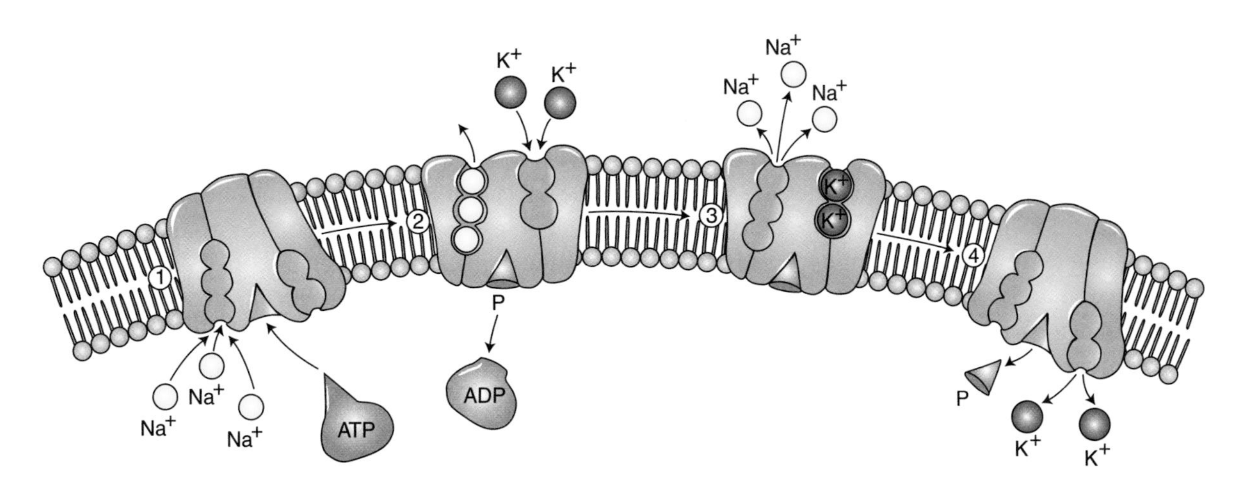

FIGURE 3-5. The protein nature of the sodium-potassium pump of the cell membrane of muscle and nerve cells.

These molecules of proteins and phospholipids are currently referred to in their arrangement as a **fluid mosaic pattern**. The phospholipid molecules are like the tiles of a mosaic but rather than being embedded in a solid cement-like material, they are embedded in a fluid and can move slightly to allow the passage of water molecules across the cell membrane and thus into the cytoplasm of the cell. This basic molecular structure of the cell membrane is the same for all other membrane-bound organelles of the cell.

Cytoplasm of the Cell

The liquid portion of a cell is called protoplasm. The protoplasm outside the nucleus is called cytoplasm; the protoplasm inside the nucleus is called nucleoplasm. The main constituent of cytoplasm is water. This water, however, has many different kinds of chemical compounds distributed among the water molecules. Some of these compounds are nucleic acids like transfer ribonucleic acid (RNA) and messsenger RNA, enzymes, hormones, and various other chemicals involved in the functioning of the cell. Some of these compounds are in solution in the water, whereas others are in a colloidal suspension. In both solutions and colloids, substances are uniformly distributed throughout the water medium. In a solution, however, *individual* atoms or ions are distributed throughout the medium. In a colloid, *clumps* of atoms rather than individual atoms are distributed throughout the medium.

The factor that determines whether a substance will go into solution or a colloidal suspension in water is the electronic interaction between the molecules of the substance and the molecules of water. Because the oxygen atom in H_2O has a stronger attraction for the electrons in the H-O bond than the hydrogen atom (it shares the electrons unequally), the oxygen atom is *slightly* negative and the two hydrogen atoms are *slightly* positive. Refer to Figure 2-5 in Chapter 2. A molecule with such an unequal electron distribution of bonding electrons is said to be **polar**. Because of this polarity of the water molecule, other polar compounds, like ionically bonded compounds such as salt (sodium chloride) are readily *soluble* in water and go into solution. The polarity of the water molecules lessens the electrostatic forces holding ionically bonded molecules together so that they dissociate into individual ions and dissolve in the water.

Other compounds such as covalently bonded molecules are made up of atoms that have equal attraction for the bonding electrons that hold them together. Thus, the bonding electrons are not attracted to one atom of the bond more than the other. Compounds with such unpolarized bonds are called **nonpolar** and do not dissolve readily in water. The organic compounds with the C-H bonds are nonpolar and thus go into a colloidal suspension in the watery medium of the cytoplasm. Proteins, carbohydrates, fats, and nucleic acids are colloidally suspended in the cytoplasm, whereas the mineral salts like sodium, potassium, calcium, chlorine, and phosphorous are in solution.

Some cellular components, such as storage granules and fat droplets, are neither dissolved nor suspended in the cytoplasm. These compounds are products of cellular functions that have collected at certain specific sites within the cytoplasm. The cytoplasm will also contain structures called **vacuoles**. A vacuole is an area within the cytoplasm that is surrounded by a vacuolar membrane. This membrane has the same structure as the cell membrane. A vacuole is generally filled with a watery mixture but can also contain stored food (food vacuole) or waste products of the cell (waste vacuole).

The Nucleus

The **nucleus** is the most prominent structure in the cell. It is clearly visible with a light compound microscope. It is a fluid-containing structure that is separated from the cytoplasm by the **nuclear membrane**, sometimes referred to as the nuclear envelope. The nucleus is the control center of the cell. Cells whose nuclei have been removed lose their functions. Cells with a nucleus transplanted from a different cell take on the characteristics of the cell from which the nucleus was taken.

Nuclear Membrane

A unique feature of the nuclear membrane or envelope is that it is composed of two membranes (Figure 3-6). The inner membrane surrounds and contains the nucleoplasm and its materials. The outer membrane is continuous with the endoplasmic reticulum (ER), an organelle discussed later. The electron microscope has revealed the presence of pores or openings in the double nuclear membrane. These pores have a very fine partition to hinder the free transport or leakage of materials of the nucleoplasm but which allow the passage of materials from the nucleoplasm, which must gain access to the cytoplasm. For example, when protein synthesis must take place, the code to make the protein is on the DNA in the nucleus but the protein is made at a ribosomal site in the cytoplasm. The code is copied from the DNA by a special molecule called messenger RNA (mRNA), which leaves the nucleus through a pore to go to the ribosome. This process is discussed in more detail later in the chapter. The structure of the nuclear membrane is the same fluid mosaic pattern as the cell or plasma membrane.

Nucleoplasm

The fluid medium of the nucleus is called the nucleoplasm. It consists of a colloidal suspension of proteins; the nucleic acids DNA, **deoxyribonucleic acid** (dee-ock-see-rye-boh-noo-**KLEE**-ik **ASS**-id), and RNA, **ribonucleic acid** (rye-boh-noo-**KLEE**-ik **ASS**-id); enzymes; and other chemicals of the nucleus. Many chemical reactions occur in the nucleoplasm and are essential to cellular function and survival, including cellular reproduction.

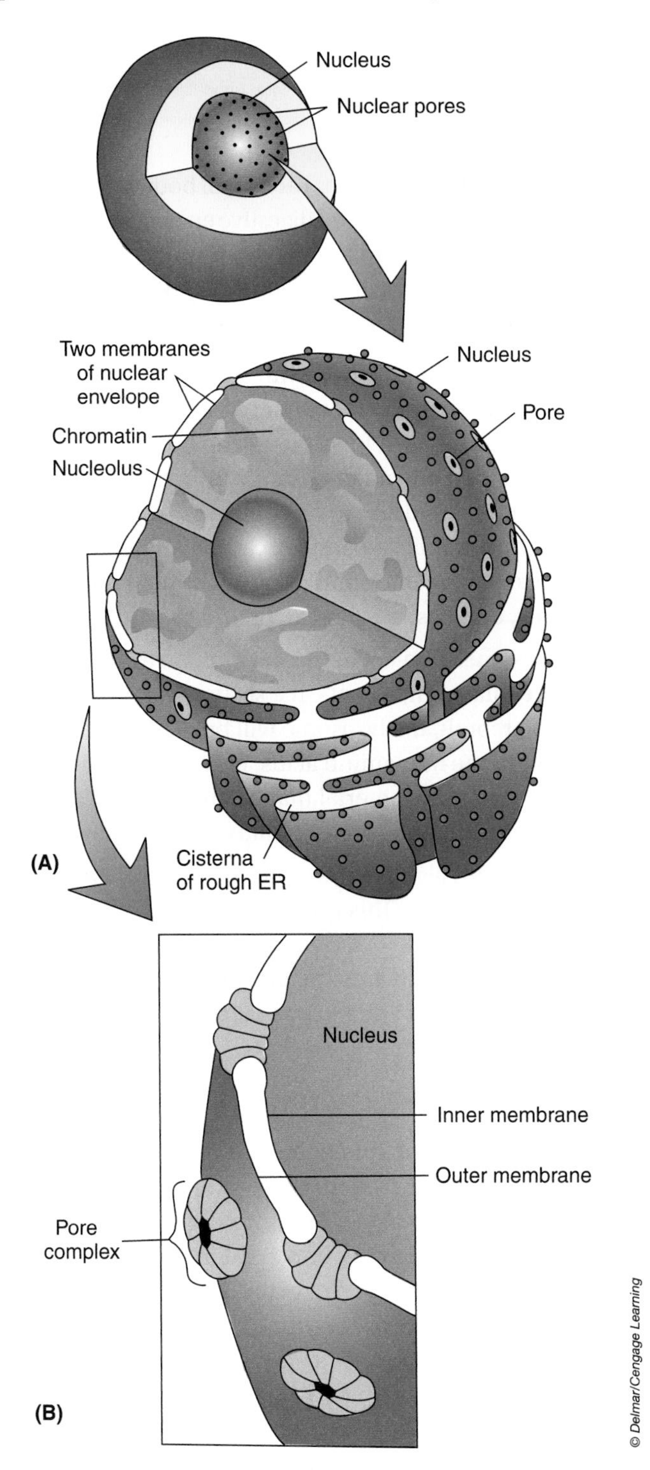

FIGURE 3-6. The structure of the nuclear membrane or envelope. (A) Diagrammatic view of the internal anatomy of the nucleus and the connection of the outer nuclear membrane with the rough endoplasmic reticulum. (B) Diagrammatic view of the pore complex.

Chromatin

When the cell is stained, fine dark threads appear in the nucleus. This material is called **chromatin** (**KROH**-mah-tin) and is the genetic material of the cell. The cells of the human body contain 46 chromosomes (22 pairs of autosomes and one pair of sex chromosomes: one member of each pair comes from the father and one member from the mother). The egg cell and the sperm cell contain one-half that number, or 23 chromosomes. Chromosomes are made of DNA molecules and proteins. When the DNA molecules duplicate during cell division, they shorten and thicken and become visible. We now call the DNA chromosomes. When the cell is not dividing, the DNA molecules are long and thin and visible only as chromatin. All of the above terms are used to describe the different levels of chromosomal organization (Figure 3-7). This is discussed in greater detail in Chapter 4. DNA controls many of the functions of the cell.

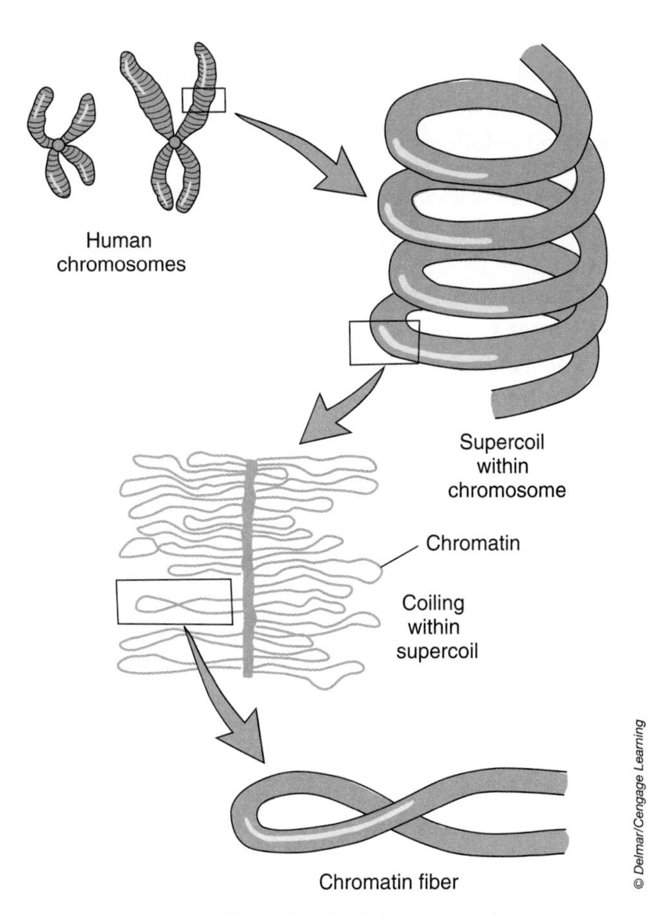

FIGURE 3-7. Some different levels of chromosomal organization.

Nucleolus

The **nucleolus** (noo-**KLEE**-oh-lus) is a spherical particle within the nucleoplasm that does not have a covering membrane around it. It is composed of primarily DNA, RNA, and proteins. A cell may have more than one nucleolus. This structure is the site of ribosomal synthesis. It is involved in protein synthesis because it makes the ribosomes and ribosomes are the sites of protein synthesis.

The Mitochondria

A **mitochondrion** (singular) or mitochondria (my-toh-**KON**-dree-ah) (plural) are small oblong-shaped structures composed of two membranes (Figure 3-8). The outer membrane gives a mitochondrion its capsule shape; the inner membrane folds on itself to provide a surface on which the energy-releasing chemical reactions of the cell occur. When viewed under a light compound microscope, mitochondria appear only as small, dark granules in the cytoplasm. It is the electron microscope that has revealed to us the true nature of the mitochondria. The folds of the inner membrane are called **cristae** (**KRIS**-tee). It is on the cristae that cellular respiration occurs, where food (chemical energy) is converted into another usable form of chemical energy, ATP. For this reason, the mitochondria are known as the powerhouses of the cell.

In its most simple expression, cellular respiration can be stated as follows:

Food (like glucose) + oxygen → energy + waste.

$$C_6H_{12}O_6 + 6O_2 \rightarrow ATP + 6CO_2 + 6H_2O.$$

This chemical reaction is discussed in greater detail in Chapter 4.

Most of the energy-producing reactions, which occur in the mitochondria, take place on the surface of the cristae. Cells with high energy requirements (like muscle cells) will have mitochondria with many folds or cristae.

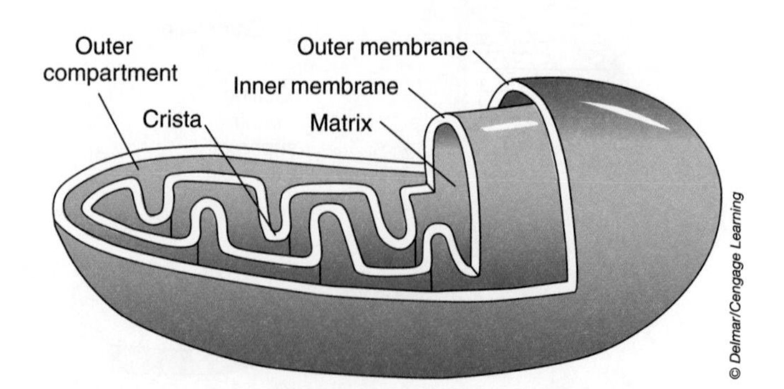

FIGURE 3-8. The membrane structure of a typical mitochondrion.

Cells with low energy requirements, like the lining of the cheek (epithelial cells), will have mitochondria with fewer folds or cristae. All cells will have approximately the same number of mitochondria. They are inherited from the mother via the egg cell. Mitochondria also contain mitochondrial DNA.

Lysosomes

Lysosomes (**LIGH**-so-sohmz) are small bodies in the cytoplasm that contain powerful digestive enzymes to enhance the breakdown of cellular components (see Figure 3-1). The structure and size of lysosomes vary but they are generally spherical. They have three general functions:

1. They act in conjunction with stored food vacuoles. When a cell needs more energy, a lysosome will fuse with a stored food vacuole to break down the stored food into a more usable form that can go to a mitochondrion to be converted into ATP. For example, starch, a complex carbohydrate, will be broken down into simple sugars, protein into amino acids, and fats into fatty acids and glycerol.
2. Lysosomes also act in the maintenance and repair of cellular components. If a section of ER needs to be rebuilt, the lysosome will break down the membrane into amino acids, fatty acids, glycerol, and so on and material that can be recycled to build new protein and phospholipids.
3. Lysosomes also act as suicide agents in old and weakened cells. This process is known as **autolysis** (aw-**TAHL**-ih-sis). The lysosome will expel all of its enzymes directly into the cytoplasm of the cell to destroy the cell and its organelles.

Endoplasmic Reticulum

The **endoplasmic reticulum** (en-doh-**PLAZ**-mik re-**TIK**-you-lum), or ER, is a complex system of membranes that forms a collection of membrane-bound cavities. These often interconnect into a membrane-bound system of channels within the cytoplasm. The shape and size of these cavities vary with the type of cell. When the cavities are sac-like or channel-like, they are called **cisternae** (sis-**TER**-nee) and are used to store and transport materials made by the cell. The ER is attached to the outer membrane of the nuclear membrane or envelope and ultimately connects with the cell membrane (Figure 3-9). With the use of the electron microscope, it was discovered that there are two types of ER: a rough ER and a smooth ER.

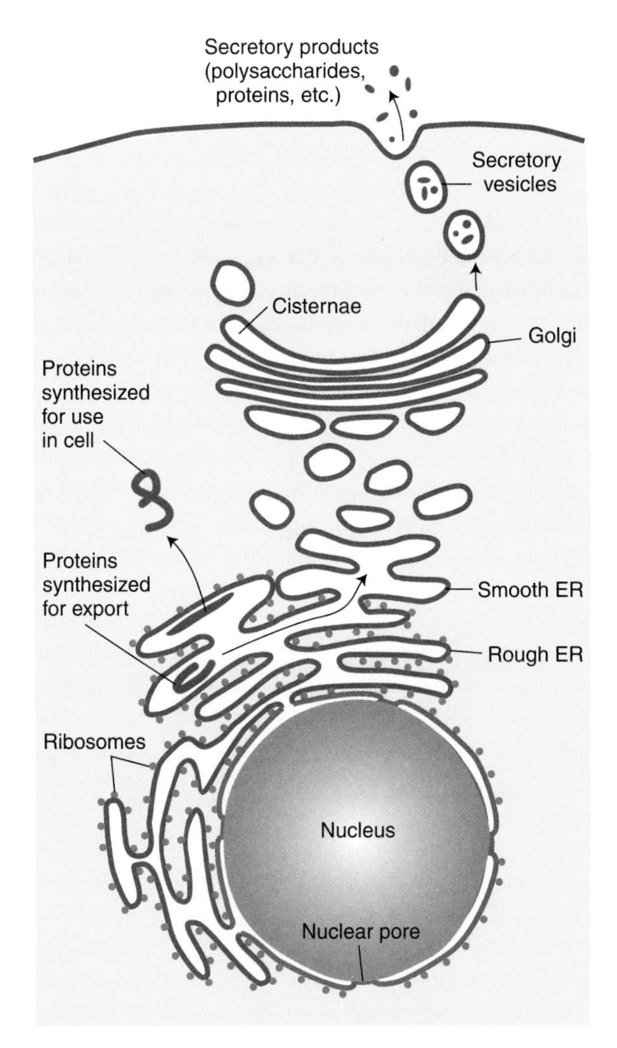

FIGURE 3-9. The structure and nature of the endoplasmic reticulum and the Golgi body.

The Rough or Granular ER

All cells will have a **rough** or **granular ER**. It is called rough or granular because it has ribosomes attached to it. These are the granules on the ER. Because of the attached ribosomes, the rough ER is a site of protein synthesis. Proteins that will be secreted by the cell are synthesized there. The cavities and vesicles of the rough ER serve in the segregation and transport of these proteins in preparation for further discharge and processing. The rough ER may also be involved in the collection of digestive enzymes to form lysosomes.

The Agranular or Smooth ER

Occasionally, a **smooth** or **agranular ER** will be attached to a granular ER (see Figure 3-9). Structurally, the agranular form differs from the rough form. It does not have attached ribosomes. It also differs in function. Only certain cells have the agranular or smooth ER. It is found in the cells of the gonads in which sex hormones are being synthesized. One function appears to be sex hormone synthesis. It is also found in the cells of the lacteals of the villi of the small intestine. Thus, it is also believed to be involved in the transportation of fats.

The Golgi Apparatus

The **Golgi** (**GOHL**-jee) **apparatus** is also called the **Golgi body.** It consists of an assembly of flat sac-like cisternae that resemble a stack of saucers or pancakes (see Figure 3-9). Golgi bodies can differ in both size and compactness. They function as the points within the cell where compounds to be secreted by the cell are collected and concentrated. They may be seen attached to the ER. When the cell's secretions are a combination of both proteins and carbohydrates, the carbohydrates will be synthesized in the Golgi apparatus and the complexes of carbohydrates and proteins are assembled there. In the pancreas, enzymes synthesized by the ribosomes are collected in the membranes of the Golgi apparatus and then are secreted. Lysosomes may also form at the Golgi body when digestive enzymes are collected there.

Ribosomes

Ribosomes (**RYE**-boh-sohmz) are tiny granules distributed throughout the cytoplasm and are attached to the rough or granular ER. They are not surrounded by a membrane. Ribosomes are composed of ribosomal RNA and proteins. Messenger RNA attaches to ribosomes during protein synthesis. There are many, many ribosomes in the cell because they are so essential to cell function. They are the sites of protein synthesis.

Protein Synthesis

Proteins are essential to cellular function and structure. Proteins are part of membrane structures (proteins are embedded in the double phospholipid layer). Enzymes are protein catalysts (all chemical reactions in the cell require enzymes), and our immune system functions through the production of antibodies (large proteins) that attack foreign proteins (antigens).

The code to make a particular protein lies on a DNA molecule in the nucleus. Genes on the DNA molecule constitute the code. However, proteins are made at the ribosomes. Therefore, this code must be copied and taken to the ribosomes. A special molecule called **messenger RNA** (mRNA) copies the code from the DNA molecule in the nucleus. This process is called **transcription** and occurs with the assistance of an enzyme called RNA polymerase. The mRNA then leaves the nucleus through a nuclear pore and goes into the cytoplasm to a ribosome or group of ribosomes. The ribosome will now assist in the assemblage of the protein because it now has the code or recipe to produce the protein. To make the protein, the ribosome now needs the ingredients, which are amino acids.

Another molecule will now go into the cytoplasm and collect the amino acids. This molecule is **transfer RNA** (tRNA). It is coded for a particular amino acid by means of three nitrogen bases at one end of the molecule known as the anticodon (Figure 3-10). These three bases will fit or match with three bases on the mRNA molecule called the codon. In this way, a series of tRNA molecules bring amino acids to certain sites on the mRNA molecule. This process is called **translation** (reading the code and bringing the appropriate amino acids in sequence along the mRNA). Now the ribosomes, with the assistance of enzymes, put the amino acids together by linking them up and forming a polypeptide chain (Figure 3-11). The numerous ribosomes found in the cell indicate the importance and significance of **protein synthesis** to the survival and function of the cell.

Centrioles

Two **centrioles** (**SEN**-tree-olz) are found only in animal cells at right angles to each other near the nuclear membrane. The pair together is referred to as a **centrosome**

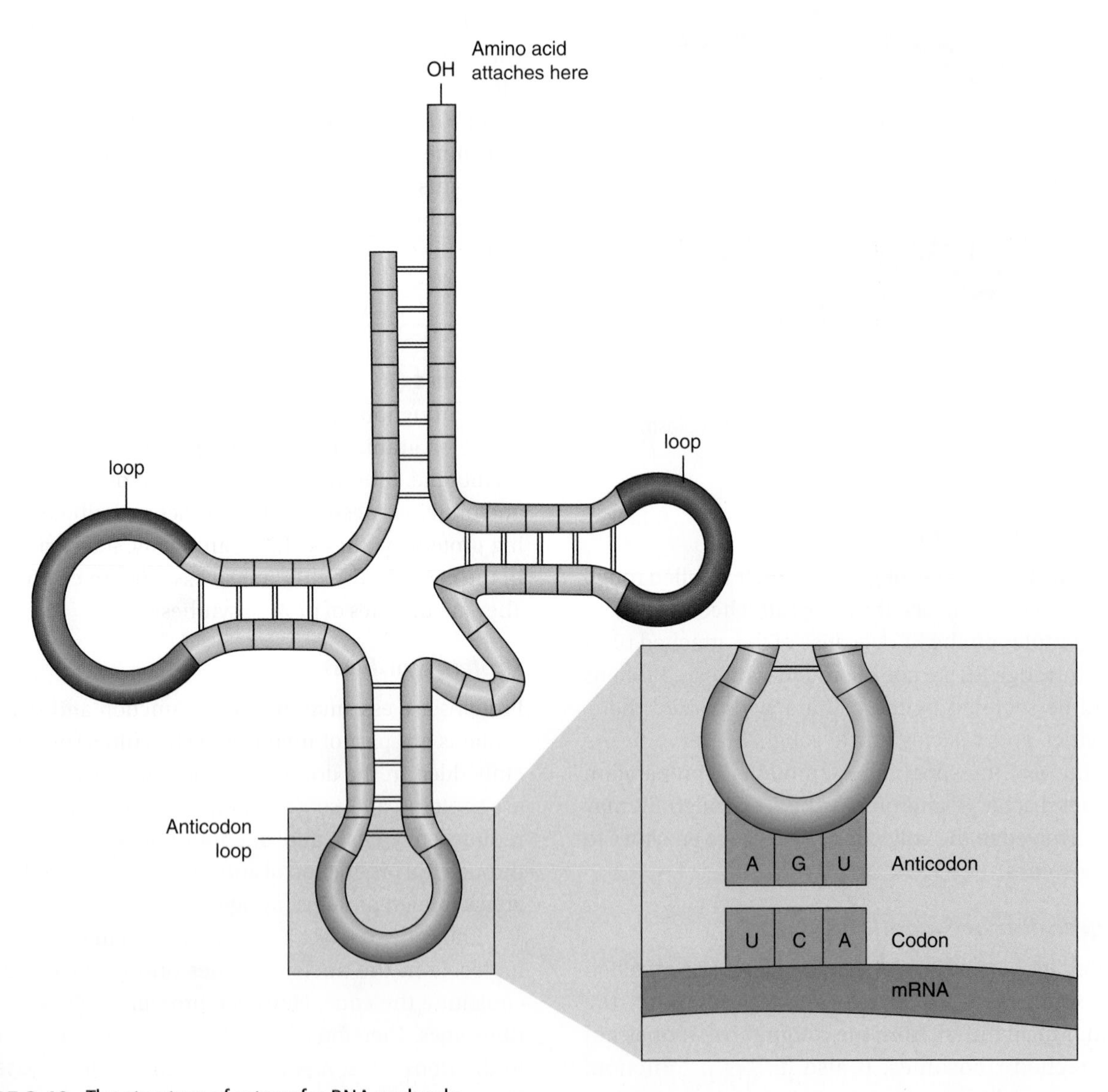

FIGURE 3-10. The structure of a transfer RNA molecule.

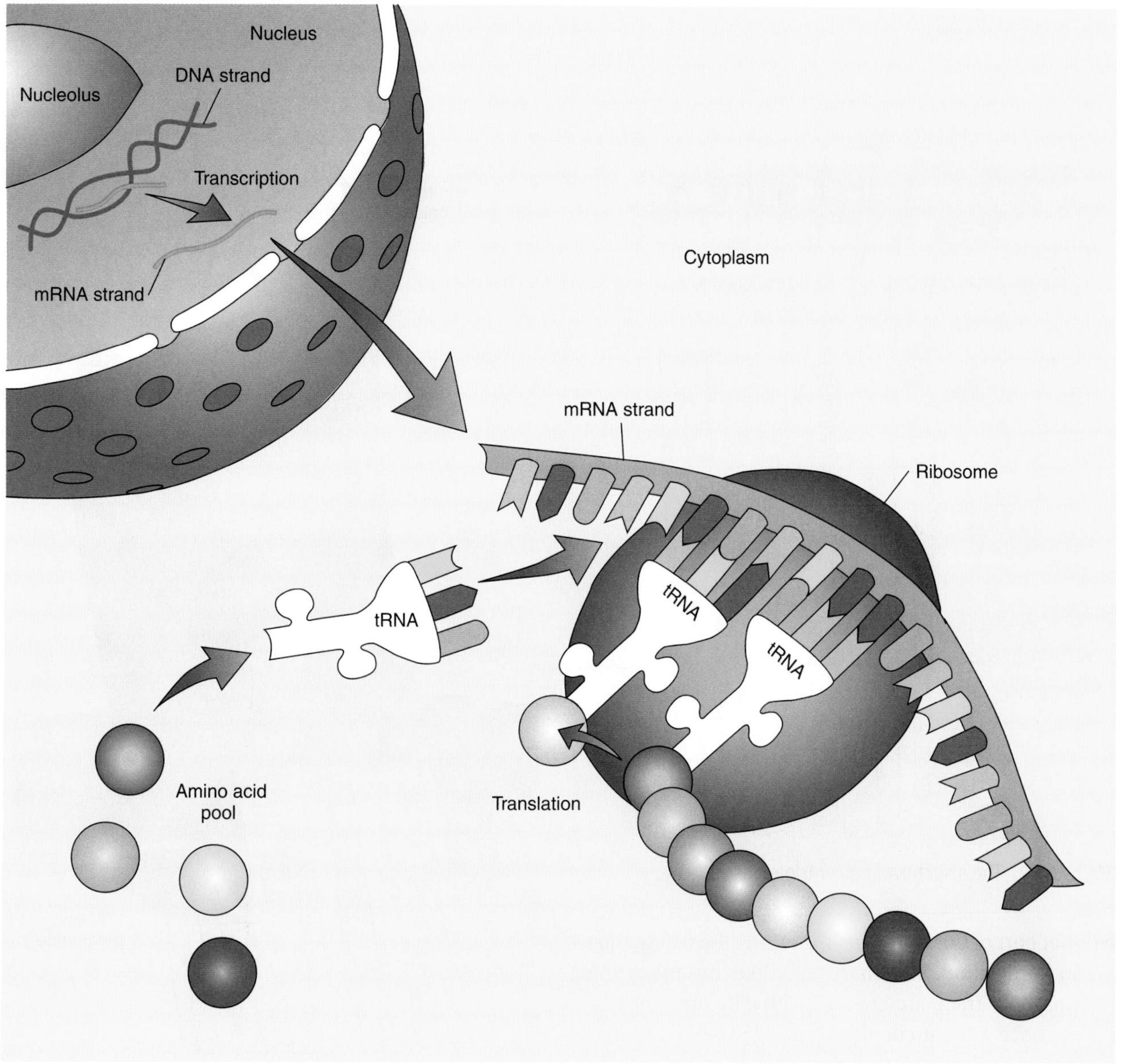

FIGURE 3-11. An overview of the process of protein synthesis.

(**SEN**-troh-sohm). They are composed of nine sets of triplet fibers (Figure 3-12). The inner fiber of each triplet is connected to the outer fiber of the adjacent triplet by a subfiber.

During cell division, the centrioles move to each side of the dividing cell and position themselves at a location called the opposite pole of the cell. They now form a system of **microtubules,** which are long, hollow cylinders made of a protein called **tubulin**. These fibers or microtubules redistribute the duplicated chromosomes during cell division into the appropriate new daughter cells.

Cilia and Flagella

Cilia (**SIL**-lee-ah) and **flagella** (fla-**JELL**-ah) are cellular organelles located on the cell surface. They are composed of fibrils that protrude from the cell and beat or vibrate. Some single-celled organisms use these structures to move through a medium. For example, *Euglena* has a flagellum that pulls it through the water, whereas a *Paramecium* is covered with row upon row of hundreds of cilia to allow it to swim in pond water. In the human body the male sperm cell is propelled by a single beating flagellum that assists it in reaching the female egg in

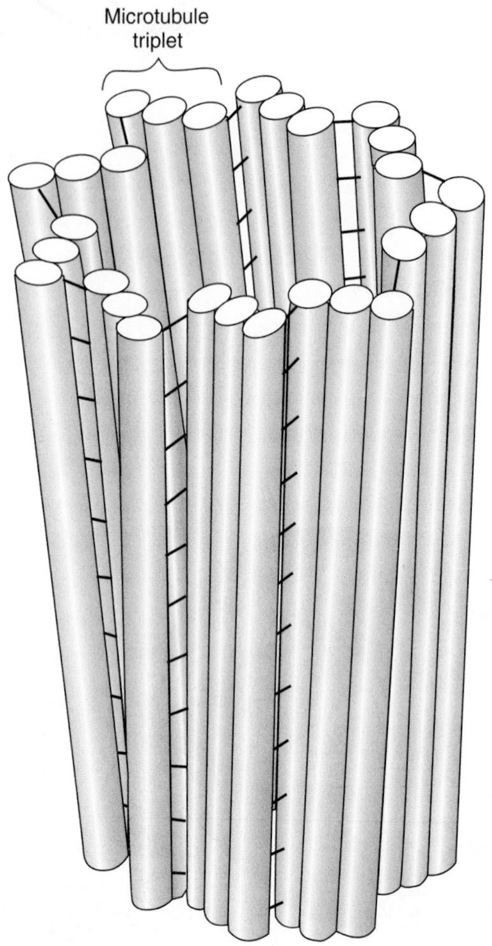

Microtubule triplet

© Delmar/Cengage Learning

FIGURE 3-12. The structure of a centriole.

the upper part of the fallopian or uterine tube where they unite in fertilization. Stationary cells, like those that line our respiratory tract, are covered with cilia on their free edge to move the mucus-dust package upward across the cell surfaces to bring this material to the throat to be swallowed and then discharged from the body.

Although cilia and flagella are similar anatomically, a flagellum is considerably longer than a cilium. A cell with cilia will have row on row of cilia, but a cell with flagella will have one (like the sperm cell) or two or four like some single-celled protozoans.

Externally, these structures are hairlike protrusions from the cell membrane. Internally, they are composed of nine double fibrils arranged in a cylindrical ring around two single, central fibrils (Figure 3-13). The microtubules or fibrils of the flagellum arise from a structure called the basal body found just below the area from which the flagellum protrudes from the surface of the cell membrane. The basal body or plate has a cylindrical structure like the centriole, that is, it is also composed of nine sets of triplet fibers.

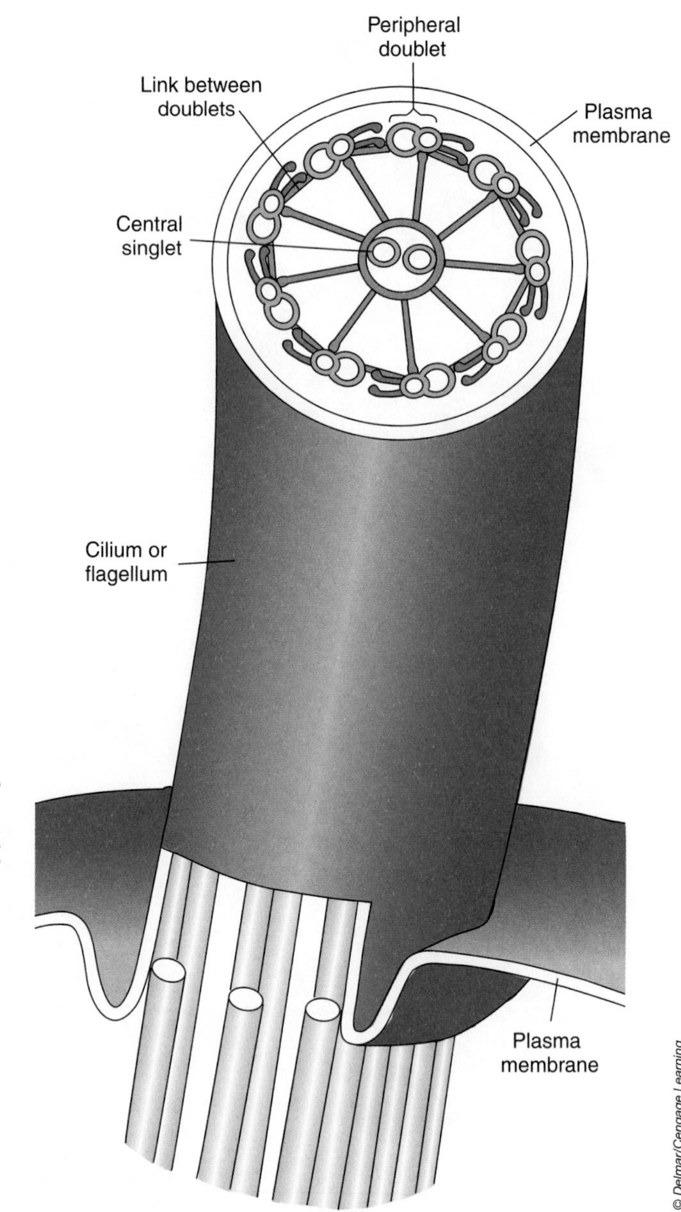

FIGURE 3-13. The internal anatomy of a cilium or flagellum.

Plastids of Plant Cells

In our laboratory exercise on cells, we will examine and compare plant cells with animal cells. Therefore, it is necessary to discuss these organelles found only in plant cells. There are three plastids found in plant cells. The most common and most numerous of these are the **chloroplasts** (**KLOR**-oh-plastz) that cause plants to look green.

Chloroplasts are large organelles found mainly in plant cells (Figure 3-14). They contain the green pigment chlorophyll. These organelles are the site of photosynthesis. It is here that the light energy of the sun is converted into chemical energy and food for use by both plants and

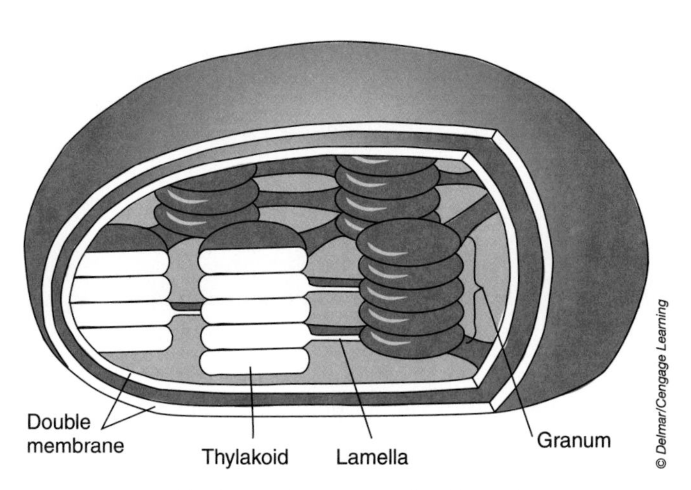

FIGURE 3-14. The internal anatomy of a chloroplast.

animals. Without plants and their chloroplasts, animals could not survive on this planet. The process of photosynthesis occurs inside the chloroplast. This chemical equation is:

$$6CO_2 + 12H_2O \rightarrow C_6H_{12}O_6 \text{ or glucose (sugar, food)} + 6O_2 \text{ (the air we breathe)} + 6H_2O$$

Chloroplasts are large enough to be easily seen with a light microscope. They are enclosed by an outer membrane but the second internal membranous structure is quite complex. The inside contains many stacks of membranes called a **granum**. A granum is made of a stack of individual double membranes called a **thylakoid** (**THIGH**-lah-koid). The grana (plural) are connected to one another by a different system of membranes called **lamellae.** The grana are made of proteins, enzymes, chlorophyll, and other pigments arranged in a layered structure.

Plant cells also have two other types of plastids. **Chromoplasts** (**KROH**-moh-plastz) are similar in structure to chloroplasts but they contain other pigments like the carotenoid pigments. The carotenoid pigments are **xanthophyll** (**ZAN**-tho-fill), which produces a yellow color (the skin on a banana), and **carotene** (**KAR**-oh-teen), which produces a red-orange color (tomatoes and carrots). These pigments also produce the colors of flower petals and fruits. Another type of plastid is the **leucoplast** (**LOO**-koh-plast). Leucoplasts do not have any pigments—they are storage plastids. An onion bulb is full of leucoplasts where sugar is stored and a potato contains leucoplasts where starch is stored.

The Cell Wall of Plant Cells

The cell membrane of plant cells is surrounded by a semirigid covering called the cell wall made of a complex carbohydrate called **cellulose** (**SELL**-you-lohs). Cellulose is synthesized by Golgi bodies by linking up glucose units. Animal cells do not have cell walls. This material is what we call fiber in our diet. It cannot be digested; thus, it keeps our stools soft. We eat fruits and vegetables to maintain a balance of fiber in our diet. This fiber may help prevent the development of colon cancer.

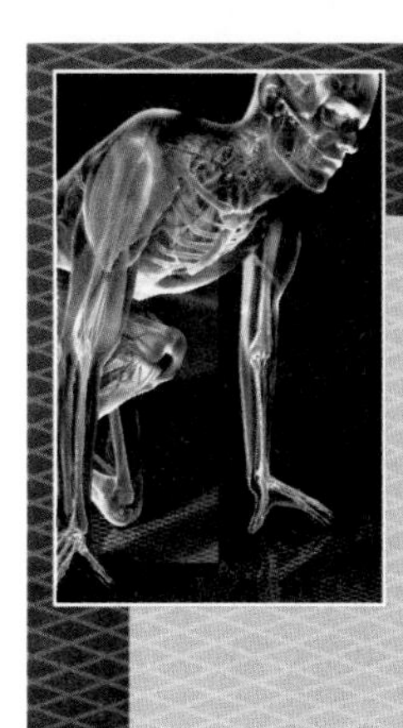

AS THE BODY AGES

Apoptosis (a-po-**TOE**-sis) is a natural process by which cells in the body die and is controlled by specific genes. This process eliminates damaged cells, cells that have outlived their natural lifetime and need to be replaced, cells that are infected with pathogens, and certain cancer cells. As we age, cellular repair and cell division to replace old cells become more difficult to achieve. Some biologists believe that there are actually "death genes" that become activated later in life, resulting in cellular deterioration and death. ■

Career FOCUS

These are careers that are available to individuals interested in cellular structure and function:

- **Cell biologists** study the anatomy and function of cellular organelles and have careers working in universities and medical centers as researchers and technicians.
- **Electron microscopists**, equipped with special training in operating an electron microscope, find employment in hospitals, research universities, industry, and pharmaceutical companies.

SUMMARY OUTLINE

INTRODUCTION

1. The cell is the basic unit of biologic organization. It is composed of a fluid medium called protoplasm surrounded by a cell or plasma membrane. Structures within this protoplasm are called organelles.
2. The protoplasm inside the nucleus is called nucleoplasm; the protoplasm outside the nucleus is called cytoplasm.

HISTORY OF THE CELL THEORY

1. Two Germans, Matthias Schleiden, a botanist, and Theodor Schwann, a zoologist, were the first biologists to propose the cell theory in the 1830s.
2. The modern cell theory states that cells are the basic units of organization of all organisms; all organisms are composed of one or more cells; cells arise from only preexisting cells through cell division; and all existing cells are the descendants of the first cells formed early in the evolutionary history of life on earth.

THE CELL MEMBRANE

1. The cell membrane or plasma membrane is made up of a double phospholipid layer with proteins embedded in the phospholipid layer.
2. The phospholipid layer allows for the free passage of water molecules through the cell membrane via osmosis, while the proteins act as channels, active transport areas, receptor sites, and identity markers for the cell.
3. This molecular arrangement of the cell membrane is referred to as a fluid mosaic pattern and is responsible for the selective permeability of the membrane. It is through the membrane that materials enter and exit the cell.

CYTOPLASM OF THE CELL

1. The main constituent of cytoplasm is water. Chemical compounds like the mineral salts are dissolved in solution in this water; chemical compounds with the C-H bond (organic molecules) are in colloidal suspension.
2. In a solution, individual atoms or ions are distributed throughout the watery medium; in a colloid, clumps of atoms rather than individual atoms are distributed throughout the watery medium.
3. The water molecule has unique properties, which determine whether molecules will go into solution or a colloidal suspension. The oxygen atom in H_2O has a stronger attraction for the electrons in the H-O bond than the hydrogen atoms; thus, the oxygen atom is slightly negative, whereas the two hydrogen atoms are slightly positive.
4. Polar compounds, such as the ionically bonded mineral salts, will dissolve in water and go into solution; nonpolar compounds, such as the covalently bonded molecules of proteins, carbohydrates, fats, and nucleic acids, will go into colloidal suspension.
5. The cytoplasm of the cell will also contain storage granules, fat droplets, and vacuoles.

THE NUCLEUS

1. The nucleus is the control center of the cell. It is surrounded by a double nuclear membrane. The inner nuclear membrane surrounds the fluid part of the nucleus, called nucleoplasm, whereas the outer nuclear membrane connects with the endoplasmic reticulum.
2. The nuclear membrane is perforated with pores that allow materials like messenger RNA to leave the nucleus and go into the cytoplasm of the cell.
3. The genetic material inside the nucleoplasm is darkly stained threads of nucleic acids called chromatin. This chromatin will duplicate, shorten, and thicken during cell division and will become visible as chromosomes.
4. The nucleolus is a spherical particle within the nucleoplasm that does not have a covering membrane. It is the site where ribosomes are made.

THE MITOCHONDRIA

1. The mitochondria are the powerhouses of the cell. Each mitochondrion is composed of two membranes. The outer membrane forms its capsular shape, and the inner membrane folds on itself to increase surface area.
2. The inner folds of the mitochondrion are called cristae. It is in the mitochondrion where the aerobic phase of cellular respiration occurs:

$$C_6H_{12}O_6 + 6O_2 \rightarrow \text{ATP (energy)} + 6CO_2 + 6H_2O$$

3. Cells with higher energy requirements, like muscle cells, will have mitochondria with many cristae; those with lower energy requirements will have fewer cristae.

LYSOSOMES

1. Lysosomes are small structures in the cytoplasm surrounded by a membrane and contain powerful digestive enzymes.
2. Lysosomes function in three different ways in the cell: they function with food vacuoles to digest stored food; they function in the maintenance and repair of cellular organelles; and they act as suicide agents in old and weakened cells.

ENDOPLASMIC RETICULUM

1. The endoplasmic reticulum or ER is a complex system of membranes that makes up a collection of membrane-bound cavities or channels. These channels are called cisternae.
2. The ER connects with the outer nuclear membrane and with the cell membrane.
3. There are two types of ER. All cells have a rough or granular ER whose membranes have attached ribosomes. The function of a rough ER is protein synthesis. Some cells also have a smooth or agranular ER, which does not have attached ribosomes. The function of a smooth ER can be the transportation of fats or the synthesis of the sex hormones.

THE GOLGI APPARATUS

1. The Golgi body or apparatus is a collection of flat sac-like cisternae that look like a stack of pancakes.
2. They function as points within the cytoplasm where compounds to be secreted by the cell are concentrated and collected. They act like storage warehouses of the cell.
3. If the cell is synthesizing carbohydrates and proteins, the carbohydrates will be synthesized in the Golgi apparatus.

RIBOSOMES

1. Ribosomes are small granules distributed throughout the cytoplasm and attached to the rough ER. They are not covered by a membrane.
2. Ribosomes are the site of protein synthesis.

PROTEIN SYNTHESIS

1. Proteins function in major and essential ways for cellular function and survival. They are part of the structure of membranes; they act as enzymes or catalysts that make chemical reactions occur in the cell; and they function in our immune response.
2. The code to make a particular protein is a gene on a DNA molecule. The DNA is found in the nucleus, whereas the protein is made at a ribosomal site in the cytoplasm.
3. A special molecule called messenger RNA copies the code from the DNA molecule in a process called transcription. The mRNA then leaves the nucleus through a nuclear pore and takes the code to a ribosome or group of ribosomes.

4. Other molecules called transfer RNAs go into the cytoplasm and pick up particular amino acids. Each tRNA molecule is coded for a particular amino acid by its anticodon loop at the end of the molecule. The anticodon loop will only match a particular site on the mRNA molecule called the codon. This process is called translation.
5. The ribosomes will now link up the amino acids brought to the mRNA molecule by the tRNA molecules and will construct the protein with the assistance of enzymes.

CENTRIOLES

1. Two centrioles are found at right angles to each other near the nuclear membrane. The pair is referred to as a centrosome.
2. Each centriole is composed of nine sets of triplet fibers.
3. The centrioles form the spindle fibers during cell division and guide the duplicated chromosomes to their daughter cells.

CILIA AND FLAGELLA

1. Cilia are short and flagella are long hairlike protrusions from the cell membrane. Internally, they are composed of nine double fibrils arranged in a ring around two, single central fibrils.
2. A cell with cilia will have row upon row of cilia. Cilia will move materials across the free surface of a cell, like respiratory tract cells, which move the mucus-dust package to our throat.
3. A cell with a flagellum, like the sperm cell, will propel the cell through a medium.

PLASTIDS OF PLANT CELLS

1. The most common plastid of plant cells is the chloroplast, which contains the green pigment chlorophyll that allows plant cells to perform photosynthesis.
2. Photosynthesis is the conversion of light energy (the sun) into chemical energy (food like glucose).
3. Chromoplasts are plastids that contain the carotenoid pigments, xanthophyll (yellow) and carotene (orange-red).
4. Leucoplasts are plastids that store food (e.g., sugar and starch). They contain no pigment and are colorless.

THE CELL WALL OF PLANT CELLS

1. The cell membrane of plant cells is surrounded by a semirigid covering called the cell wall. It is composed of cellulose, synthesized by the Golgi apparatus, and secreted through the cell membrane.
2. Cellulose is the material in our diet that we call fiber.

REVIEW QUESTIONS

1. Define a cell.
2. List the major points of the modern cell theory.
3. *Why is the molecular structure of a cell membrane referred to as a fluid mosaic pattern?
4. *Explain why some chemical compounds go into solution and others go into a colloidal suspension in the cytoplasm of a cell.
5. *Why is the nucleus considered the control center of the cell?
6. *Based on their structure, why are mitochondria called the powerhouses of the cell?
7. List three functions of lysosomes.
8. What are the two types of endoplasmic reticula and what are their functions in the cell?
9. What is the function of a Golgi apparatus?
10. *Why are ribosomes so numerous in the cytoplasm of a cell?
11. List three important functions of proteins in a cell.
12. What is a centrosome?
13. List the functions of cilia and flagella.
14. Name and define the three types of plastids found in plant cells.
15. *Why should plant cells be studied in a human anatomy and physiology class?

***Critical Thinking Questions**

FILL IN THE BLANK

Fill in the blank with the most appropriate term.

1. The cell theory was first proposed in the 1830s by ________________ and ________________.
2. Modern details of cellular structure have been extensively studied because of the invention of the ________________ microscope.

3. Cell membranes are made of layers, a double ________________ with ________________ embedded in this double layer.
4. The main component of cytoplasm is ________________.
5. In a solution, ________________ atoms or ions of a substance are distributed throughout the medium.
6. In a colloid, ________________ of atoms are distributed throughout the medium.
7. Because the oxygen atom in H_2O has a stronger attraction for the electrons in the H-O bond than the hydrogen atom do, the oxygen atom has a slightly ________________ charge and the hydrogen atoms have a slightly ________________ charge.
8. Ionically bonded molecules are called ________________, whereas covalently bonded molecules are called ________________ molecules.
9. The nuclear membrane or envelope is composed of two membranes; the outer membrane is generally continuous with the ________________, which often has ribosomes attached to it.
10. The fluid medium of the nucleus is specifically called ________________.
11. During cell division, chromatin condenses into thick rodlike structures called ________________, which become visible with a light microscope.
12. A spherical particle within the nucleus that does not have a covering membrane is the ________________; it is the site of ribosome synthesis.
13. The convolutions of the inner membrane of the mitochondrion are called ________________.
14. ________________ are small bodies in the cytoplasm that contain enzymes that enhance the breakdown of cellular components.
15. When the membrane-bound cavities of the endoplasmic reticulum are channel- or sac-like, they are called ________________.
16. The ________________ seems to function as a point within the cell where compounds to be secreted by the cell are collected and concentrated and where carbohydrates are synthesized.
17. Protein synthesis occurs at the ________________, which are composed of RNA and protein and are not surrounded by a membrane.
18. The interior of chloroplasts consists of many stacks of membranes called ________________.
19. Internally, a flagellum is composed of ________________ double fibrils arranged in a cylindrical ring around ________________ central single fibrils.
20. Centrioles form the ________________, which distribute the daughter chromosomes during cell division to the daughter cells.

MATCHING

Place the most appropriate number in the blank provided.

____	Mitochondrion	1. Lysosomes
____	Muscle cells	2. Lacteals
____	Enzymes	3. Protein synthesis
____	Sex hormone synthesis	4. Redistribute chromosomes
____	Ribosomes	5. Powerhouse of the cell
____	Chloroplasts	6. Locomotion
____	Centriole	7. Carotenoid pigments
____	Flagella	8. Cisternae
____	Cell wall	9. Many cristae
____	Channels of the ER	10. Agranular endoplasmic reticulum
		11. Cellulose
		12. Site of photosynthesis

Search and Explore

Search the Internet using "cell biologists" as a key word and discover cell biology experts and their research. Give an oral presentation on your findings.

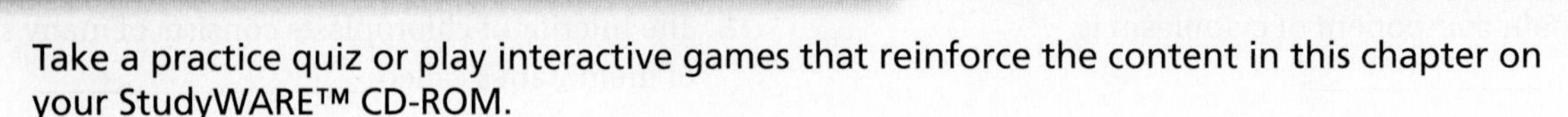

StudyWARE™ Connection

Take a practice quiz or play interactive games that reinforce the content in this chapter on your StudyWARE™ CD-ROM.

Study Guide Practice

Go to your **Study Guide** for more practice questions, labeling and coloring exercises, and crossword puzzles to help you learn the content in this chapter.

LABORATORY EXERCISE: CELL STRUCTURE

Materials needed: Compound light microscope, prepared microscope slides of the letter e, colored threads, living *Elodea* or *Cabomba* plant and an onion bulb, living culture of *Paramecium*, flat-edged toothpicks and methylene blue stain, dissecting microscope and a moss plant, videotape or CD-ROM on "How to Use a Microscope"

I. USING A COMPOUND LIGHT MICROSCOPE

Your compound microscope is an expensive and delicate piece of equipment and must be handled carefully. Review the videotape or CD-ROM, provided by your instructor, on the operation and parts of your microscope. Figure 3-15 shows the parts of a compound light microscope.

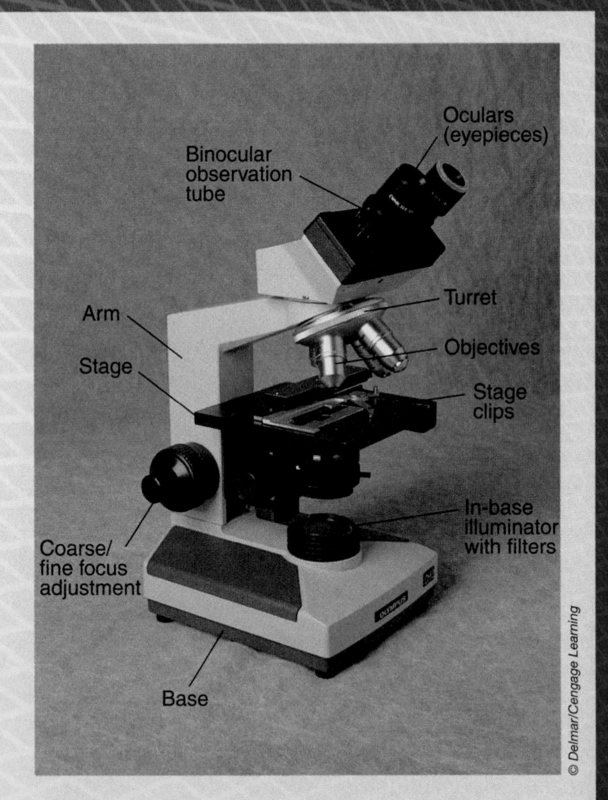

FIGURE 3-15. The parts of a compound light microscope.

A. PARTS OF A COMPOUND MICROSCOPE

1. Remove your assigned microscope from its storage area using two hands. Grab the *arm* with one hand and support the *base* with your other hand. Bring the microscope to your laboratory station and place it down gently. Unwind the electrical cord and plug it in.
2. Identify the *body tube*. At the top of the body tube is the *ocular lens* usually with a magnification of 10×. At the end of the body tube are the other magnifying elements screwed into a *revolving nosepiece*. These elements are called the *objective lenses*. Although the number of objective lenses varies, there will usually be a low-power objective (10× magnification) and a high-power objective (40× magnification).

(continues)

CELL STRUCTURE (Continued)

3. Underneath the body tube is the *stage*, a flat piece on which microscope slides are placed. It may be a mechanical, movable stage. The stage will have *stage clips* to hold the slide in place. There will be a hole in the stage to allow light to be reflected from its built in substage *lamp* through the stage opening. Light then passes through the specimen on the microscope slide into the body tube resulting in an image on the retina of your eye.
4. The importance of light makes necessary its careful adjustment. Your microscope may or may not have a condenser, which concentrates light. Just below the stage can be found the iris diaphragm. Practice moving the iris diaphragm lever to observe the changes in light by looking through the ocular lens. If you have an iris diaphragm plate or disc, practice locking in the different size holes to observe the changes in light intensity.
5. When viewing an object with a microscope, you are required to have the lens a certain distance from the object. This is called the working distance. At the correct working distance from an object, the object is *in focus*. Changes and adjustments in the focus are accomplished by using the *coarse* (larger knob) and *fine* (smaller knob) *adjustment knobs*, located on the arm.

B. MAGNIFICATION

1. The magnifying power of most objectives and oculars is usually engraved. The ocular lens will be marked at the top edge. The objectives are engraved on the side of the objective cylinder.
2. The low-power objective will be engraved 10×, which means that the objective marked will produce an image 10 times larger than the object on the microscope slide.
3. When using a compound microscope, you are using two sets of magnifiers. The ocular lens has a magnification of 10×, and the lower power objective has a magnification of 10×. Under low power, the objective forms an image in the body tube 10 times larger than the object on the microscope slide; the 10× ocular lens then magnifies that image another 10 times. Thus, the image that finally reaches your eye has a total magnification of 100 times.

C. USING THE MICROSCOPE

1. You are now ready to use the microscope, having reviewed its parts. Be sure the microscope is plugged into an electrical outlet; place the microscope at your lab table with the oculars toward you. Turn on the illuminator lamp. Clean the ocular lens and the objectives with *lens paper* provided by your instructor.
2. Get a microscope slide with a letter e and place it on the stage over the stage opening. Be sure your letter e covered with a cover slip is in the center of the stage opening. Secure the slide with the stage clips. Place the low-power objective in place. It will click and lock into position. Lower the objective into place by turning the coarse adjustment knob. Your microscope should have an automatic stop so you will not crack the slide. Look through the eyepiece. Notice the appearance of the letter e under low power. Bring it into focus by slowly adjusting the coarse adjustment knob by raising it above the stopping point. Now look at the letter e on your microscope slide. It is upside down but when you look through the microscope it is right side up. Now move the slide to the right; the letter e moves to the left; now move the slide to the left; the letter e moves to the right. Move the slide away from you (up), the letter e moves toward you (down). Move the slide towards you (down); the letter e moves away from you (up). These phenomena are called *inversions*. Practice moving the slide around to get used to the inversions.
3. If you want to see more detail, you must switch to the high-power objective. Using your thumb and index finger unlock the low-power objective and move the high-power objective into position by rotating the nosepiece. Now use *only* the fine focus adjustment knob to bring that part of the letter e you are viewing

(continues)

CELL STRUCTURE (Continued)

into focus. Notice how the field of vision got smaller as the object to be viewed got bigger. Your magnification now under high power is 10 × (ocular lens) × 40 × high-power objective = 400. Remember to adjust the light if necessary.

D. IMPORTANT CAUTIONS

1. Use only the fine focus adjustment knob to focus under high power; *never* use the coarse adjustment knob under high power.
2. Use only lens paper to clean the ocular lens and objectives; *never* use kimwipes or other material. Other materials may scratch the delicate lens. Always clean your lenses before lab and at the end of lab.
3. Keep the stage of the microscope dry at all times to prevent corrosion of metal parts.

E. DEPTH OF FOCUS

1. Get a microscope slide with three colored threads mounted together. Under low power (100×), focus where the three threads cross one over another. Now using the fine adjustment knob, slowly focus up and down. As you do this, notice how different parts of the threads and different threads become distinct. When one thread is in focus, the others above and below are blurred. By continually fine focusing up and down through the threads, you can perceive the depth dimension that is not evident when the focus is resting at one point.
2. Now turn to high power and notice that you can see much less depth than under low power. In fact you may not be able to distinguish one whole thread completely clearly under high power.
3. The vertical distance that will remain in focus at any particular time is called the depth of focus or depth of field. The medium in which the threads are embedded between the cover slip and the microscope slide is like the water depth in a swimming pool. When the specimen is near the top close to the cover slip it is in focus; it will go out of focus if it swims down to the bottom of the microscope slide (if it is a live specimen like a *Paramecium* that you will observe later on).

II. THE DISSECTING MICROSCOPE

1. Another common microscope in use in anatomy and other biology courses is the binocular dissecting microscope. Your instructor has set one up on demonstration at the front of the lab. This microscope has a pair of oculars, one for each eye. The distance between the oculars can be adjusted by pulling them together and by pushing them apart until you have adjusted them for the correct position of your eyes. The ocular pair has a magnification of 10×. The movable large single objective lens usually goes from 1× to 2×.
2. Place a small moss plant on the stage of the microscope (usually a large round glass plate). Move the plant around and change the objective lens to observe the arrangement of the small whorl of leaves on the stem.

III. THE ELECTRON MICROSCOPE

Light microscopes can only magnify to about 2000×. Most of our knowledge of cellular fine structure has been derived from pictures taken with an electron microscope. Electron microscopy uses a beam of electrons rather than light and magnets instead of glass lenses. Electron beams have a much shorter wavelength than visible light, and refracting them with magnets yields resolutions thousands of times greater than light. Special training is required to use an electron microscope.

Observe photographs taken of cellular details with an electron microscope. Your instructor will put on demonstration a number of electron micrographs.

IV. PREPARING YOUR OWN WET MOUNTS

Wet mounts are prepared by placing a drop of material on a *clean* microscope glass slide. If the material is dry, then place it in a drop of water on the center of the slide. A cover slip is then placed on top of the material by holding it at a 45° angle until the edge of the cover slip touches the drop of water. Then gently drop the cover slip on top of the material. The water will push air in front of it to prevent air bubbles.

A. HUMAN EPIDERMAL CELLS

1. Gently scrape the inside of your cheek with a clean flat toothpick. Prepare a microscope

(continues)

CELL STRUCTURE (Continued)

slide by placing a drop of water in the center of the slide. Place the scrapings in the drop of water in the center of the slide. Add a small amount of methylene blue stain to the drop of water by touching an eye droplet full of stain to the drop of water. Mix the stain, water, and cheek scrapings with your toothpick. Cover with a cover slip. This is the only slide we will use a stain on.

2. You have removed some of the protective epithelial cells that line your mouth. These cells are constantly being worn off and replaced by new cells. Therefore, these cells usually come off the cheek in masses.
3. Under low power of the microscope, scan your slide until you find individual cells. Now turn to high power. Notice the cells are flat and irregular in shape. Locate the nucleus, the nucleolus, the nuclear membrane, the cytoplasm, and the cell membrane. The dark granules in the cytoplasm are probably mitochondria. You may also see some vacuoles.

B. PLANT CELLS

1. Cut an onion into quarters; with forceps strip a piece of thin epidermis from the inside of one of the leaves of the onion bulb. Make a wet mount by placing it in a drop of water; try to avoid folding of the epidermal tissue.
2. Compare this to the human epithelial cell. These plant cells have rigid cell walls, easily observable under low power. Switch to high power and observe the light nucleus with a number of nucleoli. Notice the large cell vacuole and clear cytoplasm. There are no chloroplasts in the onion bulb.
3. Take a young leaf from the top of an aquatic *Elodea* or *Cabomba* plant and make a wet mount of it to observe large green chloroplasts in the cells of this plant. Notice the large number of chloroplasts in the cytoplasm of the cells under low power. There are so many chloroplasts that any other cellular contents are obscured. In young leaves, the cytoplasm streams; thus, the chloroplasts will be moved in a current within the cell. Again note the rigid cell wall made of cellulose. The cell membrane is pushed up against the cell wall so it is not visible as a separate entity.

C. LIVING SINGLE-CELLED ANIMAL: *PARAMECIUM*

Make a wet mount of *Paramecium* by putting a drop of the culture (supplied by your instructor) onto a microscope slide covered with a glass cover slip. *Paramecium* is a single-celled ciliate animal (Figure 3-16). Examine your slide under low power because the *Paramecia* are large and actively swim in the watery medium. Observe the animal's structures, especially the beating cilia seen around the edge of the cell membrane, the large macronucleus, and the oral groove.

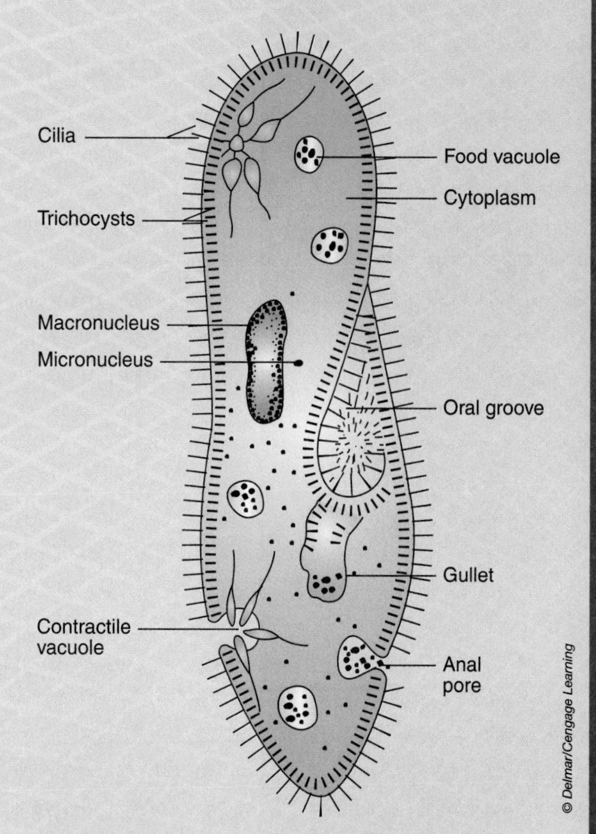

FIGURE 3-16. The structure of a *Paramecium*.

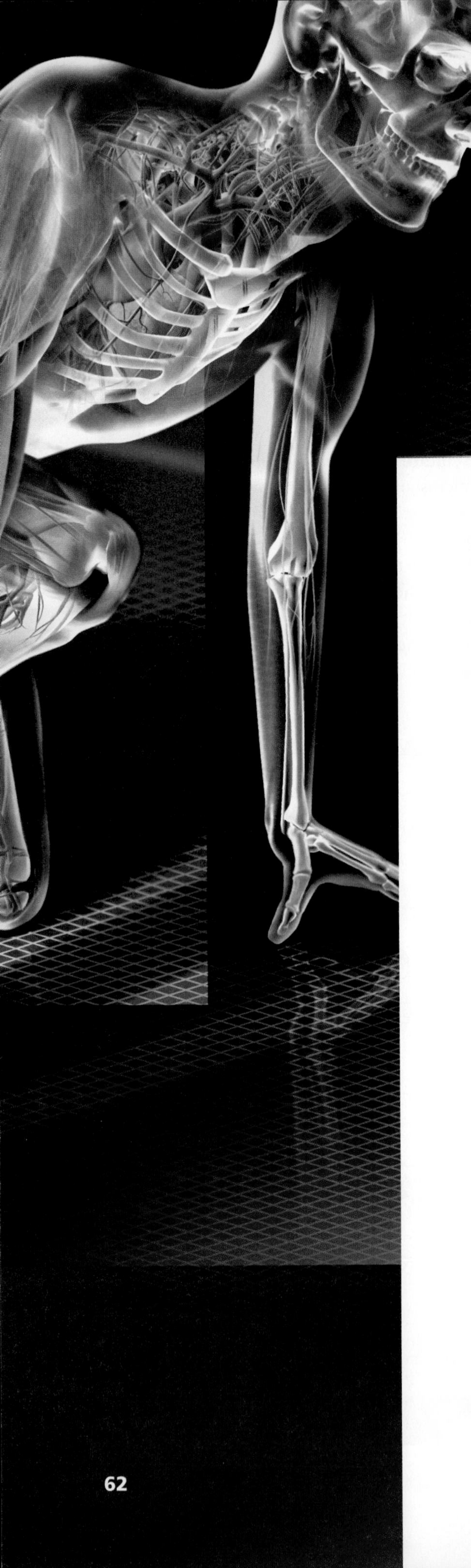

Cellular Metabolism and Reproduction: Mitosis and Meiosis

CHAPTER OBJECTIVES

After studying this chapter, you should be able to:

1. Define *metabolism*.
2. Describe the basic steps in glycolysis and indicate the major products and ATP production.
3. Describe the Krebs citric acid cycle and its major products and ATP production.
4. Describe the electron transport system and how ATP is produced.
5. Compare glycolysis with anaerobic production of ATP in muscle cells and fermentation.
6. Explain how other food compounds besides glucose are used as energy sources.
7. Name the discoverers of the anatomy of the DNA molecule.
8. Know the basic structure of the DNA molecule.
9. Name the nitrogen base pairs and how they pair up in the DNA molecule.
10. Define the stages of the cell cycle.
11. Explain the significance of mitosis in the survival of the cell and growth in the human body.
12. Understand the significance of meiosis as a reduction of the genetic material and for the formation of the sex cells.

4

KEY TERMS

Acetaldehyde 69
Acetic acid 66
Acetyl-CoA. 66
Adenine 73
Aerobic. 64
Alpha-ketoglutaric acid 66
Anabolism 64
Anaerobic respiration...... 64
Anaphase. 76
Anaphase I. 78
Anaphase II 78
Aster. 77
Calories 64
Carcinogens. 85
Carcinomas 85
Catabolism. 64
Cell cycle 76
Cell plate 78
Cellular respiration/ metabolism 64
Centromere 76
Chiasmata 81
Chromatids 76
Chromatin 76
Citric acid. 66
Cleavage furrow 78
Clones 76
Co-enzyme A. 66
Crossing-over 81
Cytochrome system 66
Cytokinesis 70
Cytosine. 73
Diploid 78
DNA (deoxyribonucleic acid) 73
Down syndrome 87
Electron transfer/ transport system 64
Ethyl alcohol 69
Fermentation. 68
Flavin adenine dinucleotide (FAD). 66
Gametogenesis 82
Gene. 73
Glucose 64
Glycolysis. 64
Guanine 73
Haploid 78
Interphase 76
Kinetochore. 77
Klinefelter's syndrome 87
Krebs citric acid cycle 64
Lactic acid 69
Malic acid. 66
Meiosis. 71
Metabolism 64
Metaphase. 76
Metaphase I. 78
Metaphase II 78
Metastases. 85
Metastasize 85
Mitosis 70
Mutation 85
Nicotinamide adenine dinucleotide (NDA) 65
Nucleic acid 71
Nucleotides 73
Oogenesis 82
Oxaloacetic acid 66
Phosphoglyceraldehyde (PGAL). 65
Phosphoglyceric acid (PGA) 65
Phosphorylation 64
Polar bodies. 82
Prophase 76
Prophase I 78
Prophase II. 78
Purines. 73
Pyrimidines 73
Pyruvic acid 64
Quinone. 66
Sarcomas 85
Spermatogenesis. 82
Spindle fibers 77
Succinic acid 66
Synapsis. 81
Tay-Sachs disease 87
Telophase. 76
Telophase I. 78
Telophase II 78
Tetrad. 81
Thymine. 73
Tubulin. 76
Tumor. 85
Zygote/fertilized egg 78

INTRODUCTION TO CELLULAR METABOLISM

For cells to maintain their structure and function, chemical reactions must occur inside the cell. These chemical reactions require an input of biologically usable energy. The most common and available form of energy within a cell is the chemical energy found within the structure of an ATP (adenosine triphosphate) molecule. We use the term **metabolism** (meh-**TAB**-oh-lizm) to describe the total chemical changes that occur inside a cell. There are two subcategories of metabolism: **anabolism** (an-**AB**-oh-lizm) is an energy-requiring process that builds larger molecules from combining smaller molecules and **catabolism** (ka-**TAB**-oh-lizm), which is an energy-releasing process that breaks down large molecules into smaller ones. These cellular metabolic processes are often called **cellular respiration** or **cellular metabolism**.

Molecules of ATP are made within the cell during a stepwise decomposition (catabolism) of organic molecules (carbohydrates, fats, and proteins). We measure the energy contained in food as **calories**. This decomposition releases the chemical energy (calories) stored in these organic foodstuffs and this energy is used to synthesize ATP (another form of chemical energy) from ADP (adenosine diphosphate) and PO_4 (inorganic phosphate). Thus, ATP is the energy source available to the cell to be used for all cell processes: chemical reactions use ATP as an energy source to maintain cellular structure and function.

Photosynthesis by plant cells is the ultimate source of the organic molecules (foodstuffs) that will be decomposed to form ATP. Photosynthesis requires $6CO_2$ + $12H_2O$ in the presence of light and chlorophyll to produce $C_6H_{12}O_6$ (glucose) an organic molecule + $6O_2$ (oxygen) as a waste product + $6H_2O$ (water) as a waste product. The formation of ATP is the final step in the transformation of light energy into the chemical energy of a biologically usable form. This explains the significance of our dependence on plants to convert sun or light energy into food or chemical energy.

The most efficient cellular process by which ATP is formed during the breakdown of organic molecules requires molecular oxygen (O_2). This process is called cellular or biochemical respiration or cellular metabolism. The overall chemical equation is:

$$C_6H_{12}O_6 + 6O_2 \rightarrow 6CO_2 + 6H_2O + \text{energy in the form of ATP.}$$

Respiration, therefore, requires an exchange of gases between the cell and its surroundings to allow the inflow of O_2 to the cell and the outflow of CO_2. Biochemical respiration is strictly the oxygen-requiring or **aerobic** process of ATP production. This biochemical meaning of respiration should not be confused with the everyday meaning of breathing. The most common substance decomposed aerobically in cells to produce ATP is glucose, $C_6H_{12}O_6$.

The breakdown of a glucose molecule into carbon dioxide gas and water is a continuous process. However, we will discuss this process in three steps. The first step is called **glycolysis** (gligh-**KOL**-ih-sis). Because it does not require oxygen, it is also occasionally called **anaerobic** (without oxygen) **respiration**. This step occurs in the cytoplasm of the cell. The next two steps are called the **Krebs citric acid cycle** and the **electron transfer** or **transport system**. These two steps require oxygen and they occur in the matrix and on the folds or cristae of the mitochondria of the cell.

CELLULAR METABOLISM OR BIOCHEMICAL RESPIRATION

Glycolysis

The first step in the biochemical respiration process is glycolysis. It is common to the aerobic breakdown of **glucose** and to the two different types of anaerobic breakdown of glucose molecules. One type of anaerobic glucose decomposition occurs in yeast cells (a type of fungus) and is called fermentation. The other type occurs in our muscle cells when we exercise and experience muscle fatigue and cannot get enough oxygen to our muscle cells. In the overall process of glycolysis, the C_6 (backbone chain of six carbon atoms) sugar glucose is slowly broken down by various enzymatic steps to two C_3 units of **pyruvic** (pye-**ROO**-vik) **acid**. Refer to Figure 4-1 as we discuss glycolysis.

The first step in glycolysis (which takes place in the cytoplasm of the cell) is the addition of a phosphate to the glucose. This process is called **phosphorylation** (fos-for-ih-**LAY**-shun). The phosphate comes from the breakdown of an ATP molecule into ADP and PO_4, releasing the energy required to add the phosphate to the glucose. The glucose phosphate quickly changes to another C_6 sugar phosphate called fructose phosphate. In another ATP requiring reaction, the fructose phosphate is phosphorylated by breaking down another ATP into ADP and PO_4. This phosphate is added to the fructose phosphate, creating fructose diphosphate. So far we have not made any ATP but rather we have used up two ATP and these must be paid back from our final ATP production at the end of glycolysis.

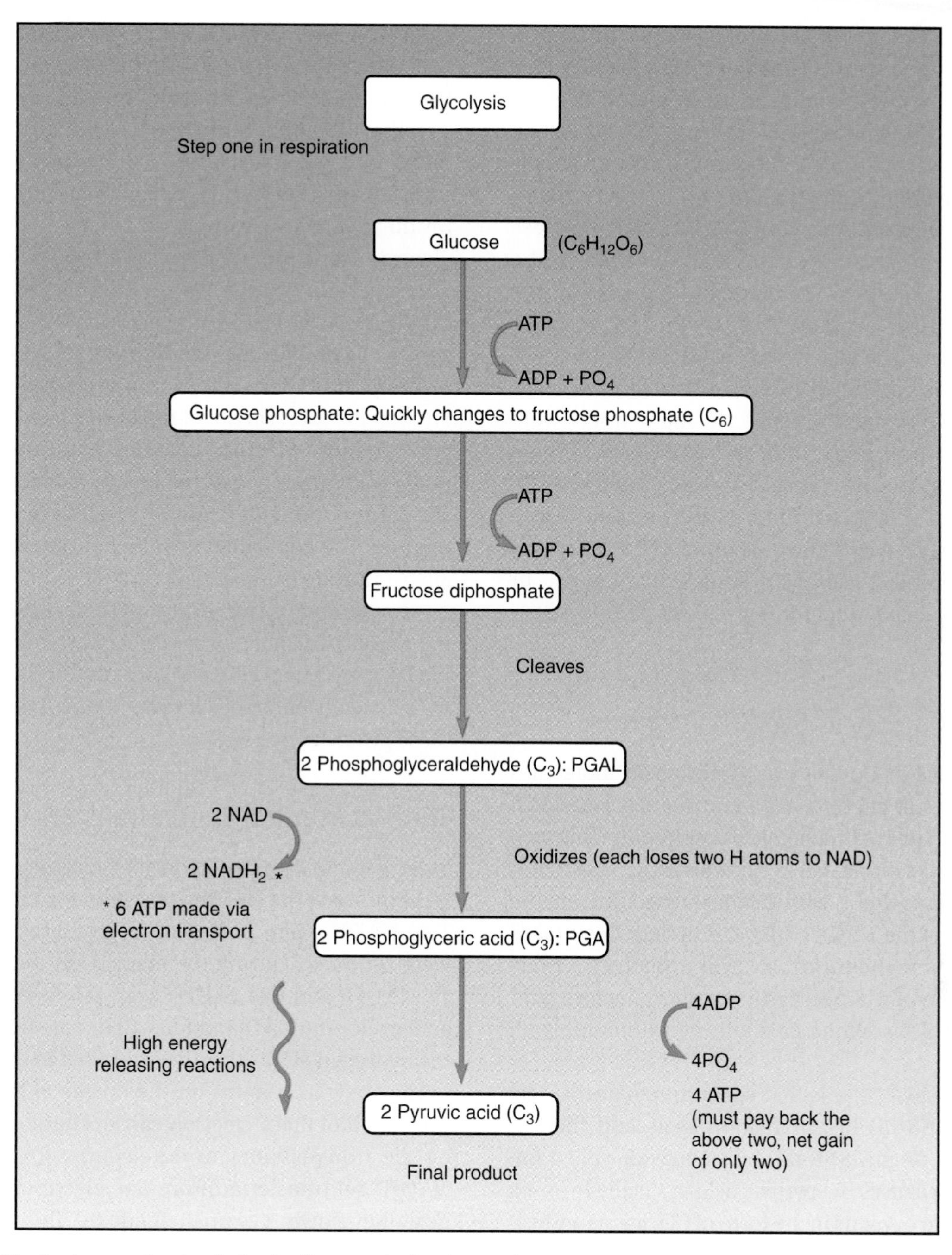

FIGURE 4-1. The basic steps in glycolysis, the first step in biochemical respiration.

In the next step of glycolysis, the fructose diphosphate splits or cleaves into two C_3 molecules of **phosphoglyceraldehyde** (fos-foh-**GLISS**-er-AL-deh-hyde), abbreviated as **PGAL**. The PGAL is now oxidized (loses electrons) by the removal of two electrons and two H+ ions to form two **phosphoglyceric** (**fos**-foh-**GLISS**-er-ik) **acids**, abbreviated as **PGA**. The two hydrogen atoms that come off each of the two PGALs go to the electron transport system and are taken up by the electron carrier molecule **nicotinamide adenine dinucleotide** (**nik**-oh-**TIN**-ah-mide **ADD**-eh-neen dye-**noo**-klee-oh-tide) abbreviated, as NAD. This step is actually part of the electron transport system and will result in the production of six ATP molecules. However, this step occurs only if oxygen is present. In this process, NAD gets reduced (gains electrons) to $NADH_2$. Because there were two PGALs, it happens twice. Each time an NAD gets reduced to $NADH_2$ and the electron transport system functions, three ATP molecules are made. Again, because it happened twice, a total of six ATP are made in this aerobic step.

Next, the two PGAs get broken down through a series of high-energy releasing enzymatic steps to two C_3 molecules of pyruvic acid. So much energy is given off in these steps that four ADP and four PO_4 get added to form four ATP molecules. The energy in the PGA molecules is converted to the high-energy four ATP molecules. In this step, we make four ATP but it is from these ATP that we must pay back the two ATP used in the beginning of glycolysis. Therefore, our net gain of ATP is only two ATP.

In summary, the glycolytic breakdown of one molecule of glucose produces two pyruvic acid molecules. It took two ATP to start the sequence and four ATP were produced. However, to pay back the two ATP, our net gain is only two ATP. However, we also produced two $NADH_2$, which are part of the electron transport system. When oxygen is present, we produce six more ATP via electron transport. Aerobic glycolysis produces six plus two or eight ATP molecules. Anaerobic glycolysis produces only two ATP.

The Krebs Citric Acid Cycle

In the presence of O_2, the two pyruvic acid molecules formed as a result of glycolysis are further broken down in the second step of biochemical respiration. This step is named after its discoverer, a German-born British biochemist, Sir Hans Krebs, who first postulated the scheme in 1937. This is the Krebs citric acid cycle (which takes place in the mitochondria). We will explain this cycle using only one of the two pyruvic acid molecules produced in glycolysis. When finished, we will multiply all products by 2.

The C_3 pyruvic acid is first converted to **acetic acid** (ah-**SEE**-tic **ASS**-id) in a transition stage and then to the C_2 **acetyl-CoA** (ah-**SEE**-tal) by an enzyme called **Coenzyme A**. This causes the pyruvic acid molecule to lose a carbon and two oxygens in the form of CO_2 gas as a waste product. It also loses two hydrogens to NAD, producing $NADH_2$ (thus, via electron transport three ATP molecules are made in this step). The acetyl-CoA now enters the Krebs citric acid cycle. This occurs on the cristae of the mitochondria (Figure 4-2).

The C_2 acetyl-CoA reacts with a C_4 molecule **oxaloacetic** (**ok**-sah-low-ah-**SEE**-tik) **acid** to form the C_6 molecule **citric acid**, hence the name of the cycle. No ATP is produced in this step but an important event occurs. CoA enzyme is regenerated to react with another acetic acid to continue the cycle. Another enzyme now converts the citric acid to the C_5 **alpha-ketoglutaric** (**AL**-fah **KEY**-toh gluh-**TAYR**-ik) **acid**. This causes the citric acid to lose a carbon and two oxygens as CO_2 gas (waste product) and two hydrogens to NAD. Thus, NAD gets reduced via electron transport to $NADH_2$ and three ATP are made.

The C_5 alpha-ketoglutaric acid now gets broken down into the first C_4 molecule **succinic** (suk-**SIN**-ik) **acid**. It loses a carbon and two oxygens as CO_2 gas (waste product) and two hydrogens twice to NAD. Thus, via electron transport six more ATP molecules are made. Succinic acid changes to another C_4 molecule, **malic** (**MAH**-lik) **acid**. Finally, the malic acid loses two hydrogens to **flavin adenine dinucleotide** (**FLAY**-vin **ADD**-eh-neen dye-**NOO**-klee-oh-tide), abbreviated as **FAD**. This is another electron carrier of the electron transport system and two more ATP molecules are made in this step. The malic acid now is converted to the oxaloacetic acid. Also going from alpha-ketoglutaric acid to oxaloacetic acid another ATP equivalent is made. This molecule is actually guanosine triphosphate (GTP).

In summary, for every pyruvic acid that enters the Krebs citric acid cycle, three CO_2, four $NADH_2$, one $FADH_2$, and one ATP (GTP) are produced. Because two pyruvic acids entered the cycle, we must multiply all of these products by 2.

The Electron Transport (Transfer) System

Most of the ATP produced during biochemical respiration is produced in the electron transport system (Figure 4-3). Two $NADH_2$ were produced in glycolysis. Two $NADH_2$ were produced during the acetyl-CoA formation. Then six $NADH_2$ and two $FADH_2$ were produced in the citric acid cycle. The NAD and FAD all donate the electrons of the hydrogen atoms that they captured in these reactions to the enzyme systems on the cristae of the mitochondria. Each of these electron carriers has a slightly different electron potential. As the electrons from the cofactor $NADH_2$ get transferred from one electron carrier to the next, they slowly give up their energy. This energy is used in the energy-requiring synthesis of ATP from ADP and inorganic phosphate.

The electron transport system functions as a series of reduction/oxidation reactions. When NAD accepts the two hydrogens, it gets reduced to $NADH_2$. When it gives up the two hydrogens to FAD, NAD gets oxidized while FAD becomes $FADH_2$ and gets reduced. This series of redox reactions continues until the electrons of the hydrogen atoms get ultimately donated to oxygen. Several kinds of electron carriers participate in this process: the cofactor NAD, the cofactor FAD, **quinone**, and the **cytochrome system**. There is some debate as to whether the hydrogen protons ($2H^+$) are transferred along with

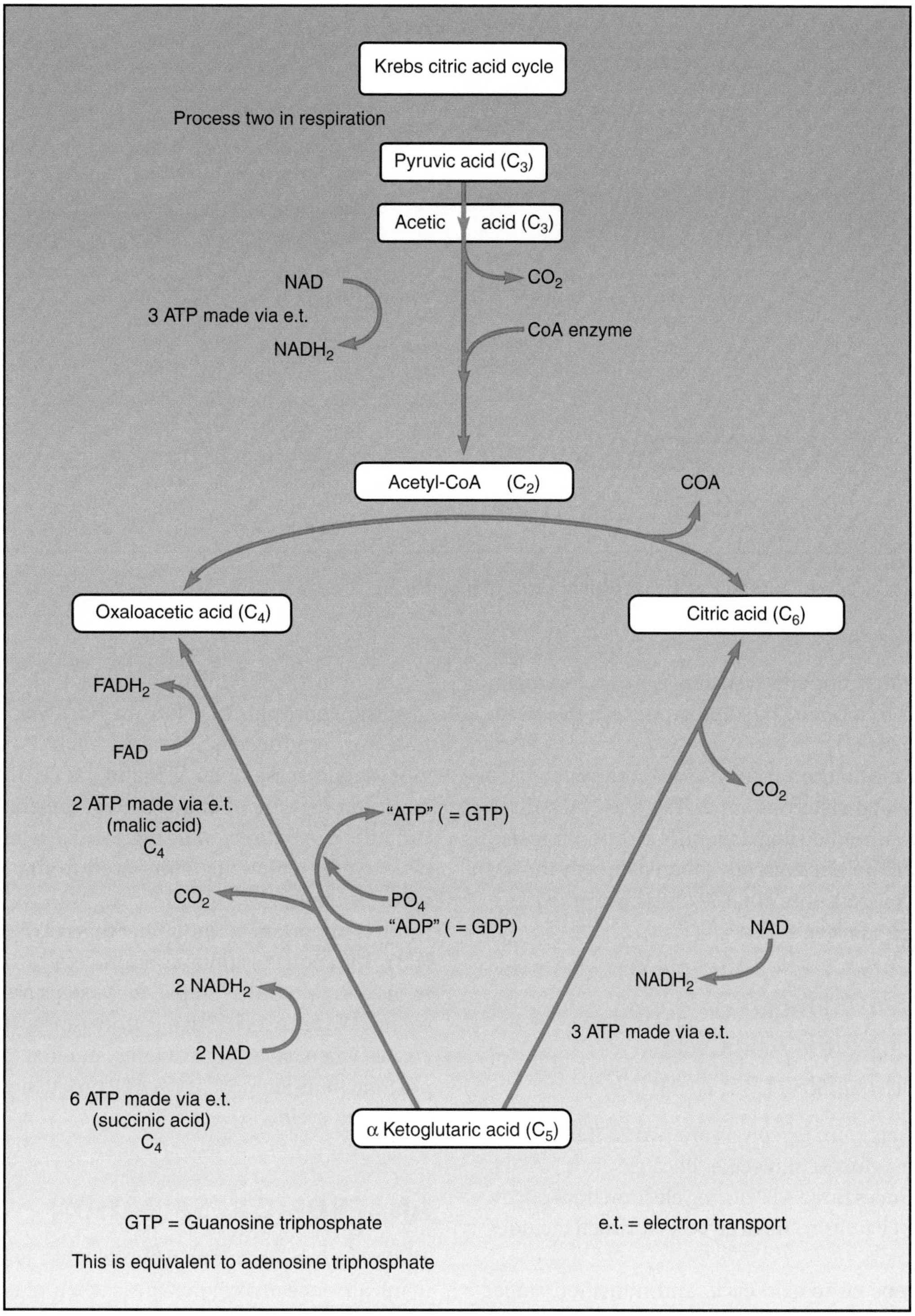

FIGURE 4-2. The Krebs citric acid cycle and its products.

the electrons ($2e^-$) in this transport or not. A currently accepted scheme is shown in Figure 4-3.

This scheme illustrates why the breakdown of glucose requires oxygen (O_2). Oxygen is the ultimate electron acceptor for the electrons captured by the cofactors during glucose decomposition. One ATP is formed during the first step of electron transfer from $NADH_2$ to FAD. During the following transfer from $FADH_2$ to quinone H_2 to the cytochrome system to O_2 (or ½ O_2 = O), two more units of ATP are formed. You will notice that the cytochrome system only accepts the two electrons and then transfers them to oxygen (O). Therefore,

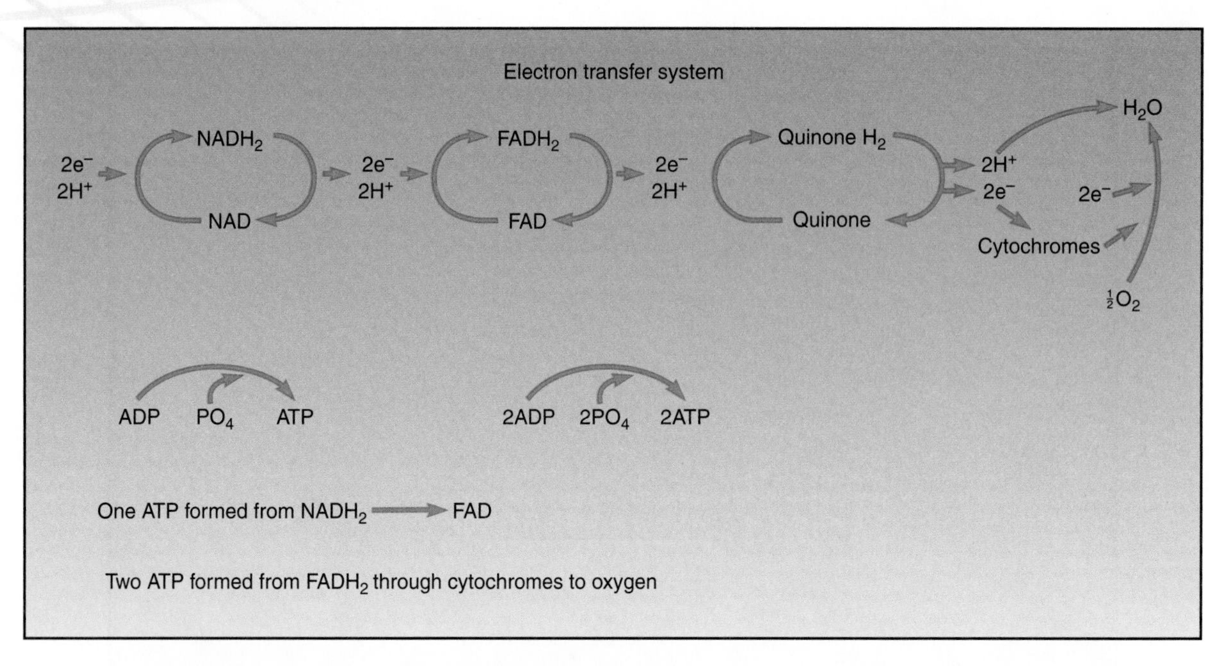

FIGURE 4-3. The electron transport or transfer system and ATP production.

quinone H_2 must directly transfer the two hydrogen protons ($2H^+$) to oxygen (O), thus producing the waste product water (H_2O).

As we examine the electron transport system, we observe that when electrons are donated to NAD, three ATP units are formed during the entire electron transfer. However, when the electrons are donated directly to FAD and NAD is bypassed, only two ATP units are formed during the electron transfer.

Summary of ATP Production during Glycolysis, the Citric Acid Cycle, and Electron Transport

The net products from glycolysis are two ATP units and two $NADH_2$ per glucose molecule. Because each $NADH_2$ molecule produces three ATP during electron transport, a total of eight ATP units result in glycolysis, which includes electron transport.

In the Krebs citric acid cycle and transition stage, four $NADH_2$, one $FADH_2$, and one ATP (or GTP) are formed during the breakdown of each pyruvic acid. However, because each glucose molecule produces two pyruvic acid molecules, we actually form eight $NADH_2$, two $FADH_2$, and two ATP (or GTP) units. The number of ATP units formed during the citric acid cycle and electron transport then is $24 + 4 + 2 +$ or 30 ATP or $24 + 4 = 28$ ATP and 2 GTP.

In total, 30 ATP from the citric acid cycle and electron transport plus 8 ATP from glycolysis and electron transport produced a net gain of 38 ATP units per each glucose molecule or 36 ATP and 2 GTP. This represents a cellular capture of about 60% of the energy available from the breakdown of a single glucose molecule. This is very high efficiency compared to that of any man-made machine.

It is important to remember that cellular or biochemical respiration is a continuous process. Although we tend to discuss it in three "steps," these steps are not separate events. We have seen that electron transport is part of glycolysis when oxygen is available and that electron transport accounts for most of the ATP production in the Krebs citric acid cycle.

ANAEROBIC RESPIRATION

There are two situations when glucose is broken down in the absence of oxygen. One is when yeast cells (a type of fungus) feed on glucose, and this process is called **fermentation**. The other situation occurs in our muscle cells when we overexercise and experience muscle fatigue and cannot get enough oxygen to the muscle cells. Then the muscle cells begin to break down glucose in the absence of oxygen, a much less efficient breakdown with less ATP produced. We will now look at these two anaerobic processes.

Fermentation

Fermentation is the process by which yeast breaks down glucose anaerobically (in the absence of oxygen). The final products of fermentation are: carbon dioxide gas (CO_2), **ethyl alcohol** (CH_3CH_2OH), and ATP. In yeast cells, glucose breaks down, as in glycolysis, to produce two molecules of pyruvic acid, a net gain of two ATP and two $NADH_2$. However, because oxygen is not used, the pyruvic acid molecules do not proceed to the citric acid cycle. Instead a yeast enzyme called a decarboxylase breaks down the pyruvic acid to CO_2 and a C_2 compound, **acetaldehyde** (ass-et-**AL**-deh-hyde) (CH_3CHO). It is the CO_2 gas that causes bread to rise and is the reason we add yeast to our flour (glucose), water, and eggs (which makes dough) when we bake bread. Because this process occurs without oxygen, the $NADH_2$ does not give its electrons to oxygen through the electron transport system as it does in aerobic respiration. Instead the $NADH_2$ donates its two hydrogen atoms to the acetaldehyde through the action of another yeast enzyme called an alcoholic dehydrogenase. This reaction regenerates the NAD and forms the final product ethyl alcohol. This product is what is produced in the beer, wine, and liquor industries to convert the sugars in grapes and the sugars in grains to alcohol.

In conclusion, the fermentation process produces only two ATP per glucose molecule. Obviously, this energy-capturing mechanism is much less efficient than aerobic respiration.

Anaerobic Production of ATP by Muscles

The second situation that can occur in anaerobic respiration is the breakdown of glucose in human muscle cells when not enough oxygen becomes available due to muscle fatigue such as when an athlete sprints. Again this process starts with glycolysis. However, the pyruvic acid formed undergoes a different fate. Again glycolysis yields two pyruvic acid molecules, a net gain of two ATP molecules, and two $NADH_2$ per glucose molecule. As it was in fermentation, the two $NADH_2$ cannot donate their electrons to oxygen. Instead the $NADH_2$ donates them to pyruvic acid to form **lactic** (**LAK**-tik) **acid**. It is the accumulation of lactic acid that causes the momentary fatigue in muscles that are overexercised. When muscles are overworked, the muscle cells need to produce extra energy in the form of ATP. Aerobic respiration produces much of this energy. However, if the muscle is worked more rapidly than oxygen (O_2) can be supplied to it from the bloodstream, the muscle cells will begin to produce the ATP anaerobically and lactic acid accumulates. Once oxygen gets to the muscle, the fatigue diminishes as lactic acid is broken down.

When we overexercise and our muscles get sore and we experience muscle fatigue, we notice that our heartbeat and breathing rates are accelerated. We sit down, breathe faster (to get more O_2 into our bodies), and the fatigue slowly diminishes. When O_2 again becomes available, the lactic acid is converted back to pyruvic acid and aerobic respiration proceeds as normal. We note that anaerobic formation of ATP by muscles is much less efficient than aerobic respiration. Only two molecules of ATP are produced per glucose molecule.

PRODUCTION OF ATP FROM GENERAL FOOD COMPOUNDS

Obviously, we do not only eat glucose. So where do the other food compounds in our diet fit into the respiration cycle to produce ATP? If we think of the steps in biochemical respiration as parts of a very efficient cellular furnace where fuel (food) is converted to another form of chemical energy, ATP, then we can grasp a better understanding of how other food molecules are burned to produce ATP (Figure 4-4).

Glucose is a simple carbohydrate. Other carbohydrates such as starch (plant carbohydrate) and glycogen (animal starch) as well as other types of sugars such as monosaccharides and disaccharides fit into the cellular furnace at the level where glucose enters the glycolytic sequence. If after digestion the food molecules are not needed immediately, they can be stored in the body (in food vacuoles or the liver, or converted to fat cells) until needed later to produce more ATP.

Digestion decomposes fat into fatty acids and glycerol. They, too, will enter the cellular furnace at a stage related to their chemical structure. Glycerol, a C_3 molecule, is similar to PGA and will enter at the PGA stage of glycolysis. Fatty acids enter the Krebs citric acid cycle. Proteins are broken down by digestion into amino acids. Again, they will enter the cellular furnace at a level related to their chemical structure. Alanine, a C_3 amino acid, and lactic acid enter at the pyruvic acid stage. Glutamic acid, a C_5 amino acid, is similar to alpha-ketogluteric acid. Aspartic acid, a C_4 amino acid, resembles oxaloacetic acid. These amino acids enter into the citric acid cycle at different stages. So when you put that piece of candy into your mouth during

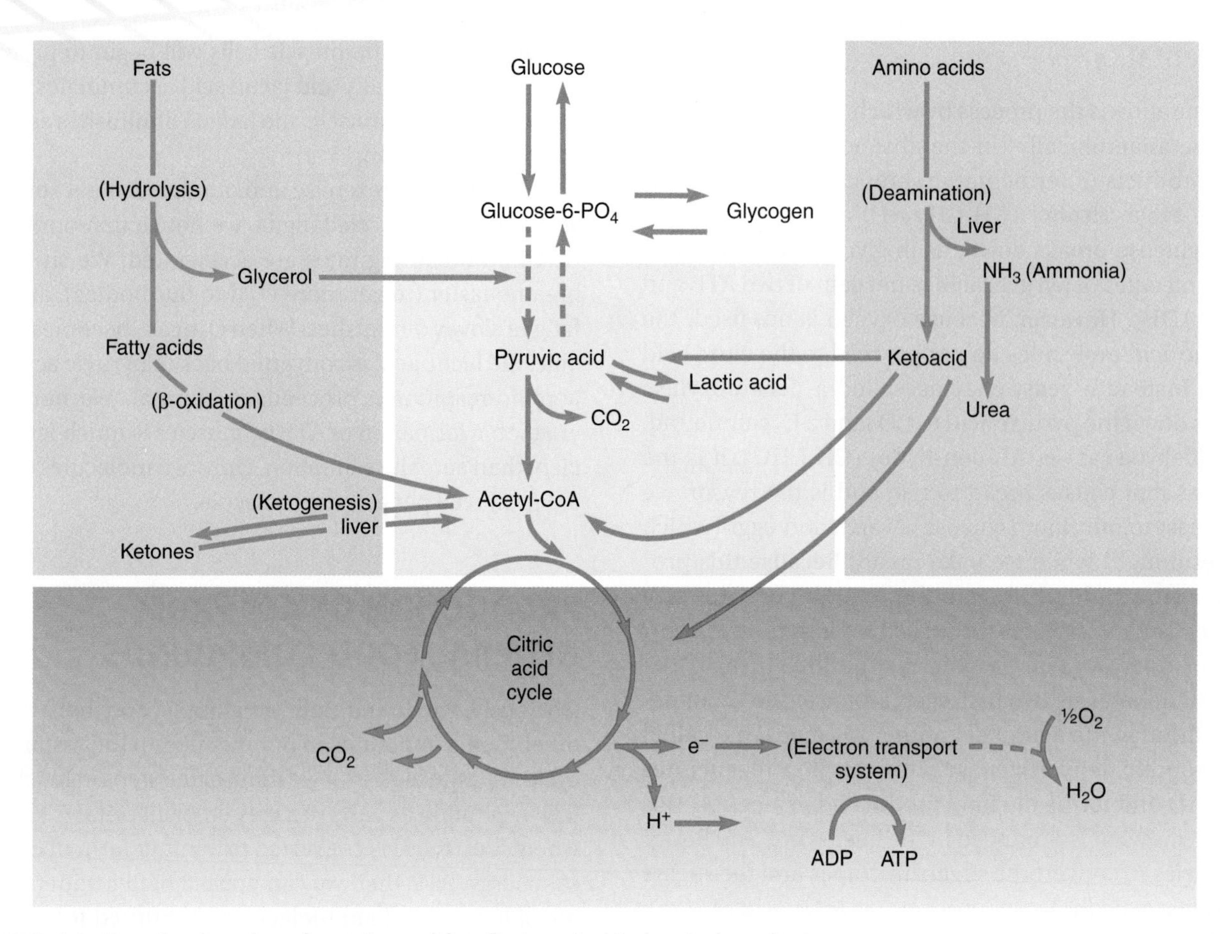

FIGURE 4-4. How the digestion of proteins and fats fits into the biochemical respiration process.

class break to get some extra energy to finish class, you now will have a better understanding of how that carbohydrate is converted to ATP (Figure 4-5), the fuel that runs our cells. The simplest way of describing cellular or biochemical respiration is to begin the process with a glucose molecule (Figure 4-6). Glycolysis occurs in the cytoplasm of the cell and produces pyruvic acid. If oxygen is available, the pyruvic acid is eventually converted to acetyl-CoA, which then enters the citric acid cycle, eventually being converted to CO_2, H_2O, and 38 ATP. If oxygen is not available, the pyruvic acid is converted to lactic acid and only two ATP molecules are produced.

SUMMARY OF ATP PRODUCTION FROM ONE GLUCOSE MOLECULE

Table 4-1 summarizes products produced and the total ATP produced in the individual stages of the cellular metabolism of one glucose molecule. The stages are broken down into glycolysis, acetyl-CoA production, and the citric acid cycle.

INTRODUCTION TO CELLULAR REPRODUCTION

Cellular reproduction is the process by which a single cell duplicates itself. In this process the genetic material in the nucleus is duplicated during interphase of the cell cycle followed by the process called **mitosis** (my-**TOH**-sis) when the nuclear material is replicated. This is followed by duplication of the cellular organelles in the cytoplasm called **cytokinesis** (sigh-toh-kye-**NEE**-sis), which is the final event of mitosis leading to two new daughter cells. These processes, part of the cell cycle, allow our bodies to grow, repair themselves, and maintain our structures and functions. In other words, these processes allow us to maintain our life.

However, cellular reproduction is also the process by which our genetic material is passed on to our offspring from one generation to the next. In this process of cellular reproduction, special cells called sex cells, the egg and the sperm, are produced. In this type of cellular reproduction, the genetic material must not only be duplicated, but it must also be reduced in half so that the female egg carries half of the genetic material or 23 chromosomes and the

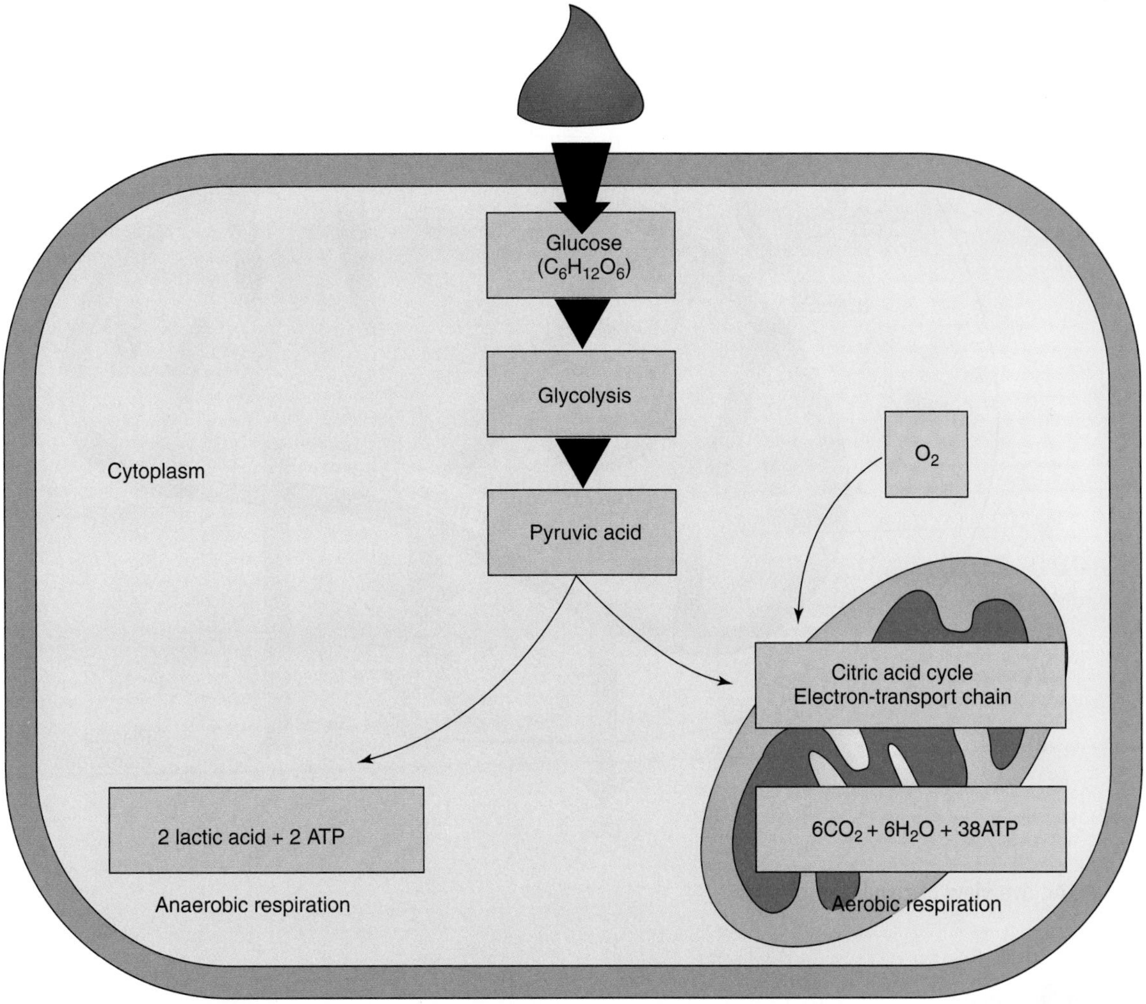

FIGURE 4-5. How a piece of candy gets metabolized to ATP.

male sperm carries the other one half of the genetic material or the other 23 chromosomes. A special kind of cellular reduction division called **meiosis** (my-**OH**-sis), which occurs only in the gonads, allows this to occur. When a sperm and and egg unite in fertilization, the genetic material is returned to its full complement of 46 chromosomes.

Before we study these processes of cellular division, it is necessary to understand the basic structure of the DNA molecule, which constitutes the genetic material. We shall also examine the history of the discovery of the structure of DNA.

THE STRUCTURE OF THE DNA MOLECULE

To better understand the structure of the DNA molecule, the history of the discovery of DNA and the anatomy of the DNA molecule will be discussed.

The History of the Discovery of DNA

One of the most significant discoveries in biology of the 20th century was the discovery of the three-dimensional structure of the DNA molecule. A number of scientists made various contributions to our knowledge of the DNA molecule. The molecule itself was discovered in 1869 by a German chemist, Friedrich Miescher. He extracted a substance from the nucleus of human cells and the sperm of fish. He called it nuclein because it came from the nucleus. Because this material was slightly acidic, it became known as **nucleic acid**.

It was not until the 1920s that any further discoveries were made. A biochemist, P. A. Levine, discovered that DNA contained three main components: phosphate (PO_4) groups, five carbon sugars, and nitrogen-containing bases called purines (adenine and guanine) and pyrimidines (thymine and cytosine).

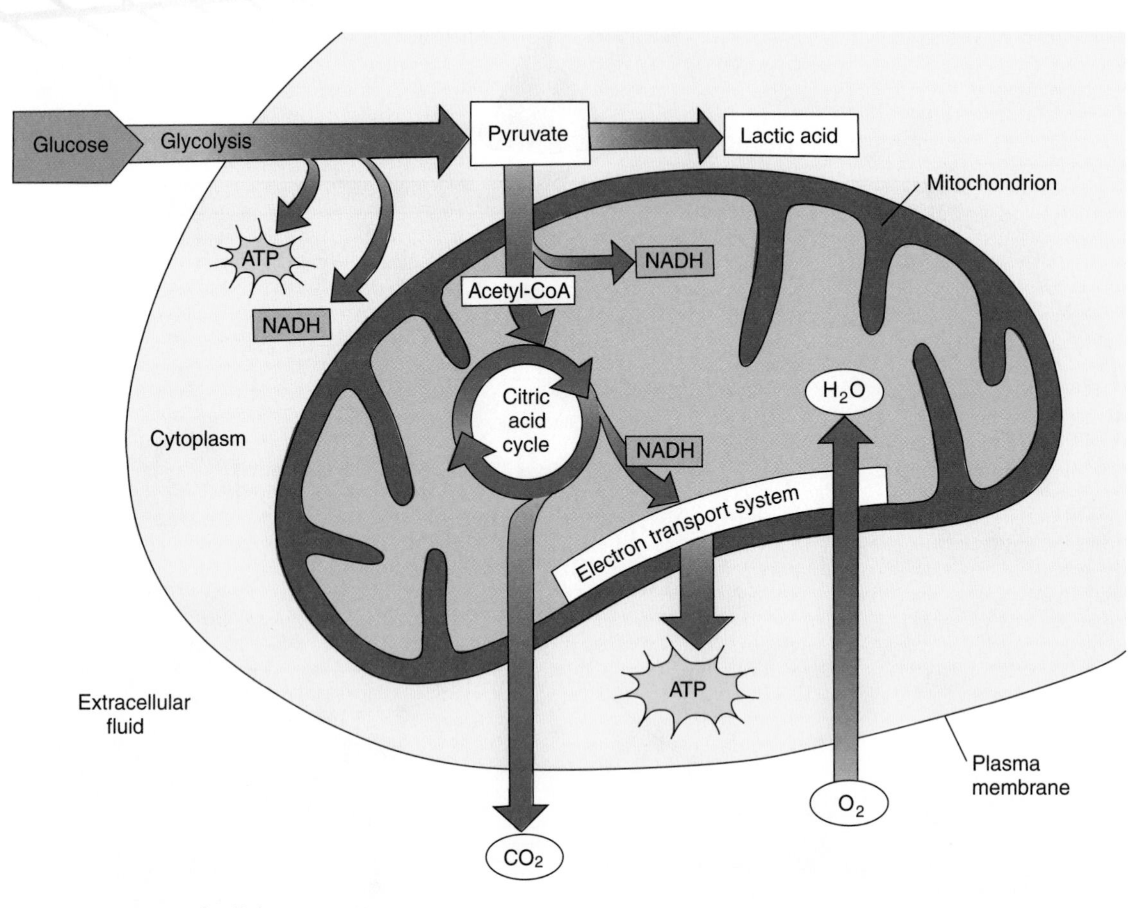

FIGURE 4-6. An overview of cellular respiration.

Table 4-1 ATP Production by Cellular Respiration

Step	Product	Total ATP Produced
Glycolysis	4 ATP	2 ATP (4 ATP produced minus 2 ATP to start cycle)
	2 $NADH_2$	6 ATP
Acetyl-CoA production	2 $NADH_2$	6 ATP
Citric acid cycle	2 ATP or 2 GTP	2 ATP or 2 GTP
	6 $NADH_2$	18 ATP
	2 $FADH_2$	4 ATP
		38 ATP or 36 ATP and 2 GTP

The actual three-dimensional structure of DNA was discovered in the 1950s by three scientists. It was a British chemist, Rosalind Franklin, who discovered that the molecule had a helical structure similar to a winding staircase. This was accomplished when she conducted an X-ray crystallographic analysis of DNA. Her photograph was made in 1953 in the laboratory of another British biochemist, Maurice Wilkins. Two other researchers were also studying the DNA molecule at this time: James Watson, an American postdoctoral student, and an English scientist, Francis Crick, at Cambridge University in England. After learning informally of Rosalind Franklin's discovery, they worked out the three-dimensional structure of the DNA molecule. Rosalind Franklin's discovery of the helical nature of DNA was published in 1953, but Watson and Crick learned of her results before they were published.

James Watson and Francis Crick won the Nobel Prize in 1962 after publishing their results. Rosalind Franklin, meanwhile, had tragically died of cancer prior to this event. Today, however, these three are given credit for discovering the structure of DNA, the molecule that contains all the hereditary information of an individual. An interesting account of the discovery of the nature of the molecule was published in 1968 by James Watson in his book *The Double Helix*. This discovery opened up whole new fields of research for the 20th century: recombinant DNA, the Human Genome Project, and genetic engineering.

The Anatomy of the DNA Molecule

DNA (deoxyribonucleic acid) is the hereditary material of the cell. It not only determines the traits an organism exhibits, but it is exactly duplicated during reproduction so that offspring exhibit their parents' basic characteristics. An organism's characteristics are due to chemical reactions occurring inside our cells. DNA governs these chemical reactions by the chemical mechanism of controlling what proteins are made.

Every DNA molecule is a double helical chain of **nucleotides** (**NOO**-klee-oh-tides; Figure 4-7). A nucleotide consists of a phosphate group (PO_4), a five-carbon sugar (deoxyribose), and an organic nitrogen-containing base, either a purine or a pyrimidine. There are two **purines** (**PYOO**-reenz), **adenine** (**ADD**-eh-neen) and **guanine** (**GWAHN**-een), and two **pyrimidines** (pih-**RIM**-ih-deenz), **thymine** (**THYE**-meen) and **cytosine** (**SYE**-toh-seen). Adenine always pairs up with thymine and guanine always pairs up with cytosine. Bonds form between the phosphate group of one nucleotide and the sugar of the next nucleotide in a chain. The organic nitrogen base extends out from the sugar of the nucleotide. It is easier to visualize the double helical nature of the DNA if we think of it as a spiral staircase. The handrails of the staircase are composed of the phosphate-sugar chain, and the stairs of the staircase are the nitrogen base pairs.

If we look at Figure 4-8, we see that a pyrimidine always pairs with a purine. A pyrimidine is a single ring of six atoms (thymine and cytosine); a purine is a fused double ring of nine atoms (adenine and guanine). These organic nitrogen bases are a complex ring structure of carbon and nitrogen atoms. Because we know how the bases pair up in the double chain of nucleotides, if we only know one side of the helix, we can figure out the second side by matching bases. The two chains of the helix are held together by weak hydrogen bonds between the base pairs. There are two hydrogen bonds between the pyrimidine thymine and the purine adenine, whereas there are three hydrogen bonds between the pyrimidine cytosine and the purine guanine. Because of the specific pairing of bases, the sequence of bases in one chain determines the sequence of bases in the other. We therefore refer to the two chains as complements of each other. A **gene** is a sequence of organic nitrogen base pairs that codes for a polypetide or a protein.

A major project of the 20th century that developed from Watson, Crick, and Franklin's discovery of DNA structure was the Human Genome Project. The objective of this project was to identify all the genes on all 46 chromosomes (DNA molecules). We know now that there are approximately 3 billion organic base pairs that code over 30,000 genes. We can think of the bases adenine (A), thymine (T), cytosine (C), and guanine (G) as the four letters of the alphabet of life. These base pairs determine all the characteristics of all the life we know on our planet—the basic structure of the DNA molecule is the same for all living organisms.

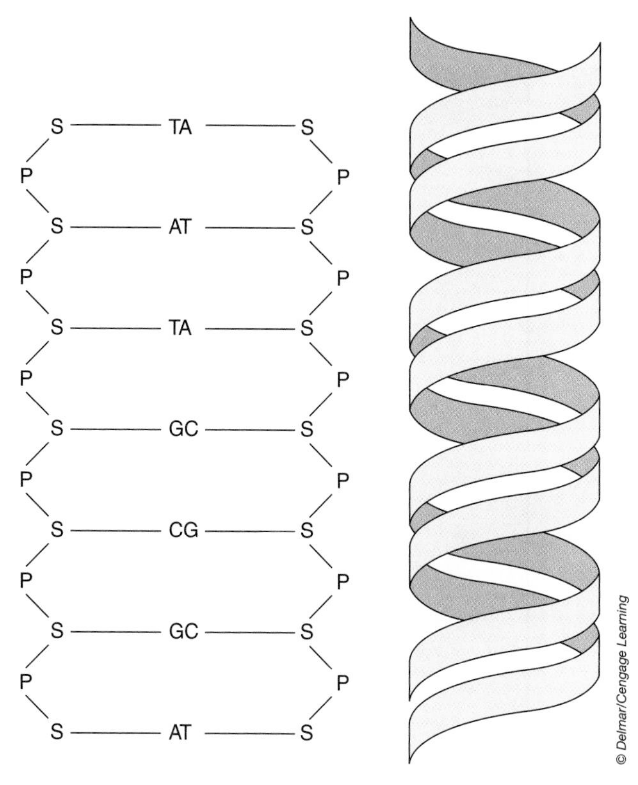

FIGURE 4-7. The double helical chain of nucleotides of a DNA molecule (a very short section).

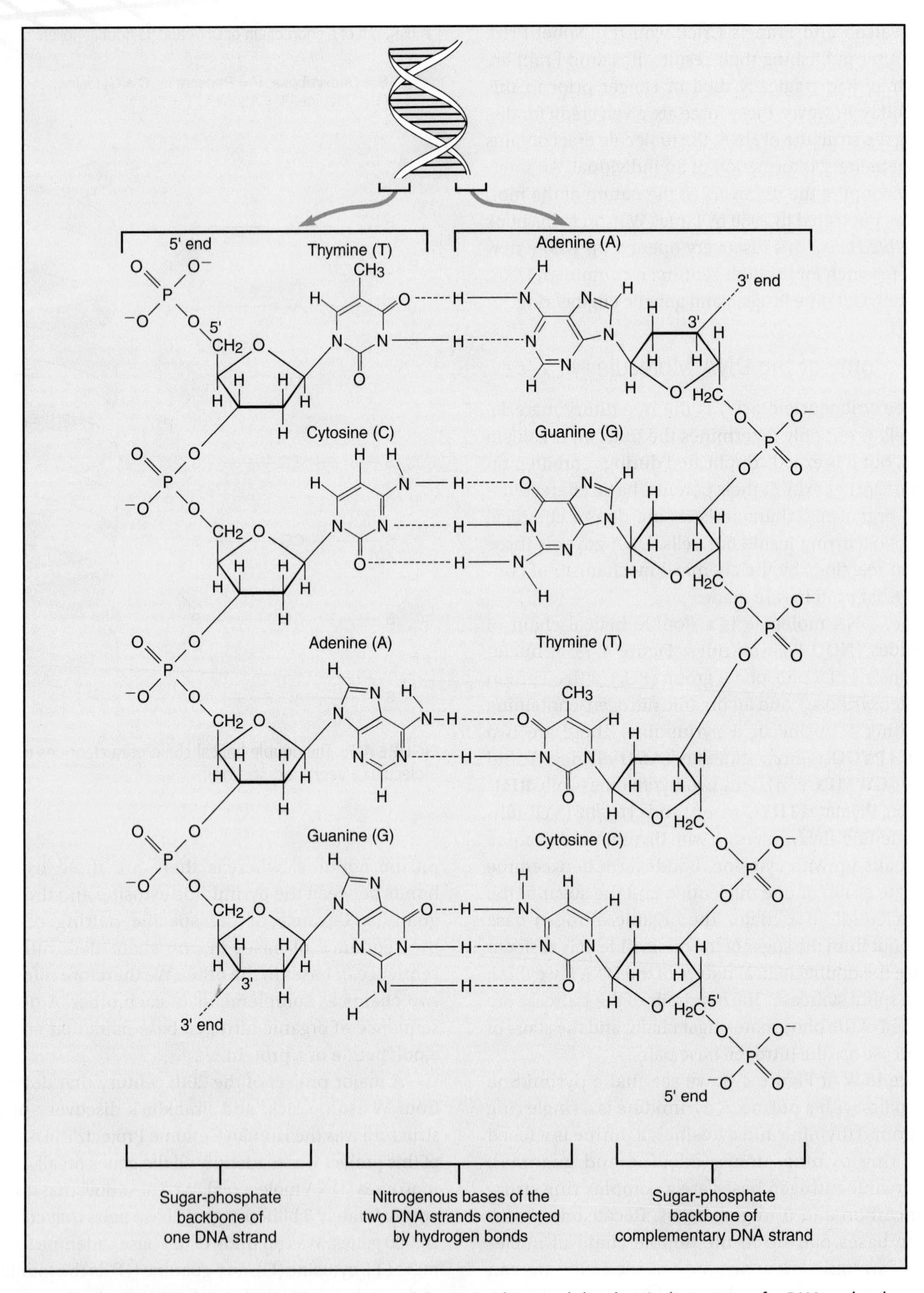

FIGURE 4-8. The hydrogen bonds between the purines and the pyrimidines and the chemical structure of a DNA molecule.

The DNA molecule must be duplicated before cell division. The molecule separates where the hydrogen bonds hold the two chains of nucleotides together and a new copy of the DNA chain is constructed (Figure 4-9). The first step is the unwinding of the molecule. This is accomplished by helicase enzymes that separate the hydrogen bonds between the base pairs and stabilize the nucleotide chains of the double helix. Then new nucleotides are added to the separated chains by DNA polymerase, another enzyme. In this way, a copy of the DNA molecule is constructed.

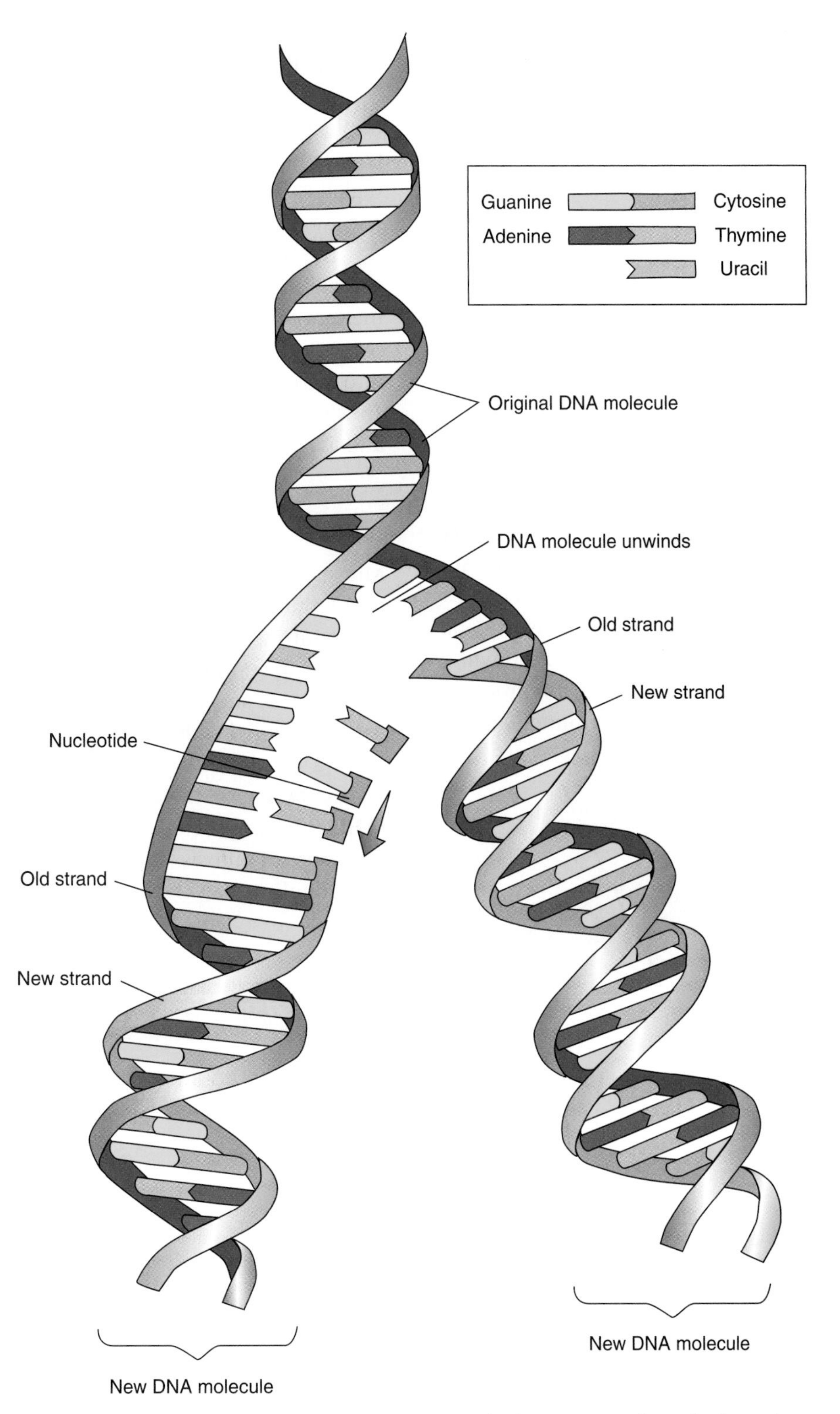

FIGURE 4-9. DNA replication. The two chains of nucleotides separate at the base pairs and are duplicated.

THE CELL CYCLE

All reproduction begins at the cellular level. The process by which a cell divides into two and duplicates its genetic material is called the **cell cycle** (Figure 4-10). The cell cycle is divided into three main stages: interphase (the stage in which great activity is occurring but this activity is not visible; thus, this stage used to be called a "resting stage"), mitosis, and cytokinesis. Two of these three stages have substages. We shall discuss all stages in detail. The time to complete a cell cycle will vary greatly among different organisms. Cells in a developing embryo will complete the cell cycle in less than 20 minutes. A dividing mammalian cell will complete the cycle in approximately 24 hours. Other cells in our bodies rarely duplicate and undergo the cell cycle, such as nerve cells and muscle cells. Human liver cells will divide only if damaged. They usually have cell cycles lasting a full year.

Interphase

Refer to Figure 4-10 for illustrations of the stages of the cell cycle discussed below. A cell spends most of its time in the stage of the cell cycle known as **interphase**. This phase, the longest and most dynamic part of a cell's life, is not part of cell division. In fact, interphase means between phases. Yet during this time the cell is growing, metabolizing, and maintaining itself. During this time, the nucleus is seen as a distinct structure surrounded by its nuclear membrane. Inside the nucleoplasm the unwound strands of chromosomes are only visible as dark threads called **chromatin** (**KRO**-mah-tin). Interphase has three subphases: growth one (G_1), synthesis (S), and growth two (G2). Some authors called the G phases gap one and gap two.

G_1 is the primary growth phase of the cell. It occupies the major portion of the life span of the cell.

The synthesis or S phase is when the strands of DNA duplicate themselves. Each chromosome now consists of two sister **chromatids** attached to each other at a central region called the **centromere** but are not yet visible. Most chromosomes consist of 60% protein and 40% DNA.

The G_2 phase is the final phase for the preparation of cell division. In animal cells, the centrioles begin movement to the opposite poles of the cell. Mitochondria are replicated as the chromosomes now condense and coil into tightly compacted bodies. **Tubulin** is synthesized. This is the protein material that forms the microtubules and assembles at the spindle.

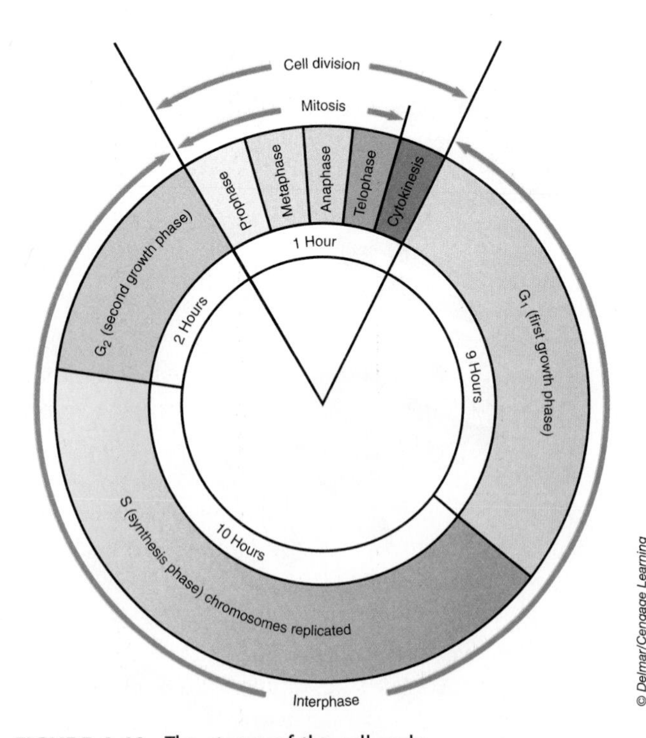

FIGURE 4-10. The stages of the cell cycle.

Mitosis

Mitosis is the process of cellular reproduction that occurs in the nucleus and forms two identical nuclei. Because of the intricate movement of daughter chromosomes as they separate, this phase of the cell cycle has received a great deal of study by biologists. This phase can also be easily observed with a light microscope. Although mitosis is a continuous process, it is subdivided into four stages: **prophase**, **metaphase**, **anaphase**, and **telophase** (**TELL**-oh-faze). Refer to Figure 4-11 for an illustration of the stages of mitosis. The cells resulting from mitosis are exact duplicates or **clones** of the parent cell.

StudyWARE™ Connection

Watch an animation on mitosis on your StudyWARE™ CD-ROM.

Prophase

The coiled, duplicated chromosomes have shortened and thickened and are now visible. Each chromosome consists of two sister or daughter chromatids. The sister chromatids remain attached to one another at the **centromere** (**SIN**-troh-meer). The centromere is a constricted or pinched-in area of the chromosome where a disk of protein called the

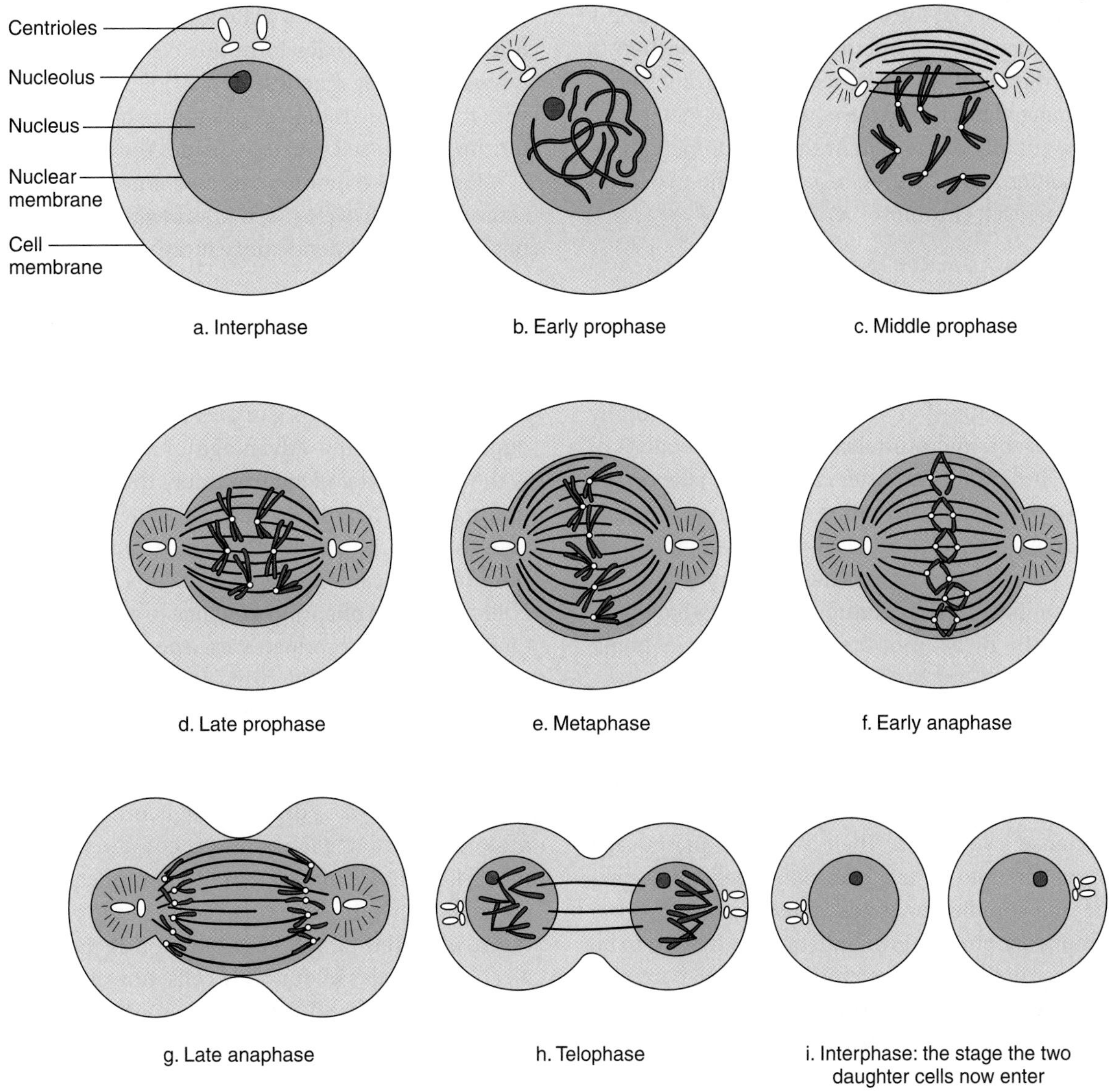

Centrioles
Nucleolus
Nucleus
Nuclear membrane
Cell membrane

a. Interphase
b. Early prophase
c. Middle prophase
d. Late prophase
e. Metaphase
f. Early anaphase
g. Late anaphase
h. Telophase
i. Interphase: the stage the two daughter cells now enter

FIGURE 4-11. An illustration of interphase and the stages of mitosis.

kinetochore (kye-**NEE**-toh-kor) is found. In animal cells, the centriole pair begins to move apart to the opposite poles of the cell forming a group of microtubules between them called the **spindle fibers**. In plant cells, a similar group of spindle fibers forms even though there are no centrioles. As the centrioles move apart to the opposite poles of the cell, they become surrounded by a cluster of microtubules of tubulin that radiate outward looking like a starburst. This starburst form is called the **aster**. The spindle fibers form between the asters. As these fibers form, they push the centrioles to the opposite ends of the cell and brace the centrioles against the cell membrane. At this time the nuclear membrane breaks down and its components become part of the endoplasmic reticulum. The nucleolus is no longer visible.

Each chromosome has two kinetochores, one for each sister chromatid. As prophase continues, a group of microtubules grow from the poles to the centromeres of the chromosomes. The microtubules attach the kinetochores to the poles of the spindle. Because these microtubules coming from the two poles attach to opposite sides of the centromere, they attach one sister chromatid to one pole and the other sister chromatid to the other, thus ensuring separation of the sister chromatids, each one going to a different daughter cell.

Metaphase

Metaphase is the second stage of mitosis and begins when the sister chromatids align themselves at the center of the cell. The chromosomes are lined up in a circle along the inner circumference of the cell called the equator of the cell. Held in place by the microtubules and attached to the kinetochore of their centromeres, the chromosomes become arranged in a ring at the equatorial or metaphase plate in the middle of the cell. At the end of metaphase each centromere now divides, separating the two sister chromatids of each chromosome.

Anaphase

Anaphase is the shortest stage of mitosis and is one of the most dynamic stages to observe. The divided centromere, each with a sister chromatid, moves toward the opposite poles of the spindle. The motion is caused by the pulling of the microtubules on the kinetochore of each sister chromatid. The sister chromatids take on a V shape as they are drawn to their respective poles. At this time, the poles also move apart by microtubular sliding and the sister chromatids are drawn to opposite poles by the shortening of the microtubules attached to them. Cytokinesis, the division of the cytoplasm, may begin in anaphase.

Telophase

The final stage of mitosis is telophase. The sister chromatids, which now can be called chromosomes, begin to decondense and uncoil. Their V-shaped or sausage form disappears into diffuse chromatin, becoming long and thin. The spindle apparatus is disassembled as the microtubules are broken down into units of tubulin to be used to construct the cytoskeleton of the new daughter cells. A nuclear membrane forms around each group of daughter chromosomes. Cytokinesis is nearly complete. In animal cells the centrioles duplicate. Plant cells do not have centrioles.

Cytokinesis

The process of cell division is not yet complete because the actual separation of the cell into two new daughter cells has yet to occur. The phase of the cell cycle in which actual cell division occurs is called cytokinesis.

In animal cells, cytokinesis occurs as the cells separate by a furrowing in or pinching in of the cell membrane referred to as a **cleavage furrow**. The cell membrane indents to form a valley outside the spindle equator. This furrow first appears in late anaphase, and in telophase, it is drawn in more deeply by the contraction of a ring of actin filaments that lie in the cytoplasm beneath the constriction points. As constriction proceeds, the furrow extends into the center of the cell and thus the cell is divided into two.

In plant cells, a **cell plate** forms at the equator. Small membranous vesicles form this cell plate, which grows outward until it reaches the cell membrane and fuses with it. Cellulose is then deposited in this new membrane forming a new cell wall that divides the cell in two.

Each new daughter cell now enters the interphase stage of the cell cycle. Each now begins its growth phase until it is ready to divide once more.

MEIOSIS: A REDUCTION DIVISION

In sexual reproduction, two specialized cells (the sperm and the egg) known as gametes unite to form a **fertilized egg** or **zygote**. The advantage of sexual reproduction is the increased genetic variability that results from the uniting of the hereditary material of two different organisms (humans). This results in a new individual, similar to but not identical to either parent. This new genetic variability gives the offspring a chance to adapt to a changing environment. To produce these special cells or gametes, a special kind of cellular division must occur. This special kind of division is called **meiosis** (mye-**OH**-sis) and it occurs only in special organs of the body—in the female gonads or ovaries and in the male gonads or testes.

Meiosis is a reduction division of the nuclear material so that each gamete contains only half as much hereditary material as the parent cell. When two gametes unite, the resulting zygote has the full complement of hereditary or DNA material. Humans have 46 chromosomes in our body cells; however, the human egg has only 23 and the human sperm has only 23 as a result of meiosis. This reduced number is called the **haploid** (**HAP**-loyd) (Greek haploos = one) or n number and the total or full complement of chromosomes is referred to as the 2n or **diploid** (**DIP**-loyd) (Greek di = two) number. Figure 4-12 illustrates the sexual cycle. We inherited 23 chromosomes from our mother through the egg fertilized at conception and 23 from our father's sperm.

Meiosis consists of two separate divisions where chromosomes are separated from one another but the DNA is duplicated only once. The first meiotic division is broken down into four substages: **prophase I**, **metaphase I**, **anaphase I**, and **telophase I**. It is in this first meiotic division that the chromosomes are reduced in half. The second meiotic division is also broken down into four substages: **prophase II**, **metaphase II**, **anaphase II**, and **telophase II**. In meiosis we end up with four daughter cells each containing only half the genetic material, whereas in mitosis we end up with two daughter cells each containing the full complement of genetic material.

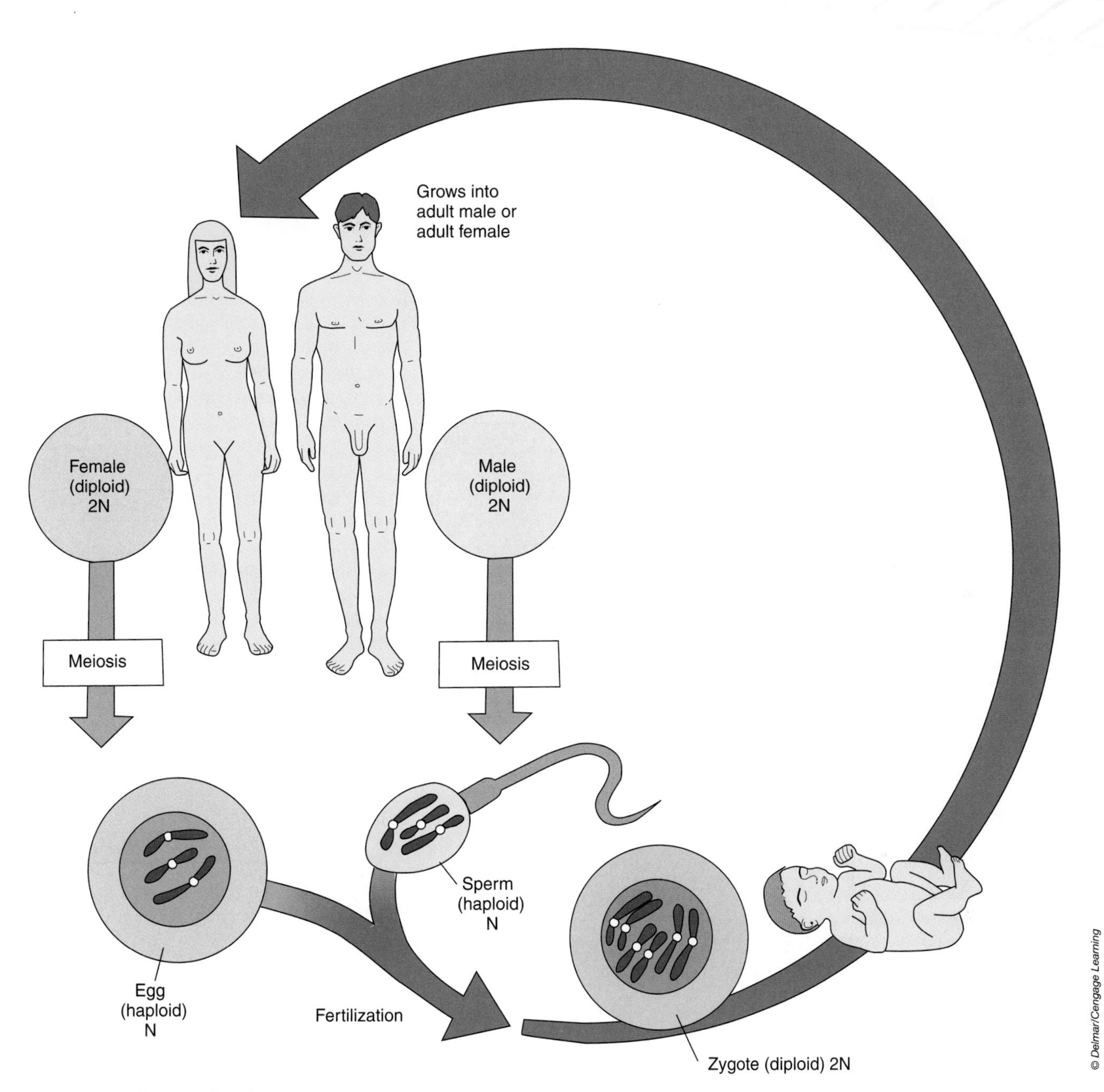

FIGURE 4-12. The sexual cycle.

The Stages of Meiosis

The stages of meiosis are discussed next. Refer to Figure 4-13 to follow the discussion.

*Study*WARE™ Connection

Watch an animation on meiosis on your StudyWARE™ CD-ROM.

Prophase I

The DNA has already duplicated before the onset of meiosis. Therefore, just like at the beginning of mitosis in interphase, each thread of DNA consists of two sister chromatids joined at their centromere. In prophase I, the duplicated chromosomes shorten, coil, thicken, and become visible. It is here in meiosis that now something very different occurs. Each chromosome pairs up with its homologue. Remember that our 46 chromosomes exist as 23 pairs. One member of each pair was inherited from our mother and the other member of each pair

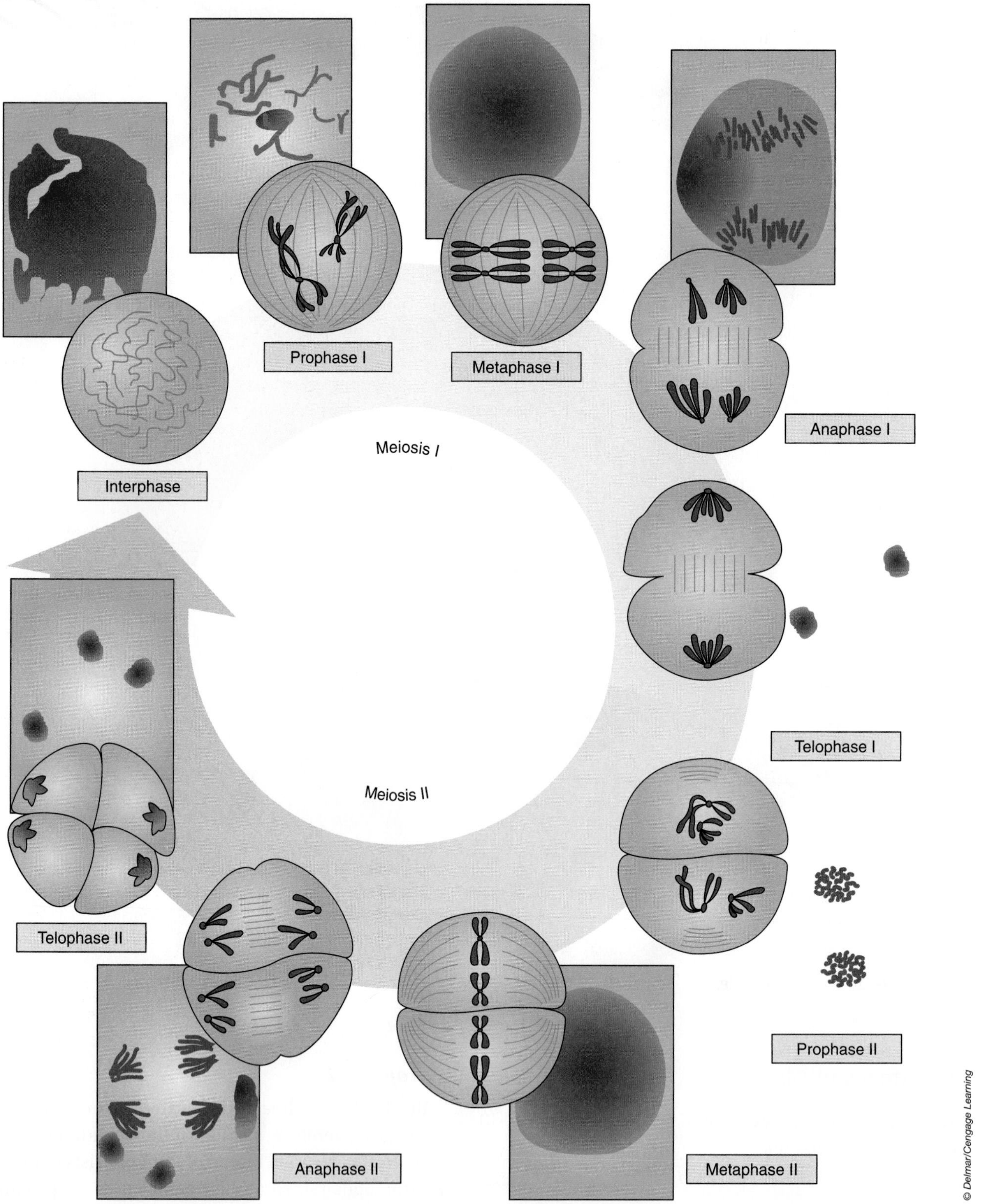

FIGURE 4-13. The stages of meiosis and the behavior of the chromosomes.

from our father. In mitosis, look-alike chromosomes did *not* pair up with one another. In meiosis, homologous chromosomes are brought so close together that they line up side by side in a process called **synapsis**. We now have a pair of homologous chromosomes each with two sister chromatids. The visible pair of chromosomes is called a **tetrad** (Figure 4-14). The chromosomes are so close together that they may actually exchange genetic material in a process called **crossing-over**. Actual segments of DNA are exchanged between the sister chromatids of the homologous chromosomes. Crossing-over is a common but random event and it occurs only in meiosis. Evidence of crossing-over can be seen with a light microscope as an X-shaped structure known as a chiasma or **chiasmata** (key-**AZZ**-mah-tah) (plural). The spindle forms from microtubules just as in mitosis; paired chromosomes separate slightly and orient themselves on the spindle attached by their centromere.

Metaphase I

Spindle microtubules attach to the kinetochore only on the outside of each centromere, and the centromeres of the two homologous chromosomes are attached to microtubules originating from *opposite* poles. This one-sided attachment in meiosis is in contrast to mitosis whose kinetochore on *both* sides of a centromere are held by microtubules. This ensures that the homologous chromosomes will be pulled to opposite poles of the cell. The homologous chromosomes line up on the equatorial plate. The centromeres of each pair lie opposite one another. The orientation on the spindle is random; thus, either homologue might be oriented to either pole.

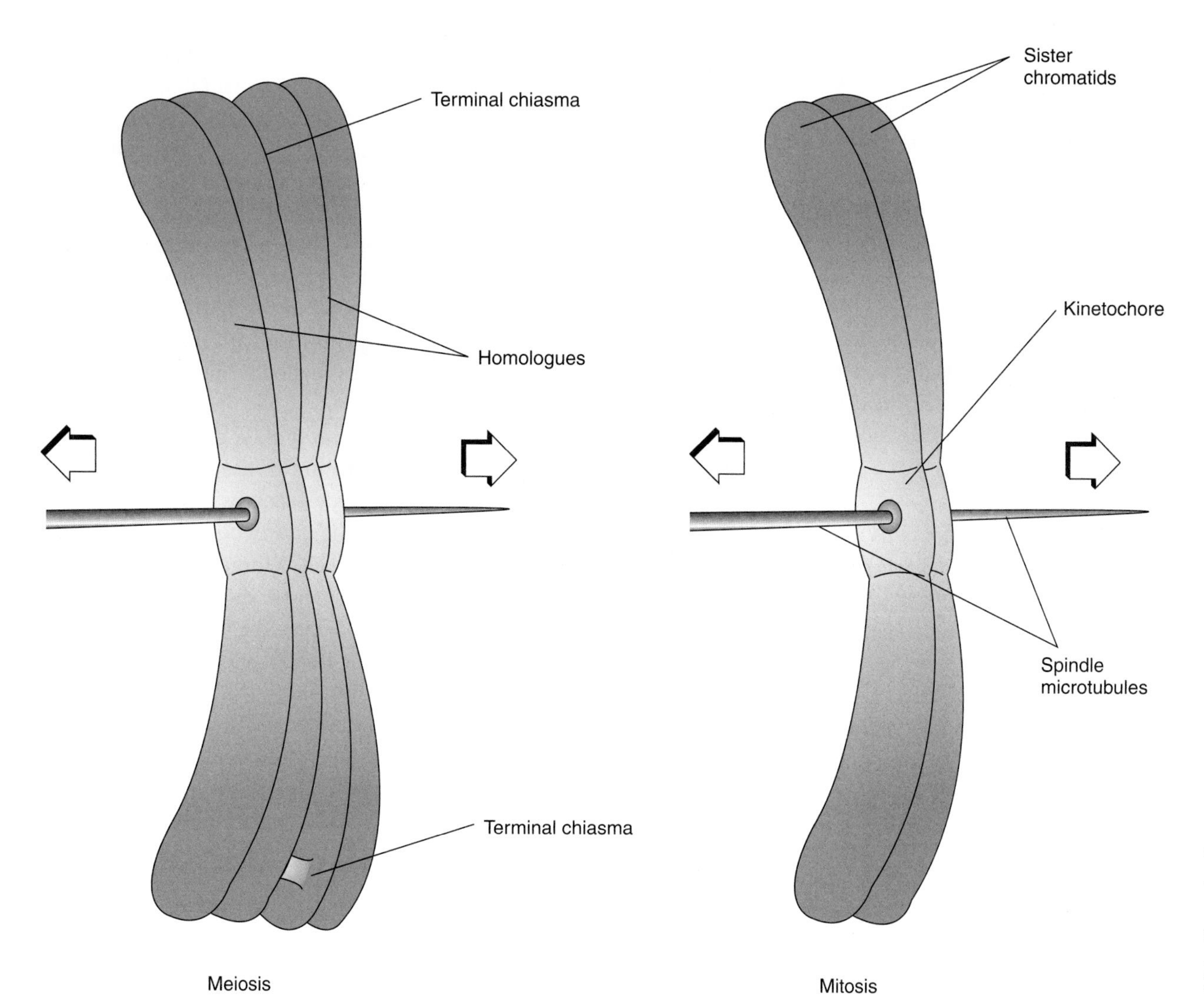

FIGURE 4-14. The pairing up of homologous chromosomes in prophase I of meiosis produces figures called tetrads. When crossing-over occurs, it is visible and it is called chiasmata.

Anaphase I

The microtubules of the spindle shorten and pull the centromeres toward the poles, dragging both sister chromatids with it. Thus, unlike mitosis, the *centromere does not divide* in this stage. Because of the random orientation of the homologous chromosomes on the equatorial plate, a pole may receive either homologue of each pair. Thus, the genes on different chromosomes assort independently.

Telophase I

The homologous chromosome pairs have separated and now a member of each pair is at the opposite ends of the spindle. Now at each pole is a cluster of "haploid" chromosomes. The number has been reduced from 46 to 23 at each pole. However, each chromosome still consists of two sister chromatids attached by a common centromere. This "duplication condition" will be corrected in the second meiotic division. Now the spindle disappears, the chromosomes uncoil and become long and thin, and a new nuclear membrane forms around each cluster of chromosomes at the opposite poles. Cytokinesis occurs and we have two new cells formed at the end of the first meiotic division. The second meiotic division closely resembles the occurrences in mitosis.

Prophase II

In each of the two daughter cells produced in the first meiotic division, a spindle forms, and the chromosomes shorten, coil, and thicken. The nuclear membrane disappears but *no duplication of DNA occurs.*

Metaphase II

In each of the two daughter cells, the chromosomes line up on the equatorial plate. Spindle fibers bind to *both* sides of the centromere. Each chromosome consists of two sister chromatids and one centromere.

Anaphase II

The *centromeres* of the chromosomes *divide.* The spindle fibers contract, pulling the sister chromatids apart and moving each one to an opposite pole. Now each chromosome is truly haploid, consisting of one chromatid and one centromere.

Telophase II

New nuclear membranes form around the separated chromatids, the spindle disappears, and the chromosomes uncoil and decondense. The result is four haploid daughter cells each containing one-half the genetic material of the original parent cell, or, in our case, each cell having 23 chromosomes instead of 46.

StudyWARE™ Connection

Play an interactive game lebeling the stages of meiosis on your StudyWARE™ CD-ROM.

GAMETOGENESIS: THE FORMATION OF THE SEX CELLS

The four haploid cells produced by meiosis are not yet mature sex cells. Further differentiation must now occur. This is known as **gametogenesis** (**gam**-eh-toh-**JEN**-eh-sis). The process occurring in the seminiferous tubules of the testes is called **spermatogenesis** (**sper**-mat-oh-**JEN**-eh-sis) (Figure 4-15). The cytoplasm of each of the four cells produced, called spermatids, becomes modified into a tail-like flagellum. A concentration of mitochondria collects in the middle piece or collar. The mitochondria will produce the ATP necessary to propel the flagellum, which causes the sperm to swim. The nucleus of each cell becomes the head of the sperm. The genetic material is concentrated in the head of the sperm. The sperm cell will penetrate an egg and fuse with the genetic material of the egg in the process called fertilization, producing a fertilized egg or zygote.

The formation of the female egg, called **oogenesis** (**oh**-oh-**JEN**-eh-sis), occurs in the ovary (Figure 4-16). However, only one functional egg is produced. In the first meiotic division, there is an unequal distribution of the cytoplasm so that one cell is larger than the other. The larger cell in the second meiotic division also has unequal distribution of the cytoplasm. The three smaller cells produced are called **polar bodies** and eventually die. They have contributed cytoplasm to the single larger cell that will become the functional egg. The union of sperm and egg is called fertilization and restores the diploid number of chromosomes to 46.

A COMPARISON OF MITOSIS AND MEIOSIS

The two types of cellular division consisting of mitosis and meiosis are easy to confuse. They have similarities but they also have differences. In both mitosis and meiosis, the chromosomes duplicate or replicate in the phase

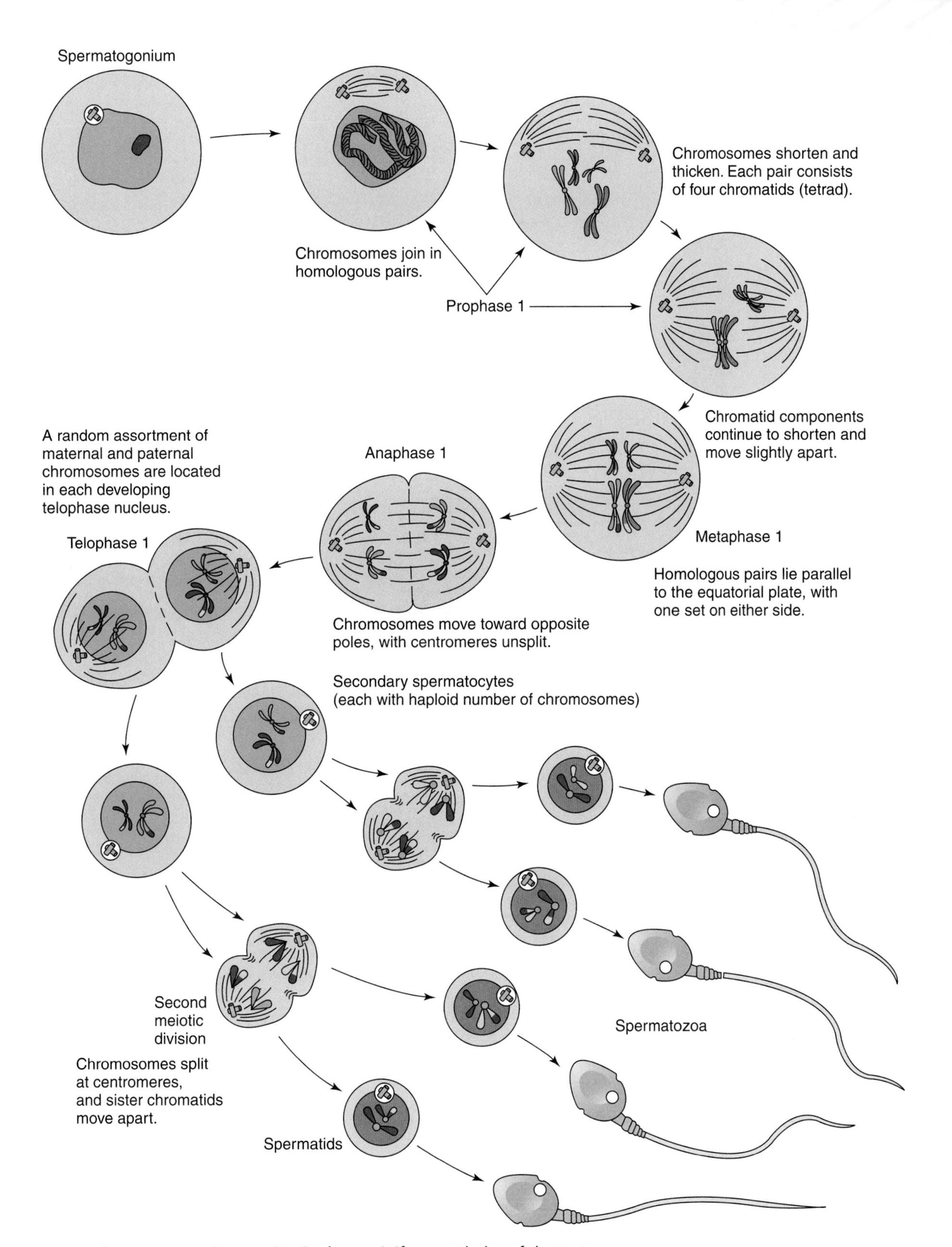

FIGURE 4-15. Spermatogenesis occurring in the seminiferous tubules of the testes.

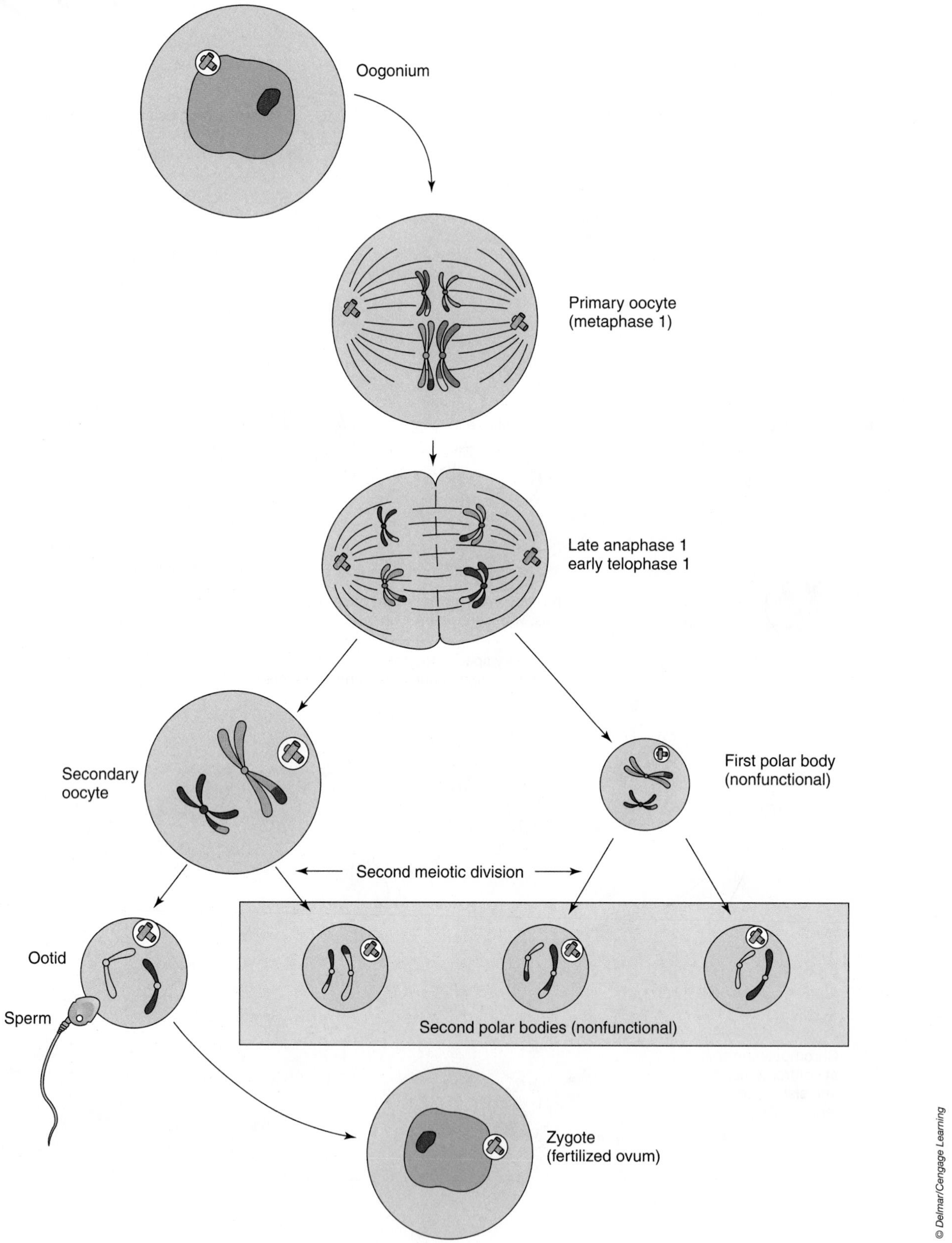

FIGURE 4-16. Oogenesis occurring in the ovary.

of the cell cycle called interphase. However, in mitosis the end result is *two* daughter cells each with exactly the same number of chromosomes as the parent cell, whereas in meiosis the end result is *four* daughter cells each with only half the number of chromosomes as the parent cell. Mitosis consists of one division, whereas meiosis consists of two divisions.

In mitosis, when the genetic material duplicates, the homologous chromosomes are scattered in the nucleus and do *not* seek one another out. In meiosis, after duplication, the homologous pairs of chromosomes line up together and come so close that they entwine. Crossing-over or exchange of segments of DNA may occur. Crossing-over occurs only in prophase I of meiosis. This results in a recombination of existing genes, thus producing new genetic characteristics. In mitosis the centromere divides in anaphase. In meiosis the centromere does not divide in anaphase I. The centromere divides only in anaphase II. Figure 4-17 provides a summary comparison of mitosis and meiosis.

Mitosis occurs in all cells of our bodies on a regular basis (except nerve, muscle, and liver cells). After the egg is fertilized, the embryo develops by mitosis. After birth, we grow and mature by mitosis. When we cut our finger or bruise our tissues, the cells are repaired and replaced by mitosis. Liver cells divide only if damaged; muscle cells and nerve cells rarely divide by mitosis. Cells produced by mitosis can live on their own. These cells all contain the same genetic information as the parent cell. In special sections of our gonads, the seminiferous tubules of the testes of the male and in the ovaries of the female, another kind of cell division occurs. Meiosis occurs only in these special cells of the gonads. It is a reduction division. The genetic material is reduced in half. This process begins at puberty in the male and in the embryo of the female. It will continue at puberty in the female. The cells produced cannot live on their own. They live for only a short time and eventually die unless they fuse in fertilization inside the female reproductive tract.

In each type of cellular division, the genetic material is exactly duplicated during interphase. Sometimes, however, the genetic material may be damaged by X-rays, radiation, or certain chemicals. When this happens, the cells' damaged genetic material may cause them not to go into interphase. They divide continuously, forming masses of tissues. This is cancer.

COMMON DISEASE, DISORDER, OR CONDITION

CANCER

Cancer has a number of causes. One of these is incorrect information in the genetic material of a cell. When a cell duplicates its genetic material during the interphase S (synthesis) stage, rarely, a mistake can occur. The exact copying of the code can be disrupted by external factors such as excessive exposure to certain chemicals (smoke, asbestos), radiation (radioactive materials), X-rays, and even some viruses. This is called a **mutation**.

When the genetic code governing cell division is affected, cells continue to grow and do not go into interphase. They keep dividing uncontrollably. This results in a cluster of cells called a **tumor**. All the cells of the tumor contain the same genetic misinformation, and some leave the tumor and travel to other parts of the body. They go to other sites and produce more tumors. We call these other tumors at different sites **metastases** (meh-**TASS**-tah-seez) and when the defective cells spread they **metastasize** (meh-**TASS**-tah-size) to other parts of the body.

Cancer can occur in any tissue and a number of terms have developed to define these cancers. Tumors developing from epithelial tissue are called **carcinomas** (kar-sin-**NOH**-mahz); those coming from connective tissue, like bone, are called **sarcomas** (sar-KOM-**ahz**). Cancer-causing agents are called **carcinogens** (kar-**SIN**-oh-jenz). Some of the most serious cancers in humans are lung cancer resulting from cigarette smoking, colorectal cancers caused by excessive red meat in the diet, and breast cancer. The cause of breast cancer appears to be related to a gene located on chromosome number 17, which is responsible for hereditary susceptibility to the disease.

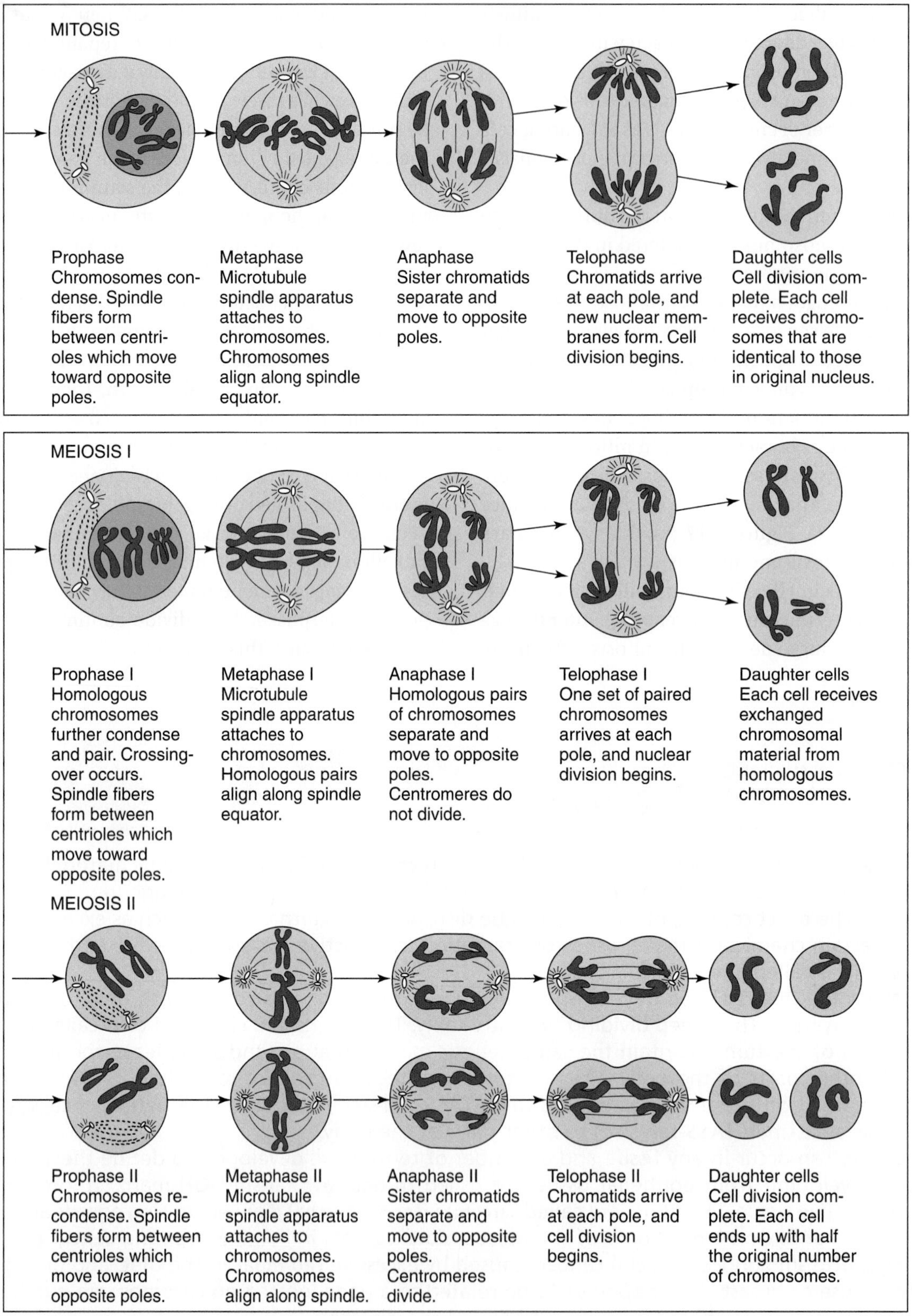

MITOSIS

Prophase
Chromosomes condense. Spindle fibers form between centrioles which move toward opposite poles.

Metaphase
Microtubule spindle apparatus attaches to chromosomes. Chromosomes align along spindle equator.

Anaphase
Sister chromatids separate and move to opposite poles.

Telophase
Chromatids arrive at each pole, and new nuclear membranes form. Cell division begins.

Daughter cells
Cell division complete. Each cell receives chromosomes that are identical to those in original nucleus.

MEIOSIS I

Prophase I
Homologous chromosomes further condense and pair. Crossing-over occurs. Spindle fibers form between centrioles which move toward opposite poles.

Metaphase I
Microtubule spindle apparatus attaches to chromosomes. Homologous pairs align along spindle equator.

Anaphase I
Homologous pairs of chromosomes separate and move to opposite poles. Centromeres do not divide.

Telophase I
One set of paired chromosomes arrives at each pole, and nuclear division begins.

Daughter cells
Each cell receives exchanged chromosomal material from homologous chromosomes.

MEIOSIS II

Prophase II
Chromosomes recondense. Spindle fibers form between centrioles which move toward opposite poles.

Metaphase II
Microtubule spindle apparatus attaches to chromosomes. Chromosomes align along spindle.

Anaphase II
Sister chromatids separate and move to opposite poles. Centromeres divide.

Telophase II
Chromatids arrive at each pole, and cell division begins.

Daughter cells
Cell division complete. Each cell ends up with half the original number of chromosomes.

FIGURE 4-17. A comparison of mitosis and meiosis.

COMMON DISEASE, DISORDER, OR CONDITION

GENETIC DISORDERS

TAY-SACHS DISEASE

Tay-Sachs disease (**TAY SAKS**) is caused by a genetic recessive trait found mainly in Jewish families of eastern European origin, especially the Ashkenazic Jews. It is a neurodegenerative disorder of lipid metabolism caused by a lack of the enzyme hexosaminidase A, resulting in a buildup of sphingolipids in the brain. This causes progressive physical and mental retardation resulting in an early death of the affected individual. Symptoms first appear by 6 months of age with convulsions, followed by blindness due to atrophy of the optic nerve, dementia, and paralysis. Children usually die between the ages of 2 and 4. There is no cure.

KLINEFELTER'S SYNDROME

Klinefelter's syndrome (**KLINE**-fell-ters) occurs in males who have reached puberty. It is caused by males who have extra X chromosomes. The most common condition is an XXY chromosome. This condition results in long legs, small and firm testes, an abnormal enlargement of the breasts, varicose veins, below normal intelligence, poor social skills, and chronic pulmonary disease. Males with an XXXXY chromosome have severe congenital abnormalities and serious mental retardation.

DOWN SYNDROME

Down syndrome is a congenital defect commonly caused by an extra chromosome 21. It occurs in about one in every 600 to 650 live births in mothers who are 35 years of age or older. It results in both mental retardation and physical defects. It was formerly referred to as mongolism. The condition can be diagnosed by an amniocentesis during pregnancy. Some of the physical defects include a slant to the eyes, a low nasal bridge, short broad hands with stubby fingers, broad stubby feet with a space between the first and second toes, and abnormal tooth development. The most prominent feature of this syndrome is mental retardation with IQ's between 50 and 60. In addition there can be heart disease, chronic pulmonary infections, and visual problems. Those children who survive their first few years tend to be stocky in build and short with incomplete or delayed sexual development.

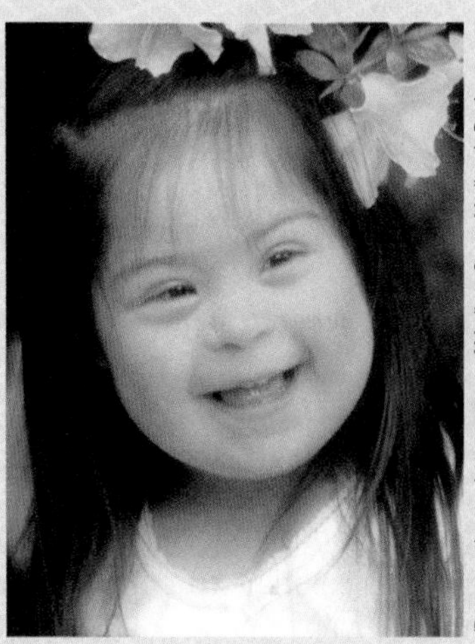

(Down syndrome photo courtesy of Marijane Scott, Marijane's Designer Portraits, Down Right Beautiful 1996 Calendar)

AS THE BODY AGES

As we age, our metabolism slows down considerably because we do not need as much energy for growth and maintenance as we did in our youth. Hence, we do not need the same caloric intake as we did in the past. We should reduce the amount of food we take in daily to prevent the "inevitable" weight gains we experience from our early 40s onward. Daily exercise also helps reduce this weight gain.

Women stop ovulating and producing eggs by meiosis after menopause. However, men continue to produce sperm cells well into their 80s. It is therefore not uncommon for a male to father a child when in his 70s. ■

Career FOCUS

These are careers that are available to individuals interested in metabolism:

- **Nutritionists** have completed advanced degrees in foods and nutrition and work as professionals in hospitals, childcare centers, school systems, and nursing homes.
- **Dieticians** establish daily balanced dietary intakes of fluids and foods for people from early childhood through old age.

There are also careers for individuals interested in cellular reproduction.

- **Genetic engineers** work with recombinant DNA to study altering and controlling the phenotype (appearance) and genotype (genes) of organisms. In the future, this work may offer the possibility of eliminating or controlling genetic disorders in humans.
- **Gerontologists** study the aging process and its sociological, economic, psychological, and clinical effects experienced by the elderly.
- **Consultants in planned parenthood** assist in family planning programs, providing information on how to become pregnant or how to prevent an unwanted pregnancy.

SUMMARY OUTLINE

INTRODUCTION TO CELLULAR METABOLISM

1. The most common form of chemical energy that maintains cellular structure and function is the molecule ATP (adenosine triphosphate).
2. Metabolism is a general term that describes the total chemical changes that occur in cells. There are two subcategories: anabolism is the energy-requiring process that builds larger molecules by combining smaller molecules, and catabolism is the energy-releasing process that breaks down larger molecules into smaller ones. These cellular metabolic processes are often called cellular respiration.
3. ATP is made in a stepwise catabolism (decomposition) of food molecules like glucose. The chemical energy in food (calories) is released and used to put together ADP (adenosine diphosphate) and PO_4 (phosphate) to make ATP.
4. The overall chemical equation for cellular respiration is: $C_6H_{12}O_6 + 6O_2 \rightarrow 6CO_2 + 6H_2O + 38$ ATP or 36 ATP and 2 GTP.
5. Cellular respiration consists of three processes or steps: glycolysis, Krebs citric acid cycle, and electron transport.

GLYCOLYSIS

1. Glycolysis occurs in the cytoplasm of the cell and does not require oxygen.
2. We must use two ATP molecules to start glycolysis and these must be "paid back" from our production of ATP.
3. The main products of glycolysis are fructose diphosphate, which splits into two phosphoglyceraldehyde molecules that oxidize to two phosphoglyceric acids, which convert to the final product of two pyruvic acid molecules.
4. When oxygen is present, two hydrogens from each of the two phosphoglyceraldehydes go to the electron transport system beginning with the cofactor NAD. This produces six ATP.
5. When phosphoglyceric acid is decomposed to pyruvic acid four ATP are made, but we must pay back the two that started the sequence. Thus, we produce only a net gain of two ATP.

6. In glycolysis we produce a total net gain of eight ATP. In anaerobic glycolysis in muscle cells and in fermentation only two ATP are produced.

THE KREBS CITRIC ACID CYCLE

1. The citric acid cycle is named for the British biochemist Sir Hans Krebs who first proposed the scheme in 1937.
2. The two pyruvic acid molecules produced in glycolysis are converted to acetic acid and then to acetyl-CoA through the action of CoA enzyme. Acetyl-CoA now enters the cristae of the mitochondria to go through the citric acid cycle.
3. The major chemical products in the cycle are citric acid, alpha-ketogluteric acid, succinic acid, malic acid, and oxaloacetic acid.
4. Most of the ATP is made by electron transport. For each of the two pyruvic acid molecules broken down, 14 ATP are made via electron transport for a total production of 28 ATP. In addition, 2 ATP or GTP are produced in the citric acid cycle, for a total of 30 ATP or 28 ATP and 2 GTP.
5. For each pyruvic acid molecule broken down, three CO_2 are given off as waste products for a total of six CO_2 molecules produced.

THE ELECTRON TRANSPORT SYSTEM

1. The electron transport system functions as a series of redox (reduction-oxidation) reactions.
2. There are several kinds of electron carriers in the electron transport system: NAD (nicotinamide adenine dinucleotide), FAD (flavin adenine dinucleotide), quinone, and the cytochrome system.
3. If the system begins with NAD, then three ATP are produced. If the system begins with FAD, then only two ATP are produced.
4. Oxygen is necessary for respiration because oxygen is the ultimate electron acceptor in the system.
5. When oxygen accepts the electrons from the two hydrogen atoms and the two hydrogen protons, water (H_2O) is produced as a waste product.

SUMMARY OF ATP PRODUCTION

1. Glycolysis produces a total net gain of eight ATP.
2. The Krebs citric acid cycle produces for each of the two pyruvic acid molecules 14 ATP via electron transport and one ATP or GTP. Thus, the total ATP production in the cycle is 28 ATP and two GTP or 30 ATP.
3. The eight ATP from glycolysis and the 30 ATP from the citric acid cycle yield a total of 38 ATP from each glucose molecule.

ANAEROBIC RESPIRATION

1. Fermentation is the process by which yeast cells break down glucose in the absence of oxygen. This process produces only two ATP and is much less efficient than glycolysis.
2. The other products of fermentation are carbon dioxide gas, which causes bread dough to rise, and ethyl alcohol, which is used in the beer, wine, and liquor industries.
3. When we overwork our muscles and cannot get enough oxygen to the muscle cells, they begin to break down glucose in the absence of oxygen. The total net gain of ATP is only two ATP molecules and pyruvic acid is converted to lactic acid.
4. The buildup of lactic acid in the muscle cells is what causes the fatigue in overworked muscles. Our breathing and heartbeat rates accelerate to get more O_2 to the cells. Eventually, the fatigue goes away as lactic acid is converted back to pyruvic acid when oxygen again becomes available.

PRODUCTION OF ATP FROM GENERAL FOOD COMPOUNDS

1. The cellular furnace that "burns" food to produce ATP consists of glycolysis, the Krebs citric acid cycle, and electron transport.
2. Carbohydrates feed into the furnace at the level of glucose in glycolysis.
3. Fats are digested into glycerol, which feeds into the furnace at the phosphoglyceric acid stage of glycolysis, and fatty acids, which feed into the citric acid cycle.
4. Proteins are digested into amino acids. They feed into the furnace at different stages of glycolysis and the citric acid cycle based on their chemical structure.
5. Carbohydrates, fats, and proteins are all potential sources of cellular energy because they can all be broken down and their chemical energy can be converted into another form of chemical energy, ATP, which runs the cell's machinery.

INTRODUCTION TO CELLULAR REPRODUCTION

1. Cellular reproduction is the process by which a single cell duplicates itself. Mitosis is duplication of the genetic material in the nucleus. Cytokinesis is the duplication of the organelles in the cytoplasm. Meiosis is a special kind of reduction division that occurs only in the gonads.

THE STRUCTURE OF THE DNA MOLECULE

The History of the Discovery of DNA

1. DNA was first discovered in 1869 by a German chemist, Friedrich Miescher.
2. In the 1920s, P. A. Levine discovered that DNA contained phosphates, five-carbon sugars, and nitrogen-containing bases.
3. A British citizen, Rosalind Franklin, discovered the helical structure of DNA via X-ray crystallography studies.
4. James Watson, an American, and British Francis Crick won the 1962 Nobel Prize for working out the three-dimensional structure of the molecule.

The Anatomy of the DNA Molecule

1. A DNA molecule is a double helical chain of nucleotides.
2. A nucleotide is a complex combination of a phosphate group (PO_4), a five-carbon sugar (deoxyribose), and a nitrogen-containing base, either a purine or a pyrimidine.
3. A pyrimidine consists of a single ring of six atoms of carbon and nitrogen. There are two pyrimidines in the molecule: thymine and cytosine.
4. A purine consists of a fused double ring of nine atoms of carbon and nitrogen. There are two purines in the molecule: adenine and guanine.
5. In the chain of nucleotides, bonds form between the phosphate group of one nucleotide and the sugar of the next nucleotide. The base extends out from the sugar.
6. Adenine of one chain always pairs with thymine of the other chain. Cytosine of one chain always pairs with guanine of the other chain. The bases are held together by hydrogen bonds.
7. A gene is a sequence of organic nitrogen base pairs that codes for a polypetide or protein.
8. In our 46 chromosomes there are billions of organic base pairs that encode over 30,000 genes.

THE CELL CYCLE

1. The cell cycle is the process by which a cell divides into two and duplicates its genetic material.
2. A cell cycle is divided into three stages: interphase, mitosis, and cytokinesis.

Interphase

1. Interphase is the time between divisions. It is divided into three substages: G_1 (growth one), S (synthesis), and G_2 (growth two).
2. The major portion of the life of the cell is spent in G_1.
3. During the S phase, the genetic material or DNA duplicates itself.
4. During the G_2 phase, mitochondria replicate and the chromosomes condense and coil. Tubulin is synthesized.

Mitosis

1. Mitosis, the cellular division in the nucleus, has four stages: prophase, metaphase, anaphase, and telophase.

Prophase

1. The duplicated chromosomes shorten, thicken, and become visible as two sister chromatids held together at a middle area called the centromere.
2. The two kinetochores are found at the centromere.
3. The centrioles move to opposite poles of the cell and form the spindle and asters in animal cells.
4. The nuclear membrane breaks down and the nucleolus disappears.
5. The microtubules attach the kinetochores to the spindle.

Metaphase

1. The sister chromatids align themselves in a circle at the equator of the cell held in place by the microtubules attached to the kinetochores of the centromere.
2. The centromere divides.

Anaphase

1. Each divided centromere pulls a sister chromatid to an opposite pole.
2. Cytokinesis begins.

Telophase

1. The chromosomes begin to uncoil and decondense.
2. The spindle apparatus breaks down.
3. A new nuclear membrane forms around the cluster of chromosomes at each pole.
4. Cytokinesis is nearly complete.

Cytokinesis

1. In animal cells, a cleavage furrow forms by a pinching in of the cell membrane, resulting in two daughter cells.
2. In plant cells, a cell plate forms at the equator and grows outward, effectively dividing the cell in two. The cell plate becomes a new cell wall.

MEIOSIS: A REDUCTION DIVISION

1. Meiosis is a reduction division of the nuclear material; it occurs only in the gonads. It reduces the genetic material from 46 (diploid or 2n) to 23 (haploid or n) chromosomes.
2. Meiosis consists of two divisions, resulting in four cells. The first meiotic division reduces the number of chromosomes in half. The second meiotic division corrects their duplicated nature.

The Stages of Meiosis

Prophase I

1. Homologous chromosomes pair and crossing-over may occur.
2. Spindle fibers form, the nuclear membrane breaks down, and the chromosomes are attached to the spindle by their centromeres.

Metaphase I

1. Microtubules attach to the kinetochore on one side of the centromere. Homologous pairs of chromosomes align along the equator of the spindle.

Anaphase I

1. The centromeres do *not* divide. The microtubules of the spindle shorten and pull the centromeres of the chromosomes to opposite poles, one member of each pair to a pole.

Telophase I

1. A member of each pair of homologous chromosomes is at each pole. The number of chromosomes has been reduced in half. They are now haploid but still duplicated.
2. The spindle disappears and a new nuclear membrane forms around each group of chromosomes at the pole.
3. The chromosomes uncoil and decondense.
4. Cytokinesis occurs and two new daughter cells are formed.

Prophase II

1. In each daughter cell a spindle forms, centrioles move to opposite poles, and the chromosomes coil and thicken.
2. The nuclear membrane disappears.

Metaphase II

1. The chromosomes line up at the equator of the cell attached by the microtubules of the spindle.
2. Microtubules bind to *both* kinetochores of the centromere.

Anaphase II

1. The centromeres divide.
2. The spindle fibers contract pulling the sister chromatids apart, one to each pole of the spindle.

Telophase II

1. The chromatids arrive at each pole, where they uncoil and decondense.
2. A new nuclear membrane forms around the chromatids, and the spindle disappears.
3. Four haploid cells are formed as cytokinesis is completed.

GAMETOGENESIS: THE FORMATION OF THE SEX CELLS

1. Spermatogenesis occurs in the seminiferous tubules of the testes. Each of the four cells produced by meiosis develops into sperm.
2. The cytoplasm of each cell develops into a tail-like flagellum and a concentration of mitochondria forms the collar or middle piece. The head of the sperm is formed by the nucleus of the cell.
3. Oogenesis occurs in the ovary. Of the four cells produced, only one becomes the functional egg. The other three are called polar bodies and contribute their cytoplasm to the functional egg.

A Comparison of Mitosis and Meiosis

1. Mitosis produces two daughter cells with the exact same genetic material as the parent cell.
2. Meiosis produces four daughter cells, each with half the genetic material of the parent cell.
3. In meiosis, homologous pairs of chromosomes line up in prophase I and exchange genetic material in a process called crossing-over.
4. Mitosis is a process of cellular division by which we grow, maintain, and repair ourselves. Most cells of our body undergo mitosis. Nerve and muscle cells rarely divide, and liver cells divide only when damaged.

5. Meiosis occurs only in the gonads. Unlike the cells produced by mitosis, meiotic cells cannot live on their own. They must unite in fertilization in the female reproductive tract.
6. Cancer is uncontrolled cellular growth caused by damaged genetic material. Cells never go into interphase; they divide continuously.

REVIEW QUESTIONS

1. Diagram glycolysis, the first step in the biochemical decomposition of a glucose molecule. Indicate where and how many ATP are produced.
2. Diagram the Krebs citric acid cycle, beginning with one molecule of pyruvic acid. Indicate where and how many ATP are produced.
3. Diagram the electron transport system with its electron carriers. Indicate where and how many ATP are produced.
*4. Why are fermentation and anaerobic production of ATP by muscle cells less efficient than glycolysis?
*5. In addition to beer and wine, name some other practical applications of the fermentation process for human advancement.
6. Name the four kinds of organic bases found in a DNA molecule, and indicate how they pair up in linking the two helical chains of the molecule. Include a linear diagrammatic drawing of a short segment of DNA.
7. Name the three main stages of the cell cycle.
8. Name and explain what happens during the three substages of interphase.
*9. Why was interphase once called a "resting stage"?
10. Name and briefly describe the stages of the two meiotic divisions of meiosis.
*11. Why is meiosis called a reduction division?
*12. Compare the major differences between mitosis and meiosis.

***Critical Thinking Questions**

TRUE OR FALSE

T F 1. The breakdown of glucose requires oxygen, because oxygen is the ultimate electron donor for the electrons used by the cofactors in the electron transport system.

T F 2. The anaerobic decomposition of glucose by yeast cells yields one ATP, ethyl alcohol, CO_2 and NAD.

T F 3. Lactic acid produced in the muscles during anaerobic respiration accumulates there and is never broken down or ever converted to other products.

T F 4. The formation of lactic acid regenerates NAD for use in the glycolytic sequence.

T F 5. Each molecule of glucose undergoing glycolysis yields one molecule of pyruvic acid.

T F 6. One molecule of glucose undergoing aerobic respiration yields 34 molecules of ATP.

T F 7. The overall chemical equation for respiration is $C_6H_{12}O_6 + 6O_2 \rightarrow 6CO_2 + 6H_2O +$ energy in the form of ATP.

T F 8. When nicotine adenine dinucleotide gets reduced in electron transport, two molecules of ATP are produced.

T F 9. Fermentation is as effective as glycolysis.

T F 10. All foods taken into the body must be broken down to glucose before respiration can proceed.

FILL IN THE BLANK

Fill in the blank with the most appropriate term.

1. ________________ discovered the helical nature of the DNA molecule via X-ray crystallography studies in England.
2. ________________ and ________________ won the Nobel Prize in 1962 for determining the three-dimensional structure of the DNA molecule.
3. The DNA molecule consists of a double helical chain of ________________.
4. In the DNA molecule, every phosphate group is bonded to a ________________.
5. The two types of nitrogen bases in the DNA molecule are: ________________, a single ring of six atoms, and ________________, a fused double ring of nine atoms of carbon and nitrogen.
6. Adenine always pairs with ________________, and cytosine always pairs with ________________.
7. The nitrogen base pairs of the two chains of the DNA molecule are held together by ________________ bonds.
8. Cellular division in the nucleus producing two identical nuclei is known as ________________.
9. ________________ is a reduction division of the nuclear material so that each gamete contains only half as much genetic material as the parent.

10. Exchange of genetic material between homologous chromosomes occurring in prophase I of meiosis is called ____________________.
11. The number of cells produced after a mitotic division is ____________________, whereas the number of cells produced after meiosis is ____________________.
12. Meiosis occurs only in the ____________________ of the human body.

Search and Explore

Search the Internet to explore metabolism changes that occur as we age. Share with the class something you learned from your research.

*Study*WARE™ Connection

Take a quiz or play interactive games that reinforce the content in this chapter on your StudyWARE™ CD-ROM.

Study Guide Practice

Go to your **Study Guide** for more practice questions, labeling and coloring exercises, and crossword puzzles to help you learn the content in this chapter.

LABORATORY EXERCISE: CELLULAR METABOLISM

The author recommends learners view the videotape "Cellular Respiration: Energy for Life" in lab. This videotape is produced by Human Relations Media, 175 Tompkins Ave., Pleasantville, NY 10570-3156. It runs 22 minutes and comes with a teacher's guide and student worksheets.

LABORATORY EXERCISE: CELLULAR REPRODUCTION

1. Your instructor will show you a videotape or a CD-ROM on cell division. A suggestion is the The Center for Humanities videotape "Mitosis and Meiosis: How Cells Divide."
2. Set up your compound light microscope and observe the stages of mitosis by looking at slides of the whitefish blastula (animal) and an onion root tip (plant). Draw and label cells showing the following stages: interphase, prophase, metaphase, anaphase, and telophase.
3. Examine a prepared slide of human sperm. Draw and label the parts of a sperm cell.
4. Examine a prepared slide of the chromosomes of a fruit fly, *Drosophila,* from a smear of the fly's salivary gland.
5. Construct a portion of a DNA molecule from a kit supplied by your instructor.
6. Utilizing a chromosome simulation biokit construct chromosomes with colored beads representing genes and replicate the stages of the mitotic cell cycle.

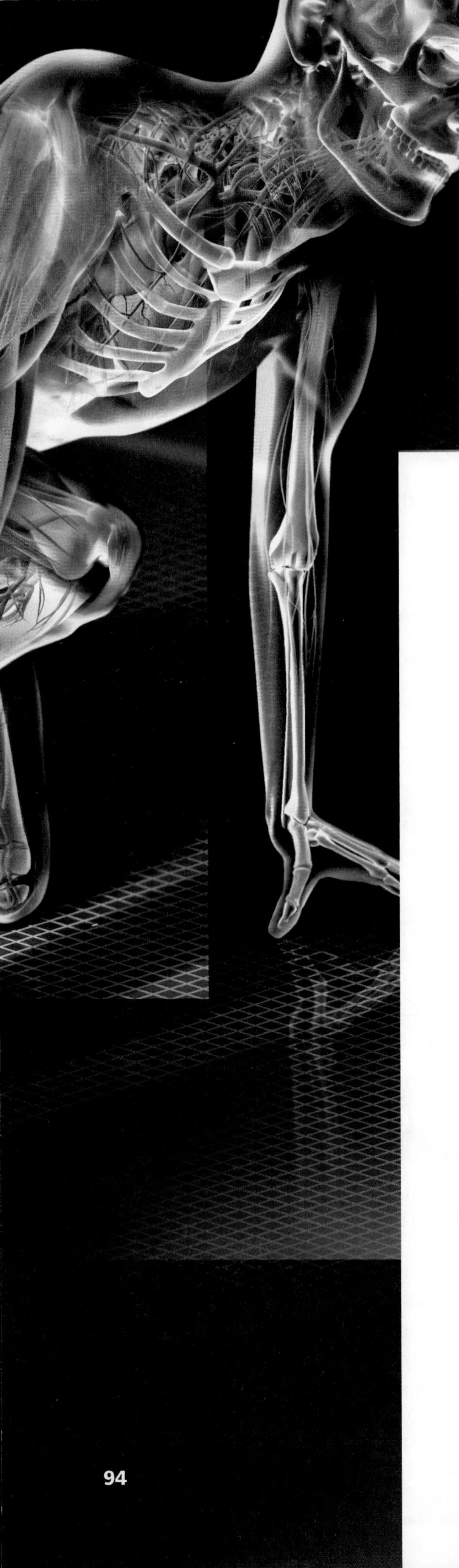

Tissues

CHAPTER OBJECTIVES

After studying this chapter, you should be able to:

1. Classify epithelial tissue based on shape and arrangement and give examples.
2. Name the types of glands in the body and give examples.
3. Name the functions of connective tissue.
4. Compare epithelial tissue with connective tissue in terms of cell arrangement and interstitial materials.
5. Name the three major types of connective tissue and give examples.
6. List the functions of epithelial tissue.
7. List the three types of muscle and describe each based on structure and function.
8. Describe the anatomy of a neuron and the function of nervous tissue.

5

KEY TERMS

INTRODUCTION

The basic units of **tissue** (**TISH**-you) are groups of cells. These cells will have a similar function and a similar structure. Tissues are classified based on how these cells are arranged and what kind and how much material is found between the cells. Cells are either tightly packed or separated by interstitial material. The study of tissue is called **histology** (hiss-**TALL**-oh-jee).

The four basic types of tissue are epithelial, connective, muscle, and nervous. Each type is further subdivided into specific examples. These tissues combine to form organs. The various organs make up the systems of the body that allow us to function and survive in our complex world.

EPITHELIAL TISSUE

Epithelial (ep-ih-**THEE**-lee-al) **tissue** functions in four major ways.

1. **It protects underlying tissues:** Our skin is epithelial tissue and protects us from the harmful rays of the sun and certain chemicals. The lining of our digestive tract is made of epithelial tissue and protects underlying tissue from abrasion as food moves through the tract.
2. **It absorbs:** In the lining of the small intestine, nutrients from our digested food enter blood capillaries and get carried to the cells of our body.
3. **It secretes:** All glands are made of epithelial tissue; the endocrine glands secrete hormones, the mucous glands secrete mucus, and our intestinal tract contains cells that secrete digestive enzymes in addition to the pancreas and the liver, which secrete the major portions of digestive enzymes.
4. **Epithelial tissue excretes:** Sweat glands excrete waste products such as urea.

When epithelial tissue has a protective or absorbing function, it is found in sheets covering a surface, like the skin or intestinal lining. When it has a secreting function, the cells involute from the surface into the underlying tissues to form glandular structures. Only a minimal, if any, amount of intercellular material is found in epithelial tissue. The cells are very tightly packed together and thus this tissue is not as easily penetrated as other tissues.

Epithelial cells are anchored to each other and to underlying tissues by a specialized membrane called the **basement membrane**. This membrane acts like the adhesive on a tile floor, the tiles being the epithelial cells. It is very important because it acts as an anchor for the attached side of the epithelial cells and it provides protection for other underlying tissue like connective tissue.

Epithelial tissue can be named according to shape and structures that might be on the free or outer edge of the cells. This surface can be plain or it can have rows of cilia (those that line the respiratory tract), a flagellum (the sperm cell), microvilli (folds), and secretory vesicles (those that line the small intestine). Epithelial tissue can be one layer or several layers thick.

Classification Based on Shape

Epithelial cells are classified as either squamous, cuboidal, or columnar. **Squamous** (**SKWAY**-mus) cells are flat and slightly irregular in shape (Figure 5-1). They serve as a protective layer. They line our mouth, blood and lymph vessels, parts of kidney tubules, our throat and esophagus, the anus, and our skin. If exposed to repeated irritation like the linings of ducts in glands, other epithelial cells can become squamous in appearance.

Cuboidal (kyoo-**BOY**-dal) cells look like small cubes (Figure 5-2). They are found in glands and the lining tissue of gland ducts (sweat and salivary), the germinal coverings of the ovaries, and the pigmented layer of the retina of the eye. Their function can be secretion and protection. In areas of the kidney tubules, they function in absorption.

Columnar cells are tall and rectangular looking (Figure 5-3). They are found lining the ducts of certain glands (e.g., mammary glands) and the bile duct of the liver. They are also found in mucus-secreting tissues such as the mucosa of the stomach, the villi of the small intestine, the uterine tubes, and the upper respiratory tract. Many of these cells are ciliated.

Classification Based on Arrangement

The four most common arrangements of epithelial cells are simple, stratified, pseudostratified, and transitional. As epithelial cells are named, a combination of the classification of both shape and arrangement is used. The **simple** arrangement is one cell layer thick. It is found in the lining of blood capillaries, the alveoli of the lungs, and in the loop of Henle in the kidney tubules. Refer to Figure 5-2 for simple cuboidal epithelium found in the lining of glandular ducts. Refer to Figure 5-3 for simple columnar epithelium found in the villi of the small intestine and the lining of the uterus. The **stratified** arrangement is several layers of cells thick. Refer to Figure 5-1B for stratified squamous epithelium found lining our mouth and throat and as the outer surface of

Characteristics and Location | **Morphology**

Squamous epithelial cells
These are flat, irregularly shaped cells. They line the heart, blood and lymphatic vessels, body cavities, and alveoli (air sacs) of lungs. The outer layer of the skin is composed of stratified and keratinized squamous epithelial cells. The stratified squamous epithelial cells on the outer skin layer protect the body against microbial invasion.

(A)

(B)

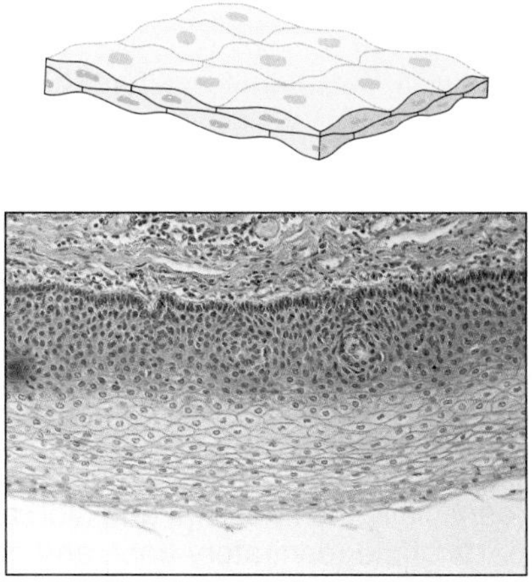

Photo (B) Courtesy of Fred Hossler, Visuals Unlimited, Inc.

FIGURE 5-1. Views of squamous epithelium.

Characteristics and Location | **Morphology**

Cuboidal epithelial cells
These are the cube-shaped cells that line the kidney tubules and cover the ovaries and secretory parts of certain glands.

(A)

(B)

Photo (B) Courtesy of Cabisco, Visuals Unlimited, Inc.

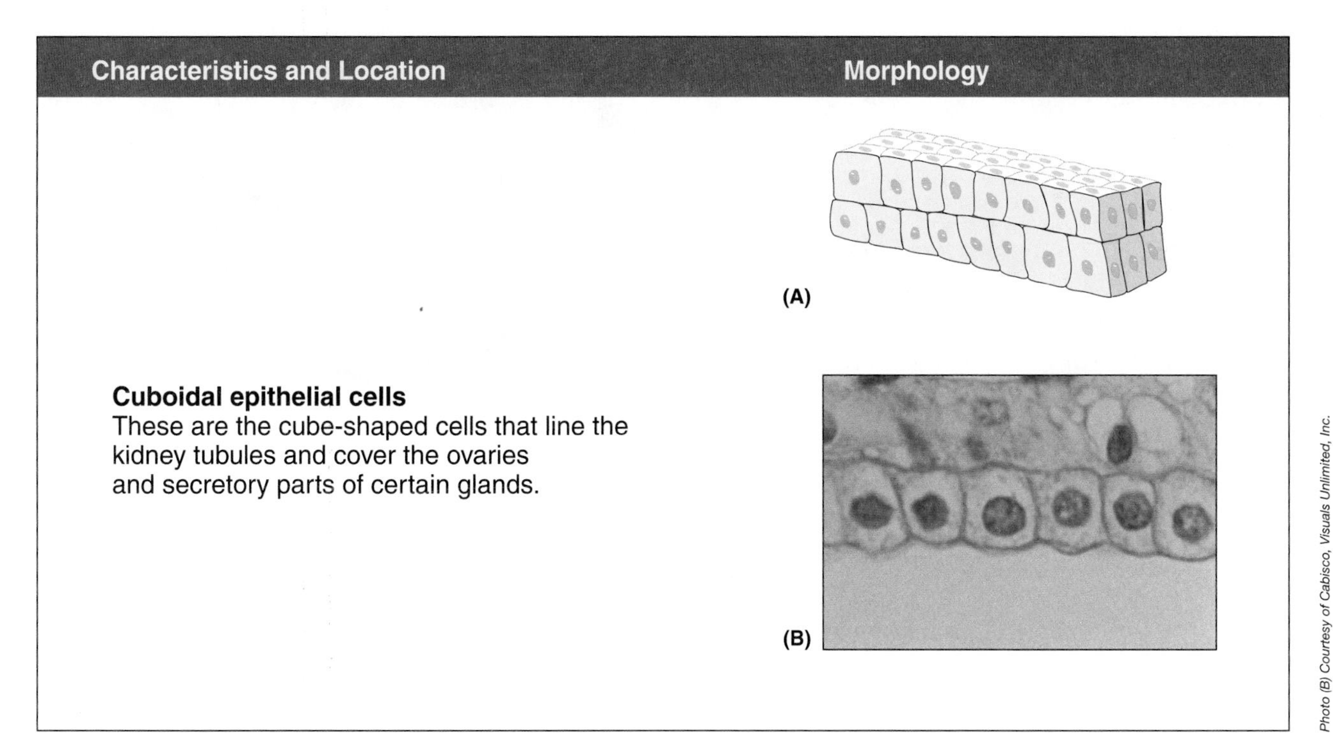

FIGURE 5-2. Views of cuboidal epithelium.

our skin. Stratified cuboidal epithelium is found lining our sweat gland ducts and salivary gland ducts. Stratified columnar epithelium is found as the lining of the ducts of the mammary glands and in parts of the male urethra.

The **pseudostratified** arrangement appears to consist of several layers due to nuclei variously positioned in the cell, but, in actuality, all cells extend from the basement membrane to the outer or free surface of the cells. This arrangement is usually seen with columnar cells.

Characteristics and Location	Morphology
Columnar epithelial cells Elongated, with the nucleus generally near the bottom and often ciliated on the outer surface; they line the ducts, digestive tract (especially the intestinal and stomach lining), parts of the respiratory tract, and glands.	(A) (B)

Photo (B) Courtesy of Richard Kessel, Visuals Unlimited, Inc.

FIGURE 5-3. Views of columnar epithelium.

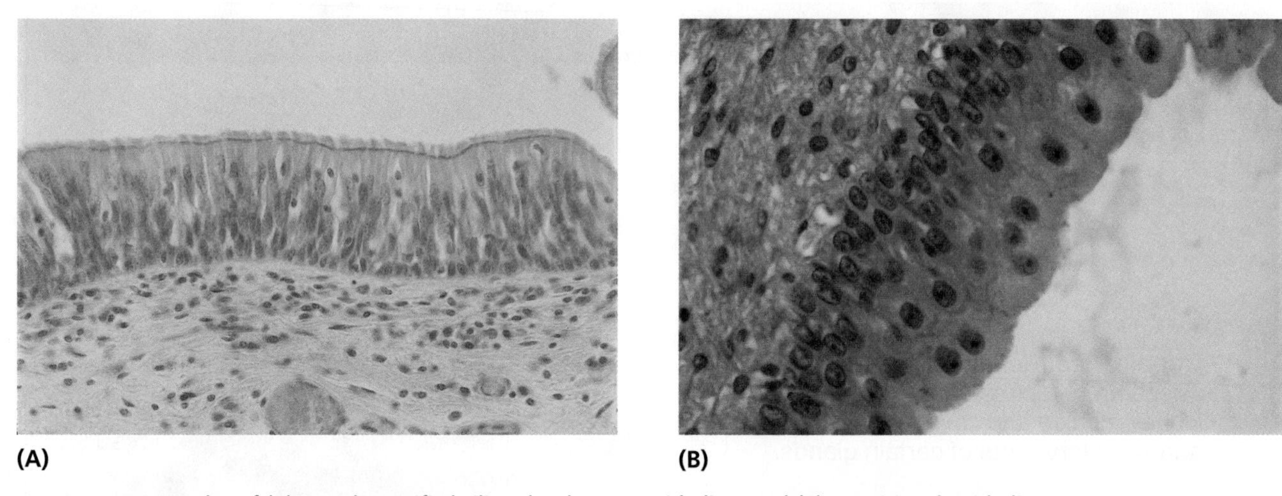

(A) (B)

Courtesy of John D. Cunningham (photo A), Richard Kessel (photo B), Visuals Unlimited, Inc.

FIGURE 5-4. Examples of (A) pseudostratified ciliated, columnar epithelium and (B) transitional epithelium.

Figure 5-4A is an example of pseudostratified ciliated, columnar epithelium. We find this tissue in the throat, trachea, and bronchi of the lungs. **Transitional** epithelium consists of several layers of closely packed, flexible, and easily stretched cells (Figure 5-4B). When the surfaces of the cells are stretched, as in a full bladder, the cells appear squamous or flat but when the tissue is relaxed, as in an empty bladder, the layers of cells look ragged like the teeth of a saw. This type of epithelium lines the pelvis of the kidney, the ureters, the urinary bladder, and the upper part of the urethra.

Classification Based on Function

Epithelial tissue can also be named or classified based on its function. The terms *mucous membrane, glands, endothelium,* and *mesothelium* all refer to epithelial tissue.

Mucous (**MYOO**-kus) **membrane** lines the digestive, respiratory, urinary, and reproductive tracts. It lines all body cavities that open to the outside. It is usually ciliated. Its most obvious function is to produce mucus, but it also concentrates bile in the gallbladder. In the intestine it secretes enzymes for the digestion of food and nutrients before absorption. Mucous membrane protects, absorbs nutrients, and secretes mucus, enzymes, and bile salts.

Glandular epithelium forms glands. Glands are involutions of epithelial cells specialized for synthesizing special compounds. The body has two types of multicellular glands. **Exocrine** (**EKS**-oh-krin) **glands** have excretory *ducts* that lead the secreted material from the gland to the surface of a lumen (passageway) on the skin. There are two types of exocrine glands. **Simple exocrine glands** have single unbranching ducts. Some examples of simple exocrine glands are the sweat glands, most of the glands of the digestive tract, and the sebaceous glands. The other type of exocrine gland is the **compound exocrine gland**. These glands are made of several component lobules each with ducts that join other ducts. Thus, the ducts are branching. Examples of compound exocrine glands are the mammary glands and the large salivary glands. **Endocrine glands** are the second type of multicellular glands in the body. They are ductless and secrete hormones; examples are the thyroid and pituitary glands. **Goblet cells** are unicellular glands that secrete mucus. They are interspersed among the epithelial cells that make up mucous membranes.

Endothelium (**en**-doh-**THEE**-lee-um) is a special name given to the epithelium that lines the circulatory system. This system is lined with a single layer of squamous-type cells. Endothelium lines the blood vessels and the lymphatic vessels. The endothelium that lines the heart gets another special name and is called **endocardium**. A blood capillary consists of only one layer of endothelium. It is through this single layer of cells that oxygen, carbon dioxide, nutrients, and waste are transported by the blood cells to the various cells of our bodies.

Our final type of epithelial tissue based on function is **mesothelium** (mezo-**THEE**-lee-um). This tissue is also called **serous** (**SEER**-us) **tissue.** It is the tissue that lines the great cavities of the body that have no openings to the outside. These membranes consists of a simple squamous cell layer overlying a sheet of connective tissue. Special names are associated with this type of epithelial tissue also. The **pleura** (**PLOO**-rah) is the serous membrane or mesothelial tissue that lines the thoracic cavity. The **pericardium** is the serous membrane that covers the heart; the **peritoneum** (**pair**-ih-toh-**NEE**-um) is the serous membrane lining the abdominal cavity. This tissue protects, reduces friction between organs, and secretes fluid. The term **parietal** refers to the walls of a cavity and **visceral** refers to the covering on an organ.

CONNECTIVE TISSUE

The second major type of tissue is **connective tissue**. This type of tissue allows movement and provides support for other types of tissue. In this tissue, unlike epithelial, there is an abundance of intercellular material called **matrix** (**MAY**-trikz). This matrix is variable in both type and amount. It is one of the main sources of differences between the different types of connective tissue. There are also fibers of **collagen** (**KOL**-ah-jen) and **elastin** (ee-**LASS**-tin) embedded in this matrix. Sometimes the fibers are very apparent under the microscope, as in a tendon, whereas in other tissues the fibers are not very apparent as in certain cartilage. We can classify connective tissue into three subgroups: loose connective tissue, dense connective tissue, and specialized connective tissue.

Loose Connective Tissue

As the name implies, the fibers of loose connective tissue are not tightly woven among themselves. There are three types of loose connective tissue: **areolar** (ah-**REE**-oh-lah), **adipose** (**ADD**-ih-pohz), and **reticular** (reh-**TIK**-you-lar). Loose connective tissue fills spaces between and penetrates into organs.

Areolar is the most widely distributed of the loose connective tissue. It is easily stretched yet resists tearing. This tissue has three main types of cells distributed among its delicate fibers: fibroblasts, histiocytes, and mast cells. **Fibroblasts** (**FYR**-broh-blastz) are small flattened cells with large nuclei and reduced cytoplasm; they are also somewhat irregular in shape. The term *fibroblast* (blast meaning germinal or embryonic) refers to the ability of these cells to form fibrils (small fibers). They are active in the repair of injury. **Histiocytes** (**HISS**-tee-oh-sightz) are large, stationary **phagocytic** (fag-oh-**SIH**-tik) cells that eat up (phago = to eat) debris and microorganisms outside the blood circulatory system. They were originally monocytes in the circulating blood. When they are motile in tissue, they are called **macrophages** (**MACK**-roh-fay-jez). A macrophage of loose connective tissue is specifically called a histiocyte. Histiocytes are stationary or fixed in tissue. **Mast cells** are roundish or polygonal in shape and are found close to small blood vessels. Mast cells function in the production of **heparin** (an anticoagulant) and **histamine** (an inflammatory substance produced in response to allergies). Areolar tissue is the basic support tissue around organs, muscles, blood vessels, and nerves (Figure 5-5). It forms the delicate membranes around the spinal cord and brain. It attaches the skin to its underlying tissues.

Adipose tissue is the second type of loose connective tissue (Figure 5-6). It is loaded with fat cells. Fat cells are so full of stored fat that their nuclei and cytoplasm are pushed up against the cell membrane. In a histologic

Function	Characteristics and Location	Morphology
Areolar (loose) connective This tissue surrounds various organs and supports both nerve cells and blood vessels, which transport nutrient material (to cells) and wastes (away from cells). Areolar tissue also (temporarily) stores glucose, salts, and water.	It is composed of a large, semifluid matrix, with many different types of cells and fibers embedded in it. These include fibroblasts (fibrocytes), plasma cells, macrophages, mast cells, and various white blood cells. The fibers are bundles of strong, flexible white fibrous protein called collagen, and elastic single fibers of elastin. It is found in the epidermis of the skin and in the subcutaneous layer with adipose cells.	(A) Mast cell; Reticular fibers; Collagen fibers; Fibroblast cell; Plasma cell; Elastic fiber; Matrix; Macrophage cell (B)

Photo (B) Courtesy of John D. Cunningham, Visuals Unlimited, Inc.

FIGURE 5-5. Views of areolar or loose connective tissue.

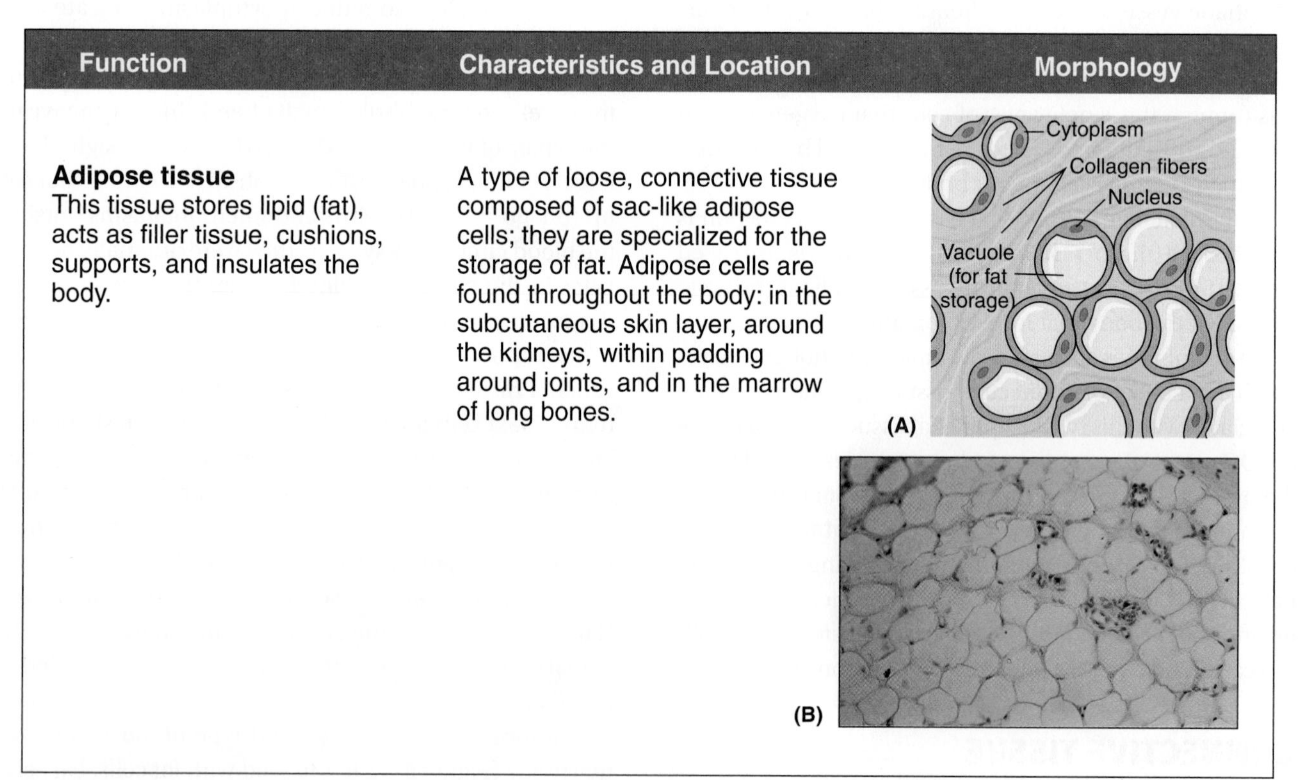

Function	Characteristics and Location	Morphology
Adipose tissue This tissue stores lipid (fat), acts as filler tissue, cushions, supports, and insulates the body.	A type of loose, connective tissue composed of sac-like adipose cells; they are specialized for the storage of fat. Adipose cells are found throughout the body: in the subcutaneous skin layer, around the kidneys, within padding around joints, and in the marrow of long bones.	(A) (B)

Photo (B) Courtesy of Fred Hossler, Visuals Unlimited, Inc.

FIGURE 5-6. Views of adipose tissue. It surrounds the lobules of mammary glands. The amount of adipose tissue determines a woman's breast size.

section under a microscope, they look like large soap bubbles and are very easy to recognize. Adipose tissue acts as a firm, protective packing around and between organs, bundles of muscle fibers, and nerves, and it supports blood vessels. The kidneys have a surrounding layer of adipose tissue to protect them from hard blows or jolts. In addition, because fat is a poor conductor of heat, adipose tissue acts as insulation for the body, protecting us from excessive heat losses or excessive heat increases in temperature. Think of the animals in the Arctic and Antarctic. They can live there because of their layers of blubber, which is adipose tissue. The camel's hump is not a water storage organ but a thick hump of fat containing adipose tissue to protect the animal's internal organs from the heat of the desert.

Reticular tissue is the third type of loose connective tissue. It consists of a fine network of fibers that form the framework of the liver, bone marrow, and lymphoid organs such as the spleen and lymph nodes.

Dense Connective Tissue

Again as the name implies, dense connective tissue is composed of tightly packed protein fibers. It is further divided into two subgroups based on how the fibers are arranged and the proportions of the tough collagen and the flexible elastin fibers. Examples of dense connective tissue having a *regular* arrangement of fibers are **tendons**, which attach muscle to bone; **ligaments**, which attach bone to bone; and **aponeuroses** (**ap**-oh-noo-**ROH**-seas), which are wide flat tendons (Figure 5-7). Tendons have a majority of tough collagen fibers, whereas ligaments (e.g., the vocal cords) have a combination of tough collagen and elastic elastin fibers.

Examples of dense connective tissue having an *irregular* arrangement of these fibers are muscle sheaths, the dermis layer of the skin, and the outer coverings of body tubes like arteries. Capsules that are part of a joint structure also have dense irregular connective tissue as do **fascia** (**FASH**-ee-ah), the connective tissue covering a whole muscle.

Specialized Connective Tissue

A number of types of connective tissue have specialized functions. Cartilage is one of these special kinds of tissues. The three types of cartilage found in the body are hyaline, fibrous, and elastic. Cells of cartilage are called **chondrocytes** (**KON**-droh-sightz); they are large round cells with spherical nuclei. When we view cartilage under

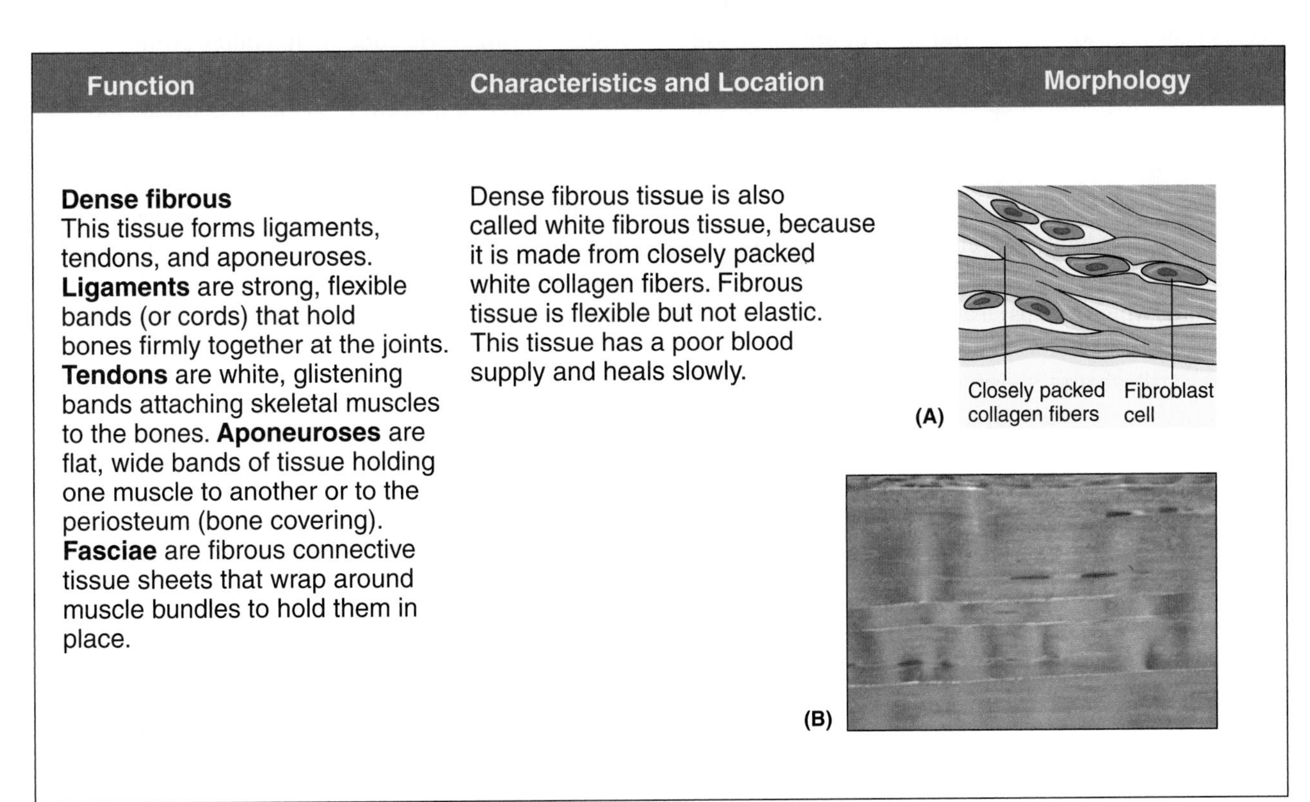

Function	Characteristics and Location	Morphology
Dense fibrous This tissue forms ligaments, tendons, and aponeuroses. **Ligaments** are strong, flexible bands (or cords) that hold bones firmly together at the joints. **Tendons** are white, glistening bands attaching skeletal muscles to the bones. **Aponeuroses** are flat, wide bands of tissue holding one muscle to another or to the periosteum (bone covering). **Fasciae** are fibrous connective tissue sheets that wrap around muscle bundles to hold them in place.	Dense fibrous tissue is also called white fibrous tissue, because it is made from closely packed white collagen fibers. Fibrous tissue is flexible but not elastic. This tissue has a poor blood supply and heals slowly.	

FIGURE 5-7. Examples of dense connective tissue with a regular arrangement of fibers are tendons, ligaments, and aponeuroses. Those with an irregular arrangement of fibers are fascia.

Photo (B) Courtesy of Carolina Biological, Visuals Unlimited, Inc.

Function	Characteristics and Location	Morphology
Cartilage Provides firm but flexible support for the embryonic skeleton and part of the adult skeleton. **Hyaline** Forms the skeleton of the embryo.	Hyaline cartilage is found on articular bone surfaces and also at the nose tip, bronchi, and bronchial tubes. Ribs are joined to the sternum (breastbone) by the costal cartilage. It is also found in the larynx and the rings in the trachea.	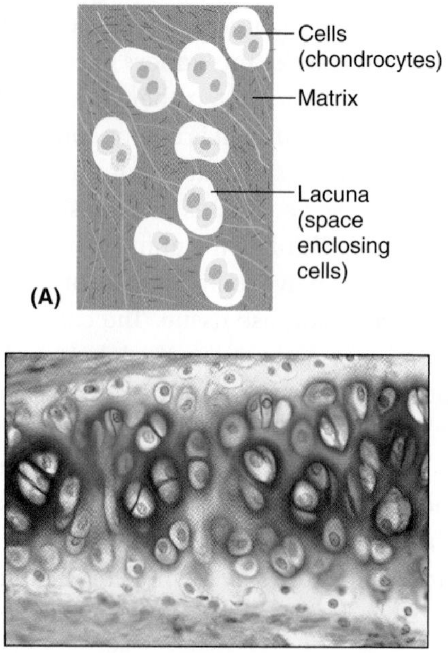 (A) Cells (chondrocytes); Matrix; Lacuna (space enclosing cells) (B)

Photo (B) Courtesy of Fred Hossler, Visuals Unlimited, Inc.

FIGURE 5-8. The anatomy of hyaline cartilage.

the microscope, these chondrocytes are found in cavities called **lacunae** (lah-**KOO**-nee). The lacunae are cavities in a firm matrix composed of protein and polysaccharides. Depending on the type of cartilage, various amounts of collagen and elastin fibers are embedded in the matrix, causing the cartilage to be either flexible or very strong and resistant.

Hyaline cartilage, when viewed under the microscope, has a matrix with no visible fibers in it, hence the name hyaline, which means clear (Figure 5-8). As the fetus forms in the womb, the skeletal system is made entirely of hyaline cartilage and is visible after the first 3 months of pregnancy. Most of this hyaline cartilage is gradually replaced by bone over the next 6 months through a process called ossification. However, some hyaline cartilage remains as a covering on the surfaces of the bones at joints. In our bodies, the costal cartilages that attach the anterior ends of our upper seven pair of ribs to the sternum is hyaline cartilage. The trachea and bronchi are kept open by incomplete rings of hyaline cartilage. The septum of our nose is also made of hyaline cartilage.

Fibrocartilage has a majority of tough collagenous fibers embedded in the matrix (Figure 5-9). These fibers make this type of cartilage dense and very resistant to stretching. The intervertebral disks that surround our spinal cord and act as shock absorbers between our vertebrae are made of this strong cartilage. It also connects our two pelvic bones at the pubic symphysis. Thus, we can flex our vertebral column and bend within a particular range of movement. During delivery, a minimal range of expansion of the birth canal can occur at the pubic symphysis due to the fibrocartilage.

The third type of cartilage is **elastic cartilage**. This type of cartilage has a predominance of elastin fibers embedded in the matrix. These fibers must be specially stained to view under a microscope (Figure 5-10). These fibers permit this type of cartilage to be easily stretched and flexible while being capable of returning to its original shape. Elastic cartilage makes up our external ear or auricle, our ear canals or auditory tubes, and our epiglottis.

Bone is very firm specialized connective tissue. Bone is covered in great detail in Chapter 7. If we section a bone, we see that it is composed of two types of bone tissue: **compact bone**, which forms the dense outer layer of bone and looks solid, and **cancellous bone**, which forms the inner spongy-looking tissue underneath the compact bone. When viewed under a microscope, the bone cells called **osteocytes** (**OSS**-tee-oh-sightz) are also found in cavities or lacunae as we saw in cartilage. However, the matrix of bone is impregnated with mineral salts, particularly calcium and phosphorous, which give bone its firm, hard appearance (Figure 5-11).

Function	Characteristics and Location	Morphology
Fibrocartilage A strong, flexible, supportive substance found between bones and wherever great strength (and a degree of rigidity) is needed.	Fibrocartilage is located within intervertebral disks and the pubic symphysis between the pubic bones.	(A) 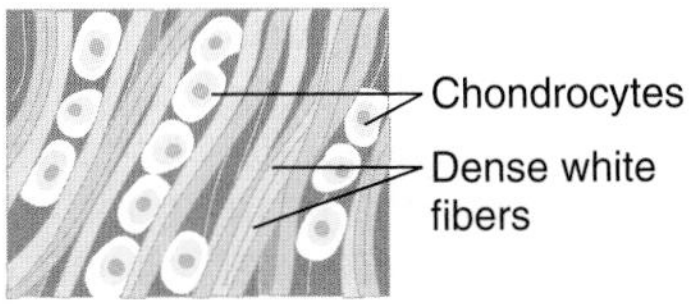 Chondrocytes; Dense white fibers (B) 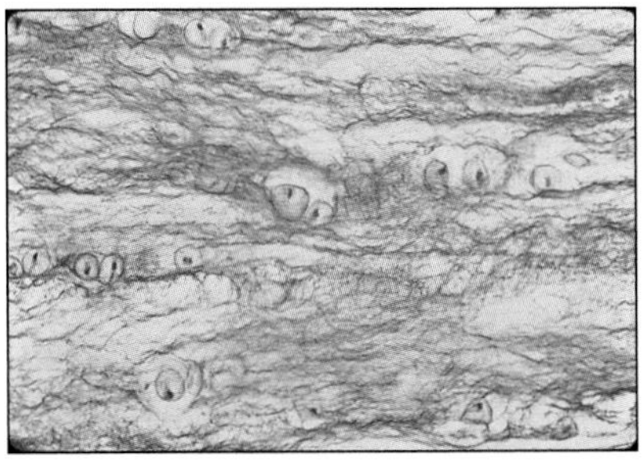

Photo (B) Courtesy of Cabisco, Visuals Unlimited, Inc.

FIGURE 5-9. The anatomy of fibrocartilage.

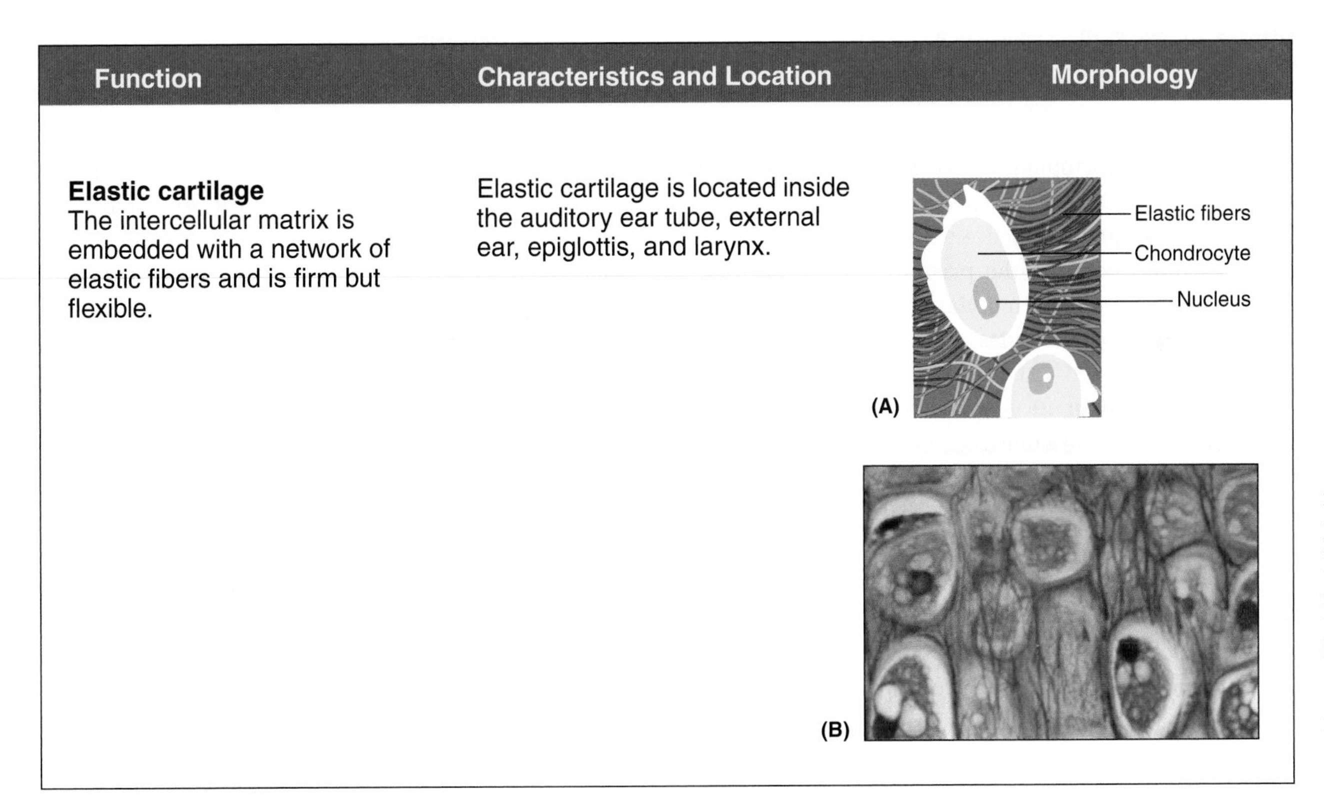

Photo (B) Courtesy of Triarch, Visuals Unlimited, Inc.

FIGURE 5-10. The anatomy of elastic cartilage.

Another specialized connective tissue is **dentin**, the material that forms our teeth. Dentin is closely related to bone in structure but is harder and denser. The crown of the tooth is covered with another material, enamel, which is white in appearance. Dentin is light brown. If you ever chipped a tooth, the brown material you saw under the white enamel was the dentin. The enamel is secreted onto the dentin of a tooth by special epithelial cells that make up the enamel organ. This secretion occurs just before the teeth break their way through the gums.

Function	Characteristics and Location	Morphology
Bone (osseus) tissue Comprises the skeleton of the body, which supports and protects underlying soft tissue parts and organs, and also serves as attachments for skeletal muscles.	Connective tissue's intercellular matrix is calcified by the deposition of mineral salts (like calcium carbonate and calcium phosphate). Calcification of bone imparts great strength. The entire skeleton is composed of bone tissue.	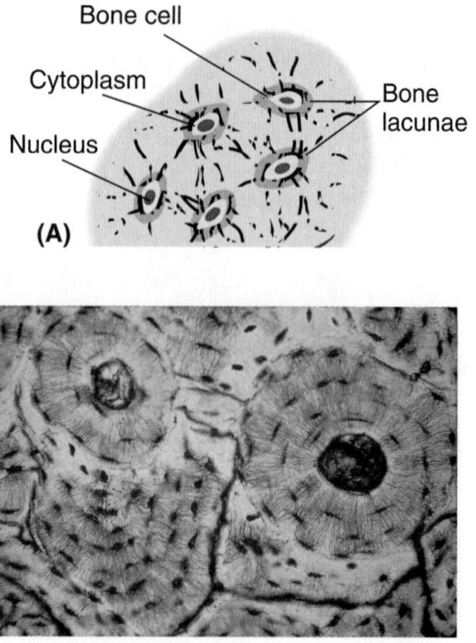 Bone cell; Cytoplasm; Nucleus; Bone lacunae (A); (B)

Photo (B) Courtesy of Fred Hossler, Visuals Unlimited, Inc.

FIGURE 5-11. Views of compact bone.

Blood and **hematopoietic** (hee-**MAT**-oh-poy-**eh**-tik) **tissue** are other examples of specialized connective tissue. Blood is unique connective tissue in that it is composed of a fluid portion (the plasma) and the formed elements of blood: the **erythrocytes** (eh-**RITH**-roh-sightz) or red blood cells and **leukocytes** (**LOO**-koh-sightz) or white blood cells (Figure 5-12). We will discuss blood in more detail in Chapter 13. Blood cells are formed in red bone marrow, and some white blood cells are also formed in lymphoid organs. Marrow and lymphoid organs are referred to as hematopoietic tissue. Blood is liquid tissue circulating through the body. It transports oxygen, nutrients, hormones, enzymes, and waste products such as carbon dioxide gas and urea. It also protects the body through its white blood cells and helps to regulate body temperature.

Lymphoid tissue is another specialized connective tissue (Figure 5-13). Lymphoid tissue is found in the lymph glands or nodes, the thymus gland, the spleen, the tonsils, and the adenoids. Lymph tissue manufactures plasma cells like the B lymphocytes. This tissue's main role is antibody production and protects us from disease and foreign microorganisms.

The **reticuloendothelial** (reh-**tik**-you-loh-**in**-doh-**THEE**-lee-al) or **RE system** consists of those specialized connective tissue cells that do phagocytosis. Three types of cells fit into this category. The first type are the RE cells that line the liver (they get another special name: **Kupffer's cells**) and those that line the spleen and bone marrow. The second type are the macrophages. These cells are also referred to as histiocytes or "resting-wandering" cells, because they are fixed in tissue until they must wander to an invader and devour it. Any phagocytic cell of the RE system can be called a macrophage. The third type of cell is a **neuroglia** (noo-**ROH**-glee-ah) which does support and a **microglia** (my-**KROG**-lee-ah) cell. This is a phagocytic cell found in the central nervous system. Other types of neuroglia cells do support.

Synovial membranes line the cavities of freely moving joints and are also classified as specialized connective tissue. These membranes also line bursae, which are small sacs containing synovial fluid found between muscles, tendons, bones, and skin and underlying structures. They prevent friction.

Connective Tissue Functions

Connective tissue has many and varied functions:

1. **Support:** Bones support other tissues of the body. On top of bones we find muscle, nerves, blood vessels, fat, and skin. Cartilage supports our nose and forms the bulk of the structure of our ear.

Function	Characteristics and Location	Morphology
Vascular (liquid blood tissue) Blood Transports nutrient and oxygen molecules to cells and metabolic wastes away from cells (can be considered as a liquid tissue). Contains cells that function in the body's defense and in blood clotting.	Blood is composed of two major parts: a liquid called plasma, and a solid cellular portion known as blood cells (or corpuscles). The plasma suspends corpuscles, of which there are two major types: red blood cells (erythrocytes) and white blood cells (leukocytes). A third cellular component (really a cell fragment) is called platelets (thrombocytes). Blood circulates within the blood vessels (arteries, veins, and capillaries) and through the heart.	Erythrocytes, Thrombocytes (platelets), Lymphocyte, Neutrophil, Monocyte, Basophil, Eosinophil (A) (B)

FIGURE 5-12. Views of blood, a unique fluid connective tissue.

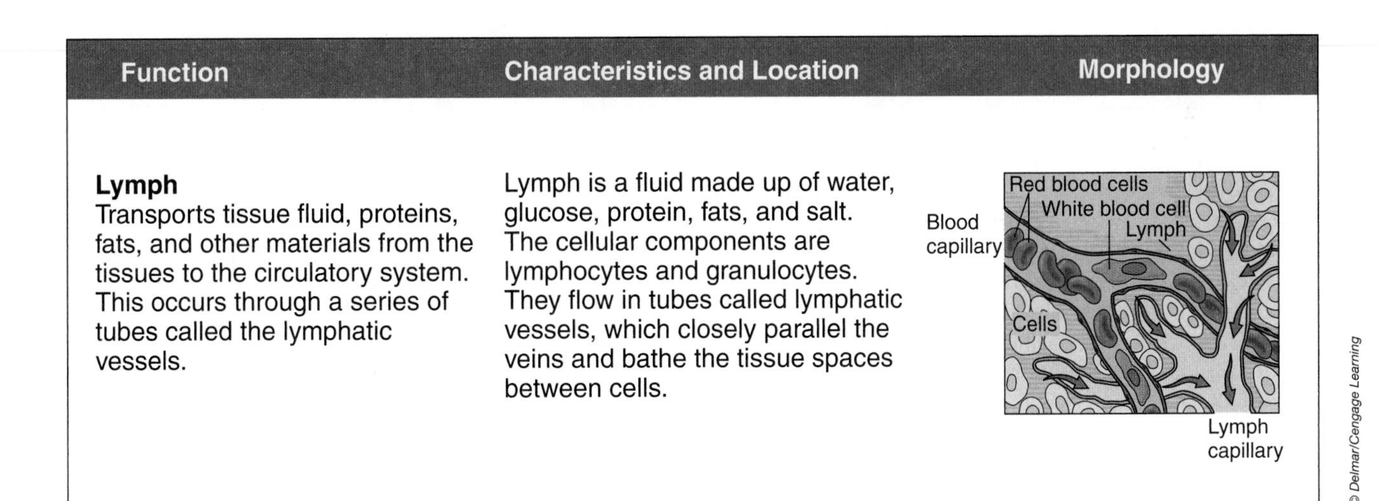

Function	Characteristics and Location	Morphology
Lymph Transports tissue fluid, proteins, fats, and other materials from the tissues to the circulatory system. This occurs through a series of tubes called the lymphatic vessels.	Lymph is a fluid made up of water, glucose, protein, fats, and salt. The cellular components are lymphocytes and granulocytes. They flow in tubes called lymphatic vessels, which closely parallel the veins and bathe the tissue spaces between cells.	

FIGURE 5-13. A diagram of a lymph capillary.

2. **Nourishment:** Blood carries nutrients to the cells of our body. Synovial membranes in joint capsules nourish the cartilage found on top of bones.
3. **Transportation:** Blood transports gases, enzymes, and hormones to cells.
4. **Connection:** Tendons connect muscles to bone, and ligaments connect bone to bone.
5. **Movement:** Muscles through tendons pull on bones, and bones move our bodies through our environment.

6. **Protection:** Bones protect vital organs of the body like the heart, lungs, brain, and spinal cord. Blood cells, especially the white blood cells, protect us from foreign microorganisms and tissue injury.
7. **Insulation:** Adipose tissue (fat) insulates us from excessive heat loss and excessive increases in temperature.
8. **Storage:** Bone stores the mineral salts calcium and phosphorous. Adipose tissue stores the high-energy molecules of fat to be used and converted to adenosine triphosphate when necessary.
9. **Attachment and separation:** Connective tissue attaches skin to underlying muscle. It also forms layers around and between organs.

MUSCLE TISSUE

The basic characteristic of **muscle tissue** is its ability to shorten and thicken or contract. This is due to the interaction of two proteins in the muscle cell: actin and myosin. Muscle cell contractility is discussed in greater detail in Chapter 9. Because a muscle cell's length is much greater than its width, muscle cells are frequently referred to as **muscle fibers**. The three types of muscle tissue are smooth, striated or skeletal, and cardiac.

Smooth muscle cells are spindle-shaped with a single nucleus (Figure 5-14). They are not striated (**STRYE**-ate-ed), that is, you do not see alternating dark and light bands when viewed under the microscope. This muscle tissue is involuntary, meaning we do not control its contraction. It is controlled by the autonomic nervous system. We find smooth muscle in the walls of hollow organs like those of the digestive tract, arteries, and veins. The muscle cells are arranged in layers: an outer longitudinal layer and an inner circular layer. Simultaneous contraction of the two layers pushes materials inside the hollow organs in one direction. Hence, food is pushed by contraction of the smooth muscles along the digestive tract, called **peristalsis** (pair-ih-**STALL**-sis), and blood is pushed along in arteries and veins. Urine is also pushed down the ureters from the kidneys by contraction of smooth muscle.

Striated or **skeletal muscle** is the muscle we normally think about when we mention muscle (Figure 5-15). It is the tissue that causes movement of our body by pulling on bones, hence the name skeletal muscle. The long thin cells of skeletal muscle are multinucleated and striated. We can see alternating light bands of the thin protein filaments of actin and dark bands of the thick protein filaments of

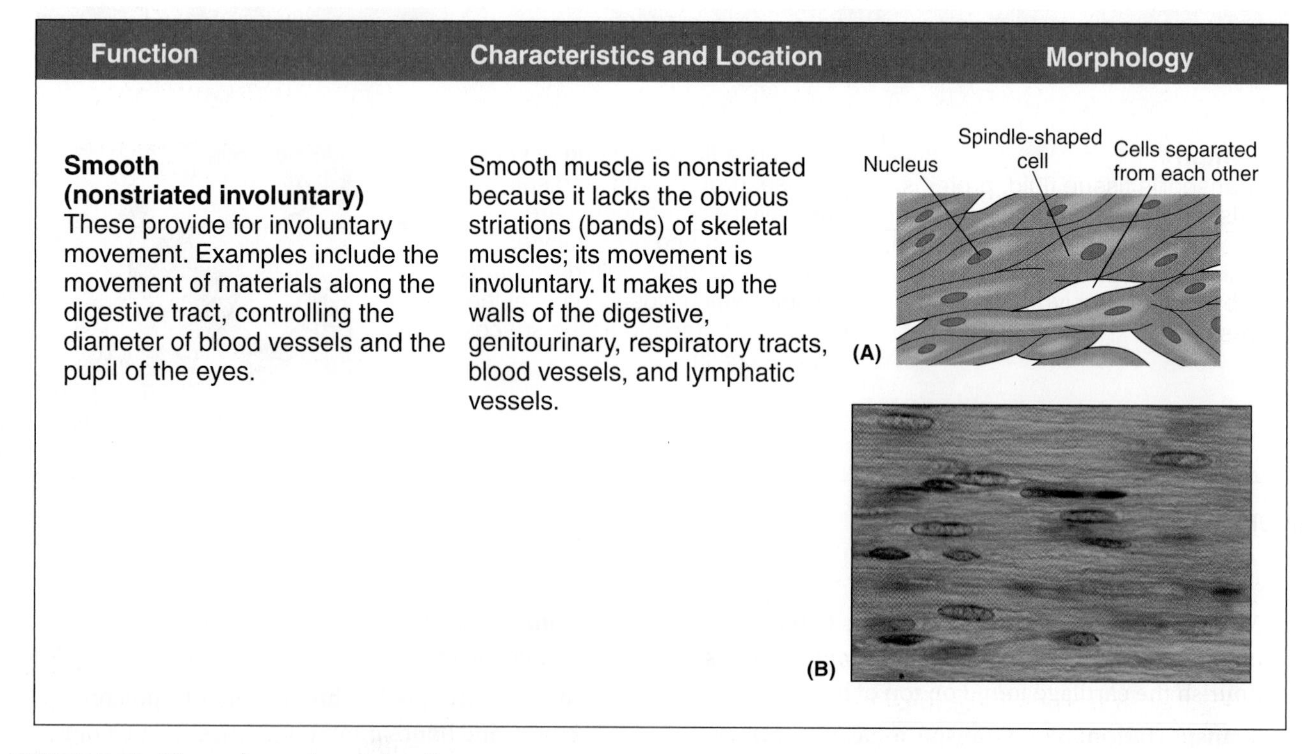

Photo (B) Courtesy of R. Calentine, Visuals Unlimited, Inc.

FIGURE 5-14. Views of smooth muscle cells.

Function	Characteristics and Location	Morphology
Skeletal (striated voluntary) These muscles are attached to the movable parts of the skeleton. They are capable of rapid, powerful contractions and long states of partially sustained contractions, allowing for voluntary movement.	Skeletal muscle is striated (having transverse bands that run down the length of muscle fiber); voluntary, because the muscle is under conscious control; and *skeletal*, because these muscles are attached to the skeleton (bones, tendons, and other muscles).	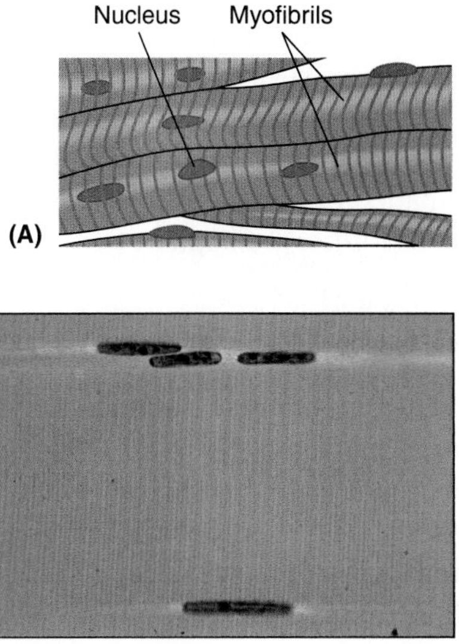 Nucleus, Myofibrils (A); (B)

Photo (B) Courtesy of R. Calentine, Visuals Unlimited, Inc.

FIGURE 5-15. Views of skeletal muscle cells.

Function	Characteristics and Location	Morphology
Cardiac These cells help the heart contract to pump blood through and out of the heart.	Cardiac muscle is a striated (having a cross-banding pattern), involuntary (not under conscious control) muscle. It makes up the walls of the heart.	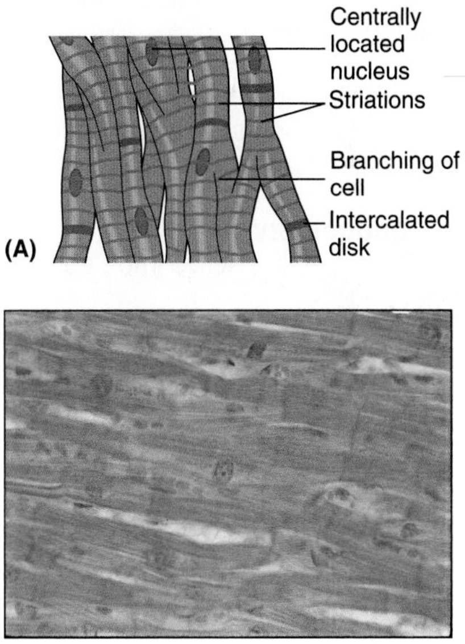 Centrally located nucleus, Striations, Branching of cell, Intercalated disk (A); (B)

Photo (B) Courtesy of John Cunningham, Visuals Unlimited, Inc.

FIGURE 5-16. Views of cardiac muscle cells.

myosin. When we eat "meat" of animals and fish, it is usually muscle that we are consuming. Muscle makes up about 40% of our total weight and mass. Striated muscle is voluntary and is under the control of the central nervous system.

Cardiac muscle is found only in the heart. Like skeletal muscle it is striated and like smooth muscle it is uninucleated and under the control of the autonomic nervous system (Figure 5-16). The cells of cardiac muscle are cylindrical in shape with branches that connect to other cardiac cells. These branches connect with one another through special areas called **intercalated** (in-**TER**-kah-**lay**-ted) **disks**. The cells are much shorter than either skeletal or smooth muscle cells. This is the muscle that causes contraction or beating of the heart; thus, it pumps the blood through our body. The interconnected branches of cardiac muscle cells guarantee coordination of the pumping action of the heart (to be discussed further in Chapter 14).

NERVOUS TISSUE

The basic unit of organization of nervous tissue is the nerve cell or **neuron** (**NOO**-ron) (Figure 5-17). Actually, the neuron is a conducting cell, whereas other cells of the system called neuroglia are supporting cells. These different types of nerve cells are discussed in greater detail in Chapter 10. Neurons are very long cells, so like muscle cells, they are called nerve fibers. It is basically impossible to view an entire neuron even under low power of the microscope due to their length. However, we can view the parts of a neuron as we scan a microscope slide. The **cell body** contains the nucleus. It also has rootlike extensions called **dendrites** (**DEN**-drytz) that receive stimuli and conduct them to the cell body. **Axons** (**AK**-sonz) are long, thin extensions of the cell body that transmit the impulse toward the **axon endings**.

Nervous tissue makes up the brain, spinal cord, and various nerves of the body. It is the most highly organized tissue of the body. It controls and coordinates body activities. It allows us to perceive our environment and to adapt to changing conditions. It coordinates our skeletal muscles. Its special senses include sight, taste, smell, and hearing. It controls our emotions and our reasoning capabilities. It allows us to learn through the memory process.

Function	Characteristics and Location	Morphology
Neurons (nerve cells) These cells have the ability to react to stimuli. **1. Irritability—** Ability of nerve tissue to respond to environmental changes. **2. Conductivity—** Ability to carry a nerve impulse (message).	Nerve tissue is composed of *neurons* (nerve cells). Neurons have branches through which various parts of the body are connected and their activities coordinated. They are found in the brain, spinal cord, and nerves.	Dendrites Nucleus Axon Sheath Myelin Axon terminal branches **(A)** Cell body **(B)**

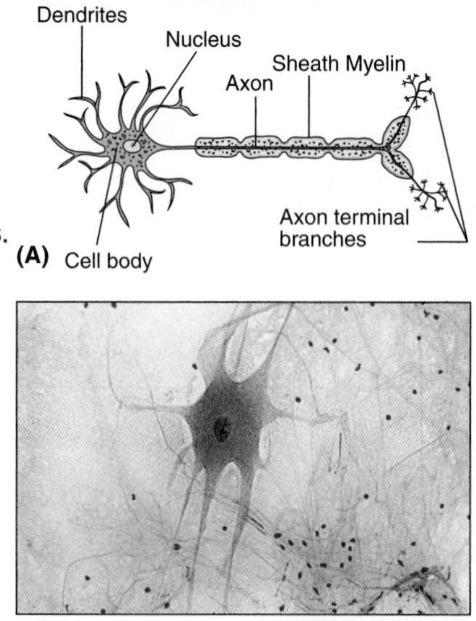

Photo (B) Courtesy of Triarch, Visuals Unlimited, Inc.

FIGURE 5-17. Views of a neuron (multipolar, motor neuron).

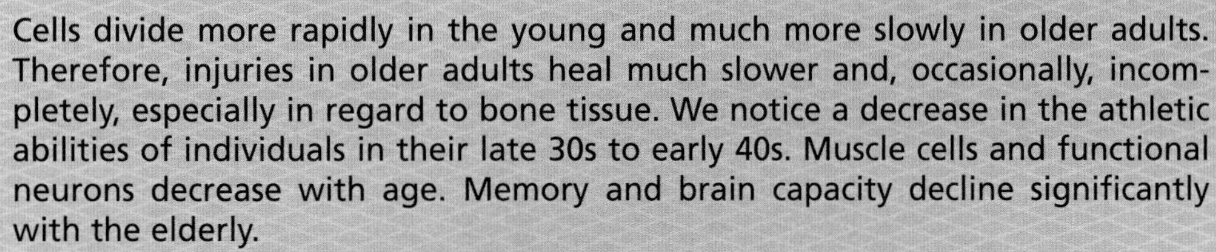

Cells divide more rapidly in the young and much more slowly in older adults. Therefore, injuries in older adults heal much slower and, occasionally, incompletely, especially in regard to bone tissue. We notice a decrease in the athletic abilities of individuals in their late 30s to early 40s. Muscle cells and functional neurons decrease with age. Memory and brain capacity decline significantly with the elderly.

As we age, the tough collagen fibers become structurally irregular, resulting in ligaments and tendons being more fragile and less flexible. Elastic fibers become less elastic. We see gradual increases over time in the wrinkling of the skin due to these connective tissue fiber changes. Thus, the older we get, the more we tend to use moisturizing and fiber-containing cosmetic creams and lotions. ■

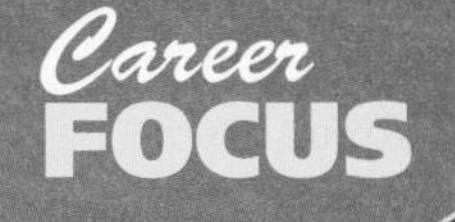

These are careers that are available to individuals interested in the study of tissues. They require training in microscopy to examine cellular structures.

- **Histologists** are medical professionals specializing in the study of the structure and function of tissues. They are employed in universities and medical centers.
- **Forensic scientists** specialize in analyzing tissue samples, looking for clues that help solve crimes. They are employed by law enforcement agencies.

SUMMARY OUTLINE

INTRODUCTION

1. Histology is the study of tissues.
2. The four kinds of tissue are epithelial, connective, muscle, and nervous.

EPITHELIAL TISSUE

1. Epithelial tissue functions in four ways: it protects underlying tissues; it absorbs nutrients; it secretes hormones, mucus, and enzymes; and it excretes waste like urea in sweat.
2. The basement membrane acts as an anchor and adhesive for epithelial cells.
3. Epithelial tissue can be named according to shape, arrangement, or function.
4. Epithelial tissue is made of cells closely packed together with very little intercellular material.

Classification Based on Shape

1. Squamous epithelial cells are flat and serve a protective function like the lining of our mouths and our skin.
2. Cuboidal epithelial cells are shaped like cubes and function in protection and secretion.
3. Columnar epithelial cells are tall and rectangular. They function in secretion and absorption.

Classification Based on Arrangement

1. Simple epithelium is one cell layer thick.
2. Stratified epithelium is several layers of cells thick.
3. Pseudostratified epithelium looks like it is several layers thick, but, in reality, all cells extend from the basement membrane to the outer surface.
4. Transitional epithelium consists of several layers of closely packed, easily stretched cells. When stretched they appear flat; when relaxed they look ragged or saw-toothed.

Classification Based on Function

1. Mucous membrane produces mucus. It protects, absorbs nutrients, and secretes enzymes and bile salts in addition to mucus.
2. Simple exocrine glands such as sweat and sebaceous glands have single unbranching ducts.
3. Compound exocrine glands are made of several branching lobules with branching ducts. Examples are the mammary glands and the large salivary glands.
4. Endocrine glands are ductless and secrete hormones directly into the bloodstream.
5. Endothelium lines the blood and lymphatic vessels. The endothelium of the heart is called the endocardium.
6. Mesothelium or serous tissue lines the great cavities of the body. The pleura lines the thoracic cavity. The peritoneum lines the abdominal cavity. The pericardium covers the heart.

CONNECTIVE TISSUE

1. Connective tissue is made of cells with lots of intercellular material called matrix.
2. Fibers of tough collagen or fibers of flexible elastin can be embedded in this matrix.
3. The three subgroups of connective tissue are loose connective tissue, dense connective tissue, and specialized connective tissue.

Loose Connective Tissue

1. The three types of loose connective tissue are areolar, adipose, and reticular.
2. Loose connective tissue fills space between and penetrates into organs.
3. Areolar is the most widely distributed type of loose connective tissue. It contains three types of cells: fibroblasts, which make fibrils for repair; histiocytes or macrophages, which do phagocytosis; and mast cells, which produce the anticoagulant heparin and histamine, an inflammatory substance.
4. Adipose tissue is loose connective tissue with fat stored in its cells. It protects and insulates.
5. Reticular tissue forms the framework of the liver, spleen, lymph nodes, and bone marrow.

Dense Connective Tissue

1. Dense connective tissue having a regular arrangement of embedded fibers are tendons, ligaments, and aponeuroses.
2. Dense connective tissue having an irregular arrangement of embedded fibers are muscle sheaths, joint capsules, and fascia.

Specialized Connective Tissue

1. The cells of cartilage are called chondrocytes. The three types of cartilage tissue are hyaline, fibrocartilage, and elastic.
2. Hyaline cartilage is found in the costal cartilages that attach the ribs to the sternum, in the septum of our nose, and in the rings that keep our trachea and bronchi open.
3. Fibrocartilage is very strong; intervertebral disks are made of fibrocartilage.
4. Elastic cartilage is easily stretched and flexible. It is found in the ears, epiglottis, and auditory tubes.
5. The two types of bone tissue are compact or dense and cancellous or spongy. Bone cells are called osteocytes. They are embedded in a matrix of calcium and phosphorous, the mineral salts responsible for the hardness of bone.
6. Our teeth are made of dentin; the crown of the tooth is covered with enamel.
7. Blood is composed of a liquid portion called plasma and the blood cells. Blood cells are formed in red bone marrow, a hematopoietic tissue.
8. Lymphoid tissue makes up our lymph glands, thymus, spleen, tonsils, and adenoids. This tissue produces the plasma cells or B lymphocytes that produce antibodies.
9. The reticuloendothelial (RE) system is involved in phagocytosis in connective tissue. Kupffer's cells line the liver; RE cells also line the spleen and bone marrow. Macrophage is a term for any phagocytic cell of the RE system. Microglia cells do phagocytosis in the nervous system; other neuroglia cells do support.
10. Synovial membranes line joints and bursae. They produce synovial fluid, which lubricates joints and nourishes cartilage.

Connective Tissue Functions

1. It supports other tissues.
2. It provides nourishment: blood carries nutrients.
3. It transports: blood transports enzymes and hormones.
4. It connects various tissues to one another.
5. It provides movement via bones.

6. It protects vital organs (bones of skull and thorax) and provides immunity (lymphoid tissue and white blood cells).
7. It insulates and maintains temperature (adipose tissue).
8. It provides storage areas: bone stores calcium and phosphorous, adipose tissue stores fat.
9. It attaches and separates other tissues of the body.

MUSCLE TISSUE

1. The three types of muscle tissue are smooth, striated or skeletal, and cardiac.
2. Due to the interaction of two proteins, actin and myosin, muscle cells can shorten their length or contract. Some pull on bones through tendons and bring about movement.
3. Smooth muscle cells are long, unicellular and nonstriated. They are involuntary and are arranged in two layers around hollow organs: an outer longitudinal layer and an inner circular one. They are found in the digestive tract, arteries and veins, and the ureters of the kidney.
4. Striated or skeletal muscle cells are long, multinucleated and striated. They are voluntary and pull on bone, causing movement.
5. Cardiac muscle cells are found only in the heart. They are striated, uninucleated and cylindrical in shape with branches that connect to branches of other cardiac cells via intercalated disks. These cells are responsible for pumping blood through the heart.

NERVOUS TISSUE

1. Nervous tissue is composed of two types of nerve cells: neurons are conducting cells and neuroglia are supporting and protecting cells.
2. A neuron is composed of a cell body with a nucleus, extensions of the cell body called dendrites, and a long axon with axon endings.
3. Nervous tissue controls and coordinates the activities of the body.

REVIEW QUESTIONS

1. Name the three cell shapes of epithelial tissue.
2. Name the two types of exocrine glands and give an example of each.
3. Name four functions of epithelial tissue.
4. *Compare the structure of epithelial tissue with that of connective tissue.
5. *Why is adipose tissue considered a good insulator?
6. Name five functions of connective tissue and give examples.
7. Name the three types of muscle tissue.
8. Name the two types of nerve cells found in nervous tissue.

*Critical Thinking Questions

FILL IN THE BLANK

Fill in the blank with the most appropriate term.

1. The three types of loose connective tissue are: ____________, ____________, and ____________.
2. ____________ are cells that form fibrils and are active in the repair of injury.
3. ____________ are phagocytic cells that operate outside the vascular system—they are often fixed and are found in areolar tissue.
4. ____________ cells function in the production of heparin and histamine.
5. ____________ tissue is loose connective tissue with fat-containing cells.
6. Examples of dense connective tissue having a regular arrangement of fibers are ____________, ____________, and ____________.
7. Examples of dense connective tissue having an irregular arrangement of fibers are ____________, ____________, ____________, and ____________ sheaths.
8. The three types of cartilage are ____________, ____________, and ____________.
9. The two common types of bone tissue are ____________ and ____________.
10. A tooth is made up of ____________; the crown of the tooth is covered by ____________, the hardest substance in the body.
11. ____________ membranes line the cavities of the freely moving joints and bursae.
12. The two types of protein fibers that can be found in the matrix of connective tissue are ____________ and ____________.

MATCHING

Place the most appropriate number in the blank provided.

_____ Heparin	1.	Cells of cartilage
_____ Histamine	2.	Blood platelets
_____ Chondrocytes	3.	Hardest substance in the body
_____ Ossification	4.	White blood cells
_____ Enamel	5.	Blood-forming tissue
_____ Marrow	6.	Anticoagulant
_____ Erythrocytes	7.	Red blood cells
_____ Leukocytes	8.	Fat cells
_____ Neuron	9.	Supports the neuron
_____ Neuroglia	10.	Conducting cell
	11.	Inflammatory substance
	12.	Formation of bone

Search and Explore

Search the Internet with key words from the chapter to discover additional information and interactive exercises. Key words might include anatomy of a neuron, muscle tissue, epithelial tissue, or connective tissue.

*Study***WARE**™ Connection

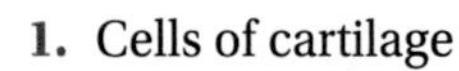

Take a quiz or play a concentration game on your StudyWARE™ CD-ROM.

Study Guide Practice

Go to your **Study Guide** for more practice questions, labeling exercises, and a crossword puzzle to help you learn the content in this chapter.

LABORATORY EXERCISE: TISSUES

Materials needed: A compound microscope and prepared microscope slides

A. EPITHELIAL TISSUE

1. Examine a prepared slide of stratified squamous epithelial tissue. Two types of slides will be available. Slides made from the internal tissues of a human like the epiglottis will be nonkeratinized, that is, the cells will have nuclei. If the slide is of the skin, it will be keratinized, that is, the cells will not have nuclei. Refer to Figure 5-1 in the text as you do your microscope examination of this tissue.
2. Examine a prepared slide of simple cuboidal epithelium. This slide will be from a duct of a gland. Refer to Figure 5-2 in the text as you examine these cells shaped like cubes attached to a basement membrane.
3. Examine a prepared slide of simple columnar epithelium. The best slide of this type of tissue comes from the intestine. Refer to

(continues)

TISSUES (Continued)

Figure 5-3 in the text as you examine these tall rectangular cells.

4. Examine a prepared slide of pseudostratified cilated, columnar epithelium. Notice how the cilia on the free edge of the cells look like flames or waves. Try to see that each cell will extend from the basement membrane. Refer to Figure 5-4A in the text.

B. CONNECTIVE TISSUE

1. Examine a prepared slide of hyaline cartilage. Notice how the chondrocytes are in a cavity or lacuna and that the matrix appears clear. No fibers are visible in the matrix. Refer to Figure 5-8 in the text as you view your slide.
2. Examine a prepared slide of elastic cartilage. This slide has been specially stained to show the splinter-like elastin fibers embedded in the matrix surrounding the chondrocytes in their lacuna. Refer to Figure 5-10 in the text as you view this slide.
3. Examine a prepared slide of fibrocartilage. Notice how thick and wavy the fibers of collagen are arranged in the matrix. Note the fewer chondocytes in their lacunae compared to the other two types of cartilage. Refer to Figure 5-9 as you view this slide.
4. Examine a prepared slide of compact bone. Refer to Figure 5-11 of the text as you view this slide. Identify the central canal surrounded by rings of bone. The central canal contains a blood capillary. The rings of bone are formed by the mineralized matrix and are called lamellae. Note the lacunae, which contain the osteocytes.
5. Examine a prepared slide of human blood stained with Wright's stain. First examine under low power. Look for an area where you see some dark-stained cells. These will be leukocytes with stained nuclei. Switch to high power. You will be able to identify the many erythrocytes without nuclei and various leukocytes with their stained nucleus that usually appears folded. Also notice the tiny stained specks in the plasma; these are thrombocytes or platelets. Refer to Figure 5-12 in the text.

C. MUSCLE TISSUE

1. Examine a prepared slide of smooth muscle. Notice the spindle tapering cells of smooth muscle with no cross striations. They are uninucleated and under the microscope look like flowing water in a stream. Refer to Figure 5-14 in the text.
2. Examine a prepared slide of skeletal muscle. Notice that the cells are large, multinucleated with visible cross striations. They look like thick poles under the microscope. The cross striations are alternating bands of thick myosin protein filaments (dark) and thin actin protein filaments (light). Refer to Figure 5-15 in the text as you view this tissue.
3. Examine a prepared slide of cardiac muscle. Refer to Figure 5-16 in the text. Notice that the cells are striated and uninucleated. The cells have branches that look like splits in a pole. Notice the thick intercalated disks that connect the branches of the cardiac cells.

D. NERVOUS TISSUE

1. Examine a prepared slide of a multipolar neuron. This slide comes from the spinal cord of an ox. Search under low power to identify a cell body with nucleus and dendrite extensions. Notice the very long axons and at the ends of axons the axon endings. Refer to Figure 5-17 in the text.

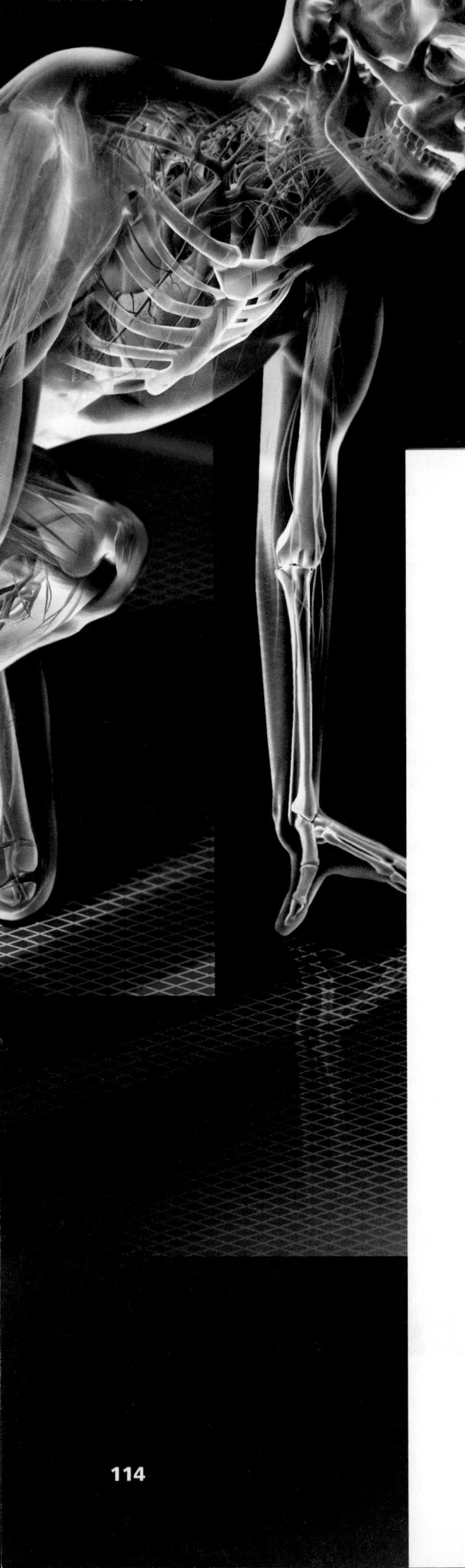

The Integumentary System

CHAPTER OBJECTIVES

After studying this chapter, you should be able to:

1. Name the layers of the epidermis.
2. Define *keratinization*.
3. Explain why there are skin color differences among people.
4. Describe the anatomic parts of a hair.
5. Compare the two kinds of glands in the skin based on structure and secretion.
6. Explain why sweating is important to survival.
7. Explain how the skin helps regulate body temperature.
8. Name the functions of the skin.

6

KEY TERMS

INTRODUCTION

The integumentary system is made up of the skin and its appendages. See Concept Map 6-1: Integumentary System. The appendages or modifications of the skin are hair, nails, sebaceous, ceruminous, and sweat glands. The word *integument* means a covering, and the skin of an average adult covers well over 3000 square inches of surface area of the body. The skin weighs about 6 pounds (this is nearly twice the weight of the brain or the liver). It receives approximately one-third of all the blood circulating through the body. It is flexible yet rugged and under normal conditions can repair and regenerate itself. Our skin is almost entirely waterproof. It protects us from the harmful ultraviolet rays of the sun through special pigment-producing cells. It is an effective barrier to most harmful chemicals, keeping them from entering our internal environment. It participates in the dissipation of water through sweating and helps regulate our body temperature.

THE LAYERS OF THE SKIN

Our skin consists of two main layers (Figure 6-1). The **epidermis** (ep-ih-**DER**-mis) is a layer of epithelial tissue that can further be divided into sublayers. It is found on top of the second layer of the skin called the **dermis**. This is a layer of dense connective tissue that connects the skin to tissues below it, like fat and muscle. Beneath the dermis is the subcutaneous layer, sometimes called the hypodermis.

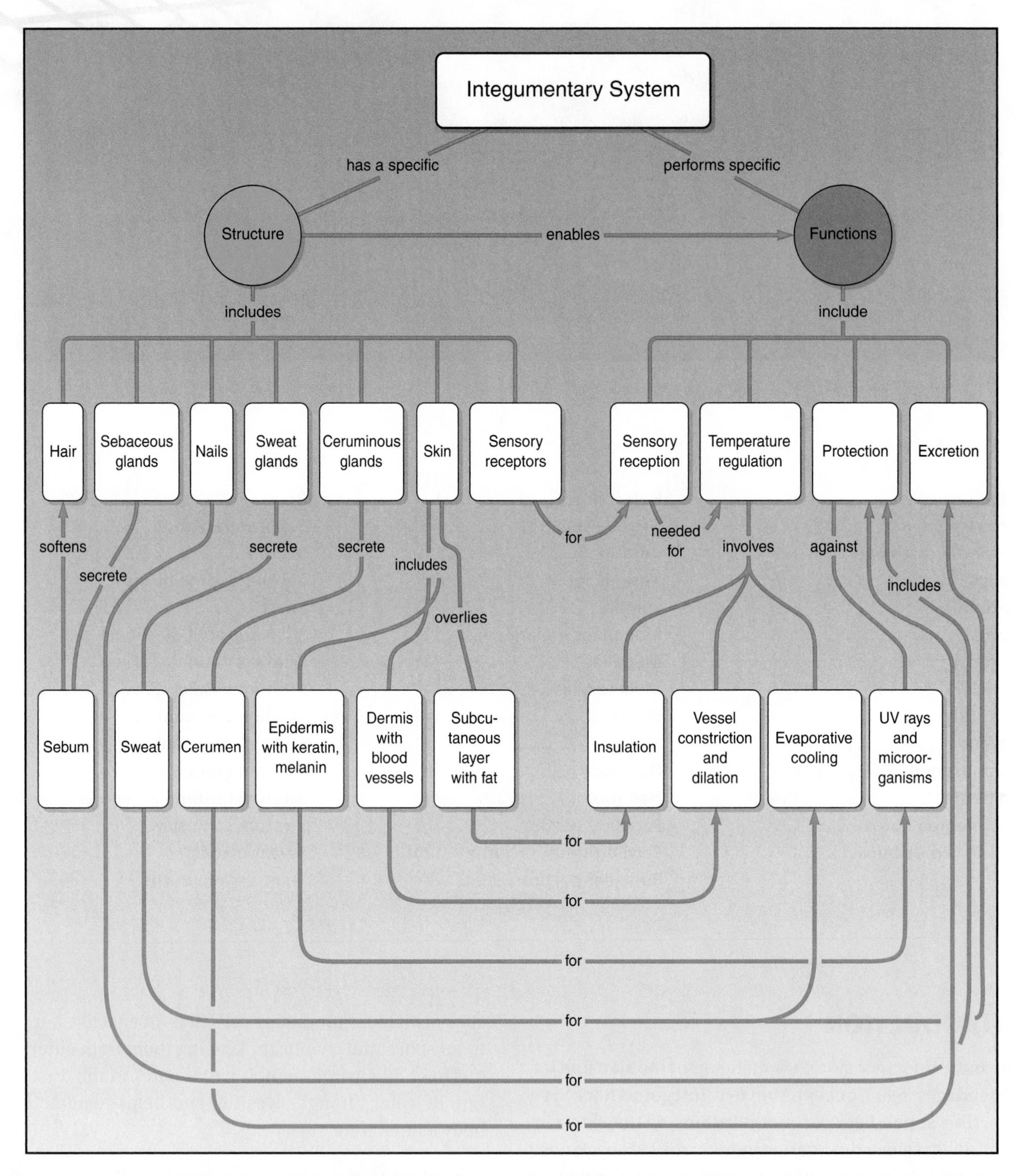

CONCEPT MAP 6-1. Integumentary System. This is the first introduction of a concept map. Each system of the body will have a concept map to introduce you to the relationship between various structures of the system and how these structures enable the system to perform its functions. A concept map is basically an outline that allows major topics in a chapter to be observed on one page, and the reader sees an overview of those topics.

The concept map breaks up a topic into its major components. First, the system of the body will be divided by connecting lines into its *structures* and *functions.* Then each of these two major topics will be further subdivided into smaller components. Connecting lines or arrows will be used to join related divisions for the purpose of showing relationships. Occasionally, a brief description or word may be used along with the connecting lines to reinforce the explanation of the relationship. Spend a few minutes examining the concept map before you read further in the chapter. It will set the stage for what you are about to read. When you go back to study the information in the chapter, referring back to the concept map will help you remember what you have read.

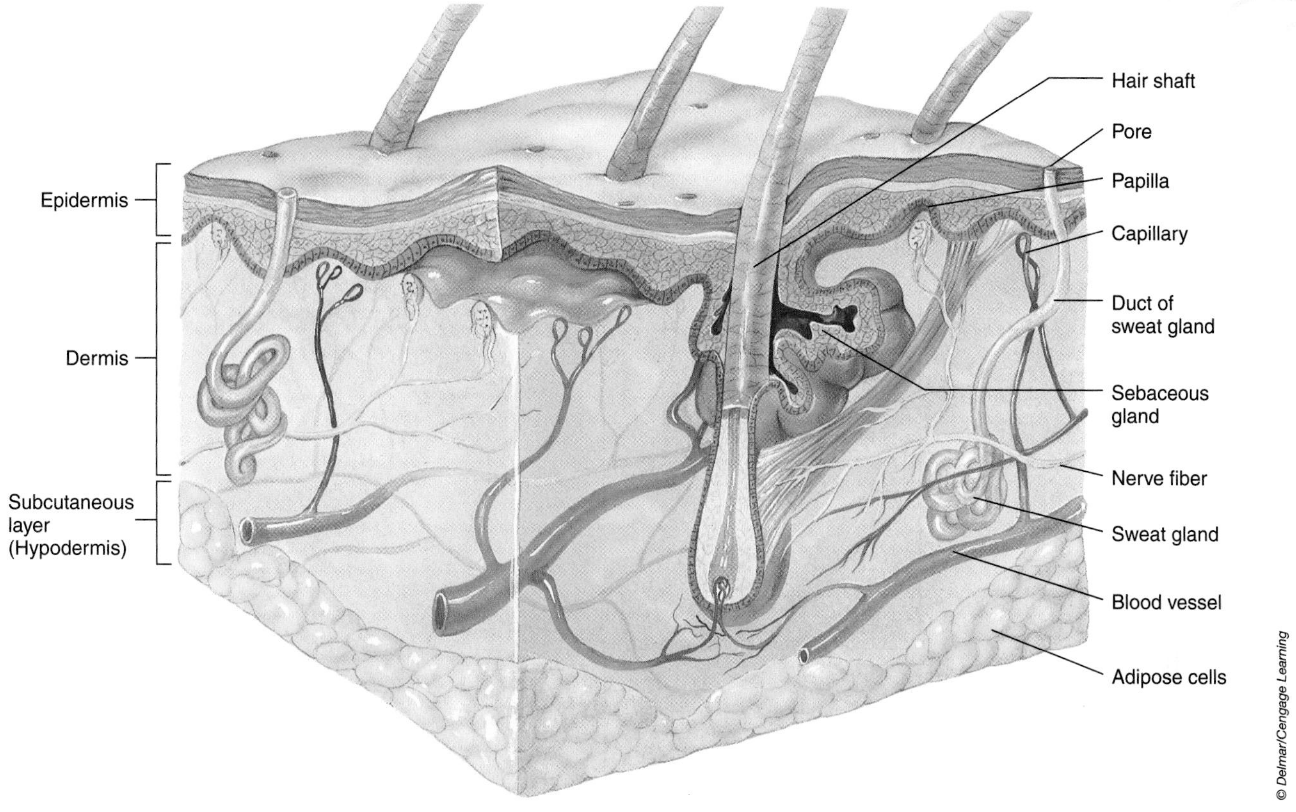

FIGURE 6-1. The layers of the skin and some of its appendages.

> ***StudyWARE™ Connection***
>
> Watch an animation on skin on your StudyWARE™ CD-ROM.

The Epidermis

The outermost or epidermal layer of the skin is composed of stratified, squamous, keratinized (no nucleus) epithelial cells. These cells are held together by highly convoluted, interlocking cellular links called **desmosomes** (**DEZ**-meh-somz). These desmosomes are responsible for the unique flexibility, entirety and whole continuous structure of the skin. The epidermis is thickest where it receives the most abrasion and weight—on the palms of the hands and the soles of the feet. It is much thinner over the ventral surface of the trunk.

The epidermis, which is not vascularized, rests on a basement membrane of connective tissue. The lowermost cells on this membrane divide by mitosis, so new cells push older cells up toward the surface. As they move up, they change shape and chemical composition because they lose most of their water and eventually die. This process is called **keratinization** (**kair**-ah-**tin**-ih-**ZAY**-shun) because the cells become filled with **keratin** (**KAIR**-ah-tin), a protein material. These dead, outermost cells are constantly being shed. This outermost layer forms an effective barrier to substances that would penetrate the skin, and this layer is very resistant to abrasion.

The process of keratinization produces distinctive layers of the epidermis called **strata** (plural) or **stratum** (**STRAT**-um) [singular] (Figure 6-2). There are five layers from outermost to deep. They are the stratum corneum (**STRAT**-um **COR**-nee-um), commonly called the horny or leathery layer; the stratum lucidum (**STRAT**-um **LOO**-sid-um), commonly called the clear layer; the stratum granulosum (**STRAT**-um **gran**-you-**LOH**-sum), commonly called the granular layer; and the stratum spinosum (**STRAT**-um spye-**NOH**-sum), commonly called the spiny or prickly layer. The innermost layer and the most important is the stratum germinativum (**STRAT**-um jer-mih-**NAY**-tih-vum) or the regenerative layer, also called the stratum basale.

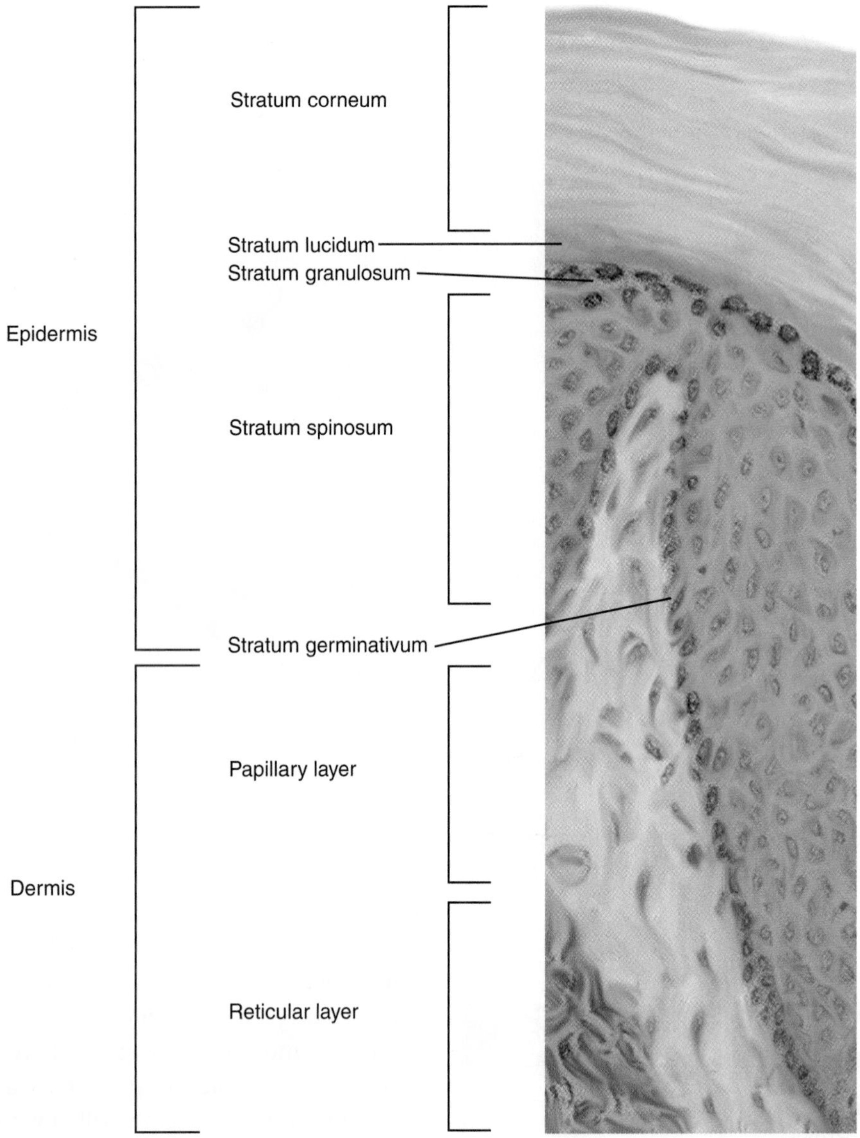

Stratum corneum
Stratum lucidum
Stratum granulosum
Epidermis
Stratum spinosum
Stratum germinativum
Papillary layer
Dermis
Reticular layer

FIGURE 6-2. The epidermal and dermal layers of the skin.

The Stratum Corneum

The **stratum corneum** forms the outermost layer of the epidermis. It consists of dead cells converted to protein. They are called keratinized cells because they have lost most of their fluid. The organelles of the cell are now just masses of the hard protein keratin that gives this layer its structural strength. These cells are also covered and surrounded with lipids to prevent any passage of fluids through this layer. These cells have only about 20% water as compared to cells in the lowermost layer that have about 70%. The cells resemble scales in shape and can consist of up to 25 layers. By the time cells reach this layer, the desmosomes have broken apart and, therefore, these cells are constantly being sloughed off. The shedding of these cells from the scalp produces what we call dandruff.

This layer also functions as a physical barrier to light and heat waves, microorganisms (e.g., bacteria, fungi, protozoa, and viruses) and most chemicals. The thickness of this layer is determined by the amount of stimulation on the surface by abrasion or weight bearing, hence the thickened palms of the hands and soles of the feet. When skin is subjected to an excessive amount of abrasion or friction, a thickened area called a **callus** (**KAL**-us) will develop. Learners who do a lot of writing will develop small calluses on their fingers that hold their pens. Abrasion on the bony prominences on the foot can produce structures we call **corns**.

The Stratum Lucidum

The **stratum lucidum** lies directly beneath the stratum corneum but is difficult to see in thinner skin. It is only one or two cell layers thick. Its cells are transparent and flat.

The Stratum Granulosum

The **stratum granulosum** consists of two or three layers of flattened cells. Because granules tend to accumulate in these cells, it was named the granular layer. These granules have nothing to do with skin color. This layer is very active in keratinization. In this layer the cells lose their nuclei and become compact and brittle.

The Stratum Spinosum

The **stratum spinosum** consists of several layers of prickly or spiny-shaped cells that are polyhedron in structure. In this layer, desmosomes are still quite prevalent. The outline caused by the polyhedral shapes causes the cell's outlines to look spiny, hence the name. In some classification schemes, this layer is included with the stratum germinativum.

The Stratum Germinativum

The **stratum germinativum** is the deepest and most important layer of the skin because it contains the only cells of the epidermis that are capable of dividing by mitosis. When new cells are formed they undergo morphologic and nuclear changes as they get pushed upward by the dividing cells beneath them. Therefore, these cells give rise to all the other upper layers of the epidermis. The epidermis will regenerate itself only so long as the stratum germinativum remains intact. Its basal layer, called the **stratum basale** (**STRAT**-um **BAY**-sil), rests on the basement membrane.

*Study***WARE**™ Connection

Watch an animation on tissue repair and healing on your StudyWARE™ CD-ROM.

The stratum germinativum also contains cells called **melanocytes** (**MEL**-ah-no-sightz), which are responsible for producing skin color. Melanocytes are irregularly shaped with long processes that extend between the other epithelial cells of this layer. They produce a pigment called **melanin** (**MEL**-ah-nin), which is responsible for variations in skin pigmentation. All races have the same number of melanocytes, but the different races have specific genes that determine the amount of melanin produced by the melanocytes. Darker-skinned individuals have more active melanocytes that produce more melanin. Melanocytes are activated to produce melanin by exposure to sunlight. We darken when we expose ourselves to the sun. All races get darker after exposure to the sun over a period of time. We call this getting a suntan.

Based on the discoveries and research done in anthropology by the Leakey family in Olduvai Gorge, Tanzania, scientists believe humans evolved in Africa. The first humanoids were probably very dark to protect themselves from the harmful ultraviolet rays of the sun. They had very active melanocytes like today's Africans. Over time, some humans migrated away from the equator and genetic recombinations and mutations governing the activity of their melanocytes occurred. Over long periods, this led to the evolution of the different races, whose variations in skin color are determined by the amount of melanin produced and its distribution. The strongest factor in increasing pigmentation in the skin is the sun's stimulating effect on melanocytes. Melanin cross-links with protein to form a tough resistant compound. Hence, heavily pigmented skin is more resistant to external irritation. People who live closer to the equator, where there is maximum exposure to sunlight, will be darker than people who live in the north like the Baltic States of Norway, Sweden, Finland, and Denmark. This variation in melanin content is the principal factor responsible for the color differences among races. Individuals of darker-skinned races have more active melanocytes, while individuals of lighter-skinned races have less active melanocytes.

Larger amounts of melanin can occur in certain areas of the body, producing the darkened areola area of the nipples, freckles, and moles, although other areas of the body's skin have less melanin, like the palms of the hands and the soles of the feet. Even though many genes are responsible for skin color, one mutation can cause the absence of skin color by preventing the production of melanin. This condition is called **albinism** (**AL**-bih-nizm) and results from a recessive gene that causes the absence of melanin. Albinos have no pigment in their skin and appendages of the skin. Their hair is white, their eyes pink, and their skin very fair. These individuals must be very careful to avoid overexposure to the sun.

The Dermis

The dermis is also known as the **corium** (**KOH**-ree-um). It lies directly beneath the epidermis and is often referred to as the true skin. It is composed of dense connective tissue

COMMON DISEASE, DISORDER, OR CONDITION

BURNS

Burns are classified into three major categories: first-degree, second-degree, and third-degree. First- and second-degree burns can also be categorized as **partial-thickness burns**. These burns do not completely destroy the stratum germinativum's basal layer and regeneration of the epidermis will occur from both within the burn area or from the edges of the burn.

First-degree burns involve just the epidermis (Figure 6-3A). They can be caused by brief contact with very hot or very cold objects. They can also be caused by sunburn, being overexposed to the harmful rays of the sun. Sunscreen should always be used to protect the skin from sunburn. Symptoms of first-degree burns are redness and pain. There may also be slight swelling or edema. These burns can heal in about 7 days with no scarring.

Second-degree burns involve both the epidermis and the dermis (Figure 6-3B). With minor dermal involvement, symptoms will include redness, pain, swelling, and blisters. Healing can take up to 2 weeks with no scarring. If there is major dermal involvement, the burn can take several months to heal and the wound might appear white. Scar tissue may develop.

In **third-degree burns**, also called **full-thickness burns**, the epidermis and the dermis are completely destroyed (Figure 6-3C). Recovery can only occur from the edge of the burn wound. Interestingly third-degree burns are usually painless because the sensory receptors in the skin have been destroyed. The pain usually comes from the area around the third-degree burn where first- and second-degree burns surround the area. Third-degree burns usually require skin grafts because they take a long time to heal and will form disfiguring scar tissue. Skin grafts will prevent these complications and speed healing. Skin grafts use the epidermis and part of the dermis from another part of the body, usually the buttocks or thighs, and the graft is then placed on the burn. Interstitial fluid from the burn helps heal the area.

with tough white collagenous fibers and yellow elastin fibers. Blood vessels, nerves, lymph vessels, smooth muscles, sweat glands, hair follicles, and sebaceous glands are all embedded in the dermis.

The dermis can be divided into two portions (see Figure 6-2). The **papillary portion** is the area adjacent to the epidermis, and the **reticular portion** is found between the papillary portion and the fatty subcutaneous tissue beneath. A sheet of areolar tissue, usually containing fat (adipose tissue), is known as the subcutaneous tissue or superficial fascia and attaches the dermis to underlying structures like muscle or bone. This subcutaneous tissue is sometimes referred to as the **hypodermis**. It is into this area that hypodermic injections are given. The pink tint of light-skinned individuals is due to blood vessels in the dermis. There are no blood vessels in the epidermis. When an individual is embarrassed, blood vessels in the dermis dilate. This causes "blushing" or the reddish tint seen in the facial area.

When a light-skinned individual suffocates or drowns, carbon dioxide in the blood causes the blood to take on a bluish tinge. This results in the bluish discoloration of skin or **cyanosis** (sigh-ah-**NOH**-sis) caused by lack of oxygen in the blood. When a dark-skinned individual suffocates or drowns, the same condition occurs but results in a grayish or ashy tinge to the skin rather than a bluish tinge.

THE ACCESSORY STRUCTURES OF THE SKIN

The structures associated with the skin include hair, nails, sebaceous glands, ceruminous glands, or wax glands in the ear canal and sweat glands.

Hair

Hair, in addition to mammary glands, is a main characteristic of all mammals. When the hair is very thick and covers most of the surface of the body, as on a dog or cat, it is called fur. Even on humans, hair covers the entire body except the palms of the hands, the soles of the feet, and certain portions of the external genitalia (e.g., the head of the penis). In some parts of the body, the hair is so small that it appears invisible, yet in other places it is very obvious as on the head, in the armpits, and around the genitalia. The amount of hair a person develops is related to complex genetic factors.

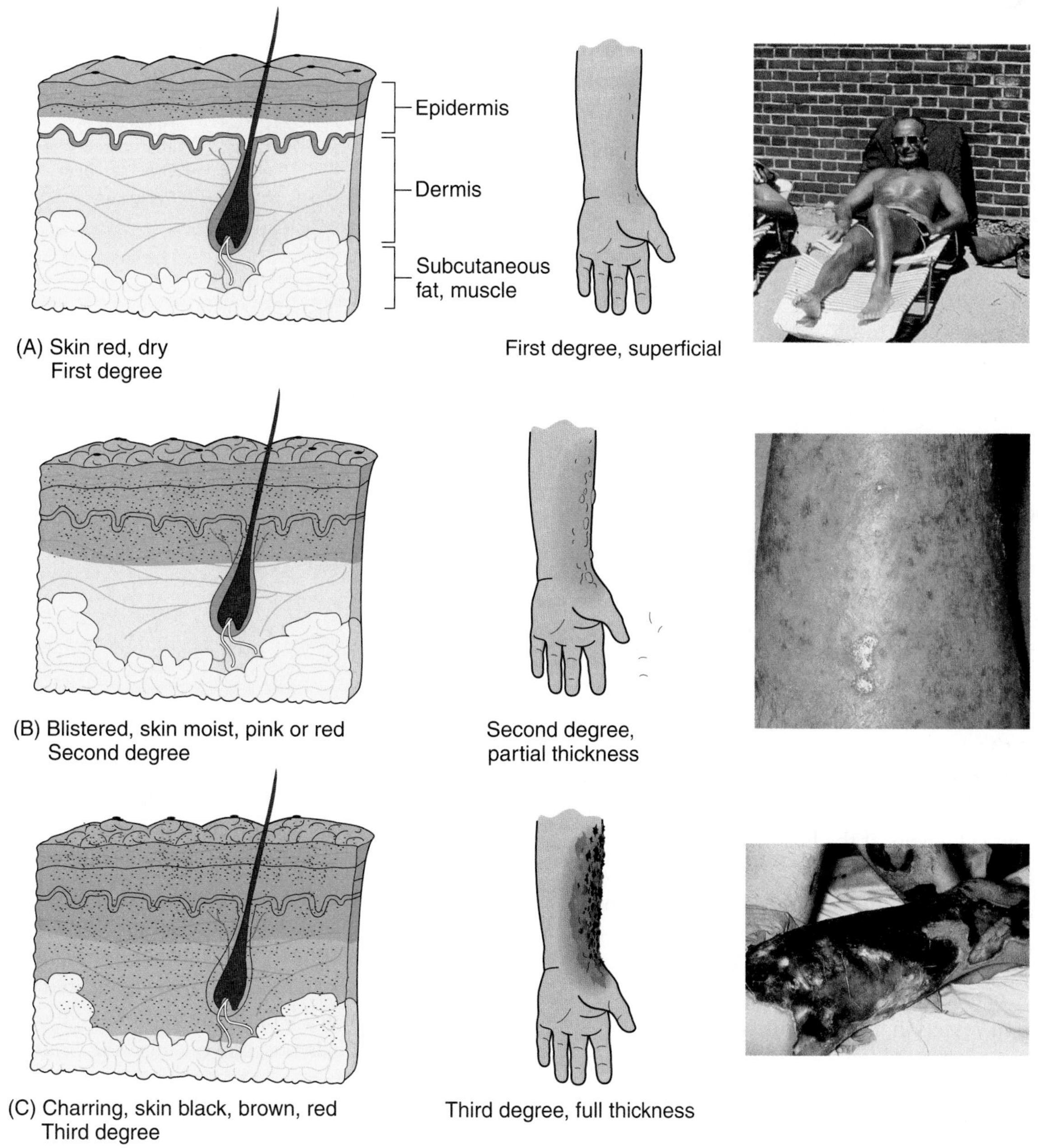

Epidermis
Dermis
Subcutaneous fat, muscle

(A) Skin red, dry
First degree

First degree, superficial

(B) Blistered, skin moist, pink or red
Second degree

Second degree, partial thickness

(C) Charring, skin black, brown, red
Third degree

Third degree, full thickness

Photos courtesy of The Phoneix Society for Burn Surviors, Inc.

FIGURE 6-3. Burns are usually referred to as (A) first, (B) second, or (C) third degree.

Each individual hair is composed of three parts (Figure 6-4): the **cuticle**, the **cortex**, and the **medulla**. The outermost portion is the cuticle, which consists of several layers of overlapping scalelike cells. The cortex is the principal portion of the hair. Its cells are elongated and united to form flattened fibers. In people with dark hair, these fibers contain pigment granules. The middle or central part of the hair is called the medulla. It is composed of cells with many sides. These cells frequently contain air spaces. There are other parts to the anatomy of a hair. The **shaft** is the visible portion of the hair. The **root** is found in an epidermal tube called the **hair follicle**. The follicle is made of an outer connective tissue sheath and an inner epithelial membrane continuous with the stratum germinativum.

Attached to the hair follicle is a bundle of smooth muscle fibers that make up the **arrector** (ah-**REK**-tohr) **pili** (**PIH**-lye) **muscle**. This muscle causes the goose flesh appearance on our skin when we get scared or when we get a chill. The muscle is involuntary and when it contracts it pulls on the hair follicle, causing the hair to "stand on its end." We see the goose flesh appearance where hair is scarce. When dogs or cats get angry, their hairs stand up on the nape of their necks. This is all the result of contraction of the arrector pili muscles.

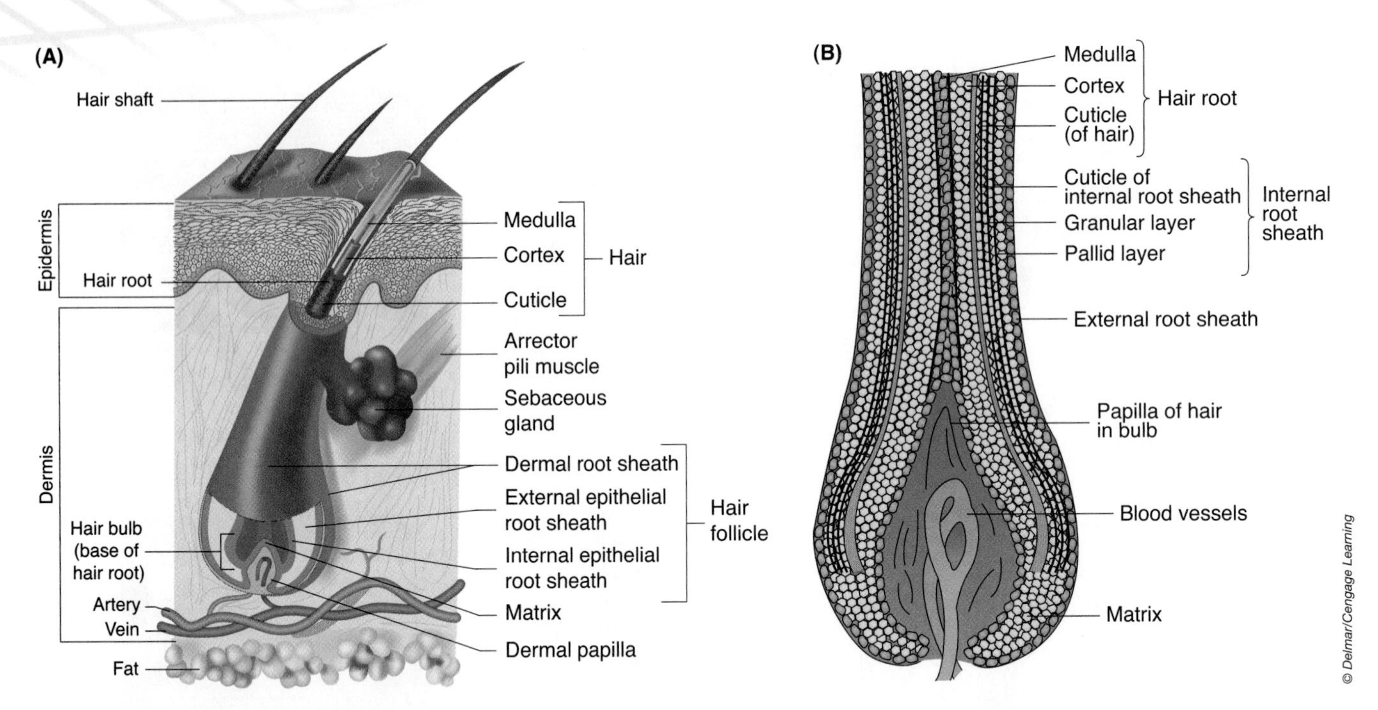

FIGURE 6-4. The anatomy of an individual hair.

Hair Growth

Hair growth is similar to the growth of the epidermis. Note that the hair follicle is an involution of the epidermis. The deeper cell layers at the base of the hair follicle are responsible for the production of new cells by mitosis. The epithelial cells of the hair follicle divide by mitosis and get pushed upward because of the basement membrane. As the cells move upward, they keratinize and form the layers of the hair shaft. Hair growth begins in the hair bulb. Blood vessels in the hair bulb provide the nourishment to produce the hair. Hair grows in cycles. The duration of the cycle depends on the hair. Scalp hair grows for 3 years and rests for 1 or 2 years. Hair loss normally means the hair is being replaced because the old hair falls out of the follicle when a new hair begins to form. Some people, particularly men, have a genetic predisposition for what is called pattern baldness. These men suffer a permanent loss of hair because the hair follicles are also lost. This occurs because male sex hormones affect the hair follicles of men with this genetic trait and they become bald.

Hair Texture

We classify hair texture as straight, curly, or tightly curly sometimes referred to as kinky. This is due to genetic factors controlling the nature of the keratin of the hair. The keratin of the cortex of the hair is polymerized and cross-linked chemically in a characteristic folded configuration called alpha keratin, making the fibers elastic. The alpha keratin chain in some individuals produces straight hair, in others curly, and in still others tightly curled. When stretched, the keratin chain gets drawn or pulled out into a more linear form called beta keratin. Unless the hair is greatly distended or altered by chemical agents, it will return immediately to its normal alpha configuration. When you wash your hair, it can be elongated to one and one-half its normal length due to the weight of the water on the hair. This is possible because the protein keratin can be readily stretched in the direction of the long axis of the molecular chains of amino acids.

Permanent waves act on this principle and people can change the texture of their hair by going to a beauty salon for treatment. The hair stylist will stretch and mold the hair into the desired new wave: big rollers for straighter hair, small tight rollers for curlier hair. Then a chemical reducing agent is placed on the hair to rupture the old disulfide bonds of the alpha keratin chain. Next a new chemical oxidizing agent is placed on the hair to reestablish new stabilizing cross-links in the new position of the beta chain. Remember, the chemicals only affected the visible portion of the hair or the shaft. The new cells growing from the hair bulb will not have the new texture, and the permanent wave or new style will eventually "grow out." Another visit to the beauty salon must occur in a few months to redo the process.

Hair Color

Hair color is also determined by complex genetic factors. For example, some people turn gray in their youth, yet others turn gray in their 40s, 50s, or even as late as their 60s. We do know that gray hair occurs when pigment is absent in the cortex of the hair. White hair results from both the absence of pigment in the cortex plus the formation of air bubbles in the shaft.

Hereditary and other unknown factors determine the graying of hair. An interesting research project was done with black cats and gray hair. The hair of a black cat turned gray when its diet was deficient in pantothenic acid (an amino acid). Restoring this substance to its diet caused the gray hair to return to black. Unfortunately, this only works with cats. So the hair coloring industry is still secure.

Great frights, like being in a serious plane or car accident, can cause people's hair to change color and go gray or white. This occurs quite rarely. We do not know what physiologic processes are triggered that cause this to occur, other than the trauma of such an experience.

Nails

At the ends of fingers and toes, we have nails (Figure 6-5). Other animals have claws (birds, reptiles, cats, and dogs) or hooves (horses, cows, deer, and elk). The nail is a modification of horny (leathery) epidermal cells composed of very hard keratin. Air mixed in the keratin matrix forms the white crescent at the proximal end of each nail called the **lunula** (**LOO**-noo-lah) and the white as the free edge of the nail. Again the size of the lunula will vary from person to person and sometimes from nail to nail due to genetic factors. The **nail body** is the visible part of the nail. The **nail root** is the part of the nail body attached to the **nail bed** from which the nail grows approximately 1 mm per week unless inhibited by disease. The cuticle or eponychium is stratum corneum that extends out over the proximal end of the nail body.

Our fingernails grow faster than our toenails. Regeneration of a lost fingernail occurs in 3½ to 5½ months. Regeneration of a lost toenail occurs in 6 to 8 months as long as the nail bed remains intact. As we age, the rate of growth of nails slows.

Sebaceous Glands

Sebaceous (see-**BAY**-shus) **glands** (see Figure 6-1) develop along the walls of hair follicles and produce **sebum** (**SEE**-bum). This is an oily substance that is responsible for lubricating the surface of our skin, giving it a glossy appearance. Sebaceous secretions consist of entire cells containing the sebum. As the cells disintegrate, the sebum moves along the hair shaft to the surface of the skin where it produces a cosmetic gloss. Brushing hair causes the sebum to cover the shaft of our hair, making hair shiny. Remember how good your dog or cat looks after a good brushing. The coat of fur glistens and shines due to the sebum.

Sebaceous secretion is under the control of the endocrine system. It increases at puberty, resulting in acne problems in adolescents, and it decreases in later life, resulting in dry skin problems. It also increases in late pregnancy.

Sweat Glands

Sweat glands (see Figure 6-1) are simple tubular glands found in most parts of the body. They are not found on the margins of the lips or the head of the penis. They are most numerous in the palms of our hands and in the soles of our feet. It has been estimated that there are 3000 sweat glands per square inch on the palms of our hands. When you get nervous, think about which area of your body gets sweaty first—your hands!

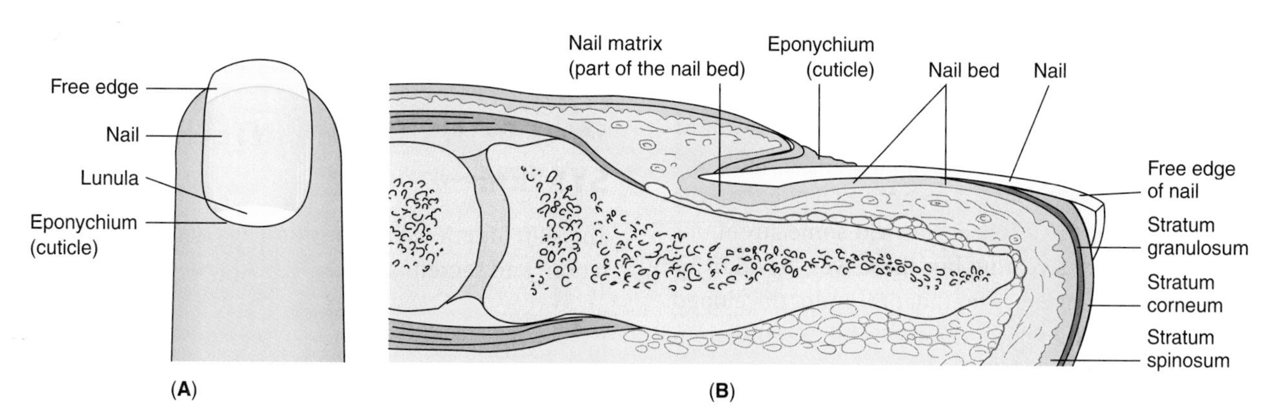

FIGURE 6-5. The anatomy of a nail. (A) Posterior view of finger and (B) fingernail and underlying structures.

HEALTH ALERT

ACNE IN ADOLESCENTS

During puberty the sebaceous glands secrete an excessive amount of sebum. The opening of the gland can become clogged with this oily substance. Because so much is being secreted, it cannot make its way onto the surface of the skin. The fatty oil oxidizes in the presence of air and becomes discolored, producing a "blackhead." The sebum retained in the gland can provide a growth medium for pus-producing bacteria. If the clog is near the surface of the skin, this results in a pimple. If the clog is deep in the gland along the shaft of the hair, the result is a boil. This must be lanced by a physician so that it can be drained of excess fluid and the bacterial infection.

To prevent blackheads and pimples, adolescents should frequently wash their faces on a regular daily basis. They may use astringents like alcohols to dry the surface of their skin, depending on skin type. ■

Each sweat gland consists of a secretory portion and an excretory duct. The secretory portion is located in the deep dermis, occasionally in the subcutaneous tissue, and is a blind tube twisted and coiled on itself. A blind tube is one that has only one opening, in this case at the top. From the coiled secretory portion that produces the sweat, the excretory duct spirals up through the dermis into the epidermis and finally opens on the surface of the skin.

The most common and most numerous type of sweat glands are referred to as eccrine (**EK**-rin) sweat glands. Another smaller group of sweat glands are called apocrine (**AP**-oh-krin) sweat glands. These are found only in the armpits, in the scrotum of males, in the labia majora of females, and around the anus. They are not involved in regulating body temperature. They usually open into hair follicles just above the sebaceous gland. They become active at puberty and contribute to the development of body odor.

Sweat contains the same inorganic materials as blood but in a much lower concentration. Its chief salt is sodium chloride, which is the reason sweat tastes salty. Its organic constituents include urea, uric acid, amino acids, ammonia, sugar, lactic acid, and ascorbic acid. Sweat itself is practically odorless. That may surprise you because many of us have been in a locker room at a gym. Actually the odor is produced by the action of bacteria feeding on the sweat. Remember the last time you did some strenuous exercise? You were sweating but there was no odor for the first 10 or 15 minutes. After that time, odor developed because it took that long for the bacterial population to grow in the sweat and their effects to be smelled.

Sweating is also an important physiologic process that cools the body. Sweating leads to loss of heat in the body because heat is required to evaporate the water in sweat. Therefore, sweating helps lower body temperature. Some people are born without sweat glands—they have a congenital absence of these glands. These individuals can easily die of heat stroke if exposed to high temperatures even if only for a brief period of time. Other individuals have overactive sweat glands and must use stronger deodorants and antiperspirants. Due to hair in the armpits, sweat accumulates there. Because our armpits are usually covered while our arms are at our sides, the environment is ideal for bacteria to feed on the sweat, hence the need for deodorants.

*Study***WARE**™ Connection

Play an interactive game labeling the layers of the skin and some of its appendages on your StudyWARE™ CD-ROM.

FUNCTIONS OF THE INTEGUMENTARY SYSTEM

The skin functions in sensation, protection, thermoregulation, and secretion.

Sensation

Receptor sites in the skin detect changes in the external environment for temperature and pressure. Receptor

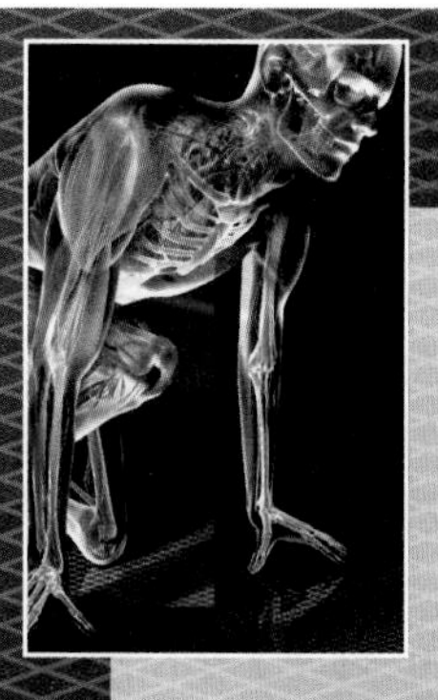

AS THE BODY AGES

As we age, many visible changes occur in our skin such as crow's feet around the eyes and mouth. A decrease in secretions of sebum results in drier skin. We hear and see advertisements for skin moisturizers, a big source of income for the cosmetic industry. Loss of collagen and elastin fibers in the dermis results in the skin sagging and wrinkling. Blood flow to the skin decreases with age, and the skin becomes thinner and looks more translucent. The number of melanocytes decreases in certain areas, causing the darkening appearance of other areas of the skin referred to as age spots or liver spots. Lack of melanin also affects the graying and production of white hair with increased age.

As we age, hair does not grow as regularly on our heads as it did in our youth, and many people, females as well as males, experience thinning hair or baldness. This is also related to genetic factors. Some individuals experience baldness as early as their 20s whereas others maintain a full head of hair into their 70s.

Nails, especially fingernails, may become brittle and break easily, while the toenails can become susceptible to fungal infections and become discolored and thicker than normal.

Older adults may become sensitive to the cold due to poor blood circulation in the dermis and fewer fat deposits underneath the skin for insulation in the subcutaneous layer. In addition, due to the skin not being able to repair and maintain itself as easily as it did in youth, infections of the skin are more likely to develop in older adults.

There is less sebum secretion from the sebaceous glands later in life, resulting in drier skin and hence the need for skin moisturizers. During the winter months in north temperate climates, this condition is further exaggerated by drying winds, resulting in cracked skin with small, open, and painful wounds around the tips of fingers. During these times daily applications of hand cream moisturizers are especially helpful. ■

sites are in contact with sensory neurons that transmit the impulse to the brain and spinal cord for interpretation (see Chapters 10 and 11). Temperature receptors produce the sensations of hot and cold. Pressure receptor sites allow us to interpret excessive pressure that results in the sensation of pain as when we get pinched. They also detect mild pressure that results in the sensation of pleasure as from a gentle massage or a petting stroke. Combinations of varying degrees of those stimulations at these receptor sites produce other sensations that we call burning, itching, or tickling. These receptor sites allow us to react to external stimuli and to interpret what is occurring in the outside world.

Protection

The skin is an elastic, resistant covering. It prevents passage of harmful physical and chemical agents. The melanin produced by the melanocytes in the stratum germinativum darkens our skin and protects us from the damaging ultraviolet rays of sunlight. Most chemicals cannot gain entry into the body through the skin, but the chemicals that cause poison ivy and poison oak can penetrate this barrier. Fat-soluble chemicals like DDT, a chlorinated hydrocarbon pesticide, can also get through the skin. If you put your hand in a can of gasoline, you will not be poisoned. However, if you put your hand in a container of DDT, you could be poisoned. People who work with certain insecticides must wear protective clothing to prevent the penetration of these chemicals through their skin.

The lipid content of the skin inhibits the excessive loss of water and electrolytes through the skin. Normal skin is impermeable to water, carbohydrates, fat, and protein. However, all true gases and certain volatile substances will pass through the epidermis like the organic pesticide just mentioned. The numerous openings around hair follicles can act as channels for absorption of these materials.

Skin also has an "acid mantle." This acidity kills most bacteria and other microorganisms that make contact with our skin. Soaps and shampoos will often be labeled as pH balanced, which indicates that these cleansers will not destroy the acid mantle of the skin. Some skin diseases will destroy the acidity of certain areas of the skin,

COMMON DISEASE, DISORDER, OR CONDITION

SKIN CANCER

Skin cancer usually develops from excessive exposure to the ultraviolet rays of the sun, which is the reason sunscreens with an SPF of 15 or above should be used when engaging in outdoor recreational activities, or when one's occupation requires outdoor work. Common areas of development of skin cancers are the face, neck, and hands. There are three types of skin cancers, two of which can be lethal.

The most common type of skin cancer is **basal cell carcinoma**. This cancer produces an open ulcer and can be easily treated with radiation therapy or surgical removal. This cancer rarely spreads. The first dangerous type of skin cancer is **squamous cell carcinoma**, which produces a nodular tumor that is highly keratinized and found in the epidermis (Figure 6-6). If untreated it can spread to the dermis, metastasize, and cause death. The most dangerous type of skin cancer is **malignant melanoma** (Figure 6-7). It is rare and associated with a mole on the skin. A mole is a grouping of melanocytes. They become cancerous and metastasis is common. The melanoma can appear as a dark nodule or a spreading flat lesion. Unless treated early, this cancer is fatal.

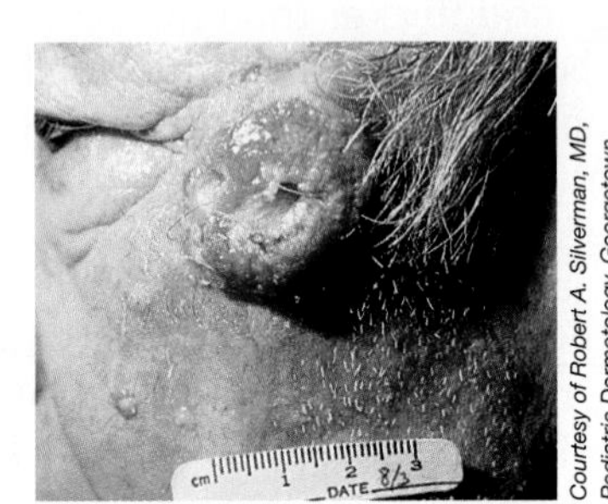

Courtesy of Robert A. Silverman, MD, Pediatric Dermotology, Georgetown University

FIGURE 6-6. Squamus cell carcinoma on the face.

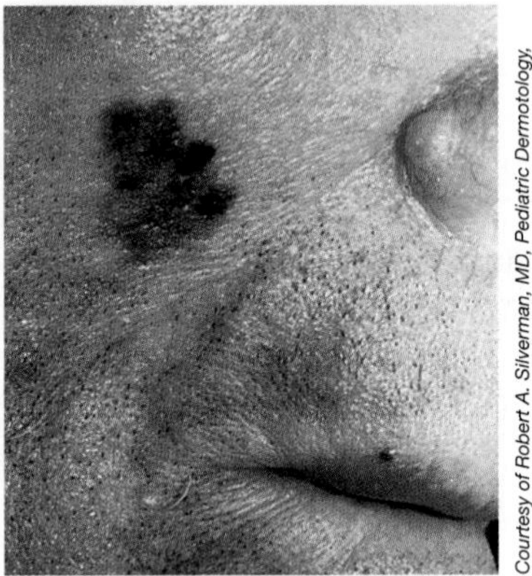

Courtesy of Robert A. Silverman, MD, Pediatric Dermotology, Georgetown University

FIGURE 6-7. Facial malignant melanoma.

impairing the self-sterilizing capabilities of our skin. These diseases make the skin prone to bacterial infections.

Nails protect the ends of our digits. Fingernails can also be used in defense. Hair on our head acts as an insulator and helps prevent heat loss. Hairs in our nose filter out large foreign particles like soot. Eyelashes protect our eyes from foreign objects.

Thermoregulation

Normal body temperature is maintained at approximately 98.6°F (37°C).Temperature regulation is critical to our survival because changes in temperature affect the functioning of enzymes. The presence of enzymes is critical for normal chemical reactions to occur in our cells. When people get high fevers, they can die because the heat of a fever destroys the enzymes by breaking up their chemical structure. Without enzymes, chemical reactions cannot occur and our cellular machinery breaks down and death results.

When external temperatures increase, blood vessels in the dermis dilate to bring more blood flow to the surface of the body from deeper tissue beneath. In the skin the blood with its temperature or heat is then lost by radiation, convection, conduction, and evaporation. When we sweat, the water in sweat evaporates, which requires energy and thus carries away heat to reduce body temperature.

When external temperatures decrease, the first response is for blood vessels in the dermis to dilate to bring heat to the surface to warm our extremities. Light-skinned individuals will have rosy cheeks when they first go out during a cold wintry day. Excessive exposure to the cold cannot be maintained for long, so blood vessels then constrict to bring the heat inside to preserve the vital organs of the body. Frostbite occurs when the skin of the extremities no longer gets a blood supply due to the maintained constriction of the blood vessels in the dermis to conserve heat. The tissues in the tips of these extremities die and turn black.

COMMON DISEASE, DISORDER, OR CONDITION

DISORDERS OF THE INTEGUMENTARY SYSTEM

RINGWORM

Ringworm is caused by several species of fungus. Its symptoms include itchy, patchy scalelike lesions with raised edges. In earlier times, it was believed that this condition was caused by worms, hence its other name, tinea, which is Latin for worm. Ringworm on the feet is called athlete's foot; in the groin area it is called jock itch. Ringworm of the scalp is called tinea capitus and is most common in children. Untreated, it can lead to hair loss and secondary bacterial infections.

PSORIASIS

Psoriasis (soh-**RYE**-ah-sis) is a common chronic skin disorder that may be genetic in orgin. The actual cause is unknown. It is characterized by red patches covered with thick, dry, and silvery scales that develop from excessive production of epithelial cells through hyperactivity of the stratum germinativum. These patches can develop anywhere on the body. When the scales are scraped away, bleeding usually results. There is no known cure for this disease, but it can be controlled with corticosteroids, ultraviolet light, and tar solution creams and shampoos.

WARTS

Warts are caused by human papillomavirus. The virus causes uncontrolled growth of epidermal tissue. The virus is transmitted by direct contract with an infected individual. The growths are usually benign and disappear spontaneously. They can also be removed surgically or with topical applications.

COLD SORES

Cold sores, also known as fever blisters, are caused by Type I herpes simplex virus. Initial infections show no symptoms but the virus can remain dormant in the skin around the mouth and in the mucous membrane of the mouth. When activated it produces small, fluid-filled blisters that can be both painful and irritating. Stress seems to activate the virus.

IMPETIGO

Impetigo (im-peh-**TEYE**-go) is a highly contagious skin disease of children caused by the bacterium *Staphylococcus aureus* (Figure 6-8). The skin erupts with small blisters containing pus that rupture easily, producing a honey-colored crust. The blisters usually develop on the face and can spread. The bacteria are spread by direct contact and enter the skin through abrasions. Treatment includes cleansing with antibacterial soaps and antibiotics.

SHINGLES

Shingles is caused by the herpes zoster or chickenpox virus that develops after the childhood infection. The virus remains dormant within cranial or spinal nerves. Trauma or stress somehow activates the virus to travel through the nerve paths to the skin where it produces very painful, vesicular skin eruptions. Treatment is symptomatic with lotions to relieve itching and analgesics to control pain.

VITILIGO

Vitiligo (vit-ill-**EYE**-go) is an acquired skin disease resulting in irregular patches of skin of various sizes completely lacking in any pigmentation. The depigmented white patches are often located on exposed areas of skin. The cause of the disease is unknown.

MOLES

Moles are produced by groupings of melanocytes that develop during the first years of life. They are common disorders of the skin that are usually benign and developed by most people. Moles reach maximum size and elevation at puberty. They vary in size and may have hair associated with them. If

(continues)

COMMON DISEASE, DISORDER, OR CONDITION

DISORDERS OF THE INTEGUMENTARY SYSTEM (continued)

they enlarge and darken later in life, moles may be a first indication of skin cancer. Moles should be regularly monitored for changes, beginning in midlife at around age 30. Moles that are consistently irritated or become infected on a normal basis should be removed surgically and sent to a pathology lab for examination.

ALOPECIA

Alopecia (al-o-**PE**-she-ah) is commonly known as baldness and can be caused by a number of factors in both men and women. Male pattern baldness (common baldness) is influenced by genetic factors and aging. Some individuals may begin losing scalp hair as early as in their 20s, while others may have a full head of hair well into their senior years. Baldness is also influenced by male sex hormones. The cosmetic industry has marketed a number of drugs, such as minoxidil, to regrow and counter the effects of pattern baldness. Alopecia may also be caused by malnutrition, diabetes, certain endocrine disorders, chemotherapy for cancer, and drug interactions. Other forms of alopecia include alopecia universalis, which is a total loss of all body hair; alopecia areata, which results in bald spots on the face and scalp caused by an autoimmune disorder; and alopecia capitis totalis, an uncommon disorder, which results in the complete loss of all scalp hair.

CHICKENPOX

Chickenpox is caused by the virus *Varicella zoster* (Figure 6-9). It develops in young children producing many highly itchy, vesicular eruptions all over the skin. The fluid of the eruptions and their scabs are highly contagious, except when completely dry. Transmission occurs through contact with the skin lesions but can also occur through droplets sneezed or coughed up from infected individuals. A vaccine is available for children 12 months or older to prevent the disease.

POISON IVY, OAK, AND SUMAC DERMATITIS

Poison ivy dermatitis is caused by contact with a chemical toxicodendrol, present in the leaves of the poison ivy, climbing vine, plant *Rhus*, which is characterized by three pointed shiny leaves. It is characterized by itching and burning vesicular eruptions. It can be treated with topical applications of corticosteroid creams or lotions. Poison oak and poison sumac are caused by contact with species of the shrub plant also of the genus *Rhus* (Figure 6-10).

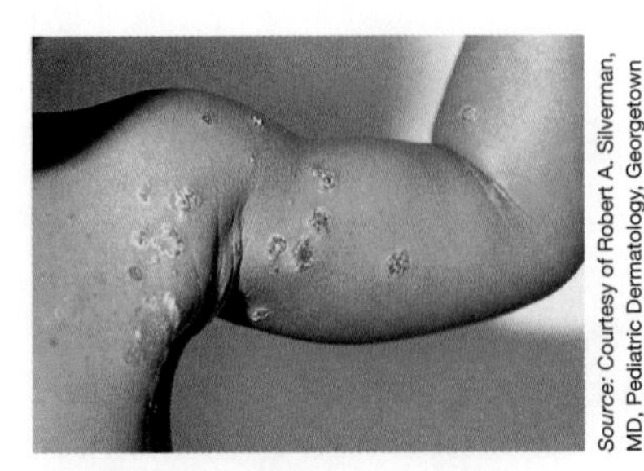

Source: Courtesy of Robert A. Silverman, MD, Pediatric Dermatology, Georgetown University

FIGURE 6-8. Impetigo pustules on the arm and trunk caused by *Staphylococcus*.

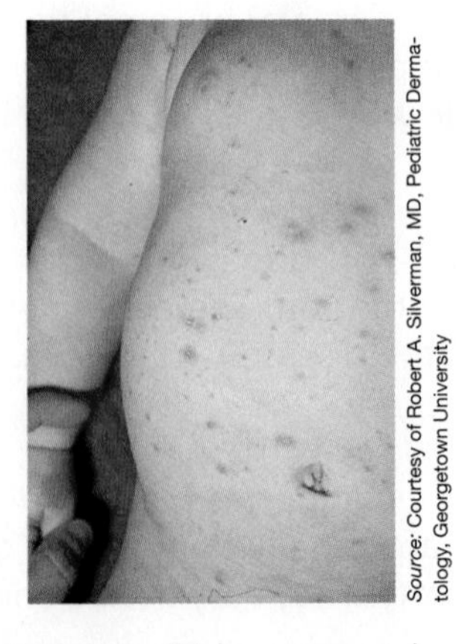

Source: Courtesy of Robert A. Silverman, MD, Pediatric Dermatology, Georgetown University

FIGURE 6-9. Chickenpox pustules on the trunk of a young child.

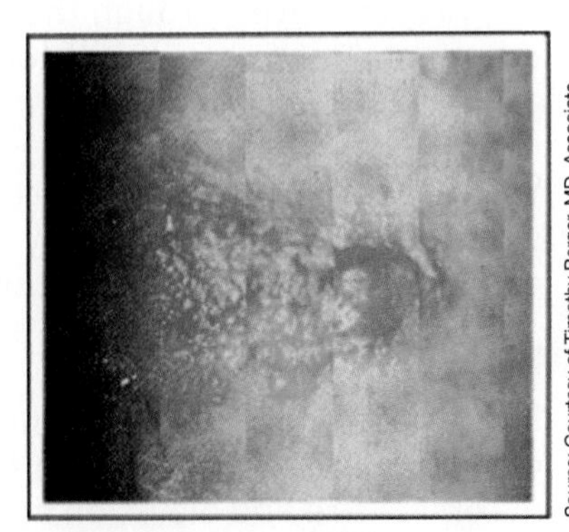

Source: Courtesy of Timothy Berger, MD, Associate Clinical Professor, University of California, San Francisco

FIGURE 6-10. Contact dermatitis caused by poison oak.

COMMON DISEASE, DISORDER, OR CONDITION

DISORDERS OF THE INTEGUMENTARY SYSTEM (continued)

PARONYCHIA, ONYCHOMYCOSIS, AND ONYCHOCRYPTOSIS

Paronychia (par-oh-**NIK**-e-ah) occurs when the fold of skin at the edge of the nail becomes infected (Figure 6-11A).

Onychomycosis (on-e-koh-my-**KOH**-sis) is a fungal infection of the nails resulting in dry, thickened, and brittle nails, usually accompanied by a yellowish discoloration (Figure 6-11B). It is very difficult to treat as topical applications are ineffective.

Onychocryptosis (on-e-koh-crypt-**TOE**-sis) is known as an ingrown toenail (Figure 6-11C). It occurs when the lateral distal margin of the nail grows into or is pressed into the skin of a toe causing inflammation. Prevention of the condition is accomplished by wearing proper shoes and correctly aligned trimming of the toe nails.

SKIN LESIONS

Various surface lesions can develop on the skin (Figure 6-12). A **papule** is a skin lesion that is a solid, small elevation less than one centimeter in diameter. A **macule** is a flat, small discoloration of the skin that is even with the skin surface. A small rash can be considered a macule. A **wheal** is a pale or reddened elevation produced by a localized edema or swelling, like that caused by a mosquito bite. A **crust** is a hard, solid layer on the surface of the skin caused by dried blood, serum, or pus. It is also referred to as a scab. A **furuncle** or **boil** is a *Staphylococcus* infection of a hair follicle or gland with pus formation. It is characterized by swelling, pain, and redness. It is treated with antibiotics and surgical removal usually in a doctor's office. A **bulla** or **vesicle** is a thin blister of the skin containing clear, serous fluid. It is usually larger than one centimeter in diameter. A **pustule** is a small elevation of the skin, similar to a vesicle or bulla, but is filled with pus. They are usually less than one centimeter in diameter. A **cyst** is an encapsulated sac in the dermis or under the skin in the subcutaneous tissue. It is lined with epithelium and can contain either fluid or a semisolid mass.

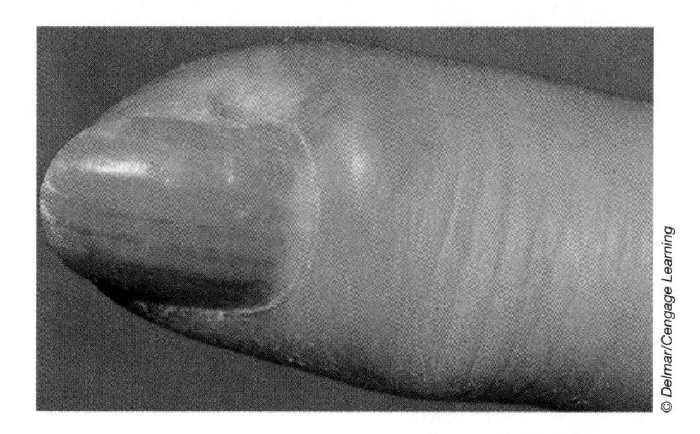

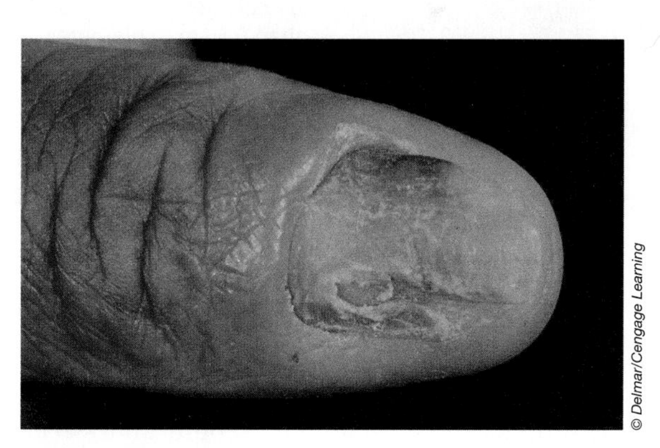

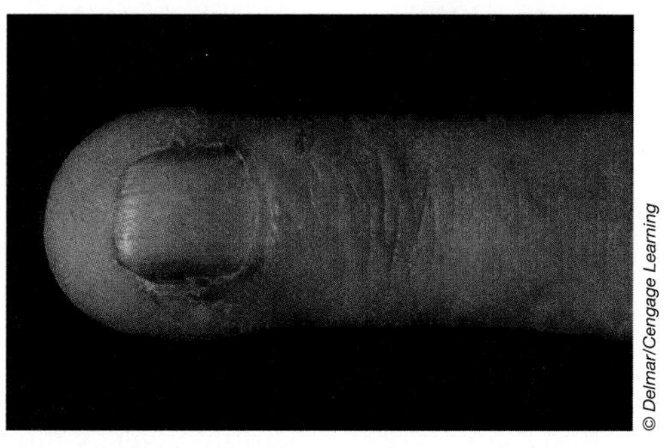

FIGURE 6-11. (A) Paronychia, (B) onychomycosis, (C) onychocryptosis.

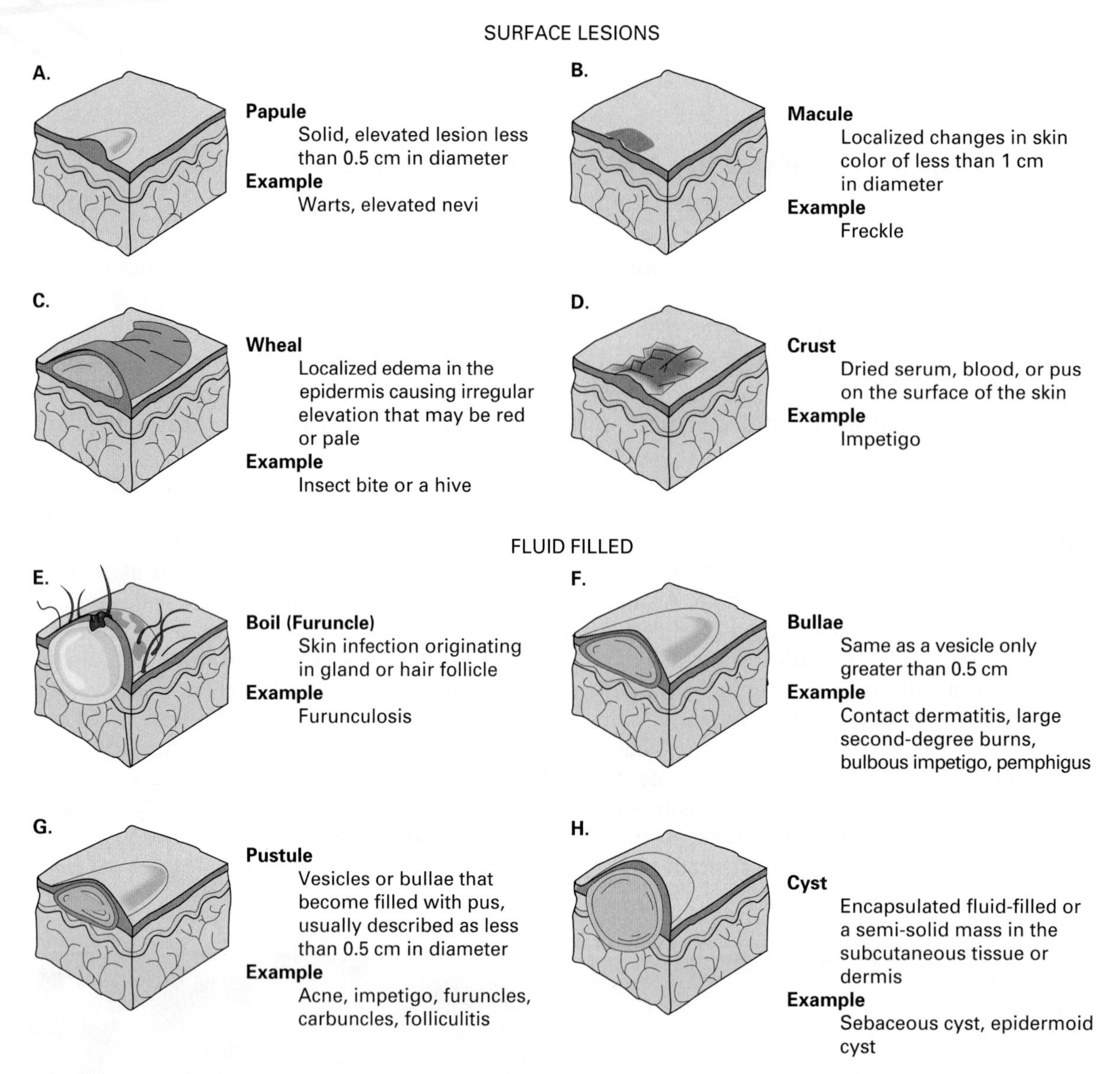

FIGURE 6-12. Various skin lesions.

Secretion

The skin produces two secretions: sebum and sweat. Sebum is secreted by the sebaceous glands. In addition to imparting a cosmetic gloss to our skin and moisturizing our skin, sebum has both antifungal and antibacterial properties. It helps prevent infection and maintains the texture and integrity of the skin. Sweat is produced by the sweat glands and is essential in the cooling process of the body. Sweat also contains waste products such as urea, ammonia, and uric acid and so can also be considered an excretion. A secretion is something beneficial, whereas an excretion is something the body does not need and could be harmful.

The skin is actively involved in the production of vitamin D. Exposure to the ultraviolet rays of the sun stimulates our skin to produce a precursor molecule of vitamin D that then goes to the liver and kidneys to become mature vitamin D. Vitamin D is necessary for our bodies because it stimulates the intake of calcium and phosphorus in our intestines. Calcium is necessary for muscle contraction and bone development. Phosphorus is an essential part of adenosine triphosphate. Because we live indoors, and in colder climates wear heavy clothing, we sometimes do not get enough exposure to the sun to adequately produce enough vitamin D. We also should ingest vitamin D through our diets. Good sources of vitamin D are milk, other dairy products, and fish oils.

These are careers that are available to individuals interested in the integumentary system:

- **Cosmetologists** have careers as makeup artists and hair stylists with positions in television, the cinema, and local hair salons.
- **Dermatologists** are physicians who specialize in the diseases and disorders of the skin.
- **Allergists** are physicians who specialize in the inflammatory responses of the skin and reactions of the immune system.
- **Plastic surgeons** are physicians who specialize in cosmetic surgery to correct birth defects and to counteract the effects of aging.
- **Nurses** can also have careers that specialize in skin care.

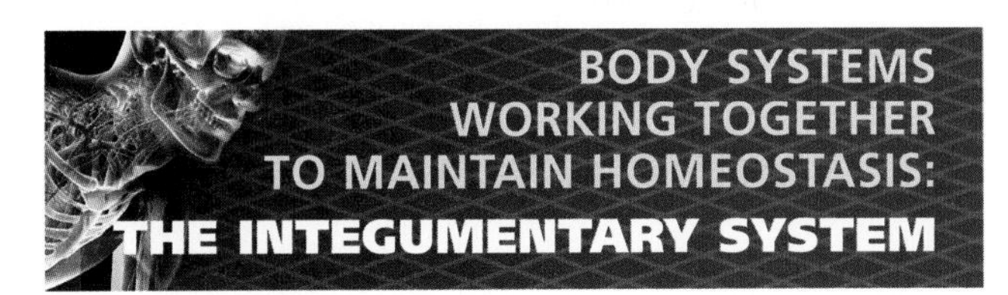

Skeletal System
- Vitamin D manufactured by the skin helps provide calcium for strong bones.

Muscular System
- Vitamin D helps provide calcium for muscle contraction.
- Facial muscles produce facial expressions of body language.
- Shivering helps control body temperature by warming the body.

Nervous System
- Receptor sites for temperature and pressure changes in the skin provide information to the nervous system so we can cope with our external environment.
- Nerves activate sweat glands.

Endocrine System
- Hormones control the secretion of sebum from the sebaceous glands.
- Hormones increase blood flow to the skin.
- Hormones control the amount of fat in subcutaneous tissue.

Cardiovascular System
- Blood vessels in the dermis help regulate body temperature by dilating or constricting.
- Dilation of blood vessels in light skin produces blushing during embarrassing moments.

Lymphatic System
- Skin is an effective barrier against invasion by microorganisms providing a first defense for the immune system.
- Sebum has antifungal and antibacterial properties.
- The acid mantle of the skin helps prevent most bacterial infections.

Digestive System
- Vitamin D produced by the skin causes calcium and phosphorus to be absorbed in the intestine.
- Excess calories can be stored as fat in subcutaneous tissue.

Respiratory System
- Receptor sites in the skin can bring about changes in breathing rates.

Urinary System
- Kidneys can restore water and electrolytes lost during sweating.

Reproductive System

- Stimulation of receptor sites in the skin can bring about sexual interest.
- Sucking on the nipple causes the postnatal female to produce milk from her mammary glands.

SUMMARY OUTLINE

INTRODUCTION

1. The integumentary system is made up of the skin, hair, nails, sebaceous glands, ceruminous or wax glands in the ear canal, and sweat glands.
2. The skin is waterproof, it protects us from ultraviolet radiation, and through sweating it dissipates water and helps regulate body temperature.

THE LAYERS OF THE SKIN

1. The skin is composed of two layers: the upper epidermis and the lower dermis or corium.

The Epidermis

1. The epidermis is composed of stratified, squamous, keratinized epithelium.
2. As cells move up to the surface of the epidermis they lose water, and their nuclei change chemically, a process called keratinization.
3. The five layers of the epidermis are from outermost to deep:
   stratum corneum or horny layer
   stratum lucidum or clear layer
   stratum granulosum or granular layer
   stratum spinosum or spiny layer
   stratum germinativum or regenerative layer.

The Stratum Corneum

1. It consists of dead cells converted to protein or keratinized cells that constantly are being shed.
2. It is a barrier to light and heat waves, most chemicals, and microorganisms.

The Stratum Lucidum

1. This layer is only one or two flat and transparent layers of cells thick. It is difficult to see.

The Stratum Granulosum

1. This layer is two or three layers of cells very active in keratinization.

The Stratum Spinosum

1. This layer consists of several layers of spiny-shaped polyhedron-like cells.
2. The interlocking cellular bridges or desmosomes are found in this layer.

The Stratum Germinativum or Basale

1. This layer rests on the basement membrane. Its lowermost layer of cells is called the stratum basale.
2. This layer is the layer that produces new epidermal cells by mitosis.
3. Melanocytes of this layer produce melanin. This pigment is responsible for skin color and protection from the harmful ultraviolet rays of the sun.
4. Dark-skinned individuals have more active melanocytes.
5. Albinism is a genetic condition that results from the absence of melanin.

The Dermis

1. The dermis is also called the corium or true skin and is composed of dense connective tissue.
2. Blood and lymph vessels, nerves, muscles, glands, and hair follicles are found in the dermis.
3. It is divided into two portions: the papillary portion below the epidermis and the reticular portion above the subcutaneous tissue.
4. The subcutaneous tissue can be called the hypodermis.

THE APPENDAGES OF THE SKIN

1. The appendages of the skin include hair, nails, sebaceous glands, ceruminous or wax glands, and sweat glands.

Hair

1. Hair covers the entire body except the palms of the hands, the soles of the feet, and parts of the external genitalia.
2. Each individual hair is made of three parts: the outer cuticle, the cortex, which is the principal portion with pigment granules, and the inner medulla with air spaces.
3. The visible portion of a hair is called the shaft.
4. The root of a hair is in a hair follicle.

5. When the arrector pili smooth muscle contracts, it causes a hair to stand on end and produces "goose flesh."

Hair Growth

1. Hair growth begins with the cells deep in the hair follicle at the hair bulb growing by mitosis and nourished by blood vessels.
2. Hair grows in cycles and rests between cycles.

Hair Texture

1. Hair texture can be classified as straight, curly, or tightly curly, and is due to genetic factors.
2. Hair in the alpha keratin chain is elastic; when stretched it is in the beta keratin chain.

Hair Color

1. Hair color is determined by complex genetic factors.
2. Gray hair occurs when pigment is absent in the cortex.
3. White hair results from both the absence of pigment and air bubbles in the shaft.
4. Heredity and other unknown factors cause hair to turn gray.

Nails

1. A nail is a modification of epidermal cells made of very hard keratin.
2. The lunula is the white crescent at the proximal end of a nail caused by air mixed with the keratin.
3. The nail body is the visible portion of a nail. The nail root is the part covered by skin.
4. The nail grows from the nail bed.
5. The cuticle is stratum corneum that extends over the nail body.

Sebaceous Glands

1. Sebaceous glands produce sebum and are found along the walls of hair follicles.
2. Sebum, an oil, gives a cosmetic gloss to skin and moisturizes it.
3. Sebaceous secretion is controlled by the endocrine system, increasing during puberty and late pregnancy and decreasing with age.

Sweat Glands

1. Sweat glands are most numerous in the palms of our hands and in the soles of our feet.
2. The secretory, blind tube portion of a sweat gland is in the subcutaneous tissue. The excretory portion goes through the dermis to the surface.
3. The odor of sweat is produced by the action of bacteria feeding on the sweat.
4. Sweating is an important physiologic process that helps cool the body.

FUNCTIONS OF THE INTEGUMENTARY SYSTEM

1. The skin functions in sensation, protection, thermoregulation, and secretion.

Sensation

1. Receptor sites for changes in temperature (hot and cold) and pressure (pleasure and pain) are found in the skin.
2. Combinations of stimulations result in the sensations of itching, burning and tickling.

Protection

1. The skin prevents the entrance of harmful physical and chemical agents into the body.
2. Melanin protects us from the harmful ultraviolet rays of the sun.
3. The lipid content of skin prevents excessive water and electrolyte loss.
4. The acidic pH of skin kills most bacteria and microorganisms that come in contact with our skin.
5. Hair acts as an insulator, protects our eyes, and filters out foreign particles in our nose.

Thermoregulation

1. Normal body temperature is regulated by blood vessel dilation and constriction in the dermis of the skin.
2. Sweating is an evaporation process that cools the body.

Secretion

1. Sebum has antifungal and antibacterial properties.
2. Sweat contains waste products such as urea, uric acid, and ammonia, so it is also an excretion.
3. The skin helps manufacture vitamin D through exposure to ultraviolet rays of the sun.

REVIEW QUESTIONS

1. List the five layers of the epidermis with their common names.
2. List and describe four functions of the integumentary system.
*3. If all people have the same number of melanocytes in their skin, how do we explain the differences in color among the various races?
*4. Why would a person born with an absence of sweat glands be very susceptible to death by exposure to heat?

*Critical Thinking Questions

Search and Explore

- Visit the Skin Cancer Foundation website at http://www.skincancer.org and research one of the types of skin cancer. Write a short paper of one to two paragraphs about what you learned.
- Visit the MedlinePlus website at http://www.nlm.nih.gov/medlineplus and search on any of the skin conditions mentioned in this chapter.

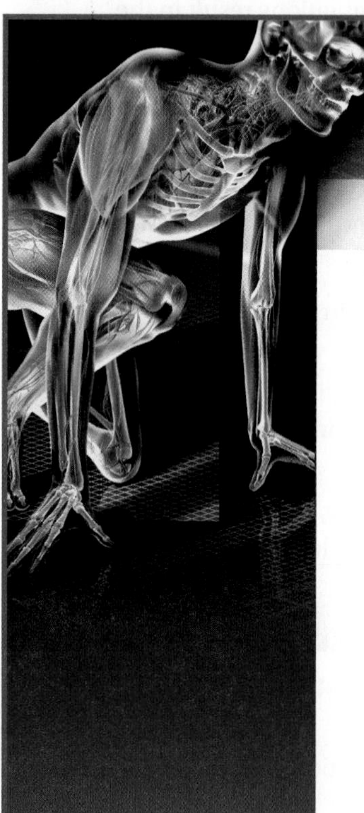

CASE STUDY

Siesha, a 21-year-old woman, is visiting her dermatologist following a referral from her primary health care provider. Siesha is worried about a mole that has developed on her right thigh. The health care provider notes that Siesha has a dark tan. During history taking, Siesha says that she and her friends play beach volleyball every weekend during the summer, and she uses tanning beds during the winter. After examination of the mole, the health care provider believes that Siesha might have a form of skin cancer.

Questions

1. What type of skin cancer do you think Siesha might have?
2. What is a second type of skin cancer that is potentially fatal?
3. Which is the least dangerous type of skin cancer?
4. What aspects of Siesha's history indicate that she is at risk for skin cancer?
5. What precautions can individuals take to help prevent skin cancer?

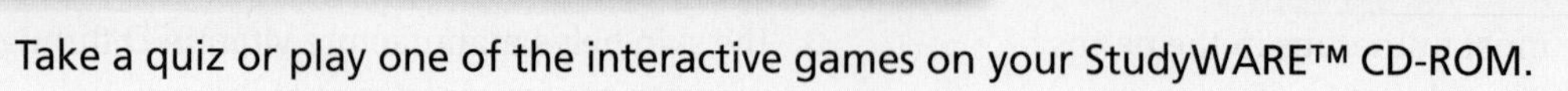

*Study*WARE™ Connection

Take a quiz or play one of the interactive games on your StudyWARE™ CD-ROM.

Study Guide Practice

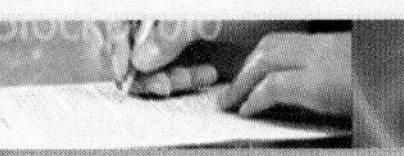

Go to your **Study Guide** for more practice questions, labeling and coloring exercises, and crossword puzzles to help you learn the content in this chapter.

MATCHING

Place the most appropriate number in the blank provided.

_____ Melanin	**1.** Central axis of hair
_____ Keratinization	**2.** Skin pigment
_____ Dermis	**3.** Visible portion of hair
_____ Cuticle	**4.** White crescent of nail
_____ Medulla	**5.** Desmosomes
_____ Shaft	**6.** Produces brittle dry cells without nuclei
_____ Cortex	**7.** Lubricates surface of skin
_____ Lunula	**8.** Blue or ashy color of skin
_____ Sebum	**9.** Arrector pili muscle
_____ Cyanosis	**10.** Principal portion of hair, cells are elongated
	11. Corium
	12. Outermost layer of hair, scalelike cells

LABORATORY EXERCISE: THE INTEGUMENTARY SYSTEM

Materials needed: A compound microscope and prepared microscope slides

1. Examine a prepared slide of a section of skin from the palm of the hand. Try to distinguish the different layers of the epidermis. Note the basement membrane. Notice that the cells closest to the basement membrane are nonkeratinized and cuboidal, but those farthest away in the stratum corneum are keratinized and squamous. Note the two layers of the dermis: the papillary portion and the reticular portion.
2. Examine a prepared slide of hair shafts from the human scalp. Note the structure of the hair follicle surrounding the shaft of hair.
3. Your instructor will show you a videotape or CD-ROM on "The Skin."

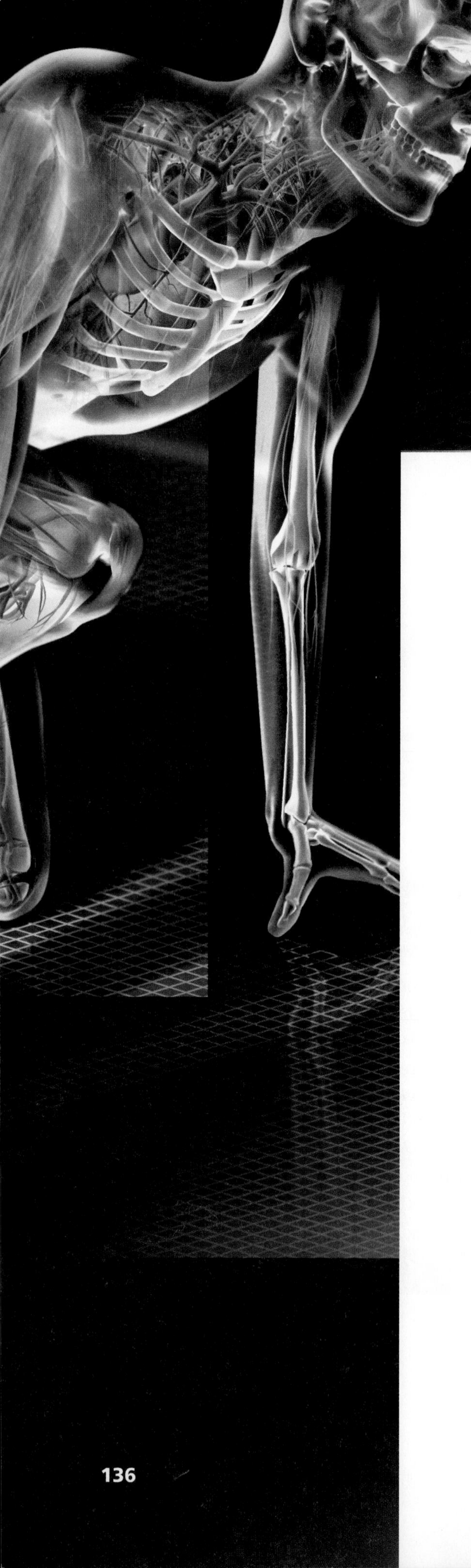

The Skeletal System

CHAPTER OBJECTIVES

After studying this chapter, you should be able to:

1. Name the functions of the skeletal system.
2. Name the two types of ossification.
3. Describe why diet can affect bone development in children and bone maintenance in older adults.
4. Describe the histology of compact bone.
5. Define and give examples of bone markings.
6. Name the cranial and facial bones.
7. Name the bones of the axial and appendicular skeleton.

7

KEY TERMS

Acetabulum 159
Acromial process 155
Alveolus 151
Atlas 153
Auditory ossicles 149
Axis 153
Calcaneus 161
Canaliculi 142
Cancellous or spongy bone 142
Capitate 157
Carpals 157
Cartilage 138
Cervical vertebrae 152
Clavicle 155
Coccygeal vertebrae/coccyx 152
Compact or dense bone . . 142
Condyle 145
Coracoid process 155
Coronal suture 146
Costae 155
Crest 146
Cuboid 161
Cuneiforms 161
Diaphysis 144
Endochondral ossification 140
Endosteum 140
Epiphyseal line 144
Epiphysis 144
Ethmoid bone 149
External occipital crest . . 146
External occipital protuberance 146
Femur 159
Fibula 160
Fontanelle 140
Foramen 146
Foramen magnum 146
Fossae 145
Fracture 144
Frontal bone 146
Gladiolus 155
Glenoid fossa 155
Hamate 157
Haversian/central canals 142
Head 146
Hematopoiesis 138
Humerus 155
Hyoid bone 152
Ilium 158
Incus 149
Intramembranous ossification 140
Ischium 158
Kyphosis 154
Lacrimal bones 151
Lacunae 142
Lamella 142
Lambdoid suture 146
Ligaments 138
Line 146
Lordosis 154
Lumbar vertebrae 152
Lunate 157
Malleus 149
Mandible bone 151
Manubrium 155
Mastoid portion of temporal bone 146
Maxillary bones 150
Meatus/canal 146
Medullary cavity 145
Metacarpal bones 157
Metaphysis 144
Metatarsals 161
Nasal bones 150
Navicular/scaphoid 157
Neck 146
Obturator foramen 159
Occipital bone 146
Occipital condyle 146
Olecranon process 157
Orbital margin 146
Ossification 140
Osteoblasts 140
Osteoclasts 140
Osteomalacia 144
Osteon 142
Osteoprogenitor cell 140
Palatine bones 150
Parietal bones 147
Patella 159
Pelvic girdle 158
Periosteum 140
Petrous part of temporal bone 146

(continues)

KEY TERMS (*continued*)

INTRODUCTION

The supporting structure of the body is the framework of joined bones that we refer to as the skeleton. It enables us to stand erect, to move in our environment, to accomplish extraordinary feats of artistic grace like ballet moves and athletic endeavors like the high jump as well as normal physical endurance. The skeletal system allows us to move a pen and write and aids us in breathing. It is closely associated with the muscular system. The skeletal system includes all the bones of the body and their associated cartilage, tendons, and ligaments. Despite the appearance of the bones, they are indeed composed of living tissue. The hard, "dead" stonelike appearance of bones is due to mineral salts like calcium phosphate embedded in the inorganic matrix of the bone tissue. Leondardo da Vinci (1452–1519), the famous Italian Renaissance artist and scientist, is credited as the first anatomist to correctly illustrate the skeleton with its 206 bones. See Concept Map 7-1: Skeletal System.

THE FUNCTIONS OF THE SKELETAL SYSTEM

The skeleton has five general functions:

1. It supports and stabilizes surrounding tissues such as muscles, blood and lymphatic vessels, nerves, fat, and skin.
2. It protects vital organs of the body such as the brain, spinal cord, the heart, and lungs, and it protects other soft tissues of the body.
3. It assists in body movement by providing attachments for muscles that pull on the bones that act as levers.
4. It manufactures blood cells. This process is called **hematopoiesis** (**hem**-ah-toh-poy-**EE**-sis) and occurs chiefly in red bone marrow.
5. It is a storage area for mineral salts, especially phosphorus and calcium, and fats.

Associated with the bones are cartilage, tendons, and ligaments. **Cartilage**, a connective tissue, is the environment in which bone develops in a fetus. It is also found at the ends of certain bones and in joints in adults, providing a smooth surface for adjacent bones to move against each other. **Ligaments** are tough connective tissue structures that attach bones to bones like the ligament that attaches the head of the femur to the acetabulum of the pelvic bone in the hip joint. **Tendons** are similar structures that attach muscle to bone.

THE GROWTH AND FORMATION OF BONE

The skeleton of a developing fetus is completely formed by the end of the third month of pregnancy.

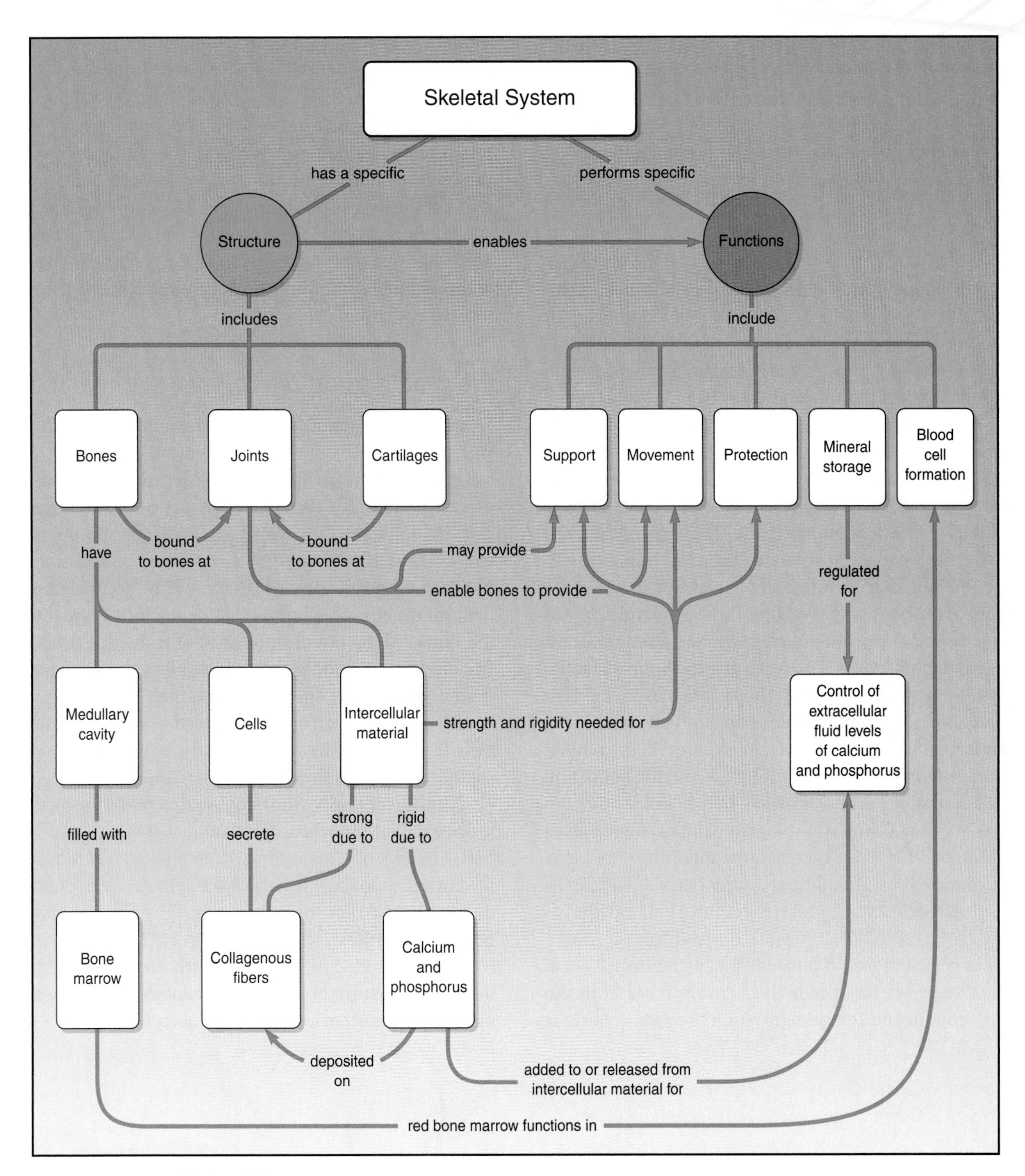

CONCEPT MAP 7-1. Skeletal System.

However, at this time, the skeleton is predominantly cartilage. During the subsequent months of pregnancy, ossification, the formation of bone, and growth occur. The osteoblasts invade the cartilage and begin the process of ossification. Longitudinal growth of bones continues until approximately 15 years of age in girls and 16 years of age in boys. This takes place at the epiphyseal line or plate. Bone maturation and remodeling continue until the age of 21 in both sexes. It would be incorrect to state that cartilage actually turns into bone. Rather cartilage is the environment in which the bone develops.

The strong protein matrix is responsible for a bone's resilience or "elasticity" when tension is applied to the bone so that it gives a little under pressure. The mineral salts deposited into this protein matrix are responsible for the strength of the bone so that it does not get crushed when pressure is applied to the bone.

Deposition of Bone

Bone develops from spindle-shaped cells called **osteoblasts** that develop from undifferentiated bone cells called **osteoprogenitor** (**oss**-tee-oh-pro-**JEN**-ih-tohr) **cells** (Figure 7-1). These osteoblasts are formed beneath the fibrovascular membrane that covers a bone called the **periosteum** (pair-ee-**AHS**-tee-um) (Figure 7-2). These osteoblasts are also found in the **endosteum** (en-**DOS**-tee-um), which lines the bone marrow or medullary cavity. Deposition of bone is controlled by the amount of strain or pressure on the bone. The more strain, the greater the deposition of bone. The heel bone, or calcaneum, is a large strong bone because it receives the weight of the body when walking. Bones (and muscles) in casts will waste away or atrophy, whereas continued and excessive strain via exercise will cause the bone and muscles to grow thick and strong. This is the reason children are told to run and play to develop strong bones during their formative years.When a cast is removed, the patient participates in physical therapy to build up the bone (and muscles) that became weak while in the cast.

A break in a bone will stimulate injured osteocytes to proliferate. They then secrete large quantities of matrix to form new bone. In addition, other types of bone cells called **osteoclasts** are present in almost all cavities of bone (see Figure 7-1). They are derived from immune system cells and are responsible for the reabsorption of bone. These are large cells that remove bone from the inner side during remodeling, such as when a bone is broken. These cells are also responsible for the ability of a crooked bone to become straight. If a young child is detected to be bow-legged, the physician will apply braces to the legs. Periodic tightening of the braces puts pressure on the bone so that new bone is deposited by osteocytes (mature osteoblasts), or mature bone cells, while the osteoclasts remove the old bone during this remodeling process. This process can cause a broken bone that was set improperly to heal incorrectly. To correct this, the bone must be broken again and correctly reset to straighten properly.

Types of Ossification

There are two types of **ossification** (**oss**-sih-fih-**KAY**-shun) (the formation of bone by osteoblasts). The first type is **intramembranous ossification**, in which dense connective tissue membranes are replaced by deposits of inorganic calcium salts, thus forming bone. The membrane itself will eventually become the periosteum of the mature bone. Underneath the periosteum will be compact bone with an inner core of spongy or cancellous bone. Only the bones of the cranium or skull form by this process. Because complete ossification in this way does not occur until a few months after birth, one can feel these membranes on the top of a baby's skull as the *soft spot* or **fontanelle** (fon-tah-**NELL**). This allows the baby's skull to give slightly as it moves through the birth canal.

The other bones of the body are formed by the second process called **endochondral** (en-doh-**KON**-dral) **ossification** (Figure 7-3). This is the process in which cartilage is the environment in which the bone cells develop (endo = inside, chondro = cartilage). As the organic matrix becomes synthesized, the osteoblast becomes completely surrounded by the bone matrix and the osteoblast becomes a mature bone cell or osteocyte. Both types of ossification result in compact and cancellous bone.

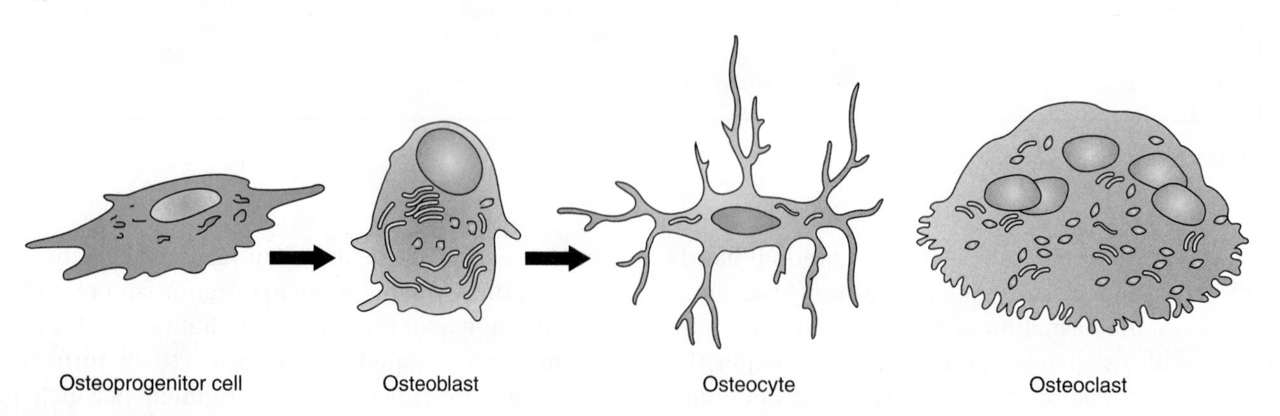

FIGURE 7-1. The different types of bone cells.

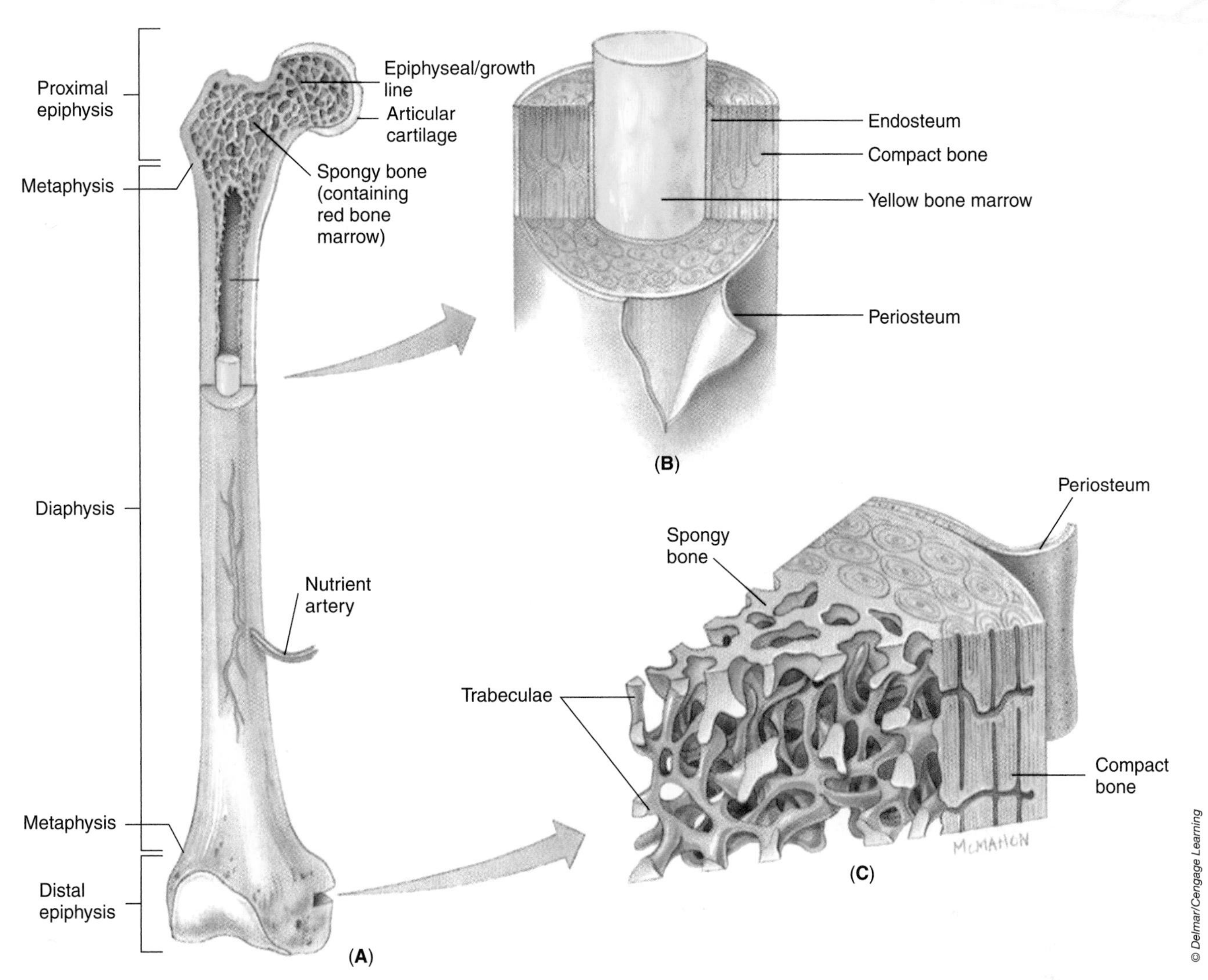

FIGURE 7-2. The structure of a typical long bone. (A) Diaphysis, epiphysis, and medullary cavity. (B) Compact bone surrounding yellow bone marrow in the medullary cavity. (C) Spongy bone and compact bone in the epiphysis.

HEALTH ALERT

STRONG BONES

In order to maintain strong and healthy bones throughout our lives, it is important to maintain a balanced diet with a daily intake of calcium. We can do this by consuming dairy products such as milk, yogurt, and cheeses. In addition to diet, a regular regimen of exercise is also important. As bones are developing in children and adolescents, it is important to increase calcium intake and exercise more rigorously. As we mature we still require calcium; however, we require it in smaller amounts. Daily exercise, as simplistic as walking in older age and running or playing sports in middle age, will help maintain a healthy skeletal system. When playing sports, walking, or running, it is crucial to wear proper foot attire with arch supports and a good fit. This will prevent future problems with bones of the feet. Proper posture during walking and sitting will also maintain healthy and strong bones. ■

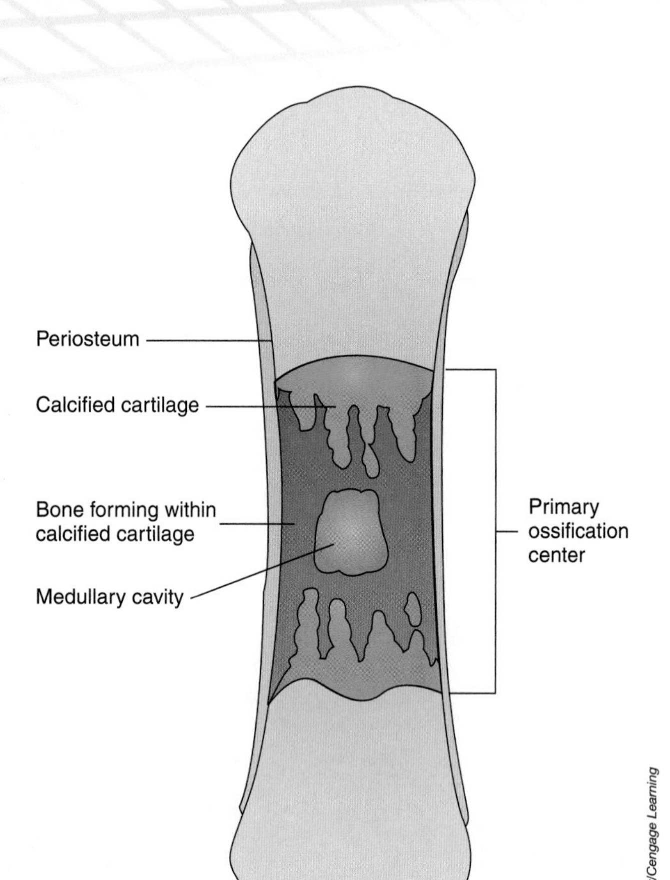

FIGURE 7-3. Endochondral ossification where cartilage is the environment in which bone develops.

Maintaining Bone

In a healthy body, a balance must exist between the amount of calcium stored in the bones, the calcium in the blood, and the excess calcium excreted by the kidneys and via the digestive system. The proper calcium ion concentration in the blood and bones is controlled by the endocrine system. Two hormones, calcitonin and parathormone, control the calcium concentration in our bodies. Calcitonin causes calcium to be stored in the bones; parathormone causes it to be released into the bloodstream.

THE HISTOLOGY OF BONE

There are two types of bone tissue: **compact** or **dense bone** and **cancellous** or **spongy bone** (see Figure 7-2C). In both types of tissue, the osteocytes are the same, but the arrangement of how the blood supply reaches the bone cells is different. The two types of tissue have different functions. Compact bone is dense and strong, whereas cancellous bone has many open spaces, giving it a spongy appearance. It is in these spaces that bone marrow can be found.

The Haversian System of Compact Bone

The **haversian** (hah-**VER**-shan) **canal**, also called an **osteon**, was named for an English physician, Clopton Havers (1650–1702), who first described it as a prominent feature of compact bone (Figure 7-4). This system allows for the effective metabolism of bone cells surrounded by rings of mineral salts. It has several components. Running parallel to the surface of the bone are many small canals containing blood vessels (capillaries, arterioles, venules) that bring in oxygen and nutrients and remove waste products and carbon dioxide. These canals are called haversian or **central canals** and are surrounded by concentric rings of bone, each layer of which is called a **lamella** (lah-**MELL**-ah). Between two lamellae or rings of bone are several tiny cavities called **lacunae** (lah-**KOO**-nee). Each lacuna contains an osteocyte or bone cell suspended in tissue fluid. The lacunae are all connected to each other and ultimately to the larger haversian or central canals by much smaller canals called **canaliculi** (**kan**-ah-**LIK**-you-lye). Canals running horizontally to the haversian (central) canals, also containing blood vessels, are called **Volkmann's** or **perforating canals.** It is tissue fluid that circulates through all these canals and bathes the osteocyte, bringing in oxygen and food and carrying away waste products and carbon dioxide, keeping the osteocytes alive and healthy.

> *Study***WARE**™ **Connection**
>
> Watch an animation of a fracture as a result of direct force on your StudyWARE™ CD-ROM.

Cancellous Bone

Cancellous or spongy bone is located at the ends of long bones and forms the center of all other bones. It consists of a meshwork of interconnecting sections of bone called **trabeculae** (trah-**BEK**-you-lee), creating the spongelike appearance of cancellous bone (Figure 7-2C). The trabeculae give strength to the bone without the added weight of being solid. Each trabecula consists of

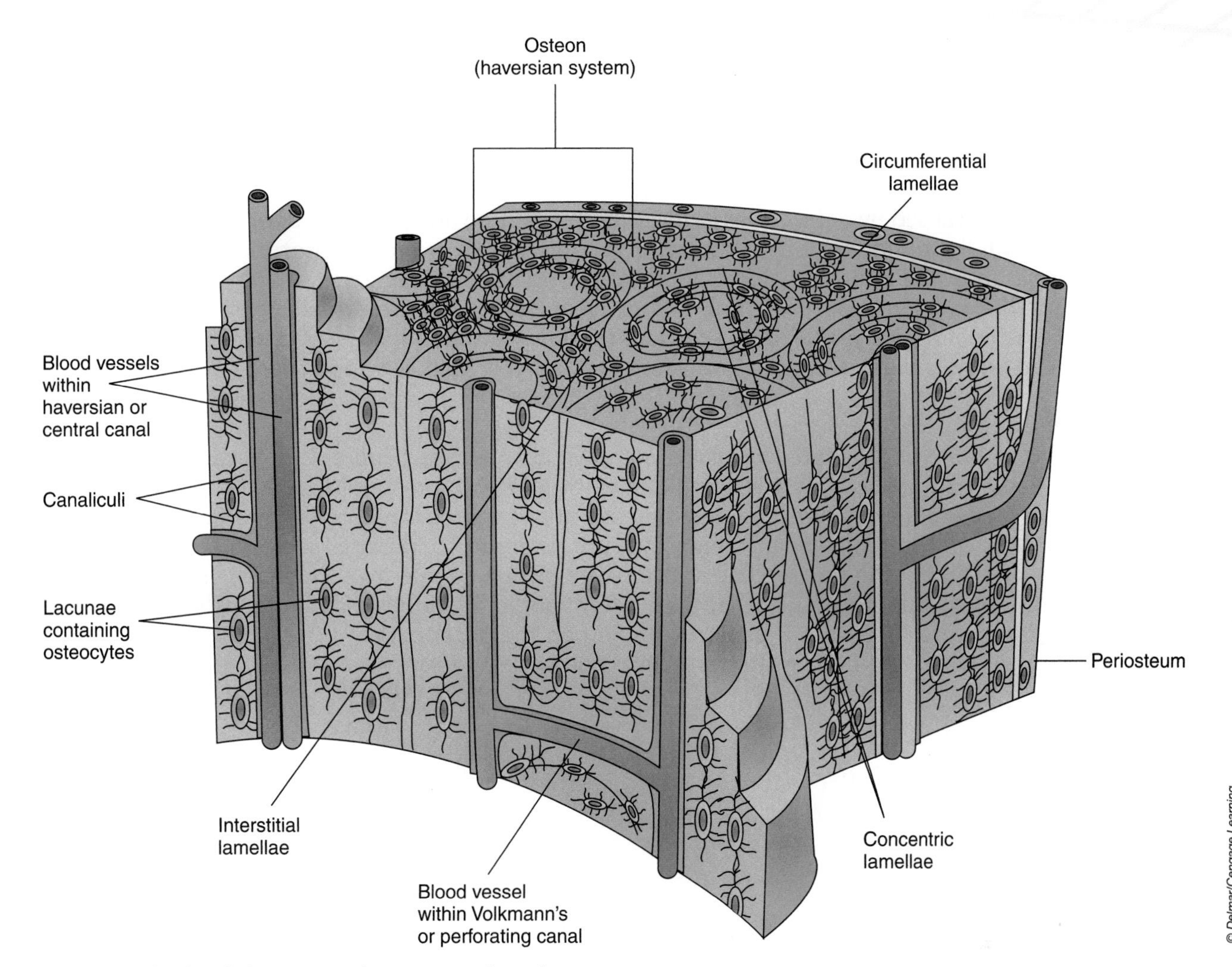

FIGURE 7-4. The detailed structure of compact or dense bone.

several lamellae with osteocytes between the lamellae just as in compact bone. The spaces between the trabeculae are filled with bone marrow. Nutrients exit blood vessels in the marrow and pass by diffusion through the canaliculi of the lamellae to the osteocytes in the lacunae.

Bone Marrow

The many spaces within certain cancellous bone are filled with **red bone marrow.** This marrow is richly supplied with blood and consists of blood cells and their precursors. The function of red bone marrow is hematopoiesis, or the formation of red and white blood cells and blood platelets. Therefore, blood cells in all stages of development will be found in red bone marrow. We shall discuss in more detail the different stages of blood cell development in Chapter 13.

In an adult, the ribs, vertebrae, sternum, and bones of the pelvis all contain red bone marrow in their cancellous tissue. These bones produce blood cells in adults. Red bone marrow within the ends of the humerus or upper arm and the femur or thigh is plentiful at birth but gradually decreases in amount as we age.

Yellow bone marrow is connective tissue consisting chiefly of fat cells. It is found primarily in the shafts of long bones within the medullary cavity, the central area of the bone shaft (see Figure 7-2B). Yellow bone marrow extends into the osteons or haversian systems, replacing red bone marrow when it becomes depleted.

COMMON DISEASE, DISORDER, OR CONDITION

DISORDERS OF THE SKELETAL SYSTEM

RICKETS

Rickets is a disease caused by deficiencies in the minerals calcium and phosphorus or by deficiencies in vitamin D and sunlight. Vitamin D is necessary for calcium and phosphorus absorption. The condition causes changes in bones known as rickets in children and **osteomalacia** (**oss**-tee-oh-mah-**LAY**-she-ah) in adults. The bones fail to ossify, resulting in soft, weak bones that are easily broken. Rickets occurs in children who do not receive adequate exposure to sunlight (sunlight is necessary for vitamin D production in the body) or whose diets are deficient in vitamin D (milk is a food source of vitamin D).

FRACTURES

The breaking of bone or associated cartilage is known as a **fracture** (Figure 7-5). Because bone supports other tissues, a fracture is usually accompanied by injury to surrounding soft tissues like muscle or connective tissue. Bone fractures are classified as either open or compound if the bone protrudes through the skin, or closed or simple if this skin is not perforated. Fractures can also be categorized based on the direction of the fracture line as transverse (at right angles to a long axis), linear (parallel to a long axis), or oblique (an angle other than a right angle to a long axis). A greenstick fracture is an incomplete fracture. The bone is bent but broken on the outer part of the bone. A comminuted fracture is one in which the bone is shattered into numerous pieces.

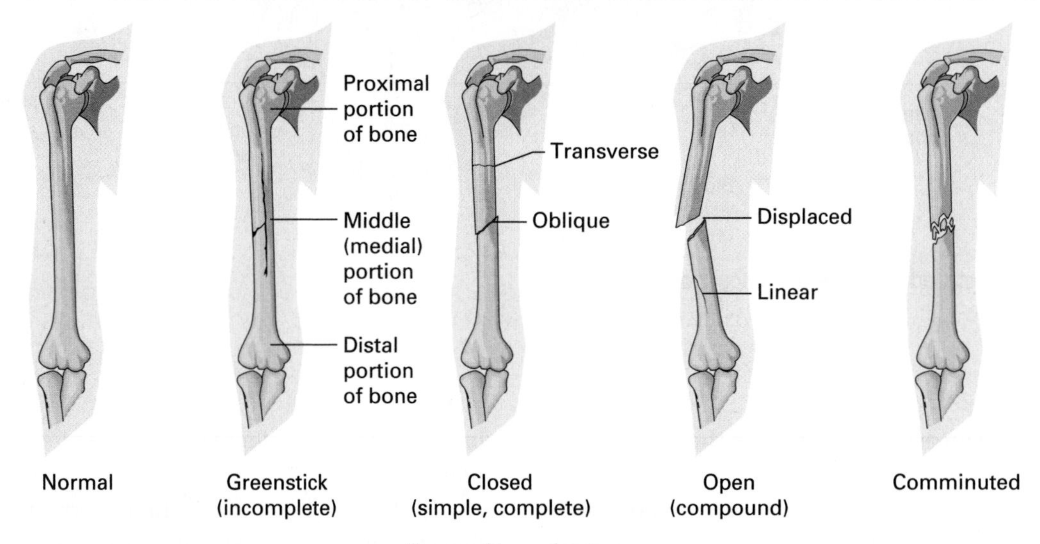

FIGURE 7-5. Types of bone fractures.

THE CLASSIFICATION OF BONES BASED ON SHAPE

The individual bones of the body can be divided by shape into five categories: long, short, flat, irregular, and sesamoid (Figure 7-6).

Long Bones

Long bones (see Figure 7-2) are bones whose length exceeds their width and consist of a **diaphysis** (dye-**AFF**-ih-sis) or shaft composed mainly of compact bone, a **metaphysis** (meh-**TAFF**-ih-sis) or flared portion at each end of the diaphysis consisting mainly of cancellous or spongy bone, and two extremities, each called an **epiphysis** (eh-**PIFF**-ih-sis), separated from the metaphysis by the **epiphyseal or growth line** where longitudinal growth of the bone occurs. The shaft consists mainly of compact bone. It is thickest toward the middle of the bone because strain on the bone is greatest at that point. The strength of a long bone is also ensured by the slight curvature of the shaft, a good

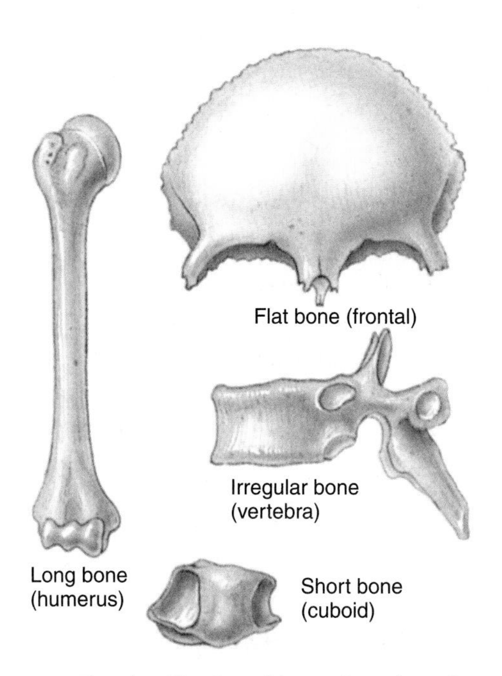

FIGURE 7-6. The classification of bones based on shape.

engineering design to distribute weight. The interior of the shaft is the **medullary cavity** filled with yellow bone marrow. The extremities or the epiphyses of the long bone have a thin covering of compact tissue overlying a majority of cancellous tissue, which usually contains red marrow. The epiphyses are usually broad and expanded for articulation with other bones and to provide a large surface for muscle attachment. Examples of obvious long bones are the clavicle, humerus, radius, ulna, femur, tibia, and fibula. Not so obvious are those short versions of a long bone, the metacarpals of the hand, the metatarsals of the foot, and the phalanges of the fingers and toes.

Short Bones

Short bones are not merely shorter versions of long bones. They lack a long axis. They have a somewhat irregular shape.They consist of a thin layer of compact tissue over a majority of spongy or cancellous bone. Examples of short bones of the body are the carpal bones of the wrist and the tarsal bones of the foot.

Flat Bones

Flat bones are thin bones found whenever there is a need for extensive muscle attachment or protection for soft or vital parts of the body. These bones, usually curved, consist of two flat plates of compact bone tissue enclosing a layer of cancellous bone. Examples of flat bones are the sternum, ribs, scapula, parts of the pelvic bones, and some of the bones of the skull.

Irregular Bones

Irregular bones are bones of a very peculiar and different or irregular shape. They consist of spongy bone enclosed by thin layers of compact bone. Examples of irregular bones are the vertebrae and the ossicles of the ears.

Sesamoid Bones

Sesamoid (SESS-ah-moyd) bones are small rounded bones. These bones are enclosed in tendon and fascial tissue and are located adjacent to joints. They assist in the functioning of muscles. The kneecap, or patella, is the largest of the sesamoid bones. Some of the bones of the wrist and ankle could also be classified as sesamoid bones as well as short bones.

BONE MARKINGS

The surface of any typical bone will exhibit certain projections called **processes** or certain depressions called **fossae (FOSS-ee)**, or both. These markings are functional in that they can help join one bone to another, provide a surface for the attachments of muscles, or serve as a passageway into the bone for blood vessels and nerves. The following is a list of some terms and definitions regarding bone markings.

Processes

A process is a general term referring to any obvious bony prominence. The following is a list of specific examples of processes.

1. **Spine:** any sharp, slender projection such as the spinous process of a vertebra (see Figure 7-14).
2. **Condyle (KON-dial):** a rounded or knuckle-like prominence usually found at the point of articulation with another bone such as the lateral and medial condyles of the femur (see Figure 7-23).
3. **Tubercle (TOO-ber-kl):** a small round process like the lesser tubercle of the humerus (see Figure 7-19).
4. **Trochlea (TROK-lee-ah):** a process shaped like a pulley as in the trochlea of the humerus (see Figure 7-19).
5. **Trochanter (tro-KAN-ter):** a very large projection like the greater and lesser trochanter of the femur (see Figure 7-23).

6. **Crest**: a narrow ridge of bone like the iliac crest of the hip bone (see Figure 7-22).
7. **Line**: a less prominent ridge of bone than a crest.
8. **Head**: a terminal enlargement like the head of the humerus and the head of the femur (see Figures 7-19 and 7-23).
9. **Neck**: that part of a bone that connects the head or terminal enlargement to the rest of the bone, like the neck of the femur (see Figures 7-19 and 7-23).

Fossae

A fossa is a general term for any depression or cavity in or on a bone. The following is a list of specific examples of fossae.

1. **Suture**: a narrow junction often found between two bones like the sutures of the skull bones (see Figure 7-9).
2. **Foramen**: an opening through which blood vessels, nerves, and ligaments pass like the foramen magnum of the occipital bone of the skull or the obturator foramen of the pelvic bone (see Figure 7-22).
3. **Meatus** or **canal**: a long tube-like passage, like the auditory meatus or canal (see Figure 7-9).
4. **Sinus** or **antrum**: a cavity within a bone like the nasal sinuses or frontal sinus (see Figure 7-8A).
5. **Sulcus**: a furrow or groove like the intertubercular sulcus or groove of the humerus (see Figure 7-19).

DIVISIONS OF THE SKELETON

The skeleton typically has 206 named bones. The *axial* part consists of the skull (28 bones, including the cranial and facial bones), the hyoid bone, the vertebrae (26 bones), the ribs (24 bones), and the sternum. The *appendicular* part of the skeleton consists of the bones of the upper extremities or arms (64 bones, including the shoulder girdle bones) and the bones of the lower extremities or legs (62 bones, including the bones of the pelvic girdle) (Figure 7-7).

THE AXIAL SKELETON

The skull, in the correct use of the term, includes the cranial and the facial bones. We will discuss the cranial bones first.

The Cranial Bones

The bones of the cranium have a number of important functions. They protect and enclose the brain and special sense organs like the eyes and ears. Muscles for mastication or chewing and muscles for head movement attach to certain cranial bones. At certain locations, air sinuses or cavities are present that connect with the nasal cavities (Figure 7-8). All of the individual bones of the cranium are united by immovable junction lines called sutures.

The **frontal bone** is a single bone that forms the forehead, the roof of the nasal cavity, and the orbits, which are the bony sockets that contain the eyes (Figure 7-9). Important bone markings are the **orbital margin,** a definite ridge above each orbit located where eyebrows are found, and the **supraorbital ridge**, which overlies the frontal sinus and can be felt in the middle of your forehead. The **coronal suture** is found where the frontal bone joins the two parietal bones.

The two **parietal** (pah-**RYE**-eh-tal) **bones** form the upper sides and roof of the cranium. They are joined at the **sagittal suture** in the midline.

The **occipital bone** is a single bone that forms the back and base of the cranium (see Figure 7-9) and joins the parietal bones superiorly at the **lambdoid suture.** The inferior portion of this bone has a large opening called the **foramen magnum** through which the spinal cord connects with the brain. On each lower side of the occipital bone is a process called the **occipital condyle.** These processes are significant because they articulate with depressions in the first cervical vertebra (atlas), thus allowing the head to connect with and rest on the vertebrae. Other notable markings are the **external occipital crest** and the **external occipital protuberance,** which can be felt through the scalp at the base of the neck. Several ligaments and muscles attach to these regions.

The two **temporal bones** help form the lower sides and base of the cranium (see Figure 7-9). Each temporal bone encloses an ear and bears a fossa for articulation with the lower jaw or mandible. The temporal bones are irregular in shape and each consists of four parts: the squamous, petrous, mastoid, and tympanic parts. The **squamous portion** is the largest and most superior of the four parts. It is a thin flat plate of bone that forms the temple. Projecting from its lower part is the zygomatic process that forms the lateral part of the zygomatic arch or cheek bone. The **petrous part** is found deep within the base of the skull where it protects and surrounds the inner ear. The **mastoid portion** is located behind and below the auditory meatus or opening of the ear. The mastoid process is a rounded projection of the mastoid portion of the temporal bone easily felt behind the ear. Several muscles of the neck attach

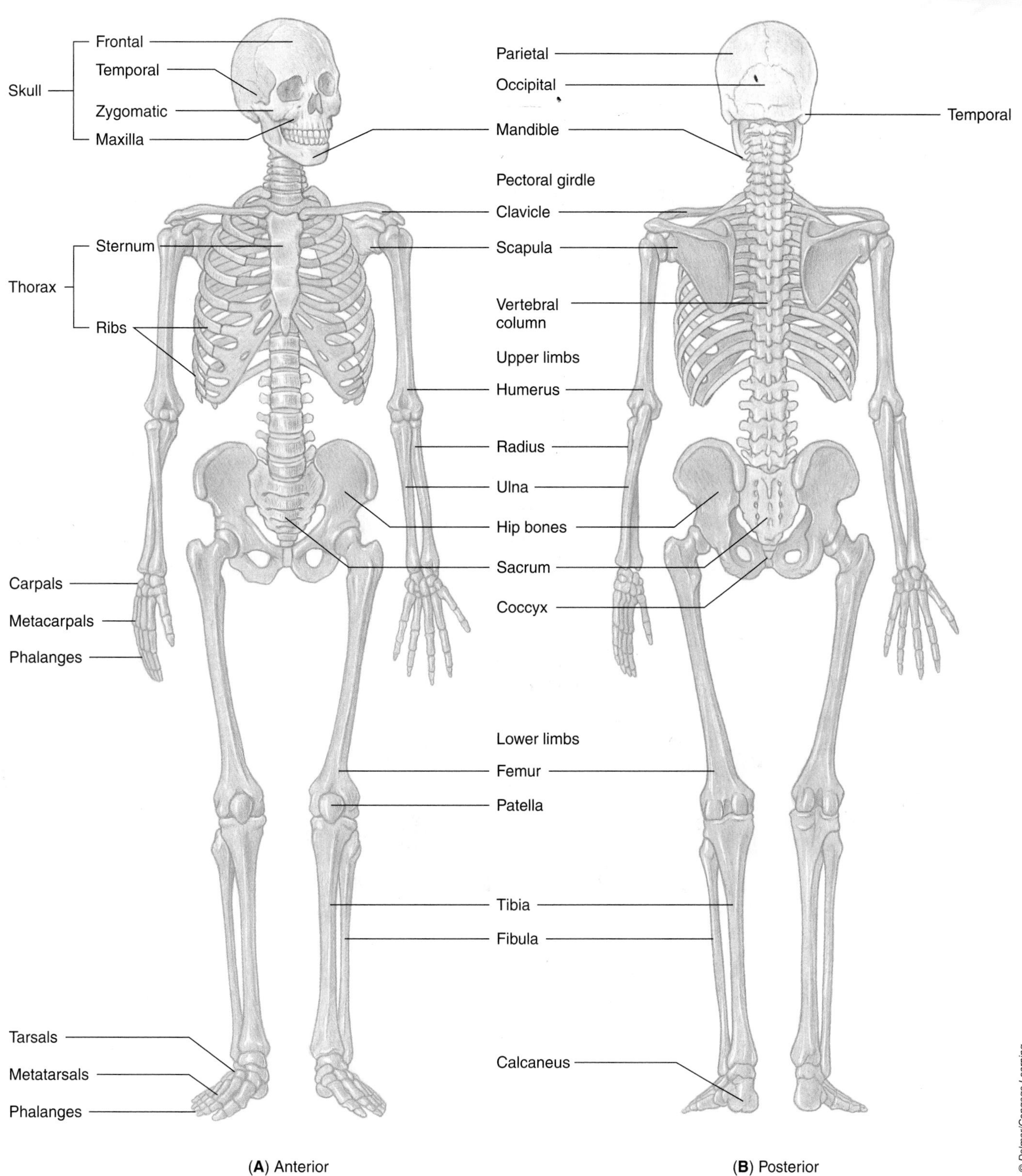

FIGURE 7-7. The human skeletal system. (A) Anterior view. (B) Posterior view.

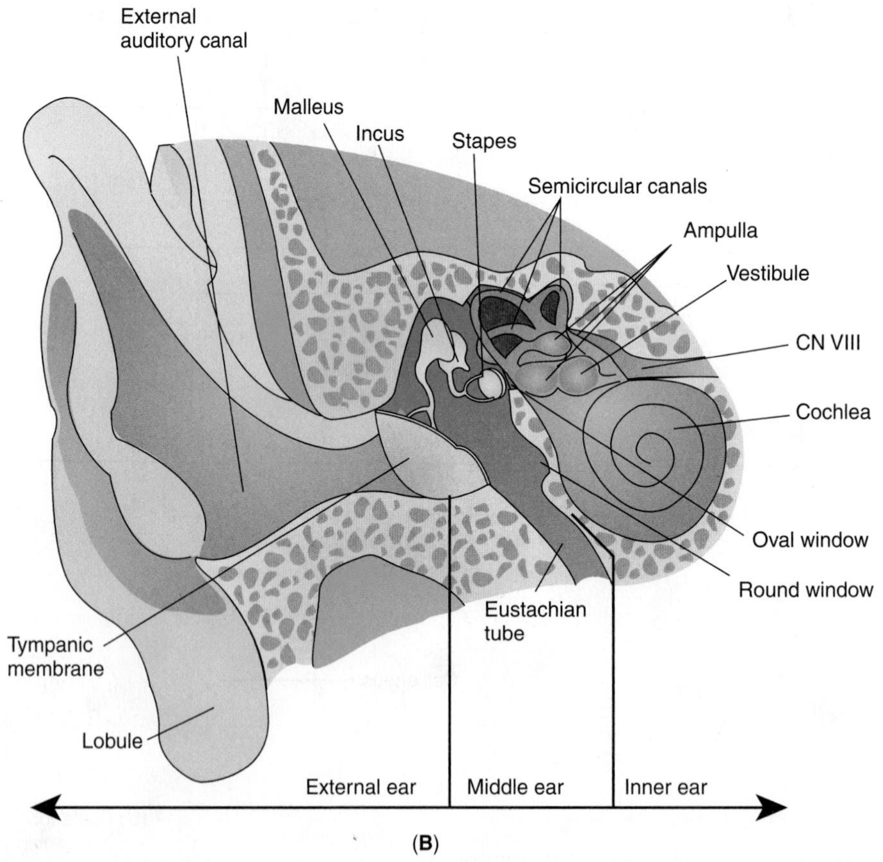

FIGURE 7-8. (A) The paranasal sinuses. (B) Cross section of the ear showing the ossicles.

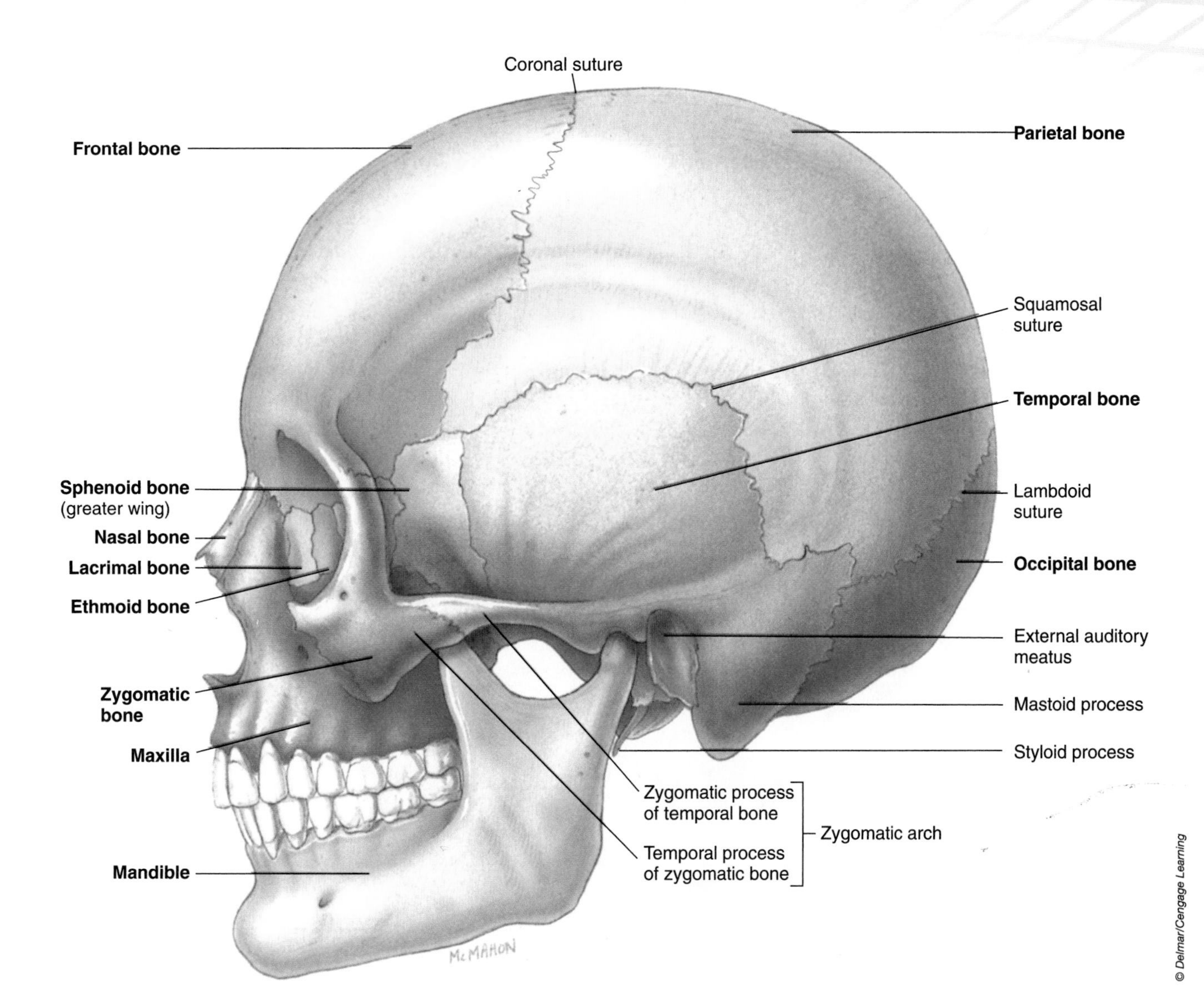

FIGURE 7-9. The cranial bones.

to this mastoid process and assist in moving your head. Finally, the **tympanic plate** forms the floor and anterior wall of the external auditory meatus. A long and slender styloid process can be seen extending from the undersurface of this plate. Ligaments that hold the hyoid bone in place (which supports the tongue) attach to this styloid process of the tympanic plate of the temporal bone.

The single **sphenoid bone** forms the anterior portion of the base of the cranium (Figures 7-9 and 7-10). When viewed from below it looks like a butterfly. It acts as an anchor binding all of the cranial bones together.

The single **ethmoid bone** is the principal supporting structure of the nasal cavities and helps form part of the orbits. It is the lightest of the cranial bones (see Figures 7-9 and 7-10).

The six **auditory ossicles** are the three bones found in each ear (see Figure 7-8B): the **malleus** or hammer, the **stapes** (**STAY**-peez) or stirrup, and the **incus** or anvil. These tiny bones are highly specialized in both structure and function and are involved in exciting the hearing receptors.

The **wormian bones** or **sutural bones** are located within the sutures of the cranial bones. They vary in number, are small and irregular in shape, and are never included in the total number of bones in the body. They form as a result of intramembranous ossification of the cranial bones.

StudyWARE™ Connection

Play an interactive game labeling the cranial bones on your StudyWARE™ CD-ROM.

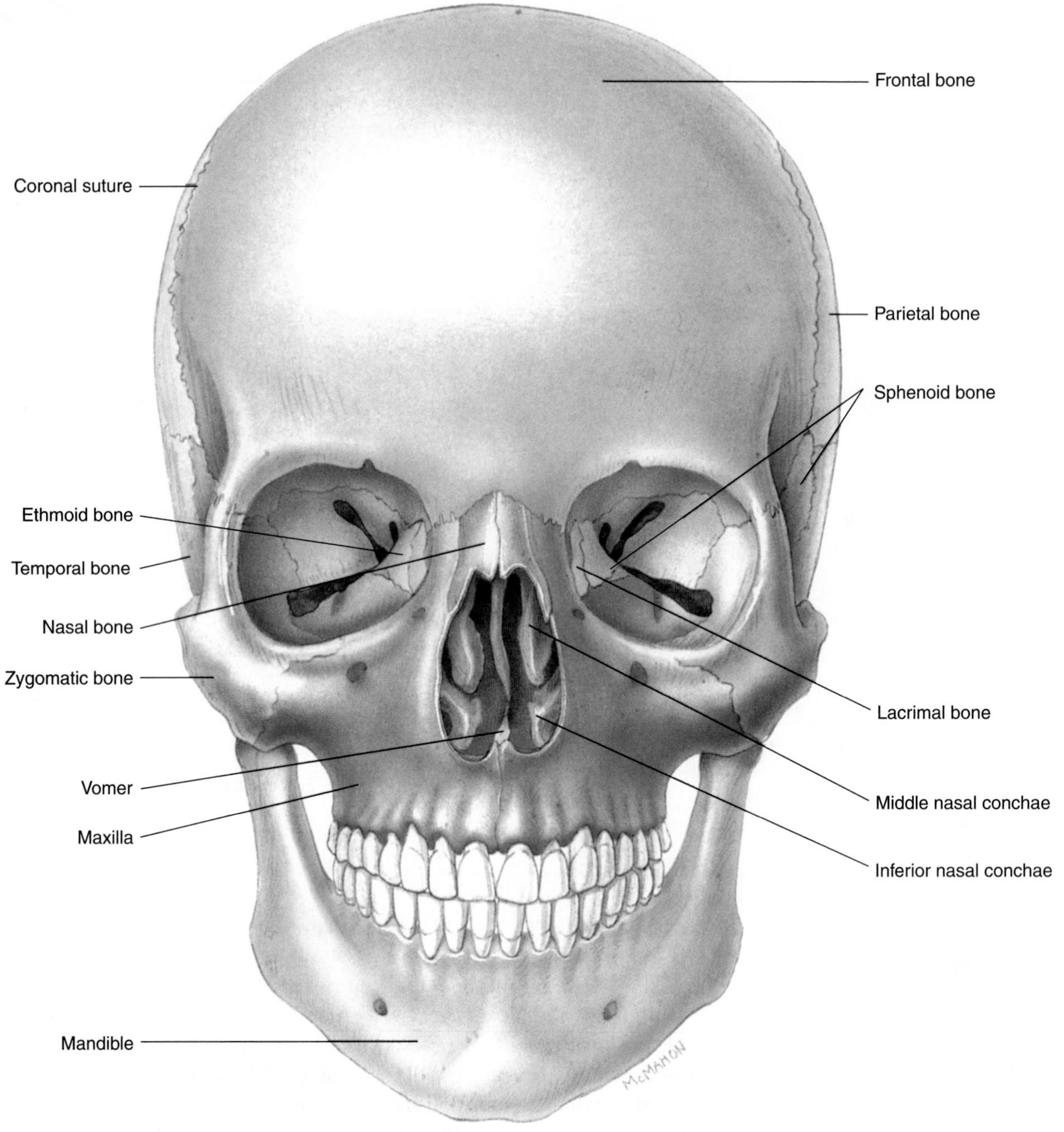

FIGURE 7-10. The facial bones.

The Facial Bones

Like the bones of the cranium, the facial bones are also united by immovable sutures, with one exception: the lower jawbone or mandible. This bone is capable of movement in a number of directions. It can be elevated and depressed as in talking, and it can protract and retract and move from side to side as in chewing.

The two **nasal bones** are thin and delicate bones that join in a suture to form the bridge of the nose (see Figure 7-10).

The two **palatine bones** form the posterior part of the roof of your mouth or part of the hard palate. This region is the same as the floor of the nasal cavity. Upward extensions of the palatine bones help form the outer walls of the nasal cavity.

The two **maxillary bones** make up the upper jaw (see Figure 7-10). Each maxillary bone consists of five parts: a body, a zygomatic process, a frontal process, a palatine process, and an alveolar process. The large body of the maxilla forms part of the floor and outer wall of the nasal cavity, the greater part of the floor of the orbit, and much of the anterior face below the temple. The body is covered by a number of facial muscles and contains a large maxillary sinus located lateral to the nose. The zygomatic process extends laterally to participate in the formation of the cheek. (Processes are named according to the bone they go to; thus, the zygomatic process of the maxillary bone goes toward and joins the zygomatic or cheekbone.) The frontal process extends upward to the frontal bone or forehead. The palatine process extends posteriorly in a horizontal plane to join or articulate with the palatine bone and actually forms the greater anterior portion of the hard palate or roof of the mouth. The alveolar processes bear the teeth of the upper jaw, and each tooth is embedded in an **alveolus** (al-**VEE**-oh-lus) or socket.The two maxillary bones join at the intermaxillary suture. This fusion is usually completed just before birth. If the two bones do not unite to form a continuous structure, the resulting defect is called a cleft palate and is usually associated with a cleft lip. With today's surgical techniques, the defect can be repaired early in the development of the child.

The two **zygomatic bones**, also known as the **malar bones**, form the prominence of the cheek and rest on the maxillae (see Figure 7-10). Its maxillary process joins the maxillary bone by connecting with the maxillary bone's zygomatic process. Each zygomatic bone has a frontal process extending upward to articulate with the frontal bone and a smaller temporal process that joins laterally with the temporal bone, thus forming the easily identified zygomatic arch.

The two **lacrimal** (**LAK**-rim-al) **bones** make up part of the orbit at the inner angle of the eye (see Figure 7-10). These very small and thin bones lie directly behind the frontal process of the maxilla. Their lateral surface has a depression or fossa that holds the lacrimal sac or tear sac and provides a canal for the lacrimal duct. Tears are directed from this point to the inferior meatus of the nasal cavity after they have cleansed and lubricated the eye.

The two **turbinates** or **nasal conchae bones** are very thin and fragile (see Figure 7-10). There is one in each nostril on the lateral side. They extend to but do not quite reach the bony portion of the nasal septum. They help form a series of shelves in the nasal cavity where air is moistened, warmed, and filtered.

The single **vomer bone** is a flat bone that makes up the lower posterior portion of the nasal septum (see Figure 7-10).

The single **mandible bone** develops in two parts. The intervening cartilage ossifies in early childhood, and the bone becomes fused into a single continuous structure. It is the strongest and longest bone of the face (see Figure 7-10). It consists of a U-shaped body with alveolar processes to bear the teeth of the lower jaw (just like the maxillary bone's alveolar processes that bear the teeth of the upper jaw). On each side of the body are the rami that extend perpendicularly upward. Each ramus has a condyle for articulation with the mandibular fossa of the temporal bone, thus allowing for the wide range of movement of the lower jawbone (see Figure 7-9).

The Orbits

The orbits are the two deep cavities in the upper portion of the face that protect the eyes. A number of bones of the skull contribute to their formation. Refer to Figure 7-10 to view these bones. Each orbit consists of the following bones:

Area of Orbit	Participating Bones
Roof	Frontal, sphenoid
Floor	Maxilla, zygomatic
Lateral wall	Zygomatic, greater wing of sphenoid
Medial wall	Maxilla, lacrimal, ethmoid

The Nasal Cavities

The framework of the nose surrounding the two nasal fossae is located in the middle of the face between the hard palate inferiorly and the frontal bone superiorly.

The nose is formed by the following bones (see Figure 7-10):

Area of Nose	Participating Bones
Roof	Ethmoid
Floor	Maxilla, palatine
Lateral wall	Maxilla, palatine
Septum of medial wall	Ethmoid, vomer, nasal
Bridge	Nasal

The Foramina of the Skull

If one views the skull inferiorly and observes the floor of the cranial cavity, one can observe the largest foramen of the skull, the foramen magnum. One can also observe a number of much smaller foramina or openings that penetrate the individual bones of the skull. They all have names and are passageways for blood vessels and nerves entering and exiting the various organs of the skull.

The Hyoid Bone

The single **hyoid bone** is a unique component of the axial skeleton because it has no articulations with other bones (Figure 7-11). It is rarely seen as part of an articulated skeleton in a lab. Rather, it is suspended from the styloid process of the temporal bone by two styloid ligaments. Externally, you can detect its position in the neck just above the larynx or voice box a fair distance from the mandible. It is shaped like a horseshoe consisting of a central body with two lateral projections. The larger projections are the greater cornu, and the smaller lateral projections are the lesser cornu. The hyoid bone acts as a support for the tongue and its associated muscles. It also helps elevate the larynx during swallowing and speech.

How to Study the Bones of the Skull

When learning the different bones of the skull, one of the best methods is to first refer to the colored plates in your textbook where each individual bone is portrayed in a different color. Refer to Figure 7-10, the anterior view of the skull, and Figure 7-9, the lateral view of the skull. Once you get a sense of where these bones are located, use a model of a human skull (either real bone or a good plastic reproduction) and search for sutures as a guide. Remember that in a real skull the older the skull, the less obvious the sutures become. As we age, the sutures tend to disappear or become very faint. The colored plates will greatly assist you in learning where the bones of the skull are found.

The Torso or Trunk

The sternum, ribs, and vertebrae make up the trunk or torso of the axial skeleton. The vertebrae are rigid and provide support for the body but the fibrocartilaginous disks between the vertebrae allow for a high degree of flexibility. The disks and vertebrae protect the delicate spinal cord contained within their articulated channels formed from successive foramina.

The spinal column is formed from a series of 26 irregular bones called vertebrae, separated and cushioned by the intervertebral disks of cartilage. A typical vertebra has the following parts or features (Figure 7-12): the *body* is a thick disk-shaped anterior portion pierced with numerous small holes for nerves and blood vessels that nurture the bone. The *neural arch* encloses a space, the *neural or vertebral foramen*, for passage of the spinal cord. The arch has three processes for muscle attachment: the *spinous process*, quite large on the thoracic vertebrae, directed backward, and two *transverse processes*, one on each side of the vertebra. The articular processes are used for articulating with the vertebra immediately above by the *two superior articular processes* and with the vertebra immediately below by the *two inferior articular processes.* The vertebral arch is composed of two portions on each side, the *pedicles* notched above and below for passage of nerves to and from the spinal cord, and the *laminae*, which form the posterior wall of the vertebral column.

Refer to Figure 7-13 for views of the structure of the vertebral column. There are 7 **cervical vertebrae**, 12 **thoracic vertebrae**, and 5 **lumbar vertebrae**. These all remain separate throughout life and are referred to as movable. In addition there are five **sacral vertebrae** that become fused by adult life and form the single **sacrum**. There are also four **coccygeal vertebrae** that unite firmly to form the single **coccyx** or tailbone. These last two, the sacrum and coccyx, are called fixed; hence the vertebrae are referred to in number as 26 rather than 33.

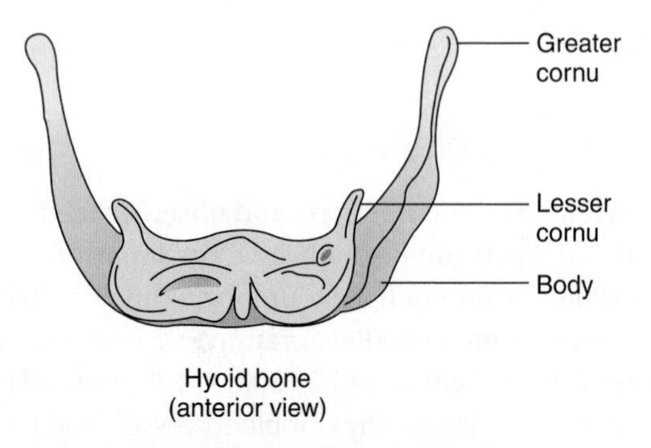

FIGURE 7-11. The hyoid bone (anterior view).

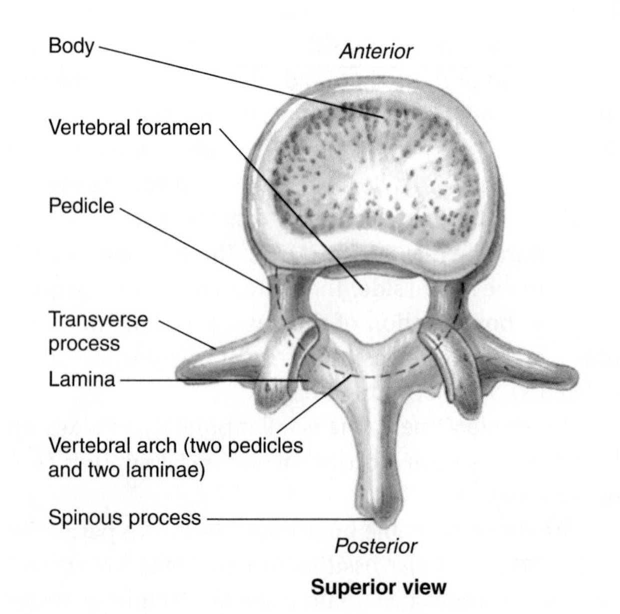

FIGURE 7-12. The characteristics of a typical vertebra.

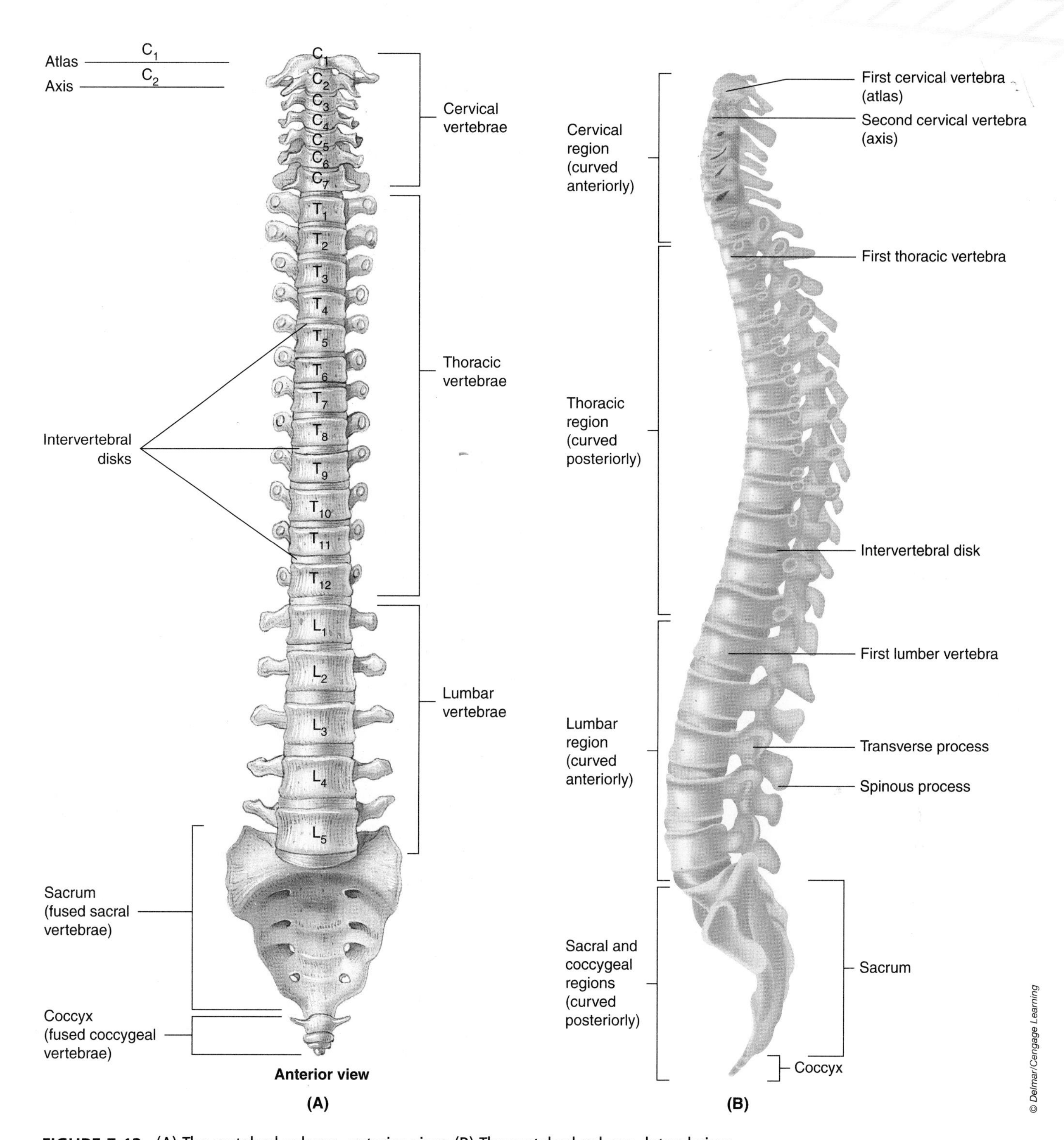

FIGURE 7-13. (A) The vertebral column, anterior view. (B) The vertebral column, lateral view.

The cervical vertebrae are the smallest vertebrae. The first two have been given special names (Figure 7-14). The first is called the **atlas** (named after Atlas in Greek mythology who held up the world); it supports the head by articulation with the condyles of the occipital bone. The second vertebra is the **axis**; it acts as the pivot on which the atlas and head rotate. The thoracic vertebrae have two distinguishing characteristics: the long spinous process pointing downward and six facets, three on each side for articulation with a rib. The lumbar vertebrae are the largest and the strongest. They are modified for the attachment of the powerful back muscles. The sacrum is a triangular and slightly curved bone. The curving coccyx can move slightly to increase the size of the birth canal during delivery in the female.

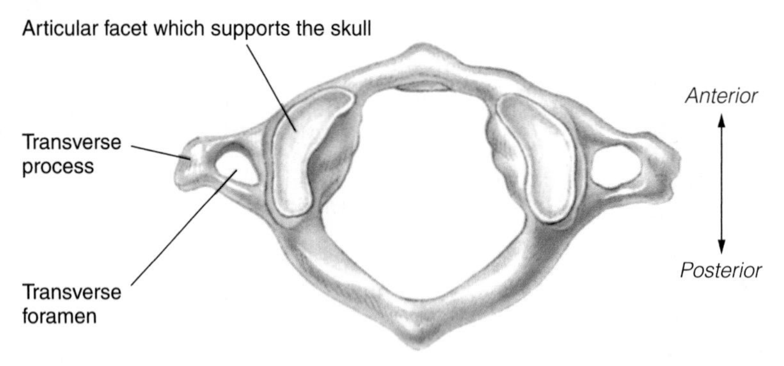

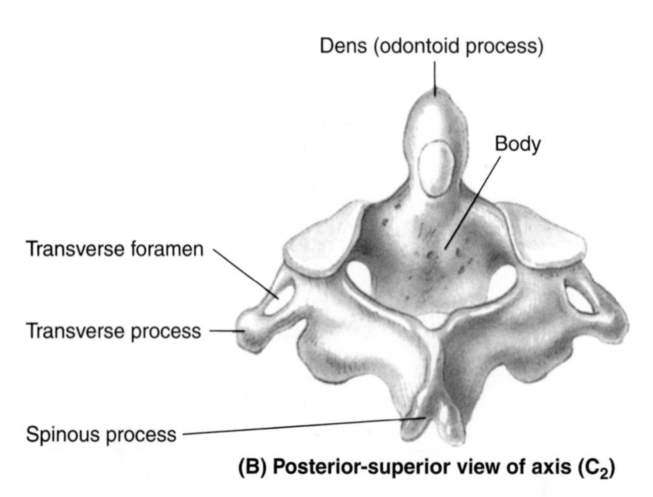

FIGURE 7-14. (A) Superior view of atlas (C_1) and (B) posterior-superior view of axis (C_2).

COMMON DISEASE, DISORDER, OR CONDITION

DISORDERS OF THE SPINE

The normal curvatures of the spine can become exaggerated as a result of injury, poor body posture, or disease (Figure 7-15). When the posterior curvature of the spine is accentuated in the upper thoracic region, the condition is called **kyphosis**. This results in the commonly referred to condition called hunchback. It is particularly common in older individuals due to osteoporosis. It can also be caused by tuberculosis of the spine, osteomalacia, or rickets. **Lordosis**, or swayback, is an abnormal accentuated lumbar curvature. It also can result from rickets or spinal tuberculosis. Temporary lordosis is common in men with potbellies and pregnant women who throw back their shoulders to preserve their center of gravity, thus accentuating the lumbar curvature. **Scoliosis** (skoh-lee-**OH**-sis), meaning twisted condition, is an abnormal lateral curvature of the spine that occurs most often in the thoracic region. It can be common in late childhood for girls, but the most severe conditions result from abnormal vertebral structure, lower limbs of unequal length, or muscle paralysis on one side of the body. Severe cases can be treated with body braces or surgically before bone growth ceases.

In addition to providing protection for the spinal cord and support for the body, the vertebral column is also built to withstand forces of compression many times the weight of the body. The fibrocartilaginous intervertebral disks act as cushions so that landing on your feet after a jump or a fall will help prevent the vertebrae from fracturing. Epidural anesthetics are commonly injected into the lower lumbar region during labor and birth.

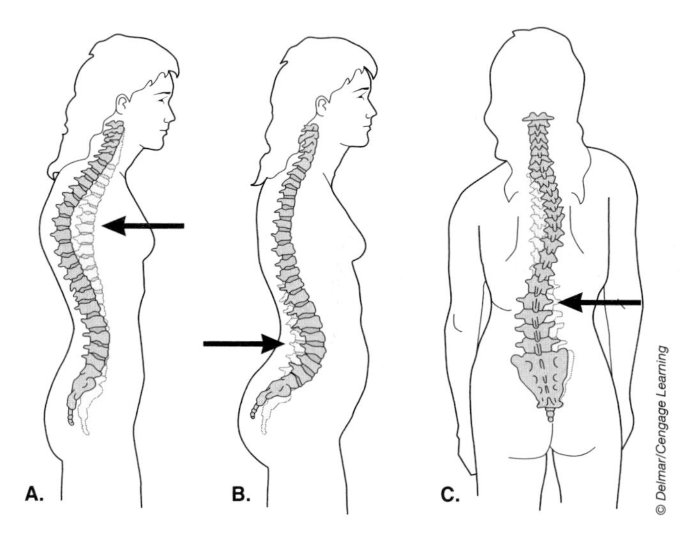

FIGURE 7-15. Abnormal curvatures of the spine: (A) kyphosis; (B) lordosis; (C) scoliosis.

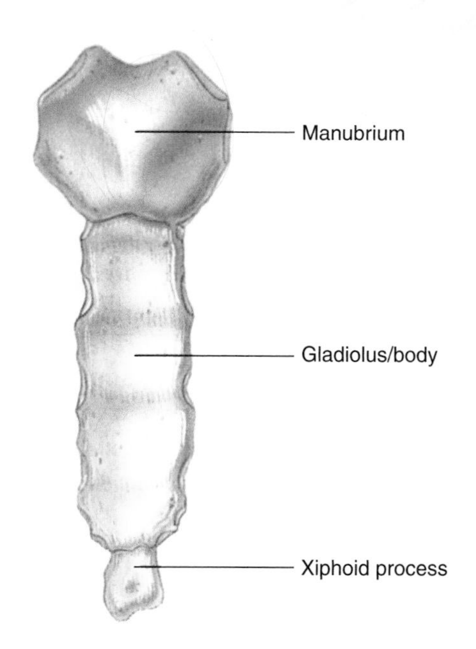

FIGURE 7-16. The sternum, anterior view.

The Thorax

The thorax or the rib cage of the body is made up of the sternum, the costal cartilages, the ribs, and the bodies of the thoracic vertebrae. This bony cage encloses and protects the heart and lungs. It also supports the bones of the shoulder girdle and the bones of the upper extremities.

The Sternum

The **sternum** is also known as the breastbone (Figure 7-16). It develops in three parts: the **manubrium,** the **gladiolus,** and the **xiphoid** (**ZIFF**oyd) **process.** The sternum resembles a sword, with the manubrium resembling the handle of the sword, the gladiolus or body forming the blade, and the xiphoid process forming the tip of the sword. No ribs are attached to the xiphoid, but the manubrium and gladiolus have notches on each side for attachment of the first seven costal (rib) cartilages. The manubrium articulates with the clavicle or collarbone (Figure 7-17). Between these two points of attachment is the suprasternal or jugular notch easily felt through the skin. The diaphragm and the rectus abdominis muscles attach to the xiphoid.

The Ribs

The 12 pairs of ribs are also referred to as the **costae** (Figure 7-17). They are named according to their anterior attachments. Because the upper seven pairs articulate directly with the sternum, they are called *true ribs.* The lower five pairs are called false ribs. The costal cartilages of the 8th, 9th, and 10th rib pairs are attached to the cartilage of the 7th rib so they join the sternum only indirectly. Because the 11th and 12th pairs of ribs have no cartilage and do not attach at all anteriorly, these "false" ribs have another name, *floating ribs.* Of course, all ribs attach posteriorly to the thoracic vertebrae.

THE APPENDICULAR SKELETON

The Bones of the Upper Extremities

The bones of the upper extremities include the bones of the shoulder girdle, the arm, the forearm, the wrist, the hand, and the fingers.

The bones of the shoulder girdle are the **clavicle** (**KLAV**-ih-kl) and the **scapula** (**SKAP**-you-lah). The clavicle or collarbone is a long slim bone located at the root of the neck just below the skin and anterior to the first rib. The medial end articulates with the manubrium of the sternum and the lateral end with the **acromial** (ah-**KRO**-mee-al) **process** of the scapula. The scapula or shoulder blade is a large, flat, triangular bone located on the dorsal portion of the thorax, covering the area from the second to the seventh rib (Figure 7-18). Two other prominent bony projections on the scapula are the **coracoid process**, which functions as an attachment for muscles that move the arm, and the **glenoid fossa**, which receives the head of the humerus and helps form the shoulder joint.

The **humerus** (**HYOO**-mehr-us) is the largest and longest bone of the upper arm (Figure 7-19). Its head is rounded and joined to the rest of the bone by its

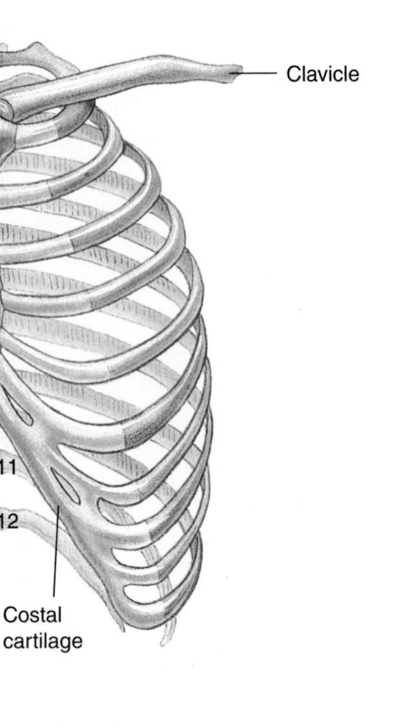

FIGURE 7-17. Thoracic cage, anterior view.

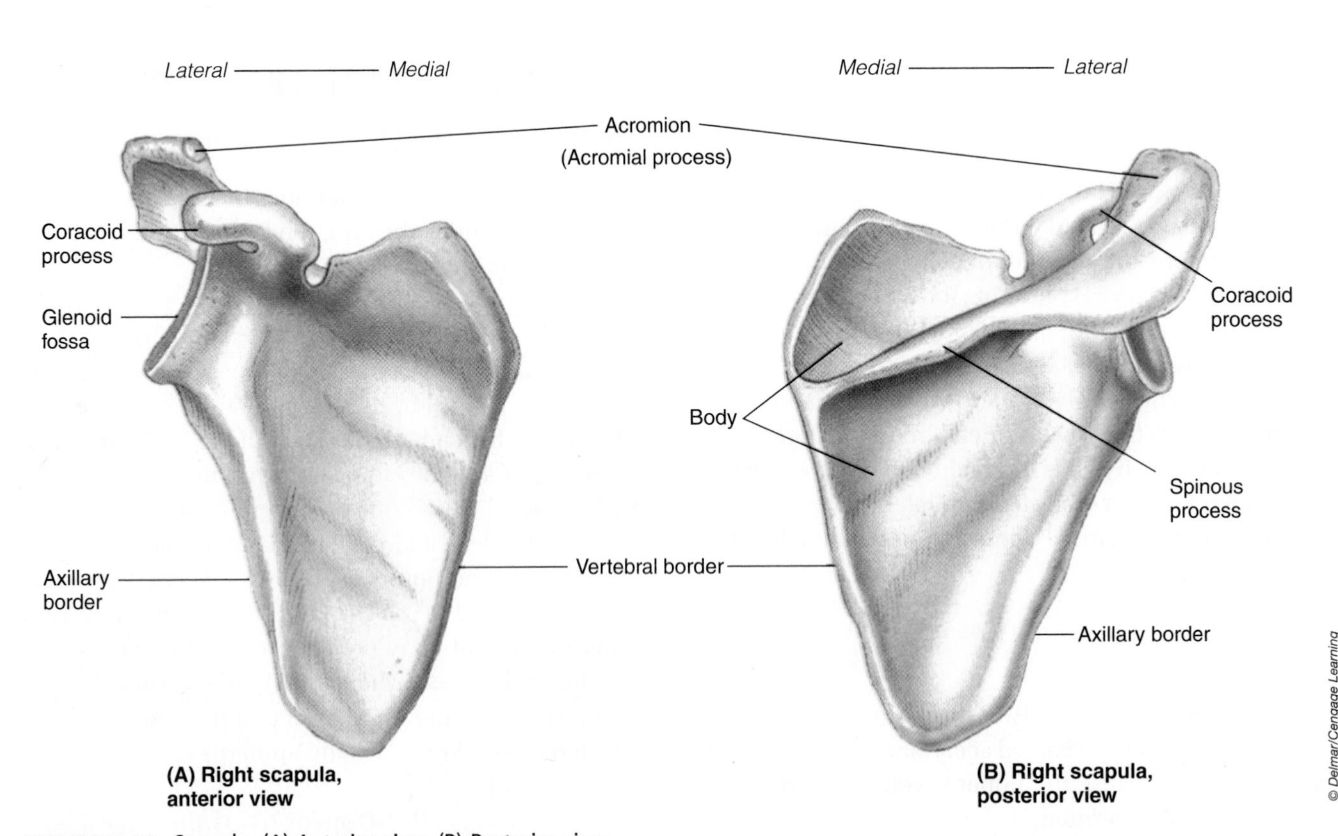

FIGURE 7-18. Scapula. (A) Anterior view. (B) Posterior view.

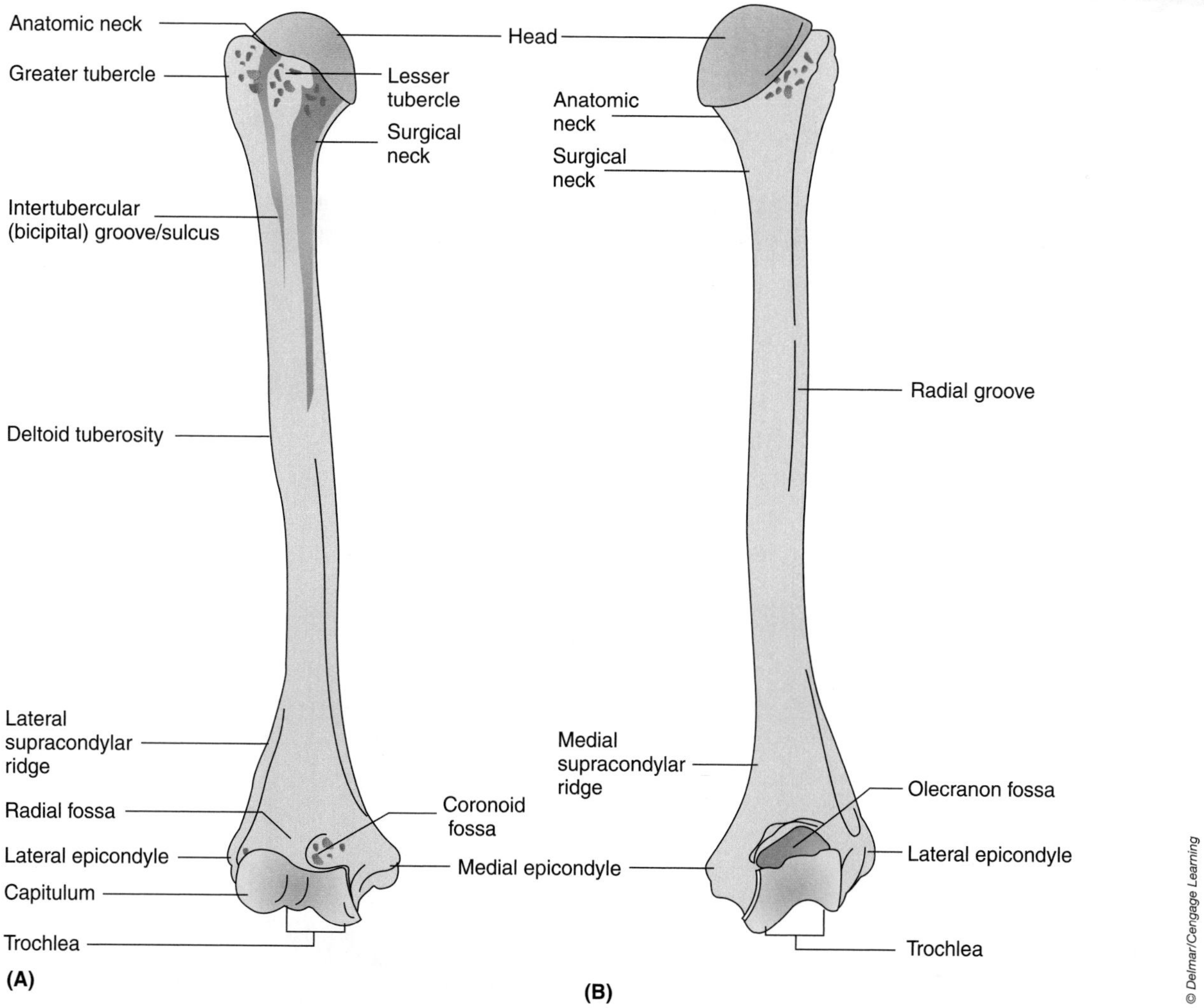

FIGURE 7-19. The right humerus. (A) Anterior view. (B) Posterior view.

anatomic neck. The upper part of the bone has two prominences, the greater and lesser tubercles, which function as insertions for many of the muscles of the upper extremity.

The **ulna** is the *longer*, medial bone of the forearm (Figure 7-20). Its shaft is triangular, and the distal, or lower, end is called the head. At its proximal end is the **olecranon** (oh-**LEK**-rah-non) **process** or elbow. When banged, nerves are pressed causing the tingling sensation, which gives it the common name of "funny bone."

The **radius** is the *shorter*, lateral bone of the forearm. It is joined to the ulna by an interosseus membrane traversing the area between the shafts of the two bones. They move as one. The styloid process of the radius articulates with some of the bones of the wrist.

The bones of the wrist are called **carpals** (Figure 7-21). They are arranged in two rows of four each. In the proximal row from medial to lateral they are the **pisiform** (**PYE**-zih-form), **triquetral** (try-**KWEE**-tral), **lunate** (**LOO**-nate), and **scaphoid** (**SKAFF**-oyd), also known as the **navicular** (nah-**VIK**-you-lahr). In the distal row from medial to lateral are the **hamate**, **capitate** (**KAP**-ih-tate), **trapezoid** (**TRAP**-eh-zoyd), or lesser multiangular, and the **trapezium** (trah-**PEE**-zee-um), or greater multiangular.

The palm of the hand is made up of the five **metacarpal bones** (see Figure 7-21). These are small, long bones, each with a base, a shaft, and a head. They radiate out from the wrist bones like the spokes of a wheel rather than being parallel. They each articulate with a proximal **phalanx** (**FAY**-langks) of a finger. Each finger, except the thumb, has three **phalanges** (fah-**LAN**-jeez): a proximal, a middle, and a terminal, or distal, phalanx. The thumb has only a proximal and distal phalanx.

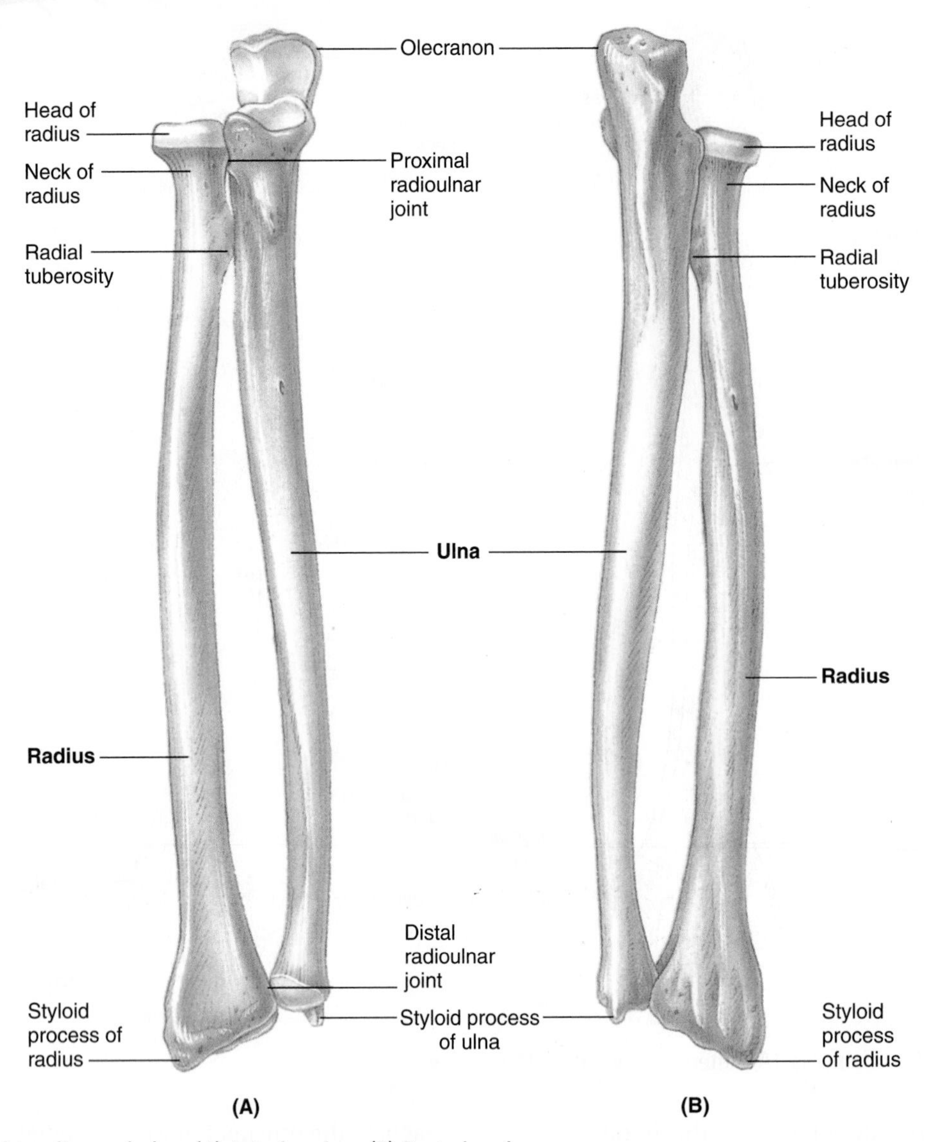

FIGURE 7-20. Right radius and ulna. (A) Anterior view. (B) Posterior view.

The Bones of the Lower Extremities

The bones of the lower extremities include the pelvic girdle, which supports the trunk and provides attachment for the legs. It consists of the paired hip bones or coxal bones. Each hipbone consists of three fused parts: the **ilium** (**ILL**-ee-um), the **ischium** (**ISS**-kee-um), and the **pubis** (**PYOO**-bis). Other bones of the lower extremity include the thigh, the kneecap, the shin, the calf, the ankle bones, the foot, and the toes.

The **pelvic girdle** is actually made up of two hip or coxal bones that articulate with one another anteriorly at the pubic symphysis. Posteriorly, they articulate with the sacrum. This ring of bone is known as the pelvis.

The ilium is the uppermost and largest portion of the hipbone. It forms the expanded prominence of the upper hip or iliac crest. It is usually wider and broader in females and smaller and narrower in males. Its crest is projected into the anterior superior iliac spine and the anterior inferior iliac spine (Figure 7-22). The ischium is the strongest portion of a hipbone and is directed slightly posteriorly. Its curved edge is viewed from the front as the lowermost margin of the pelvis. It has the rounded and thick ischial tuberosity, which you sit on, and thus bears the weight of the body in the sitting position. The pubis is superior and slightly anterior to the ischium. Between the pubis and the ischium is the large

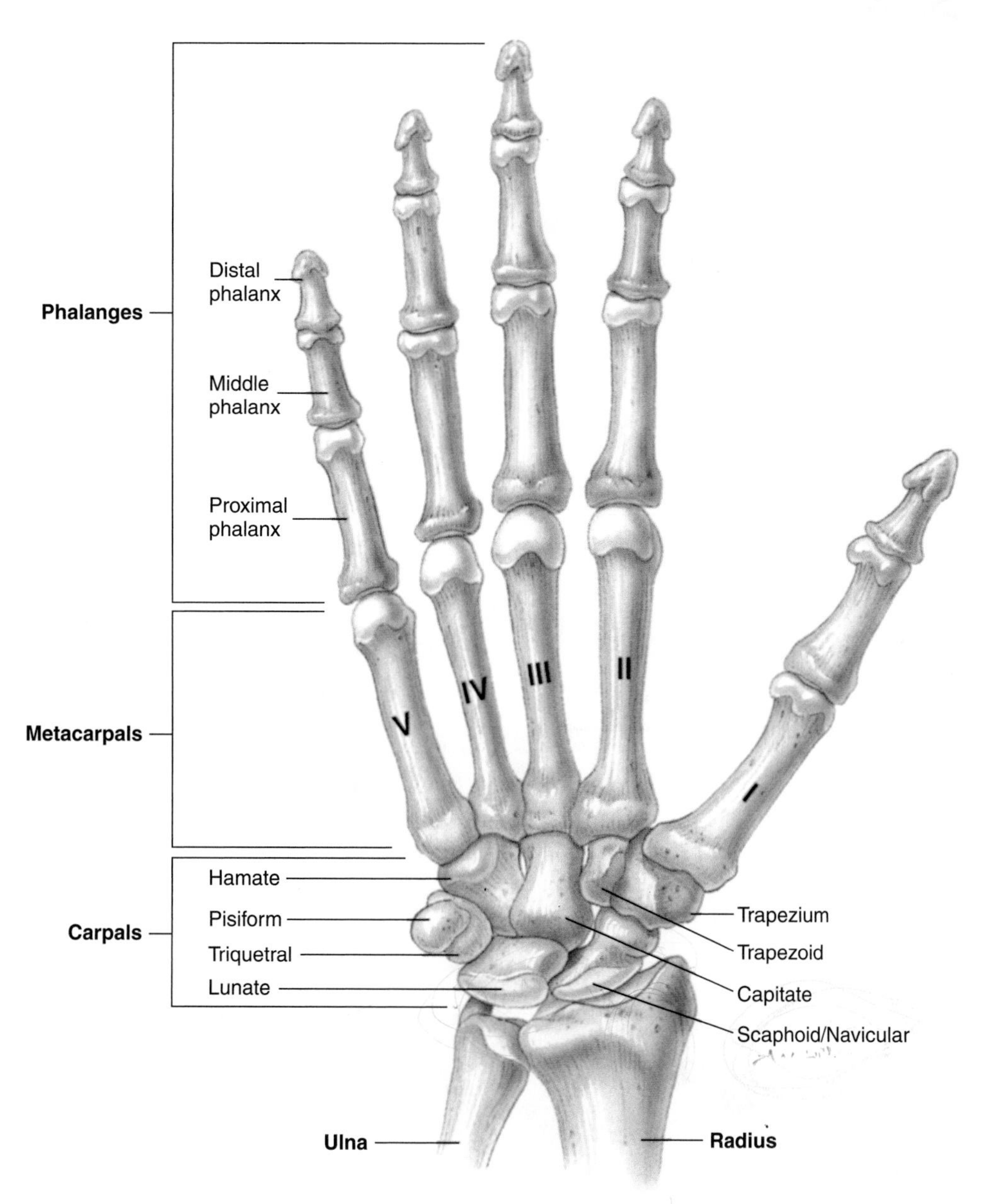

FIGURE 7-21. The bones of the wrist and hand.

obturator (**OB**-tuh-**ray**-tohr) **foramen**. This is the largest foramen in the body and allows for the passage of nerves, blood vessels, and tendons. On the lateral side of the hip just above the obturator foramen is the deep socket called the **acetabulum** (ass-eh-**TAB**-you-lum). All three parts of the pelvic bone meet and unite in this socket. It also receives the head of the femur to help form the hip joint.

The **femur** (**FEE**-mehr), or thigh, is the largest and heaviest bone of the body (Figure 7-23). This single large bone of the upper leg is not in a vertical line with the axis of the erect body. Rather it has a unique engineering design that allows it to bear and distribute the weight of the body. It is positioned at an angle, slanting downward and inward so that the two femurs appear as a large letter V (Figure 7-6). Its upper extremity bears a large head that fits into the acetabulum of the pelvic bone, with an anatomic neck. Its lower portion is widened into a large lateral condyle and an even larger medial condyle. It articulates distally with the tibia.

The **patella** (pah-**TELL**-ah), or kneecap, is the largest of the sesamoid bones. It is somewhat flat and triangular, lying right in front of the knee joint, and is enveloped within the tendon of the quadriceps femoris muscle. Its

Iliac crest
Ilium
Anterior superior iliac spine
Iliac fossa
Anterior inferior iliac spine
Pubis
Ischial spine
Body of pubis
Ischium
Acetabulum
Ischial tuberosity
Obturator foramen
(A)
Right coxal bone, medial view

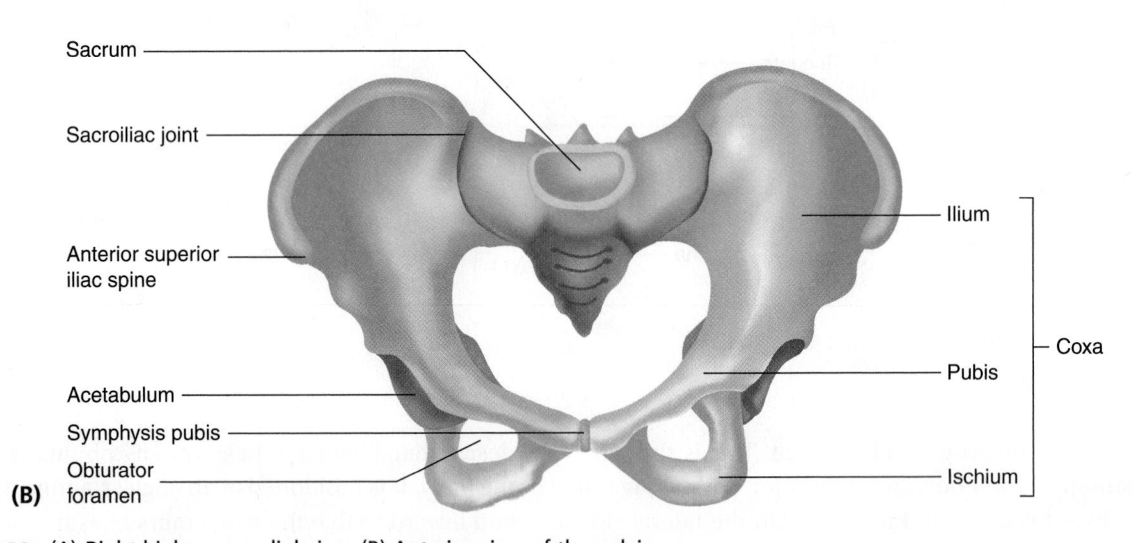

FIGURE 7-22. (A) Right hipbone, medial view. (B) Anterior view of the pelvis.

only articulation is with the femur. It is a movable bone, and it increases the leverage of the muscles that straighten out the knee.

The **tibia** (**TIB**-ee-ah) is the larger of the two bones forming the lower leg (Figure 7-24). It is also known as the shin bone. The rounded condyles of the femur rest on the flat condyle at the proximal end of the tibia.

The **fibula** (**FIB**-you-lah) is also known as the calf-bone. In proportion to its length, it is the most slender bone of the body. It lies parallel with and on the lateral side of the tibia. It does not articulate with the femur but attaches to the proximal end of the tibia via its head.

The bones of the ankle are known as the **tarsal bones** (Figure 7-25). The seven short tarsal bones

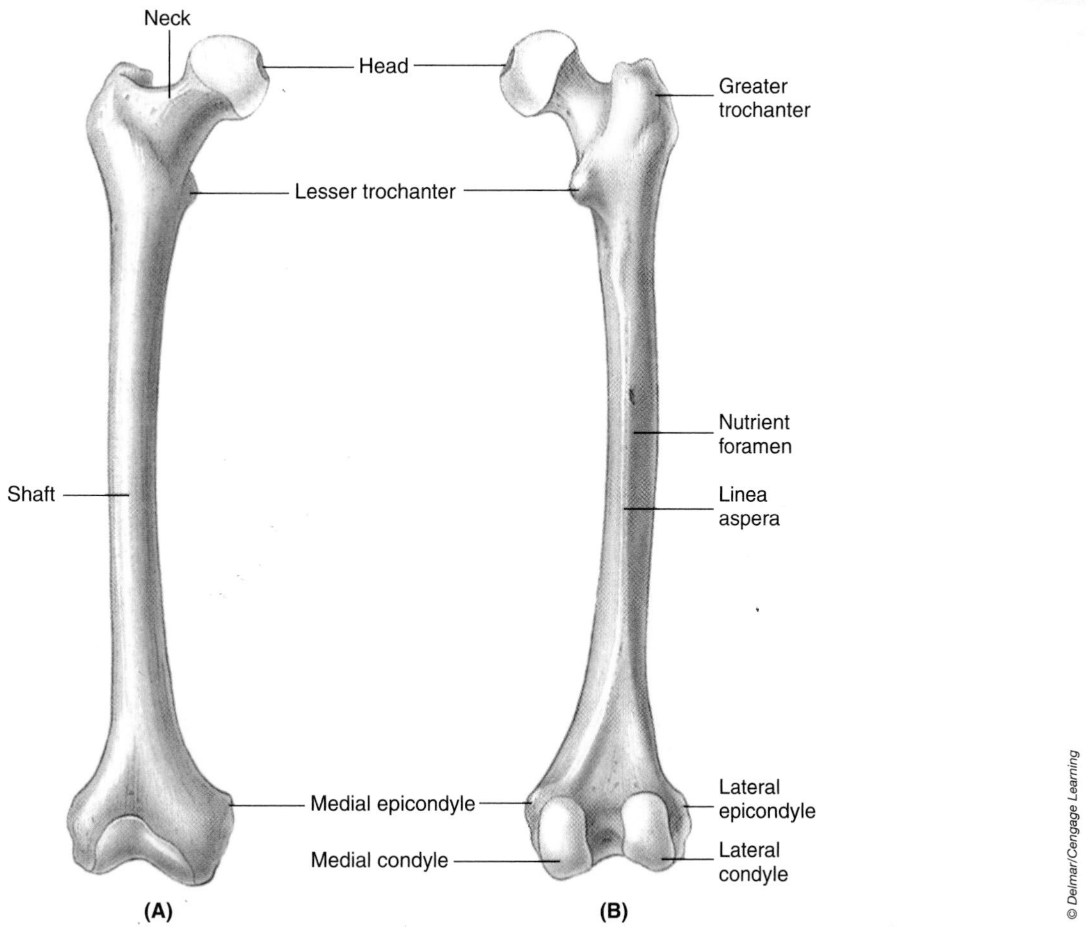

FIGURE 7-23. The femur. (A) Anterior view. (B) Posterior view.

resemble the carpal bones of the wrist but are larger. They are arranged in the hindfoot and forefoot. The tarsal bones of the hindfoot are the **calcaneus** (kal-**KAY**-nee-us), sometimes called the calcaneum, which is the largest of the tarsal bones and forms the heel, the **talus** or ankle bone, the **navicular** (nah-**VIK**-you-lar), and the **cuboid** (**KYOO**-boyd). Because the calcaneus or heel bone receives the weight of the body when walking, it has developed as the largest of the tarsal bones. The tarsal bones of the forefoot are the medial (I), intermediate (II), and lateral (III) **cuneiforms** (kyoo-**NEE**-ih-formz).

The rest of the forefoot bones are the **metatarsals** and phalanges. There are five metatarsal bones in the forefoot. Each is classified as a long bone based on shape and each has a base, shaft, and a head. The heads formed at the distal ends of the metatarsals form what we call the ball of the foot. The bases of the first, second, and third metatarsals articulate with the three cuneiforms; the fourth and fifth metatarsals articulate with the cuboid. The intrinsic muscles of the toes are attached to the shafts of the metatarsals. The first metatarsal is the largest due to its weight-bearing function during walking.

The phalanges of the toes are classified as long bones despite their short length because again they have a base, shaft, and head (see Figure 7-25B). They have the same arrangement as the phalanges of the fingers. There are two phalanges in the great toe, proximal and distal. The proximal one is large due to its weight-bearing function when walking. The other four toes have three each, proximal, middle, and distal phalanges.

THE ARCHES OF THE FOOT

The bones of the foot are arranged in a series of arches that enable the foot to bear weight while standing and to provide leverage while walking. There are two longitudinal arches and one transverse arch. The *medial*

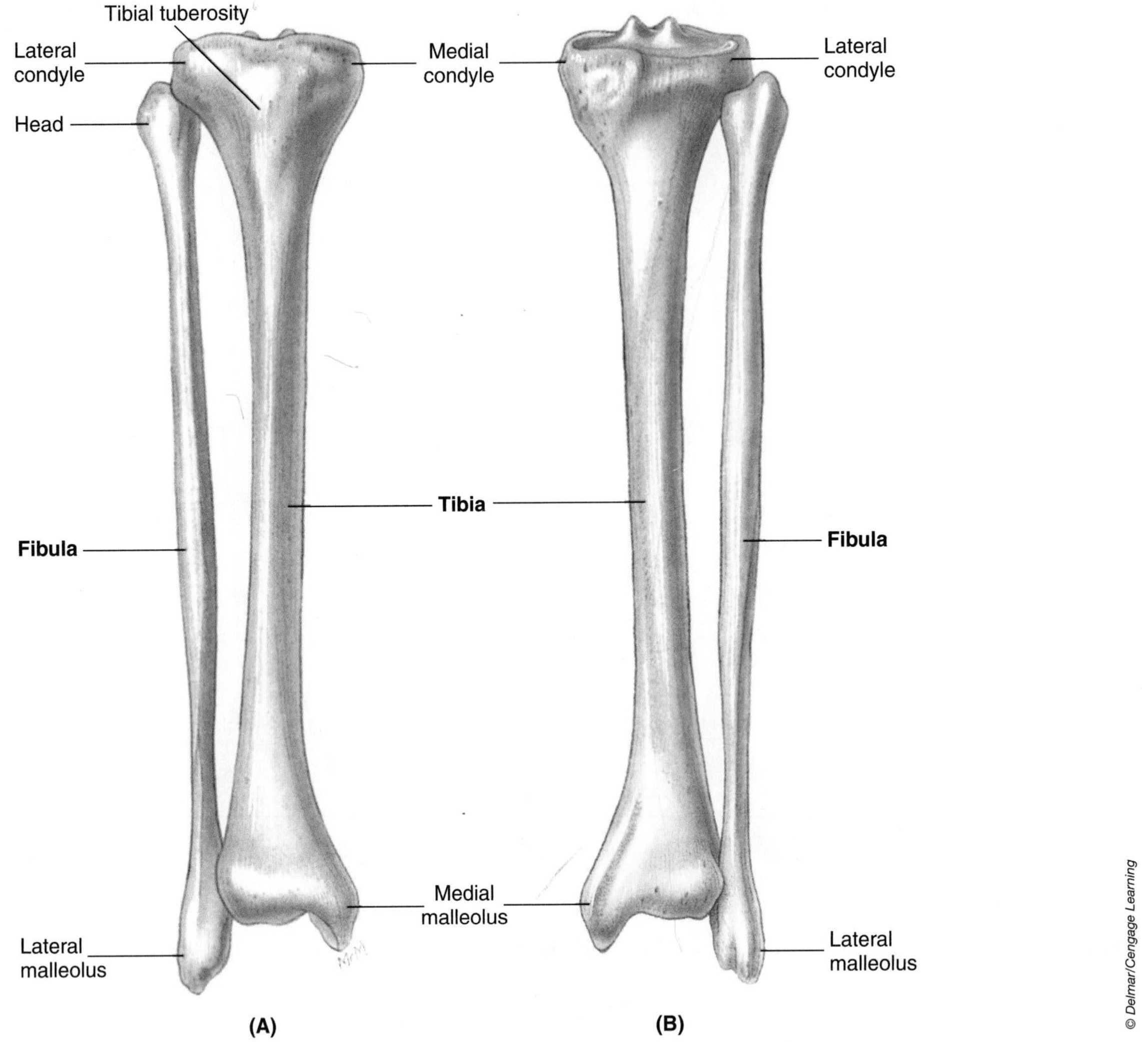

FIGURE 7-24. The tibia and fibula. (A) Anterior view. (B) Posterior view.

longitudinal arch is formed by the calcaneus, talus, navicular, the three cuneiforms, and the three medial metatarsals. This is the highest arch of the foot and can easily be noted. The *lateral longitudinal arch* is much lower and is formed by the calcaneus, the cuboid, and the two lateral metatarsals. The *transverse arch* is perpendicular to the longitudinal arches and is most pronounced at the base of the metatarsals.

The term *pes planus*, or flatfoot, indicates a decreased height of the longitudinal arches. It rarely causes any pain and can be inherited or result from muscle weakness in the foot.

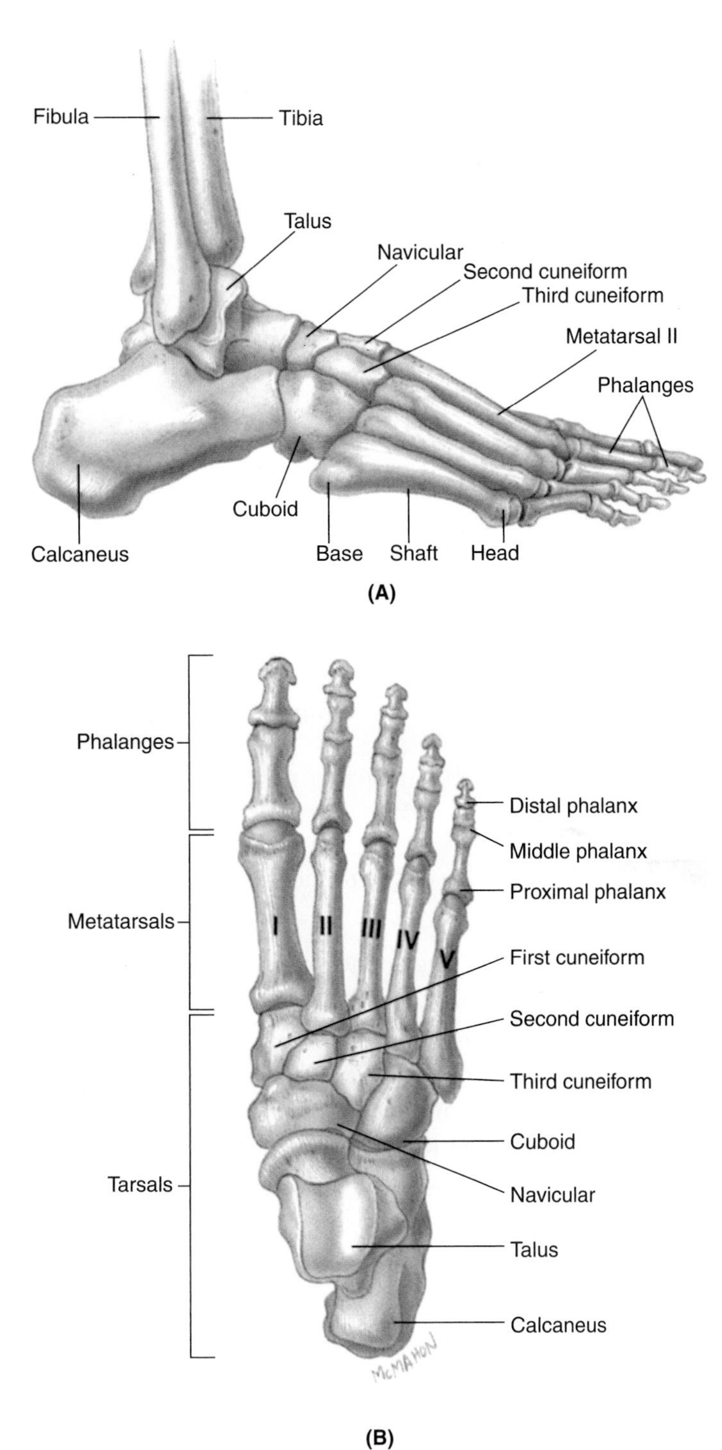

FIGURE 7-25. (A) Right ankle and foot, lateral view. (B) Right ankle and foot, superior view.

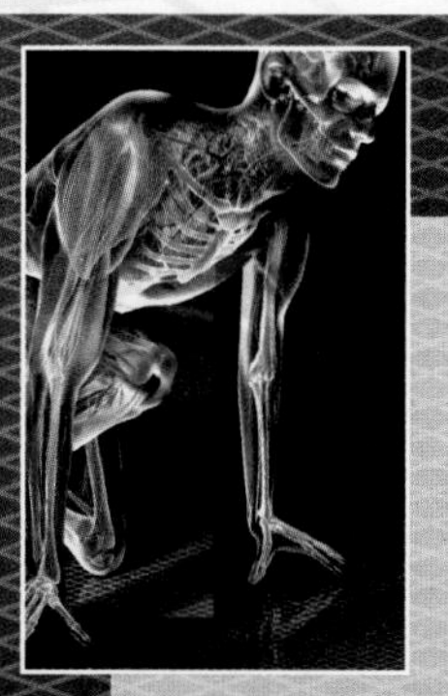

AS THE BODY AGES

As we age, less protein matrix is formed in our bone tissue accompanied by a loss of calcium salts. Bones become more fragile and tend to break more easily in older adults. Older adults also develop stiffness and less flexibility of the skeleton due to a decrease in the protein collagen found in the tendon connective tissue that connects bone to muscle, and in ligaments that connect bone to bone. Hence, as we age we should be more conscious of our diet and include more foods that contain calcium. Regular exercise can also help maintain healthy bone tissues. Walking is an excellent way to exercise both bones and muscles.

Do we really "shrink" as we grow older? Shrinking is caused by a thinning of the intervertebral disks in the spinal column. Starting at around age 40, individuals can lose about one-half inch in height every 20 years due to the loss of disk protein. ■

Career FOCUS

There are many careers available to individuals who are interested in the skeletal system.

- **Athletic trainers** provide guidance to develop muscles and bones for agility, good looks, and sports training.
- **Chiropractors** or **doctors of chiropractic** complete at least two years of premedical studies, followed by four years of study in an approved chiropractic school, learning mechanical manipulation of the spinal column as a method to maintain a healthy nervous system.
- **Prosthetists** are individuals who create artificial limbs.
- **Orthopedists** are physicians specializing in preventing and correcting disorders of the skeleton, joints, and muscles. There are also careers in orthopedic nursing.
- **Orthotists** are individuals who design, make, and fit braces or other orthopedic devices prescribed by a physician.
- **Paramedics** and **emergency medical technicians** can also further train and specialize in the treatment of skeletal system disorders like broken bones and fractures.

BODY SYSTEMS WORKING TOGETHER TO MAINTAIN HOMEOSTASIS: THE SKELETAL SYSTEM

Integumentary System
- Vitamin D is produced in the skin by UV light.
- It enhances the absorption of calcium in bones for bone and tooth formation.

Muscular System
- Through their tendons, muscles pull on bones, bringing about movement.
- Calcium from bones is necessary for muscle contraction to occur.

Nervous System
- The cranial bones protect the brain, and the vertebrae and intervertebral disks protect the spinal cord.
- Receptors for pain monitor trauma to bones.
- Calcium from bones is necessary for nerve transmission.

Endocrine System
- The hormone calcitonin causes calcium to be stored in bones.
- The hormone parathormone causes calcium to be released from bones.
- Growth hormone from the anterior pituitary gland affects bone development.

Cardiovascular System
- Blood cells transport oxygen and nutrients to bone cells and take away carbon dioxide and waste products.
- Calcium from bones is necessary for blood clotting and normal heart functions.

Lymphatic System
- Red bone marrow produces lymphocytes, which function in our immune response.

Digestive System
- Calcium, necessary for bone matrix development, is absorbed in the intestine from our daily food intake.
- Excess calcium can be eliminated via the bowels.

Respiratory System
- Oxygen is brought into the body via the respiratory system and transported by the blood to bone cells for biochemical respiration.
- The ribs along with the intercostal muscles and diaphragm bring about breathing.

Urinary System
- The kidneys help regulate blood calcium levels.
- Excess calcium can also be eliminated via the kidneys.

Reproductive System
- Bones are a source of calcium during breastfeeding.
- The pelvis aids in supporting the uterus and developing fetus during pregnancy in the female.

COMMON DISEASE, DISORDER, OR CONDITION

DISORDERS OF THE SKELETAL SYSTEM

OSTEOPOROSIS

Osteoporosis (oss-tee-oh-poh-**ROH**-sis) is a disorder of the skeletal system characterized by a decrease in bone mass with accompanying increased susceptibility to bone fractures. This results from decreased levels of estrogens that occur after menopause in women and in both men and women in old age. Estrogens help maintain bone tissue by stimulating osteoblasts to form new bone.

Osteoporosis occurs more often in middle-aged and elderly women, but it can also affect teenagers who do not eat a proper balanced diet, people allergic to dairy products, and anyone with eating disorders. The bone mass becomes depleted in such a way that the skeleton cannot withstand everyday mechanical stress. Bone fractures are common, even from normal daily activities. This disease is responsible for height loss, hunched backs, and pain in older individuals. Adequate diet and exercise can prevent osteoporosis.

(continues)

COMMON DISEASE, DISORDER, OR CONDITION

DISORDERS OF THE SKELETAL SYSTEM (continued)

PAGET'S DISEASE

Paget's disease is a common nonmetabolic disease of bone whose cause is unknown. It usually affects middle-aged and elderly individuals. Symptoms include an irregular thickening and softening of the bones. There is excessive bone destruction and unorganized bone repair. Areas of the body affected are the skull, pelvis, and limbs. Treatment includes a high-protein and high-calcium diet with mild but regular exercise.

GIGANTISM

Gigantism (**JYE**-gan-tizm) is the result of excessive endochondral ossification at the epiphyseal plates of long bones. This results in abnormally large limbs, giving the affected individual the appearance of a very tall "giant."

DWARFISM

Dwarfism is the opposite condition of gigantism and results from inadequate ossification occurring at the epiphyseal plates of long bones. This results in an individual being abnormally small. This condition is not to be confused with a genetic dwarf.

SPINA BIFIDA

Spina bifida (**SPY**-nah **BIFF**-ih-dah) is a congenital defect in the development of the posterior vertebral arch in which the laminae do not unite at the midline. It is a relatively common disorder. It may occur with only a small deformed lamina, or it may be associated with the complete absence of laminae, causing the contents of the spinal canal to protrude posteriorly. If the condition does not involve herniation of the meninges or contents of the spinal canal, treatment is not required.

HERNIATED DISK

A **herniated disk** is a rupture of the fibrocartilage surrounding an intervertebral disk that cushions the vertebrae above and below. This produces pressure on spinal nerve roots, causing severe pain and nerve damage. The condition occurs most often in the lumbar region and is also known as a slipped disk. Treatment can include prolonged bed rest to promote healing or surgery to remove the damaged disk.

CLEFT PALATE AND CLEFT LIP

A **cleft palate**, more common in females, occurs when the palatine processes of the maxillary bones do not fuse properly, resulting in an opening between the oral and nasal cavities (Figure 7-26). A person with this condition cannot speak clearly and has difficulty eating or drinking. A child born with this condition can have it surgically repaired and corrected; hence, we rarely see this condition in developed countries. A **cleft lip**, more common in males, occurs when the maxillary bones do not form normally, producing a cleft in the upper lip. Treatment is by surgical repair in infancy. A child could be born with both a cleft palate and a cleft lip.

BLACK EYE

A **black eye** is caused by a blow to the supraorbital ridge, which overlies the frontal sinus. This results in a skin laceration with bleeding. Tissue fluid and blood accumulate in the connective tissue around the eyelid. This swelling, bruising, and discoloration produces a black eye, also called a periorbital ecchymosis or bruise.

DEVIATED SEPTUM

A **deviated septum** develops when the nasal septum shifts to the left (usually) during normal growth. It can be aggravated by a severe blow to the nasal area. Serious deflections can interfere with nasal flow, can cause frequent nose bleeding, or may result in headaches or shortness of breath. Severe deviation can be corrected through surgical procedures.

COMMON DISEASE, DISORDER, OR CONDITION

DISORDERS OF THE SKELETAL SYSTEM (continued)

SINUSITIS

Sinusitis is an inflammation of any one or more of the paranasal sinuses: the frontal, ethmoidal, sphenoidal, and/or the maxillary sinus. This inflammation and swelling of the mucous membrane blocks drainage from the sinuses to the nose resulting in accumulation of drainage materials causing pressure, pain, and headache. It is caused by a number of factors including infections, allergic reactions, and changes in atmospheric pressure as when flying in a plane or underwater swimming. Treatment includes antibiotics, decongestants, analgesics, and surgery to aid drainage in individuals with chronic sinusitis.

WHIPLASH

A **whiplash** injury affects the cervical vertebrae and their associated muscles and ligaments. It is caused by a violent back-and-forth movement of the neck and head as experienced in a rear-end car collision or by athletic injuries. It results in severe pain and stiffness to the neck region and can produce fractures to the spinous process of the cervical vertebrae and/or torn ligaments, tendons, and muscles in this area.

ACROMEGALY

Acromegaly is a chronic condition caused by overactivity of the anterior pituitary gland, resulting in excessive secretions of growth hormone (Figure 7-27). It produces widening and thickening of the bones of the hands, face, jaws, and feet with accompanying tissue enlargement. Developing complications over time include heart disease, hypertension, excess blood sugar, and atherosclerosis (cholesterol containing plague in the arteries). Treatment can include radiation or surgical removal of part of the pituitary gland.

FRACTURED CLAVICLE

A **fractured clavicle** is the most commonly broken bone in the body. It can occur from falling when you use your outstretched arm to soften the fall or from excessive force on the anterior thorax as during an automobile accident when using a shoulder seat belt harness. The most common treatment for a fractured clavicle is the use of a shoulder/arm sling to keep the arm stationary and not allowing it to move outward. This allows the osteoclasts to reabsorb the damaged bone and the osteoblasts to lay down new bone for repair.

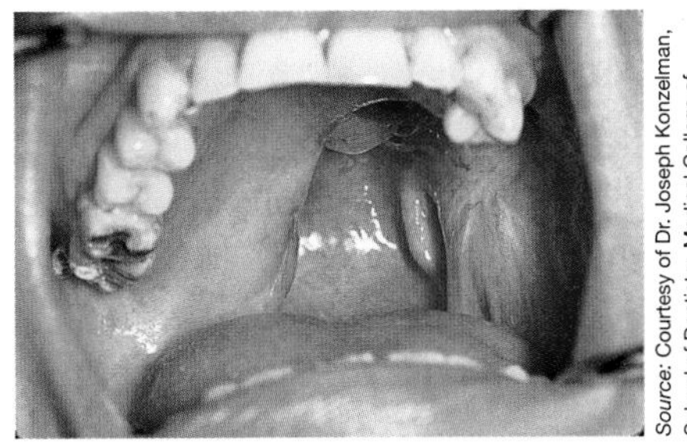

Source: Courtesy of Dr. Joseph Konzelman, School of Dentistry, Medical College of Georgia

FIGURE 7-26. Cleft palate.

FIGURE 7-27. Comparison of a normal individual with an individual with acromegaly.

SUMMARY OUTLINE

INTRODUCTION

1. The skeleton is the supporting structure of the body; it allows muscles to bring about movement and breathing.
2. The solid appearance of bone is due to mineral salts that form the inorganic matrix surrounding the living bone cells.
3. Leonardo da Vinci was the first to correctly illustrate the 206 bones of the body.

THE FUNCTIONS OF THE SKELETAL SYSTEM

The skeleton has five functions:

1. Support surrounding tissues
2. Protect vital organs and soft tissues
3. Provide levers for muscles to pull on
4. Manufacture blood cells in red bone marrow by hematopoiesis
5. Act as a storage area for mineral salts, especially calcium and phosphorus, and fat in yellow marrow

THE GROWTH AND FORMATION OF BONE

1. After 3 months, the fetal skeleton is completely formed and made primarily of hyaline cartilage. Ossification and growth then develop.
2. Longitudinal growth of bone continues until approximately 15 years of age in girls and 16 in boys.
3. Bone maturation continues until 21 years of age in both sexes.

Deposition of the Bone

1. Bone develops from spindle-shaped embryonic bone cells called osteoblasts.
2. Osteoblasts develop into mature bone cells called osteocytes. They form under the fibrovascular membrane covering bone, called the periosteum, and under the membrane lining of the medullary cavity, called the endosteum.
3. The more strain or pressure on a bone, the more the bone will develop.
4. Osteoclasts are large cells that are responsible for the reabsorption of injured bone. They also reabsorb bone during remodeling.

Types of Ossification

The two types of ossification are:

1. Intramembranous ossification: a process in which dense connective membranes are replaced by deposits of inorganic calcium salts. The bones of the cranium form in this way.
2. Endochondral ossification: the process whereby cartilage is the environment in which the bone cells develop. All other bones of the body develop in this way.

Maintaining the Bone

1. The correct amount of calcium stored in the bones, the proper amount of calcium in the blood, and the excretion of excess calcium are controlled by the endocrine system.
2. The parathyroid glands secrete parathormone, which causes calcium to be released into the bloodstream. Another hormone, calcitonin, causes calcium to be stored in the bones.

THE HISTOLOGY OF BONE

There are two types of bone tissue:

1. Compact or dense bone is strong and solid.
2. Cancellous or spongy bone has many open spaces filled with bone marrow.

The Haversian System of Compact Bone

1. An English physician, Clopton Havers (1650–1702), first described the histologic features of compact bone.
2. Haversian canals or osteons are small canals containing blood vessels running parallel to the surface of compact bone and are surrounded by concentric rings of solid bone called lamellae.
3. In these rings of bone are cavities called lacunae; each lacuna contains an osteocyte bathed in fluid.
4. Lacunae are connected to one another and eventually to the osteons by smaller canals called canaliculi.
5. The tissue fluid that circulates in these canals carries nutrients and oxygen to and waste away from the bone cells.

Cancellous Bone

1. Cancellous bone consists of a meshwork of bone called trabeculae.
2. Trabeculae create the spongy appearance of cancellous bone.
3. The spaces between the trabeculae are filled with bone marrow.

Bone Marrow

There are two types of bone marrow:

1. Red bone marrow's function is hematopoiesis, the formation of blood cells.
2. In an adult the ribs, vertebrae, sternum, and pelvis contain red bone marrow in their cancellous tissue.
3. Yellow bone marrow is found in the shafts of long bones within their cancellous tissue.
4. Yellow bone marrow stores fat cells.

THE CLASSIFICATION OF BONES

The bones of the body can be classified, based on shape, into five categories.

1. Long bones consist of a shaft or diaphysis, a flared portion at the end of the diaphysis called a metaphysis, and two extremities called epiphyses. Examples are the clavicle, humerus, radius, ulna, femur, tibia, and fibula as well as the phalanges, metacarpals, and metatarsals.
2. Short bones have a somewhat irregular shape. Examples are the tarsal bones of the foot and the carpal bones of the hand.
3. Flat bones are flat and serve to protect or provide extensive muscle attachment. Examples are some bones of the cranium, the ribs, scapula, and part of the hipbone.
4. Irregular bones have a very peculiar or irregular shape. Examples are the vertebrae and the ossicles of the ear.
5. Sesamoid bones are small rounded bones enclosed in tendon and fascial tissue near joints. One example is the largest sesamoid bone, the patella.

BONE MARKINGS

1. Bones exhibit certain projections called processes. Examples of processes are the spine, condyle, tubercle, trochlea, trochanter, crest, line, head, and neck.
2. Bones also exhibit certain depressions called fossae. Examples of fossae are suture, foramen, meatus or canal, sinus or antrum and sulcus.
3. These markings are functional to help join bones to one another, to provide a surface for muscle attachment, or to serve as a passageway for blood vessels and nerves into and out of the bone.

DIVISIONS OF THE SKELETON

1. The human skeleton has 206 bones.
2. The skeleton can be divided into the axial skeleton (skull, hyoid, vertebrae, ribs, and sternum) and the appendicular skeleton (bones of the upper and lower extremities).

THE AXIAL SKELETON

1. The cranial bones consist of the frontal bone, the two parietal bones, the occipital bone, the two temporal bones, the sphenoid bone, the ethmoid bone, the six auditory ossicles (malleus, incus, stapes in each ear), and the varying wormian or sutural bones.
2. The facial bones consist of the two nasal bones, the two palatine bones, the two maxillary bones (upper jaw), the two zygomatic or malar bones (cheekbones), the two lacrimal bones, the two turbinates or nasal conchae, the single vomer bone, and the single lower jawbone, the mandible.

The Orbits, Nasal Cavities, and Foramina

1. The orbits are the two deep cavities that enclose and protect the eyes. A number of bones of the skull contribute to their formation.
2. The framework of the nose surrounds the two nasal cavities made by a number of bones of the skull.
3. Foramina are passageways for blood vessels and nerves. The largest foramen of the skull is the foramen magnum for passage of the spinal cord.

The Hyoid Bone

1. The hyoid bone does not articulate with any other bones. It is suspended by ligaments from the styloid process of the temporal bone.
2. Its function is to support the tongue.

The Torso or Trunk

1. The sternum, ribs, and vertebrae make up the torso or trunk.
2. A typical vertebra has a number of characteristics: a disk-shaped body, an arch that encloses the spinal foramen, a spinous process and two transverse processes for muscle attachment, and two superior articular processes and two inferior articular processes for articulation with the vertebrae immediately above and below.
3. There are seven cervical vertebrae: the first is called the atlas and the second the axis.

4. There are 12 thoracic vertebrae that articulate with the ribs.
5. There are five lumbar vertebrae, the strongest.
6. The single sacrum is made of five fused sacral vertebrae.
7. The single coccyx or tailbone is made up of four fused coccygeal vertebrae.
8. The sternum or breastbone develops in three parts; it looks like a sword: the manubrium or handle, the gladiolus or body that looks like the blade, and the xiphoid process that resembles the tip of the sword.
9. There are 12 pairs of ribs: the upper seven pairs articulate directly with the sternum through their costal cartilages and are called true ribs; the lower five pairs are called false ribs; because the 11th and 12th pairs have no costal cartilage to articulate indirectly with the sternum like the 8th, 9th, and 10th pairs, they are called floating ribs.

THE APPENDICULAR SKELETON

The Bones of the Upper Extremities

1. The bones of the shoulder girdle are the clavicle or collarbone and the scapula or shoulder blade.
2. The humerus is the bone of the upper arm.
3. The forearm bones are the ulna, the longer of the two bones, with its proximal olecranon process or funny bone of the elbow, and the radius, the shorter bone that articulates with some of the wrist or carpal bones.
4. The carpal bones of the wrist are the pisiform, triquetral, lunate, and scaphoid (in the proximal row); the hamate, capitate, trapezoid or lesser multiangular; and the trapezium (type) or greater multiangular (in the distal row).
5. The bones of the palm of the hand are the five metacarpals.
6. The bones of the fingers are the 14 phalanges in each hand.

The Bones of the Lower Extremities

1. Each hip or pelvic bone consists of three fused bones: the ischium, ilium, and pubis. They form the pelvic girdle. The female ilium is wider than the male's, and we all sit on our ischial tuberosity.
2. The femur or thighbone is the largest bone in the body.
3. The patella or kneecap is the largest of the sesamoid bones; it is wrapped in the tendon of the quadriceps femoris muscle.
4. The tibia or shinbone is the largest bone of the lower leg.
5. The fibula of the lower leg is the most slender bone in the body. It is also known as the calfbone.
6. The tarsal bones of the foot are the calcaneus or heel, the talus or ankle, the navicular, and the three cuneiforms.
7. The metatarsals make up the rest of the foot bones along with the 14 phalanges of the toes.

THE ARCHES OF THE FOOT

1. The foot has three arches: the medial longitudinal arch is the highest, the lateral longitudinal arch, and the transverse arch.
2. Pes planus or flatfoot results from decreased height in the longitudinal arches.

REVIEW QUESTIONS

1. Name five functions of the skeleton.
*2. Why should parents make sure that their young child drinks milk, exercises, and plays in the sunlight on a daily basis?
3. Name the cranial bones.
4. Name the facial bones.
5. Name the carpal bones of the wrist.
6. Name the tarsal bones of the foot.

***Critical Thinking Question**

FILL IN THE BLANK

Fill in the blank with the most appropriate term.

1. The two common types of bone tissue are ____________________ and ____________________.
2. Bone develops from spindle-shaped cells called ____________________ found beneath the periosteum.
3. ____________________ are large cells, present in the cavities of bone, which function in the reabsorption of bone.
4. ____________________ ossification is a process in which dense connective tissue or membranes are replaced by deposits of inorganic calcium.
5. The "replacement" of cartilaginous structures with bone is called ____________________ ossification.
6. A disease of bone in children caused by a deficiency of vitamin D and sunlight is

_______________; in adults it is called _______________.

7. Haversian canals are surrounded by concentric rings of bone, each layer of which is called a _______________; between these are tiny cavities called _______________, each containing an osteocyte.
8. _______________ bone marrow's function is hematopoiesis.
9. _______________ bone marrow consists chiefly of fat cells.
10. The bridge of the nose is made up of the paired _______________ bones.
11. The hard palate of the roof of the mouth is made up of the two _______________ bones.
12. The _______________ bone, found in the axial skeleton, has no articulations with other bones and functions as a support for the tongue.
13. The first cervical vertebra is called the _______________ and supports the head; the second cervical vertebra is called the _______________.
14. The sternum or breastbone develops in three parts: the _______________, the body or _______________, and the _______________.
15. There are 12 pairs of ribs. The upper seven pairs articulate directly with the sternum and are called _______________ ribs; the lower five pairs do not directly join the sternum and are called _______________ ribs.

MATCHING

Place the most appropriate number in the blank provided.

_____ Periosteum	1. Rounded or knuckle-like prominence
_____ Osteomalacia	2. Depression in or on a bone
_____ Epiphysis	3. Canal, tubelike passage
_____ Process	4. Orifice through which vessels and nerves pass
_____ Condyle	5. Fibrovascular membrane covering bone
_____ Fossa	6. Zygomatic bone
_____ Fissure/suture	7. Any marked, bony prominence
_____ Foramen	8. Frontal bone
_____ Meatus	9. Cavity within a bone
_____ Sinus	10. Two extremities of a long bone
_____ Sulcus	11. Furrow or groove
_____ Forehead	12. Alveolus
_____ Cheekbone	13. Narrow ridge of bone
_____ Tooth socket	14. Narrow slit between two bones
	15. Rickets in adults
	16. Temporal bone
	17. Parietal bone

TRUE OR FALSE

T F 1. Cartilage actually turns into bone during ossification.

T F 2. The protein matrix of bone is responsible for its elasticity and the salts deposited in the matrix prevent crushing.

T F 3. The more strain on a bone, the less the bone will develop.

T F 4. It is possible for crooked bones in children to become straight due to the continued process of reabsorption.

T F 5. The proper calcium ion concentration of the blood is controlled and maintained by the parathyroid glands.

T F 6. The foramen magnum is the largest orifice in the skeleton and is found at the base of the parietal bone.

T F 7. Like the bones of the cranium, all the facial bones are united by immovable sutures.

T F 8. If the thoracic vertebrae become excessively curved, a condition known as kyphosis develops.

T F 9. The 11th and 12th ribs have another name, floating ribs, because they do not articulate at all with other parts of the skeleton.

T F 10. The scapula is the bone whose common name is the collarbone.

Search and Explore

- Search the Internet for any one of the following famous people: Andre the Giant, Richard Kiel, or Carel Struycken. Give an oral presentation on the individual you chose and describe their disease.

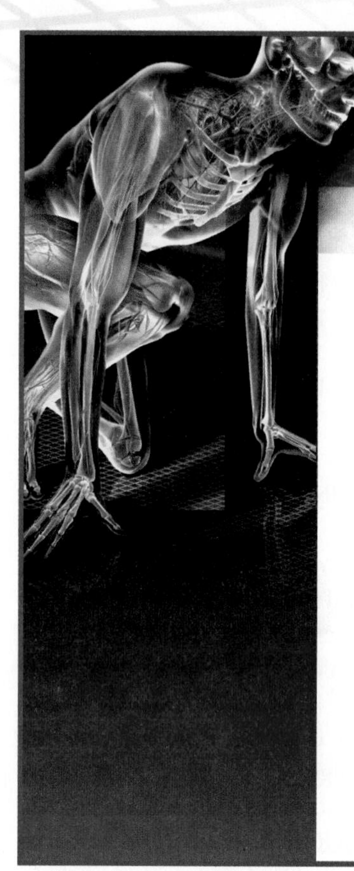

CASE STUDY

Lorette, a 70-year-old woman, is being evaluated by an orthopedist at a local clinic following a recent wrist fracture. A year ago, Lorette fell and experienced a vertebral fracture, which is still creating serious back pain. The health care provider notes that Lorette has a stooped posture and a hunched back. Her medical record reveals a history of small fractures, as well as an abnormal yearly loss of height.

Questions

1. What condition might be responsible for Lorette's medical history and current problems?
2. Which individuals are most likely to develop this condition?
3. What hormone protects women from developing this problem prior to menopause?
4. What is the medical term for hunchback, and what causes it to develop?
5. Why might Lorette have lost height over the years?
6. What measures can individuals take to prevent this skeletal disorder?

*Study*WARE™ Connection

Take a quiz or play a championship game on your StudyWARE™ CD-ROM.

Study Guide Practice

Go to your **Study Guide** for more practice questions, labeling and coloring exercises, and crossword puzzles to help you learn the content in this chapter.

LABORATORY EXERCISE: THE SKELETAL SYSTEM

Materials needed: An articulated human skeleton, either real bone, if possible, or a good plastic reproduction; a number of skulls (one skull per 4–5 learners); disarticulated examples of human bones, an articulated foot, and an articulated hand; a microscope slide of compact bone

1. Break into groups of 4 to 5 learners. Use the colored plates from your textbook and identify the cranial and facial bones with their major sutures by working with the skulls provided by your instructor.
2. Move to the articulated skeleton and identify the other bones of the body.
3. Look at an articulated hand and a foot and identify the carpal bones of the wrist and the tarsal bones of the foot.
4. Examine a hyoid bone and identify its parts.
5. Try to identify various bone markings mentioned in your text.
6. Examine a long bone that has been split open to view compact and cancellous tissue.
7. Review the histology of compact bone by viewing a microscope slide of compact bone. Identify all the parts of the haversian system or osteon.

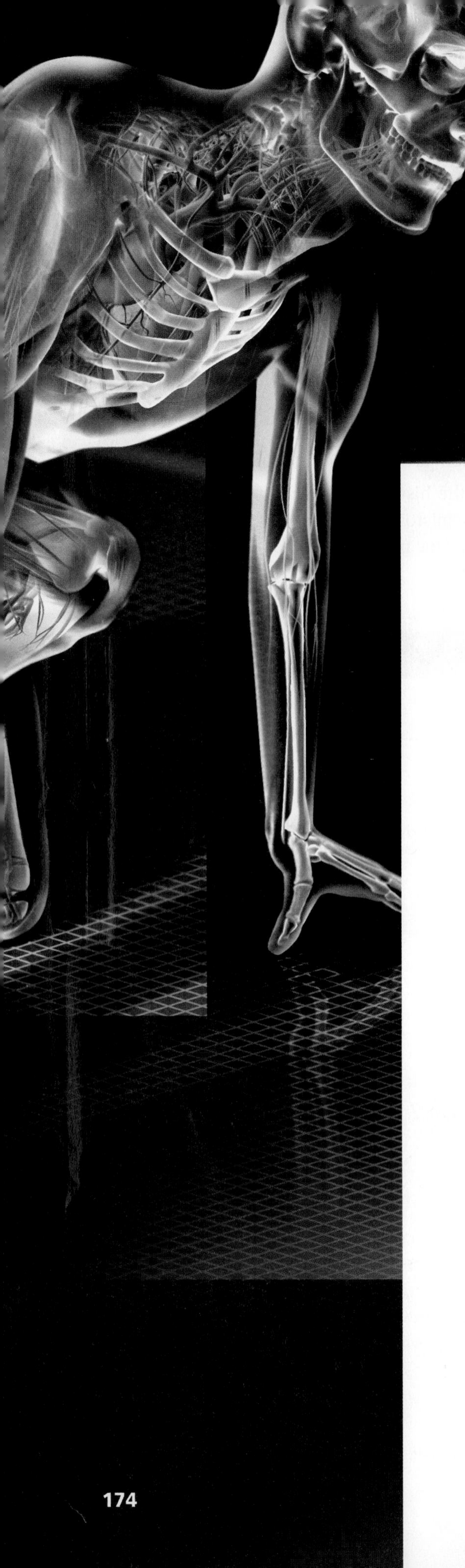

The Articular System

CHAPTER OBJECTIVES

After studying this chapter, you should be able to:

1. Name and describe the three types of joints.
2. Name examples of the two types of synarthroses joints.
3. Name examples of the two types of amphiarthroses joints.
4. Describe and give examples of the six types of diarthroses or synovial joints.
5. Describe the capsular nature of a synovial joint.
6. Describe the three types of bursae.
7. Name some of the disorders of joints.
8. Describe the possible movements at synovial joints.

KEY TERMS

INTRODUCTION

An **articulation** is a place of union or junction between two or more bones, regardless of the degree of movement allowed by this union. The sutures between various bones of the skull are considered as much a part of the articular system as the knee or elbow joint. When we think of a joint, we tend to think of the freely moving joints such as the shoulder or hip joint, but other types of joints have limited or no movement at all occurring at their site.

THE CLASSIFICATION OF JOINTS: STRUCTURE AND FUNCTION

Joints are classified into three major groups according to the degree of movement they allow (function) and the type of material that holds the bones of the joint together (structure).

Synarthroses

Synarthroses (**sin**-ahr-**THRO**-seez) are joints or unions between bones that do not allow movement. *Syn* as a prefix means joined together. There are three examples of synarthroses or immovable joints.

The first type is a **suture** (**SOO**-chur). A suture is an articulation in which the bones are united by a thin layer of fibrous tissue. The suture joints of the skull are examples. Recall from Chapter 7 that the bones of the skull are formed by intramembranous ossification. The fibrous tissue in the suture is the remnant of that process and helps form the suture.

The second example is a **syndesmosis** (sin-dez-**MOH**-sis). Syndesmoses (plural) are joints in which the bones are connected by ligaments between the bones. Examples are where the radius articulates with the ulna and where the fibula articulates with the tibia. These bones move as one when we pronate and supinate the forearm or rotate the lower leg. Some authors consider syndesmosis as an example of an amphiarthrosis (little movement) joint.

The third example is a **gomphosis** (gohm-**FOH**-sis). Gomphoses (plural) are joints in which a conical process fits into a socket and is held in place by ligaments. An example is a tooth in its alveolus (socket), held in place by the periodontal ligament.

Amphiarthroses

Amphiarthroses (**am**-fee-ahr-**THRO**-seez) are joints that allow only slight movement. There are two examples of amphiarthroses.

The first example of an amphiarthrosis is a **symphysis** (**SIM**-fah-sis). Symphyses (plural) are joints in which the bones are connected by a disk of fibrocartilage. An example of a symphysis is the pubic symphysis where the two pelvic bones at the pubis are joined. During delivery this joint allows the pelvic bone slight movement to increase the size of the birth canal.

The second example of an amphiarthrosis is a **synchondrosis** (**sin**-kon-**DRO**-sis). Synchondroses (plural) are joints in which two bony surfaces are connected by hyaline cartilage. The cartilage is replaced by permanent bone later in life. An example of a synchondrosis is the joint between the epiphyses (flared portions) and the diaphysis (shaft) of a long bone. Remember from Chapter 7 that this is the location of the growth plate and where long bones develop longitudinally by endochondral ossification. Some authors consider a synchondrosis as an example of a synarthrosis (no movement). Another example is the hyaline cartilage connection of the ribs to the sternum.

Diarthroses or Synovial Joints

Diarthroses (**dye**-ahr-**THRO**-seez) or **synovial joints** are freely moving joints or articulations (Figure 8-1). They are

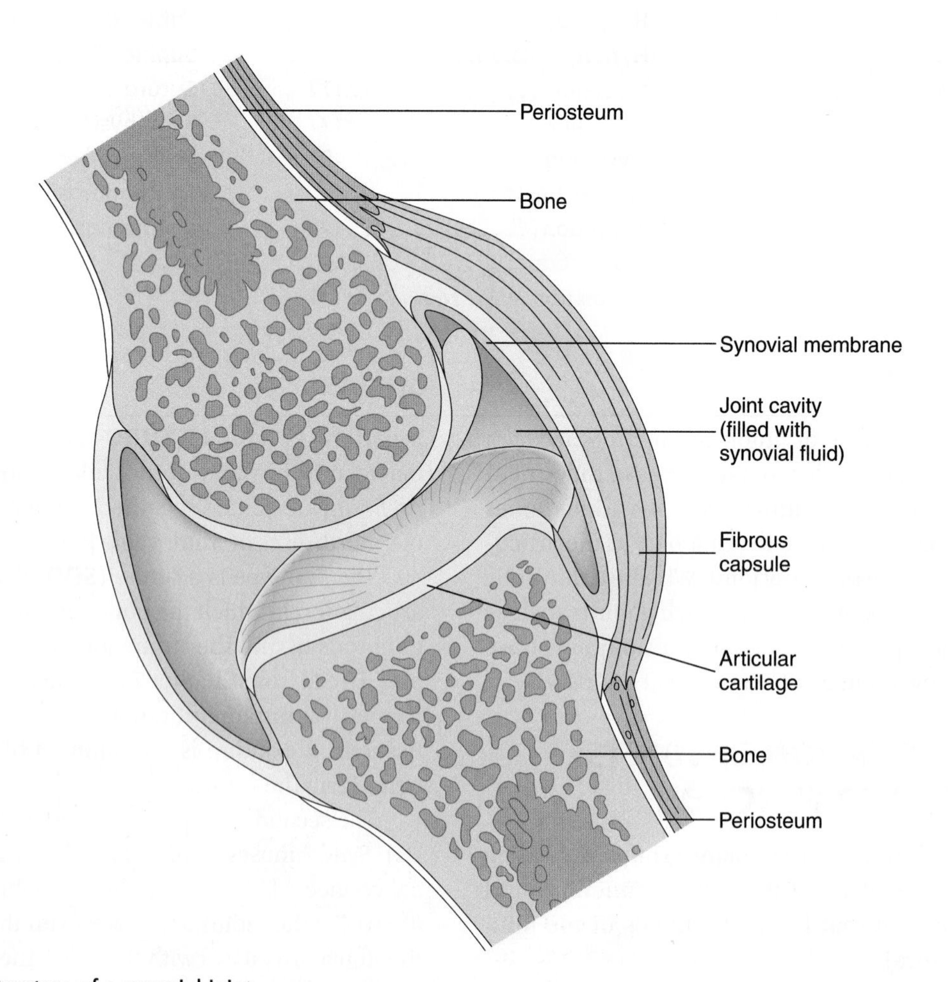

FIGURE 8-1. The structure of a synovial joint.

always characterizied by the presence of a cavity enclosed by a capsule. This cavity may contain various amounts and concentrations of a number of tissues. The cavity may be enclosed by a capsule of fibrous articular cartilage. Ligaments can reinforce the capsule, and cartilage will cover the ends of the opposing bones. This capsule will be lined on the inside with synovial membrane, which produces synovial fluid. Most joints of the upper and lower limbs are diarthroses.

The articular cartilage in the joint provides a smooth, gliding surface for opposing bone. This is made possible because of the lubrication caused by the synovial fluid. The opposing bones do not wear or erode over time due to the constant friction caused by movement at the joint. Articular cartilage has a limited blood supply. It receives its nourishment from the synovial fluid and from a small number of subsynovial blood vessels at the junction of the cartilage and the joint capsule. Synovial fluid has two functions: creating a smooth gliding surface for opposing bones and nourishing the articular cartilage. Cartilage also functions as a buffer between the vertebrae in the spinal column to minimize the forces of weight and shock from running, walking, or jumping.

Collagenous fibers connecting one bone to another in the synovial joint form the capsule enclosing the joint. The range of motion of the joint is related to the laxity or looseness of the joint. This is directly related to the structure of the capsule and how it is formed over the opposing bones. In the shoulder joint, which has the greatest range of movement, the capsule is loose enough to permit the head of the humerus to be drawn away from the glenoid fossa of the scapula. However, in the hip joint the range of motion is much more restricted, because the capsule is thicker and shorter and the head of the femur sits deeply in the acetabulum of the pelvic bone. The femur is also connected to the acetabulum by a series of strong ligaments. This structure is necessary because of the need for greater strength in this joint.

In addition to the above tissues that make up the capsule, muscles and their tendons can also be found as the outermost layer of the capsule. They provide an important mechanism for maintaining the stability of a diarthrosis or synovial joint. They have advantages over ligaments because during both relaxation and contraction they maintain the joint surfaces in firm contact at every position of the joint.

In summary, synovial joints have a number of functions. First, they bear weight and allow movement; second, their construction in the form of a capsule made of ligaments, tendons, muscles, and articular cartilage provides stability; and third, synovial fluid lubricates the joint and nourishes the cartilage.

MOVEMENTS AT SYNOVIAL JOINTS

The following movements can occur at diarthroses or synovial joints.

Flexion (**FLEK**-shun) is the act of bending or decreasing the angle between bones.

Extension (eks-**TEN**-shun) is the act of increasing the angle between bones and is the opposite of flexion. Refer to Figure 8-2A for flexion/extension and hyperextension.

Hyperextension increases the joint angle beyond the anatomic position.

Abduction (ab-**DUCK**-shun) is moving the bones or limb away from the midline of the body while the opposite is **adduction** (add-**DUCK**-shun), which is moving the bone or limb toward the midline of the body (see Figure 8-2B).

Rotation (row-**TAY**-shun) is the act of moving the bone around a central axis; the plane of rotational motion is perpendicular to the axis, as when rotating our head.

Circumduction (sir-kum-**DUCK**-shun) is moving the bone in such a way that the end of the bone or limb describes a circle in the air and the sides of the bone describe a cone in the air (see Figure 8-2C).

Supination (soo-pin-**NAY**-shun) and **pronation** (proh-**NAY**-shun) refer to the movement of the forearm and hand (Figure 8-3A). Supination is moving the bones of the forearm so that the radius and ulna are parallel. If the arm is at the side of the body, the palm is moved from a posterior to an anterior position; if the arm is extended, the palm faces up as in carrying a bowl of soup. Pronation is moving the bones of the forearm so that the radius and ulna are not parallel. If the arm is at the side of the body, the palm is moved from an anterior to a posterior position; if the arm is extended, the palm faces down.

Eversion (ee-**VER**-zhun) and **inversion** (in-**VER**-zhun) refer to movements of the foot (see Figure 8-3B). Eversion is moving the sole of the foot outward at the ankle while inversion is moving the sole of the foot inward at the ankle.

Protraction (pro-**TRACK**-shun) is moving a part of the body forward on a plane parallel to the ground.

Retraction (rih-**TRACK**-shun) is moving a part of the body backward on a plane parallel to the ground. Refer to Figure 8-3C for protraction and retraction of the lower jaw.

Elevation is raising a part of the body; **depression** is lowering a part of the body. Refer to Figure 8-3D for elevation and depression of the shoulder.

Opposition is movement that occurs only with the thumb and is unique to primates. It occurs when the tip of the thumb and the fingers are brought together. The action allows us to use tools as when writing with a pen.

Reposition occurs when the digits return to their normal positions.

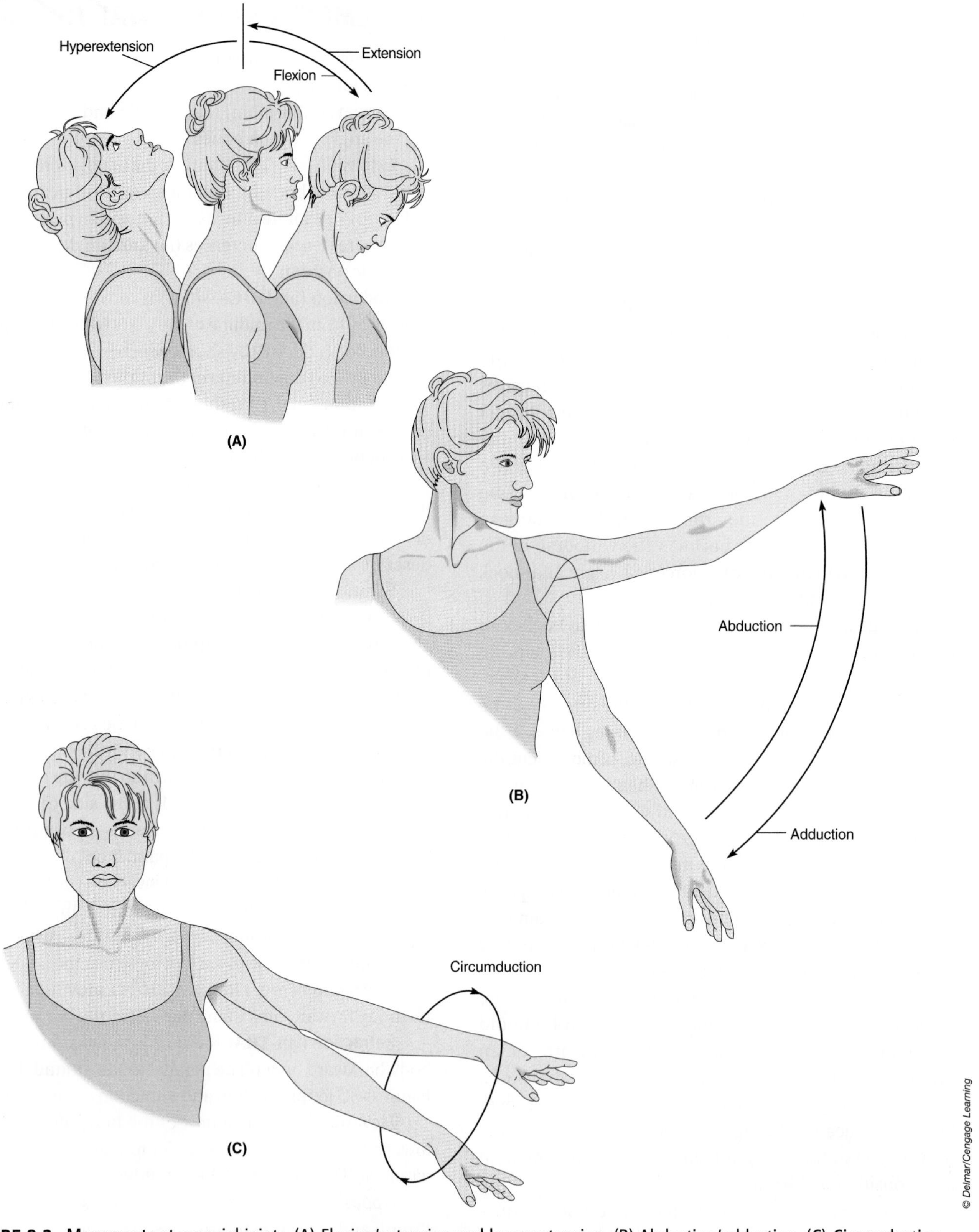

Hyperextension
Extension
Flexion
(A)
Abduction
(B)
Adduction
Circumduction
(C)

FIGURE 8-2. Movements at synovial joints. (A) Flexion/extension and hyperextension. (B) Abduction/adduction. (C) Circumduction.

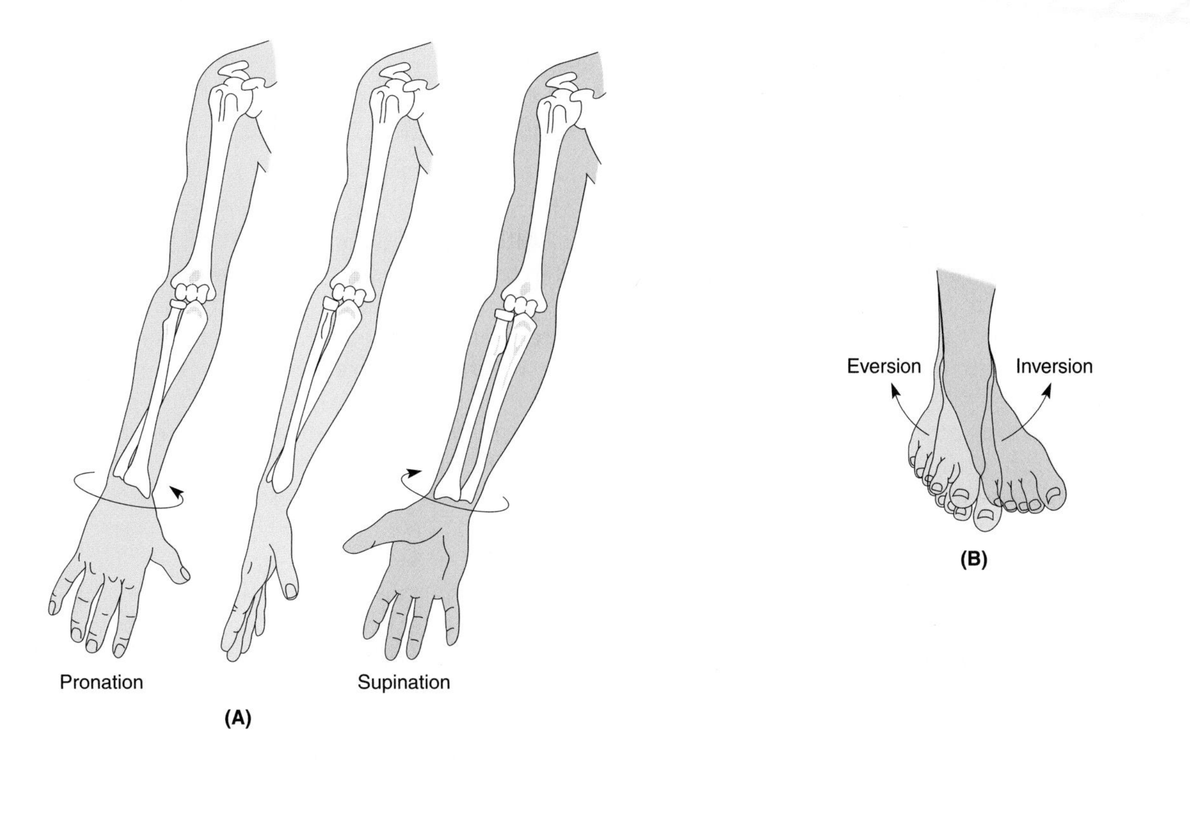

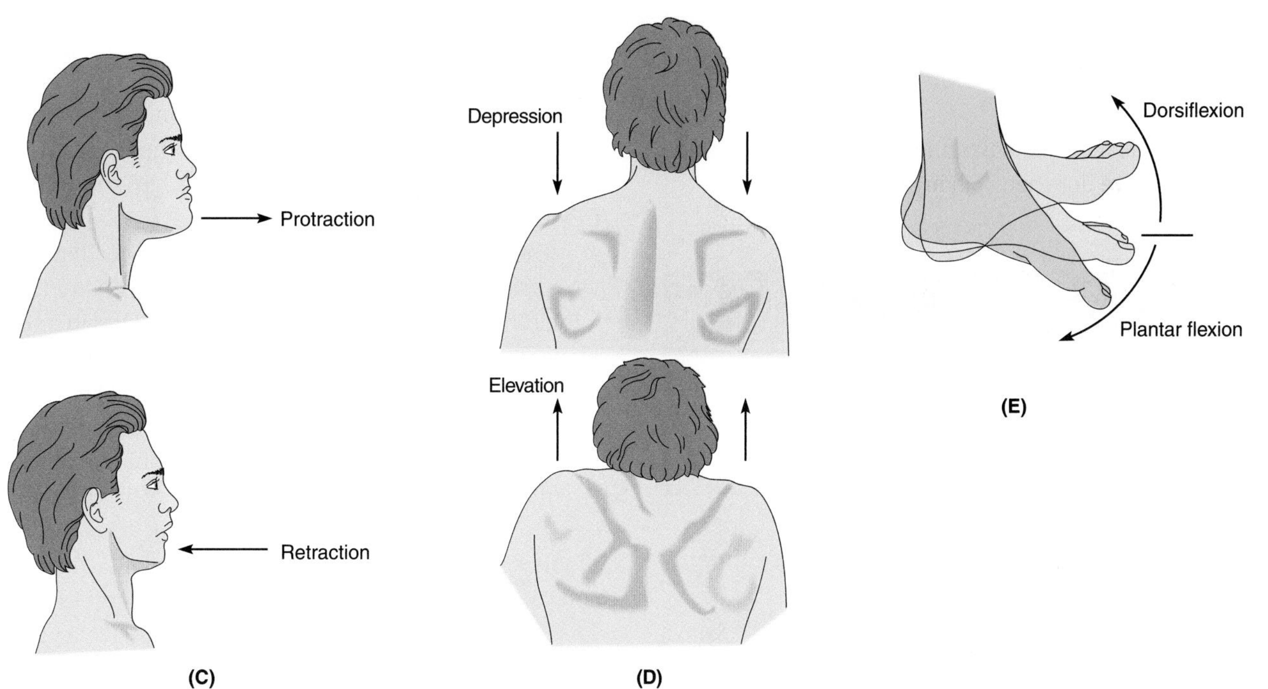

FIGURE 8-3. Movements at synovial joints. (A) Pronation/supination. (B) Eversion/inversion. (C) Protraction/retraction. (D) Depression/elevation. (E) Dorsiflexion/plantar flexion.

HEALTH ALERT

HEALTHY JOINTS

When we think of our joints, it is the freely moving diathroses or synovial joints that come to mind, such as the shoulder, elbow, hip, and knee joints. Although the construction of these joints permits a wide range of movements, consistent or excessive movements can cause injuries to them. These injuries are referred to as repetitive motion injuries and can affect the associated structure of the joint such as muscles, nerves, tendons, ligaments, and the bursae.

Injuries that can develop quickly due to excessive mechanical stress include "tennis elbow" or "canoeist elbow." Tennis athletes in competition frequently develop this type of overuse injury. Those of us who insist on paddling down a river in a canoe for a weekend, 8 hours a day, frequently develop canoeist elbow. Water skiers who consistently ski barefoot can, over time, cause major damage to the meniscus cartilage in the knee. These injuries are acute and can be temporary if the athlete gives the elbow or knee time to recuperate and return to normal motions.

Other types of injuries that develop over a long period are other repetitive motion injuries. In our technological age of computers, carpal tunnel syndrome, affecting the wrist, develops in individuals who regularly use the keyboard for long periods. Early symptoms include mild discomfort in the joint, tingling sensations, and muscle fatigue. If caught early and treated, this syndrome can be prevented. To maintain healthy joints, moderate exercise and movement is essential to maintain joint stability and lubrication. If, however, an occupation requires repetitive motion, the joint should be given frequent rest and good body posture and positioning should be maintained. This can help relieve stress on a constantly working joint. ■

Dorsiflexion is raising the foot up at the ankle joint and **plantar flexion** is pushing the foot down at the ankle joint, actions we do when walking (see Figure 8-3E).

*Study***WARE**™ Connection

Play an interactive game labeling movements of synovial joints on your StudyWARE™ CD-ROM.

THE SIX TYPES OF DIARTHROSES OR SYNOVIAL JOINTS

There are six types of freely moving or synovial joints. Refer to Figure 8-4 for the geometric structure and examples of these joints that permit certain types of movements.

A **ball-and-socket joint** is an example of a multiaxial joint. In this type of joint, a ball-shaped head fits into a concave socket. Two examples are the ball-shaped head of the femur fitting into the concave socket of the acetabulum of the pelvic bone and the head of the humerus fitting into the glenoid fossa of the scapula. This type of joint provides the widest range of motion. Movement can occur in all planes and directions. Of the two ball-and-socket joints, the hip and the shoulder, the shoulder has the widest range of movement.

The **hinge joint** is structured in such a way that a convex surface fits into a concave surface. In this type of a joint, motion is limited to flexion and extension in a single plane. Two examples are the elbow and knee joint. Because motion is restricted to one plane these joints are also called uniaxial hinge joints. Refer to Figure 8-5 to see the structure of the uniaxial knee joint. Other uniaxial hinge joints are the middle and distal phalanges of the fingers and toes.

The **pivot joint** is another uniaxial joint because motion is limited to rotation in a single plane. The joint is constructed in such a way that a pivot-like process rotates within a bony fossa around a longitudinal axis. One example is the joint between the atlas vertebra (the pivot process) that rotates within the bony fossa of the axis vertebra.

The **condyloid** (**KON**-dih-loyd) **joint**, sometimes called an ellipsoidal joint, is a biaxial joint that consists of an oval-shaped condyle that fits into an elliptical cavity. Motion is possible in two planes at right angles to each other. The wrist joint between the radius of the forearm and some of the carpal bones of the wrist is a condyloid joint. The hand can be flexed and extended in one plane

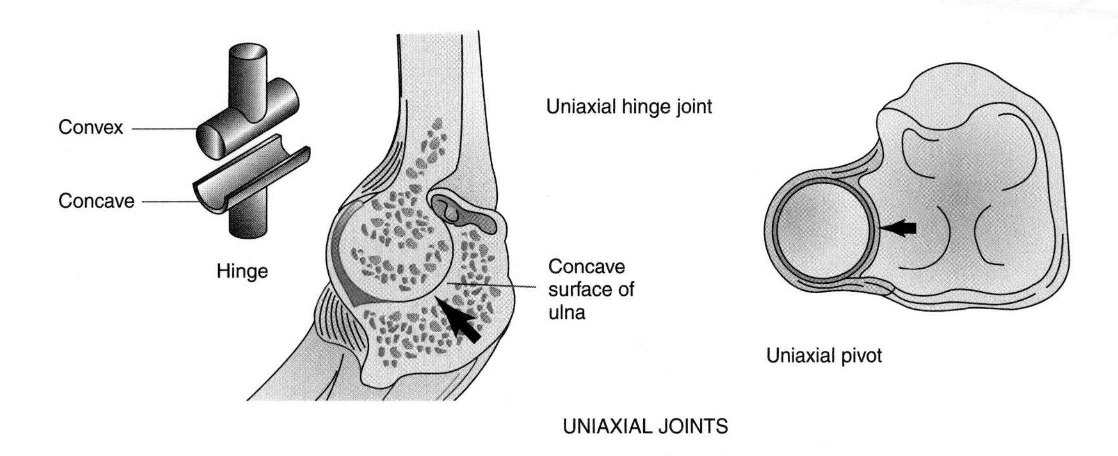

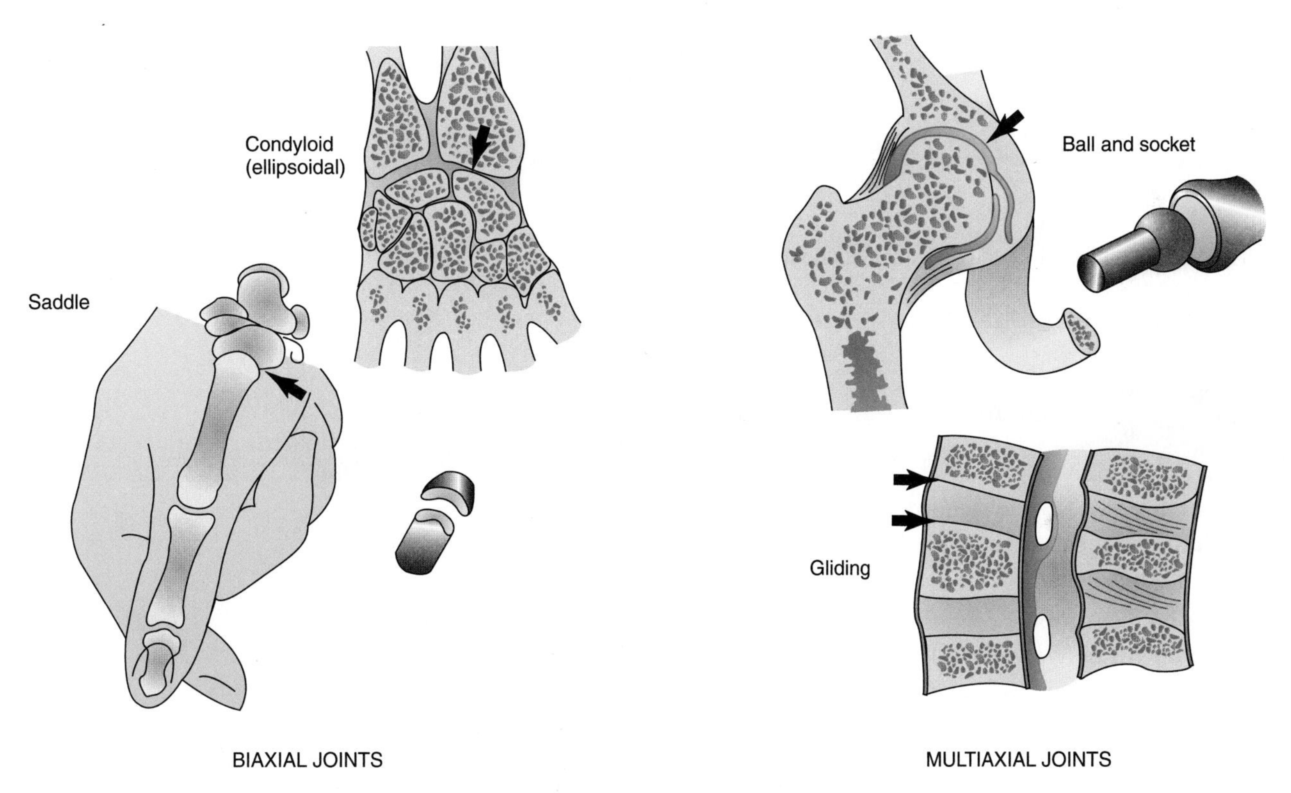

FIGURE 8-4. The six types of freely moving diarthroses or synovial joints.

like raising your hand in a sign to stop and returning it to a downward position. It can also be abducted and adducted like waving good-bye when moving the hand from side to side.

The **saddle joint**, another biaxial joint, is a bit more complex in its structure. In this type of a joint, one articular surface is concave in one direction and convex in the other (the trapezium, a carpal bone of the wrist), while the other articular surface is reciprocally convex and concave (the metacarpal bone in the thumb). Thus, the two bones fit together. Refer to Figure 8-4 to study its construction. Movement is possible in two planes at right angles to each other: flexion and extension plus abduction and adduction. This construction also permits opposition of the thumb, an evolutionary advancement allowing phenomenal dexterity of the hand to grasp and use tools.

The **gliding joint** is the last type of synovial joint and is a multiaxial joint. This type of joint is formed by either opposing plane surfaces or slightly convex and concave surfaces. This type of joint only allows gliding movement. Examples of gliding joints are those between the superior and inferior articular processes of the vertebrae in the spine.

Table 8-1 shows the classification of the three types of joints, including examples of each.

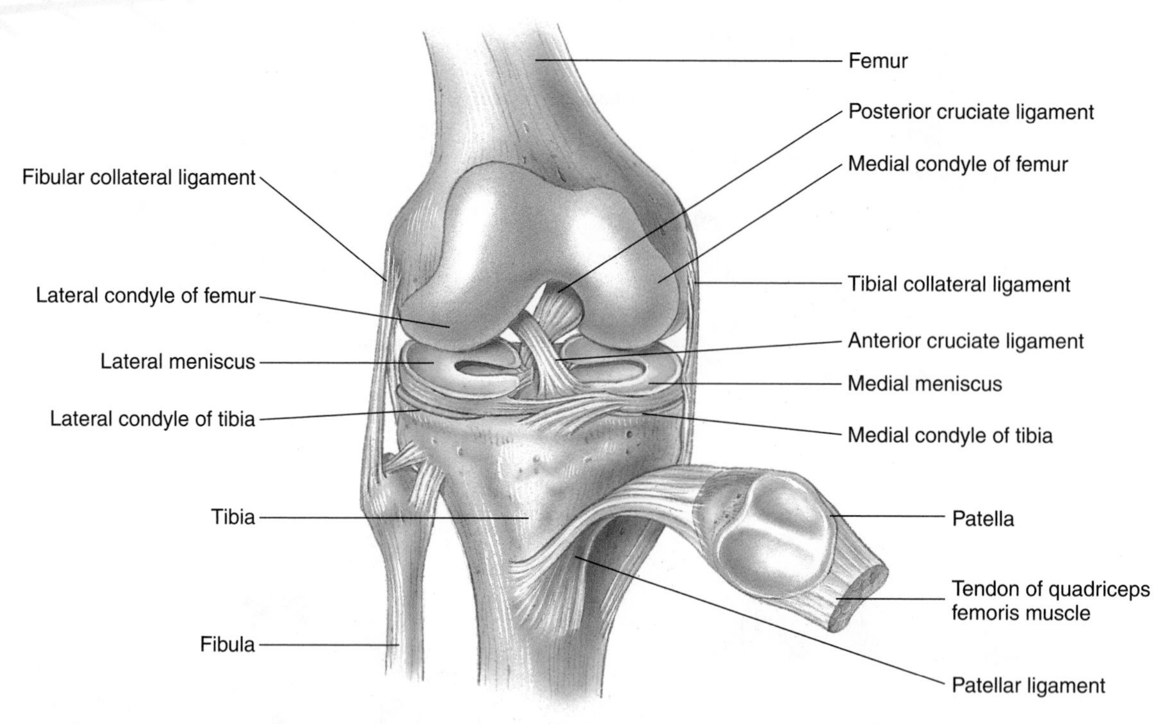

FIGURE 8-5. The structure of the uniaxial hinge joint of the knee.

Table 8-1 Classification of the Three Types of Joints with Examples

Type of Joint	Examples
1. Synarthroses: No movement possible	
a. Suture (bones united by a thin layer of fibrous tissue)	Sutures of the skull
b. Syndesmosis (bones connected by ligaments between bones)	Borders of the radius and ulna; tibia and fibula articulations
c. Gomphosis (joints in which a conical process fits into a socket held in place by ligaments)	Tooth in its alveolus
2. Amphiarthroses: Slightly movable articulations	
a. Symphysis (bones connected by a disk of fibrocartilage)	Pubic symphysis
b. Synchondrosis (two bony surfaces connected by cartilage)	Joint between the epiphyses (flared portion) and the diaphysis (shaft) of a long bone; pubic synchondroses; sternocostal synchondrosis; sternal synchondrosis
3. Diarthroses or synovial: Freely moving	
a. Ball-and-socket	Shoulder, hip
b. Hinge	Knee, elbow, fingers, toes
c. Pivot	Neck
d. Condyloid	Wrist joint between the radius and carpal bones
e. Saddle	Carpal-metacarpal articulation in the thumb
f. Gliding	Intervertebral joints

*Study*WARE™ Connection

Watch an animation about synovial joints on your StudyWARE™ CD-ROM.

BURSAE

Bursae (burr-**SEE**) are closed sacs with a synovial membrane lining. Bursae can be found in the spaces of connective tissue between tendons, ligaments, and bones. Bursae are found wherever friction could develop during movement. They facilitate the gliding of either muscle over muscle or tendons over bony ligamentous surfaces. Bursae are classified into three types based on where they are found.

Subcutaneous bursae are found under (sub) the skin (cutaneous) wherever the skin is on top of an underlying bony process (e.g., the knee joint). Between the patella or kneecap and its overlying skin is a subcutaneous bursa preventing friction between bone and skin. See Figure 8-6 for bursae of the knee joint.

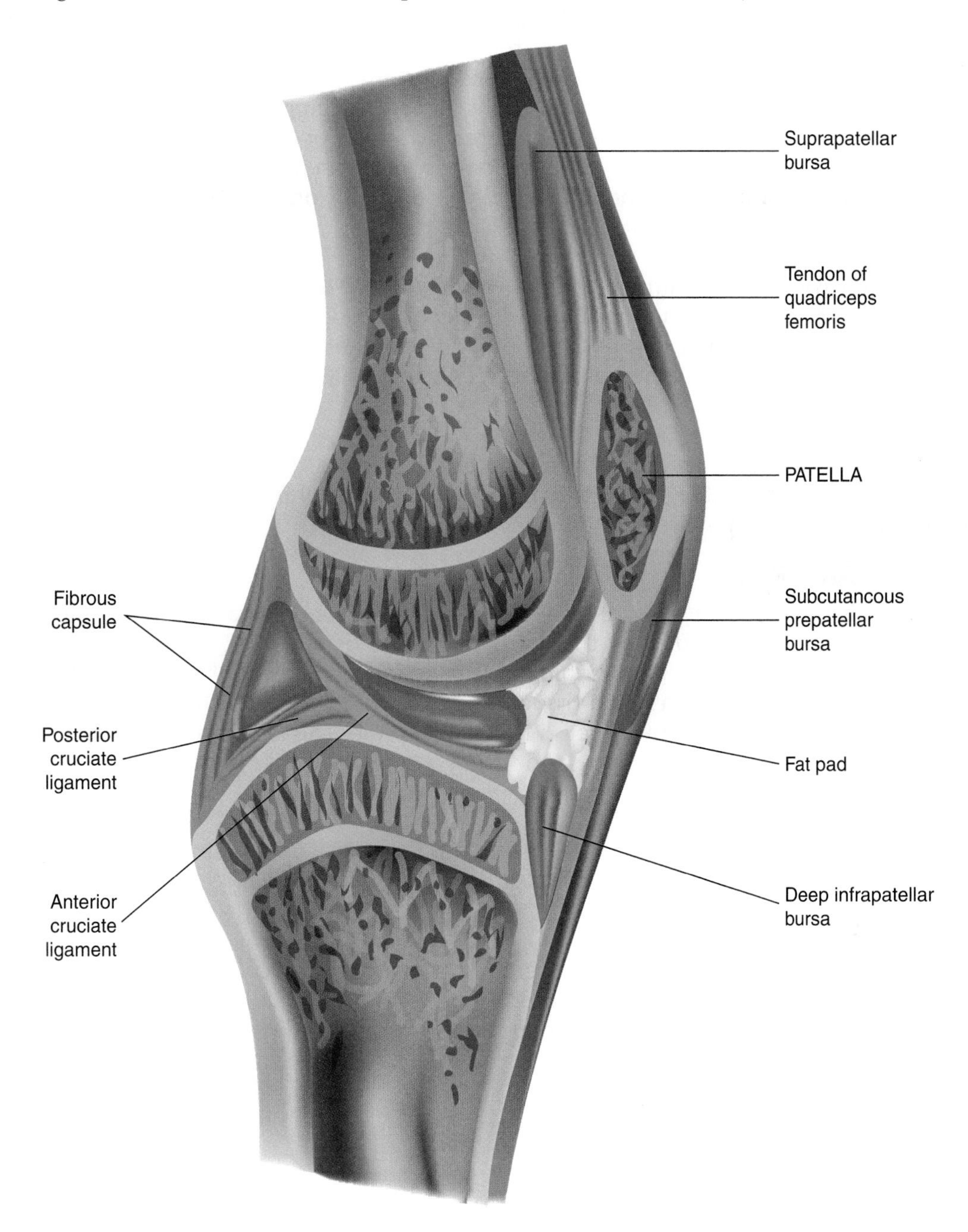

FIGURE 8-6. Lateral view of the knee joint showing bursae.

Subfascial (sub-**FASH**-ee-al) **bursae** are located between muscles. They are found above the **fascia** (**FASH**-ee-ah) of one muscle and below the fascia of another. The fascia is the fibrous connective tissue that covers the epimysium of a muscle bundle. We will discuss fascia in Chapter 9.

Subtendinous bursae are found where one tendon overlies another tendon or where one tendon overlies some bony projection, as in the shoulder.

COMMON DISEASE, DISORDER, OR CONDITION

DISORDERS OF JOINTS

BURSITIS

Bursitis (burr-**SIGH**-tis) is an inflammation of the synovial bursa that can be caused by excessive stress or tension placed on the bursa. Playing tennis for long periods of time causes tennis elbow. It is an example of bursitis in the elbow joint caused by excessive stress. You may experience canoeist elbow if you go canoeing and paddle for long hours. This is, of course, temporary. The elbow and the shoulder are common sites of bursitis. It can also be caused by a local or systemic inflammatory process. If bursitis persists, as in chronic bursitis, the muscles in the joint can eventually degenerate or atrophy and the joint can become stiff even though the joint itself is not diseased.

ARTHRITIS

Arthritis (ahr-**THRY**-tis) is an inflammation of the whole joint. It usually involves all the tissues of the joint: cartilage, bone, muscles, tendons, ligaments, nerves, blood supply, and so on. There are well over 100 varieties of arthritis, and 10% of the population experiences this disorder, which has no cure. Pain relief is common through analgesics but these only relieve a symptom of arthritis, the pain.

RHEUMATIC FEVER

Rheumatic fever is a disease involving a mild bacterial infection. If undetected in childhood, the bacterium can be carried by the bloodstream to the joints, resulting in possible development of rheumatoid arthritis later on in life.

RHEUMATOID ARTHRITIS

Rheumatoid arthritis is a connective tissue disorder resulting in severe inflammation of small joints. It is severely debilitating and can destroy the joints of the hands and feet. The cause is unknown. A genetic factor may be involved, or an autoimmune reaction may be involved in which an immune reaction develops against a person's own tissues. The synovial membranes of the joints and connective tissues grow abnormally to form a layer in the joint capsule. This layer grows into the articulating surfaces of the bones, destroying cartilage and fusing the bones of the joint.

PRIMARY FIBROSITIS

Primary fibrositis is an inflammation of the fibrous connective tissue in a joint. It is commonly called rheumatism by the layman. If it is in the lower back, it is commonly called lumbago.

OSTEOARTHRITIS

Osteoarthritis, sometimes referred to as degenerative joint disease, occurs with advancing age especially in people in their 70s. It is more common in overweight individuals and affects the weight-bearing joints. Mild exercising can prevent joint deterioration and increases the ability to maintain movement at joints.

(continues)

COMMON DISEASE, DISORDER, OR CONDITION

DISORDERS OF JOINTS

GOUT

Gout (GOWT) is an accumulation of uric acid crystals in the joint at the base of the large toe and other joints of the feet and legs. It is more common in men than in women. These waste product crystals can also accumulate in the kidneys, causing kidney damage.

SPRAIN

A sprain occurs when a twisting or turning action tears ligaments associated with a joint. The most common sites for sprains are the ankle and wrist.

SLIPPED DISK

A slipped disk, also referred to as a ruptured or herniated disk, develops when the fibro-cartilagenous intervertebral disk protrudes or moves out of place and puts pressure on the spinal cord. It can occur anywhere, but the most common areas are the lumbar and sacral regions of the spine. This results in severe pain. If a slipped disk does not respond to physical therapy and medications, a laminectomy, the surgical removal of the protruding disk, may have to be performed.

DISLOCATION

A dislocation is a temporary displaced bone from a joint caused by excessive strain on the joint. Dislocations occur most frequently in the shoulder and hip but also can occur in the fingers and knees. A physician treats this condition by forcing the bone back into the joint and then immobilizing it with a cast or splint while it heals.

GINGIVITIS

Gingivitis (jin-ja-VI-tis) is an inflammation of the tissues of the gum (the gingiva). It is caused by poor oral hygiene resulting in a bacterial infection with symptoms of swelling and reddish bleeding gums. Bacterial plaque builds up on the teeth and the infection can spread to the alveolus in the tooth socket. This will destroy the periodontal ligaments and cause the bone of the tooth sockets to degenerate, resulting in the loss of teeth. Daily brushing after meals and flossing along with visits to the dentist for semiannual cleanings will help prevent tooth loss and the development of gingivitis.

HYPEREXTENSION

Hyperextension is defined as the movement at a joint to a position well beyond the maximum normal extension of the joint. This is a forced movement like that caused by a fall and using your hands to cushion the fall resulting in a hyperextension of the wrist joint. This can cause either a severe sprain or broken bones. Hyperextension of the neck can occur in a rear-end collision in an automobile.

DISLOCATED HIP

A **dislocated hip** occurs when the head of the femur becomes displaced from the socket or acetabulum of the hip. This condition can be acquired during an automobile accident or can be congenital. It is accompanied by swelling, pain, stiffness, and loss of movement. A dislocation can also occur at any other joint of the body.

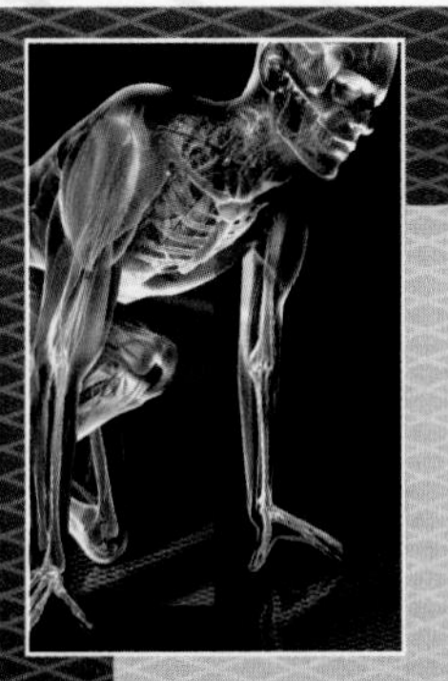

AS THE BODY AGES

Older adults experience some major changes in the articular system, particularly in the synovial joints. Elastin and collagen fibers in a joint become less flexible and tissue repair declines. The articular cartilage surfaces wear and decline because older adults are not able to replace cartilage as quickly as when they were younger. Many individuals in their 50s take glucosamine chondroitin pills to supplement cartilage buildup and help repair and lubricate stiff joints. This nutritional supplement comes from sharks. The production of synovial fluid also declines with age, as does the flexibility of tendons and ligaments, thus decreasing the range of motion in synovial joints. This is why moderate but regular exercising is so important as we age to help keep joints as flexible as possible. ■

SUMMARY OUTLINE

INTRODUCTION

An articulation or joint is a place of union between two or more bones regardless of the degree of movement allowed by the union.

THE CLASSIFICATION OF JOINTS

1. Joints are classified into three main groups based on the degree of movement they allow and their structure: synarthroses, amphiarthroses, and diarthroses.

Synarthroses

1. Synarthroses do not allow movement. The three examples of synarthroses are suture, syndesmosis, and gomphosis.
2. A suture is a joint in which the bones are joined by a thin layer of fibrous connective tissue, like the sutures of the skull.
3. A syndesmosis is a joint in which the bones are connected by ligaments between the bones, like the radius and ulna articulations and the tibia and fibula articulations. Some authors classify this as an amphiarthrosis.
4. A gomphosis consists of a conical process in a socket held together by ligaments, like a tooth in its socket.

Amphiarthroses

1. Amphiarthroses only allow slight movement. The two examples are a symphysis and a synchondrosis.
2. A symphysis is a joint in which the bones are joined by a disk of fibrocartilage, as in the pubic symphysis.
3. A synchondrosis is a joint where two bony surfaces are joined by hyaline cartilage, like the growth plate between the diaphysis and epiphysis of a long bone. Some authors classify this as a synarthrosis.

Diarthroses or Synovial Joints

1. Diarthroses or synovial joints are freely moving joints.
2. They are characterized by having a capsular structure with an internal cavity.
3. The capsule of the joint can be made up of a number of different kinds of tissue: fibrous cartilage, ligaments, tendons, muscle, and synovial membranes.
4. The diarthroses or synovial joints have several functions. They bear weight and allow movement; the ligaments, tendons, muscles, and articular cartilage provide stability; and the synovial fluid lubricates surfaces and nourishes the cartilage.

MOVEMENTS AT SYNOVIAL JOINTS

1. Flexion decreases the angle between bones.
2. Extension increases the angle between bones.
3. Hyperextension increases the joint angle beyond the anatomic position.
4. Dorsiflexion raises the foot upward at the ankle joint.
5. Plantar flexion pushes the foot down at the ankle joint.

6. Abduction moves a bone away from the midline.
7. Adduction moves a bone toward the midline.
8. Rotation moves a bone around a central axis, perpendicular to the axis.
9. Circumduction moves a bone so the end of it describes a circle and the sides of it describe a cone.
10. Supination moves the palm of the hand to an upright position or from a posterior to an anterior position if at the side of the body.
11. Pronation moves the palm of the hand to a downward position or from an anterior position to a posterior position if at the side of the body.
12. Eversion moves the sole of the foot outward at the ankle.
13. Inversion moves the sole of the foot inward at the ankle.
14. Protraction moves a part of the body forward on a plane parallel to the ground.
15. Retraction moves a part of the body backward on a plane parallel to the ground.
16. Elevation raises a part of the body.
17. Depression lowers a part of the body.
18. Opposition, unique to the thumb, allows the tip of the thumb and the fingers to be brought together.
19. Reposition is the opposite of opposition.

THE SIX TYPES OF DIARTHROSES OR SYNOVIAL JOINTS

1. The ball-and-socket joint (multiaxial) allows the widest range of movement, as in the shoulder and hip joint.
2. The hinge joint (uniaxial) limits movement to flexion and extension; examples are the knee, elbow, and the middle and distal phalanges of the fingers and toes.
3. The pivot joint (uniaxial) limits movement to rotation in one plane, such as the atlas and axis articulation in the spine.
4. The condyloid joint or ellipsoidal (biaxial) joint allows motion in two planes at right angles to each other, as in the wrist joint between the radius and carpal bones.
5. The saddle joint (biaxial), found only in the thumb, allows movement in two planes at right angles to one another and is located at the carpal-metacarpal articulation in the thumb.
6. The gliding joint (multiaxial) allows only gliding motion, as the intervertebral joints in the spine.

BURSAE

1. There are three types of bursae. Bursae are closed sacs with a synovial membrane lining that prevents friction between overlapping tissues.
2. Subcutaneous bursae are found between skin and underlying bony processes.
3. Subfascial bursae are found where muscles overlie one another.
4. Subtendinous bursae are found where one tendon overlies another or overlies a bony projection.

REVIEW QUESTIONS

1. Name and describe the three types of joints found in the human body.
2. Name two types of synarthroses and give an example of each.
3. Name two types of amphiarthroses and give an example of each.
*4. Why must diarthroses or synovial joints be constructed like a capsule for maximum function?
5. Name the six types of diarthroses and give an example of each.
6. Name and define the three types of bursae found in the human body.
*7. How can an individual try to prevent the occurrence of osteoarthritis?
8. Name the movements that can occur at the synovial joints.

*Critical Thinking Questions

Search and Explore

Write about a family member or someone you know that has one of the common diseases, disorders, or conditions introduced in this chapter, and tell about the disease.

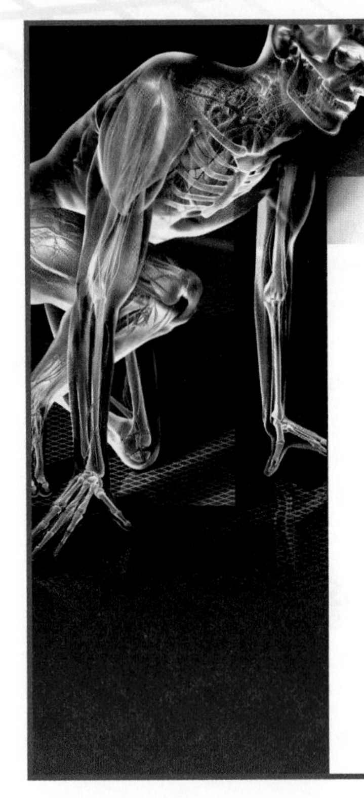

CASE STUDY

Mabel, a 42-year-old woman, is having a checkup with her arthritis specialist. She tells the specialist that she is experiencing more joint pain and stiffness than usual. The specialist examines Mabel and notes that her hands and feet are becoming more deformed in appearance due to severe joint inflammation. It is also harder for Mabel to perform activities of daily living. Mabel states that it is difficult for her to open bottles, turn doorknobs, and put on her socks and shoes. She also experiences pain and tires easily when walking short distances.

Questions

1. Given her symptoms, what type of arthritis might Mabel have?
2. What are the major characteristics of this disorder?
3. What is the cause of this condition and the resulting joint damage?
4. How widespread is arthritis?

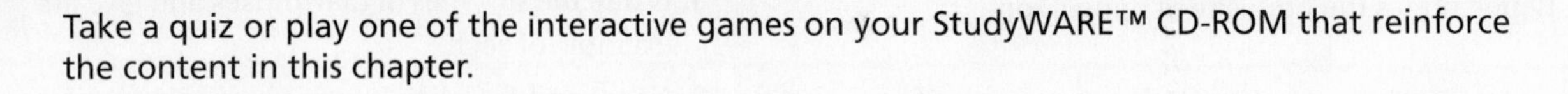

*Study*WARE™ Connection

Take a quiz or play one of the interactive games on your StudyWARE™ CD-ROM that reinforce the content in this chapter.

Study Guide Practice

Go to your **Study Guide** for more practice questions, labeling and coloring exercises, and crossword puzzles to help you learn the content in this chapter.

LABORATORY EXERCISE: THE ARTICULAR SYSTEM

Materials needed: An articulated skeleton, anatomic models of the shoulder joint and hip joint that can be disarticulated showing muscles, tendons, bones, and cartilage

1. Examine the ball-and-socket joint of the hip and shoulder. Identify the capsular nature of the joints by viewing the muscle, tendon, ligaments, and cartilage. If possible, pop out the head of the femur from the acetabulum and view the structure of the joint.
2. Examine the bones of the elbow and knee joint on the skeleton, noting how the bones fit together to allow flexion and extension.
3. Study the hand. Note the flexion and extension hinge joints of the fingers, and the saddle joint of the thumb.
4. Examine the wrist joint and the ankle joint. On the wrist, note the condyloid joint where the radius articulates with the carpal bones.
5. Your instructor will show you either a DVD or a videotape on the anatomy of human joints.

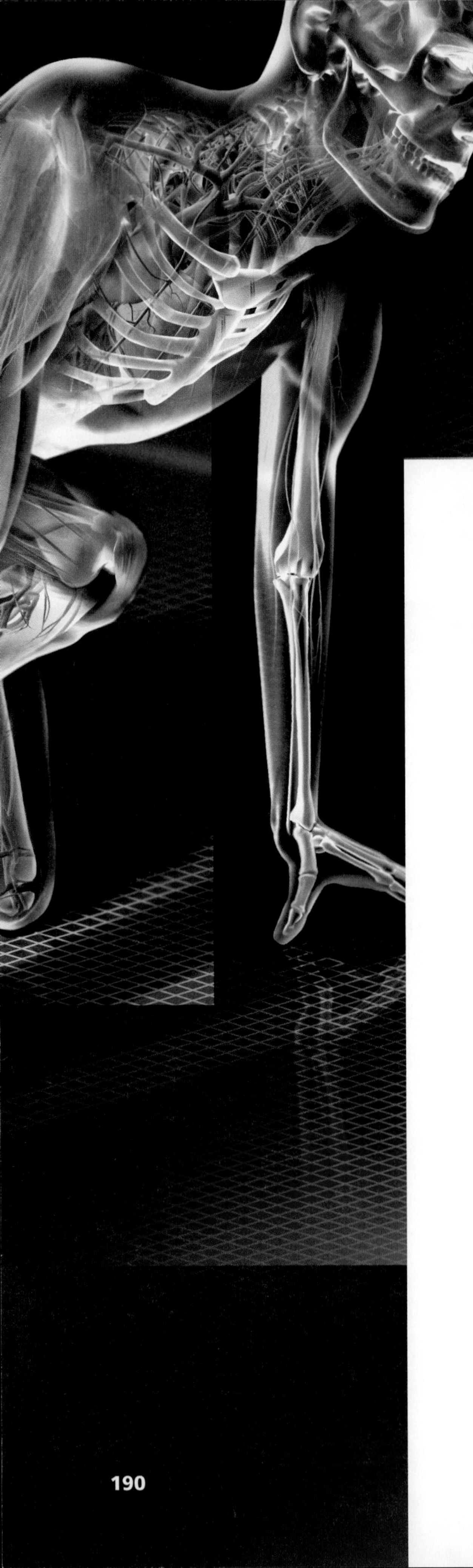

The Muscular System

CHAPTER OBJECTIVES

After studying this chapter, you should be able to:

1. Describe the gross and microscopic anatomy of skeletal muscle.
2. Describe and compare the basic differences between the anatomy of skeletal, smooth, and cardiac muscles.
3. Explain the current concept of muscle contraction based on three factors: neuroelectrical, chemical, and energy sources.
4. Define *muscle tone* and compare isotonic and isometric contractions.
5. List factors that can cause muscles to malfunction, causing various disorders.
6. Name and identify the location of major superficial muscles of the body.

9

KEY TERMS

A bands . . . 192
Abductor digiti minimi . . 213
Abductor hallucis . . . 211
Abductor pollicis . . . 211
Acetylcholine . . . 195
Actin . . . 193
Action potential . . . 195
Adductor pollicis . . . 211
Agonists . . . 202
All-or-none law . . . 200
Anconeus . . . 205
Antagonists . . . 202
Aponeurosis . . . 201
Biceps brachii . . . 205
Biceps femoris . . . 212
Brachialis . . . 205
Brachioradialis . . . 205
Buccinator . . . 202
Cardiac muscle . . . 200
Deltoid . . . 205
Diaphragm . . . 211
Electrical potential . . . 195
Endomysium . . . 192
Epimysium . . . 192
Extensor carpi . . . 208
Extensor digitorum . . . 211
Extensor hallucis . . . 213
Extensor pollicis . . . 211
External intercostals . . . 212
External oblique . . . 211
Fascia . . . 192
Fascicle . . . 192
Fasciculi . . . 192
Fibrillation . . . 201
Flexor carpi . . . 208
Flexor digitorum . . . 211
Flexor hallucis . . . 213
Flexor pollicis . . . 210
Frontalis . . . 202
Gastrocnemius . . . 213
Gluteus maximus . . . 212
Gluteus medius . . . 212
Gluteus minimus . . . 212
Gracilis . . . 212
H band or zone . . . 193
I bands . . . 193
Iliacus . . . 212
Inferior oblique . . . 202
Inferior rectus . . . 202
Infraspinatus . . . 205
Insertion . . . 201
Internal intercostals . . . 212
Internal oblique . . . 211
Interossei . . . 211
Isometric contraction . . . 200
Isotonic contraction . . . 200
Lateral rectus . . . 202
Latissimus dorsi . . . 205
Levator labii superioris . . 202
Levator scapulae . . . 202
Masseter . . . 202
Mastication . . . 202
Medial rectus . . . 202
Motor unit . . . 195
Muscle twitch . . . 199
Myosin . . . 192
Occipitalis . . . 202
Opponens pollicis . . . 211
Orbicularis oris . . . 202
Origin . . . 201
Pectoralis major . . . 205
Pectoralis minor . . . 202
Perimysium . . . 192
Peroneus longus . . . 213
Peroneus tertius . . . 213
Phosphocreatine . . . 199
Plantaris . . . 213
Popliteus . . . 212
Pronator quadratus . . . 210
Pronator teres . . . 210
Psoas . . . 212
Pterygoid . . . 202
Quadriceps femoris . . . 213
Rectus abdominis . . . 211
Rectus femoris . . . 213
Resting potential . . . 195
Rhomboids . . . 202
Sarcolemma . . . 192
Sarcomere . . . 193
Sarcoplasmic reticulum . . 195
Sarcotubular system . . . 193
Sartorius . . . 212
Semimembranosus . . . 212
Semitendinosus . . . 212
Serratus anterior . . . 202
Skeletal muscle . . . 192
Smooth muscle . . . 200
Soleus . . . 213
Sternocleidomastoid . . . 202

(*continues*)

KEY TERMS (*continued*)

INTRODUCTION

As you read this introduction, skeletal muscles are moving your eyes to read the words. Muscles allowed you to first pick up this book and open it to the correct page. You walked to your desk, and you took this book off a shelf. All of these actions allowed you to function in your environment. In addition, smooth muscle is containing the blood in your arteries and veins, food is being pushed through your digestive tract, and urine is being transported from your kidneys via the ureters to your bladder. Meanwhile, cardiac muscle is pumping the blood, carrying oxygen and nutrients to your body cells, and carrying away waste.

Muscles make up about 40% to 50% of the body's weight. They allow us to perform extraordinary physical feats of endurance (running, playing sports) and grace (ballet, figure skating). When they contract, they bring about movement of the body as a whole and cause our internal organs to function properly. Muscles of the diaphragm, chest, and abdomen allow us to breathe. See Concept Map 9-1: Muscular System.

THE TYPES OF MUSCLE

From the discussion of tissues in Chapter 5, you recall that there are three types of muscle tissue: skeletal or striated, smooth or visceral, and cardiac. Recall that **skeletal muscle** is voluntary, that is, we can control its contraction. Under the microscope, skeletal muscle cells are multinucleated and striated; we can see alternating dark and light bands. Smooth muscle, on the other hand, is involuntary, uninucleated, and nonstriated. It is found in places like the digestive tract. Cardiac muscle is involuntary, striated, and uninucleated and is found only in the heart.

THE ANATOMY OF SKELETAL OR STRIATED MUSCLE

Mature skeletal or striated muscle cells are the longest and most slender muscle fibers, ranging in size from 1 to 50 mm in length and 40 to 50 micrometers in diameter (Figure 9-1). Because of this unique structure of the cell, that is, their length being much greater than their width, skeletal muscle cells are often referred to as skeletal muscle fibers. In addition, each muscle cell or fiber is multinucleated and is surrounded by a special cell membrane. This cell membrane is electrically polarized and is called a **sarcolemma** (**sahr**-koh-**LEM**-ah). The sarcolemma is surrounded by the first of three types of connective tissue found in a muscle, the **endomysium** (**in**-do-**MISS**-ee-um), which is delicate connective tissue.

As we study Figure 9-1, we see that the entire muscle consists of a number of skeletal muscle bundles called **fasciculi** (fah-**SICK**-you-lye). Each individual bundle of muscle cells, or **fascicle** (**FASS**-ih-kl), is surrounded by another layer of connective tissue called the **perimysium** (pair-ih-**MISS**-ee-um). This is visible to the naked eye. This perimysium connects with the coarse irregular connective tissue that surrounds the whole muscle called the **epimysium** (eh-pih-**MISS**-ee-um). These three layers of connective tissue act like cement holding all of the muscle cells and their bundles together. In addition, a layer of areolar tissue covers the whole muscle trunk on top of the epimysium and is called the **fascia** (**FASH**-ee-ah).

When skeletal muscle is viewed under a microscope, the cells appear to have alternating dark and light bands referred to as cross-striations. The striations are due to an overlapping of the dark and light bands of protein on the myofibrils. The dark bands are made of the thick filaments of the protein **myosin**. Being thick, they therefore appear dark and are called the **A bands** (hint to remember:

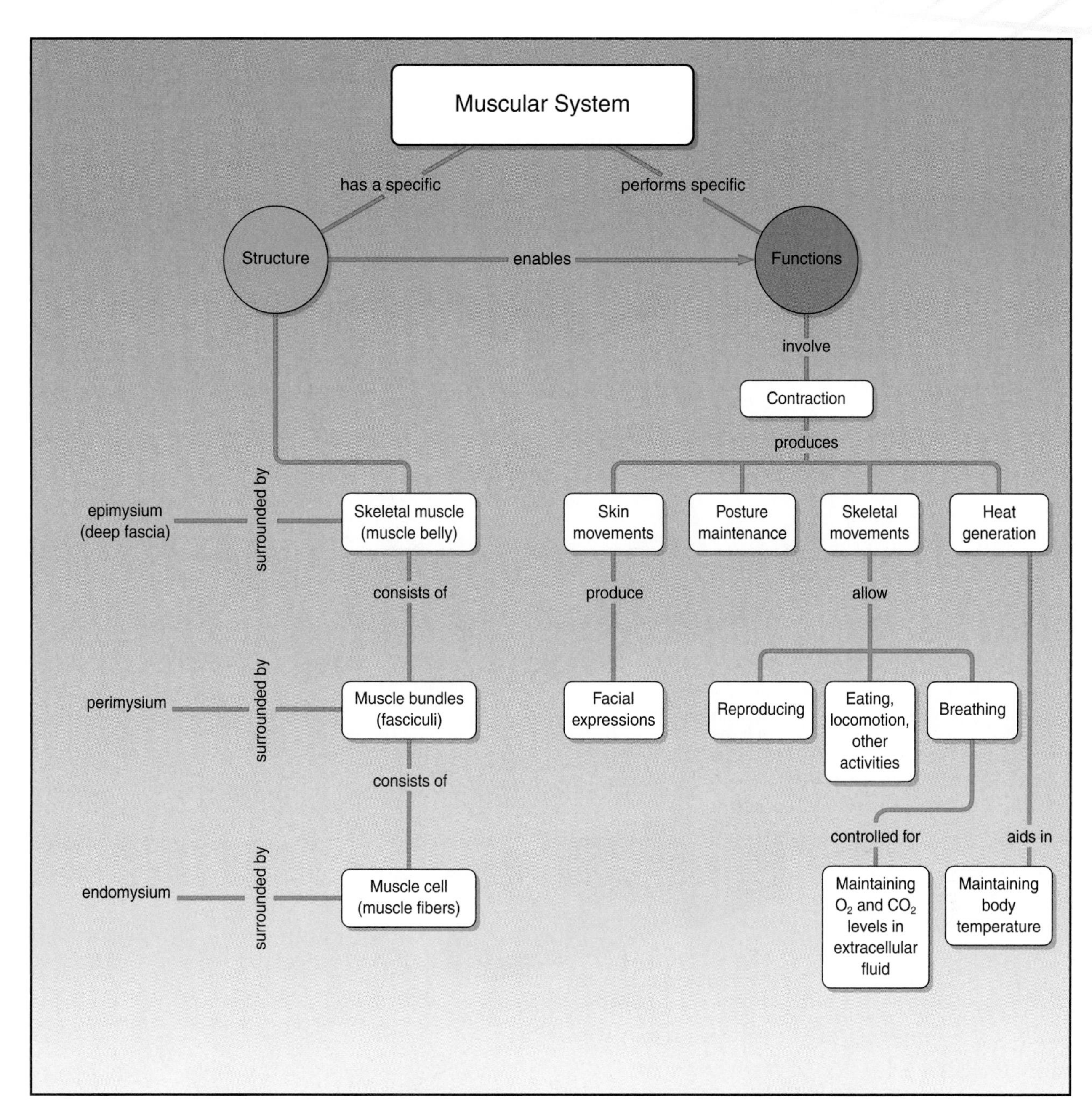

CONCEPT MAP 9-1. Muscular system.

the second letter in the word dark is A). The light bands are made of the thin filaments of the protein **actin**; being thin, they appear light and are called the **I bands** (hint to remember: the second letter in the word light is I).

A number of other markings are important to note. A narrow, dark staining band found in the central region of the I band that looks like a series of the letters Z one on top of another is called the **Z line**. A slightly darker section in the middle of the dark A band is called the **H band** or **H zone**. This is where the myosin filaments are thickest and where there are no cross-bridges on the myosin filaments. The area between two adjacent Z lines is called a **sarcomere** (**SAHR**-koh-meer). It is here at the molecular level that the actual process of contraction occurs via chemical interactions, which is discussed later.

Electron microscopy has also revealed the fact that muscle fibrils (thousands of tiny units that make up a muscle cell) are surrounded by structures made up of membranes in the form of vesicles and tubules. These structures constitute what is referred to as the **sarcotubular** (**sahr**-koh-**TYOO**-byoo-lar) **system**. The sarcotubular system is made up of two components: the **T system** or

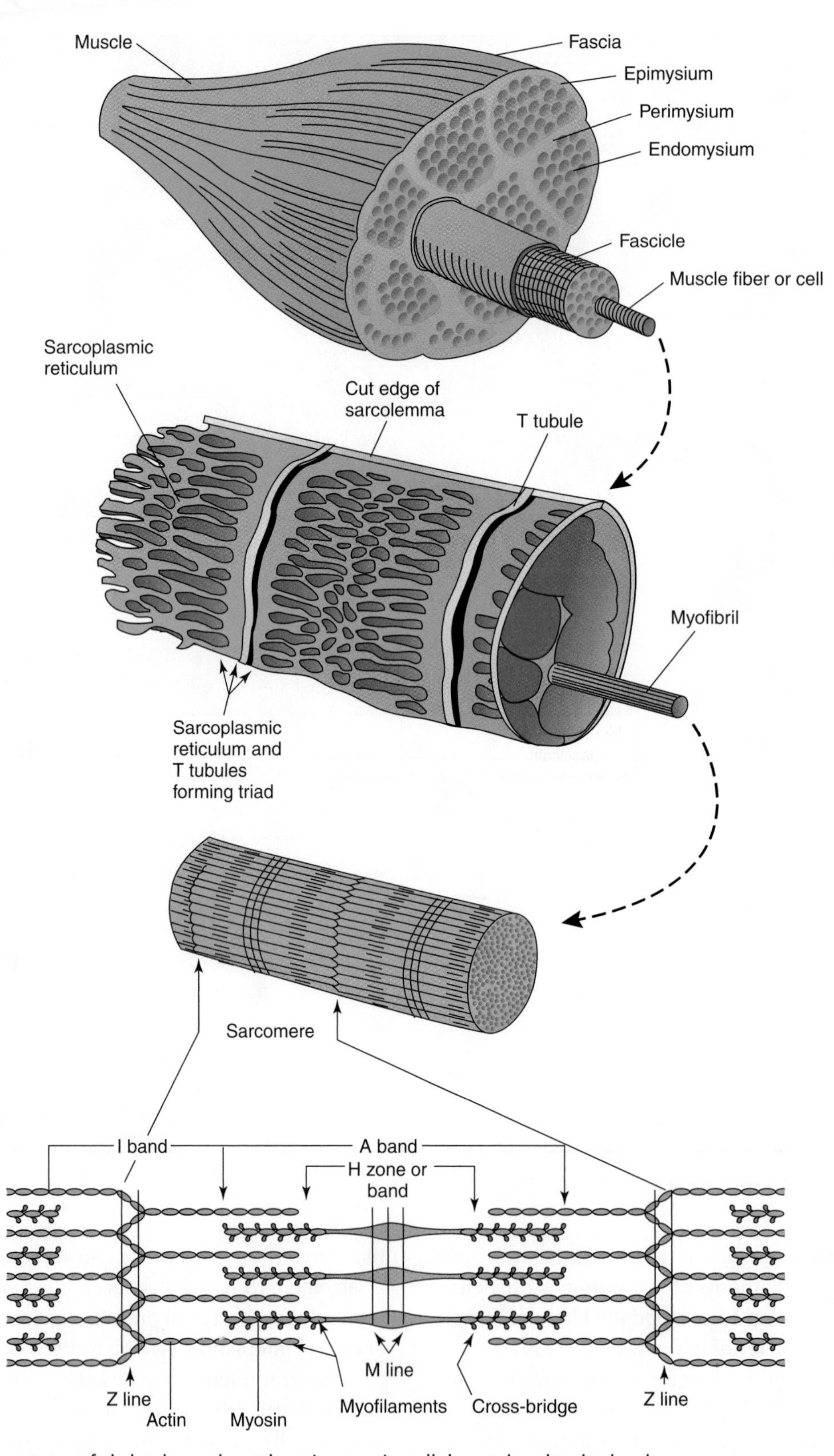

FIGURE 9-1. The anatomy of skeletal muscle at the microscopic, cellular, and molecular levels.

tubules and the **sarcoplasmic reticulum** (reh-**TIK**-you-lum). The tubules of the T system are continuous with the cell membrane or sarcolemma of the muscle fiber and form a grid perforated by individual muscle fibrils. The sarcoplasmic reticulum forms an irregular curtain around each of the fibrils. Again refer to Figure 9-1 for these complex structures. This T system functions in the rapid transmission of a nerve impulse at the cell membrane to all the thousands of fibrils that make up the muscle cell. A muscle cell could be thought of as a single thread of cloth. If you put a single thread under a microscope, you would see that it was made up of hundreds of smaller units of fiber. Hence, just like the thread, the muscle cell or fiber is made up of thousands of smaller units called myofibrils. At the molecular level, each myofibril is made up of microscopic filaments of the proteins myosin (which is thick and looks dark under the microscope) and actin (which is thin and looks light under the microscope).

THE PHYSIOLOGY OF MUSCLE CONTRACTION

To understand how a muscle contracts, it is necessary to first describe what a motor unit is and what properties muscle cells possess. Let's first discuss a motor unit.

All of the muscle cells or fibers innervated by one motor neuron are called a **motor unit** because they (the muscle cells) are always excited simultaneously and therefore contract together. It is important to remember that the terminal divisions or axon endings of a motor neuron are distributed throughout the belly of the whole muscle. Stimulation of a single motor unit causes weak but steady contractions in a broad area of the muscle rather than a strong contraction at one tiny specific point.

Muscles controlling very fine movements (like muscles that move the eye) are characterized by the presence of only a few muscle fibers in each motor unit. Another way to state this would be the ratio of nerve fibers to muscle cells is high. For example, each motor unit present in the ocular muscle contains about 10 muscle cells. However, gross movements (like lifting an object with your hand) will contain a motor unit with 200 or more muscle cells. On the average, a single motor nerve fiber innervates about 150 muscle cells.

Muscle cells possess four properties: excitability, conductivity, contractility, and elasticity. Muscle fibers can be excited by a stimulus. In our bodies this stimulus is a nerve cell. In the laboratory, we can stimulate and excite a muscle with an electrical charge. Besides the property of excitability, all protoplasm in the muscle cell possesses the property of conductivity, which allows a response to travel throughout the cell. The type of response will depend on the type of tissue that is excited. In muscle cells, the response is a contraction. Elasticity then allows the muscle cell to return to its original shape after contraction. Muscle contraction is caused by the interactions of three factors: neuroelectrical factors, chemical interactions, and energy sources.

Neuroelectrical Factors

Surrounding the muscle fiber's membrane or sarcolemma are ions. Refer to Figure 9-2 for the ionic and electrical distribution. The ionic distribution is such that there is a greater concentration of potassium ions (K^+) inside the cell than outside the cell, whereas there is a greater concentration of sodium ions (Na^+) outside the cell membrane than inside the cell. These ions are all positively charged. Because of an uneven distribution of these ions, there is an electrical distribution around the muscle cell. The inside of the cell is negatively charged and the outside of the cell is positively charged electrically. This situation is known as the muscle cell's **resting potential**.

As the nerve impulse reaches the neuromuscular junction where the axon terminals of the nerve cell are in close proximity to the muscle and its numerous cells, it triggers the axon terminals to release a neurotransmitter substance called **acetylcholine** (**ah**-seh-till-**KOH**-leen). This chemical substance affects the muscle cell membrane. It causes the sodium ions (which were kept outside during the resting potential) to rush inside the muscle cell. This rapid influx of sodium ions creates an **electrical potential** that travels in both directions along the muscle cell at a rate of 5 meters per second. This influx of Na^+ causes the inside of the cell to go from being electrically negative to being positive. This is a signal to the muscle cell to generate its own impulse called the **action potential**. This is the signal to contract. Meanwhile the potassium ions that were kept inside begin to move to the outside to restore the resting potential, but they cannot change back to the resting potential situation because so many sodium ions are rushing in.

This action potential not only travels over the surface of the muscle cell membrane but passes down into the cell by way of the T tubules and also deep into all the cells that make up the muscle. This action potential causes the sarcoplasmic reticulum to release stored calcium ions into the fluids surrounding the myofibrils of the muscle cell. Surrounding the actin myofilaments are two inhibitor substances: **troponin** (**TRO**-poh-nin) and **tropomyosin**

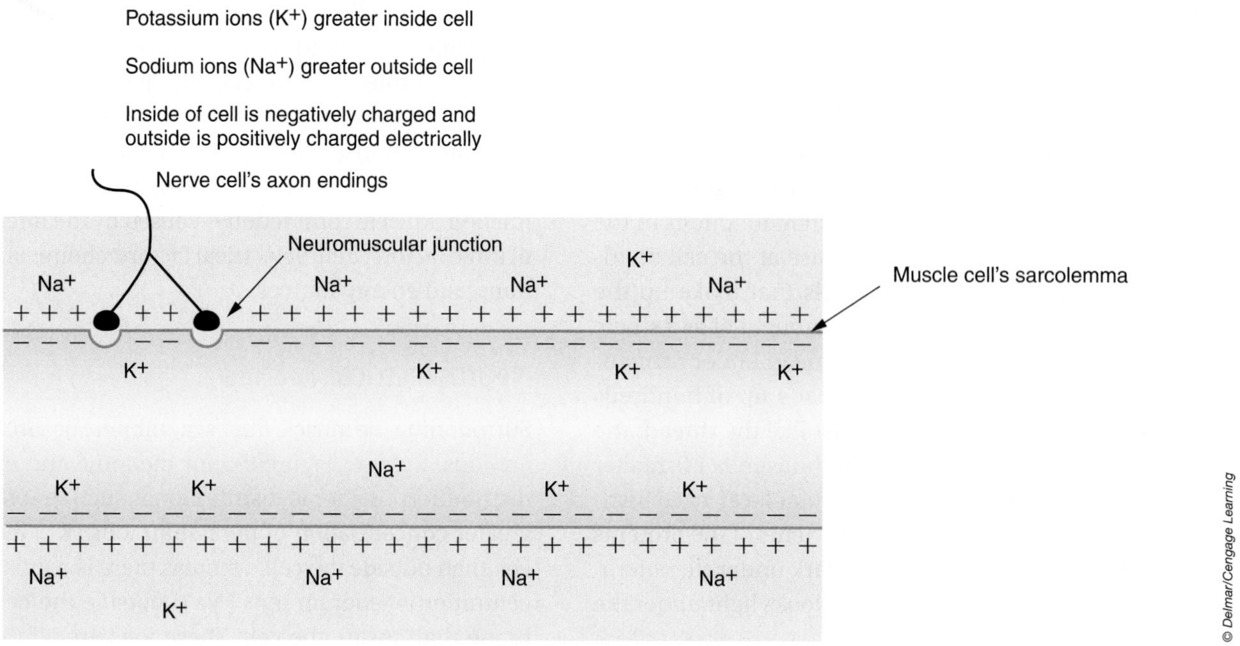

FIGURE 9-2. Ionic and neuroelectrical factors affecting the skeletal muscle cell.

(**troh**-poh-**MY**-oh-sin). Refer to Figure 9-3. These substances keep the actin and myosin protein filaments from interacting. However, when calcium ions are released by the sarcoplasmic reticulum, the action of these inhibitor substances is negated. It is the release of the calcium ions that brings about the contractile process at the molecular level in the myofilaments. When the action potential ceases to stimulate the release of the calcium ions from the reticulum, these ions begin to return and recombine with the sarcoplasmic reticulum. What causes this to happen is the sodium-potassium pump of the muscle cell membrane. As the sodium ions rushed into the cell and potassium rushed out to try to restore the original resting potential but could not do so, the sodium-potassium pump began operating to restore the ionic distribution to its normal resting potential. Contraction occurs in a few thousandths of a second and once the sodium-potassium pump restores ionic distribution, contraction ceases because the action potential is now stopped and all the calcium ions are once again bound to the reticulum. A continued series of action potentials is necessary to provide enough calcium ions to maintain a continued contraction. Now let's discuss the chemical interactions and those calcium ions.

Chemical Interactions

In 1868 a German scientist named Kuhne extracted a protein, which he called myosin, from muscle using a strong salt solution. In 1934, myosin was shown to gel in the form of threads. Shortly thereafter, it was discovered that the threads of myosin became extensible when placed near adenosine triphosphate (ATP). It was not until 1942 that scientists discovered that this myosin was not homogeneous, and that in fact there was another protein in the muscle distinct from myosin and it was called actin. In actuality, the actin unites with the myosin to form actomyosin during the contraction process.

The release of the calcium ions from the sarcoplasmic reticulum inhibits the activity of the troponin and the tropomyosin, which have kept the actin and myosin myofilaments apart. The calcium ions attach to the troponin and now cause the myosin to become activated myosin. The myosin filaments have large heads that contain ATP molecules. The activated myosin releases the energy from the ATP at the actin active site when the myosin links up and forms actomyosin. The head linkage makes a cross-bridge that pulls the actin filaments inward among the myosin filaments and breaks down the ATP into adenosine diphosphate (ADP) and PO_4 and the release of energy, which causes contraction. Refer to Figure 9-4. The shortening of the contractile elements in the muscle is brought about by the pulling of the actin filaments over the myosin filaments. The width of the A bands remains constant while the Z lines move closer together during contraction (see Figure 9-1). When the sodium-potassium pump (Figure 9-5) has restored the resting potential of the cell and sodium ions are back

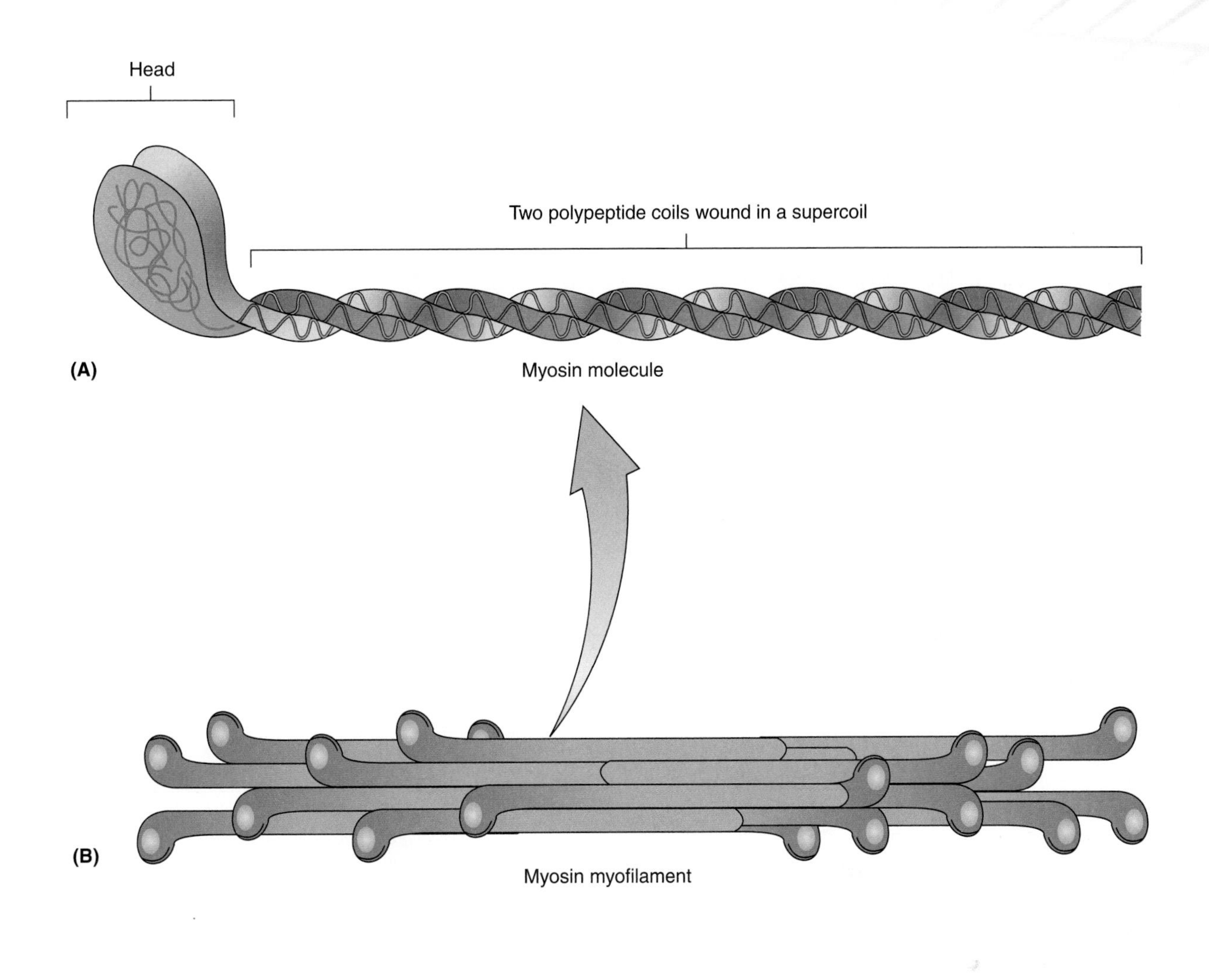

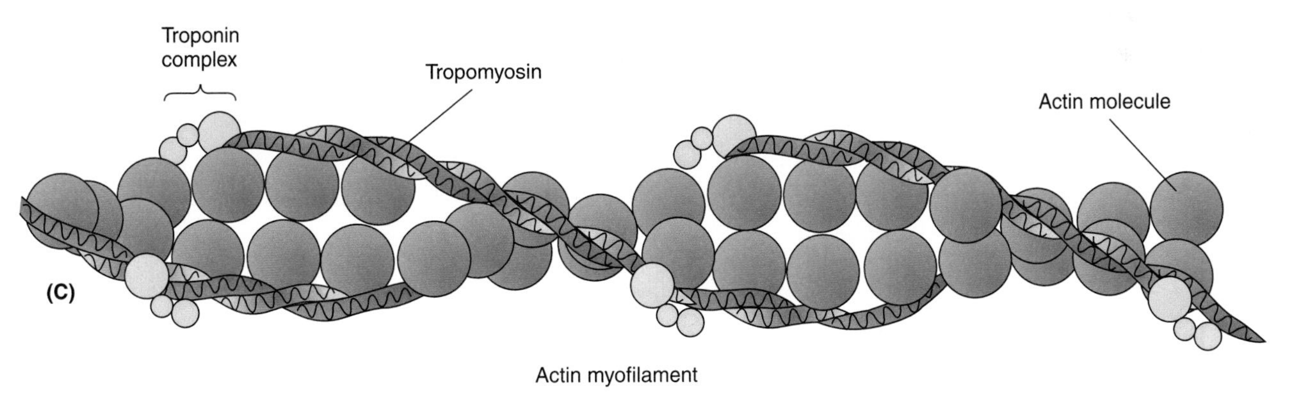

FIGURE 9-3. The structure of the actin and myosin myofilaments of a muscle cell. (A) Myosin molecule. (B) Myosin myofilament. (C) Actin myofilament.

outside and potassium ions are back inside the cell, the action potential ceases and calcium ions get reabsorbed by the sarcoplasmic reticulum. Now contraction ceases and the actin filaments get released from the myosin and the Z lines move further apart. This whole complex process occurs in 1/40 of a second. Keep in mind that we discussed only one small part of a muscle cell's filaments. There are thousands of myofilaments in a single muscle cell, and a muscle like your biceps contains hundreds of thousands of muscle cells, all interacting and coordinating together at the molecular level to bring about contraction.

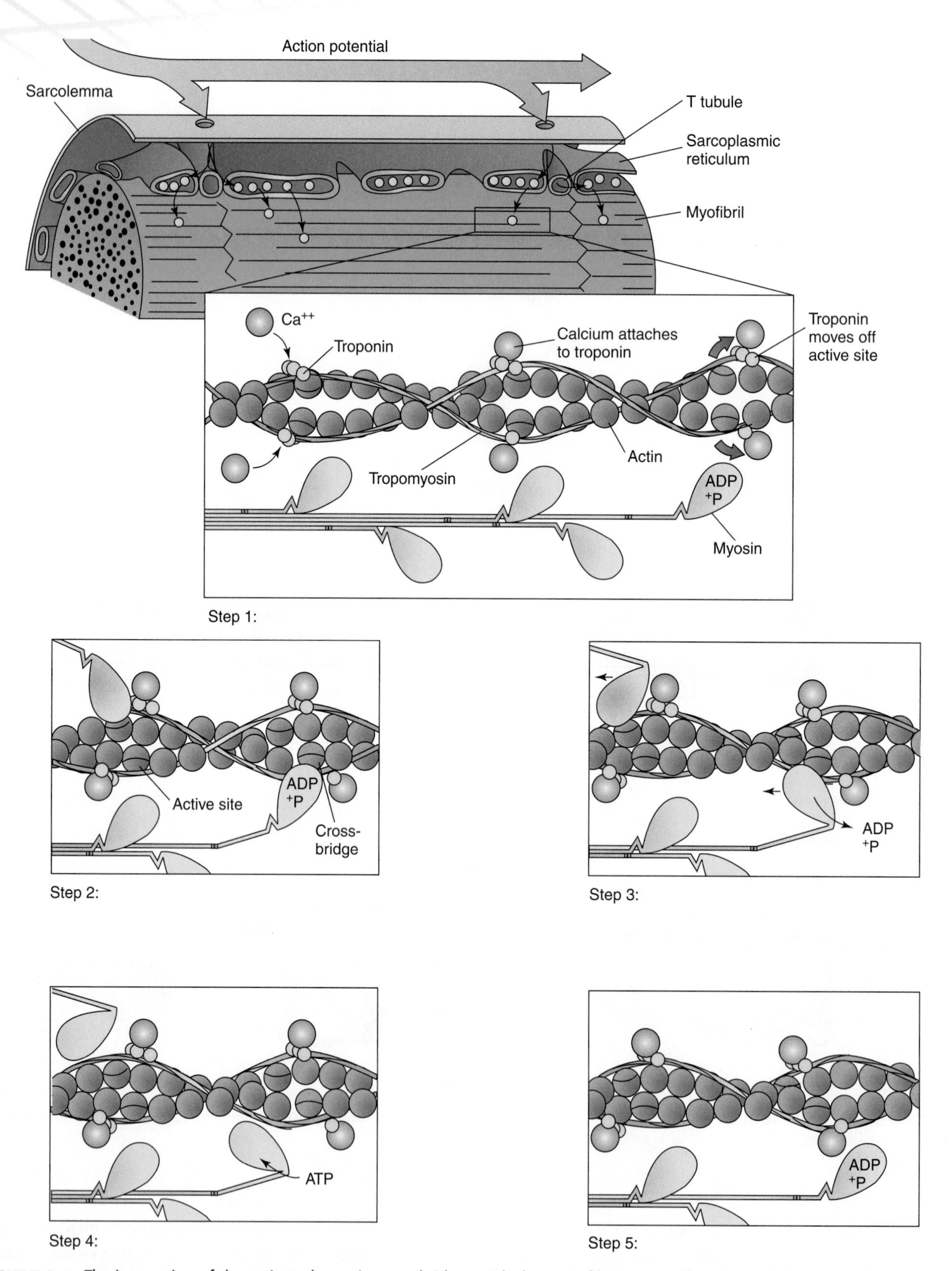

FIGURE 9-4. The interaction of the activated myosin cross-bridges with the actin filaments pulling the actin in among the myosin, resulting in contraction.

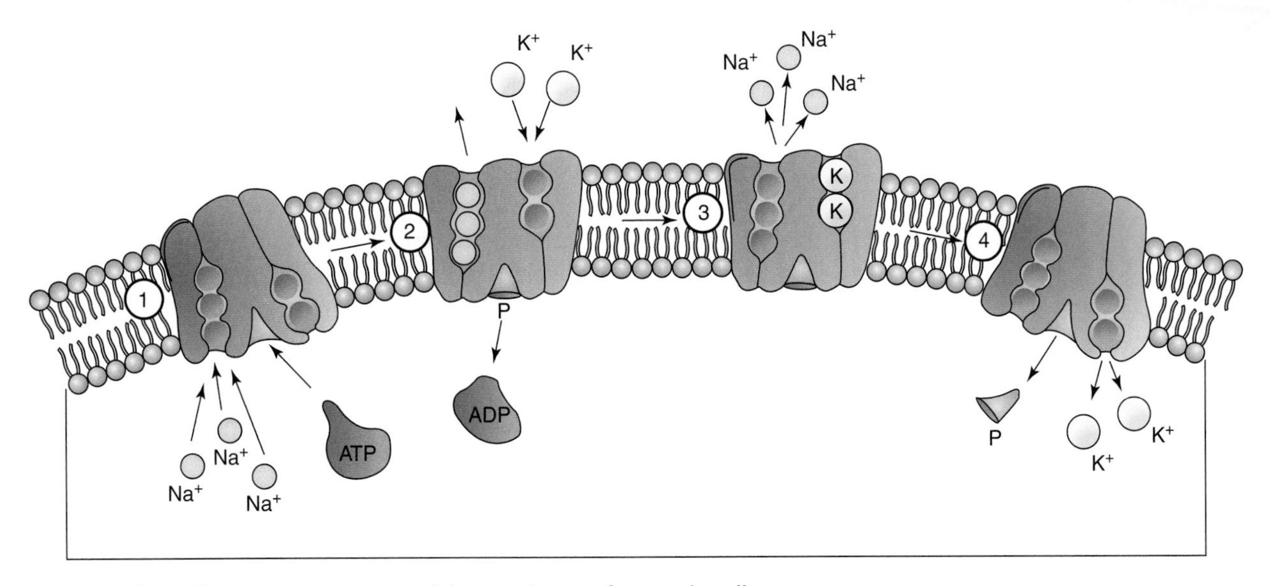

FIGURE 9-5. The sodium-potassium pump of the membrane of a muscle cell.

Energy Sources

Muscle cells convert chemical energy (ATP) into mechanical energy (contraction). This source of energy is ATP molecules (review Chapter 4). Actin + myosin + ATP → actomyosin + ADP + PO_4 + energy (causing contraction). The energy given off by the breakdown of ATP is used when the actin and myosin filaments intermesh. ATP is synthesized by glycolysis, the Krebs citric acid cycle, electron transport, and in muscle cells, by the breakdown of phosphocreatine.

In glycolysis, you will recall from Chapter 4, glucose present in the blood enters cells where it is broken down through a series of chemical reactions to pyruvic acid. A small amount of energy is released from the glucose molecule with a net gain of two molecules of ATP.

In the Krebs citric acid cycle and electron transport, if oxygen is present, the pyruvic acid is further broken down into CO_2 and H_2O and 36 more ATP molecules. If oxygen is not available to the muscle cell, the pyruvic acid changes to lactic acid and builds up in the muscle cell with only two ATP produced until oxygen again becomes available.

Muscle cells have two additional sources of ATP. **Phosphocreatine** (**fos**-foh-**KREE**-ah-tin) is found only in muscle tissue and provides a rapid source of high-energy ATP for muscle contraction. When muscles are at rest, excess ATP is not needed for contraction so phosphate is transferred to creatine to build up a reserve of phosphocreatine. During strenuous exercise, the phosphocreatine takes up ADP to release ATP and creatine, thus supplying the muscle with an additional supply of ATP. The overall reaction, which goes in both directions, is phosphocreatine + ADP ↔ creatine + ATP.

In addition, skeletal muscle cells can take up free fatty acids from the blood and break them down as another source of energy into CO_2, H_2O, and ATP. Of course, during any contraction, heat is produced as a waste product.

In summary, muscle cells have four sources of ATP for the energy of contraction:

1. Glucose + 2 ATP → CO_2 + H_2O + 38 ATP (aerobic)
2. Glucose + 2 ATP → 2 lactic acid + 2 ATP (anaerobic)
3. Phosphocreatine + ADP → creatine + ATP
4. Free fatty acids → CO_2 + H_2O + ATP

In these processes, glycolysis, the Krebs citric acid cycle, and electron transport play a vital role.

THE MUSCLE TWITCH

When the contraction of a skeletal muscle is studied in the laboratory by applying an electrical charge to the muscle, the analysis of the contraction is called a **muscle twitch** (Figure 9-6). This reveals a brief latent period directly following stimulation just before contraction begins. This latent period is followed by a period of contraction followed by a period of relaxation. This latent period occurs because the resting potential of the muscle cells must change into the electrical potential as sodium ions rush in. This is caused by the acetylcholine released by the nerve cell's axon terminals into the neuromuscular junction. The

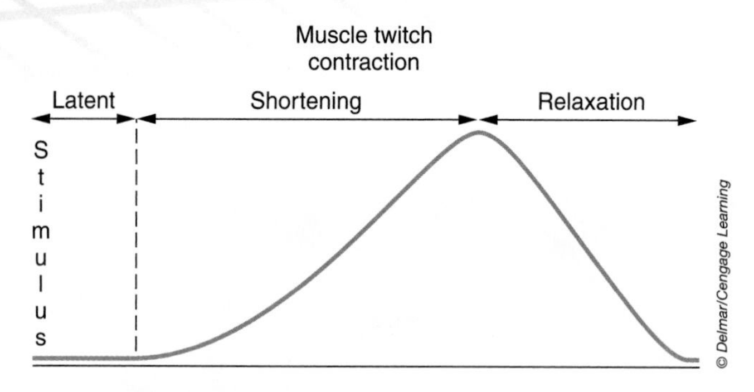

FIGURE 9-6. A laboratory analysis of a muscle twitch.

electrical potential then becomes the action potential as the signal travels down the T tubules to the sarcoplasmic reticulum. Then calcium ions get released into the fluids around the myofibrils of actin and myosin and contraction occurs. Once the sodium-potassium pump operates, calcium gets reabsorbed and relaxation occurs.

The strength of the contraction depends on a number of factors: the strength of the stimulus (a weak stimulus will not bring about contraction); the duration of the stimulus (even if the stimulus is quite strong, if it is applied for a millisecond it may not be applied long enough for it to be effective); the speed of application (a strong stimulus applied quickly and quickly pulled away may not have time enough to take effect even though it is quite strong); the weight of the load (one can pick up a waste basket with one hand but not a dining room table); and, finally, the temperature (muscles operate best at normal body temperature 37°C or 98.6°F in humans). A stimulus strong enough to elicit a response in an individual muscle cell will produce maximal contraction. The contraction either occurs or it does not. This is known as the **all-or-none law**.

MUSCLE TONE

Tone is defined as a property of muscle in which a steady or constant state of partial contraction is maintained in a muscle. Some muscle cells in a particular muscle will always be contracting while other muscle cells are at rest. Then those at rest will contract, while those that were contracting will go into relaxation. This allows us, for example, to maintain body posture for long periods of time without showing any evidence of tiring. This is accomplished because nerve stimuli alternate between various groups of muscle cells, thus allowing all to have periods of rest. Tone results in skeletal muscles exhibiting a certain degree of firmness as they maintain a slight and steady pull on attached bones. Tone maintains pressure on abdominal contents, maintains blood pressure in arteries and veins, and assists in digestion in the stomach and intestines.

There are two types of contraction. When lifting a weight, muscles become shorter and thicker. In this type of contraction, tone or tension remains the same and is referred to as **isotonic contraction**. When we push against a wall or attempt to lift a huge boulder, the muscles involved remain at a constant length while the tension against the muscle increases, and this is known as **isometric contraction**. From this fact, a whole series of exercises have been developed called isometric exercises (like locking fingers of opposite hands and pulling to develop the biceps). These exercises help develop tone or firmness in muscles.

THE ANATOMY OF SMOOTH MUSCLE

Smooth muscle is found in hollow structures of the body like the intestines, blood vessels, and urinary bladder. It cannot be controlled at will because it is under the control of the autonomic nervous system and also may be hormonally stimulated. Each smooth muscle cell contains a single large nucleus and because its fiber is more delicate than skeletal muscle, cross-striation of the myosin and actin arrangements is not visible. The cells connect by fibrils extending from one cell to another closely adjoining cell. In hollow structures like the small intestine, the smooth muscle is arranged in two layers, an outer longitudinal layer and an inner circular layer. Contraction of these two layers, with the circular layer contracting first, results in reducing both the length of the tube and the circumference of the tube. This contraction pushes whatever is in the tube in a forward direction, for example, digested food or chyme in the intestine or blood in the arteries and veins. Smooth muscle cells produce a slower contraction than skeletal muscle, but smooth muscle contraction allows greater extensibility of the muscle.

The actin and myosin fibers are not so regularly arranged in smooth muscle as in striated muscle. Therefore, contraction occurs in a similar way but without the regular rearrangement of the fibrils. The fibrils do slide together and rhythmically shorten the cell, but a slow wave of contraction passes over the entire muscle mass as the nerve impulse reaches a cell and gets transmitted to the remainder of the cells or fibers.

THE ANATOMY OF CARDIAC MUSCLE

Cardiac muscle cannot be influenced at will because it, like smooth muscle, is under the control of the autonomic nervous system. It is uninucleated, similar to smooth

HEALTH ALERT

STRONG MUSCLES

In order to maintain healthy and strong muscles, it is necessary to exercise them by stretching on a daily basis. After a good night's sleep, get out of bed and stretch. Start slowly moving your arms and legs; walk to an area where there is fresh air and take deep breaths, stretching your breathing muscles and filling your lungs to capacity. This action can set your routine of moving and stretching muscles through normal, daily activities. Walking is one of the best exercises to maintain healthy muscles. Remain relaxed when stretching, start slowly to warm up the muscles and then move to a more rigorous pace. Even when running or lifting weights, try to remain in a relaxed mode because tension puts excessive strain on muscles, and can cause damage to muscular tissue.

As children grow, instill in them the importance of exercise to maintain both healthy muscles and bones. Regular exercising should become part of our regular daily routines throughout life. Even older adults should be encouraged to take daily walks. Have you ever noticed the older "mall walkers" early in the morning before the stores open? Individuals who are confined to bed for periods of time should be realigned in body positions a number of times a day to allow stretching of muscles that normally would not be worked. Daily exercise, like walking or more rigorous jogging or weight lifting, will help maintain a healthy muscular system. ■

muscle; however, it is striated like skeletal muscle. Cardiac muscle also has another unique quality. If one muscle cell is stimulated, all the muscle cells or fibers are stimulated so all the muscle cells contract together. Also, the muscle cell that contracts the fastest will control the speed of other muscle cells, causing them all to contract at the faster rate.

The rapid rhythm of cardiac muscle is the result of a special property of this type of cell to receive an impulse, contract, immediately relax, and then receive another impulse. These events all occur about 75 times a minute. However, the period of an individual contraction is slower in cardiac (about 0.8 second) as opposed to skeletal muscle, which is much faster (about 0.09 second).

If rapid, uncontrolled contraction of individual cells in the heart occurs, this is called **fibrillation**. This results in the heart's inability to pump the blood properly and can result in death.

THE NAMING AND ACTIONS OF SKELETAL MUSCLES

Muscles can be named according to their action (adductor, flexor, extensor); according to shape (quadratus, trapezius); according to origin and insertion (sternocleidomastoid); according to location (e.g., frontalis, tibialis, radialis); according to their number of divisions (e.g., biceps, triceps, quadriceps); and, finally, according to the direction their fibers run (transverse, oblique).

The more fixed attachment of a muscle that serves as a basis for the action is the **origin**. The movable attachment, where the effects of contraction are seen, is called the **insertion**. The origin is the proximal (closer to the axial skeleton) attachment of the muscle to a bone; the insertion is the distal (farthest away from the axial skeleton) attachment to the other bone. Most voluntary or skeletal muscles do not insert directly to a bone, but rather they insert through a strong, tough, nonelastic, white collagenous fibrous cord known as a tendon. Tendons vary in their lengths from a fraction of an inch to those more than a foot in length, like the Achilles tendon in the lower leg, which inserts on the heel bone. If a tendon is wide and flat, it is called an **aponeurosis** (**ap**-oh-noo-**ROH**-sis).

Muscles are found in many shapes and sizes. Muscles that bend a limb at a joint are called flexors. Muscles that straighten a limb at a joint are called extensors. If a limb is moved away from the midline, an abductor is functioning; however, if the limb is brought in toward the midline, an adductor is functioning. The muscles rotating an involved limb are rotators. In movements of the ankle, muscles of dorsiflexion turn the foot upward, and muscles of plantar flexion bring the foot toward the ground. In movements of the hand, turning the forearm when it is extended out so that the palm of the hand faces the ground is pronation, whereas turning the forearm so that the palm faces upward is supination. Levators raise a part of the body, and depressors lower a part of the body. See Chapter 8 for a review of movements possible at synovial joints.

In performing any given movement, such as bending the leg at the knee joint, the muscles performing the actual movement are called the prime movers or **agonists**. Those muscles that will straighten the knee are the **antagonists**. The agonist or prime mover must relax for the antagonists to perform their function and vice versa. **Synergists** (**SIN**-er-jistz) are the muscles that assist the prime movers.

THE FUNCTION AND LOCATION OF SELECTED SKELETAL MUSCLES

The superficial muscles of the body are those that can be found directly under the skin (Figure 9-7). Some parts of the body, like the arms and legs, will have up to three different layers of muscles (superficial, middle, and deep layers). Other areas will have only superficial muscles, like the cranial area of the skull. These muscles can be better seen on a living human who is a bodybuilder or an athlete. These individuals exercise regularly at a gym developing their superficial muscles.

Muscles of Facial Expression

A number of muscles are involved in creating facial expressions and body language (Figure 9-8). Table 9-1 lists the muscles and functions they perform. The **occipitalis** (ok-**sip**-ih-**TAL**-is) draws the scalp backward. The **frontalis** (frohn-**TAL**-is) raises your eyebrows and wrinkles the skin of your forehead. The **zygomaticus** (**zye**-go-**MAT**-ick-us) muscles are involved in smiling and laughing. The **levator labii superioris** (leh-**VAY**-ter **LAY**-bee-eye soo-peer-ee-**OR**-is) raises your upper lip. The **orbicularis oris** (or-**BICK**-you-lah-ris **OR**-is) closes your lips and the **buccinator** (**BUCK**-sin-aye-tohr) compresses your cheek. These two muscles are involved in puckering up to kiss.

Muscles of Mastication

Mastication (mass-tih-**KAY**-shun) or chewing is caused by some very strong muscles. Table 9-2 lists the muscles of mastication and the functions they perform. The **masseter** (mass-**SEE**-ter) and the **temporalis** (tim-poh-**RAL**-is) are the main muscles that close your jaw by bringing up the mandible in a bite grip. They are assisted by the **pterygoid** (**TEHR**-ih-goyd) muscles.

Muscles of the Eye

The muscles that move the eyes are unique in that they do not insert on bone; instead they insert on the eyeball. Table 9-2 lists the muscles that move the eyes and the functions they perform. The **superior rectus** raises the eye; the **inferior rectus** lowers the eye. The **medial rectus** rolls the eye medially and the **lateral rectus** rolls the eye laterally. The **superior** and **inferior oblique** muscles rotate the eyeball on an axis.

Muscles Moving the Head

The main muscle that moves the head is the **sternocleidomastoid** (stir-noh-kyle-doh-**MASS**-toyd) muscle (see Figure 9-8). Table 9-3 lists the muscles of the head and the functions they perform. Contraction of both sternocleidomastoids causes flexion of the neck; contraction of one at a time results in rotation to the left or right. Other muscles of the neck assist the sternocleidomastoid in moving the head.

StudyWARE™ Connection

Play an interactive game labeling muscles of the head and neck on your StudyWARE™ CD-ROM.

Muscles Moving the Shoulder Girdle

The muscles that move the scapula are the **levator scapulae** (leh-**VAY**-ter **SKAP**-you-lee), the **rhomboids** (**ROM**-boydz), the **pectoralis** (peck-toh-**RAL**-is) **minor,** and the **trapezius** (trah-**PEE**-zee-us). The trapezius is seen superficially between the neck and the clavicle. Refer to Figure 9-7 to view superficial anatomy of the muscles of the trunk. The **serratus** (sir-**AYE**-tis) **anterior** muscle looks like the teeth of a saw on the lateral upper side of the trunk. These muscles all move the scapula. Table 9-3 lists the muscles that move the shoulder girdle and the functions they perform.

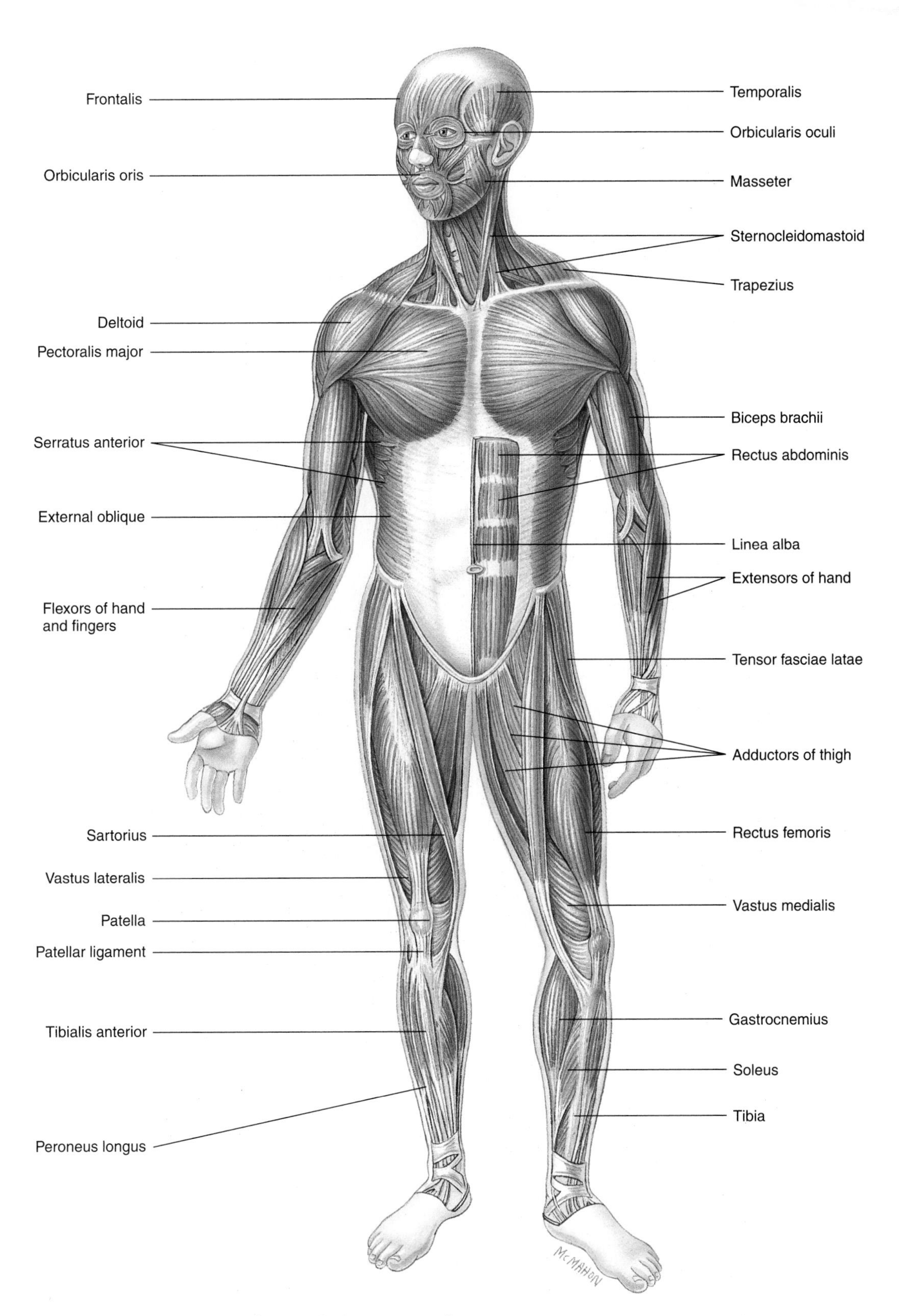

FIGURE 9-7A. The superficial muscles of the body (anterior view).

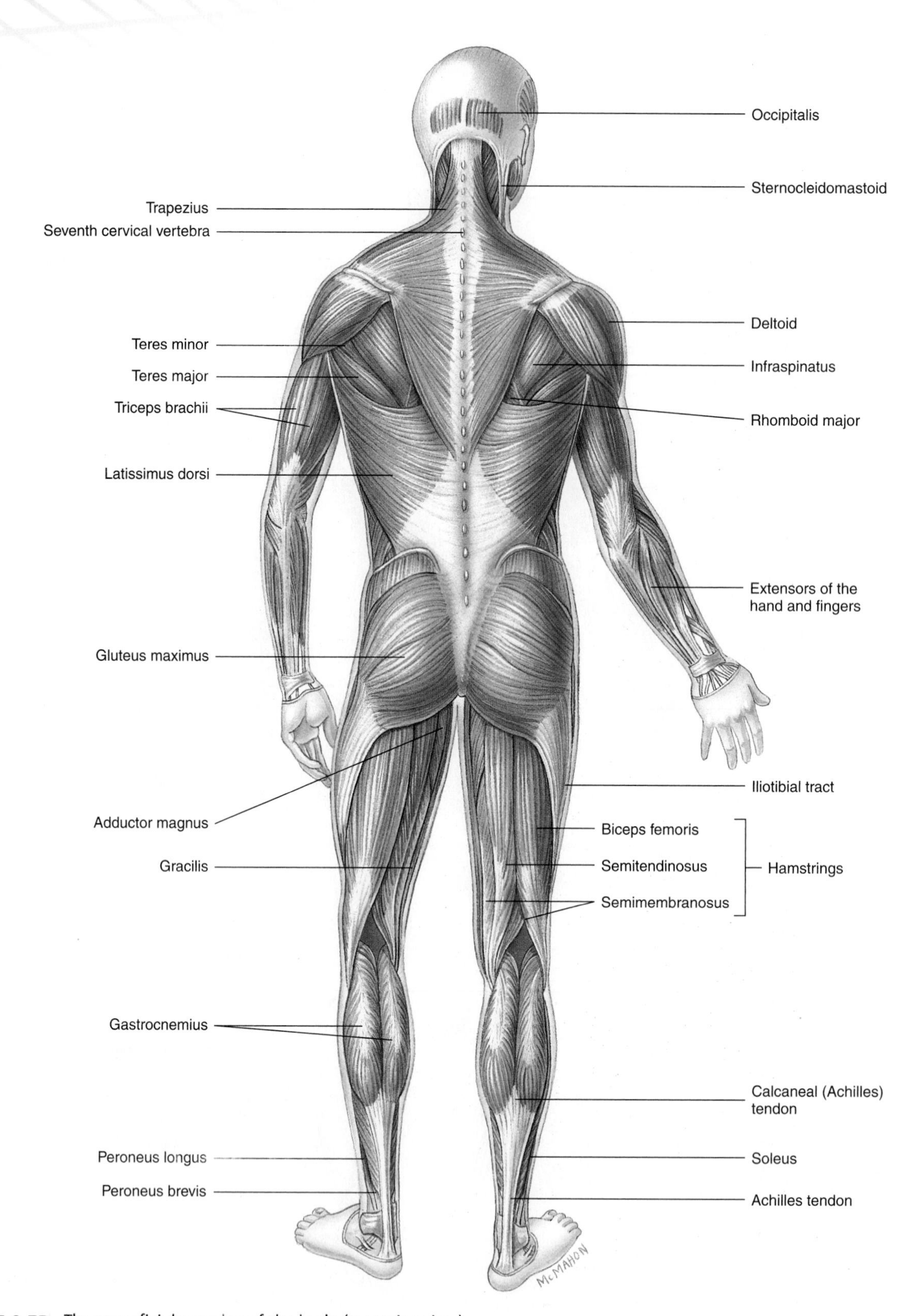

FIGURE 9-7B. The superficial muscles of the body (posterior view).

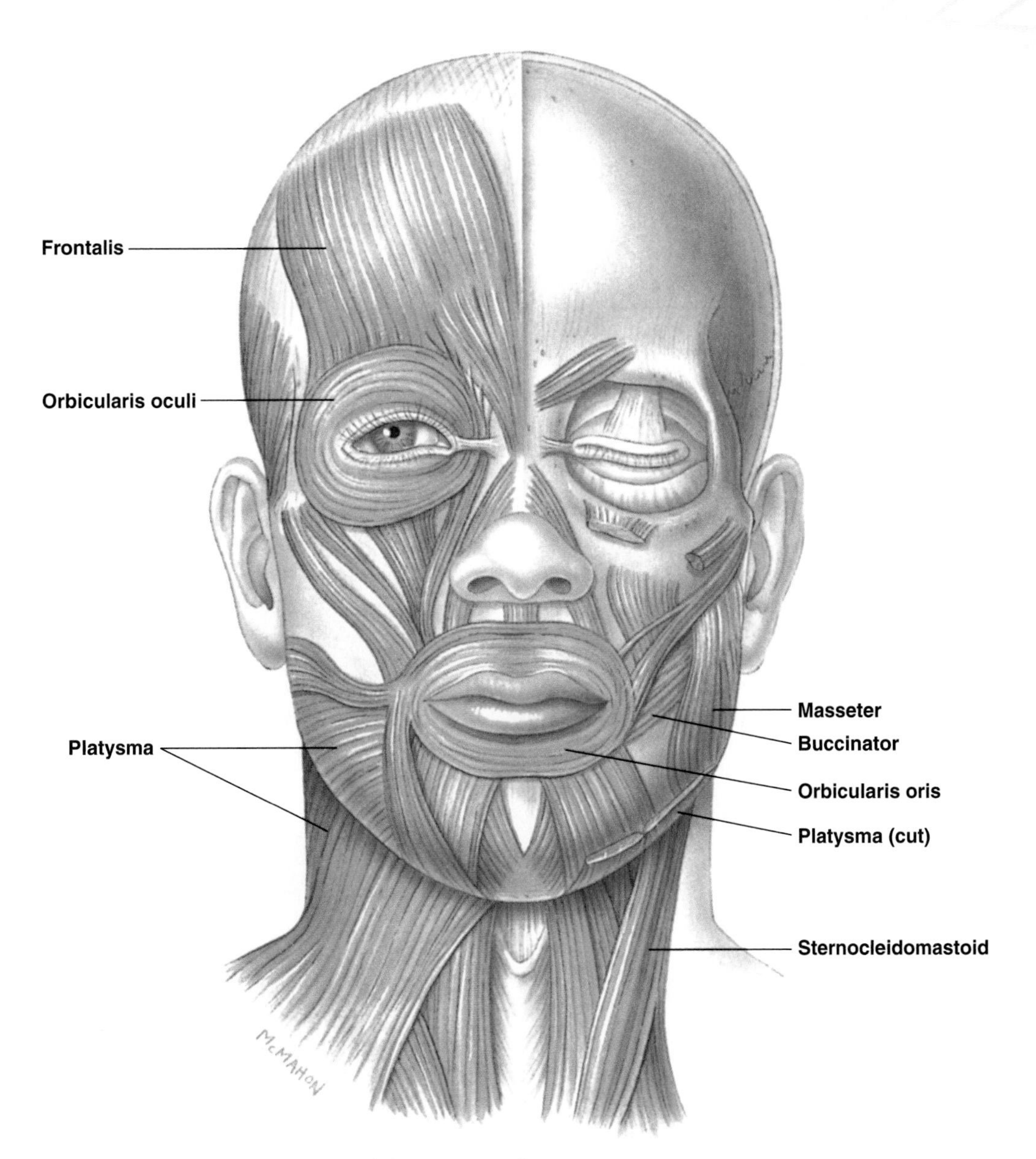

FIGURE 9-8A. Some muscles of the head and neck (anterior view).

Muscles Moving the Humerus

Most of the muscles that move the humerus originate on the bones of the shoulder girdle (Figure 9-9). Table 9-4 lists the muscles that move the humerus and the functions they perform. The **pectoralis major** flexes and adducts the arm. The **latissimus dorsi** (lah-**TISS**-ih-mus **DOR**-sigh) muscle extends, adducts, and rotates the arm medially. Because these movements are used in swimming, this muscle is often called the swimmer's muscle.

The following muscles are often referred to as the rotator cuff muscles. The **teres minor** adducts and rotates the arm. The **deltoid** (**DELL**-toyd) abducts the arm and is also the muscle that receives injections. The **supraspinatus** (**sue**-prah-spye-**NAH**-tus) also abducts the arm. The **infraspinatus** (**in**-frah-spye-**NAH**-tus) rotates the arm.

Muscles Moving the Elbow

Three muscles flex the forearm at the elbow: the **brachialis** (**bray**-kee-**AL**-us), the **biceps brachii** (**BYE**-seps **BRAY**-kee-eye), and the **brachioradialis** (**bray**-kee-oh-**ray**-dee-**AH**-lus). Table 9-5 lists the muscles that move the elbow and the functions they perform. Two muscles extend the arm: the **triceps brachii** and the **anconeus** (an-**KOH**-nee-us).

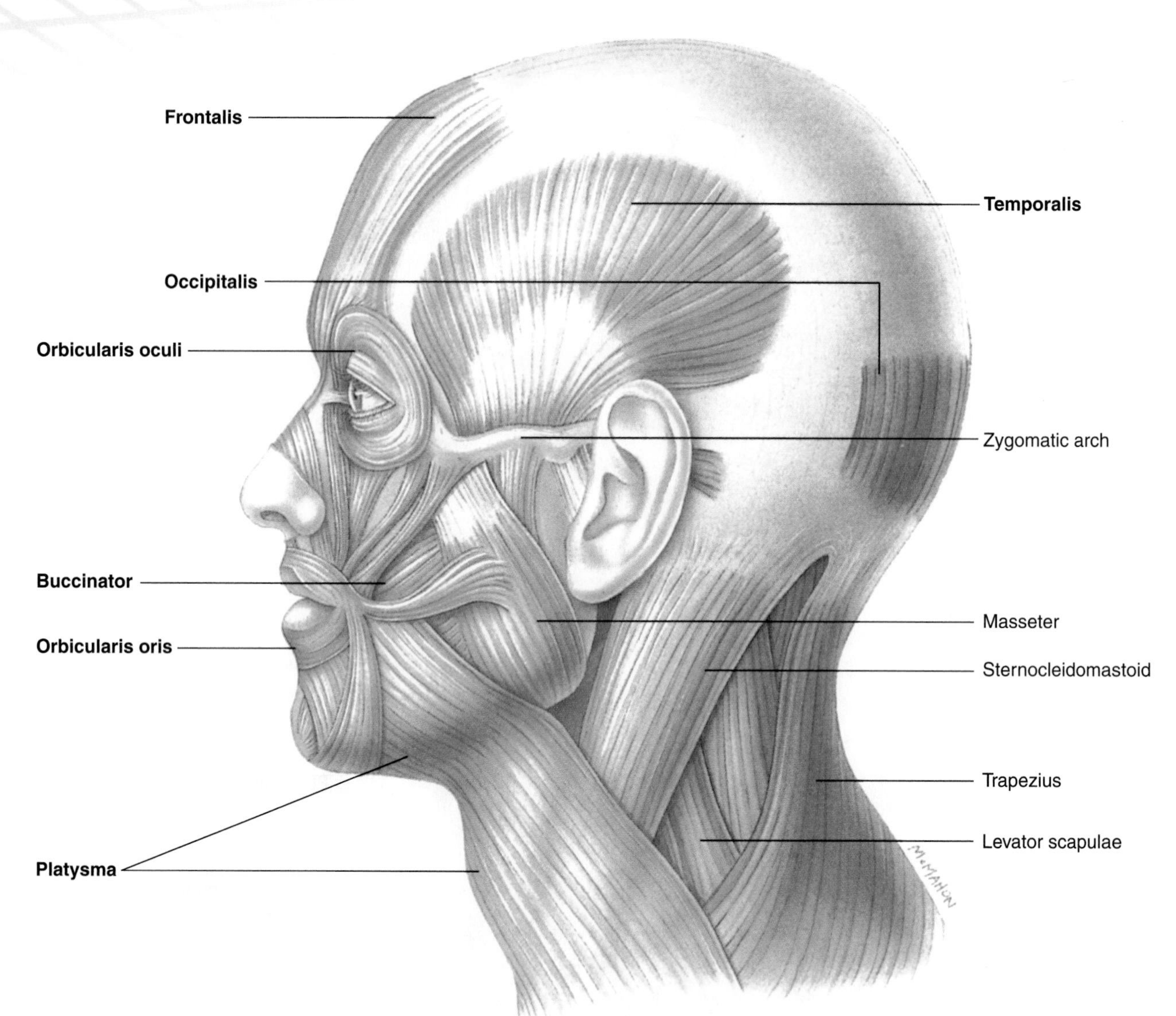

FIGURE 9-8B. Some muscles of the head and neck (lateral view).

Table 9-1 Muscles of Facial Expression

Muscle	Function
Occipitalis	Draws scalp backward
Frontalis	Elevates eyebrows, wrinkles skin of forehead
Zygomaticus minor	Draws upper lip upward and outward
Levator labii superioris	Elevates upper lip
Levator labii superioris alaeque nasi	Raises upper lip and dilates nostril
Buccinator	Compresses cheek and retracts angle
Zygomaticus major	Pulls angle of mouth upward and backward when laughing
Mentalis	Raises and protrudes lower lip as when in doubt
Orbicularis oris	Closes lips
Risoris	"Smiling" muscle

Table 9-2 Muscles of Mastication and the Muscles That Move the Eyes

Muscles of Mastication

Muscle	Function
Masseter	Closes jaw
Temporalis	Raises mandible and closes mouth; draws mandible backward
Medial pterygoid	Raises mandible; closes mouth
Lateral pterygoid (two-headed)	Brings jaw forward

Extrinsic Muscles of the Eye

Muscle	Function
Superior rectus	Rolls eyeball upward
Inferior rectus	Rolls eyeball downward
Medial rectus	Rolls eyeball medially
Lateral rectus	Rolls eyeball laterally
Superior oblique	Rotates eyeball on axis
Inferior oblique	Rotates eyeball on axis

Table 9-3 Muscles of the Head and Shoulder Girdle

Muscles Moving the Head

Muscle	Origin	Insertion	Function
Sternocleidomastoid	Two heads sternum and clavicle	Temporal bone	Flexes vertebral column; rotates head

Muscles Moving the Shoulder Girdle

Muscle	Origin	Insertion	Function
Levator scapulae	Cervical vertebrae	Scapula	Elevates scapula
Rhomboid major	2nd–5th thoracic vertebrae	Scapula	Moves scapula backward and upward; slight rotation
Rhomboid minor	Last cervical and 1st thoracic vertebrae	Scapula	Elevates and retracts scapula
Pectoralis minor	Ribs	Scapula	Depresses shoulder and rotates scapula downward
Trapezius	Occipital bone 7th cervical 12th thoracic	Clavicle	Draws head to one side; rotates scapula
Serratus anterior	8th, 9th rib	Scapula	Moves scapula forward away from spine and downward and inward toward chest wall

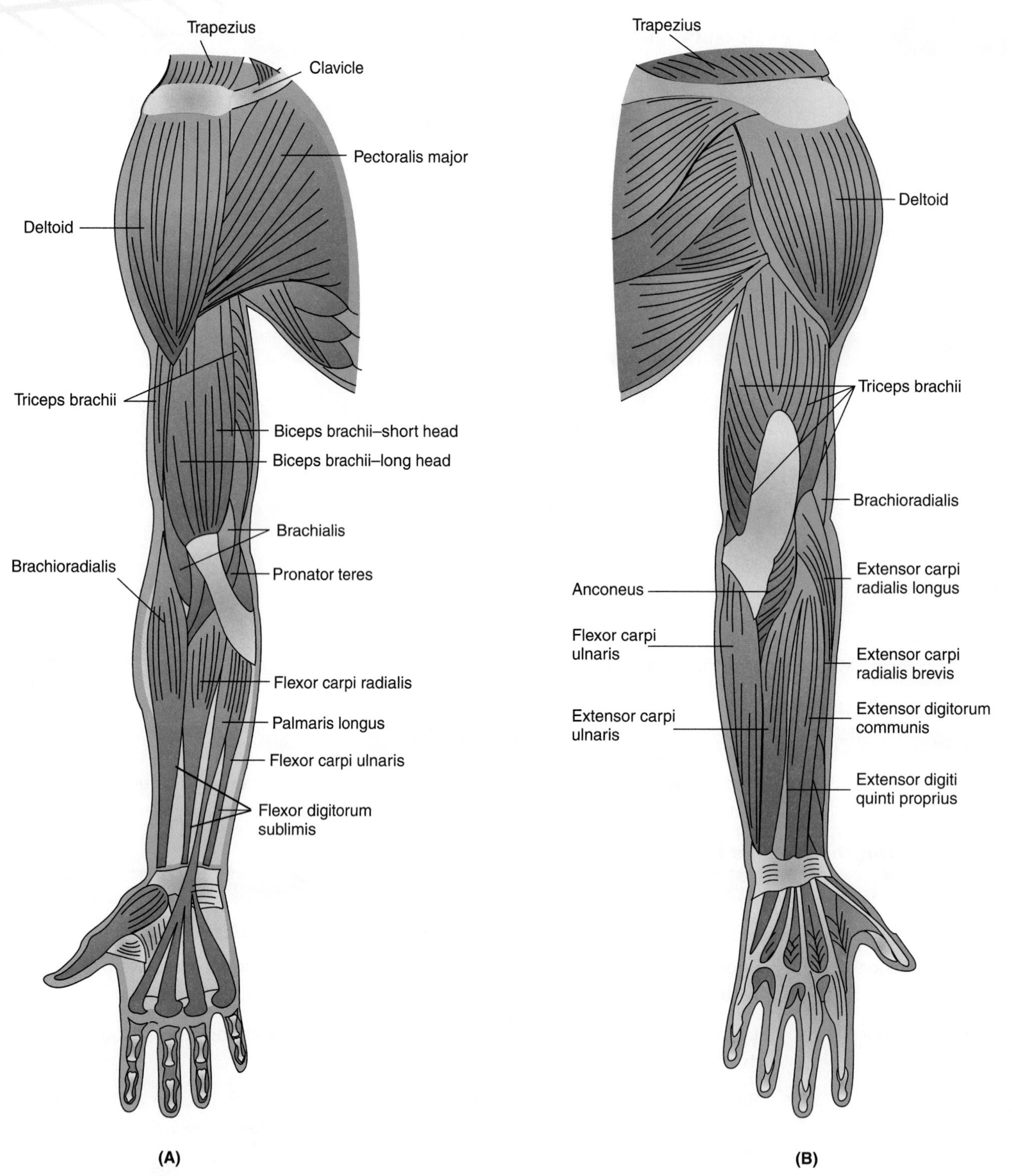

FIGURE 9-9. Muscles that move the arm and fingers: (A) anterior view, (B) posterior view.

Muscles Moving the Wrist

The two **flexor carpi** (**FLEKS**-ohr **KAHR**-pye) muscles flex the wrist and the three **extensor carpi** muscles extend the wrist with the assistance of the extensor digitorum communis. Table 9-5 lists the muscles that move the wrist and the functions they perform. These muscles are also involved in abducting and adducting the wrist. When your pulse is taken, the tendon of the flexor carpi radialis is used as the site to locate the radial pulse.

Table 9-4 Muscles Moving the Humerus

Muscle	Origin	Insertion	Function
Coracobrachialis	Scapula	Humerus	Flexes, adducts arm
Pectoralis major	Clavicle; sternum six upper ribs	Humerus	Flexes, adducts, rotates arm medially
Teres major	Scapula	Humerus	Adducts, extends, rotates arm medially
Teres minor	Scapula	Humerus	Rotates arm laterally and adducts
Deltoid	Clavicle, scapula	Humerus	Abducts arm
Supraspinatus	Scapula	Humerus	Abducts arm
Infraspinatus	Scapula	Humerus	Rotates humerus outward
Latissimus dorsi	Lower six thoracic; lumbar vertebrae; sacrum; ilium lower four ribs	Humerus	Extends, adducts, rotates arm medially, draws shoulder downward and backward

Table 9-5 Muscles Moving the Elbow and the Wrist

Muscles Moving the Elbow

Muscle	Function
Brachialis	Flexes forearm
Triceps brachii (three heads)	Extends and adducts forearm
Biceps brachii (two heads)	Flexes arm; flexes forearm; supinates hand
Anconeus	Extends forearm
Brachioradialis	Flexes forearm

Muscles Moving the Wrist

Muscle	Function
Flexor carpi radialis	Flexes, abducts wrist
Flexor carpi ulnaris	Flexes, adducts wrist
Extensor carpi radialis brevis	Extends and abducts wrist joint
Extensor carpi radialis longus	Extends and abducts wrist
Extensor carpi ulnaris	Extends, adducts wrist
Palmaris longus	Flexes wrist joint
Palmaris brevis	Tenses palm of hand
Extensor digitorum communis	Extends wrist joint

Muscles Moving the Hand

Supination of the hand so that the palm is facing upward is caused by the **supinator** (soo-pin-**NAY**-tohr) muscle. The two muscles that pronate the hand so that the palm faces downward are the **pronator teres** (pro-**NAY**-tohr **TAYR**-eez) and the **pronator quadratus** (pro-**NAY**-tohr kwod-**RAH**-tus). These muscles are found beneath the superficial muscles deep in the arm. Table 9-6 lists the muscles that move the hand, thumb, and fingers and the functions they perform.

Muscles Moving the Thumb

The thumb is capable of movement in many directions, giving the hand a unique capability that separates humans from all other animals. We can grasp and use tools because of our thumb. The two **flexor pollicis**

Table 9-6 Muscles Moving the Hand, Thumb, and Fingers

Muscles Moving the Hand

Muscle	Function
Supinator	Supinates forearm
Pronator teres	Pronates forearm
Pronator quadratus	Pronates forearm

Muscles Moving the Thumb

Muscle	Function
Flexor pollicis longus	Flexes 2nd phalanx of thumb
Flexor pollicis brevis	Flexes thumb
Extensor pollicis longus	Extends terminal phalanx
Extensor pollicis brevis	Extends thumb
Adductor pollicis	Adducts thumb
Abductor pollicis longus	Abducts, extends thumb
Abductor pollicis brevis	Abducts thumb
Opponens pollicis	Flexes and opposes thumb

Muscles Moving the Fingers

Muscle	Function
Flexor digitorum profundus	Flexes terminal phalanx
Flexor digiti minimi brevis	Flexes little finger
Interossei dorsalis	Abduct, flex proximal phalanges
Flexor digitorum superficialis	Flexes middle phalanges
Extensor indicis	Extends index finger
Interossei palmaris	Adduct, flex proximal phalanges
Abductor digiti minimi	Abducts little finger
Opponens digiti minimi	Rotates, abducts 5th metacarpal
Extensor digitorum communis	Extends the fingers

(**FLEKS**-ohr pol-**ISS**-is) muscles flex the thumb, *pollicis* coming from the Latin for "thumb." The two **extensor pollicis** muscles extend the thumb. Refer to Table 9-6. The **adductor pollicis** muscle adducts the thumb; the two **abductor pollicis** muscles abduct the thumb. The unique **opponens** (oh-**POH**-nenz) **pollicis** flexes and opposes the thumb and is used when we write.

Muscles Moving the Fingers

The **flexor digitorum** (**FLEKS**-ohr dij-ih-**TOHR**-um) muscles flex the fingers; the **extensor digitorum** muscle extends the fingers. Refer to Table 9-6. The little finger and the index finger have separate similar muscles. The **interossei** (in-tehr-**OSS**-eye) muscles, found between the metacarpals, cause abduction of the proximal phalanges of the fingers. The tendons of the extensor digitorum are visible on the surface of your hand. Extend your fingers to view these tendons.

Muscles of the Abdominal Wall

Three layers of muscles along the side of the abdomen constrict and hold the abdominal contents in place. They are from outer to inner: the **external oblique**, the **internal oblique**, and the **transversus abdominis.** In the front over your belly is the **rectus abdominis.** This is the muscle that we develop when we do sit-ups and try to get that "washboard" look. Table 9-7 lists the muscles of the abdominal wall and respiration. See Figure 9-10.

Muscles of Respiration or Breathing

The main muscle used in breathing is the **diaphragm** (**DYE**-ah-fram). Its contracting causes air to enter the

Table 9-7 Muscles of the Abdominal Wall and Respiration

Muscles of the Abdominal Wall

Muscle	Origin	Insertion	Function
External oblique	Lower eight ribs	Iliac crest Anterior rectus sheath	Compresses abdominal contents
Internal oblique	Iliac crest	Costal cartilage lower three or four ribs	Compresses abdominal contents
Transversus abdominis	Iliac crest cartilage of lower six ribs	Xiphoid cartilage linea alba	Compresses abdominal contents
Rectus abdominis	Crest of pubis, pubic symphysis	Cartilage of 5th, 6th, 7th rib	Flexes vertebral column, assists in compressing abdominal wall

Muscles of Respiration

Muscle	Origin	Insertion	Function
Diaphragm	Xiphoid process, costal cartilages, lumbar vertebrae	Central tendon	Increases vertical diameter of thorax
External intercostals	Lower border of rib	Upper border of rib below	Draws adjacent ribs together
Internal intercostals	Ridge on inner surface of rib	Upper border of rib below	Draws adjacent ribs together
Quadratus lumborum	Iliac crest	Last rib and upper four lumbar vertebrae	Flexes trunk laterally

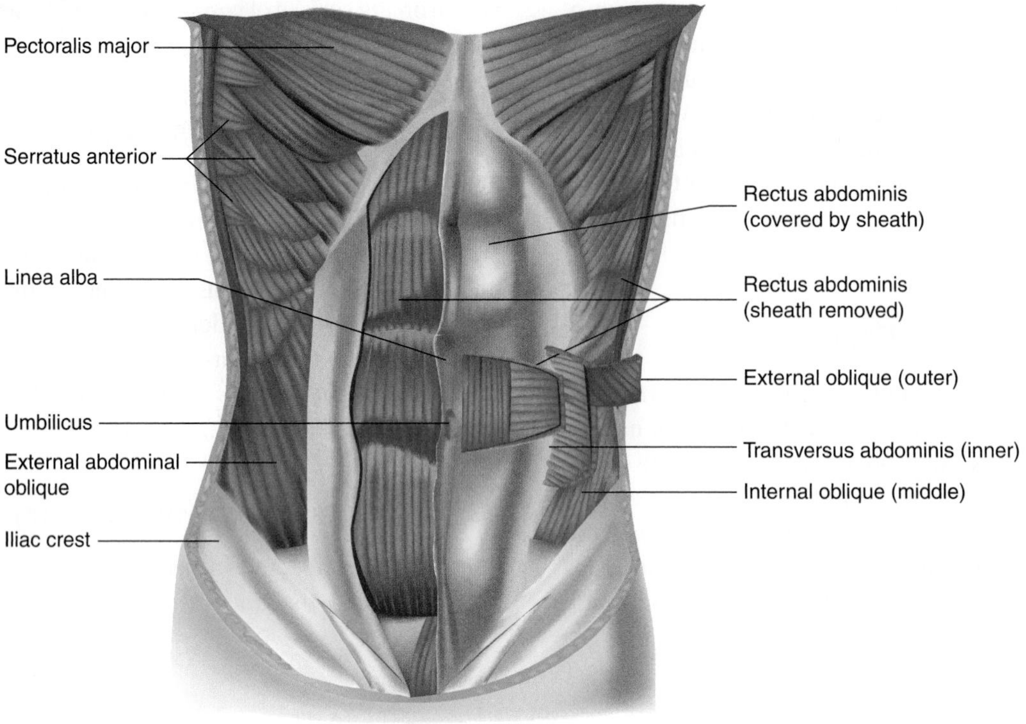

FIGURE 9-10. Muscles of the abdominal wall.

lungs. When it relaxes air is expelled from the lungs. To expand the ribs while the lungs fill with air, the **external** and **internal intercostal** muscles come into play. The external intercostals elevate the ribs when we breathe in or inspire, and the internal intercostals depress the ribs when we breathe out or expire. Refer to Table 9-7.

*Study***WARE**™ Connection

Watch an animation of accessory muscle use on your StudyWARE™ CD-ROM.

Muscles Moving the Femur

Refer to Table 9-8 for the list of muscles involved in moving the thigh or femur. The **psoas** (**SO**-us) muscles and the **iliacus** (ill-ee-**ACK**-us) muscle flex the thigh. Three gluteal muscles form the buttocks: the **gluteus** (**GLOO**-tee-us) **maximus** forms most of the buttocks; the **gluteus medius**, where injections are administered, is above and lateral to the maximus; and the **gluteus minimus**. The gluteus maximus extends the thigh. There are two adductor muscles and one abductor. The **tensor fascia lata** (**TIN**-sir **FASH**-ee-ah **LAH**-tuh) tenses the fascia lata, which is a thick band of connective tissue on the lateral side of the thigh causing abduction of the femur.

Muscles Moving the Knee Joint

Six muscles involved in flexion of the knee are found posteriorly on the thigh and four muscles involved in extension are found on the anterior surface of the thigh (Figure 9-11). Table 9-9 lists the muscles involved in flexion of the knee. The flexors of the knee are the **biceps femoris** (**BYE**-seps **FEM**-ohr-iss), the **semitendinosus** (**sim**-ee-**tin**-dih-**NO**-sus), the **semimembranosus** (**sim**-ee-**mim**-brah-**NO**-sus) (these first three are also known as the hamstrings), the **popliteus** (**pop**-lih-**TEE**-us), the **gracilis** (**GRASS**-ih-liss), and the **sartorius** (sahr-**TOHR**-ee-us). The hamstrings get their name because the tendons of these muscles in hogs or pigs were used to suspend the hams during curing or smoking. Many predators bring down their prey by biting through these hamstrings. When persons "pull

Table 9-8 Muscles Moving the Femur

Muscle	Origin	Insertion	Function
Psoas major	Transverse process of lumbar vertebrae	Femur	Flexes, rotates thigh medially
Psoas minor	Last thoracic and lumbar vertebrae	Junction of ilium and pubis	Flexes trunk
Iliacus	Last thoracic and lumbar vertebrae	Junction of ilium and pubis	Flexes, rotates thigh medially
Gluteus maximus	Ilium, sacrum, and coccyx	Fascia lata, gluteal ridge	Extends, rotates thigh laterally
Gluteus medius	Ilium	Tendon on femur	Abducts, rotates thigh medially
Gluteus minimus	Ilium	Femur	Abducts, rotates thigh medially
Tensor fascia lata	Ilium	Femur	Tenses fascia lata
Abductor brevis	Pubis	Femur	Abducts, rotates thigh
Adductor magnus	Ischium, ischiopubic ramus	Femur	Adducts, extends thigh
Obturator externus	Ischium, ischiopubic ramus	Femur	Rotates thigh laterally
Pectineus	Junction of ilium and pubis	Femur	Flexes, adducts thigh
Adductor longus	Crest and symphysis of pubis	Femur	Adducts, rotates, flexes thigh

a hamstring," they have torn the tendons of one of these muscles.

The **quadriceps femoris** muscle consists of four parts that extend the knee. They are the **rectus femoris**, the **vastus lateralis**, the **vastus medialis**, and the **vastus intermedius**. The vastus medialis and vastus lateralis are easily seen superficially on the anterior thigh (see Figure 9-11). The sartorius muscle is the longest muscle of the body and is known as the "tailors" muscle. It flexes the thigh and leg and rotates the thigh laterally for sitting cross-legged, a position some tailors sit in while hand sewing to hold their materials in their lap.

Muscles Moving the Foot

Five muscles plantar flex the foot or bring it downward. They are the **gastrocnemius** (**gas**-trok-**NEE**-mee-us) or calf muscle, the **tibialis posterior**, the **soleus** (**SO**-lee-us), the **peroneus** (**payr**-oh-**NEE**-us) **longus**, and the **plantaris** (plan-**TAH**-ris). Two muscles dorsally flex the foot or bring it upward. They are the **tibialis anterior** and the **peroneus tertius**. Table 9-10 lists the muscles involved in moving the foot and toes.

Muscles Moving the Toes

Two muscles flex the great toe: the **flexor hallucis** (**FLEKS**-ohr **HAL**-uh-kiss) brevis and longus; one muscle extends the great toe, the **extensor hallucis** (see Table 9-10). The **flexor digitorum** muscles flex the toes while the **extensor digitorum** extends the toes. The **abductor hallucis** abducts the great toe and the **abductor digiti minimi** abducts the little toe. A total of 20 intrinsic muscles of the foot move the toes to flex, extend, adduct, and abduct.

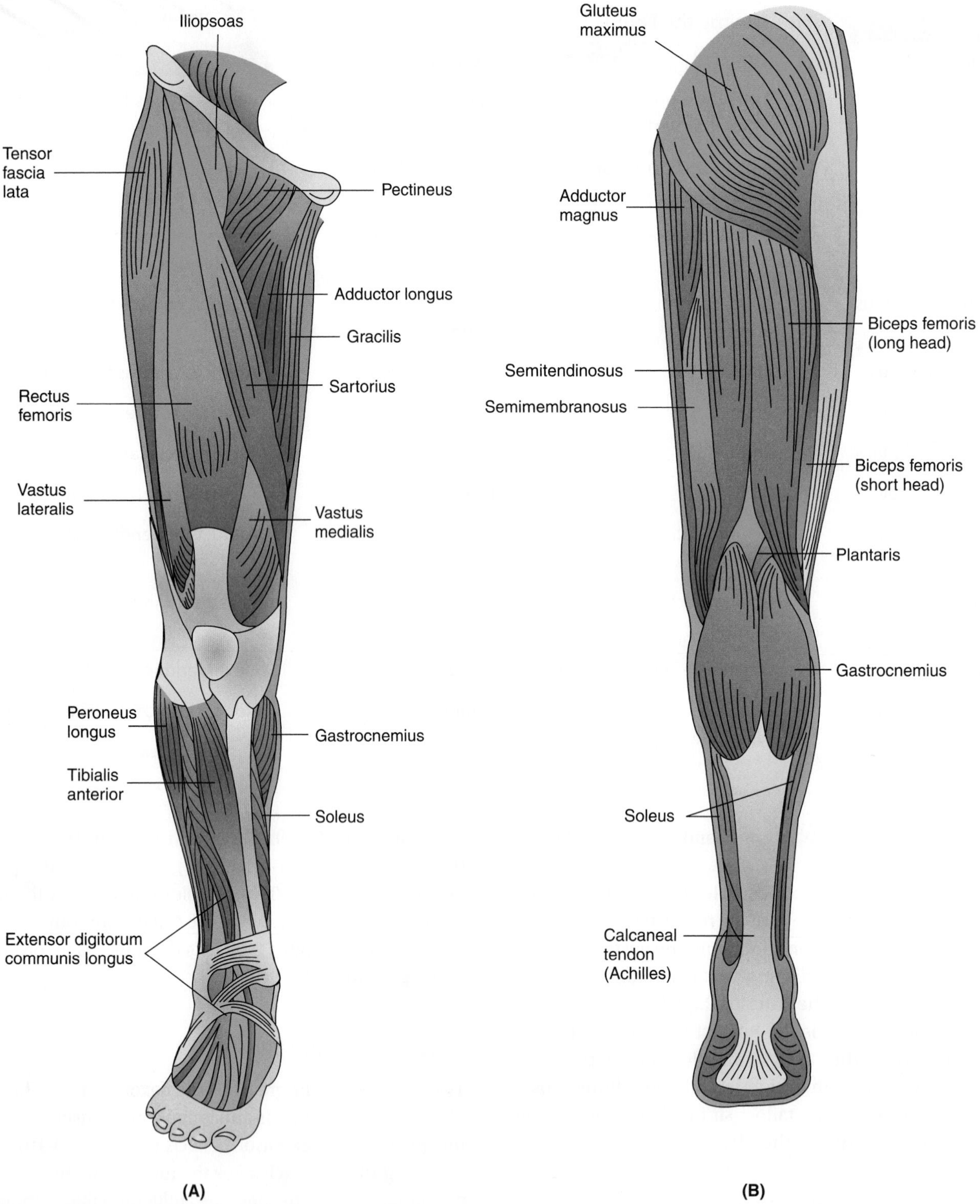

FIGURE 9-11. Superficial muscles of the leg: (A) anterior view, (B) posterior view.

Table 9-9 Muscles Moving the Knee Joint

Muscle	Function
Biceps femoris (two heads)	Flexes leg; rotates laterally after flexed
Semitendinosus	Flexes leg, extends thigh
Semimembranosus	Flexes leg, extends thigh
Popliteus	Flexes leg, rotates it
Gracilis	Adducts thigh, flexes leg
Sartorius	Flexes thigh, rotates it laterally
Quadriceps femoris: (four heads)	Extends leg and flexes the thigh
Rectus femoris	
Vastus lateralis	
Vastus medialis	
Vastus intermedius	

Table 9-10 Muscles Moving the Foot and Toes

Muscles Moving the Foot

Muscle	Function
Gastrocnemius	Plantar flexes foot, flexes leg, supinates foot
Soleus	Plantar flexes foot
Tibialis posterior	Plantar flexes foot
Tibialis anterior	Dorsally flexes foot
Peroneus tertius	Dorsally flexes foot
Peroneus longus	Everts, plantar flexes foot
Peroneus brevis	Everts foot
Plantaris	Plantar flexes foot

Muscles Moving the Toes

Muscle	Function
Flexor hallucis brevis	Flexes great toe
Flexor hallucis longus	Flexes great toe
Extensor hallucis longus	Extends great toe, dorsiflexes ankle
Interossei dorsales	Abduct, flex toes
Flexor digitorum longus	Flexes toes, extends foot
Extensor digitorum longus	Extends toes
Abductor hallucis	Abducts, flexes great toe
Abductor digiti minimi	Abducts little toe

COMMON DISEASE, DISORDER, OR CONDITION

DISORDERS OF MUSCLE

Disorders causing diseases to muscles can originate from a number of sources: the vascular supply, the nerve supply or the connective tissue sheaths around the muscle cells, or muscle bundles. The major symptoms of muscular disorders are paralysis, weakness, degeneration or atropy of the muscle, pain, and spasms.

CONTRACTURE

A **contracture** is a condition in which a muscle shortens its length in the resting state. Contractures commonly occur in individuals who are bedridden for long periods and the muscles are not properly exercised. Contractures can be prevented by keeping the body in proper alignment when resting, shifting positions periodically, and by periodically exercising the muscles. If contractures occur, they are treated by the slow and painful procedure of relengthening and exercising the muscles.

CRAMPS

Cramps are spastic and painful contractions of muscles that occur because of an irritation within the muscle such as inflammation of connective tissue or lactic acid buildup.

MYALGIA

Myalgia (my-**ALL**-jee-ah) is a term that means muscle pain.

MYOSITIS

Myositis (my-oh-**SIGH**-tis) means inflammation of muscular tissue.

ATROPHY

Atrophy (**AT**-troh-fee) is a decrease in muscle bulk due to a lack of exercise, as when a limb is in a cast for a prolonged period. Stimulation of nerves with a mild electric current can keep muscular tissue viable until full muscular activity can return. In severe cases, the muscle fibers are actually lost and replaced with connective tissue.

HYPERTROPHY

Hypertrophy (high-**PER**-troh-fee) (the opposite of atrophy) is an increase in the size of a muscle caused by an increase in the bulk of muscle cells through exercises, like weightlifting. This activity increases the amount of protein within the muscle cell. We are born with all the muscle cells we will ever have. They do not increase in numbers, only in size.

TENDINITIS

Tendinitis (tin-den-**EYE**-tis) is an inflammation of a tendon.

MUSCULAR DYSTROPHY

Muscular dystrophy (**MUSS**-kew-lehr **DIS**-troh-fee) is an inherited muscular disorder, occurring most often in males, in which the muscle tissue degenerates over time, resulting in complete helplessness.

MYASTHENIA GRAVIS

Myasthenia gravis (mye-as-**THEE**-nee-ah **GRAV**-is) is characterized by the easy tiring of muscles, or muscle weakness. It usually begins in the facial muscles. It is caused by the abnormal destruction of acetylcholine receptors at the neuromuscular junction. This is an autoimmune disorder caused by antibodies that attack acetylcholine receptors.

(continues)

COMMON DISEASE, DISORDER, OR CONDITION

DISORDERS OF MUSCLE (continued)

AMYOTROPHIC LATERAL SCLEROSIS (ALS)

Amyotrophic lateral sclerosis is also known as Lou Gehrig's disease. It is a progressively degenerative disease of the motor neurons of the body. It affects people in middle age. Around 10% of the cases can be genetically inherited, the gene causing the condition being on chromosome 21. The disease is caused by a degeneration of the motor neurons of the anterior horns of the spinal cord and the corticospinal tracts. It begins with muscle weakness and atrophy. It usually first involves the muscles of the legs, forearm, and hands. It then spreads to involve muscles of the face, affecting speech, and other muscles of the body. Within 2–5 years there is a loss of muscle control that can lead to death. There is no known cure for the disease. It also sometimes is referred to as wasting palsy.

RIGOR MORTIS

Rigor mortis occurs after death when muscles cannot contract (RIGOR = rigidity, MORTIS = of death). This occurs as calcium ions leak out of the sarcoplasmic reticulum and cause contraction to occur. Since no ATP is being produced, the myosin cross-bridges cannot detach from the actin filaments. Therefore, the muscles remain in a state of rigidity for about 24 hours. After 12 hours later, the tissues degenerate and decay and the rigidity is lost.

SNORING

Snoring is caused by the rapid vibration of the uvula and soft palate producing a harsh, rasping sound during sleep. This is caused by breathing through both the mouth and the nose. Over-the-counter nasal dilators are on the market today to help alleviate this problem.

TETANUS

Tetanus is caused by the bacterium *Clostridium tetani*. The bacterium is anaerobic, living in the absence of oxygen. Thus, stepping on an old nail, producing a deep puncture wound, will transfer the bacterium to tissues with very little oxygen. This can also occur with any deep cut or wound. The bacterium is very common in our environment. When in tissues, it releases a strong toxin that suppresses the activity of motor neurons. The motor neurons produce a sustained contraction of skeletal muscles. Because it affects the muscles of the mouth, the disease is also called lockjaw, causing difficulty in swallowing. Other symptoms include headache, muscle spasms, and muscle stiffness. In the United States we have an immunization program to control tetanus. After the initial tetanus vaccine, booster shots are recommended every 10 years to prevent the disease.

POLIO

Polio is caused by a virus that is an *Enterovirus*. It enters the spinal cord of the central nervous system and affects the peripheral nerves and muscles that they control. The virus attacks the motor neurons in the anterior horn of the spinal cord, which is gray matter (*polio* in Greek means gray matter). In the 1940's and 1950's thousands of children in the United States contracted the disease called acute paralysis poliomyelitis. Many developed paralysis of their limbs, had to wear braces, and some had to be placed in "iron lungs" because they could not breathe properly. Fortunately a vaccine was developed (Salk and Sabin vaccines) and the disease is now quite rare in the United States. However, it does occur in many third world countries where vaccines are not made available to people in spite of the Third World Health Organization's effort to eradicate the disease.

(continues)

COMMON DISEASE, DISORDER, OR CONDITION

DISORDERS OF MUSCLE (continued)

PLANTAR FASCIITIS

Plantar fasciitis is an inflammation of the connective tissue (fascia) that is part of the arches of the foot. It can be very painful and is caused by continuous stretching of the muscles and ligaments of the foot. Usually long-distance runners or individuals whose occupations require lots of walking can develop this condition.

FIBROMYALGIA

Fibromyalgia is a form of rheumatism but does not affect the joints. It is characterized by long-term tendon and muscle pain accompanied by stiffness, occasional muscle spasms, and fatigue. There is no cure. It also causes sleep disturbance. It is also known as fibrositis. It can produce pain in the shoulder and neck, arms, hands, lower back, hips, legs, knees, and feet. Muscles relaxants, anti-inflammatory drugs, and physical therapy are methods of treatment for temporary relief.

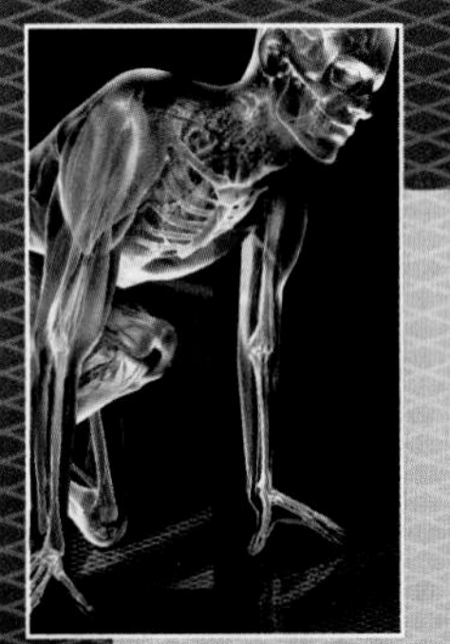

AS THE BODY AGES

As we age, sometimes beginning in our late 20s, a gradual loss of muscle cells or fibers occurs. By 40 years of age, a gradual decrease begins to occur in the size of each individual muscle. By the late 70s, 50% of our muscle mass disappears. Consistent exercising such as walking can delay and decrease this effect of aging. Resistant exercise, like working out at the gym with some weights, is an even better way to maintain muscle mass. As aging continues, the time it takes for a muscle to respond to nervous stimuli decreases, resulting in reduced stamina and a loss of power. Older adult women, in particular, may become bent over due to changes in the sacrospinalis muscle, which is found on either side of the vertebral column. Its loss of power produces the hunchback appearance often seen in the older adults. Remaining physically active can prevent many of the age-related changes that can occur in skeletal muscle. ■

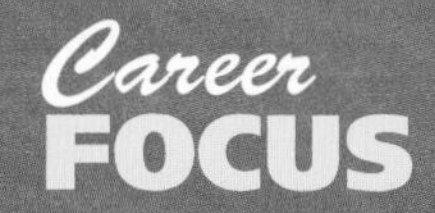

These are careers that are available to individuals who are interested in the muscular system.

- **Physicians** can specialize in sports medicine and treat sports-related problems and injuries of muscles, bones, and joints.
- **Doctors of osteopathic medicine** take a therapeutic approach to medicine by placing greater emphasis on the relationship between the organs and the musculoskeletal system. These doctors also use drugs, radiation, and surgery for medical diagnosis and therapy.
- **Massage therapists** manipulate the muscles by stroking, kneading, and rubbing to increase circulation of blood to the muscles to improve muscle tone and bring relaxation to the patient.

Integumentary System

- Sensory receptors in the skin stimulate muscle contraction in response to environmental changes in temperature or pressure.
- Skin dissipates heat during muscle contraction.

Skeletal System

- Bones provide attachments for muscles and act as levers to bring about movement.
- Bones store calcium necessary for muscular contraction.

Nervous System

- Motor neurons stimulate muscle contraction by releasing acetylcholine at their axon terminals in the neuromuscular junction.

Endocrine System

- Growth hormone from the anterior pituitary gland stimulates muscular development.
- Hormones increase blood flow to muscles during exercise.

Cardiovascular System

- The heart pumps blood to the muscle cells, carrying nutrients to and wastes away from the muscle cells.
- Red blood cells carry oxygen to and carbon dioxide gas away from the muscle cells.

Lymphatic System

- Skeletal muscle contractions push lymph through the lymphatic vessels, particularly by the action of breathing.
- Lymphocytes combat infection in the muscles and develop immunities.

Digestive System

- Skeletal muscle contraction in swallowing brings food to the system; smooth muscle contraction pushes digested food through the stomach and intestines.
- The intestines absorb digested nutrients to make them available to muscle cells for their energy source.

Respiratory System

- Breathing depends on the diaphragm and intercostal muscles.
- The lungs provide oxygen for muscle cells and eliminate the carbon dioxide waste from cellular respiration.

Urinary System

- Smooth muscles push urine from the kidneys down the ureters to the bladder.
- Skeletal muscles control urine elimination.
- In the loop of Henle in the nephrons of the kidneys, calcium levels are controlled by eliminating any excess or restoring needed calcium to the blood for muscle contraction.

Reproductive System

- Skeletal muscles are involved in kissing, erections, transferring sperm from the male to the female, and other forms of sexual behavior and activity.
- Smooth muscle contractions in the uterus bring about delivery of the newborn.

SUMMARY OUTLINE

INTRODUCTION

1. Skeletal muscles help us read by moving our eyes, allow us to move in our environment and breathe.
2. Smooth muscles push food through our intestines, contain blood in our arteries and veins, and push urine down our ureters.
3. Cardiac muscle pumps blood through our heart and blood vessels and maintains blood pressure.
4. Muscles make up 40% to 50% of our body weight.

THE TYPES OF MUSCLE

The three types of muscle tissue are skeletal, smooth or visceral, and cardiac.

1. Skeletal muscle cells are voluntary, striated, multinucleated cells that are much longer than their width, hence are also called muscle fibers.
2. Smooth muscle cells are involuntary (we cannot control them at will), nonstriated, and uninucleated fibers.
3. Cardiac muscle cells are also involuntary but are striated and uninucleated. These cells do not look like fibers but have extensions or branches.

THE ANATOMY OF SKELETAL OR STRIATED MUSCLE

1. The skeletal muscle cell or fiber is surrounded by an electrically polarized cell membrane called a sarcolemma.
2. A muscle actually consists of a number of skeletal muscle bundles called fasciculi. Each bundle or fascicle is composed of a number of muscle fibers or cells.
3. Each muscle cell in a fascicle is surrounded by delicate connective tissue called the endomysium.
4. Each bundle or fascicle is surrounded with another layer of connective tissue called the perimysium.
5. The perimysium of each fascicle is covered with another layer of connective tissue surrounding the whole muscle called the epimysium.
6. A layer of areolar tissue called the fascia is on top of the epimysium.
7. Under a microscope, skeletal muscle cells have cross-striations due to the overlap of the dark bands of the thick protein myosin (called A bands) and the light bands of the thin protein actin (called I bands).
8. In the middle of an I band is a Z line.
9. In the middle of an A band is an H line or zone.
10. The area between two adjacent Z lines is called a sarcomere.
11. Electron microscopy reveals that the muscle fibrils of actin and myosin that make up a muscle cell are surrounded by a sarcotubular system composed of T tubules and an irregular curtain called the sarcoplasmic reticulum.
12. The function of the T tubules is the rapid transmission of a nerve impulse to all the fibrils in a cell while the sarcoplasmic reticulum stores calcium ions.

THE PHYSIOLOGY OF MUSCLE CONTRACTION

1. All of the muscle cells or fibers innervated by the same motor neuron is called a motor unit.
2. Muscle cells have four properties: excitability by a stimulus; conductivity of that stimulus through their cytoplasm; contractility, which is the reaction to the stimulus; and elasticity, which allows the cell to return to its original shape after contraction.
3. Muscle contraction is caused by the interaction of three factors: neuroelectrical, chemical, and energy sources.

Neuroelectrical Factors

1. Muscle cells have positively charged sodium ions (Na^+) in greater concentration outside the muscle cell than inside the cell.
2. Muscle cells have positively charged potassium ions (K^+) in greater concentration inside the muscle cell than outside the muscle cell.
3. The outside of a muscle cell is positively charged electrically and the inside is negatively charged. This electrical distribution is known as the resting potential of the cell membrane.
4. When a motor neuron innervates the muscle cell, acetylcholine is secreted from the axon terminals into the neuromuscular junction. This causes sodium ions to rush inside the cell membrane, creating an electrical potential (changing the inside from negative to positive).
5. Potassium ions move outside the cell membrane to try to restore the resting potential but cannot do so because so many sodium ions are rushing in.
6. The influx of positive sodium ions causes the T tubules to transmit the stimulus deep into the muscle cell, creating an action potential.
7. The action potential causes the sarcoplasmic reticulum to release calcium ions into the fluids surrounding the myofibrils of actin and myosin.
8. Troponin and tropomyosin (inhibitor substance) have kept the actin and myosin filaments separate but the calcium ions inhibit the action of the troponin and tropomyosin.
9. The calcium causes the myosin to become activated myosin. The activated myosin now links up with the actin filaments.

Chemical Factors

1. The cross-bridges or heads of myosin filaments have ATP. When the cross-bridges link with the actin, the breakdown of the ATP releases energy that is used to pull the actin filaments in among the myosin filaments. The area between two Z lines gets smaller, whereas the A band remains the same. This is contraction at the molecular level.
2. Meanwhile, the sodium-potassium pump has operated. It has pumped out the sodium ions that initially rushed in and pulled back in the potassium ions that had rushed out, restoring the muscle cell's resting potential. The calcium ions get reabsorbed by the sarcoplasmic reticulum, causing the action potential to cease and restoring the resting potential. The muscle cell now relaxes as contraction ceases.
3. The whole process of contraction occurs in 1/40 of a second.

Energy Sources

1. ATP is the energy source for muscle contraction: actin + myosin + ATP $\rightarrow$ actomyosin + ADP + PO_4 + the energy of contraction.
2. ATP is produced in glycolysis, the Krebs citric acid cycle, and electron transport yielding 36 ATP.
3. ATP is produced occasionally in the absence of oxygen in muscle cells during anaerobic respiration, yielding only 2 ATP with a buildup of lactic acid during strenuous exercise.
4. Muscle cells can also take up free fatty acids from the blood and break those down into ATP.
5. Muscle cells also use phosphocreatine as a source of phosphate to produce ATP.

THE MUSCLE TWITCH

1. Laboratory analysis of a muscle contraction reveals a brief latent period immediately following the stimulus followed by actual contraction. Relaxation follows contraction. This is called a muscle twitch.
2. The strength of a contraction depends on the strength, speed, and duration of the stimulus as well as the weight of the load and the temperature.
3. The all-or-none law states that a stimulus strong enough to cause contraction in an individual muscle cell will result in maximal contraction.

MUSCLE TONE

1. Tone is that property of a muscle in which a state of partial contraction is maintained throughout a whole muscle.
2. Tone maintains pressure on the abdominal contents, helps maintain blood pressure in blood vessels, and aids in digestion. Tone gives a firm appearance to skeletal muscles.

3. There are two types of contraction: isotonic contraction occurs when muscles become shorter and thicker as when lifting a weight and tension remains the same; isometric contraction occurs when tension increases but the muscles remain at a constant length as when we push against a wall.

THE ANATOMY OF SMOOTH MUSCLE

1. Smooth muscle is found in hollow structures like the intestines, arteries, veins, and bladder. It is under the control of the autonomic nervous system.
2. Smooth muscle cells are involuntary, uninucleated, and nonstriated.
3. In hollow structures, smooth muscle is arranged in two layers: an outer longitudinal layer and an inner circular layer. This results in material being pushed forward in the tube by simultaneous contraction of both layers.

THE ANATOMY OF CARDIAC MUSCLE

1. Cardiac muscle is found only in the heart and is controlled by the autonomic nervous system.
2. Cardiac muscle cells are involuntary, uninucleated, and striated. They also have intercalated disks for coordinating contraction.
3. Cardiac muscle cells can receive an impulse, contract, immediately relax, and receive another impulse. This occurs about 75 times a minute.

THE NAMING AND ACTIONS OF SKELETAL MUSCLE

1. Muscles can be named according to their action, shape, origin and insertion, location, or the direction of their fibers.
2. The origin is the more fixed attachment; the insertion is the movable attachment of a muscle.
3. Tendons attach a muscle to a bone. A wide flat tendon is called an aponeurosis.
4. Muscles that bend a limb at a joint are called flexors; those that straighten a limb are called extensors.
5. Abductors move a limb away from the midline; adductors bring a limb toward the midline of the body.
6. Rotators revolve a limb around an axis.
7. Muscles that raise the foot are dorsiflexors; those that lower the foot are plantar flexors.
8. Muscles that turn the palm upward are supinators; those that turn the palm of the hand downward are pronators.
9. Levators raise a part of the body; those muscles that lower a part of the body are depressors.
10. Prime movers are muscles that bring about an action. Those that assist the prime movers are synergists.

THE FUNCTION AND LOCATION OF SELECTED SKELETAL MUSCLES

1. Facial muscles around the eyes and mouth assist in nonverbal communication like smiling.
2. Muscles around the upper and lower jaw assist in chewing or mastication.
3. Six muscles attach to the eye and move the eye in all directions.
4. The main muscle that moves the head is the sternocleidomastoid.
5. The upper arm is moved mainly by the deltoid, pectorals, and rotator cuff muscles.
6. The forearm can be flexed and extended; the supinators and pronators supinate and pronate the forearm and move the hand.
7. The wrist and fingers can be flexed, extended, abducted, and adducted.
8. The thumb does opposition and can grasp implements, resulting in all the unique abilities of the hand.
9. Three layers of trunk muscle compress our abdominal contents laterally, while the rectus abdominus in the front produces the washboard effect from sit-ups.
10. Breathing is accomplished by the diaphragm muscle and the intercostal muscles of the ribs.
11. Muscles of the hip flex, extend, abduct, and adduct the thigh.
12. Muscles of the thigh, like the hamstrings, flex the knee; the quadriceps femoris extends the knee.
13. Muscles of the foot and toes produce plantar flexion and dorsiflexion as in walking, eversion and inversion of the sole of the foot, and flexion and extension of the toes.

REVIEW QUESTIONS

*1. Explain muscle contraction based on neuroelectrical factors, chemical interactions, and energy sources.

*2. Compare the anatomy of a skeletal muscle cell with that of a smooth and cardiac muscle cell.

*3. Compare isometric contraction with isotonic contraction.

4. Define *muscle tone.*

5. What are some symptoms of muscle disorders?

*6. Explain why disorders of muscles can be caused by a number of problematic areas in tissue other than muscles.

***Critical Thinking Questions**

FILL IN THE BLANK

Fill in the blank with the most appropriate term.

1. Myofibrils have dark bands known as the __________________ bands composed of the protein __________________.
2. Myofibrils also have light bands known as the __________________ bands composed of the protein __________________.
3. A dark line in the light band is known as the __________________ line, and the area between two of these adjacent lines is called a __________________.
4. Electron microscopy has revealed that muscle fibrils are surrounded by the sarcotubular system. Part of that system is the __________________ system that functions in the rapid transmission of the stimulus to all fibrils in the muscle via the release of __________________ ions from the sarcoplasmic reticulum.
5. All of the muscle fibers that are innervated by the same nerve fiber are called a __________________ unit.
6. __________________ ions have a greater concentration inside the resting muscle cell, whereas __________________ ions have a greater concentration outside the resting muscle cell.
7. A nerve impulse causes __________________ to be released at the neuromuscular junction, which causes _________ ions to rush inside the muscle cell, changing its polarity.
8. Two inhibitor substances surrounding the myofilaments of actin and myosin are __________________ and __________________.
9. Smooth and cardiac muscles are under the control of the __________________ nervous system.
10. The source of energy for muscle contraction is __________________ molecules.

MATCHING

Place the most appropriate number in the blank provided.

_____	Sarcolemma	1.	Muscle bundles
_____	Fascia	2.	Muscle pain
_____	Epimysium	3.	Antagonists
_____	Prime movers	4.	Inflammation of muscular tissue
_____	Myasthenia gravis	5.	Areolar tissue covering entire muscle
_____	Myositis	6.	Steady state of contraction
_____	Extensors	7.	Easy tiring of muscles
_____	Fasciculi	8.	Electrically polarized muscle cell membrane
_____	Tone	9.	Agonists
_____	Myalgia	10.	Connective tissue covering whole muscle
		11.	Contracture
		12.	Connective tissue covering a fascicle

Search and Explore

- Visit the Muscular Dystrophy Association website at http://www.mda.org and research one of the types of muscular dystrophy. Write one to two paragraphs in your notebook on what you learned about this disease.
- Search the Internet with key words from the chapter to discover additional information and interactive exercises. Key words might include skeletal muscle, smooth muscle, cardiac muscle, or physiology of muscle contraction.

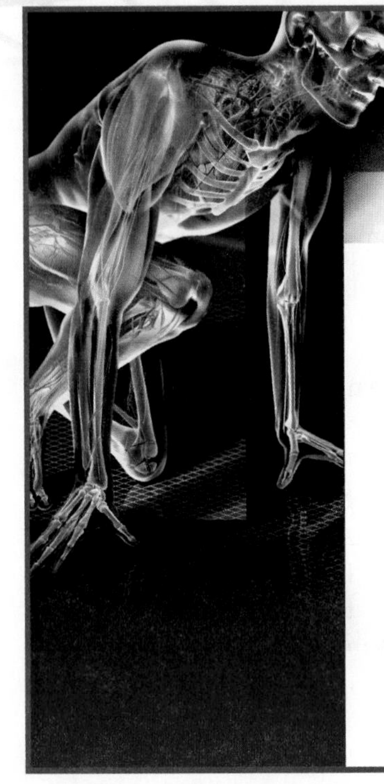

CASE STUDY

Nico Fapoulas, a 48-year-old man, is talking with his health care provider about symptoms he is experiencing. He states that when he goes for his morning jog, his legs feel weak and tired. He is having problems with simple tasks that require manual dexterity such as writing or unlocking doors. Upon examination, the health care provider observes some atrophy of the muscles of Nico's legs, forearms, and hands. He also notes that Nico is having a slight problem with his speech.

Questions

1. What disorder do you think Nico might be developing?
2. Why might Nico be experiencing muscle weakness and atrophy?
3. What is the cause of this condition?
4. What is the prognosis for an individual with this condition?

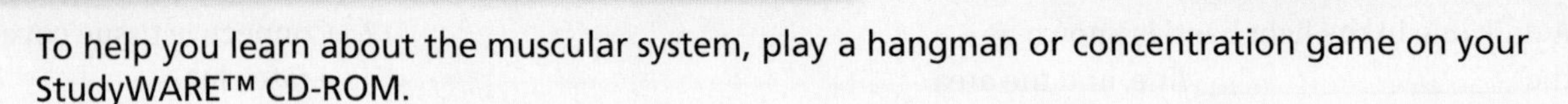

*Study*WARE™ Connection

To help you learn about the muscular system, play a hangman or concentration game on your StudyWARE™ CD-ROM.

Study Guide Practice

Go to your **Study Guide** for more practice questions, labeling and coloring exercises, and crossword puzzles to help you learn the content in this chapter.

LABORATORY EXERCISE: THE MUSCULAR SYSTEM

Materials needed: A model of the human torso with skeletal muscles; either photographs of an athlete or dancer, or a live model

1. Examine a model of a human torso with various superficial muscles.
2. Your instructor will show you a DVD or a videotape on how muscles function.
3. Through photographs or a human model (if available) with good muscle development, identify as many of the superficial muscles of the body as possible.

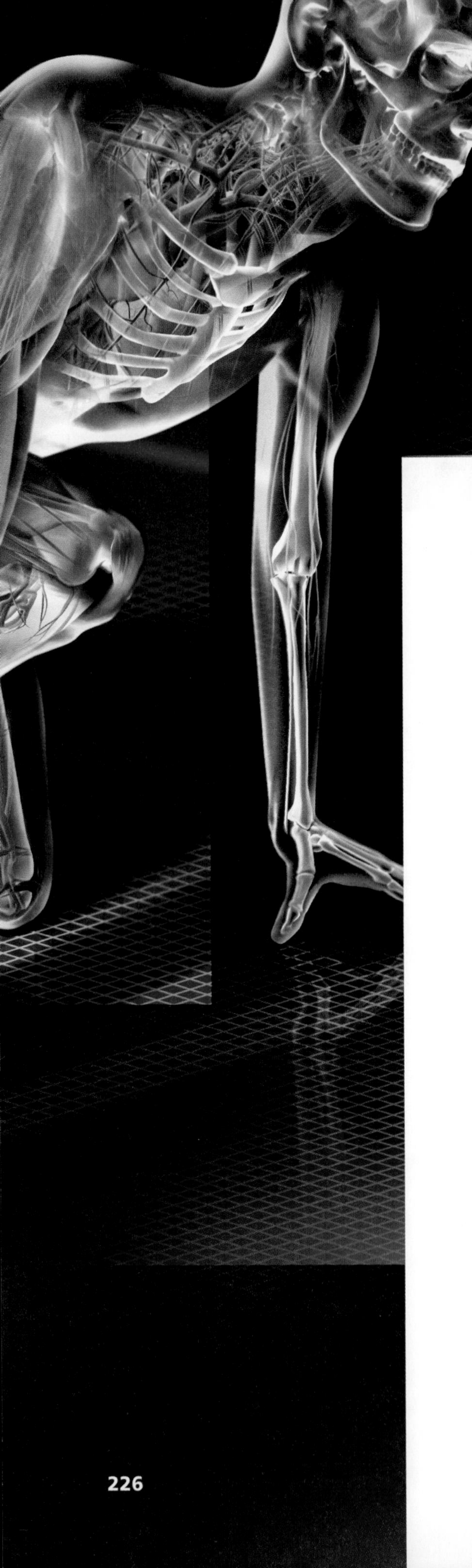

The Nervous System Introduction, Spinal Cord, and Spinal Nerves

CHAPTER OBJECTIVES

After studying this chapter, you should be able to:

1. Name the major subdivisions of the nervous system.
2. Classify the different types of neuroglia cells.
3. List the structural and functional classification of neurons.
4. Explain how a neuron transmits a nerve impulse.
5. Name the different types of neural tissues and their definitions.
6. Describe the structure of the spinal cord.
7. Name and number the spinal nerves.

10

KEY TERMS

INTRODUCTION

The nervous system is the body's control center and communication network. It directs the functions of the body's organs and systems. It allows us to interpret what is occurring in our external environment and helps us to decide how to react to any environmental change or stimulus by causing muscular contractions. It shares in the maintenance of homeostasis (the internal environment of our bodies) with the endocrine system by controlling the master endocrine gland (the pituitary) through the hypothalamus of the brain. See Concept Map 10-1: Spinal Cord and Spinal Nerves.

ORGANIZATION

The nervous system can be grouped into two major categories (Figure 10-1). The first is the **central nervous system (CNS)**, which is the control center for the whole system. It consists of the brain and spinal cord. All body sensations and changes in our external environment must be relayed from receptors and sense organs to the CNS to be interpreted (what do they mean?) and then, if necessary, acted on (such as move away from a source of pain or danger).

The second major category is the **peripheral nervous system (PNS)**, which is subdivided into several smaller units. This second category consists of all the nerves that connect the brain and spinal cord with sensory receptors, muscles, and glands.

The PNS can be divided into two subcategories: the **afferent peripheral system**, which consists of afferent or sensory neurons that convey information from receptors in the periphery of the body to the brain and spinal cord, and the **efferent peripheral system**, which consists of efferent or motor neurons that convey information from the brain and spinal cord to muscles and glands.

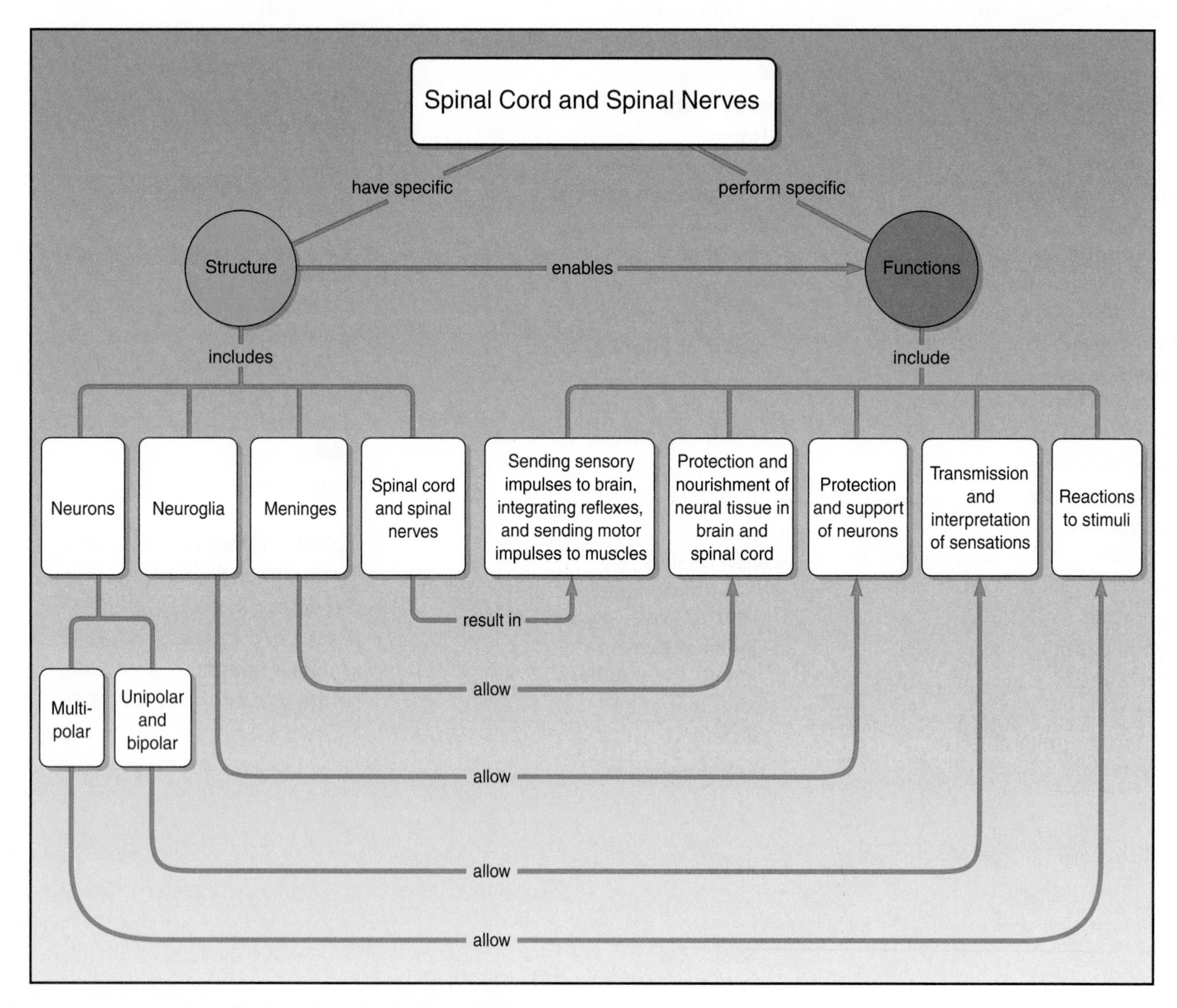

CONCEPT MAP 10-1. Spinal cord and spinal nerves.

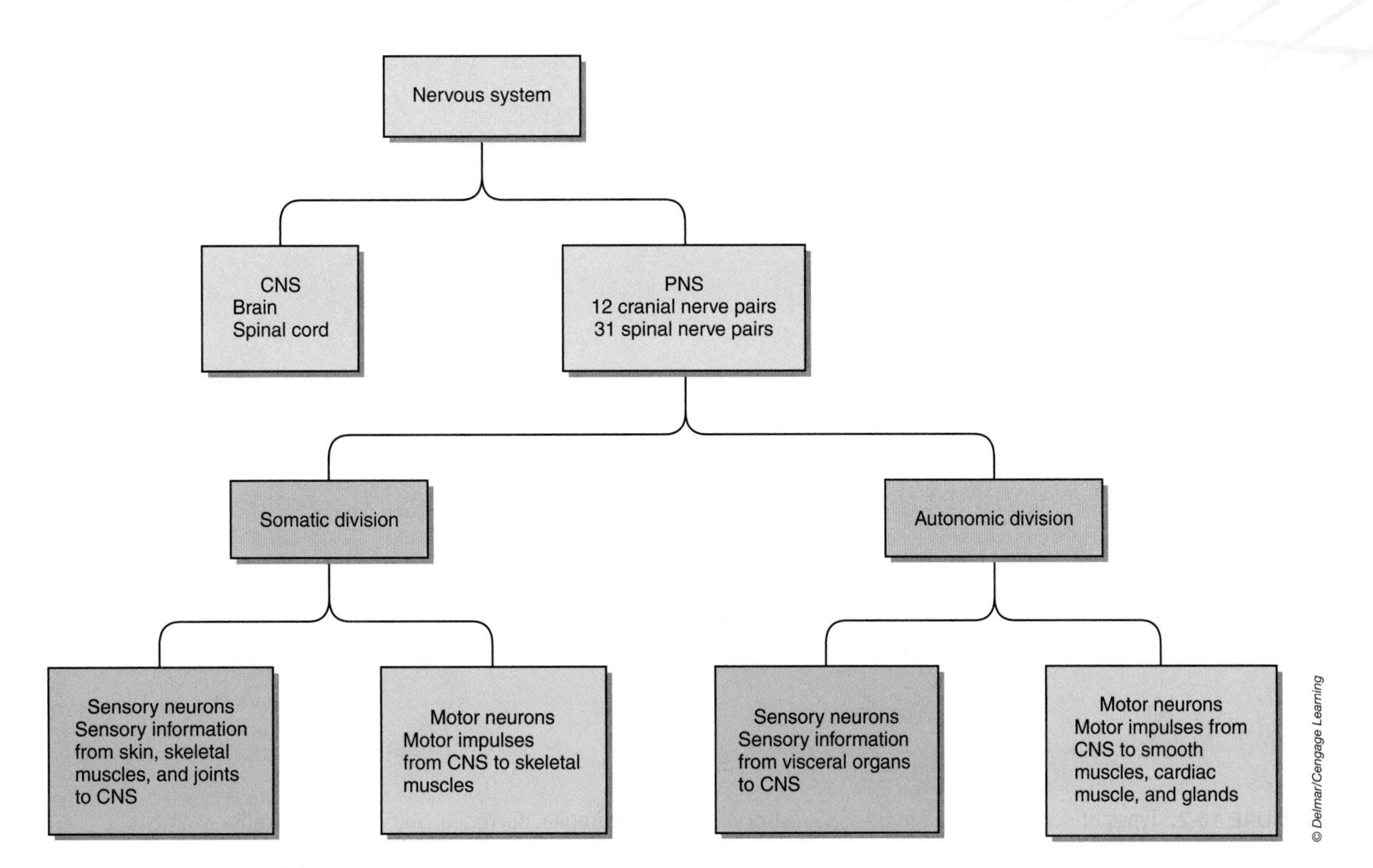

FIGURE 10-1. Divisions of the nervous system.

The efferent peripheral system can be further subdivided into two subcategories. The first is the **somatic nervous system**, which conducts impulses from the brain and spinal cord to skeletal muscle, thereby causing us to respond or react to changes in our external environment. The second is the **autonomic nervous system (ANS)**, which conducts impulses from the brain and spinal cord to smooth muscle tissue (like the smooth muscles of the intestine that push food through the digestive tract), to cardiac muscle tissue of the heart, and to glands (like the endocrine glands). The ANS is considered to be involuntary. The organs affected by this system receive nerve fibers from two divisions of the ANS: the **sympathetic division**, which stimulates or speeds up activity and thus involves energy expenditure and uses **norepinephrine** (nor-ep-ih-**NEH**-frin) as a neurotransmitter, and the **parasympathetic division**, which stimulates or speeds up the body's vegetative activities such as digestion, urination, and defecation and restores or slows down other activities. It uses **acetylcholine** (**ah**-seh-till-**KOH**-leen) as a neurotransmitter at nerve endings.

CLASSIFICATION OF NERVE CELLS

Nervous tissue consists of groupings of nerve cells or **neurons** (**NOO**-ronz) that transmit information called nerve impulses in the form of electrochemical changes. A **nerve** is a bundle of nerve cells or fibers. Nervous tissue is also composed of cells that perform support and protection. These cells are called **neuroglia** (noo-**ROWG**-lee-ah) or **glial** (**GLEE**-al) **cells** (neuroglia means nerve glue). Over 60% of all brain cells are neuroglia cells.

Neuroglia Cells

There are different kinds of neuroglia cells, and, unlike neurons, they do not conduct impulses (Figure 10-2). Table 10-1 lists the types of neuroglia. **Astrocytes** are star-shaped cells that wrap around nerve cells to form a supporting network in the brain and spinal cord. They attach neurons to their blood vessels, thus helping regulate nutrients and ions that are needed by the nerve cells. **Oligodendroglia** (all-ih-goh-**DEN**-droh-**GLEE**-ah) look like small astrocytes. They also provide support by forming semirigid connective-like tissue rows between neurons in the brain and spinal cord. They produce the fatty **myelin** (**MY**-eh-lin) **sheath** on the neurons of the brain and spinal cord of the CNS. **Microglia** (my-**KROWG**-lee-al) **cells** are small cells that protect the CNS and whose role is to engulf and destroy microbes like bacteria and cellular debris. **Ependymal** (eh-**PIN**-dih-mal) **cells** line the fluid-filled ventricles of the brain. Some

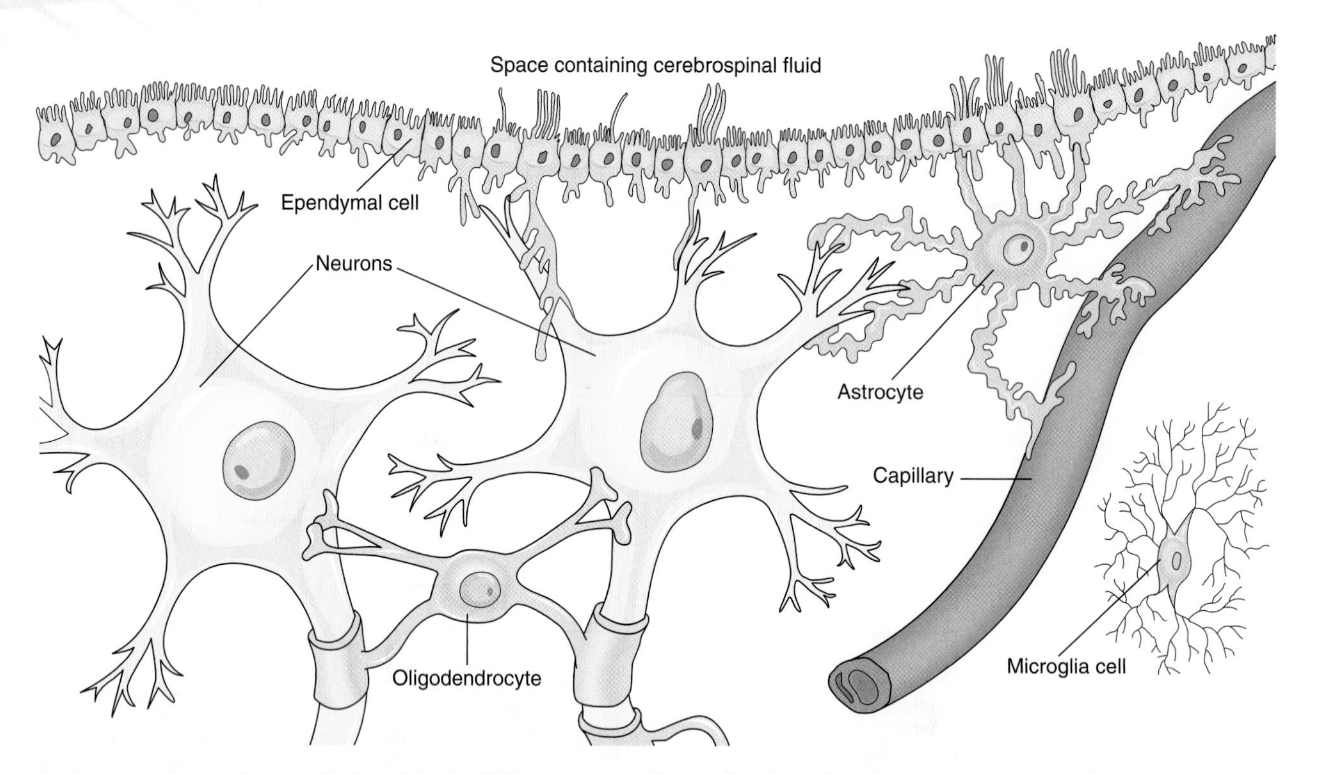

FIGURE 10-2. Types of neuroglia found in the CNS: astrocytes, oligodendroglia, microglia, and ependymal cells.

Table 10-1 Types of Neuroglia

Type	Description
Astrocytes	Star-shaped cells that function in the blood-brain barrier to prevent toxic substances from entering the brain
Oligodendroglia	Provide support and connection
Microglia	Involved in the phagocytosis of unwanted substances
Ependymal cells	Form the lining of the cavities in the brain and spinal cord
Schwann cells	Located only in the PNS and make up the neurilemma and myelin sheath

produce cerebrospinal fluid and others with cilia move the fluid through the CNS. **Schwann cells** form myelin sheaths around nerve fibers in the PNS.

The Structure of a Neuron

Each nerve cell's body contains a single nucleus (Figure 10-3). This nucleus is the control center of the cell. In the cytoplasm there are mitochondria, Golgi bodies, lysosomes, and a network of threads called neurofibrils that extend into the axon part of the cell, referred to as the fiber of the cell. In the cytoplasm of the cell body there is extensive rough endoplasmic reticulum (ER). In a neuron, the rough ER has ribosomes attached to it. These granular structures are referred to as **Nissl** (**NISS**-l) **bodies**, also called **chromatophilic substance**, and are where protein synthesis occurs.

There are two kinds of nerve fibers on the nerve cell: **dendrites** (**DEN**-drightz) and **axons**. Dendrites are short and branched, like the branches of trees. These are the receptive areas of the neuron and a multipolar neuron will have many dendrites. A nerve cell, however, will have only one axon, which begins as a slight enlargement of the cell body called the axonal hillock. The axon is a long process or fiber that begins singly but may branch and at its end has many fine extensions called **axon terminals** that contact with dendrites of other neurons. Numerous mitochondria and neurofibrils are in the axon.

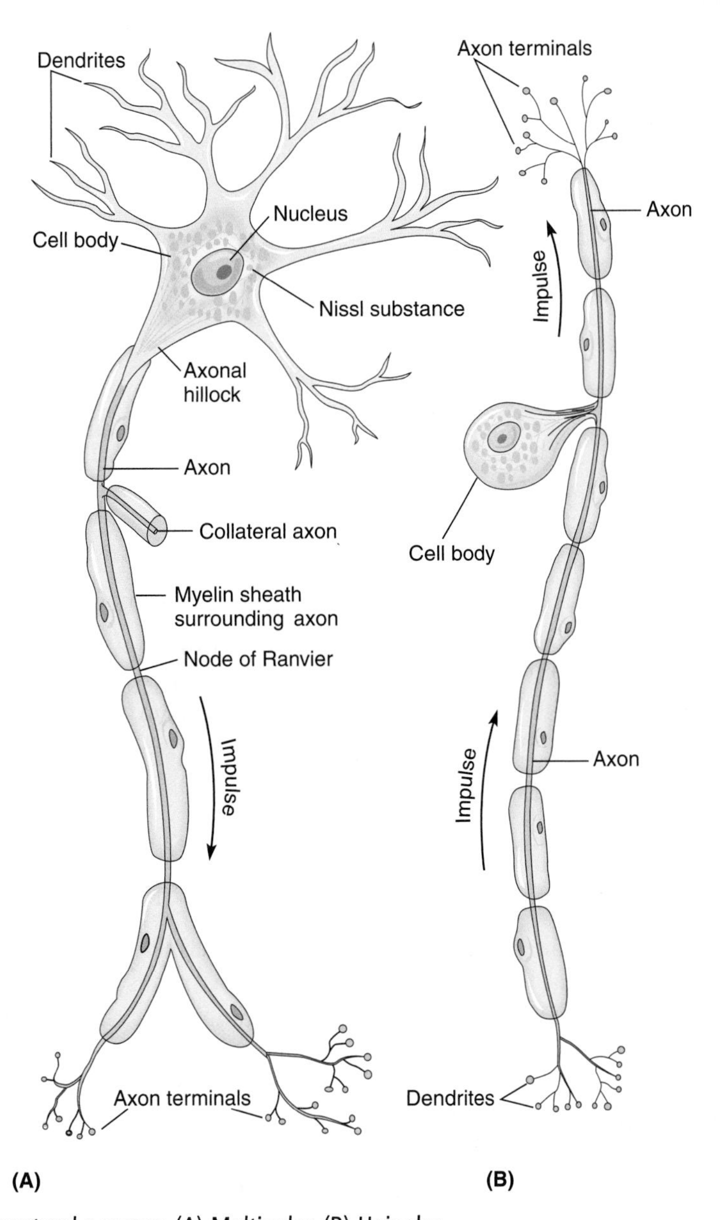

FIGURE 10-3. Two types of structural neurons. (A) Multipolar. (B) Unipolar.

The large peripheral axons are enclosed in fatty myelin sheaths produced by the Schwann cells. These are a type of neuroglial cell that wrap tightly in layers around the axon, producing fatty sheets of lipoprotein. The portions of the Schwann cell that contain most of the cytoplasm of the cell and the nucleus remain outside of the myelin sheath and make up a portion called the neurilemma. Narrow gaps in the sheath are the nodes of Ranvier.

Structural Classification of Neurons

Cells that conduct impulses from one part of the body to another are called neurons. They may be classified by both function and structure. The structural classification consists of three types of cells. **Multipolar neurons** are neurons that have several (multi) dendrites and one axon. Most neurons in the brain and spinal cord are this type. The neuron studied in Chapter 5 is this type. Recall that the part of the neuron with the nucleus is called the cell body. The smaller extensions of the cell body are the dendrites, and the single long extension is called the axon. Single cells called Schwann cells, also called **neurolemmocytes** (**noo**-row-leh-**MOH**-sights), surround the axon at specific sites and form the fatty myelin sheath around the axons in the peripheral nervous system (Figure 10-4). Gaps in the myelin sheath are called **nodes of Ranvier** (**NOHDZ** of rahn-vee-**A**), also called **neurofibral nodes**. These gaps allow ions to flow freely from the extracellular fluids to the axons, assisting in developing action potentials for nerve transmission.

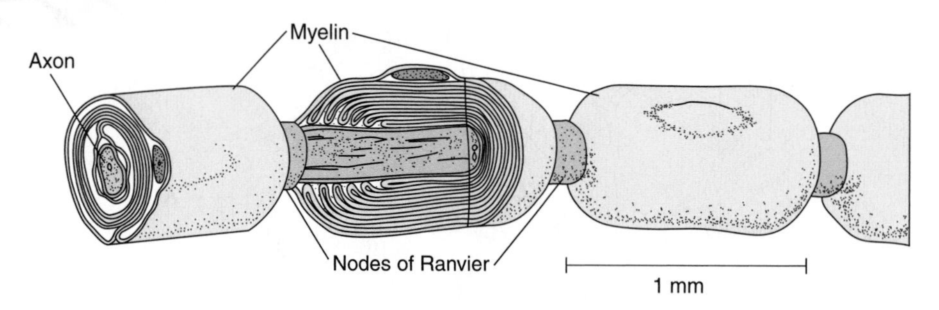

FIGURE 10-4. A section of an axon from the peripheral nervous system with its myelin sheath, a fatty substance made by Schwann cells.

Bipolar neurons (see Figure 11-11 in Chapter 11) have one dendrite and one axon. They function as receptor cells in special sense organs. Only two (bi) processes come off the cell body. They are found in only three areas of the body: the retina of the eye, the inner ear, and the olfactory area of the nose. **Unipolar neurons** have only one process extending from the cell body. This single process then branches into a central branch that functions as an axon and a peripheral branch that functions as a dendrite. Most sensory neurons are unipolar neurons (see Figure 10-3). The branch functioning as an axon enters the brain or spinal cord; the branch functioning as a dendrite connects to a peripheral part of the body.

Functional Classification of Neurons

Nerve cells pick up various changes in the environment (stimuli such as changes in temperature or pressure) from receptors. Receptors are the peripheral nerve endings of sensory nerves that respond to stimuli. There are many different types of receptors. Our skin has an enormous number of such receptors. These receptors change the energy of a stimulus, like heat, into nerve impulses. The first nerve cell receiving this impulse directly from a receptor is called a **sensory** or **afferent neuron**. These neurons are of the unipolar type. The receptors are in contact with only one end of the sensory neuron (the peripheral process in the skin), thus ensuring a one-way transmission of the impulse. The central process of the sensory neuron goes to the spinal cord.

From the sensory neuron, the impulse may pass through a number of **internuncial** or **association neurons.** These are found in the brain and the spinal cord and are of the multipolar type. They transmit the sensory impulse to the appropriate part of the brain or spinal cord for interpretation and processing.

From the association or internuncial neurons, the impulse is passed to the final nerve cell, the **motor** or **efferent neuron.** The motor neuron is of the multipolar type. This neuron brings about the reaction to the original stimulus. It is usually muscular (like pulling away from a source of heat or pain) but it can also be glandular (like salivating after smelling freshly baked cookies).

StudyWARE™ Connection

Play an interactive game labeling a the structures of a neuron on your StudyWARE™ CD-ROM.

THE PHYSIOLOGY OF THE NERVE IMPULSE

A nerve cell is similar to a muscle cell in that there are concentrations of ions on the inside and the outside of the cell membrane. Positively charged sodium (Na^+) ions are in greater concentration outside the cell than inside. There is a greater concentration of positively charged potassium (K^+) ions inside the cell than outside. This situation is maintained by the cell membrane's sodium-potassium pump (Figure 10-5). In addition to the potassium ion, the inside of the fiber has negatively charged chloride (Cl^-) ions and other negatively charged organic molecules. Thus, the nerve fiber has an electrical distribution as well, such that the outside is positively charged while the inside is negatively charged (Figure 10-6). This condition is known as the **membrane** or **resting potential.** Na^+ and K^+ ions tend to diffuse across the membrane but the cell maintains the resting potential through the channels of the sodium-potassium pump that actively extrudes Na^+ and accumulates K^+ ions.

When a nerve impulse begins, the permeability to the sodium (Na^+) ions changes. Na^+ rushes in, causing a change from a negative (-) to a positive (+) charge inside the nerve membrane. This reversal of electrical charge is called **depolarization** and creates the cell's **action potential.** The action potential moves in one direction down the nerve fiber.

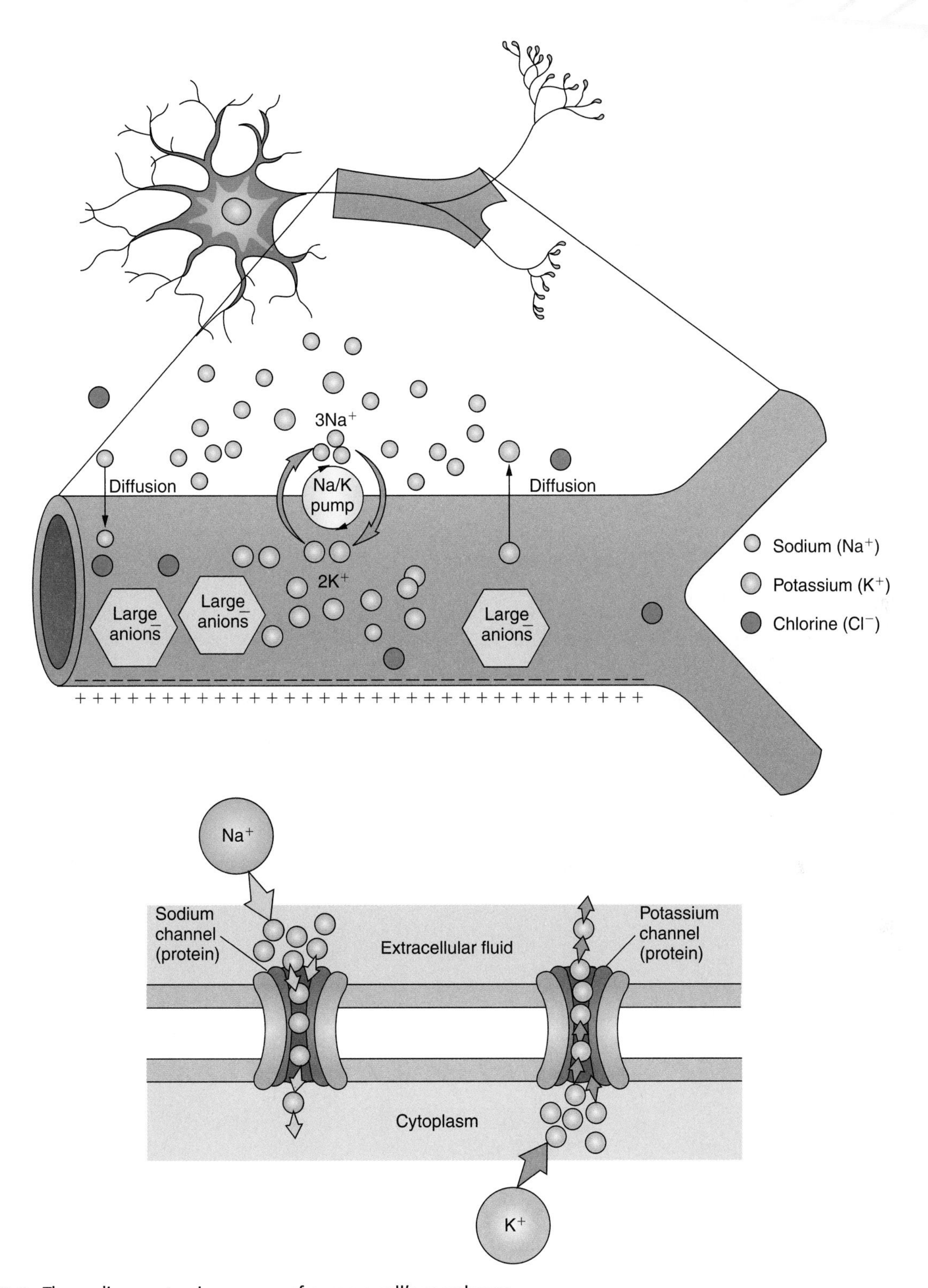

FIGURE 10-5. The sodium-potassium pump of a nerve cell's membrane.

Now the potassium ions begin to move outside to restore the resting membrane potential. The sodium-potassium pump begins to function, pumping out the sodium ions that rushed in and pulling back in the potassium ions that moved outside, thus restoring the original charges. This is called **repolarization**, as shown in Figure 10-6, and the inside of the cell again becomes negative. This process continues along the nerve fiber acting like an electrical current, carrying the nerve impulse along the fiber. The nerve impulse is a self-propagating wave of depolarization followed by repolarization moving down the nerve fiber.

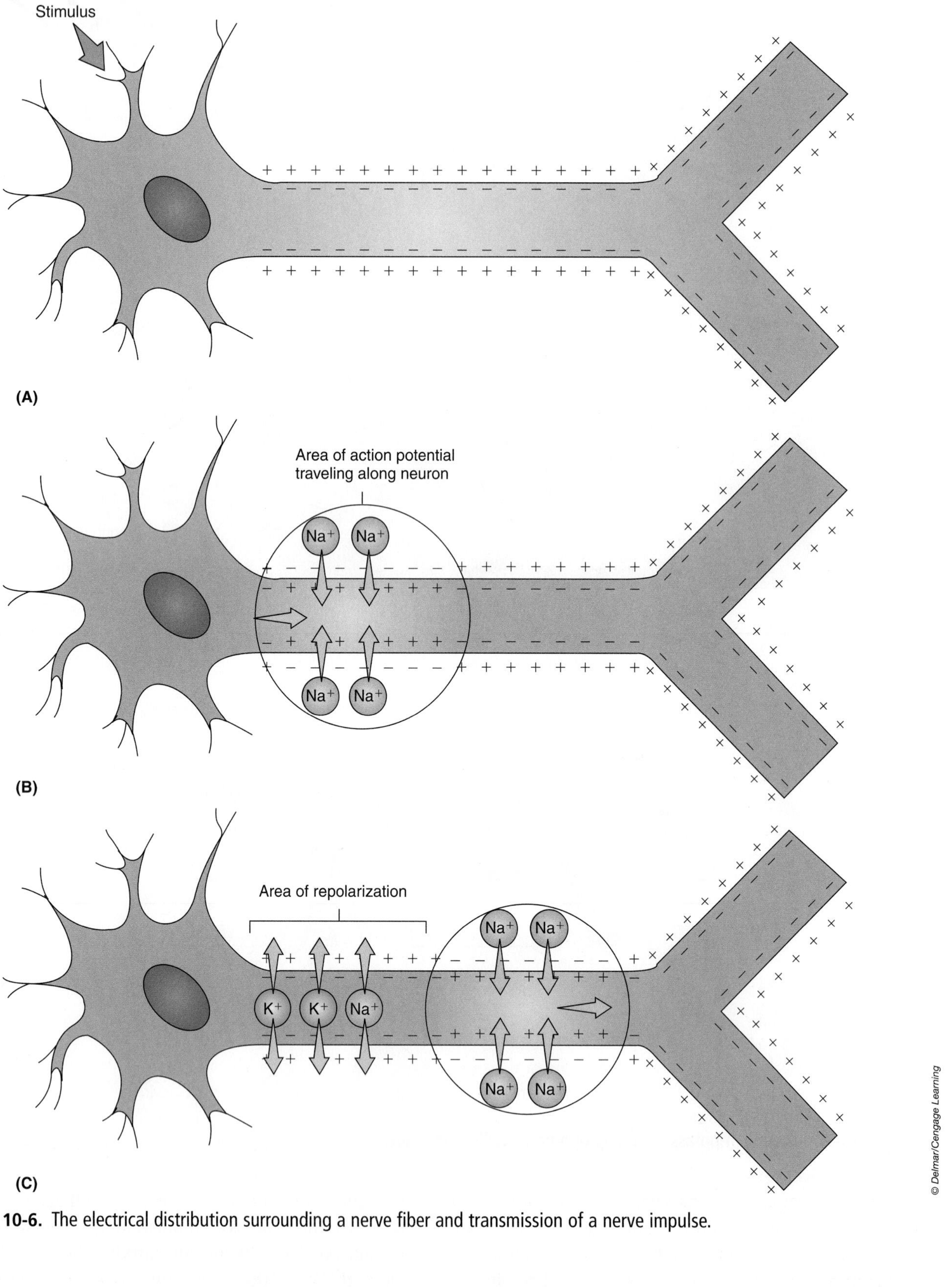

FIGURE 10-6. The electrical distribution surrounding a nerve fiber and transmission of a nerve impulse.

An unmyelinated nerve fiber conducts an impulse over its entire length, but the conduction is slower than that along a myelinated fiber. A myelinated fiber is insulated by the myelin sheath, so transmission occurs only at the nodes of Ranvier between adjacent Schwann cells. Action potentials and inflow of ions occur only at these nodes, allowing the nerve impulse to jump from node to node, and the impulse travels much faster. An impulse on a myelinated motor fiber going to a skeletal muscle could travel about 120 meters per second, while an impulse on an unmyelinated fiber would travel only 0.5 meter per second.

On any nerve fiber, the impulse will never vary in strength. If the stimulus or change in the environment is barely great enough to cause the fiber to carry the impulse, the impulse will be the same strength as one excited by a stronger stimulus. This is known as the **all-or-none law**, which states that if a nerve fiber carries any impulse, it will carry a full strength impulse.

THE SYNAPTIC TRANSMISSION

Synapses (sin-**AP**-seez) are the areas where the terminal branches of an axon (the axon terminals) are anchored close to, but not touching, the ends of the dendrites of another neuron. These synapses are one-way junctions that ensure that the nerve impulse travels in only one direction. This area is called a synaptic cleft. Other such areas of synapses are between axon endings and muscles or between axon endings and glands. An impulse continuing along a nerve pathway must cross this gap.

Transmission across synapses is brought about by the secretions of very low concentrations of chemicals called neurotransmitters that move across the gap. As the nerve impulse travels down the fiber, it causes vesicles in the axon endings of a presynaptic neuron to release the chemical neurotransmitter. Most of the synapses in our bodies use acetylcholine as the neurotransmitter. The acetylcholine allows the impulse to travel across the synaptic cleft to the postsynaptic neuron. However, it does not remain there because an enzyme in the cleft, **acetylcholinesterase**, immediately begins to break down the acetylcholine after it performs its function (Figure 10-7). The autonomic nervous system in addition uses **adrenaline** (also called **epinephrine**) as a transmitting agent. Many kinds of neurotransmitters are found in the nervous system. Some neurons produce only one type; others produce two or three. The best known neurotransmitters are acetylcholine and norepinephrine. Some others are **serotonin** (**sayr**-oh-**TOH**-nin), **dopamine** (**DOH**-pah-meen), and the **endorphins** (in-**DOHR**-finz).

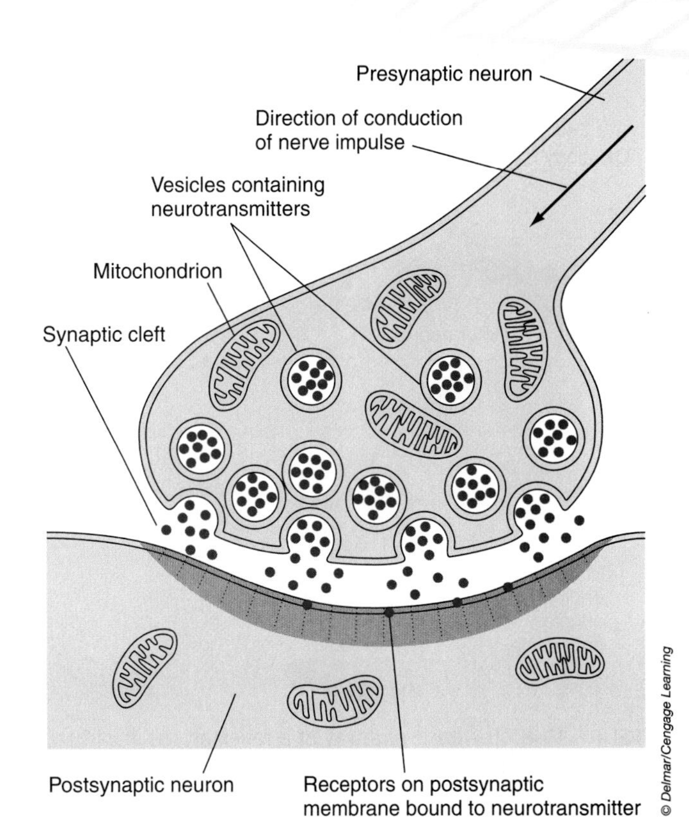

FIGURE 10-7. The release of neurotransmitter molecules by a presynaptic neuron into the synaptic cleft, transmitting the nerve impulse to the postsynaptic neuron.

StudyWARE™ Connection

Watch an animation on the firing of neurotransmitters on your StudyWARE™ CD-ROM.

THE REFLEX ARC

When we have an involuntary reaction to an external stimulus, we experience what is called a **reflex**. This is experienced when we prick our finger on a rose thorn and immediately pull away from the source of pain. The reflex allowed us to respond much more quickly than if we had to consciously think about what to do and interpret the information in the CNS. A reflex then is an involuntary reaction or response to a stimulus applied to our periphery and transmitted to the CNS.

A **reflex arc** is the pathway that results in a reflex (Figure 10-8). It is a basic unit of the nervous system and is the smallest and simplest pathway able to receive a stimulus, enter the CNS (usually the spinal cord) for immediate interpretation, and produce a response. The reflex arc has

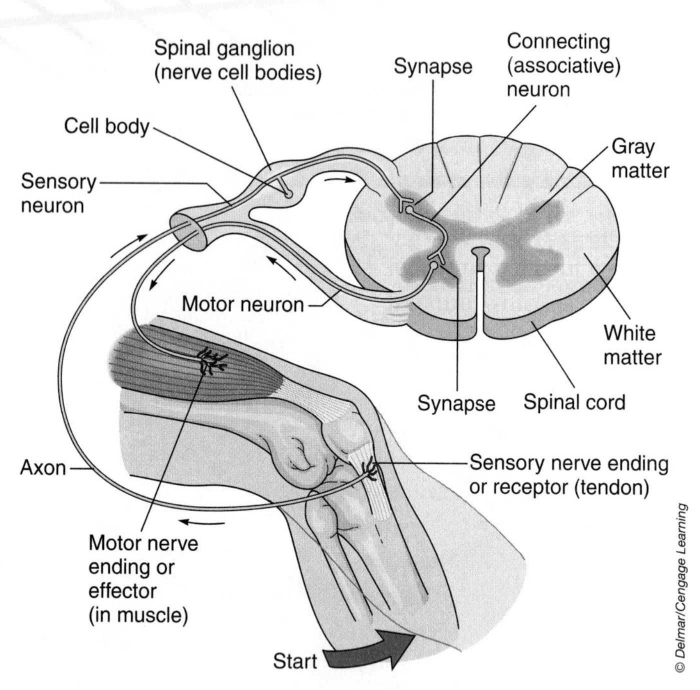

FIGURE 10-8. The basic pathway of a reflex arc, as illustrated by the knee-jerk reflex.

five components: (1) a sensory receptor in the skin; (2) a sensory or afferent neuron; (3) association or internuncial neurons within the spinal cord; (4) a motor or efferent neuron; and (5) an effector organ like a muscle. You have probably experienced a reflex arc when you had a physical examination and the doctor hit you below your knee with a rubber mallet. This is the knee-jerk reflex, also called the patellar tendon reflex. The doctor hits the patellar tendon just below the patella (the stimulus), causing the stimulation of stretch receptors within the quadriceps femoris muscle. They send the impulse via sensory neurons to the spinal cord for interpretation. The impulse then travels to a motor neuron (response) back to the muscles that contract and your leg extends.

Reflexes also occur within our bodies to help maintain homeostasis. Heartbeat rate, digestion, and breathing rates are controlled and maintained by reflexes concerned with involuntary processes. Coughing (the choke reflex), sneezing, swallowing, and vomiting are other examples of automatic subconscious reactions to changes within or outside our body.

GROUPING OF NEURAL TISSUE

In the nervous system, a number of terms are used to describe nervous tissue organization. It is important to understand the meanings of these terms. The term **white matter** refers to groups of myelinated axons (myelin has a whitish color) from many neurons supported by neuroglia. White matter forms nerve tracts in the CNS. The gray areas of the nervous system are called **gray matter**, consisting of nerve cell bodies and dendrites. It also can consist of bundles of unmyelinated axons and their neuroglia. The gray matter on the surface of the brain is called the **cortex**.

A **nerve** is a bundle of fibers located outside the CNS. Most nerves are white matter. Nerve cell bodies that are found outside the CNS are generally grouped together to form **ganglia** (**GANG**-lee-ah). Because ganglia are made up primarily of unmyelinated nerve cell bodies, they are masses of gray matter.

A **tract** is a bundle of fibers inside the CNS. Tracts can run long distances up and down the spinal cord. Tracts are also found in the brain and connect parts of the brain with each other and parts of the brain with the spinal cord. Ascending tracts conduct impulses up the cord and are concerned with sensation. Descending tracts conduct impulses down the cord and are concerned with motor functions. Tracts are made of myelinated fibers and therefore are classified as white matter. Two other terms are of note: a **nucleus** is a mass of nerve cell bodies and dendrites inside the CNS, consisting of gray matter; **horns** are the areas of gray matter in the spinal cord.

THE SPINAL CORD

The spinal cord begins as a continuation of the medulla oblongata of the brainstem. Its length is approximately 16 to 18 inches. Its diameter varies at different levels because it is surrounded and protected by bone (the vertebrae) and by disks of fibrocartilage (the intervertebral disks). It is made up of a series of 31 segments, each giving rise to a pair of spinal nerves. In addition to the above protection, the spinal cord (as well as the brain) is further protected by the **meninges** (men-**IN**-jeez), a series of connective tissue membranes. Those associated specifically with the spinal cord are called the **spinal meninges** (Figure 10-9).

The outermost spinal meninx is called the **dura mater** (**DOO**-rah-**MATE**-ehr), which means *tough mother*. It forms a tough outer tube of white fibrous connective tissue. The middle spinal meninx is called the **arachnoid mater** (ah-**RACK**-noyd **MATE**-ehr) or *spider layer*. It forms a delicate connective membranous tube inside the dura mater. The innermost spinal meninx is known as the **pia mater** (**PEE**-ah **MATE**-ehr) or *delicate mother*. It is a transparent fibrous membrane that forms a tube around and adheres to the surface of the spinal cord (and brain). It contains numerous blood vessels and nerves that nourish the underlying cells.

Between the dura matter and the arachnoid is a space called the subdural space, which contains serous

HEALTH ALERT

ILLEGAL DRUG USE

Some of the most commonly abused drugs that affect the CNS are depressants ("downers"), stimulants ("uppers"), and hallucinogens as well as the anabolic steroids.

One example of a depressant is Valium, which is prescribed in low doses to relieve tension. However, higher doses cause drowsiness, sedation, and loss of any pain sensations. Another group of depressants are the opiate drugs like codeine and heroin. Codeine can be prescribed by a doctor, but heroin has no legal use in the United States. These drugs act as sedatives and analgesics, which relieve pain in prescribed doses. However, they can cause psychological and physical dependency. They also produce a feeling of euphoria. Overuse can result in coma, convulsions, and respiratory problems that could lead to death. A drug derived from the hemp plant (*Cannabis*) is marijuana. It is not as potent as hashish. Hashish is made from the resin in the flowering tips of the hemp plant. This drug produces a state of euphoria free of anxiety and alters one's perception of time and space. Marijuana has been prescribed for individuals with advanced stages of incurable diseases (e.g., certain cancers) and glaucoma of the eye. Higher doses can lead to hallucinations and respiratory problems.

Examples of stimulants to the CNS are cocaine, hallucinogens such as LSD (lysergic acid diethylamide), and amphetamines. Hallucinogens like LSD cause distortions in the five senses. The perceptions of sight, sound, smell, and taste are heightened and exaggerated. A person may have a sense of being able to do anything physically, leading to serious injury to oneself. Cocaine produces great psychological and physical dependence on the user. When inhaled it produces a quick state of euphoria. However, it causes changes in personality, seizures, and death from stroke or abnormal rhythms of the heart. Amphetamines overstimulate postsynaptic neurons, resulting in muscle spasms, restlessness, rapid heartbeat, and hypertension. This is a high price to pay for the feeling of euphoria and elation first produced. Death can result from respiratory or heart failure.

The anabolic steroids act like the male sex hormones and are commonly used by athletes because they cause skeletal muscle cells to increase in size. Athletes, like bodybuilders, can quickly increase their bulk, although this benefit has dangerous side effects. Taken in large doses, synthetic androgens or anabolic steroids have a negative feedback effect on the hypothalamus of the brain and the pituitary gland. This causes a reduction in gonadotropin-releasing hormone, luteinizing hormone, and folliclestimulating hormone. As a result, the testes can atrophy and sterility will occur. The abuse of these steroids can also lead to liver problems, heart disease, and personality changes. ■

fluid. Between the arachnoid and the pia mater is the subarachnoid space. It is here that the clear, watery cerebrospinal fluid circulates. The meninges do not attach directly to the vertebrae. They are separated by a space called the epidural space. This space contains loose connective tissue and some adipose tissue that acts as a protective cushion around the spinal cord.

Functions of the Spinal Cord

A major function of the spinal cord is to convey sensory impulses from the periphery to the brain and to conduct motor impulses from the brain to the periphery. Ascending nerve tracts of the spinal cord carry sensory information from body parts to the brain, and descending tracts conduct motor impulses from the brain to muscles and glands. A second principal function is to provide a means of integrating reflexes. A pair of spinal nerves is connected to each segment of the spinal cord. Each pair of spinal nerves is connected to that segment of the cord by two pairs of attachments called roots (see Figure 10-9). The **posterior** or **dorsal root** is the sensory root and contains only sensory nerve fibers. It conducts impulses from the periphery (like the skin) to the spinal cord. These fibers extend into the **posterior** or **dorsal gray horn** of the spinal cord. The other point of attachment of the spinal nerve to the cord is the **anterior** or **ventral root** and this is the motor root. It contains motor nerve fibers only and conducts impulses from

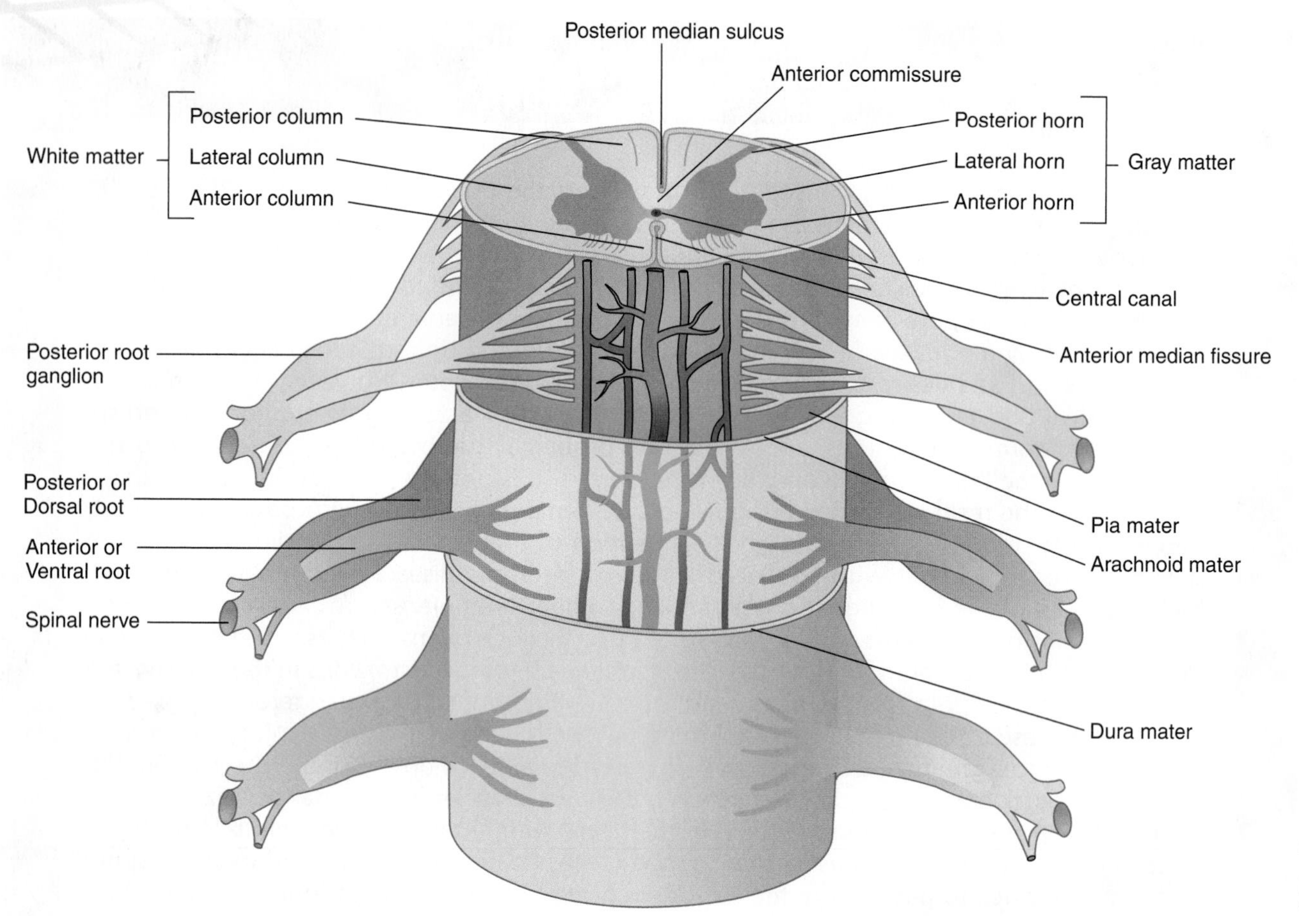

FIGURE 10-9. The anatomy of the spinal meninges and the spinal cord.

HEALTH ALERT

SPINAL TAP

The spinal cord extends down to the second lumbar vertebra. Yet spinal nerves, surrounded by the meninges, go all the way down to the end of the vertebral column. Because there is no spinal cord at the end of the vertebral canal, a needle can be inserted into the subarachnoid space in this area without damaging the spinal cord. This is done to perform a spinal tap and extract cerebrospinal fluid, which can then be examined for infectious organisms like those causing meningitis, or for detecting blood, in the case of a hemorrhage. A needle could also be inserted with an anesthetic agent to administer spinal anesthesia in this way. If a radiopaque substance is injected in this area, an X-ray can be taken of the spinal cord to detect any damage or defects in the cord. ■

the spinal cord to the periphery (like muscles). It connects with the **anterior** or **ventral gray horn** of the spinal cord.

THE SPINAL NERVES

The 31 pairs of spinal nerves arise from the union of the dorsal and ventral roots of the spinal nerves (see Figure 10-9). All the spinal nerves are mixed nerves because they consist of both motor and sensory fibers. Most of the spinal nerves exit the vertebral column between adjacent vertebrae. They are named and numbered according to the region and level of the spinal cord from which they emerge (Figure 10-10). There are 8 pairs of cervical nerves, 12 pairs of thoracic nerves, 5 pairs of

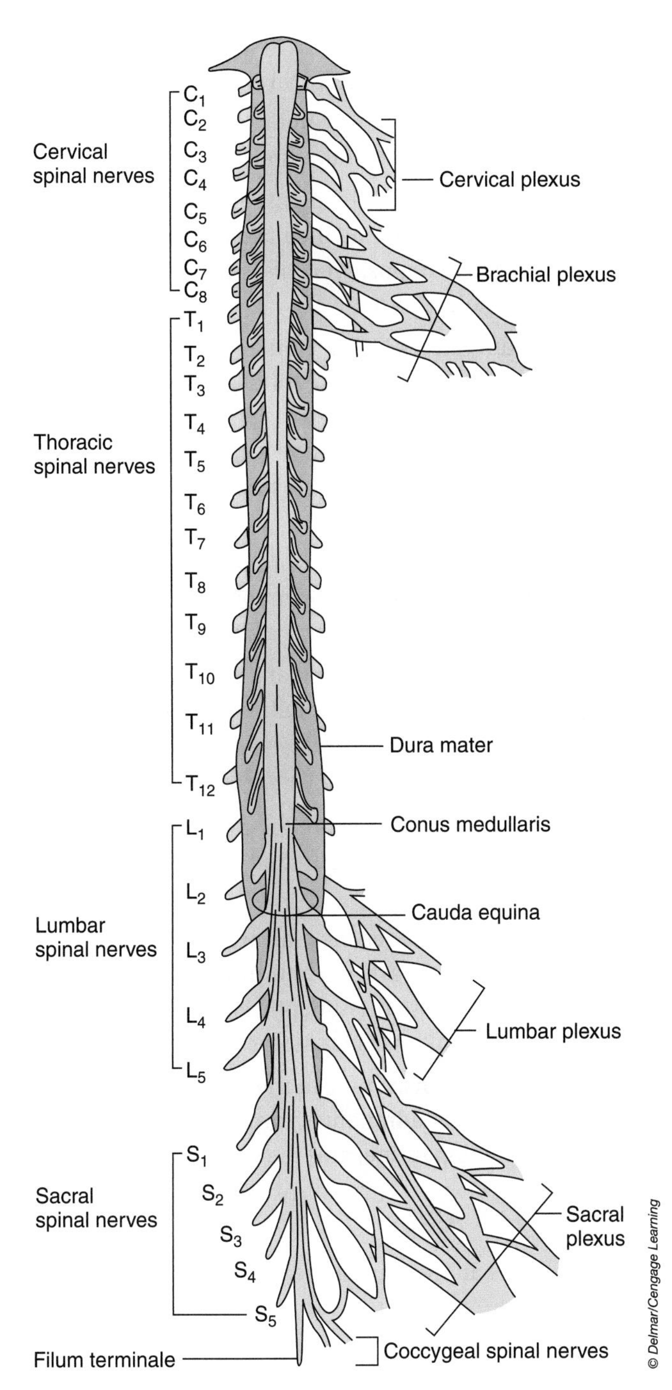

FIGURE 10-10. The names and emergent levels of the 31 spinal nerves.

lumbar nerves, 5 pairs of sacral nerves, and a single pair of coccygeal nerves. The spinal nerves are also numbered according to the order (starting superiorly) within the region. Thus, the 31 pairs are C1 through C8 (cervical), T1 through T12 (thoracic), L1 through L5 (lumbar), S1 through S5 (sacral), and Cx (coccygeal).

SUMMARY OUTLINE

INTRODUCTION

1. The nervous system is the body's control center and communication network.
2. It shares in the maintenance of homeostasis with the endocrine system.

ORGANIZATION

1. The central nervous system (CNS) consists of the brain and spinal cord.
2. The peripheral nervous system (PNS) consists of the afferent peripheral system (sensory neurons) and the efferent peripheral system (motor neurons).
3. The efferent peripheral system can be subdivided into the somatic nervous system, which sends signals to skeletal muscles, and the autonomic nervous system (ANS), which sends signals to cardiac and smooth muscles and glands.
4. The ANS has two divisions: the sympathetic division, which stimulates and speeds up activity, and the parasympathetic division, which restores or slows down certain activities but stimulates the body's vegetative activities.

CLASSIFICATION OF NERVE CELLS

1. Neurons are nerve cells that transmit nerve impulses in the form of electrochemical changes.
2. A nerve is a bundle of nerve cells.
3. Neuroglia cells are nerve cells that support and protect the neurons.

Neuroglia Cells

1. Astrocytes are star-shaped cells that wrap around neurons for support in the brain and spinal cord and connect neurons to blood vessels.
2. Oligodendroglia look like small astrocytes. They form connective-like tissue rows for support and form the fatty myelin sheath on the neurons in the brain and spinal cord.
3. Microglia are small cells that do phagocytosis of microbes and cellular debris.
4. Ependymal cells line the fluid-filled ventricles of the brain. Some produce cerebrospinal fluid and others, with cilia, move it through the CNS.
5. Schwann cells form myelin sheaths around nerve fibers in the peripheral nervous system.

The Structure of a Neuron

1. A neuron is composed of a cell body with a nucleus and other intracellular organelles.
2. Dendrites are extensions of the cell body and are the receptive areas of the neuron.
3. An axon is a single long extension of the cell body that begins as a slight enlargement, the axon hillock. The axon may branch, but at its end there are many extensions called axon terminals.
4. On large peripheral axons, a Schwann cell produces a fatty myelin sheath that surrounds and insulates the axon. Narrow gaps in the sheath are called nodes of Ranvier.

Structural Classification of Neurons

1. Multipolar neurons have several dendrites coming off the cell body and one axon. Most neurons in the brain and spinal cord are multipolar neurons.
2. Bipolar neurons have one dendrite and one axon. They are found in the retina of the eye, the inner ear, and in the olfactory area of the nose.
3. Unipolar neurons have only one process extending from the cell body, which then branches into a central branch that functions as an axon and a peripheral branch that functions as a dendrite. Most sensory neurons are unipolar neurons.

Functional Classification of Neurons

1. Receptors detect stimuli in our environment.
2. Sensory or afferent neurons receive the impulse directly from the receptor site. They are unipolar neurons.
3. Internuncial or association neurons are found in the brain and spinal cord. They transmit the impulse for interpretation and processing. They are multipolar neurons.
4. Motor or efferent neurons bring about the reaction to the stimulus. They are multipolar neurons.

THE PHYSIOLOGY OF THE NERVE IMPULSE

1. A nerve cell fiber has a higher concentration of Na^+ on the outside than inside and a higher concentration of K^+ on the inside than on the outside. This is maintained by the sodium-potassium pump.
2. The nerve fiber has a negative electrical charge on the inside and a positive electrical charge on the outside.
3. This electrical ionic distribution is called the membrane or resting potential.
4. When a nerve impulse begins, the sodium ions (Na^+) rush in changing the inside electrical charge from negative to positive. This is the action potential called depolarization.
5. Potassium ions (K^+) move out to try to restore the resting membrane potential and the sodium-potassium pump operates to restore the original charge. This is repolarization and restores the fiber's membrane to the original resting or membrane potential.
6. The nerve impulse is a self-propagating wave of depolarization followed by repolarization moving in one direction down the nerve fiber.
7. The all-or-none law states that if a nerve fiber carries any impulse, it will carry a full strength impulse.

THE SYNAPTIC TRANSMISSION

1. A synapse is an area where the terminal branches of an axon are close to but not touching the dendrites of another neuron.
2. When an impulse reaches the axon terminals, it triggers the release of a neurotransmitter like acetylcholine into the synaptic cleft, which allows the impulse to travel across the synapse.
3. Other neurotransmitters in the body are epinephrine or adrenaline, norepinephrine, serotonin, dopamine, and the endorphins.

THE REFLEX ARC

1. A reflex is an involuntary reaction to an external stimulus.
2. A reflex arc is the pathway that causes a reflex.
3. A reflex arc has five components: a sensory receptor in the skin; a sensory or afferent neuron; association or internuncial neurons in the spinal cord; a motor or efferent neuron; and an effector organ.

GROUPING OF NEURAL TISSUE

1. White matter refers to groups of myelinated axons from many neurons supported by neuroglia.

2. Gray matter consists of nerve cell bodies and dendrites, as well as groups of unmyelinated axons and their neuroglia.
3. A nerve is a bundle of fibers outside the CNS.
4. Ganglia are nerve cell bodies found outside the CNS.
5. A tract is a bundle of fibers inside the CNS.
6. A nucleus is a mass of nerve cell bodies and dendrites inside the CNS.
7. Horns are areas of gray matter in the spinal cord.

THE SPINAL CORD

1. The spinal cord is a continuation of the medulla oblongata.
2. The spinal cord is made of 31 segments, each giving rise to a pair of spinal nerves.
3. The spinal cord is protected by the spinal meninges.
4. The outermost spinal meninx is the dura mater or *tough mother*, the middle spinal meninx is the arachnoid mater or *spider layer*, and the innermost meninx is the pia mater or *delicate mother*.
5. Between the dura mater and the arachnoid is a space called the subdural space, which contains serous fluid.
6. Between the arachnoid and the pia mater is the subarachnoid space in which the cerebrospinal fluid circulates.

FUNCTIONS OF THE SPINAL CORD

1. The spinal cord conveys sensory impulses from the periphery to the brain (ascending tracts) and conducts motor impulses from the brain to the periphery (descending tracts).
2. The spinal cord also integrates reflexes.
3. Each pair of spinal nerves connects to a segment of the spinal cord by two points of attachments called the roots.
4. The posterior or dorsal root is sensory and connects with the posterior or dorsal gray horn of the spinal cord.
5. The anterior or ventral root is motor and connects with the anterior or ventral gray horn of the spinal cord.

THE SPINAL NERVES

1. There are eight pairs of cervical nerves (C1–C8).
2. There are 12 pairs of thoracic nerves (T1–T12).
3. There are five pairs of lumbar nerves (L1–L5).
4. There are five pairs of sacral nerves (S1–S5).
5. There is one pair of coccygeal nerves (Cx).

REVIEW QUESTIONS

1. Name the 31 spinal nerves and indicate how many there are of each.
*2. Discuss the factors involved in the transmission of a nerve impulse.
3. Name and describe five types of neuroglia cells.
4. Classify the organization of the nervous system.
5. Name and describe the three types of structural neurons.
*6. Explain how a reflex arc functions, and name its components.
7. Name the two functions of the spinal cord.

***Critical Thinking Questions**

FILL IN THE BLANK

Fill in the blank with the most appropriate term.

1. The central nervous system consists of the ____________________ and the ____________________.
2. The peripheral nervous system consists of the afferent system composed of ____________________ neurons, and the efferent system composed of ____________________ neurons.
3. The autonomic nervous system is divided into the ____________________ division, which stimulates, and the ____________________ division, which restores activities and stimulates vegetative functions.
4. The meninges have an outer meninx called the ____________________, a middle meninx called the ____________________ or spider layer, and an inner meninx called the ____________________.
5. A ____________________ is an area of gray matter in the spinal cord.

MATCHING

Place the most appropriate number in the blank provided.

_____ Astrocytes
_____ White matter
_____ Ganglia
_____ Nucleus
_____ Oligodendroglia
_____ Horns
_____ Meninges
_____ Tract
_____ Gray matter
_____ Microglia

1. Produce myelin sheath on neurons
2. Nerve cell bodies outside CNS
3. Engulf and destroy microbes
4. Bundles of fibers in the CNS
5. Attach neurons to their blood vessels
6. Unmyelinated axons and neuroglia
7. Coverings around brain and spinal cord
8. Gray matter in spinal cord
9. Reflex arc
10. Mass of nerve cell bodies and dendrites in CNS
11. Myelinated neurons
12. White matter in spinal cord

Search and Explore

Search the Internet with key words from the chapter to discover additional information. Key words might include central nervous system, classification of nerve cells, structure of a neuron, physiology of nerve impulse, spinal cord, and spinal nerves.

Three high school students are being evaluated by health care providers at a drug rehabilitation clinic that serves adolescents. Hector, a 15-year-old boy, appears agitated and restless. He tells his health care provider that he is having muscle spasms. Upon examination, the care provider notes that Hector has a rapid heart rate and a dangerously high blood pressure. Carolyn, a 16-year-old girl, is being reassessed following a visit to the emergency room last night. According to her ER records, Carolyn was at a party with her friends when she suddenly had a seizure. Her records also document that she had an abnormal heart rate upon admission. Dante, a 14-year-old boy, was also admitted to the ER the night before. Dante's parents found him lying on the couch in a deep sleep. When they could not arouse their son, they rushed him to the hospital where he experienced a convulsion. Later that night, Dante's parents discovered medication hidden in his room that his mother had used earlier, following major surgery.

Questions

1. What type of drug do you think Hector might be using?
2. What long-term problems can this type of drug cause?
3. What drug likely caused Carolyn's seizure and abnormal heart rhythm?
4. How do users obtain the most dramatic "high" when using this drug?
5. What drug might have caused Dante's symptoms?
6. Do you think the drug Dante used is legal or illegal, or does it depend on the circumstances?

*Study*WARE™ Connection

Take a quiz or play interactive games that help you learn about the nervous system on your StudyWARE™ CD-ROM.

Study Guide Practice

Go to your **Study Guide** for more practice questions, labeling and coloring exercises, and crossword puzzles to help you learn the content in this chapter.

LABORATORY EXERCISE: THE NERVOUS SYSTEM

Materials needed: Prepared microscope slides of neurons and neuroglia cells, anatomic model of the spinal cord, rubber mallet

1. Examine a prepared microscope slide of the crushed spinal cord of an ox. Study the parts of multipolar internuncial neurons, noting the following structures: the cell body with nucleus and dendrite extensions of the cell body; find axons with branches and axon terminals. Notice the small astrocytes and their stained nuclei scattered in the slide preparation.
2. Examine an anatomic model of a section of the spinal cord with attached spinal nerves. Study the dorsal and ventral gray horns of the cord. Note how the sensory dorsal posterior root of the spinal nerve enters the dorsal gray horn and how the motor ventral anterior root of the spinal nerve leaves the ventral gray horn of the cord.
3. Test the knee-jerk reflex on your lab partner by gently tapping the patellar tendon with a small rubber mallet. Note how the extension of the knee is completely involuntary and subconscious, illustrating the performance of the reflex arc.

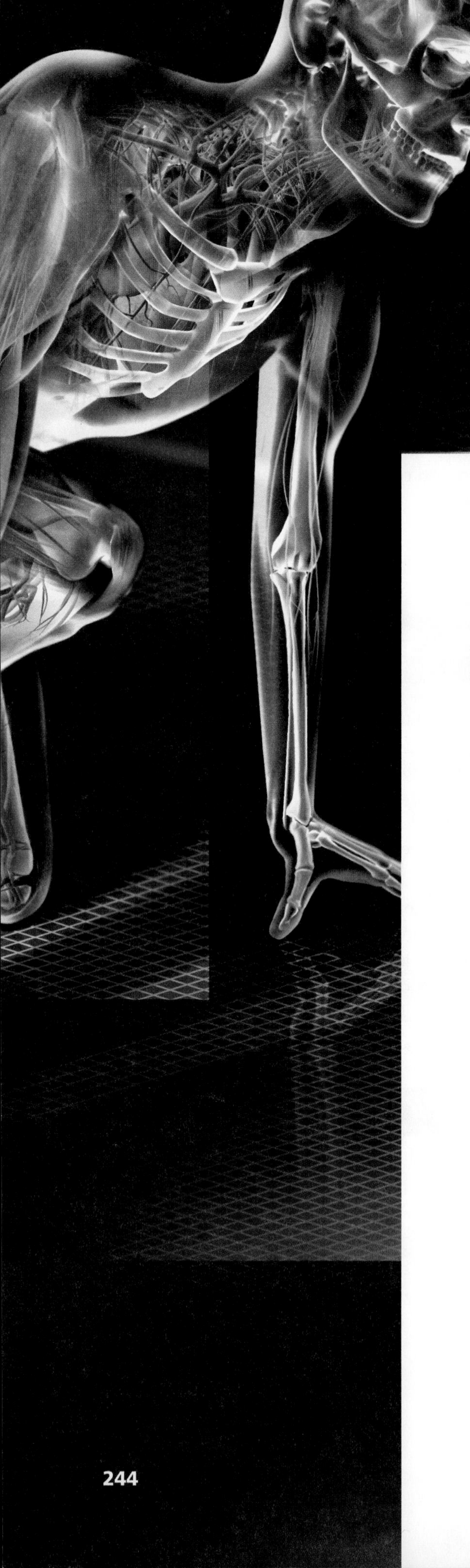

The Nervous System: The Brain, Cranial Nerves, Autonomic Nervous System, and the Special Senses

CHAPTER OBJECTIVES

After studying this chapter, you should be able to:

1. List the principal parts of the brain.
2. Name the functions of the cerebrospinal fluid.
3. List the principal functions of the major parts of the brain.
4. List the 12 cranial nerves and their functions.
5. Name the parts of the autonomic nervous system and describe how it functions.
6. Describe the basic anatomy of the sense organs and explain how they function.

KEY TERMS

INTRODUCTION

This chapter is a continuation of the discussion of the nervous system that begins in Chapter 10. The brain is divided into four main parts. The brainstem controls breathing, heartbeat rates, and reactions to visual and auditory stimuli. The diencephalon includes the thalamus and the hypothalamus, which controls many functions, including those related to homeostasis. The cerebrum controls intellectual processes and emotions, while the cerebellum maintains body posture and balance. The autonomic nervous system controls all the involuntary functions of the body such as regulating our internal organs and controlling glands.The special senses are part of the nervous system and include sight, hearing and balance, smell, and taste.

See Concept Map 11-1: The Brain, Concept Map 11-2: The Cranial Nerves, and Concept Map 11-3: The Autonomic Nervous System and Special Senses.

THE PRINCIPAL PARTS OF THE BRAIN

The brain is one of the largest organs of the body (Figure 11-1). It weighs about 3 pounds in an average adult. It is divided into four major parts: (1) the brainstem, which consists of three smaller areas, the medulla oblongata (meh-**DULL**-ah **ob**-long-**GAH**-tah), the pons

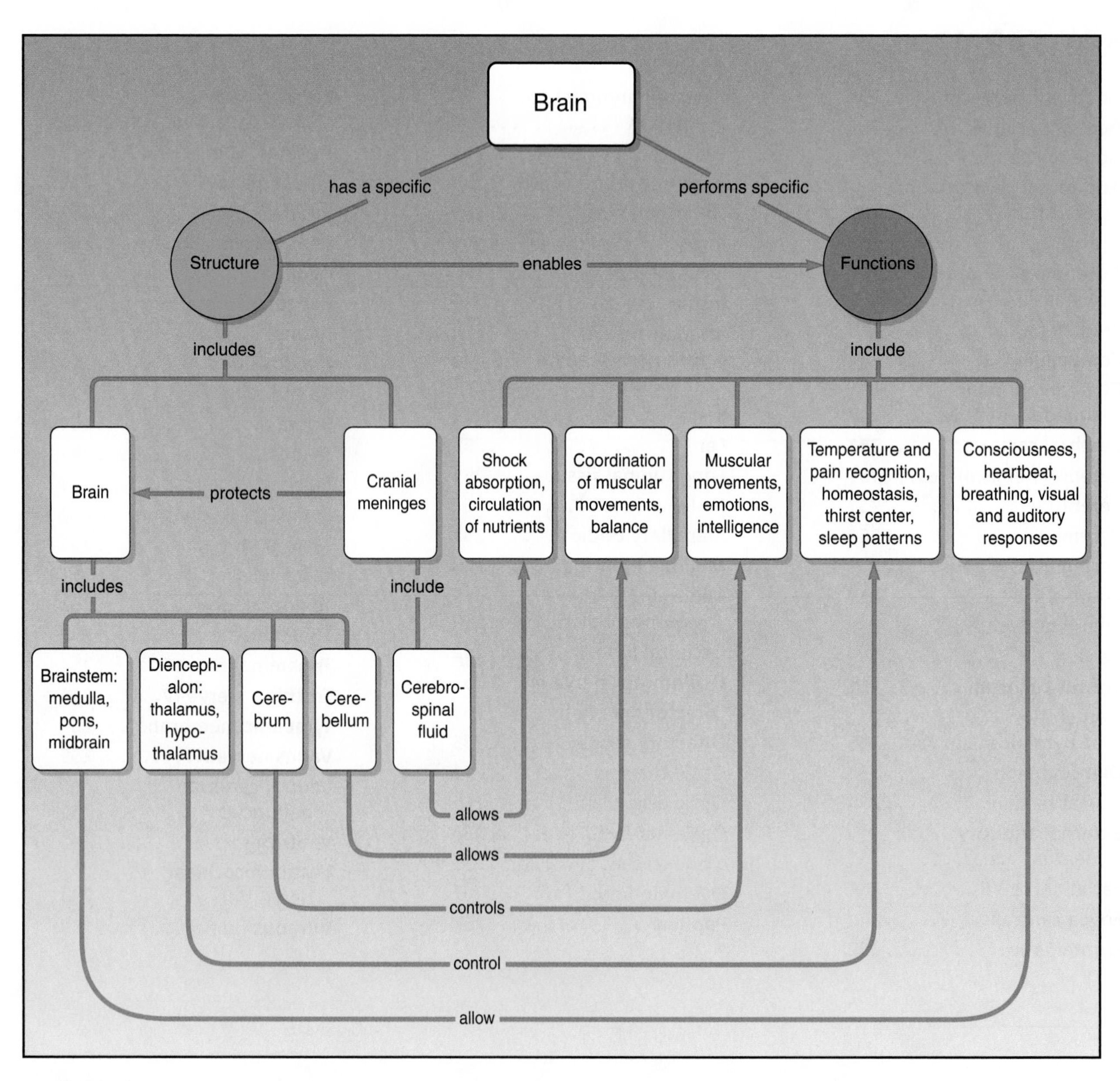

CONCEPT MAP 11-1. The brain.

varolii (**PONZ** vah-**ROH**-lee-eye) and the midbrain; (2) the diencephalon, (**dye**-en-**SEFF**-ah-lon), consisting of the thalamus (**THAL**-ah-muss) and the hypothalamus; (3) the cerebrum (seh-**REE**-brum); and (4) the cerebellum (seh-ree-**BELL**-um).

The brain is protected by the cranial bones and the meninges. The cranial meninges is the name given to the meninges that protect the brain, and they have the same structure as the spinal meninges: the outer dura mater, the middle arachnoid mater, and the inner pia mater (discussed in Chapter 10). The brain, like the spinal cord, is further protected by the cerebrospinal fluid that circulates through the subarachnoid space around the brain and spinal cord and through the ventricles of the brain. The **ventricles** are cavities within the brain that connect with each other, with the subarachnoid space of the meninges and with the central canal of the spinal cord. The cerebrospinal fluid serves as a shock absorber for the central nervous system and circulates nutrients.

The brain has four ventricles (Figure 11-2). There are two lateral ventricles in each side or hemisphere of the cerebrum under the corpus callosum (**KOR**-pus kah-**LOH**-sum). The third ventricle is a slit between and inferior to the right and left halves of the thalamus, and situated between the lateral ventricles. Each lateral ventricle connects with the third ventricle by a narrow oval opening called the **interventricular foramen** or **foramen of Monroe**. The fourth ventricle lies between the cerebellum and the lower brainstem. It connects with the third ventricle via the **cerebral aqueduct** also known as the **aqueduct of Sylvius**. The roof of this fourth ventricle has three openings through which it connects with the subarachnoid space of the brain and spinal meninges, thus allowing a flow of cerebrospinal fluid through the spinal cord, the brain, and the ventricles of the brain.

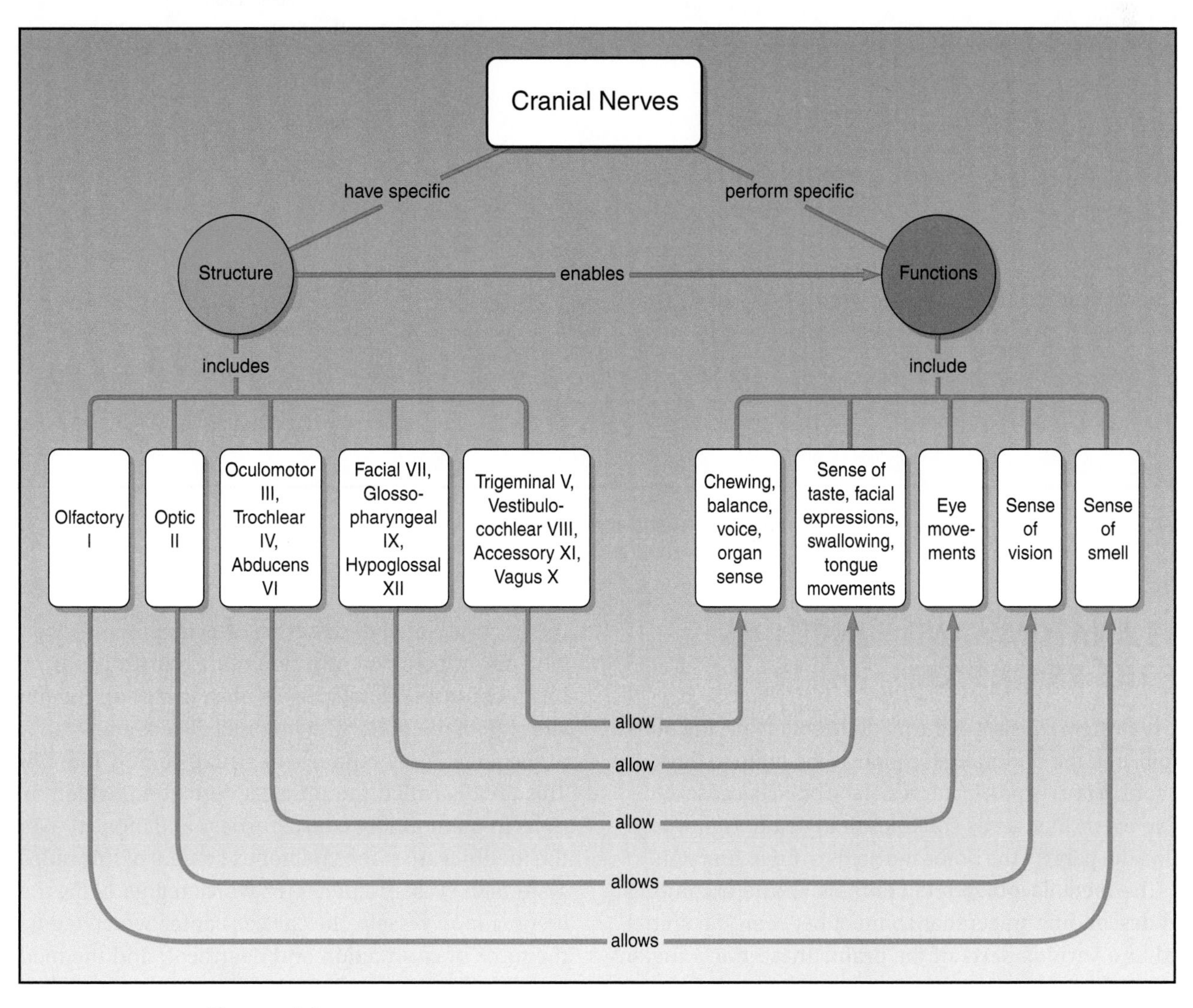

CONCEPT MAP 11-2. The cranial nerves.

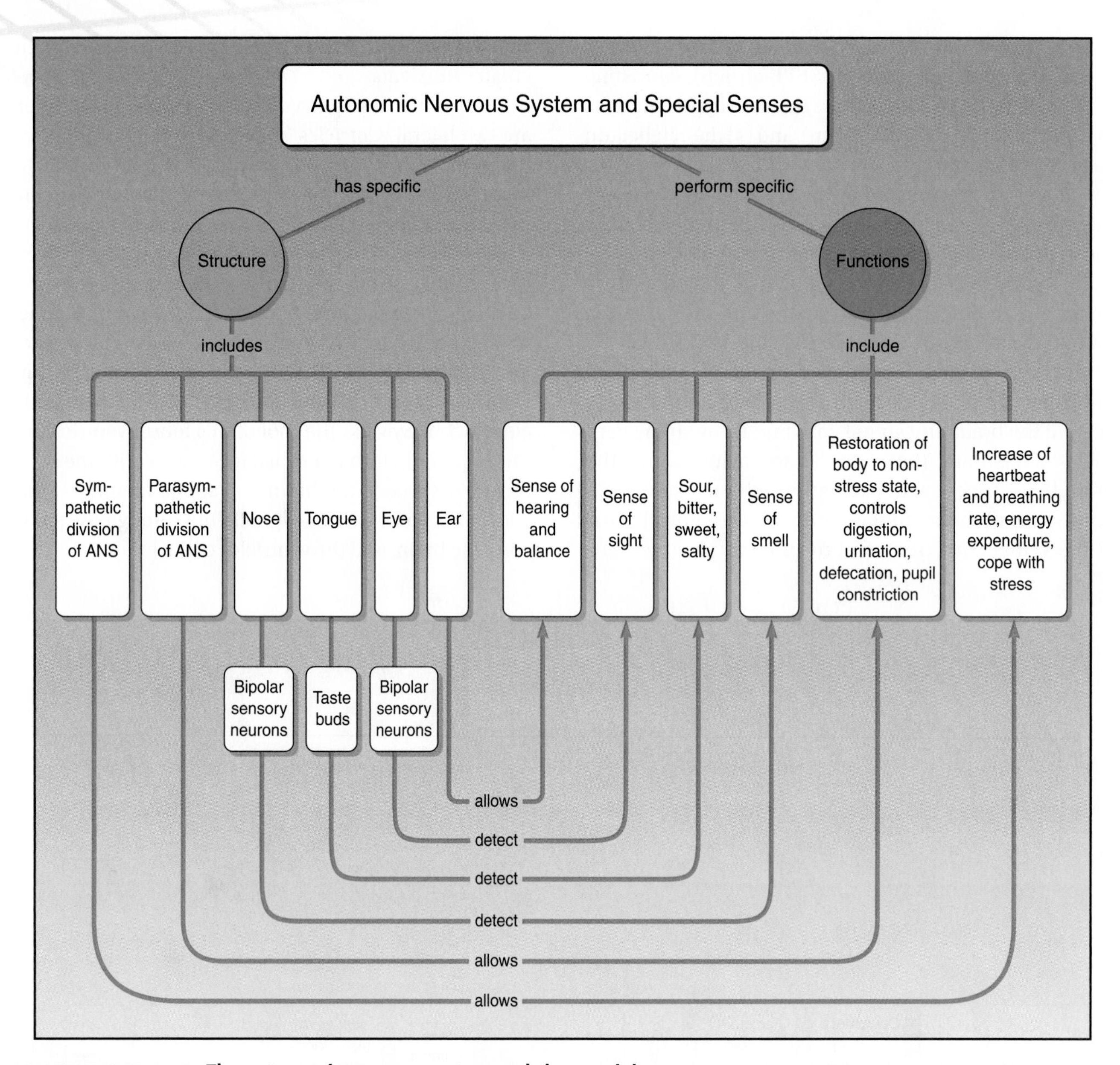

CONCEPT MAP 11-3. The autonomic nervous system and the special senses.

THE ANATOMY AND FUNCTIONS OF THE BRAINSTEM

The **brainstem** consists of the medulla oblongata, the pons varolii, and the midbrain. It connects the brain to the spinal cord. It is a very delicate area of the brain because damage to even small areas could result in death. Figure 11-3 shows the parts of the brain and areas of brain function.

The **medulla oblongata** contains all the ascending and descending tracts that connect between the spinal cord and various parts of the brain. These tracts make up the white matter of the medulla. Some motor tracts cross as they pass through the medulla. The crossing of the tracts is called **decussation of pyramids** and explains why motor areas on one side of the cortex of the cerebrum control skeletal muscle movements on the opposite side of the body. The medulla also contains an area of dispersed gray matter containing some white fibers. This area is called the **reticular formation**, which functions in maintaining consciousness and arousal. Within the medulla are three vital reflex centers of this reticular system: the vasomotor center, which regulates the diameter of blood vessels; the cardiac center, which regulates the force of contraction and heartbeat; and the medullary rhythmicity area, which adjusts your basic rhythm of breathing.

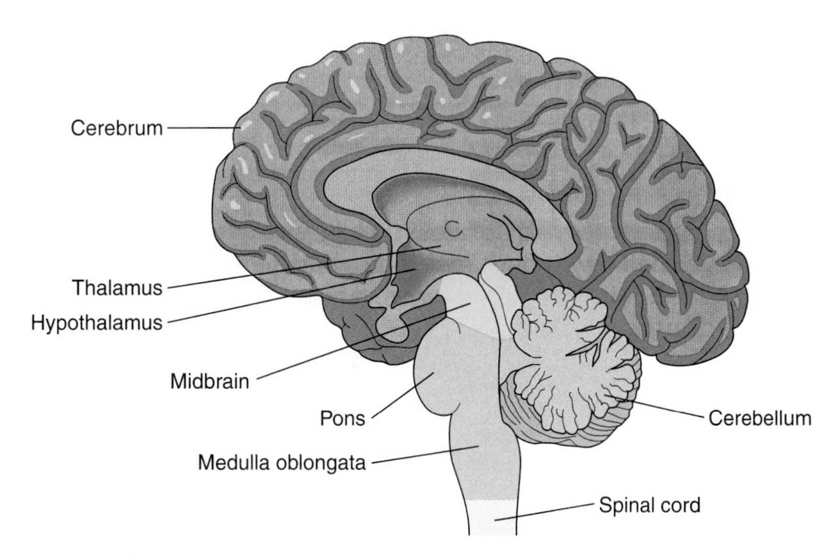

FIGURE 11-1. The principal parts of the brain.

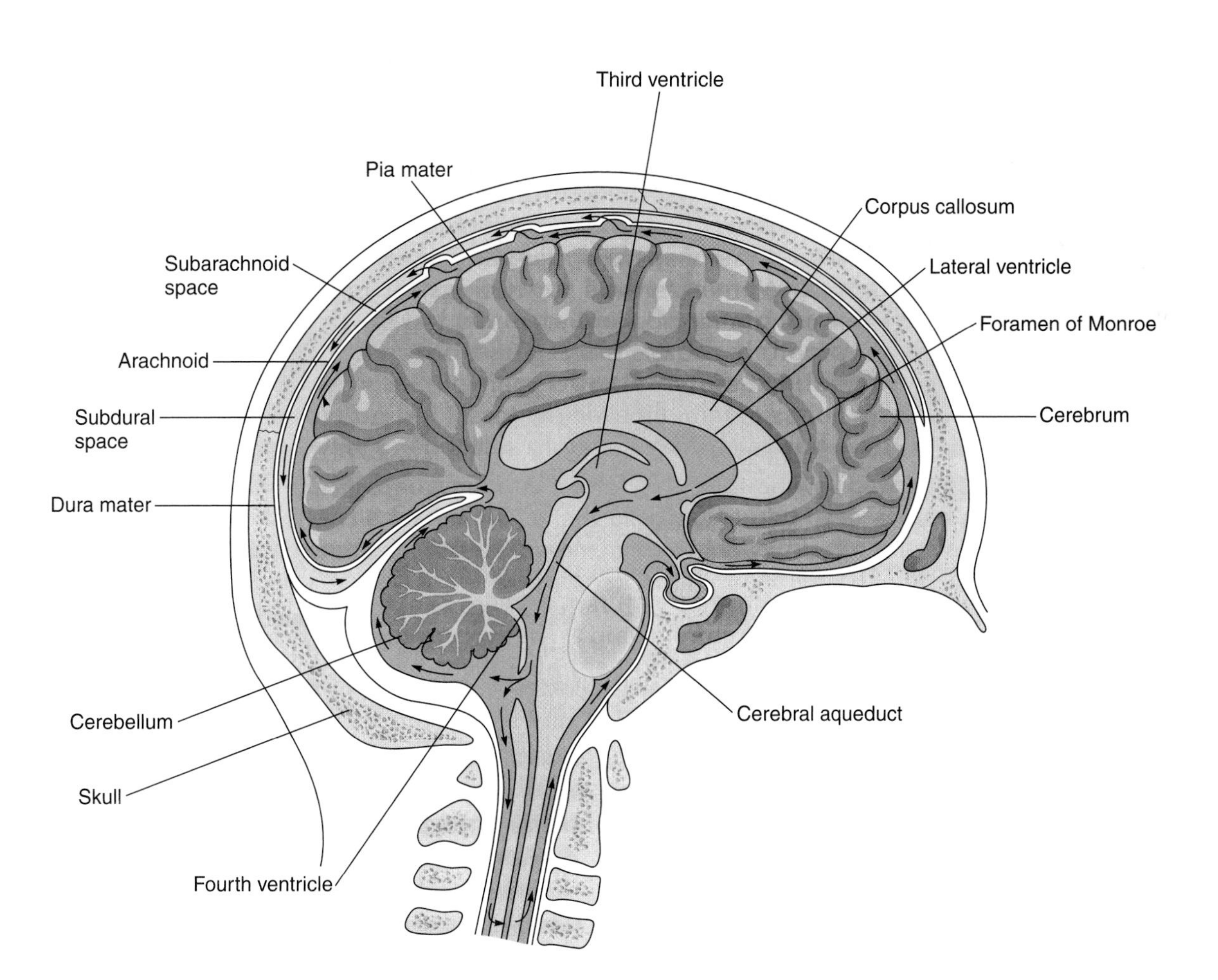

FIGURE 11-2. The ventricles of the brain, the cranial meninges, and the flow pattern of the cerebrospinal fluid.

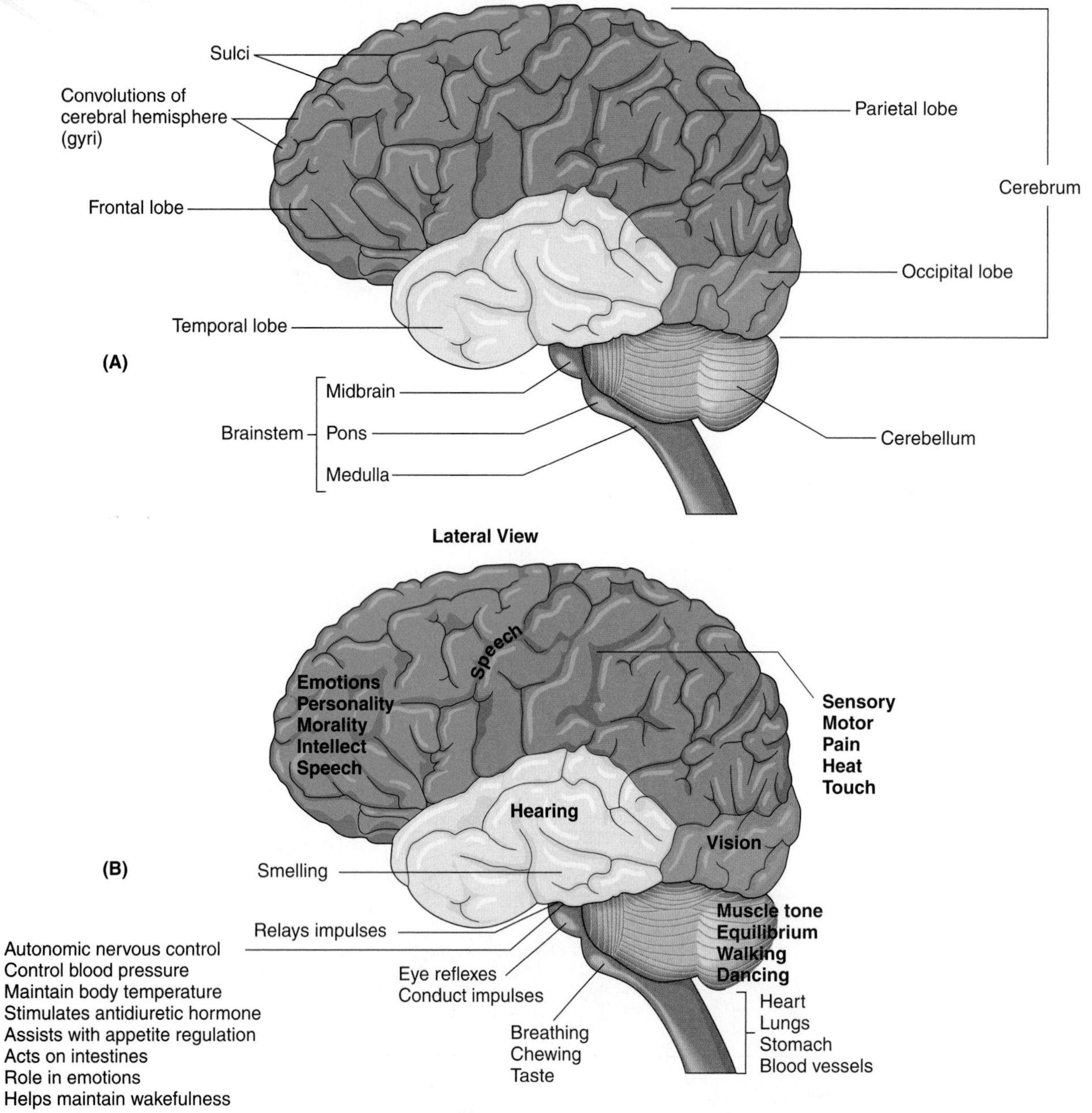

FIGURE 11-3. (A) The parts of the brain. (B) Areas of brain function.

The **pons varolii** is a bridge (*pons* is Latin for "bridge") that connects the spinal cord with the brain and parts of the brain with each other. Longitudinal fibers connect with the spinal cord or medulla with the upper parts of the brain, and transverse fibers connect with the cerebellum. Its pneumotaxic and apneustic area help control breathing.

The **midbrain**, also called the **mesencephalon** (**mess**-in-**SEFF**-ah-lon), contains the **ventral cerebral peduncles** (seh-**REE**-bral peh-**DUN**-kullz) that convey impulses from the cerebral cortex to the pons and spinal cord. It also contains the **dorsal tectum**, which is a reflex center that controls the movement of the eyeballs and head in response to visual stimuli; it also controls the movement of the head and trunk in response to auditory stimuli, such as loud noises.

THE ANATOMY AND FUNCTIONS OF THE DIENCEPHALON

The **diencephalon** is superior to the midbrain and between the two cerebral hemispheres. It also surrounds the third ventricle. It is divided into two main areas: the **thalamus** and the **hypothalamus**. It also contains the **optic tracts** and **optic chiasma** where optic nerves cross each other; the **infundibulum**, which attaches to the **pituitary gland**; the **mamillary bodies**, which are involved in memory and emotional responses to odor; and the **pineal** (**PIN**-ee-al) **gland**, which is part of the epithalamus. The pineal gland is a pinecone-shaped endocrine gland that secretes melatonin, which affects our moods and behavior. This is discussed further in Chapter 12.

The thalamus is the superior part of the diencephalon and the principal relay station for sensory impulses that reach the cerebral cortex coming from the spinal cord, brainstem, and parts of the cerebrum. It also plays an important role as an interpretation center for conscious recognition of pain and temperature and for some awareness of crude pressure and touch.

The epithalamus is a small area superior and posterior to the thalamus. It contains some small nuclei that are concerned with emotional and visceral responses to odor. It contains the pineal gland.

The hypothalamus is the inferior part of the diencephalon and, despite its small size, controls many bodily functions related to homeostasis. It controls and integrates the autonomic nervous system. It receives sensory impulses from the internal organs. It is the intermediary between the nervous system and the endocrine system because it sends signals and controls the pituitary gland. It is the center for mind-over-body phenomena. When we hear of unexplainable cures in people diagnosed with terminal illness but who refused to accept the diagnosis and recovered, the hypothalamus may have been involved in this mind controlling the body phenomenon. It is the hypothalamus that controls our feelings of rage and aggression. It controls our normal body temperature. It contains our thirst center, informing us of when and how much water we need to sustain our bodies. It maintains our waking state and sleep patterns, allowing us to adjust to different work shifts or jetlag travel problems within a day or so. It also regulates our food intake.

THE CEREBRUM: STRUCTURE AND FUNCTION

The **cerebrum** makes up the bulk of the brain. Its surface is composed of gray matter and is referred to as the **cerebral cortex**. Beneath the cortex lies the cerebral white matter. A prominent fissure, the **longitudinal fissure**, separates the cerebrum into right and left halves or **cerebral hemispheres**. On the surface of each hemisphere are numerous folds called **gyri** (**JYE**-rye) with intervening grooves called **sulci** (**SULL**-sigh). The folds increase the surface area of the cortex, which has motor areas for controlling muscular movements, sensory areas for interpreting sensory impulses, and association areas concerned with emotional and intellectual processes. A deep bridge of nerve fiber known as the **corpus callosum** connects the two cerebral hemispheres (Figure 11-4).

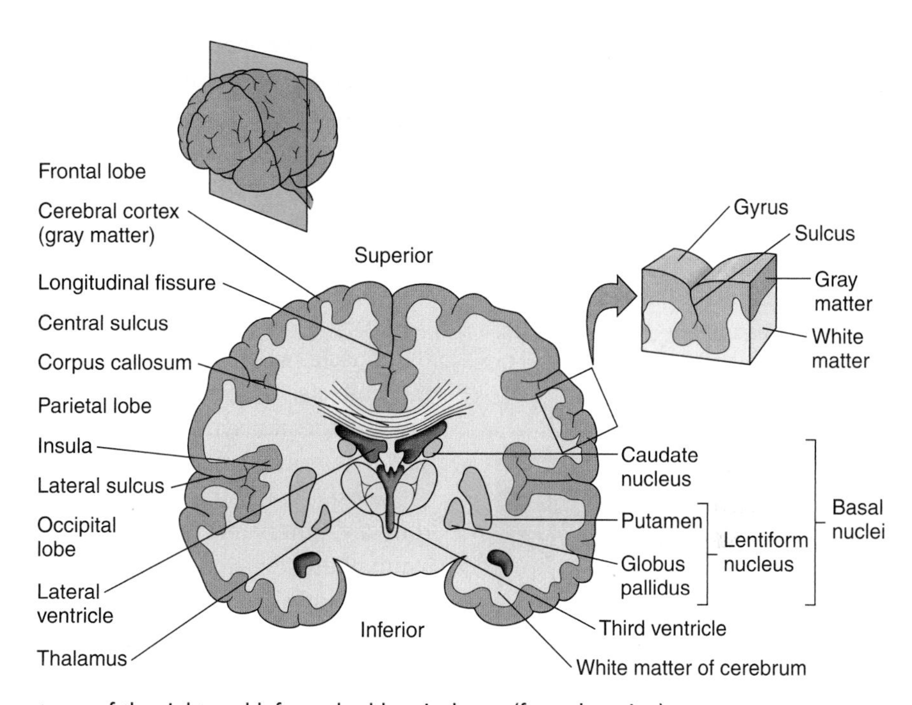

FIGURE 11-4. The anatomy of the right and left cerebral hemispheres (frontal section).

The lobes of the cerebral hemispheres are named after the bones of the skull that lie on top of them. The **frontal lobe** forms the anterior portion of each hemisphere. It controls voluntary muscular functions, moods, aggression, smell reception, and motivation. The **parietal lobe** is behind the frontal lobe and is separated from it by the central sulcus. It is the control center for evaluating sensory information of touch, pain, balance, taste, and temperature. The **temporal lobe** is beneath the frontal and parietal lobes and is separated from them by the lateral fissure. It evaluates hearing input and smell as well as being involved with memory processes. It also functions as an important center for abstract thoughts and judgment decisions. The **occipital lobe** forms the back portion of each hemisphere; its boundaries are not distinct from the other lobes. It functions in receiving and interpreting visual input (see Figures 11-1 and 11-3). A fifth lobe, the **insula**, is embedded deep in the lateral sulcus. The central sulcus separates the frontal and parietal lobes. The lateral sulcus separates the cerebrum into frontal, parietal, and temporal lobes.

THE CEREBELLUM: STRUCTURE AND FUNCTION

The **cerebellum** is the second largest portion of the brain. It is shaped somewhat like a butterfly. It is located beneath the occipital lobes of the cerebrum and behind the pons and the medulla oblongata of the brainstem (see Figure 11-3). It consists of two partially separated hemispheres connected by a centrally constricted structure called the vermis. The cerebellum is made up primarily of white matter with a thin layer of gray matter on its surface called the cerebellar cortex. It functions as a reflex center in coordinating complex skeletal muscular movements, maintaining proper body posture, and keeping the body balanced. If damaged, there can be a decrease in muscle tone, tremors, a loss of equilibrium, and difficulty in skeletal muscle movements.

*Study*WARE™ Connection

Play an interactive game labeling the parts of the brain on your StudyWARE™ CD-ROM.

THE AUTONOMIC NERVOUS SYSTEM

The **autonomic nervous system** is a subdivision of the efferent peripheral nervous system. It functions automatically without conscious effort. It regulates the functions of internal organs by controlling glands, smooth muscles, and cardiac muscle. It assists in maintaining homeostasis by regulating heartbeat and blood pressure, breathing, and body temperature. This system helps us to deal with emergency situations, emotions, and physical activities.

Receptors within organs send sensory impulses to the brain and spinal cord. Motor impulses travel along peripheral nerve fibers that lead to ganglia outside the central nervous system within cranial and spinal nerves. These ganglia are part of the autonomic nervous system.

There are two parts to the autonomic nervous system. The **sympathetic division** (Figure 11-5) prepares the body for stressful situations that require energy expenditure, such as increasing heartbeat and breathing rates to flee from a threatening situation. The fibers of the system arise from the thoracic and lumbar regions of the spinal cord. Their axons leave the cord through the ventral roots of the spinal nerves but then leave the spinal nerve and enter members of a chain of paravertebral ganglia extending longitudinally along the side of the vertebral column. Leaving the paravertebral ganglion, another neuron, the postganglionic fiber, goes to the effector organ. The sympathetic division uses acetylcholine in the preganglionic synapses as a neurotransmitter but uses norepinephrine (or noradrenaline) at the synapses of the postganglionic fibers.

The **parasympathetic division** operates under normal nonstressful conditions. It also functions in restoring the body to a restful state after a stressful experience, thus counterbalancing the effects of the sympathetic division. The preganglionic fibers of the parasympathetic division arise from the brainstem and the sacral region of the spinal cord (Figure 11-6). They lead outward in the cranial and sacral nerves to ganglia located close to the viscera. The postganglionic fibers are short and go to the muscles or glands within the viscera to bring about their effects. The preganglionic and the postganglionic fibers of the parasympathetic division use acetylcholine as the neurotransmitter into the synapses.

Most organs that receive autonomic motor neurons are innervated by both the parasympathetic and sympathetic divisions. However, there are some exceptions: blood vessels and sweat glands are innervated by sympathetic neurons, and smooth muscles associated with the lens of the eye are controlled by parasympathetic neurons.

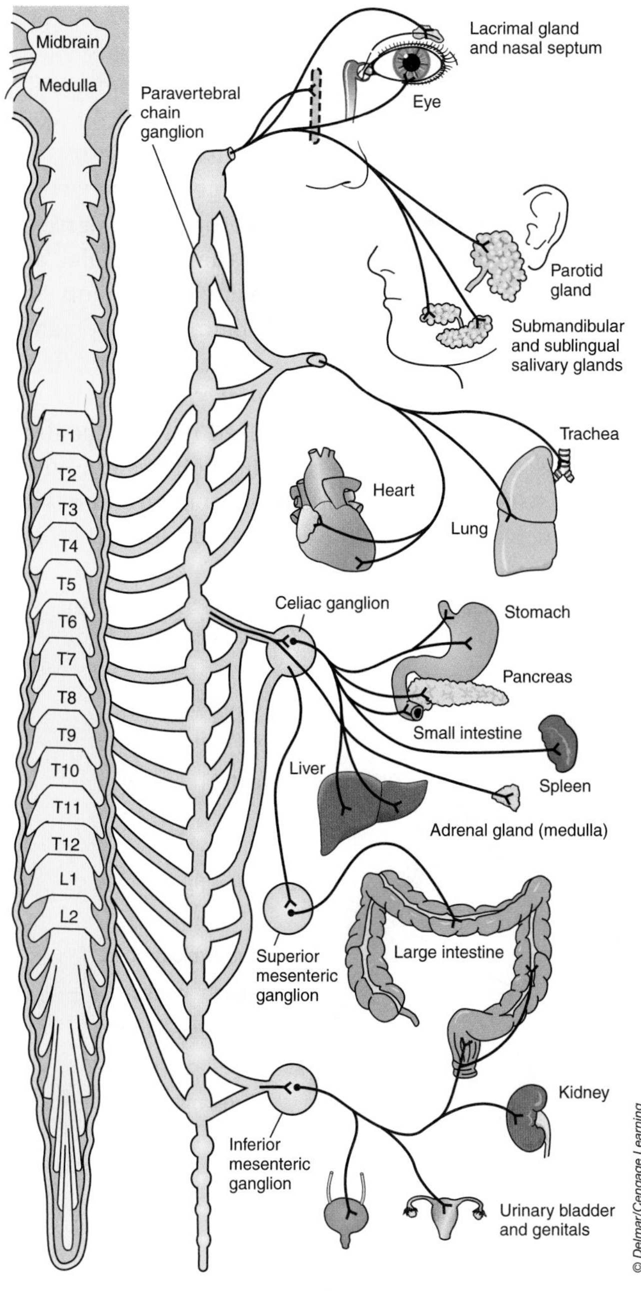

FIGURE 11-5. The nerve pathways of the sympathetic division of the autonomic nervous system.

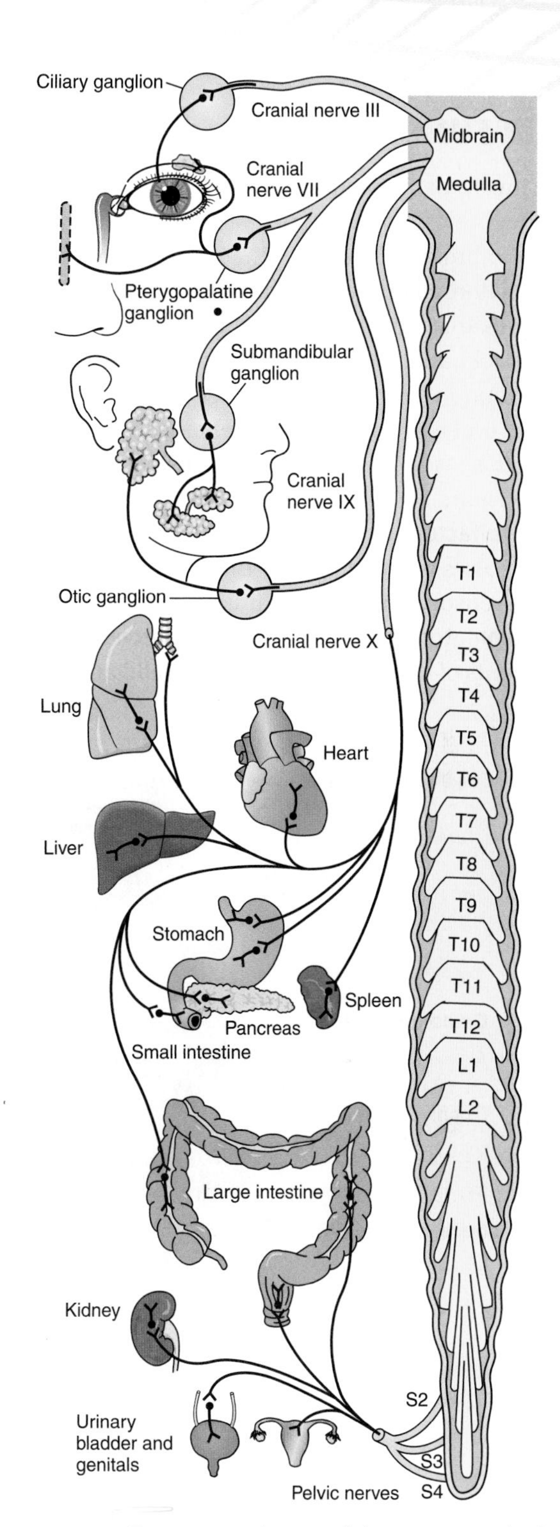

FIGURE 11-6. The nerve pathways of the parasympathetic division of the autonomic nervous system.

The sympathetic division prepares us for physical activity by increasing blood pressure and heartbeat rate, it dilates respiratory passageways for increased breathing rates, and it stimulates sweating. It also causes the release of glucose from the liver as a quick source of energy while inhibiting digestive activities. This system is occasionally called the fight-or-flight system because it prepares us to face a threat or flee quickly from it.

The parasympathetic division stimulates digestion, urination, and defecation. It also counteracts the effects of the sympathetic division by slowing down heartbeat rate, lowering blood pressure, and slowing the breathing rate. It is also responsible for the constriction of the pupil of the eye. This division is occasionally called the rest and repose system.

THE 12 CRANIAL NERVES AND THEIR FUNCTIONS

There are 12 pairs of cranial nerves. Ten pairs originate from the **brainstem**. All 12 pairs leave the skull through various foramina of the skull. They are designated in two ways: by Roman numerals indicating the order in which the nerves arise from the brain (from the front of the brain to the back) and by names that indicate their function or distribution. Some cranial nerves are only sensory or afferent; others are only motor or efferent. Cranial nerves with both sensory and motor functions are called mixed nerves (Figure 11-7).

The **olfactory nerve (I)** is entirely sensory and conveys impulses related to smell. The **optic nerve (II)** is also entirely sensory and conveys impulses related to sight. The **oculomotor nerve (III)** is a motor nerve. It controls movements of the eyeball and upper eyelid and conveys impulses related to muscle sense or position called proprioception. Its parasympathetic function causes constriction of the pupil of the eye. The **trochlear nerve (IV)** is a motor nerve. It controls movement of the eyeball and conveys impulses related to muscle sense. It is the smallest of the cranial nerves. The **trigeminal nerve (V)** is a mixed nerve and it is the largest of the cranial nerves. It has three branches: the maxillary, the mandibular, and the ophthalmic. It controls chewing movements and delivers impulses related to touch, pain, and temperature in the teeth and facial area. The **abducens nerve (VI)** is a motor nerve that controls movement of the eyeball.

The **facial nerve (VII)** is a mixed nerve. It controls the muscles of facial expression and conveys sensations related to taste. Its parasympathetic function controls the tear and salivary glands. The **vestibulocochlear nerve (VIII)** (ves-**tib**-yoo-loh-**KOK**-lee-ar) is entirely sensory. It transmits impulses related to equilibrium and hearing. The

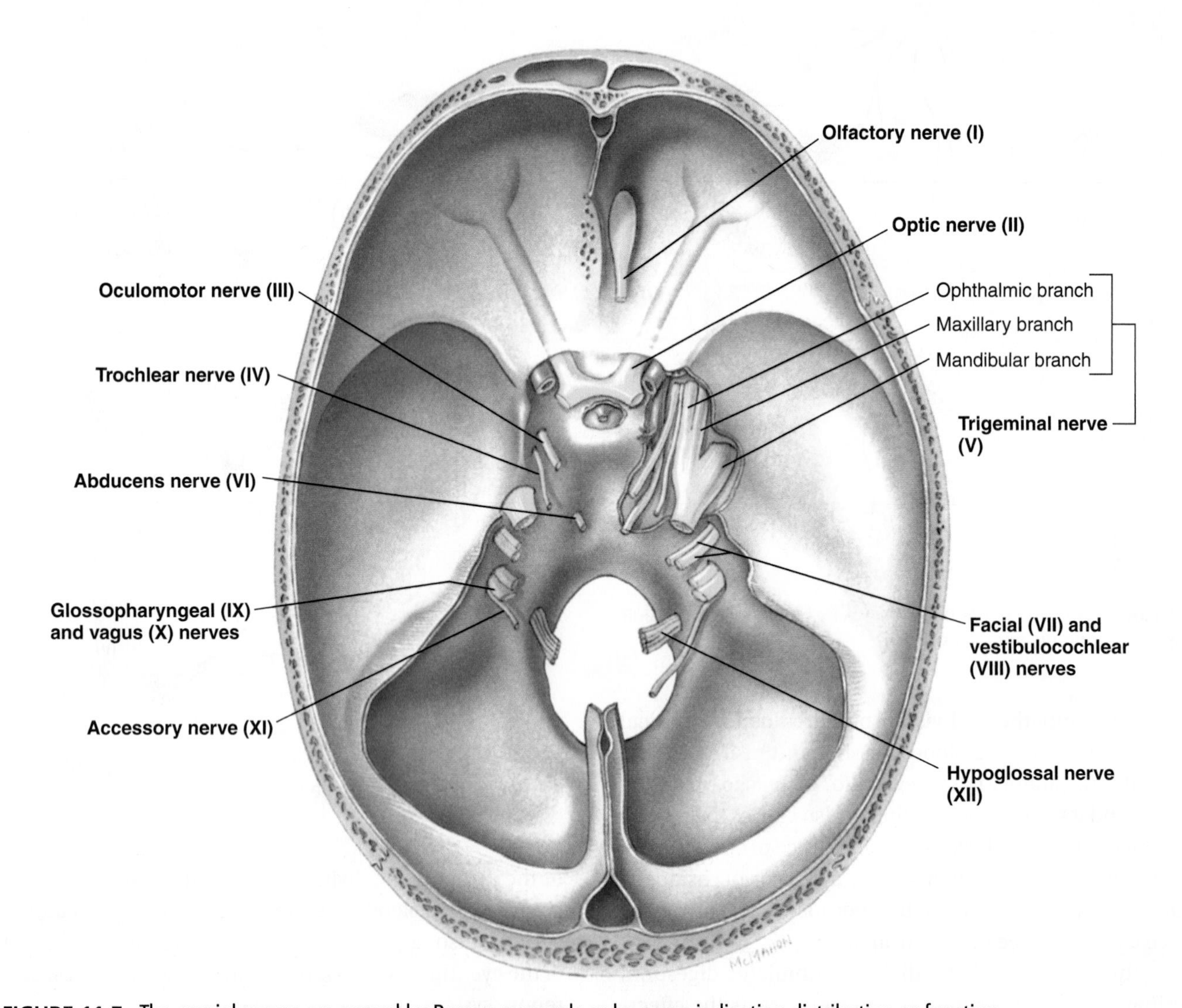

FIGURE 11-7. The cranial nerves are named by Roman numerals or by name indicating distribution or function.

HEALTH ALERT

MENTAL ALERTNESS

Consistent mental activity leads to mental alertness and a healthy brain. As children grow, toys that require mental interaction, thought, and choice help them develop mentally. Reading should become a well-developed habit throughout life. Crossword puzzles, novels, plays, and viewing a good movie are all activities that keep us mentally alert during our free time and "exercise our brain."

Diet also plays a role in maintaining good mental functions. Protein is an essential food for developing minds in young children. Many of us have heard fish referred to as "brain food." Fish is an excellent source of protein as are meats and poultry. There are also many plants that contain sources of protein such as peanut, soybean, and wheat. Peanut butter and jelly sandwiches or tuna fish sandwiches made with whole wheat bread are excellent sources of protein in children's lunch boxes. ■

HEALTH ALERT

CAFFEINE ALERT

Caffeine is found in coffee, tea, Coca-Cola, Pepsi, and other soda drinks as well as in small amounts in chocolate. It is one of our "staple" and legal drugs enjoyed by millions around the world. Caffeine functions in the same way as the sympathetic division of the autonomic nervous system. That is, it stimulates physiological activity, causing increased heartbeat rates. We view this as giving us a jump-start in the morning or keeping us alert during busy work hours. Moderate amounts of caffeine pose no serious threats to our health. However, excessive consumption can lead to high blood pressure, anxiety, irregular heartbeat rate, and difficulty falling asleep.

Caffeine levels of tolerance vary from individual to individual. We should monitor ourselves and determine how much caffeine is safe and what amounts can lead to problematic symptoms. A moderate amount of caffeine is equivalent to about two cups of coffee or cola per day. Since studies with animals have shown a link to birth defects and caffeine consumption, pregnant women should avoid or moderately reduce their caffeine consumption throughout pregnancy. ■

glossopharyngeal nerve (IX) (**GLOSS**-oh-fair-in-**GEE**-al) is a mixed nerve. It controls swallowing and senses taste. Its parasympathetic function controls salivary glands. The **vagus nerve (X)** is a mixed nerve. It controls skeletal muscle movements in the pharynx, larynx, and palate. It conveys impulses for sensations in the larynx, viscera, and ear. Its parasympathetic function controls viscera in the thorax and abdomen. The **accessory nerve (XI)** is a motor nerve. It originates from the brainstem and the spinal cord. It helps control swallowing and movements of the head. Finally, the **hypoglossal nerve (XII)** is a motor nerve. It controls the muscles involved in speech and swallowing and its sensory fibers conduct impulses for muscle sense. Table 11-1 provides a summary of the names and functions of the cranial nerves.

THE SPECIAL SENSES

The five special senses are smell, taste, vision, hearing, and balance. The senses of smell and taste are initiated by the interactions of chemicals with sensory receptors on the tongue and in the nose. Vision occurs due to the interaction of light with sensory receptors in the eye. Hearing and balance function due to the interaction of

Table 11-1 The Cranial Nerves

Number	Name	Function
I	Olfactory	Sensory: smell
II	Optic	Sensory: vision
III	Oculomotor	Motor: movement of the eyeball, regulation of the size of the pupil
IV	Trochlear	Motor: eye movements
V	Trigeminal	Sensory: sensations of head and face, muscle sense Motor: mastication Note: divided into three branches: the ophthalmic branch, the maxillary branch, and the mandibular branch
VI	Abducens	Motor: movement of the eyeball, particularly abduction
VII	Facial	Sensory: taste Motor: facial expressions, secretions of saliva
VIII	Vestibulocochlear	Sensory: balance, hearing Note: divided into two branches: the vestibular branch responsible for balance and the cochlear branch responsible for hearing
IX	Glossopharyngeal	Sensory: taste Motor: swallowing, secretion of saliva
X	Vagus	Sensory: sensation of organs supplied Motor: movement of organs supplied Note: supplies the head, pharynx, bronchus, esophagus, liver, and stomach
XI	Accessory	Motor: shoulder movement, turning of head, voice production
XII	Hypoglossal	Motor: tongue movements

mechanical stimuli (sound waves for hearing and motion for balance) with sensory receptors in the ear.

The Sense of Smell

The sense of smell is also known as the **olfactory** (**ol**-**FAK**-toh-ree) **sense.** Molecules in the air enter the nasal cavity and become dissolved in the mucous epithelial lining of the superior nasal conchae, the uppermost shelf area in the nose (Figure 11-8A). Here they come in contact with olfactory neurons modified to respond to odors. These neurons are bipolar neurons. Their dendrites are found in the epithelial surface of the uppermost shelf and contact the olfactory receptor sites in the nose. The odor molecules bind to these receptor sites. The olfactory neurons transmit the impulse along their axons whose ends become enlarged olfactory bulbs. From here, they connect with association neurons to the area of the brain called the olfactory cortex found in the temporal and frontal lobes of the cerebrum.

The receptor cells are neurons that have cilia at the distal ends of their dendrites (see Figure 11-8B). It is these cilia that function as chemoreceptors to detect odors. These molecules first become dissolved in the mucous membrane that lines the olfactory shelf in the nose and then are detected. The sense of smell is closely related to the sense of taste. We use these two senses to decide whether or not to eat a particular food. Our sense of smell is complex because a small number of receptors detect a great variety of odors. It is the brain that then interprets these receptor combinations into a type of olfactory code. The exact mechanism of how this works is still being investigated by biologists. However, we do know that olfactory receptors rapidly adapt to odors and after a short time we no longer perceive the odor as intensely as it was initally detected.

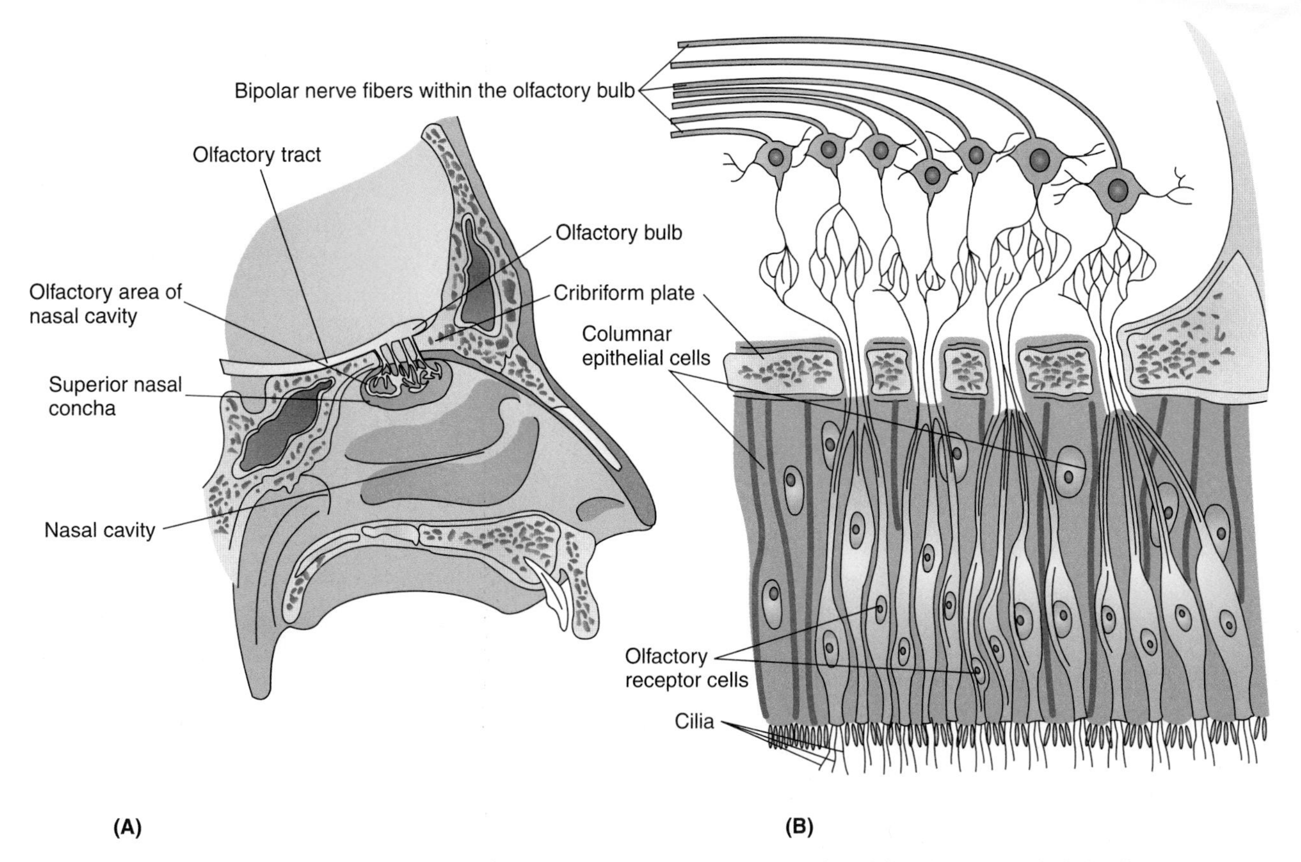

FIGURE 11-8. (A) The olfactory area in the nose formed by the superior nasal conchae. (B) Columnar epithelial cells support olfactory receptor cells with cilia at their ends.

The Sense of Taste

Taste buds are the sensory structures found on certain **papillae** (pah-**PILL**-e), which are elevations of the tongue, that detect taste stimuli (Figure 11-9). Taste buds are also found on the palate of the roof of the mouth, in certain regions of the pharynx, and on the lips of children. Each taste bud is composed of two types of cells. The first type are specialized epithelial cells that form the exterior capsule of the taste bud. The second type of cell forms the interior of the taste bud. These cells are called **taste cells** and function as the receptor sites for taste. The taste bud is spherical with an opening called the taste pore. Taste hairs are tiny projections of the taste cells that extend out of the taste pore. It is these taste hairs that actually function as the receptors of the taste cell. Cranial nerves VIII, IX, and X conduct the taste sensations to the brain, which perceives and interprets the taste.

Before a chemical can be tasted, it must first be dissolved in a fluid (just like the odors in the nose). The saliva produced by the salivary glands provides this fluid medium. Nerve fibers surrounding the taste cells transmit the impulses to the brain for interpretation. The sensory impulses travel on the facial (VIII), glossopharyngeal (IX), and vagus (X) cranial nerves to the gustatory (taste) cortex of the parietal lobe of the cerebrum for interpretation. The four major types of taste sensations are sweet, sour, salty, and bitter. Although all taste buds can detect all four sensations, taste buds at the back of the tongue react strongly to bitter, taste buds at the tip of the tongue react strongly to sweet and salty, and taste buds on the side of the tongue respond more strongly to sour tastes (see Figure 11-9). Taste sensations are also influenced by olfactory sensations. Holding one's nose while swallowing reduces the taste sensation. This is a common practice when taking bad-tasting medicine.

The Sense of Sight

The eyes are our organs of sight. They are protected by the orbits of the skull. See Chapter 7 to review the bones that make up the orbits. In addition, the eyebrows help shade the eye and keep perspiration from getting into the eye

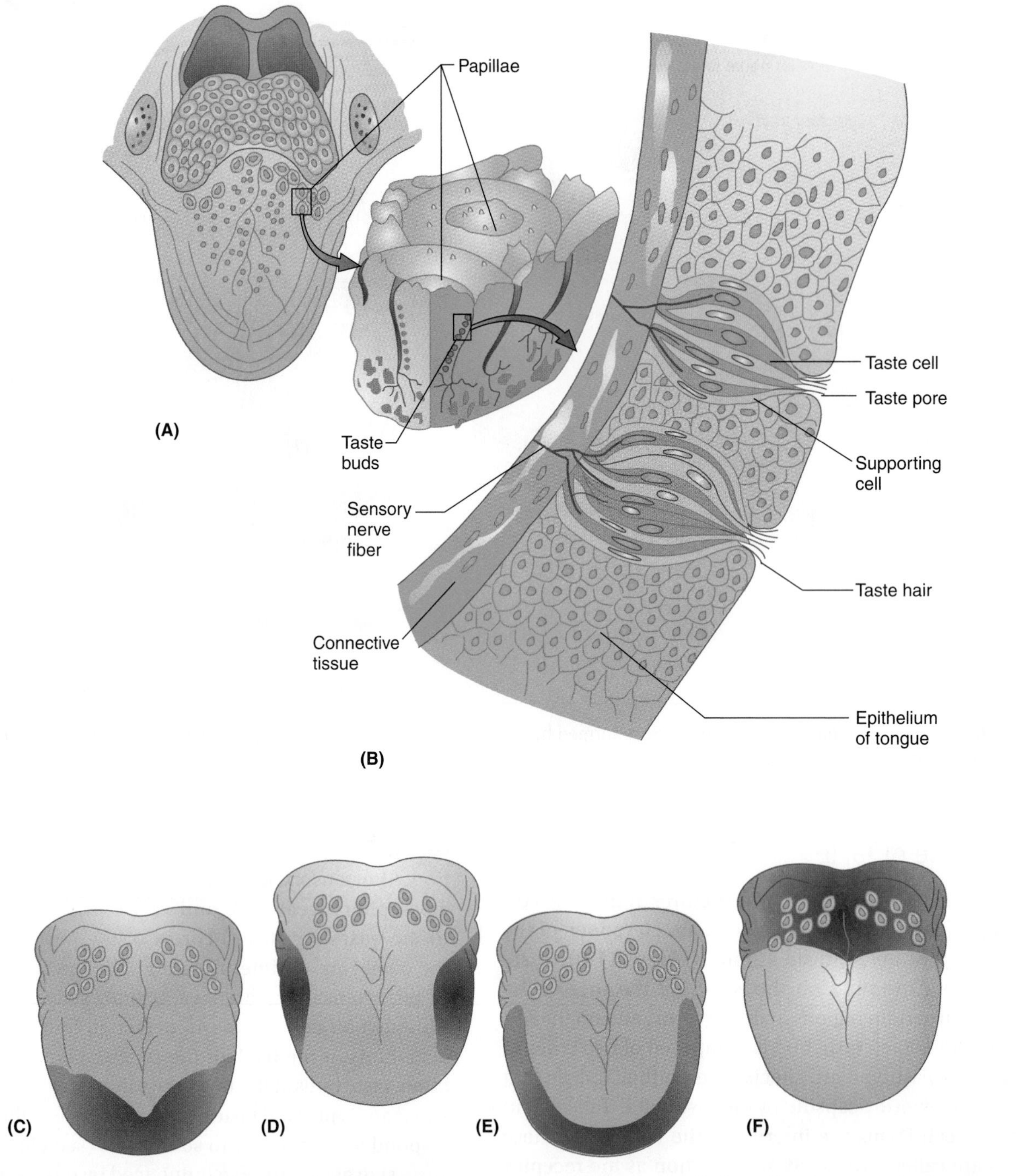

FIGURE 11-9. (A) Taste buds on the surface of the tongue are associated with elevations called papillae. (B) A taste bud contains taste cells with an opening called the taste pore at its free surface. Colored sections indicate common patterns of taste receptors: (C) sweet, (D) sour, (E) salt, (F) bitter.

and causing an irritation to the eye. Eyelids and eyelashes protect the eye from foreign objects. Blinking of the eyelids lubricates the surface of the eye by spreading tears that are produced by the lacrimal gland. The tears not only lubricate the eye but also help to combat bacterial infections through the enzyme lysozyme, salt, and gamma globulin.

StudyWARE™ Connection

Watch an animation that explains how we see on your StudyWARE™ CD-ROM.

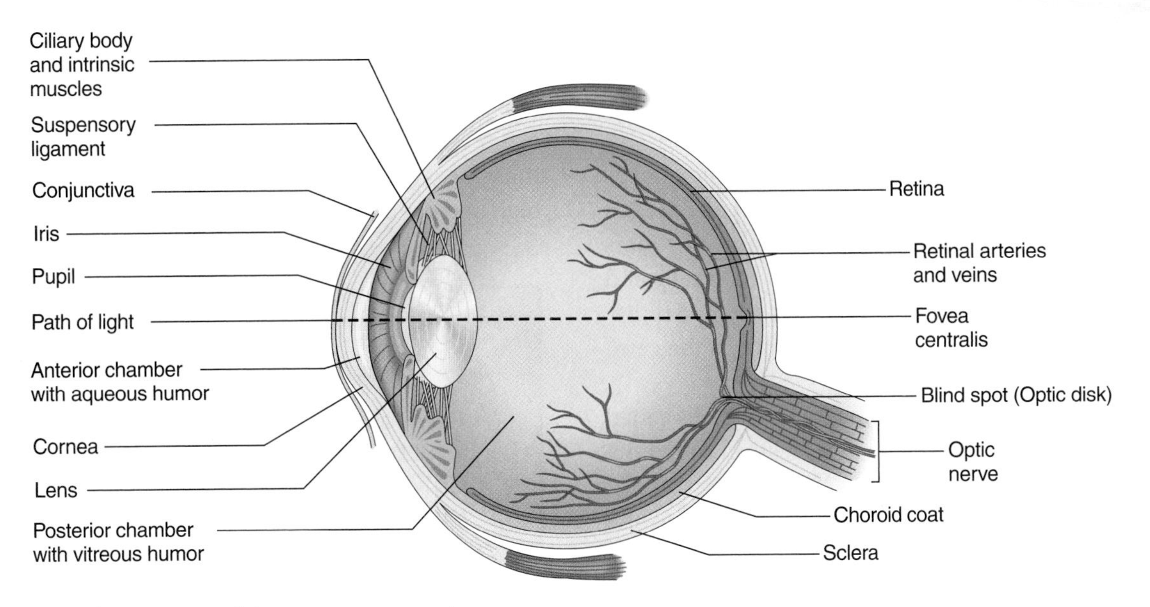

FIGURE 11-10. The anatomy of the eye, transverse view.

The Anatomy of the Eye

The eye is a sphere filled with two fluids (Figure 11-10). The skeletal muscles that move the eye are discussed in Chapter 9. They are the rectus muscles and the oblique muscles.

The wall of the eye is composed of three layers, or tunics, of tissue. The outermost layer is the **sclera** (**SKLAIR**-ah). It is white and composed of tough connective tissue. We see it as the white of the eye when looking in a mirror. The **cornea** (**COR**-nee-ah) is the transparent part of this outermost layer that permits light to enter the eye. The second layer is the **choroid** (**KOR**-oyd). It contains numerous blood vessels and pigment cells. It is black in color and absorbs light so that it does not reflect in the eye and impair vision. The innermost layer of the eye is the **retina** (**RET**-ih-nah). It is gray in color and contains the light-sensitive cells known as the rods and cones.

The **ciliary** (**SIL**-ee-**air**-ee) **body** consists of smooth muscles that hold the biconvex, transparent, and flexible **lens** in place. The **iris** is the colored part of the eye consisting of smooth muscle that surrounds the **pupil**. The iris regulates the amount of light that enters through the diameter of the pupil. When we go into a dark room, the iris opens to allow more light to enter. When we go out into strong sunlight, the iris constricts, letting less light enter the pupil.

The interior of the eye is divided into two compartments. In front of the lens is the anterior compartment that is filled with a fluid called the **aqueous humor**. This fluid helps to bend light, is a source of nutrients for the inner surface of the eye, and maintains ocular pressure. It is produced by the ciliary body. The posterior compartment of the eye is filled with **vitreous** (**VIT**-ree-us) **humor**. It too helps to maintain ocular pressure, refracts or bends light and holds the retina, and lens in place.

The retina is the innermost layer of the eye and contains the photosensitive cells (Figure 11-11). The retina has a pigmented epithelial layer that helps keep light from being reflected back into the eye. The sensory layer is made up of the rods and cones. There are more rods than cones in this layer. Rods are quite sensitive to light and function in dim light but do not produce color vision.

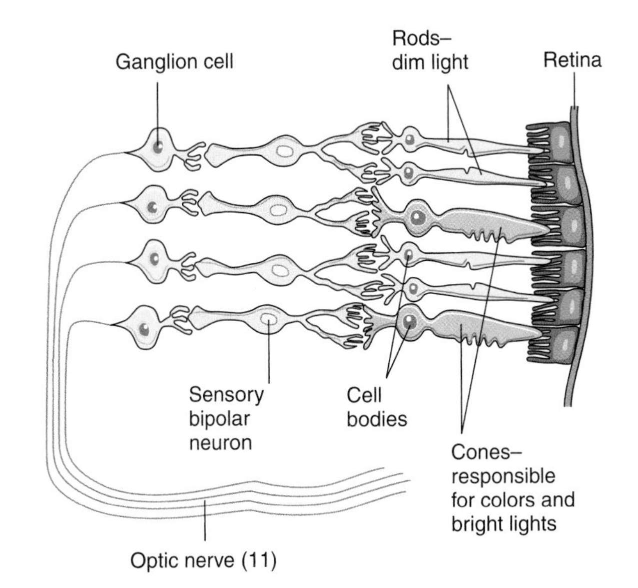

FIGURE 11-11. The layers of the retina illustrating the rods and cones and other cellular layers.

It is the cones that produce color and they require lots of light. Three different types of cones are sensitive to red, green, or blue. Combinations of these cones produce all the other colors we see.

The rod and cone cells synapse with the bipolar cells of the retina. The bipolar cells synapse with ganglia cells whose axons form the **optic nerve**. Eventually, the fibers of the optic nerve reach the thalamus of the brain and synapse at its posterior portion and enter as optic radiations to the visual cortex of the occipital lobe of the cerebrum for interpretation.

The yellowish spot in the center of the retina is called the macula lutea. In its center is a depression called the **fovea centralis**. This region produces the sharpest vision, like when we look directly at an object. Medial to the fovea centralis is the **optic disk**. It is here that nerve fibers leave the eye as the optic nerve. Because the optic disk has no receptor cells, it is called the blind spot.

Both rods and cones contain light-sensitive pigments. Rod cells contain the pigment called **rhodopsin** (roh-**DOP**-sin). Cone cells contain a slightly different pigment. When exposed to light the rhodopsin breaks down into a protein called opsin and a pigment called retinal. Manufacture of retinal requires vitamin A. Someone with a vitamin A deficiency may experience night blindness, which is difficulty seeing in dim light.

Sight is one of our most important senses. Humans depend on sight as their main sense to survive and interact with our environment. We educate ourselves via visual input through reading, color interpretations, and movement. People who lose their sight tend to develop acuity with the other senses like smell and sounds, senses that our dog and cat companions have developed to a high degree.

StudyWARE™ Connection

Play an interactive game labeling the structures of the eye on your StudyWARE™ CD-ROM.

The Sense of Hearing and Equilibrium

The external, inner, and middle ear contain the organs of balance and hearing (Figure 11-12). The external ear is that part of the ear that extends from the outside of the head to the eardrum. Medial to the eardrum is the air-filled chamber called the middle ear, which contains the auditory ossicles: the malleus, incus, and stapes. The external and middle ear are involved in hearing. The inner ear is a group of fluid-filled chambers that are involved in both balance and hearing.

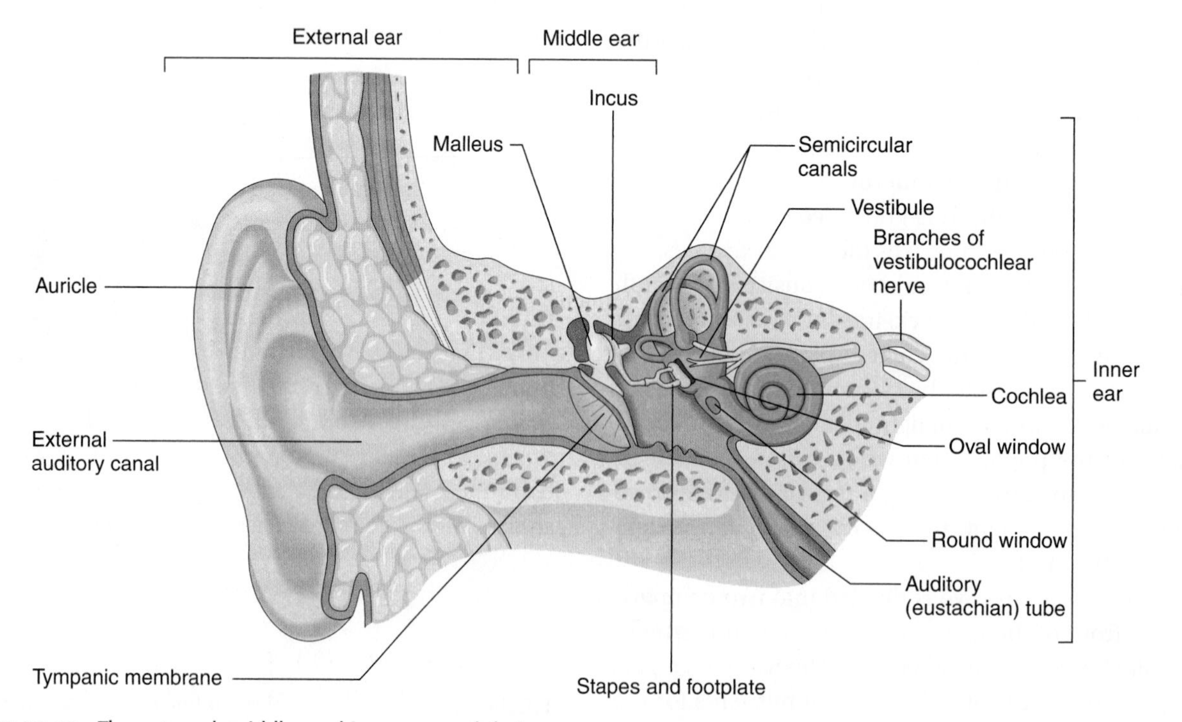

FIGURE 11-12. The external, middle, and inner ear and their organs.

The external ear consists of the flexible, visible part of our ear called the **auricle** (**AW**-rih-kl) composed mainly of elastic cartilage. This connects with our ear canal known as the **external auditory meatus** (**AW**-dih-tor-ee mee-**ATE**-us). The auricle allows sound waves to enter the ear canal, which then directs those waves to the delicate eardrum or **tympanic** (tim-**PAN**-ik) **membrane**. The ear canal is lined with hairs and modified sebaceous glands called **ceruminous** (seh-**ROO**-men-us) **glands** that produce earwax or **cerumen**. The hairs and earwax protect the eardrum from foreign objects. The thin tympanic membrane, which is silvery gray in color, is very delicate and sound waves cause it to vibrate.

The middle ear is the air-filled cavity that contains the three auditory ossicles or ear bones: the **malleus** or hammer, the **incus** or anvil, and the **stapes** or stirrup. These bones transmit the sound vibrations from the eardrum to the **oval window**. The two openings on the medial side of the middle ear are the oval window and the **round window**. They connect the middle ear to the inner ear. As the vibrations of the sound waves are transmitted from the malleus to the stapes, they are amplified in the middle ear. In the middle ear we also find the **auditory or eustachian** (yoo-**STAY**-shun) **tube**. This tube opens into the pharynx and permits air pressure to be equalized between the middle ear and the outside air, thus ensuring that hearing is not distorted. When flying in an airplane, changing altitude changes pressure. This results in muffled sounds and pain in the delicate eardrum.We can allow air to enter or exit the middle ear through the auditory tube and thus equalize the pressure by yawning, chewing, or swallowing. Sometimes we hold our nose and mouth shut and gently force air out of our lungs through the auditory tube and pop our eardrum to equalize the pressure.

The inner ear is made of interconnecting chambers and tunnels within the temporal bone. This area contains the cochlea, which is involved in hearing, and the vestibule and the semicircular canals, which are involved in balance. Balance is also called equilibrium. Static equilibrium is controlled by the vestibule and determines the position of the head in relation to gravity; kinetic equilibrium is controlled by the semicircular canals and determines the change in regard to head rotational movements.

*Study***WARE**™ Connection

- Watch an animation that explains how we hear on your StudyWARE™ CD-ROM.
- Play an interactive game labeling structures of the ear on your StudyWARE™ CD-ROM.

AS THE BODY AGES

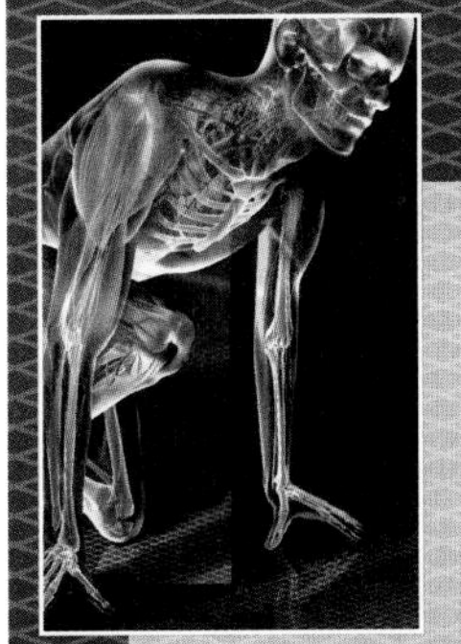

The nervous system forms very quickly in the developing embryo. By the first month, a brain and spinal cord can be seen with the beginnings of sense organs. When a child is born, the head is much larger in proportion to the rest of the body due to the developed nervous system and its neurons.

As we grow, the brain develops very rapidly during the first years of life as the neurons increase in size. The supporting cells or neuroglia grow and increase in numbers, and certain neurons develop myelinated sheaths while their dendrite branches develop and increase in number, resulting in more synapse contacts.

At maturity, the nervous system begins to undergo numerous changes. The brain actually begins to decrease in size and mass due to a loss of neurons that constitute the outer part of the cerebrum. Individuals in their mid-70s lose 7% of the weight of their brain. Accompanying this is a loss in synaptic contacts and neurotransmitters. This results in a diminished capacity to send impulses to and from the brain. Information processing is more difficult and muscular movement and responses slow down. These are all symptoms observed in older adults. A reduction in the size of the arteries supplying the brain results in less oxygen-carrying blood supplying these cells, increasing the possibilities of strokes in older adults. ■

Career FOCUS

These are careers that are available to individuals who are interested in working with the nervous system.

- **Anesthesiologists** are physicians who administer anesthesia directly to patients during surgery or supervise nurse anesthetists in the delivery of anesthesia.
- **Anesthesiologist assistants** are allied health professionals who acquire preoperative information such as a history of health-related problems and perform physical examinations such as insertion of intravenous injections and catheters as well as being involved in recovery room care.
- **Neurosurgeons** are physicians specializing in surgery of the brain, spinal cord, and the peripheral nerves.
- **Nurse anesthetists** are registered nurses who have advanced training in anesthesia who manage the care of patients during the administration of anesthesia in certain surgical situations.
- **Acupuncturists** are individuals trained in the traditional Chinese method of dulling pain by inserting fine wire needles into the skin at specific sites to produce an anesthetic effect on certain parts of the body.
- **Psychiatrists** are physicians with advanced training in the diagnosis, prevention, and treatment of mental disorders.
- **Psychologists** are individuals who specialize in the study of the function of the brain. Clinical psychologists provide testing and counseling for patients with emotional and mental disorders and have graduate training.

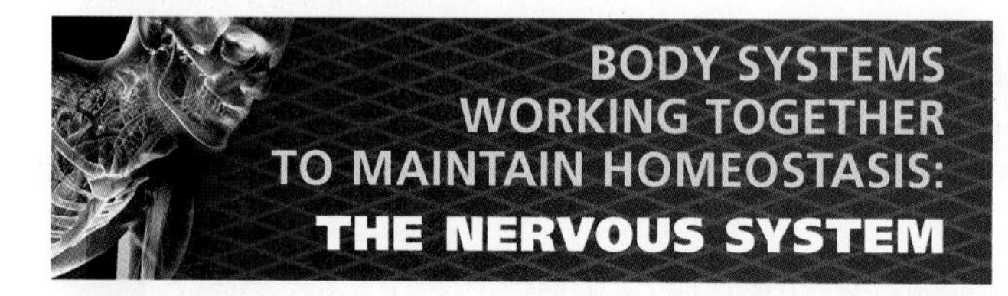

Integumentary System

- Temperature receptors in the skin detect changes in the external environment and transmit this information to the nervous system for interpretation about hot and cold sensations.
- Pressure receptors in the skin detect changes in the external environment and transmit this information to the nervous system for interpretation about pleasure and pain sensations.

Skeletal System

- The skull bones and vertebrae protect the brain and spinal cord.
- Bones store calcium for release into the blood. Calcium is necessary for nervous transmission.

Muscular System

- Muscular contraction depends on nerve stimulation.
- Muscle sense and position of body parts are controlled by sensory neurons and interpretations by the nervous system.

Endocrine System

- The hypothalamus of the brain, through neurosecretions, controls the actions of the pituitary gland, the master gland of the endocrine system, which controls the secretions of many hormones of other endocrine glands.

Cardiovascular System

- Nerve impulses control heartbeat and blood pressure.
- Nerve impulses control dilation and constriction of blood vessels, thus controlling blood flow.

Lymphatic System

- Nervous anxiety and stress can impair the immune response, a major function of the lymphatic system.

- The hypothalamus controls mind-over body phenomena and boosts the immune response, thus fighting disease.

Digestive System

- The autonomic nervous system controls peristalsis, resulting in mixing of food with digestive enzymes and moving food along the digestive tract.
- Nerve impulses inform us when to empty the tract of indigestible waste.

Respiratory System

- Respiratory rates are controlled by the nervous system, thus controlling oxygen and carbon dioxide levels in the blood.
- The phrenic nerve controls the action of the diaphragm muscle, which controls breathing rates.

Urinary System

- Nerve impulses to the kidneys control the composition and concentration of urine.
- Stretch receptors in the bladder inform us when to eliminate urine from the body.

Reproductive System

- Sperm and egg production is stimulated by the nervous system at the beginning of puberty and throughout life in men and up to menopause in women.
- Sexual pleasure is determined by sensory receptors in various parts of the body.
- Smooth muscle contractions stimulated by the nervous system initiate childbirth and delivery.
- Sucking at the breast by the newborn stimulates milk production in the mammary glands.

COMMON DISEASE, DISORDER, OR CONDITION

DISORDERS OF THE NERVOUS SYSTEM

ALZHEIMER'S DISEASE

Alzheimer's (**ALTS**-high-mers) **disease** results in severe mental deterioration. It is also known as senile dementia–Alzheimer type (SDAT). It usually affects older people but may begin in middle life with symptoms of memory loss and behavioral changes. The disease affects 10% of people older than 65 and nearly half of those 85 or older. Symptoms worsen dramatically in individuals older than 70. Symptoms include memory failure, confusion, a decrease in intellectual capacity, restlessness, disorientation, and, occasionally, speech disturbances. The disease produces a loss of neurons in the cerebral cortex of the brain, resulting in a decrease in brain size. The sulci widen and the gyri become narrowed. The temporal and frontal lobes of the cerebrum are particularly affected. Enlarged axons containing beta-amyloid protein, called plaques, form in the cortex. There is a genetic predisposition for the disease. The first symptoms of the disease usually begin with an inability to assimilate new information despite the ability to retain old knowledge, difficulty in recalling words, and a disorientation in common surroundings. Death usually occurs 8 to 12 years after the onset of symptoms. These patients should be kept comfortable and carefully observed to keep them from self-harm.

CEREBROVASCULAR ACCIDENT (CVA)

Cerebrovascular (seh-**REE**-bro-**VAS**-kyoo-lar) **accident** (**CVA**) or **stroke** can be caused by a clot or thrombus in a blood vessel, or by a piece of a clot or embolus that breaks loose and travels in the circulatory system until it lodges in a blood vessel and blocks circulation. It can also be caused by a hemorrhage in tissue or by the constriction of a cerebral blood vessel, known as a vasospasm. These situations can result in localized cellular death due to lack of blood supply to the tissue. This is known as an infarct. Symptoms are determined by the size and location of the infarct but can include paralysis or lack of feeling on the side of the body opposite the cerebral infarct, weakness, speech defects, or the inability to speak. Death may result. However, symptoms may subside in minor strokes when the resulting brain swelling subsides.

(continues)

COMMON DISEASE, DISORDER, OR CONDITION

DISORDERS OF THE NERVOUS SYSTEM (continued)

MENINGITIS

Meningitis (men-in-**JYE**-tis) is an inflammation of the meninges caused by bacterial or viral infection, which results in headache, fever, and a stiff neck. Severe cases of viral meningitis can result in paralysis, coma, and death.

ENCEPHALITIS

Encephalitis (in-**seff**-ah-**LYE**-tis) is an inflammation of brain tissue usually caused by a virus transmitted by the bite of a mosquito. It is manifested by a wide variety of symptoms, including coma, fever, and convulsions and could result in death.

TETANUS

Tetanus is caused by the introduction of the bacterium *Clostridium tetani* into an open wound. The bacterium produces a neurotoxin that affects motor neurons in the spinal cord and brainstem. It also blocks inhibitory neurotransmitters, resulting in muscle contractions. The jaw muscles are affected earliest, locking the jaw in a closed position, hence the common name *lockjaw*. Death can result from spasms of the respiratory muscles and the diaphragm.

PARKINSON'S DISEASE

Parkinson's disease is characterized by tremors of the hand when resting and a slow shuffling walk with rigidity of muscular movements. It is caused by damage to basal nuclei, resulting in deficient dopamine, an inhibitory neurotransmitter. The disease can be treated to a certain degree with L-dopa. New research uses the transplanting of fetal cells from discarded umbilical cords into the patient. These cells can produce dopamine in the individual with the disease.

CEREBRAL PALSY

Cerebral palsy (seh-**REE**-bral **PAWL**-zee) is a condition caused by brain damage during brain development or the birth process. The child's motor functions and muscular coordinations are defective. Symptoms include awkward movements, head tossing, and flailing arms. Speaking is impaired with guttural sounds, and swallowing is difficult. Body balance is poor, with spasms and tremors of muscles. Careful prenatal and obstetric care is necessary to prevent this condition.

EPILEPSY

Epilepsy is caused by a disorder in the brain where certain parts of the brain are overactive, producing convulsive seizures (involuntary muscle contractions) and possible loss of consciousness.

HEADACHE

Headache or cephalalgia can be caused by a variety of factors, from muscle tension and anxiety to swollen sinuses and toothache. Headache can also be caused by inflammation of the meninges, brain tumors, and vascular changes in the blood supply to the brain.

ANEURYSM

An **aneurysm** (**ANN**-your-**riz**-em) is an enlargement or dilation of a blood vessel wall, commonly referred to as a ballooning. This may rupture, causing bleeding or hemorrhaging in the area. Hypertension may cause an aneurysm to burst. Aneurysms are commonly developed in the aorta and on arteries that supply the brain. Hemorrhaging in the brain destroys brain tissue. Older people occasionally develop aneurysms around the popliteal artery in the leg.

MULTIPLE SCLEROSIS (MS)

Multiple sclerosis, also known as MS, is a disease caused by progressive demyelination of nerve cells in the brain and spinal cord. It is currently considered to be an autoimmune disease. It produces lesions

(continues)

COMMON DISEASE, DISORDER, OR CONDITION

DISORDERS OF THE NERVOUS SYSTEM (continued)

on the brain and spinal cord, resulting in a hardening (sclerosis) of the fatty myelin sheaths, which produces poor conduction of nerve impulses. It usually develops early in adulthood with progression and occasional bouts of remission. Symptoms of the disease are muscle weakness, double vision, vertigo, abnormal reflexes, and occasionally difficulty in urination. There is no cure for the disease. Treatments include drugs that alleviate the symptoms. Patients are encouraged to live as normal a life as possible. Some individuals with the later stages of the disease need an authorized medical scooter to assist in their mobility.

REYE'S SYNDROME

Reye's syndrome is named for Ralph Reye, an Australian pathologist. The condition usually affects individuals under 18 years of age. It usually develops after an acute viral infection like the flu, chicken pox, or an enterovirus. Symptoms include a rash, vomiting, and disorientation during the onset of the syndrome followed later by seizures, coma, and respiratory system collapse. The cause of the disease is unknown but appears to be related to the administration of aspirin. Brain cells swell and the kidneys and liver accumulate an abnormal amount of fat.

RABIES

Rabies is an acute viral, fatal disease that affects the central nervous system. It is transmitted to humans through a bite with virus-containing saliva of an infected mammal like unvaccinated dogs or cats or through the bite of wild animals such as bats, skunks, raccoons, or foxes. The virus travels to the brain and other organs. Symptoms include fever, headache, and muscle pain. If untreated, it results in encephalitis, severe muscle spasms, seizures, paralysis, coma, and eventually death. Treatment includes a series of vaccine injections administered intramuscularly. Prevention is through regular rabies shots to our domesticated cats and dogs. Since dogs with rabies are afraid of water and refuse to drink, the name hydrophobia (fear of water) is also used for the disease.

BELL'S PALSY

Bell's palsy is also known as facial palsy. It results in paralysis of the facial nerve but only on one side of the face. An affected patient may not be able to control salivation or to close one eye. The absence of muscle tone causes the face to droop. The condition is usually temporary but in severe cases it can be permanent. Expression of the symptoms can result from trauma to the nerve, compression of the nerve, or a *Herpes simplex* viral infection.

CONCUSSION

A **concussion** is caused by violent shaking or jarring to the brain as a result of a severe blow. This results in brain damage, which causes a momentary loss of consciousness. In some cases symptoms such as a headache caused by muscle tension, personality changes, or fatigue can persist for a month or more.

DEPRESSION

Depression is a condition experienced to some degree by most individuals at some time in their lives. Although described for centuries, the exact cause is neither specific nor universal for all affected individuals. Its basis is probably both psychological and physiological. By definition, depression is an abnormal emotional state with feelings of sadness, rejection, hopelessness, and worthlessness that are out of proportion to reality. Certain types of depression can be treated with either antidepressant drugs or psychotherapy. There can be certain behavioral conditions consistent with depression, such as overeating, apathy, withdrawal, anger, or even aggression.

COMMON DISEASE, DISORDER, OR CONDITION

DISORDERS OF THE SENSES

OTITIS MEDIA

Otitis media (oh-**TYE**-tis **MEE**-dee-ah) or middle ear infection is quite common in young children. It can result in a temporary loss of hearing due to fluid buildup near the tympanic membrane. Symptoms include fever and irritability, and on examination, a red eardrum.

CONJUNCTIVITIS

Conjunctivitis (kon-junk-tih-**VYE**-tis) is caused by a bacterial infection of the conjunctiva of the eye. Contagious conjunctivitis is called pinkeye and is common in children. It can be transmitted easily by hand to eye contact or by contaminated water in a swimming pool.

MYOPIA

Myopia (my-**OH**-pee-ah) is commonly called nearsightedness. It is the ability to see close objects but not distant ones.

HYPEROPIA

Hyperopia (high-per-**OH**-pee-ah) is commonly called farsightedness. It is the ability to see distant objects but not close ones. Both myopia and hyperopia can be corrected by a corrective lens (a concave lens for myopia and a convex lens for hyperopia).

PRESBYOPIA

Presbyopia (prez-bee-**OH**-pee-ah) is a decrease in the ability of the eye to accommodate for near vision. This is a normal part of aging and commonly occurs during the 40s. It can be corrected by the use of reading glasses.

COLOR BLINDNESS

Color blindness is an X chromosome inherited genetic trait occurring more frequently in males. It results in the inability to perceive one or more colors.

MOTION SICKNESS

Motion sickness is caused by a stimulation of the semicircular canals in the inner ear resulting from movements as those experienced in a boat or ship (seasickness), airplane (air sickness), or automobile (car sickness). Such actions cause the individuals to experience weakness and nausea leading to vomiting. Drugs, such as scopolamine, have been developed that can be administered via a patch placed on the skin usually behind or near the ear. It lasts up to three days to prevent motion sickness. It is usually used by individuals who take ocean cruises and those who are sensitive to motion sickness.

CATARACTS

Cataracts usually develop in older individuals. The lens of the eye becomes cloudy due to a buildup of protein materials. The aqueous humor in front of the lens supplies nutrients to the lens. Decrease or loss of nutrients leads to degeneration and cataracts, also called opacity of the lens.

GLAUCOMA

Glaucoma is caused by too much aqueous humor in front of the lens, which leads to increased pressure in the eye. Its main symptom is a narrowing of the field of vision. It occurs more often in African Americans than in Caucasians. Older individuals should be screened for developing glaucoma during their yearly eye examinations. This causes destruction of the retina or optic nerve resulting in blindness.

SUMMARY OUTLINE

THE PRINCIPAL PARTS OF THE BRAIN

1. The brain is divided into four main parts: the brainstem consisting of the medulla oblongata, the pons varolii, and the midbrain; the diencephalon consisting of the thalamus and the hypothalamus; the cerebrum consisting of two hemispheres; and the cerebellum.
2. The brain is protected by the cranial bones, the cranial meninges, and the cerebrospinal fluid.
3. Cerebrospinal fluid acts as a shock absorber for the central nervous system and circulates nutrients. In the brain, it circulates in the subarachnoid space and the four ventricles.

THE ANATOMY AND FUNCTION OF THE BRAINSTEM

1. The medulla oblongata contains all the ascending and descending tracts that connect the spinal cord with the brain. Some of these tracts cross in the medulla, known as decussation of pyramids. This explains why motor functions on one side of the cerebrum control muscular movements on the opposite side of the body.
2. The reticular formation of the medulla controls consciousness and arousal. The three vital reflex centers control the diameter of blood vessels, heartbeat, and breathing rates.
3. The pons varolii is a bridge that connects the spinal cord with the brain and parts of the brain with each other. It also helps control breathing.
4. The midbrain or mesencephalon contains the dorsal tectum, a reflex center, that controls movement of the head and eyeballs in response to visual stimulation and movement of the head and trunk in response to auditory stimuli.

THE ANATOMY AND FUNCTIONS OF THE DIENCEPHALON

1. The thalamus is a relay station for sensory impulses and an interpretation center for recognition of pain, temperature, and crude touch.
2. The hypothalamus controls functions related to homeostasis: it controls the autonomic nervous system; it receives sensory impulses from the viscera; it controls the pituitary gland; it is the center for mind-over-body phenomena; it controls our thirst center; and it maintains our waking and sleep patterns.

THE CEREBRUM: STRUCTURE AND FUNCTION

1. The surface of the cerebrum is composed of gray matter and is called the cerebral cortex. Below the cortex is the white matter.
2. A longitudinal fissure separates the cerebrum into two hemispheres. Folds on the surface of the hemispheres are called gyri with intervening grooves called sulci.
3. The corpus callosum is a bridge of nerve fibers that connects the two hemispheres.
4. The surface of the cortex has motor areas to control muscular movements, sensory areas for interpreting sensory impulses and association areas concerned with emotional and intellectual processes.
5. Each hemisphere is divided into four main lobes.
6. The frontal lobe controls voluntary muscular movements, moods, aggression, smell reception, and motivation.
7. The parietal lobe evaluates sensory information concerning touch, pain, balance, taste, and temperature.
8. The temporal lobe evaluates hearing, smell, and memory. It is a center for abstract thought and judgment decisions.
9. The occipital lobe evaluates visual input.

THE CEREBELLUM: STRUCTURE AND FUNCTION

1. The cerebellum consists of two partially separated hemispheres connected by a structure called the vermis. The cerebellum is shaped like a butterfly.
2. It functions as a center for coordinating complex muscular movements, maintaining body posture, and balance.

THE AUTONOMIC NERVOUS SYSTEM

1. The autonomic nervous system is a subdivision of the efferent peripheral nervous system.
2. It regulates internal organs by controlling glands, smooth muscle, and cardiac muscle. It maintains homeostasis by regulating heartbeat, blood pressure, breathing, and body temperature.
3. It helps us control emergency situations, emotions, and various physical activities.

4. It consists of two subdivisions: the sympathetic division and the parasympathetic division.
5. The sympathetic division deals with energy expenditure and stressful situations by increasing heartbeat rates and breathing. Its fibers arise from the thoracic and lumbar regions of the spinal cord. It uses acetylcholine as a neurotransmitter in the preganglionic synapses and norepinephrine or noradrenaline at postganglionic synapses.
6. The parasympathetic division functions in restoring the body to a nonstressful state. Its fibers arise from the brainstem and the sacral region of the spinal cord. It uses acetylcholine at both the preganglionic and postganglionic synapses as a neurotransmitter.
7. The sympathetic division prepares us for physical activity: it increases blood pressure, heart rate, breathing, and sweating; it releases glucose from the liver for quick energy. It is also known as the fight-or-flight system.
8. The parasympathetic division counteracts the effects of the sympathetic division: it slows down heart rate, lowers blood pressure, and slows breathing. It also controls digestion, urination, defecation, and constriction of the pupil. It is known as the rest or repose system.

THE 12 CRANIAL NERVES AND THEIR FUNCTIONS

1. Olfactory nerve (I) conveys impulses related to smell. It is sensory.
2. Optic nerve (II) conveys impulses related to sight. It is sensory.
3. Oculomotor nerve (III) controls movements of the eyeballs and upper eyelid. Its parasympathetic function controls constriction of the pupil. It is both sensory and motor.
4. Trochlear nerve (IV) controls movement of the eyeball. It is both sensory and motor.
5. Trigeminal nerve (V) controls chewing movements and senses touch, temperature, and pain in the teeth and facial area. It is both sensory and motor.
6. Abducens nerve (VI) also controls movement of the eyeball. It is both sensory and motor.
7. Facial nerve (VII) controls the muscles of facial expression. It also senses taste. Its parasympathetic function controls the tear and salivary glands. It is both sensory and motor.
8. Vestibulocochlear nerve (VIII) transmits impulses related to equilibrium and hearing. It is sensory.
9. Glossopharyngeal nerve (IX) controls swallowing and senses taste. Its parasympathetic function controls salivary glands. It is both sensory and motor.
10. Vagus nerve (X) controls skeletal muscle movements in the pharynx, larynx, and palate. It conveys sensory impulses in the larynx, viscera, and ear. Its parasympathetic functions control viscera in the thorax and abdomen. It is both sensory and motor.
11. Accessory nerve (XI) helps control swallowing and movement of the head. It is both sensory and motor.
12. Hypoglossal nerve (XII) controls muscles involved in swallowing and speech. It is both sensory and motor.

THE SPECIAL SENSES

1. The senses of smell and taste are initiated by the interactions of chemicals with sensory receptors on the tongue and in the nose.
2. The sense of vision occurs due to the interactions of light with sensory receptors in the eye.
3. The senses of hearing and balance occur due to the interaction of sound waves for hearing and motion for balance with sensory receptors in the ear.

The Sense of Smell

1. The sense of smell, or the olfactory sense, occurs because molecules in the air become dissolved in the mucous epithelial lining of the superior nasal conchae of the nose.
2. Bipolar sensory neurons transfer these chemical impulses to the olfactory bulbs that connect with association neurons of the olfactory cortex in the temporal and frontal lobes of the cerebrum.
3. A small number of receptors in the nose detect a great variety of odors via brain interpretation of receptor combinations.

The Sense of Taste

1. Taste buds are found on certain papillae of the tongue, on the palate of the roof of the mouth, and part of the pharynx.
2. Taste buds consist of two types of cells: epithelial cells that form the exterior capsule and taste cells that form the interior of the taste bud.

3. The taste chemical is first dissolved in the fluid of saliva. These sensory impulses are conducted by the facial, glossopharyngeal, and vagus nerves to the taste cortex of the parietal lobe of the cerebrum for interpretation.
4. There are four major types of taste sensations: bitter, strongly detected at the back of the tongue; sweet and salty, detected at the tip of the tongue; and sour, detected more strongly by the taste buds on the sides of the tongue.
5. Taste sensations are also influenced by olfactory sensations.

The Sense of Sight

1. The eyes are the organs of sight. Eyelids and eyelashes protect the eyes from foreign objects. Tears, produced by the lacrimal glands, lubricate the eyes.
2. Tears contain the bacteriolytic enzyme lysozyme.

The Anatomy of the Eye

1. The wall of the eye is composed of three layers: the sclera, the choroid, and the retina.
2. The sclera is the outermost, white, hard layer composed of tough collagenous connective tissue.
3. The cornea is the transparent part of the sclera that allows light to enter the eye.
4. The choroid is the second layer and contains blood vessels and pigment cells. It is black in color and absorbs light to prevent reflection that could impair vision.
5. The retina is the innermost layer of the eye. It contains the light-sensitive cells called rods and cones.
6. The ciliary body holds the hard, biconvex, transparent lens in place.
7. The iris is the colored part of the eye surrounding the pupil. It regulates the amount of light that can enter the pupil.
8. The interior of the eye is divided into two fluid-filled compartments. The anterior compartment is filled with aqueous humor; and the posterior compartment is filled with vitreous humor. These fluids help maintain ocular pressure, bend light, and hold the retina and lens in place.
9. There are more rods than cones in the retina. These light-sensitive cells have two functions. Rods are very sensitive to light and function in dim light; cones produce color sensations and require a lot of light.
10. The rods and cones synapse with the bipolar sensory cells of the retina. These cells synapse with the optic nerve, which reaches the thalamus of the brain to synapse with the visual cortex of the occipital lobe of the cerebrum for interpretation.

The Sense of Hearing and Equilibrium

1. The external, middle, and inner ear contain the organs of balance, or equilibrium, and hearing.
2. The visible, flexible, external ear is called the auricle. It directs sound waves to the ear canal called the external auditory meatus.
3. The ear canal is lined with hairs and ceruminous glands that produce earwax to protect the delicate eardrum, or tympanic membrane, from foreign objects.
4. The middle ear contains the auditory ossicles: the malleus or hammer, the incus or anvil, and the stapes or stirrup. These bones transmit sound vibrations from the tympanic membrane, which vibrates due to sound waves, to the oval window.
5. There are two openings on the medial side of the middle ear: the oval window and the round window, which connect the middle ear to the inner ear.
6. The middle ear also contains the auditory or eustachian tube, which connects to the pharynx and allows for equalized air pressure between the outside world and the middle ear, thus not impairing hearing.
7. The inner ear consists of fluid-filled interconnecting chambers and tunnels in the temporal bone. It contains the cochlea involved in hearing and the semicircular canals and vestibule involved in balance.

REVIEW QUESTIONS

1. Name the four principal parts of the brain and their subdivisions where appropriate.
2. Name the complex functions of the hypothalamus.
3. Name the 12 cranial nerves; include their Roman numeral designation and their functions.
4. *Explain how the hypothalamus of the brain and the autonomic nervous system allow us to fight or flee in a stressful situation.

***Critical Thinking Question**

FILL IN THE BLANK

Fill in the blank with the most appropriate term.

1. The brain is protected by the ________________ bones, the ________________, and ________________ fluid.
2. Cerebrospinal fluid acts as a ________________ and circulates ________________.
3. Cerebrospinal fluid circulates in the ________________ space and the four ________________ of the brain.
4. Crossing of tracts in the medulla oblongata is known as ________________.
5. The ________________ of the midbrain is a reflex center that controls movement of the head and eyeballs and head and trunk in response to visual and auditory stimuli.
6. Folds on the surface of the cerebrum are called ________________ and intervening grooves are called ________________.
7. The two hemispheres of the cerebrum are connected by a bridge of nerve fibers called the ________________.
8. The four main lobes of each cerebral hemisphere are: ________________, ________________, ________________, and ________________.
9. The cerebellum functions in coordinating ________________ movements and keeping the body ________________.
10. The two subdivisions of the autonomic nervous system are the ________________ system, which stimulates and involves energy expenditure, and the ________________ system, which is mainly restorative.

MATCHING

Place the most appropriate number in the blank provided.

_____ Olfactory cortex	1.	Transparent sclera
_____ Taste cortex	2.	Regulates light entering eye
_____ Tears	3.	Rods and cones
_____ Cornea	4.	Posterior compartment of eye
_____ Choroid layer	5.	Holds lens in place
_____ Retina	6.	Visible portion of the external ear
_____ Iris	7.	Temporal and olfactory lobes
_____ Ciliary body	8.	Hearing
_____ Pupil	9.	Balance
_____ Aqueous humor	10.	Earwax
_____ Vitreous humor	11.	Parietal lobe
_____ Auricle	12.	Anterior compartment of eye
_____ Ceruminous glands	13.	Colored part of eye
_____ Cochlea	14.	Blood vessels and pigment cells
_____ Semicircular canals	15.	Lacrimal gland
	16.	Outermost layer of the eye
	17.	Blind spot

Search and Explore

- Search the Internet for a famous person who was diagnosed with one of the diseases introduced in this chapter, such as Ronald Regan, 40th president of the United States, who had Alzheimer's disease.
- Visit the Human Anatomy Online website at http://www.innerbody.com and explore the nervous system.

*Study*WARE™ Connection

Take a quiz or play one of the interactive games that reinforce the content in this chapter on your StudyWARE™ CD-ROM.

Study Guide Practice

Go to your **Study Guide** for more practice questions, labeling and coloring exercises, and crossword puzzles to help you learn the content in this chapter.

LABORATORY EXERCISE: THE NERVOUS SYSTEM

Materials needed: A model of a human brain, a sheep or cow eye for dissection, a model of the external and internal ear, a dissecting pan, and a dissecting kit

1. Obtain a model of a preserved human brain showing a frontal and cross section. These are available from a biologic supply company and will be provided by your instructor. Identify the various parts of the brain, referring to Figure 11-1 in your text. In addition, identify the four ventricles of the brain.
2. Obtain a sheep or cow eye from your instructor. Make a transverse cut through the eye with your scalpel. Refer to Figure 11-10 in your text. Identify the three layers of the eye: the hard white outer sclera; the black choroid in the middle; and the innermost retina. Locate the biconvex lens. Anterior to the lens is the aqueous humor and posterior to the lens is the vitreous humor. Note the dark delicate iris surrounding the opening into the lens, the pupil. If you look carefully at the rear half of the eye, you will see a shiny greenish blue material. This is the tapetum. There is no tapetum in a human eye, but in a cow or sheep eye this area reflects light, causing the animal's eye to glow in the dark when light is shined on it.
3. Obtain an anatomic model of the ear from your instructor. Identify the auricle and external auditory meatus of the outer ear, the middle ear, and the structures of the inner ear. Refer to Figure 11-12 in your text.

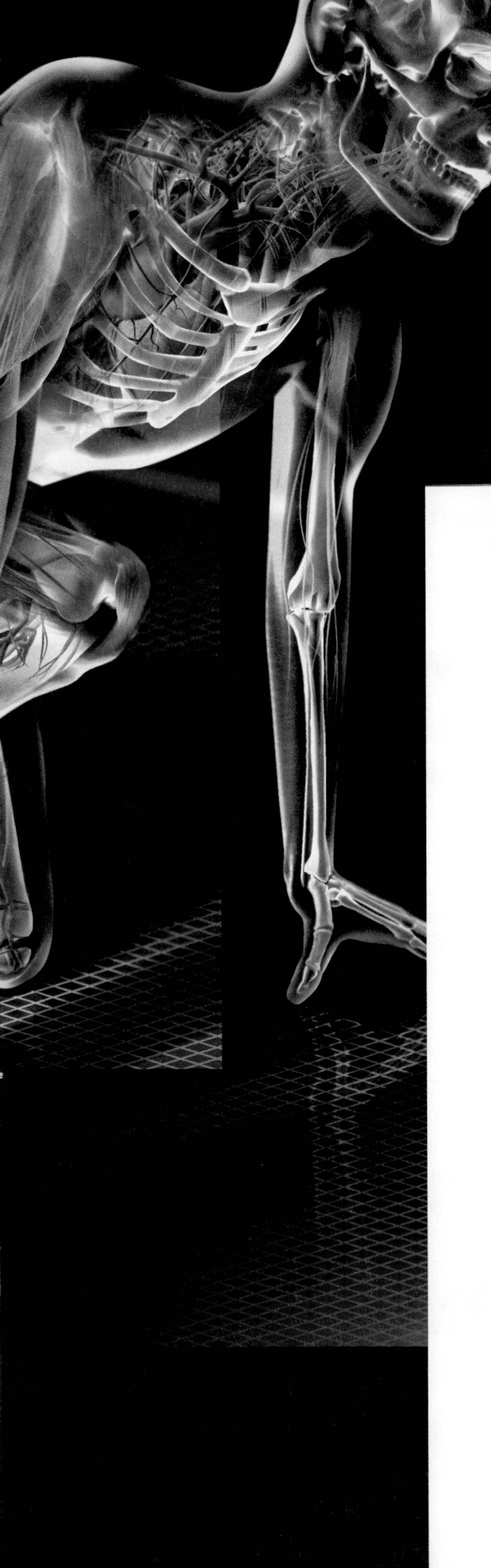

The Endocrine System

CHAPTER OBJECTIVES

After studying this chapter, you should be able to:

1. List the functions of hormones.
2. Classify hormones into their major chemical categories.
3. Describe how the hypothalamus of the brain controls the endocrine system.
4. Name the endocrine glands and state where they are located.
5. List the major hormones and their effects on the body.
6. Discuss some of the major diseases of the endocrine system and their causes.

12

KEY TERMS

Acidosis 285
Addison's disease 283
Adrenal cortex. 282
Adrenal glands/ suprarenal glands 282
Adrenal medulla 282
Adrenaline/epinephrine . 283
Adrenocorticotropic hormone (ACTH) 279
Adrenogenital syndrome 287
Aldosterone. 283
Aldosteronism. 286
Alpha cells 284
Androgens 283
Antidiuretic hormone (ADH)/vasopressin. . . . 280
Beta cells 284
Calcitonin. 281
Chief cells 282
Cortisol/hydrocortisone . . 283
Cortisone 283
Cretinism 281
Cushing's syndrome 283
Diabetes insipidus. 286
Diabetes mellitus 285
Endocrine glands 274
Estrogen. 285
Exophthalmia 281
Follicle-stimulating hormone (FSH). 279
Glucagon 284
Glycosuria 285
Goiter. 281
Graves' disease 281
Growth hormone (GH) . . 277
Homeostasis 274
Hormones 275
Hyperglycemia 285
Hyperparathyroidism . . . 282
Hyperthyroidism 281
Hypoparathyroidism. . . . 282
Hypophysis 277
Hypothalamus. 277
Hypothyroidism 281
Infundibulum. 277
Insulin 284
Lactogenic hormone (LTH)/Prolactin 279
Luteinizing hormone (LH). 279
Melanocyte-stimulating hormone (MSH). 279
Melatonin 288
Myxedema. 281
Negative feedback loop 275
Noradrenaline/ norepinephrine 283
Ovaries. 285
Oxyphil cells 282
Oxytocin (OT) 280
Pancreatic islets/islets of Langerhans 284
Parathyroid glands 281
Parathyroid hormone/ parathormone (PTH) . . 282
Pineal gland/body. 288
Pituitary gland/ hypophysis. 277
Polydipsia 285
Polyphagia. 285
Polyuria 285
Progesterone. 285
Releasing hormones 277
Releasing inhibitory hormones. 277
Seasonal affective disorder 286
Serotonin. 288
Stress 286
Testes 285
Testosterone 285
Thymosin. 288
Thymus gland 288
Thyroid gland 281
Thyroid-stimulating hormone (TSH) 279
Thyroxine or tetraiodothyronine (T_4) 281
Triiodothyronine (T_3). . . . 281

INTRODUCTION

The endocrine system exerts chemical control over the human body by maintaining the body's internal environment within certain narrow ranges. See Concept Maps 12-1 and 12-2: Endocrine System. This is known as **homeostasis** (hom-ee-oh-**STAY**-sis). This maintenance of homeostasis, which involves growth, maturation, reproduction, metabolism, and human behavior, is shared by both the endocrine system and the nervous system in a unique partnership. It is the hypothalamus of the brain (a part of the nervous system) that sends directions via chemical signals (neurotransmitters) to the pituitary gland (a part of the endocrine system). The pituitary is occasionally referred to as the master gland of the system because many of its hormones (chemical signals) stimulate the other endocrine glands to secrete their hormones.

The **endocrine glands** are ductless glands that secrete their hormones directly into the bloodstream. The blood circulatory system then carries these chemical signals

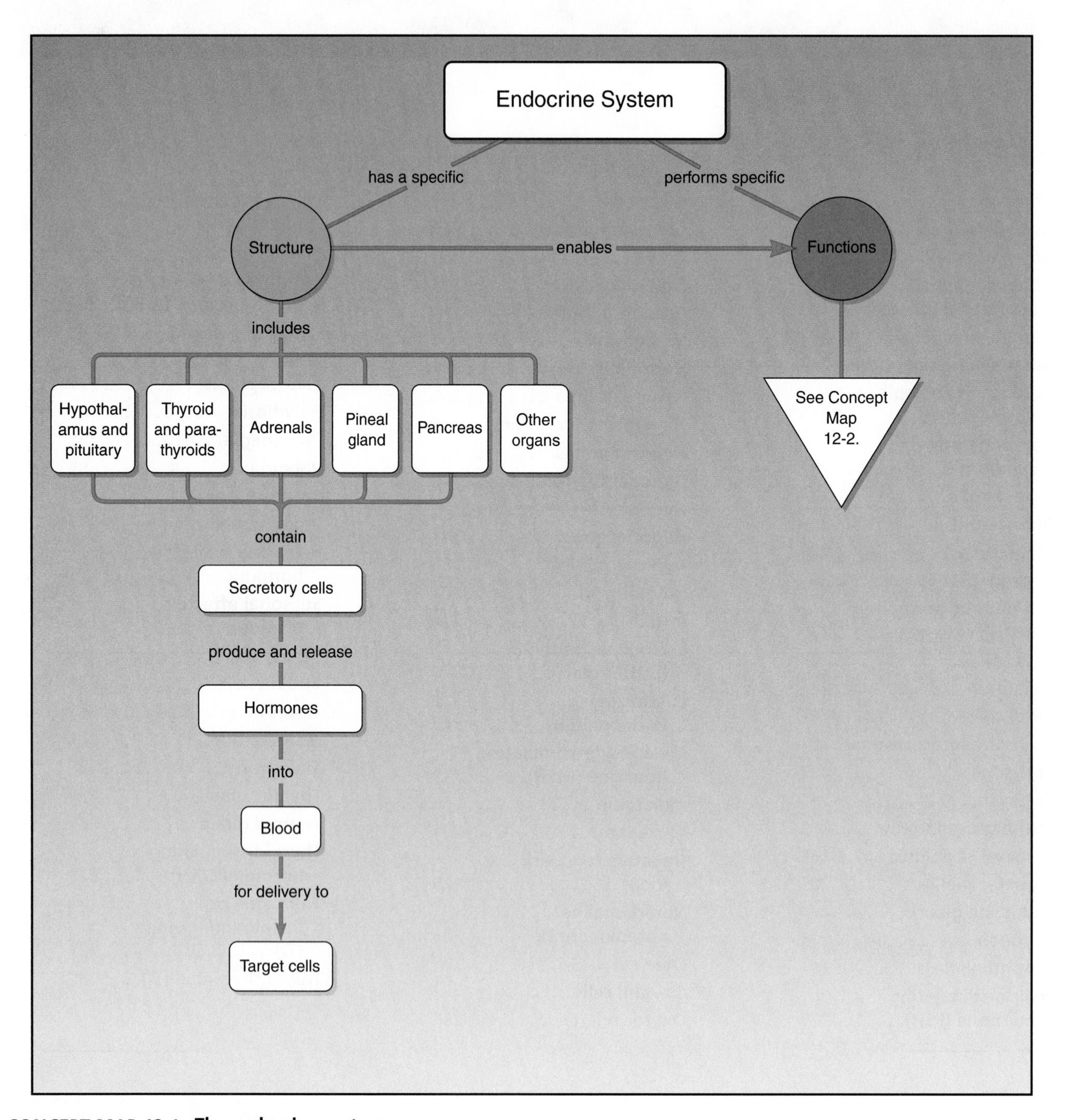

CONCEPT MAP 12-1. The endocrine system.

to target organs where their effects are seen as specific responses. These chemical signals or hormones help regulate metabolism, water and electrolyte concentrations in cells, growth, development, and the reproductive cycles. Endocrine glands are ductless, as opposed to exocrine glands, which have ducts by which their secretions are transported directly to an organ or the body surface, such as sweat glands to the surface of the body and salivary glands to the mouth.

THE FUNCTIONS OF HORMONES

Hormones control the internal environment of the body from the cellular level to the organ level of organization. They control cellular respiration, cellular growth, and cellular reproduction. They control the fluids in the body, such as water amounts and balances of electrolytes. They control the secretion of other hormones. They control our behavior patterns. They play a vital role in the reproductive cycles of men and women. In addition, they regulate our growth and development cycles.

This chemical control of the body functions primarily as a **negative feedback loop.** In our homes, our furnaces and thermostats operate as a negative feedback loop. We set our thermostat to a particular temperature and when the temperature of our home falls below that set temperature, the thermostat causes the furnace to turn on. Once the temperature inside reaches the set temperature on the thermostat, it sends another signal to the furnace to shut off. Hormonal systems function in the same way. When the concentration of a particular hormone reaches a certain level in the body, the endocrine gland that secreted that hormone is inhibited (the negative feedback) and the secretion of that hormone ceases or decreases significantly. Later as the concentration of that gland's hormone falls below normal levels, the inhibition of the gland ceases and it begins to produce and secrete the hormone once again. This kind of a negative feedback loop helps to control the concentrations of a number of hormones in our bodies.

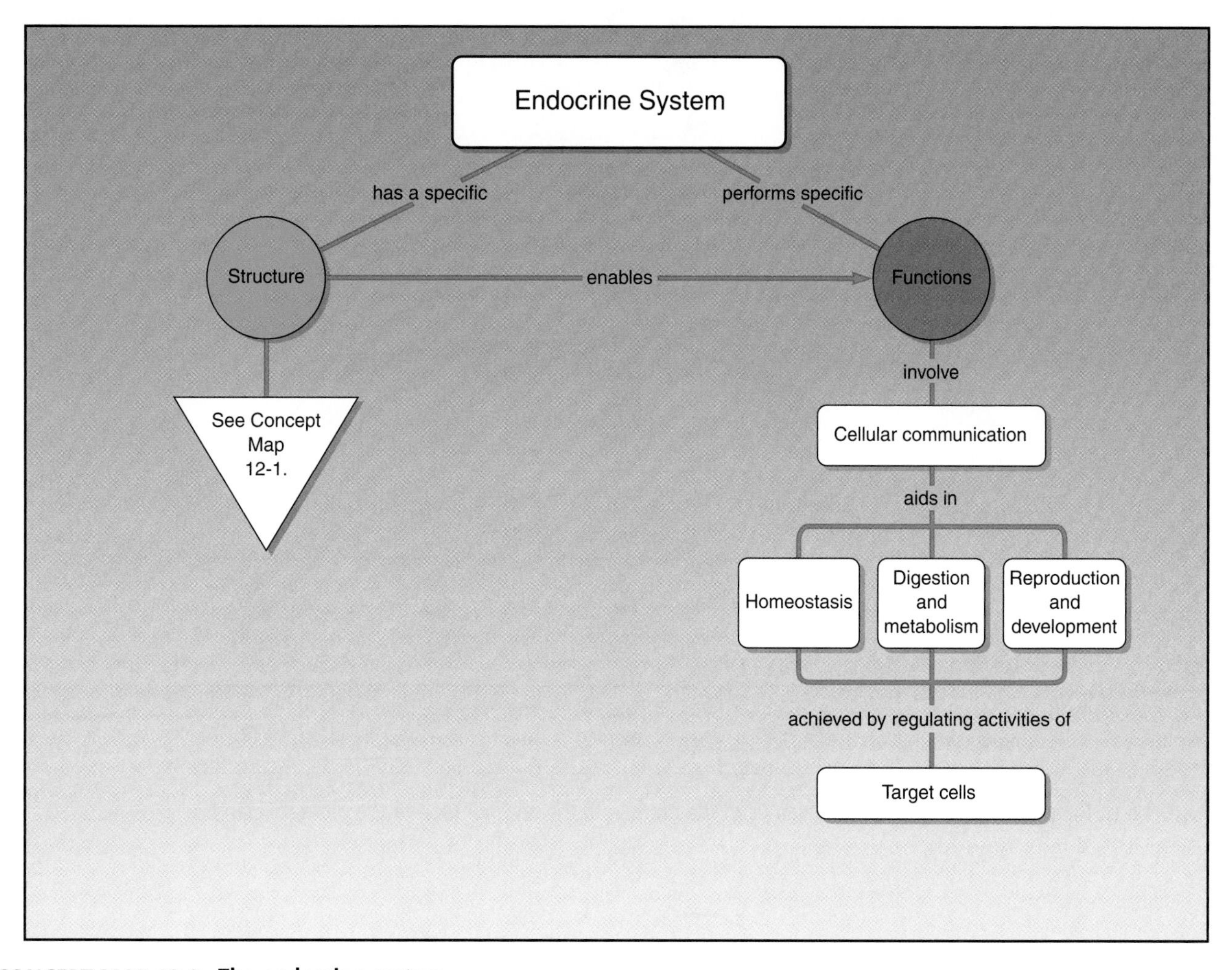

CONCEPT MAP 12-2. The endocrine system.

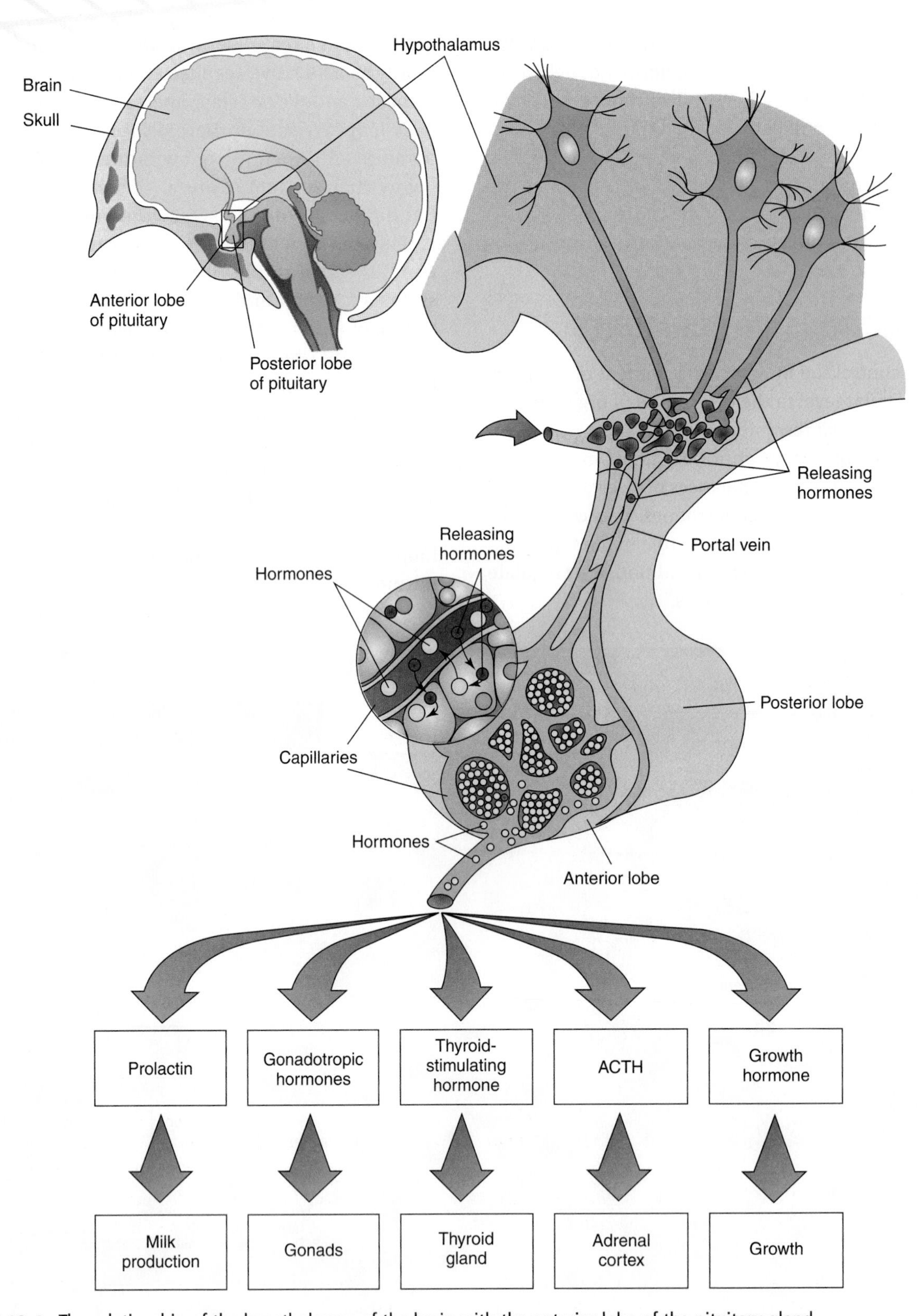

FIGURE 12-1. The relationship of the hypothalamus of the brain with the anterior lobe of the pituitary gland.

THE CLASSIFICATION OF HORMONES

Hormones can be classified into three general chemical categories. The simplest group includes hormones that are modified amino acids. Examples are the hormones secreted by the adrenal medulla: epinephrine and norepinephrine, and the hormones secreted by the posterior pituitary gland: oxytocin and vasopressin. The second category is the protein hormones: insulin from the pancreatic islets and the gonad-stimulating hormones and growth hormone from the anterior pituitary gland. The third category of hormones is the steroid hormones, which are lipids. Examples are cortisol from the adrenal cortex and estrogen and testosterone produced by the gonads.

The modified amino acid and protein hormones bind to membrane-bound receptor sites on the cells of target organs. The steroid hormones diffuse across the cell membrane and bind to intracellular (inside the cell) receptor molecules. The steroid hormones are soluble in lipids and can diffuse across the lining of the stomach and intestine and get to the circulatory system. They can be taken orally to treat illnesses. Birth control pills made of synthetic estrogen and progesterone hormones and steroids that combat inflammation are taken orally. However, the protein and modified amino acid hormones, like insulin, must be injected because they cannot diffuse across the intestinal lining because they are not soluble in lipids. They are broken down before they are transported across the lining of the digestive tract and thus their effect is destroyed. Therefore, insulin must be injected to treat diabetes mellitus. Another form of diabetes is diabetes insipidus, which is caused by a deficiency in the antidiuretic hormone (ADH).

THE HYPOTHALAMUS OF THE BRAIN

The **hypothalamus** (**high**-poh-**THAL**-ah-mus) of the brain is the inferior part of the diencephalon. It has a unique role with the endocrine system because it plays a major role in controlling secretions from the pituitary gland. There is a funnel-shaped stalk, called the **infundibulum** (in-fun-**DIB**-yoo-lum), that extends from the floor of the hypothalamus connecting it to the pituitary gland. Historically, the pituitary gland is referred to as the master gland of the endocrine system because it controls the secretions of many other endocrine glands. However, in actuality, it is the hypothalamus of the brain that sends neural and chemical signals to the pituitary gland; hence, the hypothalamus controls the pituitary gland. This relationship is akin to a concert performance. The conductor, like the pituitary gland, tells the various sections of the orchestra (the other endocrine glands) when and how to play the music. However, the conductor gets information from the sheet music or score (like the role of the hypothalamus).

Nerve cells in the hypothalamus produce chemical signals called **releasing hormones** and **releasing inhibitory hormones.** These hormones, which are actually neurosecretions, either stimulate or inhibit the release of a particular hormone from the pituitary gland (Figures 12-1 and 12-2). These releasing hormones enter a capillary bed in the hypothalamus and are transported through a portal vein in the infundibulum to a second capillary bed of the anterior pituitary gland. After leaving the capillaries, they bind to receptors controlling the regulation of hormone secretion from the pituitary gland. It is within the hypothalamus of the brain and the pituitary gland that the interactions and relationships between the endocrine and nervous systems are controlled and maintained. Conversely, due to negative feedback, the hormones of the endocrine system can influence the functions of the hypothalamus.

THE MAJOR ENDOCRINE GLANDS AND THEIR HORMONES

The endocrine glands include the pituitary gland, the pineal gland, the thyroid gland, the parathyroid glands, the thymus gland, the adrenal glands, the islets of Langerhans of the pancreas, the ovaries in women, and the testes in men (Figure 12-3).

The Anterior Pituitary Gland, Its Hormones, and Some Disorders

The **pituitary** (pih-**TYOO**-ih-**tayr**-ee) **gland** is also called the **hypophysis** (high-**POFF**-ih-sis). A small gland about the size of a pea, some of its hormones affect the functions of many other endocrine glands such as the testes, ovaries, the adrenal cortex, and the thyroid gland. It is situated in a depression of the sphenoid bone below the hypothalamus of the brain. It is divided into two lobes, a larger anterior pituitary lobe and a smaller posterior pituitary lobe.

The anterior pituitary lobe produces seven hormones (see Figure 12-1). **Growth hormone (GH)** stimulates cell metabolism in most tissues of the body, causing cells to divide and increase in size. It increases protein synthesis

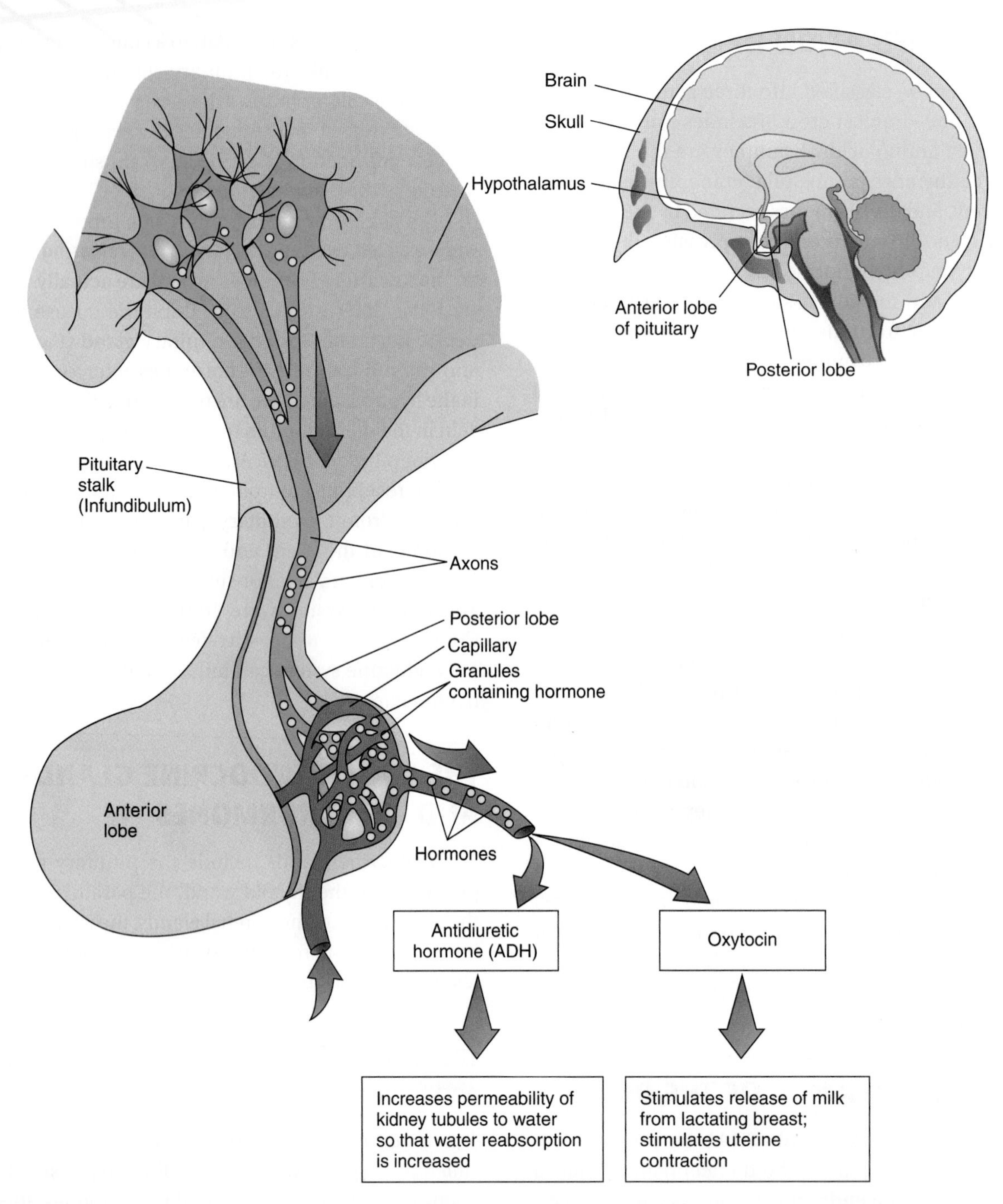

FIGURE 12-2. The relationship of the hypothalamus of the brain with the posterior lobe of the pituitary gland.

and the breakdown of fats and carbohydrates. It stimulates the growth of bones and muscles. If a young person suffers from too little GH as a result of abnormal development of the pituitary gland, a condition called pituitary dwarfism results. The person remains small, although body proportions are normal. The most famous pituitary dwarf was Charles Stratton, known as Tom Thumb, who was employed by P. T. Barnum in his circus. He died in 1888 at the age of 45 and was less than 1 meter tall. On the other hand, too much GH during childhood results in gigantism. Excess secretion of GH after childhood when bone has stopped growing results in acromegaly. Bones widen especially in the face, hands, and feet. However, in the majority of children, the anterior pituitary produces just the right amount of GH, resulting in normal growth rates. Checkups with the family doctor during childhood help to monitor the rate of growth and development. In the United States, it is now rare to see a pituitary dwarf or giant.

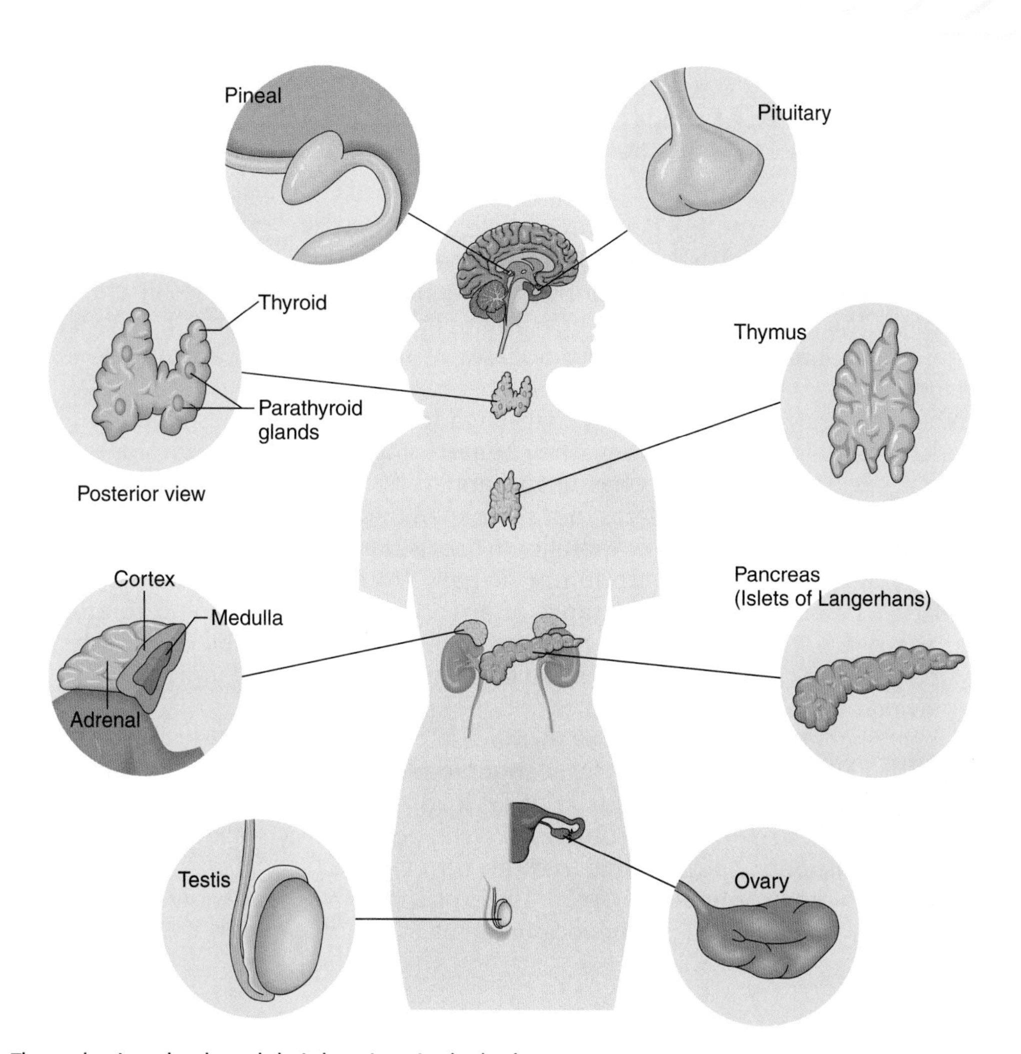

FIGURE 12-3. The endocrine glands and their locations in the body.

Secretion of GH is controlled by two releasing hormones from the hypothalamus: one stimulates secretion and the other inhibits it. Peak secretions of GH occur during periods of sleep, exercise, and fasting. Growth is also influenced by nutrition, genetics, and the sex hormones during puberty.

Thyroid-stimulating hormone (TSH) stimulates the thyroid gland to produce its hormone. The rate of TSH secretion is regulated by the hypothalamus, which produces thyrotropin-releasing hormone (TRH), which stimulates the anterior pituitary lobe to secrete TSH.

Adrenocorticotropic (ad-**ree**-noh-**KOR**-tih-koh-**TROH**-pik) **hormone (ACTH)** stimulates the adrenal cortex to secrete its hormone called cortisol. ACTH secretion is regulated by corticotropin-releasing hormone (CRH) produced by the hypothalamus. ACTH is involved with the glucose-sparing effect and helps reduce inflammation as well as stimulating the adrenal cortex.

Melanocyte-stimulating hormone (MSH) increases the production of melanin in melanocytes in the skin, thus causing a deepening pigmentation or darkening of the skin.

Follicle-stimulating hormone (FSH) stimulates development of the follicles in the ovaries of females. In males, it stimulates the production of sperm cells in the seminiferous tubules of the testes.

Luteinizing (**LOO**-tee-in-eye-zing) **hormone (LH)** stimulates ovulation in the female ovary and production of the female sex hormone progesterone. It helps maintain pregnancy. In males, it stimulates the synthesis of testosterone in the testes to maintain sperm cell production.

Lactogenic hormone (LTH), also known as **prolactin** (proh-**LACK**-tin), stimulates milk production in the mammary glands following delivery in a pregnant female. It also maintains progesterone levels following ovulation and during pregnancy in women. In males, it appears to increase sensitivity to LH and may cause a decrease in male sex hormones.

HEALTH ALERT

STEROIDS AND ATHLETES

During the 1950s, pharmaceutical companies developed anabolic steroids, which are variants of the male sex hormone testosterone. Testosterone is responsible for building muscle mass and during puberty causes bone development, deepened voice, and facial and chest hair growth in boys. The anabolic steroids were developed to treat patients who were immobile after surgeries or who had degenerating diseases of muscle, to prevent muscle atrophy in these individuals.

Bodybuilders and athletes believed that megadoses of the steroids would build muscle bulk and increase their athletic abilities. They began using the anabolic steroids in the 1960s. Although the use of these steroids has been banned by most competitions, some athletes still use them. In fact, not only athletes, but some men and women who want to add muscle bulk and increase their competitive capability in sports have also used these drugs. They can be acquired either legally by prescription or illegally from the drug market. Sports figures have admitted to their use. The advantages of anabolic steroids, according to athletes who have used them, are increased muscle bulk, greater volume of red blood cells resulting in more oxygen-carrying capabilities to muscle cells, and an increase in aggression. This results in greater athletic stamina and having a "good-looking body."

However, dangers are associated with the use of anabolic steroids. Some of the side effects are shriveled testes and infertility, changes in blood cholesterol levels that could lead to heart disease, damage to the liver that could lead to liver cancer, puffy faces (known as the cushingoid sign), and mental problems. Psychological effects range from depression, delusions, and manic personality swings that can turn violent.

Yet some athletes continue to use these drugs despite the dangers associated with them. The desire to be a winner in our society seems to cloud common sense decisions. ■

The Posterior Pituitary Gland and Its Hormones

The posterior pituitary lobe consists primarily of nerve fibers and neuroglial cells that support the nerve fibers, whereas the anterior lobe is primarily glandular epithelial cells. Special neurons in the hypothalamus produce the hormones of the posterior pituitary lobe. These hormones pass down axons through the pituitary stalk to the posterior lobe, and secretory granules near the ends of the axons store the hormones (see Figure 12-2).

Antidiuretic (an-tye-dye-yoo-**RET**-ik) **hormone (ADH)**, also known as **vasopressin** (vaz-oh-**PRES**-sin), maintains the body's water balance by promoting increased water reabsorption in the tubules of the nephrons of kidneys, resulting in less water in the urine. If secreted in large amounts, ADH can cause constriction of blood vessels, hence its other name vasopressin. A deficiency of ADH can result in a condition known as diabetes insipidus. Individuals with this condition produce 20 to 30 liters of urine daily. They can become severely dehydrated. They lose essential electrolytes, resulting in abnormal nerve and cardiac muscle functions. This condition can be treated by taking ADH as injections or in the form of a nasal spray. Again the hypothalamus regulates ADH secretion through osmoreceptors that detect changes in the osmotic pressure of body fluids. Dehydration, caused by lack of sufficient water intake, increases blood solute concentrations and these osmoreceptors signal the posterior lobe to release ADH. This causes the kidneys to conserve water. Conversely, taking in too much water or drinking too much fluid dilutes blood solutes, inhibiting ADH secretion so the kidneys excrete a more dilute (more water in it) urine until the concentration of solutes in body fluids returns to normal. In contrast, a diuretic increases urine secretion.

Oxytocin (ok-see-**TOH**-sin) (**OT**) stimulates contraction of smooth muscles in the wall of the uterus. Stretching of uterine and vaginal tissues late in pregnancy stimulates production of OT so that uterine contractions develop in the late stages of childbirth. OT also causes contraction of cells in the mammary glands causing milk ejection or lactation, forcing the milk from the glandular ducts into the nipple during breastfeeding of the newborn infant.

Occasionally, commercial preparations of OT are administered to induce labor if the uterus does not contract sufficiently on its own during childbirth. It is also given to women after childbirth to constrict blood vessels of the uterus to minimize the risk of hemorrhage.

The Thyroid Gland, Its Hormones, and Some Disorders

The **thyroid gland** consists of two lobes connected by a smaller band called the isthmus (Figure 12-4). The lobes are situated on the right and left sides of the trachea and thyroid cartilage just below the larynx. It is a highly vascular, large endocrine gland covered with a capsule of connective tissue. It is made up of spheres of cells called follicles. These follicles are composed of simple cuboidal epithelium, which produces and secretes the thyroid hormones. Thyroid output is regulated by the hypothalamus, which signals the pituitary to release TSH to increase thyroid production.

The thyroid gland requires iodine to function properly. In the United States iodized salt is used as a way to ensure the intake of adequate amounts of iodine in the diet. In countries without adequate amounts of iodine in the diet, the thyroid gland enlarges forming a **goiter** (**GOY**-ter). However, proper amounts of iodine cause the thyroid gland to effectively produce its hormones. One hormone is **thyroxine** (thigh-**ROXS**-in), also known as **tetraiodothyronine** (**teh**-trah-eye-**oh**-doh-**THIGH**-roh-neen), which contains four iodine atoms and is abbreviated as T_4. The other hormone is **triiodothyronine** (**try**-eye-**oh**-doh-**THIGH**-roh-neen), which contains three iodine atoms and is abbreviated as T_3.

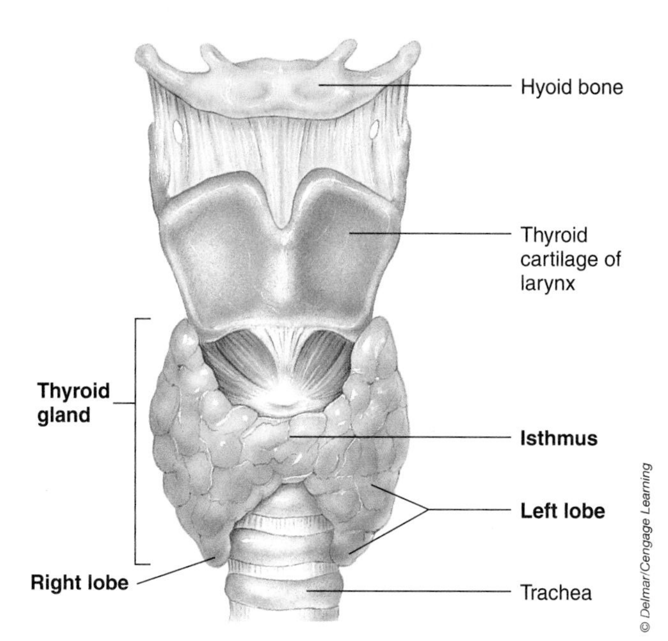

FIGURE 12-4. The thyroid gland consists of a right and left lobe joined by the isthmus.

These hormones regulate the metabolism of carbohydrates, fats, and proteins. These hormones are necessary for normal growth and development as well as for nervous system maturation. They cause an increase in the rate of carbohydrate and lipid breakdown into energy molecules as well as increasing the rate of protein synthesis. A lack of or low level of thyroid hormones is called **hypothyroidism** (**high**-poh-**THIGH**-royd-izm). In young children, this can result in a condition known as **cretinism** (**KREE**-tin-izm). The child with this condition is mentally retarded and does not grow to normal stature. In adults, this condition results in a lowered rate of metabolism, causing sluggishness, being too tired to perform normal daily tasks, and an accumulation of fluid in subcutaneous tissues called **myxedema** (miks-eh-**DEE**-mah). Too much secretion of thyroid hormones causes **hyperthyroidism** (**high**-per-**THIGH**-royd-izm). This results in extreme nervousness, fatigue, and an elevated rate of body metabolism. **Graves' disease** is a type of hyperthyroidism caused by overproduction of thyroid hormone. It is often associated with an enlarged thyroid gland or goiter and bulging of the eyeballs known as **exophthalmia** (eks-off-**THAL**-mee-ah).

Besides secreting these two thyroid hormones, the extrafollicular cells of the thyroid gland secrete a hormone called **calcitonin** (kal-sih-**TOH**-nin). This hormone lowers the calcium and phosphate ion concentration of the blood by inhibiting the release of calcium and phosphate ions from the bones and by increasing the excretion of these ions by the kidneys.

Thyroid hormone secretion is controlled by TSH produced by the anterior pituitary gland. Increased levels of thyroid hormones, through the negative feedback mechanism, inhibit the anterior pituitary gland from releasing more TSH and the hypothalamus from secreting TSH-releasing hormone. Because of negative feedback, the thyroid hormones fluctuate daily within a narrow range of concentration in the blood.

The Parathyroid Glands, Their Hormone, and Some Disorders

The **parathyroid glands** are four glands about the size of raisins that are embedded in the posterior surface of the thyroid gland (Figure 12-5). There are two in each lobe of the thyroid, a superior and an inferior gland. Each gland

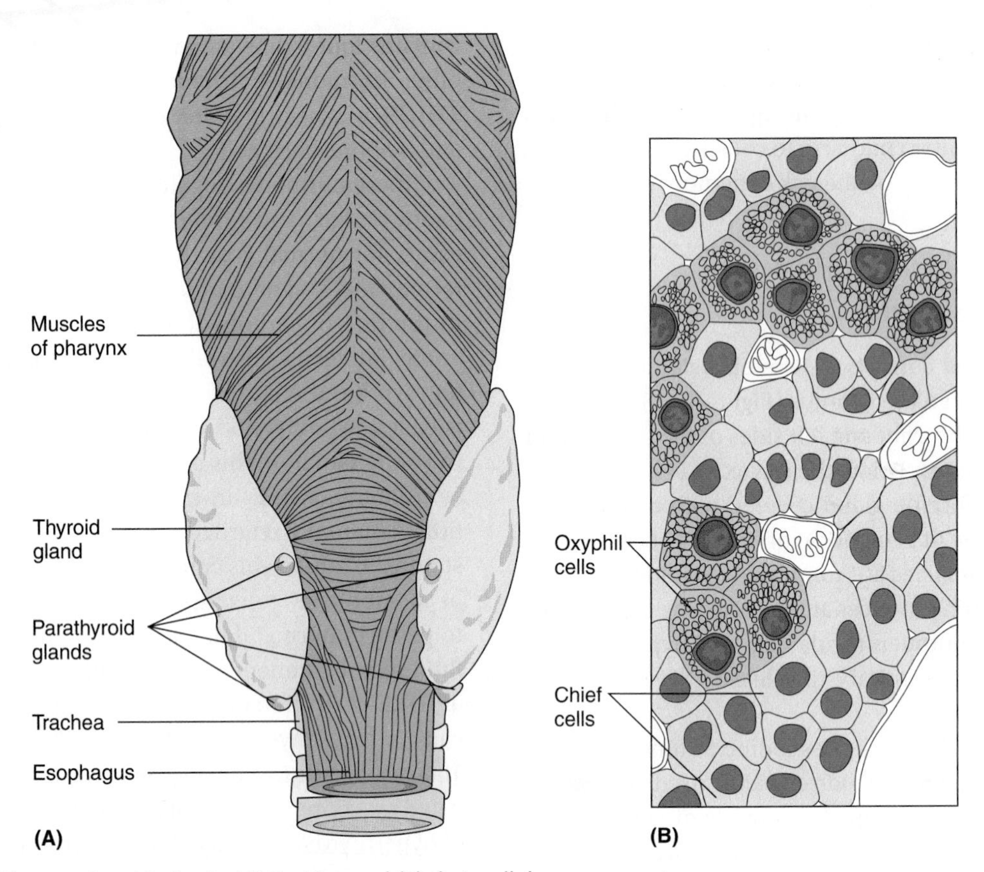

FIGURE 12-5. The parathyroid glands. (A) Position and (B) their cellular components.

consists of many tightly packed secreting cells called **chief cells** and **oxyphil cells** close to capillary networks.

The parathyroid glands secrete a single hormone called **parathyroid hormone** or **parathormone (PTH).** PTH inhibits the activity of osteoblasts and causes osteoclasts to break down bone matrix tissue, thus releasing calcium and phosphate ions into the blood. In addition, PTH causes the kidneys to conserve blood calcium and stimulates intestinal cells to absorb calcium from digested food in the intestine. This hormone raises blood calcium to normal levels.

Vitamin D also increases absorption of calcium by the intestines. Ultraviolet light from the sun acting on the skin is necessary for the first stage of vitamin D synthesis. The final stage of synthesis occurs in the kidneys and is stimulated by PTH. Vitamin D can also be supplied in the diet.

An abnormally high level of PTH secretion is known as **hyperparathyroidism** and can be caused by a tumor in the parathyroid gland. This results in breakdown of bone matrix, and bones become soft and deformed and can easily fracture. Elevated calcium levels in the blood cause muscles and nerves to become less excitable, resulting in muscle weakness and fatigue. Excess calcium and phosphate ions may become deposited in abnormal places resulting in kidney stones. An abnormally low level of PTH is called **hypoparathyroidism.** This can be caused by surgical removal of the thyroid and parathyroid glands or by injury to the glands. The decreased level of PTH reduces osteoclast activity, reduces rates of bone matrix breakdown or resorption, and reduces vitamin D formation. Bones will remain strong but the blood calcium level decreases result in nerves and muscles becoming abnormally excitable, producing spontaneous action potentials. This can cause frequent muscle cramps or tetanic contractions. If respiratory muscles are affected, breathing failure and death can occur.

The Adrenal Glands, Their Hormones, and Some Disorders

The **adrenal** (ad-**REE**-nal) **glands** are also known as the **suprarenal** (**soo**-prah-**REE**-nal) glands (Figure 12-6). They are small glands found on top of each kidney. The inner part of each gland is called the **adrenal medulla** and the outermost part is called the **adrenal cortex**. Each section functions as a separate endocrine gland.

The **adrenal medulla** produces large amounts of the hormone **adrenaline** (ad-**REN**-ih-lin), also known as

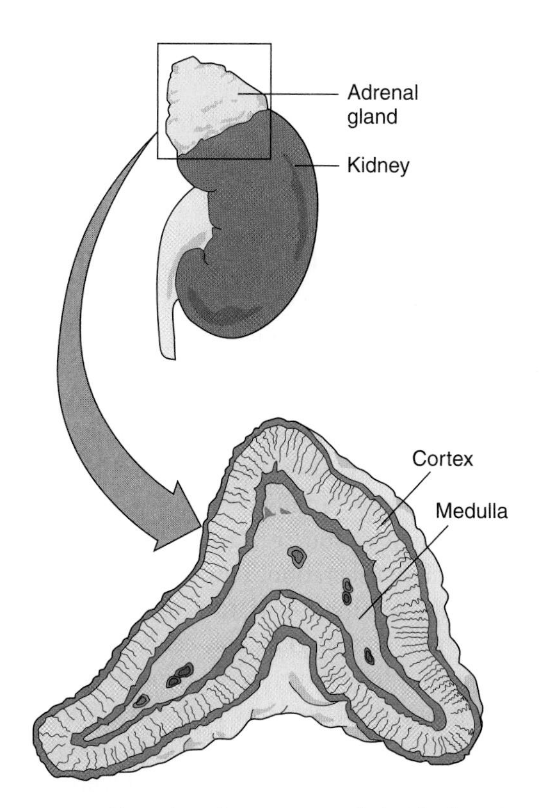

FIGURE 12-6. The adrenal or suprarenal glands, found on top of each kidney, consist of an inner adrenal medulla and an outer adrenal cortex.

epinephrine (ep-ih-**NEF**-rin), and small amounts of **norepinephrine** (nor-ep-ih-**NEF**-rin) or **noradrenaline.** These hormones are released in response to signals from the sympathetic division of the autonomic nervous system. Epinephrine and norepinephrine are commonly referred to as the fight-or-flight hormones because they get the body prepared for stressful situations that require vigorous physical activity. When a person senses danger and experiences stress, the hypothalamus of the brain triggers the adrenal gland via the sympathetic division of the autonomic nervous system to secrete its hormones.

These hormones cause the breakdown of glycogen in the liver to glucose and the release of fatty acids from stored fat cells. The glucose and fatty acids are released into the bloodstream as a quick source for synthesis of adenosine triphosphate (ATP) and increased metabolic rates. Heartbeat and blood pressure rates increase to get the glucose and fatty acids to muscle cells. Blood flow decreases to internal organs and skin but increases to muscle cells. The lungs take in more oxygen and get rid of more carbon dioxide. These changes all prepare our body to either fight or flee the stressful situation.

The adrenal cortex makes up the bulk of the adrenal gland. Its cells are organized into three densely packed layers of epithelial cells, forming an inner, a middle, and an outer region of the cortex. The outer layer of the adrenal cortex secretes a group of hormones called mineralocorticoid hormones because they regulate the concentration of mineral electrolytes. The most important of these hormones is **aldosterone** (al-**DOS**-ter-ohn), which regulates sodium reabsorption and potassium excretion by the kidneys.

The middle layer of the adrenal cortex secretes **cortisol** (**KOR**-tih-sahl), also known as **hydrocortisone** (**HIGH**-droh-**KOR**-tih-zone), which is a glucocorticoid hormone. Cortisol stimulates the liver to synthesize glucose from circulating amino acids. It causes adipose tissue to break down fat into fatty acids and causes the breakdown of protein into amino acids. These molecules are released into circulating blood to be taken up by tissues as a quick source of energy. The action of cortisol helps the body during stressful situations and helps maintain the proper glucose concentration in the blood between meals. Cortisol also helps reduce the inflammatory response. **Cortisone** (**KOR**-tih-zone), a steroid closely related to cortisol, is often given as a medication to reduce inflammation and as a treatment for arthritis. Dr. Percy Julian, an African American scientist, discovered how to synthetically produce cortisone and to use it as a treatment for the pain produced by the swelling in arthritic joints.

Cells in the inner zone of the adrenal cortex produce the adrenal sex hormones called the **androgens** (**AN**-droh-jenz). These are male sex hormones. Small amounts of androgens are secreted by the adrenal cortex in both men and women. In adult men, most androgens are secreted by the testes. Androgens stimulate the development of male sexual characteristics. In adult women, the adrenal androgens stimulate the female sex drive.

If the adrenal cortex fails to produce enough hormones, a condition known as **Addison's disease** develops. President John F. Kennedy suffered from Addison's disease and was under regular medical care for its treatment. Although President Kennedy always looked tanned and healthy, a bronzing of the skin was a symptom of the disease. In addition, other symptoms include decreased blood sodium, low blood glucose causing fatigue and listlessness, dehydration, and low blood pressure. Without treatment it can lead to death due to severe changes in electrolyte balances in the blood. Too much secretion from the adrenal cortex can lead to **Cushing's syndrome.** Blood glucose concentration remains high, lowering tissue protein. Retention of sodium causes tissue fluid increase, resulting in puffy skin. The patient exhibits obesity, a moon-shaped face, skin atrophy, and menstrual problems in women. Increases in adrenal male sex hormone production results in masculinizing changes in women, such as facial hair growth and lowering of voice pitch.

The Pancreas, Its Hormones, and Some Disorders

The pancreas has a dual role in that it is part of the digestive system where its cells, called acini, produce digestive enzymes known as pancreatic juice, and it is part of the endocrine system where its **pancreatic islets,** also known as the **islets of Langerhans,** produce the hormones **insulin** and **glucagon** (**GLOO**-kah-gon). These hormones regulate blood glucose levels. The pancreas is a flattened, elongated gland divided into head, body, and tail portions. Refer to its anatomy in Chapter 16. It is found behind the stomach and its pancreatic duct connects to the duodenum of the small intestine. This exocrine portion of the gland (the pancreatic duct) transports its digestive juices to the intestine.

Its endocrine portion consists of two main groups of cells closely associated with blood vessels. These groups of cells are known as the pancreatic islets or islets of Langerhans. **Alpha cells** secrete the hormone glucagon, and **beta cells** secrete the hormone insulin.

After a meal that consists primarily of carbohydrates like potatoes or rice, vegetables, salad, or cereals and breads, the blood glucose concentration becomes high due to the digestive processes. At this time, beta cells release insulin into the bloodstream. Insulin promotes the glucose in the blood to be transformed in the liver into glycogen, which is stored animal starch. In addition, glucose is moved into muscle cells and adipose tissue. Through negative feedback, when blood glucose levels fall, as between meals and during the night, the secretion of insulin decreases.

During the time glucose levels decrease, alpha cells in the pancreatic islets secrete the hormone glucagon. Glucagon stimulates the liver to convert the stored glycogen into glucose, thus raising blood glucose levels. Glucagon also causes the breakdown of amino acids and their conversion into glucose to raise blood sugar levels (Figure 12-7). The breakdown of the amino acids of proteins is used by the liver to synthesize more glucose. Fats are also broken down rapidly by other tissues to provide an alternative energy source. Again a negative feedback regulates glucagon secretion. Low blood sugar concentrations stimulate alpha cells to secrete glucagon. As blood sugar levels rise, glucagon secretion decreases. This mechanism helps to prevent hypoglycemia when glucose concentration gets low as during exercise and between meals.

The maintenance of blood glucose levels within a normal range is essential to body maintenance and function. A decline in blood glucose can cause nervous system malfunctions because glucose is the main source of energy for nerve cells. If the blood glucose level gets very low, the breakdown of fats releases fatty acids and

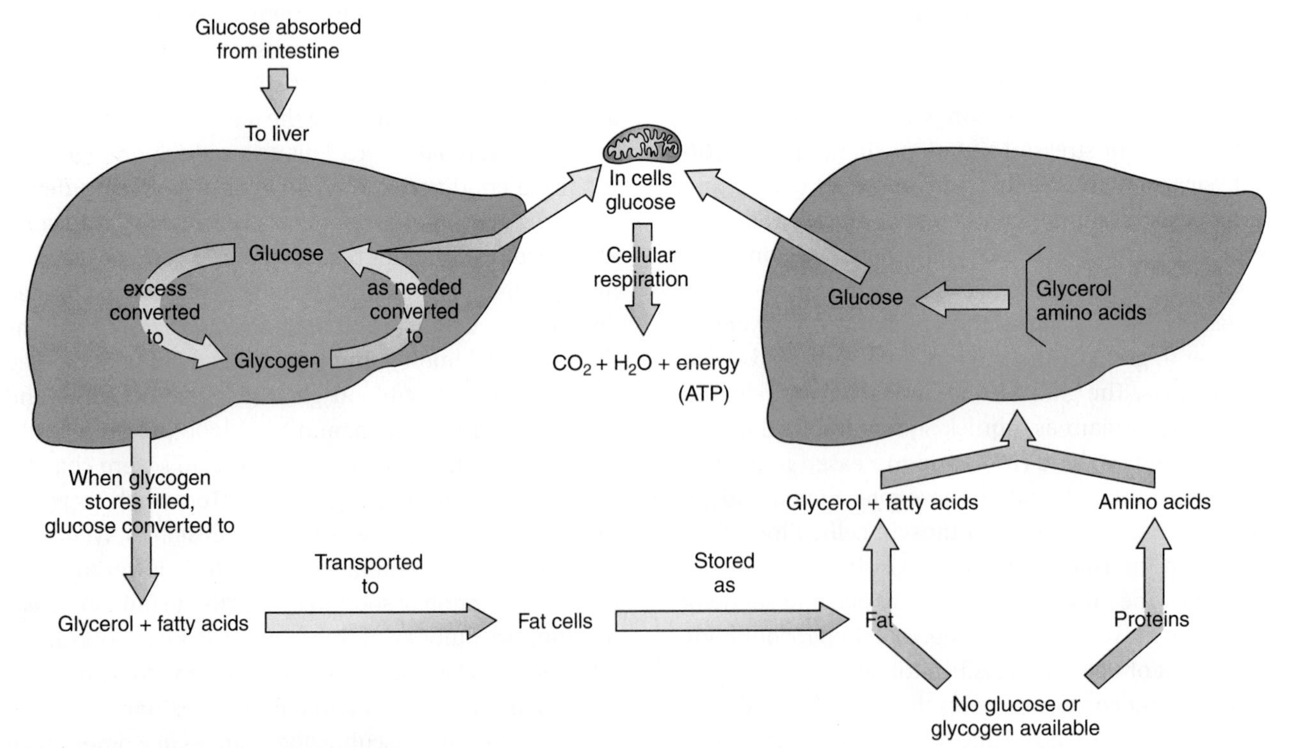

FIGURE 12-7. Glucose storage and conversion in the liver as a source of energy for the body.

ketones, causing a lowering of blood pH, a condition known as **acidosis** (**as**-ih-**DOH**-sis). If blood glucose levels are too high, the kidneys produce large amounts of urine containing high amounts of glucose, which can lead to dehydration.

The Testes and the Ovaries

The anatomy of the **testes** and the **ovaries** are discussed in detail in Chapter 19. The testes, in addition to producing sperm as exocrine glands, produce the male sex hormones as endocrine glands. The principal male sex hormone is **testosterone** (tess-**TOS**-ter-ohn). This hormone is responsible for the development of the male reproductive structures, and at puberty, the enlargement of the testes and penis. It also promotes the development of secondary male sexual characteristics, such as the growth of facial and chest hair, deepening of the voice, muscular development, bone growth resulting in broad shoulders, and narrow hips. It promotes the development of the male sexual drive and aggressiveness.

In the ovaries of the female, two groups of hormones, **estrogen** (**ESS**-troh-jen) and **progesterone** (proh-**JES**-ter-ohn), promote the development of the female reproductive structures: the uterus, vagina, and fallopian tubes. Secondary female sexual characteristics also develop such as breast enlargement, fat deposits on the hips and thighs, bone development resulting in broad hips, and a higher pitched voice. The menstrual cycle is also controlled by these hormones. Releasing hormones from the hypothalamus affect the anterior pituitary gland to produce the gonad-stimulating hormones: LH and LSH. These hormones control the secretion of hormones from the testes and ovaries. The hormones from the gonads have a negative feedback effect on the hypothalamus and the anterior pituitary gland. Thus, a constant, normal level of sex hormones is maintained in the body.

*Study*WARE™ Connection

- Watch an animation about the endocrine system on your StudyWARE™ CD-ROM.
- Play an interactive game labeling the endocrine glands on your StudyWARE™ CD-ROM.

COMMON DISEASE, DISORDER, OR CONDITION

INSULIN DEFICIENCY AND THE DISEASE DIABETES MELLITUS

Diabetes mellitus (dye-ah-**BEE**-teez **MELL**-ih-tus) is a very common disorder of the endocrine system. It is caused by a deficiency in insulin production and affects about 14 million Americans. Other individuals with diabetes have a decreased number of insulin receptors on target cells so that glucose is unable to move into cells even with normal insulin amounts. These conditions result in chronic elevations of glucose in the blood, a condition known as **hyperglycemia** (high-per-glye-**SEE**-mee-ah).

As blood sugar levels rise in diabetics, the amount of glucose filtered by the kidney tubules from the blood exceeds the ability of the tubules to reabsorb the glucose. Thus, there is a large amount of sugar in the urine, a condition known as **glycosuria** (glye-kos-**YOO**-ree-ah). This results in an increase in urine production because additional water is required to transport the extra glucose load. This is known as **polyuria**. As large amounts of fluids are lost in the urine, the diabetic individual dehydrates and craves large amounts of liquid, a condition known as **polydipsia** (pall-ee-**DIP**-see-ah), or excessive thirst. Also, because cells are not getting glucose to burn as energy, the diabetic person experiences intense food cravings or **polyphagia** (pall-ee-**FAY**-jee-ah). The diabetic person will eat ravenously but still constantly loses weight.

The disease inhibits fat and protein synthesis. Glucose-deficient cells use proteins as a source of energy, and tissues waste away. The patient is very hungry, eats yet loses body weight, and tires easily. Children will fail to grow, and both children and adults do not repair tissues well. Changes in fat metabolism build up fatty acids and ketones in the blood, resulting in low pH or acidosis. Acidosis

(continues)

COMMON DISEASE, DISORDER, OR CONDITION

INSULIN DEFICIENCY AND THE DISEASE DIABETES MELLITUS (continued)

and dehydration damage brain cells; thus, these individuals can become disoriented or may go into a diabetic coma and die.

There are two major types of diabetes mellitus: type 1 and type 2. Type 1 diabetes is also known as juvenile-onset diabetes because it usually develops between 11 and 13 years of age but before 30. It is an autoimmune disease that destroys the beta cells of the pancreas. Individuals with this type of diabetes must take daily insulin injections. This is also known as insulin-dependent diabetes mellitus (IDDM). This form of diabetes accounts for only 10% of diabetics.

Type 2 diabetes mellitus is known as noninsulin-dependent diabetes mellitus (NIDDM) and is the most common form of the disease, affecting about 90% of people with diabetes. It usually develops after 40 years of age and produces milder symptoms. Most affected persons are overweight when they develop the disease. In this situation, the beta cells still produce insulin but in reduced quantity, and insulin receptors on target cells are lost and glucose uptake diminishes. Treatment includes maintaining a balanced and controlled diet and exercise to maintain a normal body weight. Heredity and ethnic background can predispose individuals to this disease. Native Americans are at increased risk; African Americans and Hispanics are 50% more likely to develop type 2 diabetes than Caucasians. Drugs are available to treat type 2 diabetes.

Individuals with diabetes must monitor their blood glucose levels several times a day. Without monitoring and maintaining proper levels of blood glucose, nerve damage can develop. Hyperglycemia results in reduced blood flow caused by buildup of fatty materials in blood vessels, resulting in possible stroke, heart attack, and reduced circulation in the extremities. Diabetic retinopathy, causing changes in the retina of the eye, can lead to blindness. Kidney disease can be another complication of diabetes. Careful monitoring and regulation of blood sugar levels can control these symptoms. Discovery of insulin in 1921 and the development of drugs help control this disease today.

COMMON DISEASE, DISORDER, OR CONDITION

OTHER DISORDERS OF THE ENDOCRINE SYSTEM

DIABETES INSIPIDUS

Diabetes insipidus is caused by either not enough antidiuretic hormone (ADH) being produced by the posterior pituitary gland or from ADH receptors that are not functioning properly. This is not to be confused with diabetes mellitus. Individuals with diabetes insipidus excrete copious amounts of urine and thus become severely dehydrated. They also become excessively thirsty. Children with this condition often experience bedwetting. Treatment includes administration of ADH as a nasal spray.

SEASONAL AFFECTIVE DISORDER

Seasonal affective disorder occurs in individuals who are sensitive to an overproduction of melatonin that occurs in climate zones that have cloudy winter months with little bright sunshine. It produces a type of depression. Since winter months also have short days, this also contributes to more melatonin being secreted by the pineal gland (less light equals more melatonin). Individuals with this condition can be treated with daily doses of several hours of bright artificial light.

ALDOSTERONISM

Aldosteronism is caused by too much secretion of aldosterone, one of the mineralocorticoid hormones from the adrenal cortex. Symptoms of this condition include high blood pressure. This results from sodium and water retention by the kidneys, reduced levels of potassium in the blood, and an increase in blood pH.

(continues)

AS THE BODY AGES

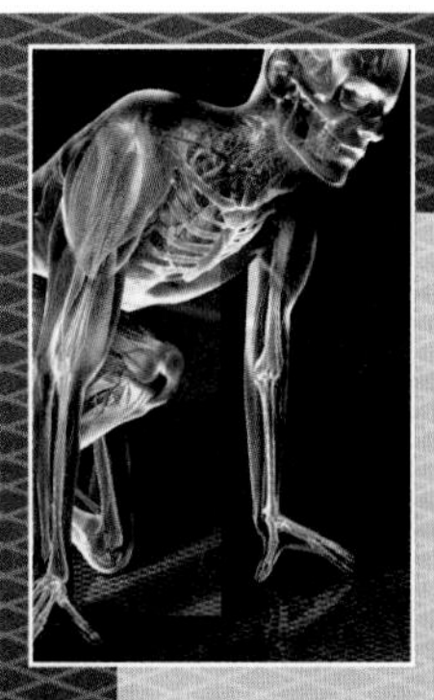

As individuals age, growth hormone decreases. This causes a decrease in bone mass, which may lead to osteoporosis. There is also a decrease in lean muscle mass with an accompanying increase in the deposition of adipose tissue. Regular exercising helps limit the decrease in growth hormone.

The production of sex hormones also declines in both men and women during the later middle years. This results in menopause in women, which is occasionally treated with sex hormone replacement therapy.

Secretion of thymosin from the thymus gland decreases with age, affecting the number of lymphocytes that can mature and provide functional immunity. This can result in susceptibility to cancers and more frequent bacterial and viral infections. Melatonin secretion from the pineal gland also decreases in older adults, resulting in changes in sleep patterns that cause tiredness during the daylight hours and require short naps during the day in addition to regular nighttime sleeping. ■

Career FOCUS

These are careers that are available to individuals who are interested in the endocrine system.

- **Nuclear medicine technologists** are individuals who administer radioactive drugs, known as radiopharmaceuticals, such as radioactive iodine for the treatment of a hyperactive thyroid gland. These radioactive drugs are also used for diagnostic imaging.
- **Endocrinologists** are physicians whose specialty is the endocrine system and the treatment of endocrine problems.
- **Diabetes dieticians** are individuals trained as dieticians who specialize in nutritional therapy, counseling, and the planning of balanced meals for patients with diabetes mellitus.

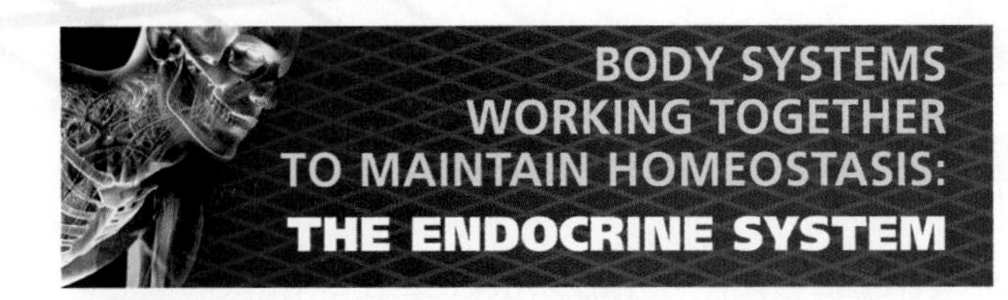

Integumentary System
- Melanocyte production in the skin is caused by melanocyte-stimulating hormone from the anterior pituitary.
- Melanocytes produce melanin, causing a darker pigmentation to skin for protection against the harmful rays of the sun.
- Androgens activate sebaceous glands; estrogen increases hydration of skin.

Skeletal System
- Calcium concentrations in bone are controlled by the hormones calcitonin and parathormone.
- Nerve cells of the hypothalamus function properly due to proper calcium ion concentration.
- Bones protect the endocrine glands in the brain, pelvis, and chest.

Muscular System
- Hormones can cause increased heartbeat rates, which help to increase the amount of blood carrying oxygen and nutrients to exercising muscle cells.
- Growth hormone stimulates muscular development.

Nervous System
- Nerve cells of the hypothalamus control the secretions of the pituitary gland.
- Through negative feedback, hormonal levels control the secretions of the hypothalamus.

Cardiovascular System
- The blood carries hormones to their target organs.
- Heart rates and the diameter of blood vessels are controlled by hormones.

Lymphatic System
- Hormones stimulate the production of T lymphocytes.
- Hormones are involved in the development of the immune system in children.
- Lymph can be a route for hormone transport.

Digestive System
- Glucose levels in the blood are controlled by hormones.
- Excess glucose is stored in the liver as glycogen and is made available to cells between meals by the combined actions of insulin and glucagon.
- Hormones also affect digestive activities, such as increased appetites during puberty caused by higher rates of metabolism.

Respiratory System
- Low levels of oxygen in the blood stimulate hormonal production of red blood cell formation in bone marrow.
- Red blood cells transport oxygen from the lungs to body cells and carbon dioxide waste from cells to the lungs.
- Epinephrine increases breathing rates.

Urinary System
- Hormones control kidney function.
- Kidneys control body water levels and balances of the electrolytes in the blood.

Reproductive System
- The sex hormones stimulate the development of the reproductive structures.
- Sex hormones also stimulate the development of secondary male and female sexual characteristics.
- Sex hormones stimulate the development of egg cells and sperm cells.

SUMMARY OUTLINE

INTRODUCTION

1. The endocrine system maintains the internal environment of the body within certain narrow limits via chemical control through its hormones. This is known as homeostasis.
2. The hypothalamus of the brain sends chemical signals that control the pituitary gland, the master gland of the system.
3. The endocrine glands are ductless glands that secrete their hormones directly into the bloodstream, which carries them to target organs.

THE FUNCTIONS OF HORMONES

1. Hormones control cellular respiration, growth, and reproduction.
2. They control body fluids and electrolyte balances.
3. They control the secretion of other hormones.
4. They control behavior patterns.

5. They regulate reproductive cycles and our growth and development.
6. Through negative feedback mechanisms, hormone levels within our bodies are maintained within normal concentrations.

THE CLASSIFICATION OF HORMONES

1. Some hormones are modified amino acids: epinephrine, norepinephrine, oxytocin, and vasopressin (ADH).
2. Other hormones are proteins: insulin and growth hormone.
3. A third category are the steroid hormones: cortisol, estrogen, and testosterone.
4. The amino acid and protein hormones bind to membrane receptor sites in the cells of target organs. When prescribed, these must be injected.
5. The steroid hormones diffuse across cell membranes and then bind to intracellular receptor molecules. When prescribed, these can be taken orally.

THE HYPOTHALAMUS OF THE BRAIN

1. The hypothalamus of the brain controls the secretions of the pituitary gland, the master gland of the endocrine system.
2. Nerve cells in the hypothalamus produce chemical signals called releasing hormones that stimulate and releasing inhibitory hormones that inhibit the release of a particular hormone from the pituitary gland.
3. The hypothalamus of the nervous system controls the secretions of the endocrine system.
4. Through negative feedback mechanisms, the endocrine system can influence the functions of the hypothalamus.

THE MAJOR ENDOCRINE GLANDS AND THEIR HORMONES

1. The endocrine glands are the anterior and posterior lobes of the pituitary gland, the pineal gland, the thyroid, the parathyroids, the thymus, the adrenal glands, the pancreatic islets, the ovaries, and the testes.

The Anterior Pituitary Gland, Its Hormones, and Some Disorders

1. The pituitary gland is also called the hypophysis. It is divided into a larger anterior lobe and a smaller posterior lobe. It is the master gland of the system.
2. The anterior pituitary lobe, made mainly of glandular epithelium, produces seven hormones.
3. Growth hormone (GH) stimulates cell metabolism and the growth of bones and muscles. Too little in childhood produces pituitary dwarfism. Too much secretion in childhood produces a condition called gigantism. Too much secretion after childhood produces enlarged hands, feet, and facial features, a condition called acromegaly.
4. Thyroid-stimulating hormone (TSH) stimulates the thyroid gland to secrete its hormones, T_3, T_4, and calcitonin.
5. Adrenocorticoid hormone (ACTH) stimulates the adrenal cortex to secrete its hormone cortisol.
6. Melanocyte-stimulating hormone (MSH) causes a darkening of the skin by stimulating melanocytes to produce melanin.
7. Follicle-stimulating hormone (FSH) stimulates the development of follicles in the ovaries of females and the production of sperm cells in males.
8. Luteinizing hormone (LH) stimulates ovulation and production of progesterone in females and the production of testosterone in males.
9. Prolactin stimulates milk production in the mammary glands of females following childbirth.

The Posterior Pituitary Gland and Its Hormones

1. The posterior lobe consists mainly of nerve fibers and neuroglial cells. It produces two hormones: antidiuretic hormone and oxytocin.
2. Antidiuretic hormone (ADH), also known as vasopressin, causes increased water reabsorption in the tubules of the kidneys, resulting in less water in the urine. A deficiency in ADH can result in a condition known as diabetes insipidus. If secreted in large amounts, it can cause constriction of blood vessels, hence its other name vasopressin.
3. Oxytocin causes contraction of uterine smooth muscles during childbirth. It also causes constriction of mammary gland cells, resulting in milk ejection or lactation during breastfeeding.

The Thyroid Gland, Its Hormones, and Some Disorders

1. The thyroid gland consists of two lobes connected by an isthmus. It is found just below the larynx on either side of the trachea. It produces three hormones, T_3, T_4, and calcitonin.

2. The thyroid gland requires iodine to function properly. This is a component of iodized salt in the United States.
3. Without sufficient iodine, the thyroid gland enlarges, forming a goiter.
4. Two thyroid hormones are thyroxine or tetraiodothyronine (T_4) and triiodothyronine (T_3). They regulate the metabolism of carbohydrates, fats, and proteins for normal growth and development and nervous system maturation.
5. Hypothyroidism (a lack of thyroid hormone) in children causes cretinism, which results in small stature and mental retardation. In adults, it results in sluggishness, fatigue, and fluid accumulation in subcutaneous tissues.
6. Hyperthyroidism (too much thyroid hormone) causes nervousness, high body metabolism, and fatigue. Graves' disease, associated with an enlarged thyroid or goiter, also has the effect of bulging eyeballs called exophthalmia.
7. The extrafollicular cells of the thyroid secrete the third hormone, calcitonin. This lowers the calcium and phosphate ion concentration in the blood by inhibiting the release of these ions from bones and increasing their excretion by the kidneys.

The Parathyroid Glands, Their Hormone, and Some Disorders

1. The four parathyroid glands are embedded in the posterior surface of the thyroid gland.
2. Their secretory cells, called chief cells, secrete the hormone parathyroid hormone or parathormone (PTH).
3. PTH causes bone cells to release calcium and phosphate into the blood, causes the kidneys to conserve blood calcium, and causes the intestinal cells to absorb calcium from digested food.
4. Vitamin D also increases absorption of calcium by the intestines.
5. High levels of PTH or hyperparathyroidism cause breakdown of bone matrix, resulting in soft, deformed, and easily fractured bones. In addition, elevated calcium affects muscle and nerves, resulting in muscle weakness and fatigue. Excess calcium can cause kidney stones.
6. Low levels of PTH or hypoparathyroidism reduce osteoclast activity, resulting in reduced rates of bone breakdown and vitamin D formation. Bones remain strong but as the blood calcium level decreases, muscle and nerves become abnormally excitable, resulting in muscle cramps and tetanic contractions. This could result in respiratory failure.

The Adrenal Glands, Their Hormones, and Some Disorders

1. The adrenal glands, also known as the suprarenal glands, are found on top of each kidney.
2. Each gland is divided into an inner part called the adrenal medulla and an outer part called the adrenal cortex.
3. The adrenal medulla produces the hormone adrenaline, also called epinephrine, in large amounts, and noradrenaline or norepinephrine in small amounts in response to signals from the sympathetic division of the autonomic nervous system.
4. Epinephrine and norepinephrine are called the fight-or-flight hormones because they prepare the body for stressful situations.
5. They cause the release of glucose from the liver and fatty acids from fat cells as a source of energy. Heart rate and blood pressure increase. Blood flow to muscle cells increases and decreases to skin and internal organs. The lungs take in more oxygen.
6. The adrenal cortex is divided into three layers. The outer layer secretes the mineralocorticoid hormones, the most important of which is aldosterone, which regulates sodium reabsorption and potassium excretion by the kidney.
7. The middle layer of the adrenal cortex secretes cortisol, also known as hydrocortisone, a glucocorticoid hormone. It causes the liver to make glucose from circulating amino acids, causes protein to be broken down into amino acids, and causes fat cells to break down fat into fatty acids as sources of energy for body cells.
8. Cortisol also inhibits the inflammatory response. Cortisone, a steroid closely related to cortisol, is given to treat arthritis and to reduce inflammation.
9. The inner layer of the adrenal cortex produces androgens, the adrenal male sex hormones. Androgens stimulate male sex characteristics. In adult men, most androgens come from the testes. In adult women, they stimulate the female sex drive.
10. Addison's disease is caused by a lack of sufficient adrenal cortex hormones. Its symptoms include a bronzing of the skin, decreased blood sodium, low

blood glucose causing fatigue, dehydration, and low blood pressure.

11. Cushing's syndrome is caused by too much secretion of the adrenal cortex. Its symptoms include high blood glucose levels and low tissue protein. Sodium retention causes tissue fluid increase, resulting in puffy skin. The patient is obese with a moon-shaped face.

The Pancreas, Its Hormones, and Some Disorders

1. The pancreatic islets or islets of Langerhans are the endocrine portion of the pancreas and produce two hormones, insulin and glucagon, which regulate blood glucose levels.
2. Beta cells of the pancreatic islets produce insulin after meals. Insulin causes excess blood glucose to be stored in the liver as animal starch or glycogen. Glucose is also moved to muscle cells and adipose tissue.
3. Alpha cells of the pancreatic islets produce glucagon between meals, when blood glucose levels are lower. Glucagon stimulates the liver to convert stored glycogen into glucose, to break down amino acids and convert them to glucose, and to break down fats in other tissues as another energy source.
4. Negative feedback mechanisms regulate the level of blood glucose concentrations.
5. A decline in blood glucose can cause nervous system malfunctions, because glucose is a main source of energy for nerve cells.
6. Low blood glucose levels cause the breakdown of fats releasing fatty acids and ketones in the blood, resulting in a lowering of blood pH, a condition called acidosis.
7. High levels of blood glucose cause the kidneys to produce large amounts of urine to dilute the excess glucose, resulting in dehydration.
8. Insufficient insulin production results in the disease diabetes mellitus.

The Testes and the Ovaries

1. The testes produce the principal male sex hormone testosterone.
2. Testosterone causes the development of the male reproductive structures and at puberty the enlargement of the testes and the penis.
3. Testosterone also causes the development of the secondary male sex characteristics like facial and chest hair, muscle development, low-pitched voice, broad shoulders, and narrow hips.
4. The ovaries produce the female sex hormones, estrogen and progesterone.
5. Estrogen and progesterone cause the development of the female reproductive organs. They also cause the development of the secondary female sexual characteristics like breast enlargement, high-pitched voice, broad hips, and fat deposits on the thighs, hips, and legs.
6. Female sex hormones also control the menstrual cycle.

The Thymus Gland and Its Hormone

1. The thymus gland produces the hormone thymosin, and the gland is crucial to the development of the immune system.
2. Thymosin causes the production of the T-lymphocyte white blood cells, which protect the body against foreign microbes.

The Pineal Gland and Its Hormone

1. The pineal gland is found in the brain near the thalamus and produces the hormone melatonin.
2. Melatonin inhibits the functions of the reproductive system and regulates body rhythms like wake and sleep patterns.
3. Bright light inhibits melatonin secretion.
4. Low levels of melatonin in bright light make us feel good and increases fertility; high levels of melatonin in dim light causes us to feel tired and depressed.
5. The pineal gland also secretes serotonin, a neurotransmitter and vasoconstrictor.

REVIEW QUESTIONS

*1. Explain how the hypothalamus of the brain controls the endocrine system.

*2. Explain how a negative feedback system functions in maintaining hormonal levels in the body.

3. Name the three chemical categories for classifying hormones and give some examples.

4. Name the major endocrine glands and their hormones.

5. Name some effects that testosterone has on the male body.

***Critical Thinking Questions**

MATCHING

Place the most appropriate number in the blank provided.

_____ Thyroid	**1.** Thyrotropic hormone
_____ Prolactin	**2.** Parathormone
_____ Adrenal medulla	**3.** Testosterone
_____ Anterior pituitary	**4.** Insulin
_____ Adrenal cortex	**5.** Vasopressin (ADH)
_____ Testis	**6.** Thyroxine
_____ Ovary	**7.** Estrogen and progesterone
_____ Parathyroid	**8.** Adrenaline/epinephrine
_____ Pancreas	**9.** Cortisol
_____ Posterior pituitary	**10.** Melatonin
	11. Milk secretion
	12. Serotonin

Search and Explore

- Visit the Endocrineweb website at http://www.endocrineweb.com and read more about the structures and diseases of the endocrine system.
- Visit the American Diabetes Association at http://www.diabetes.org to learn more about type 1 and type 2 diabetes. Write a two to three paragraph entry in your notebook comparing the differences between the two types.

CASE STUDY

Sophia, a 58-year-old moderately obese woman, is seeing her primary health care provider. Sophia is concerned because she cut her foot two weeks ago and the wound is not healing. The health care provider notes that Sophia has lost 30 pounds since her last appointment. Despite her weight loss, she states that she has been very hungry lately, and is eating much more than usual. She also reports that she is constantly thirsty, and is experiencing frequent urination. Based on her symptoms and diagnostic studies performed by her health care provider, Sophia learns she has diabetes mellitus.

Questions

1. What is the major characteristic of diabetes mellitus?
2. What type of diabetes mellitus do you think Sophia has?
3. What are the four classic symptoms of this disorder?
4. What measures will the health care provider take to control Sophia's symptoms?
5. If Sophia eventually requires insulin, why must this medication be injected?
6. What complications can individuals with uncontrolled diabetes mellitus develop?

*Study*WARE™ Connection

Take a practice quiz or play an one of the interactive games that reinforces the content in this chapter on your StudyWARE™ CD-ROM.

Study Guide Practice

Go to your **Study Guide** for more practice questions, labeling and coloring exercises, and crossword puzzles to help you learn the content in this chapter.

LABORATORY EXERCISE: THE ENDOCRINE SYSTEM

Materials needed: A dissecting kit, a fetal pig from a biologic supply company, and a dissecting pan

1. Place your fetal pig in a dissecting pan, ventral side up. Using your scalpel and pulling with your forceps, remove the skin from a square area from the middle of the lower jaw, where the ears attach, posteriorly to the chest near the sternum. Refer to Figure 12-9 cut number 2. Before you make your horizontal cut, feel for the larynx or voice box. Make your horizontal superior cut just above the larynx being careful not to cut too deeply. Then cut laterally on both sides, down to the thoracic region. As you make your horizontal posterior cut, the sternum

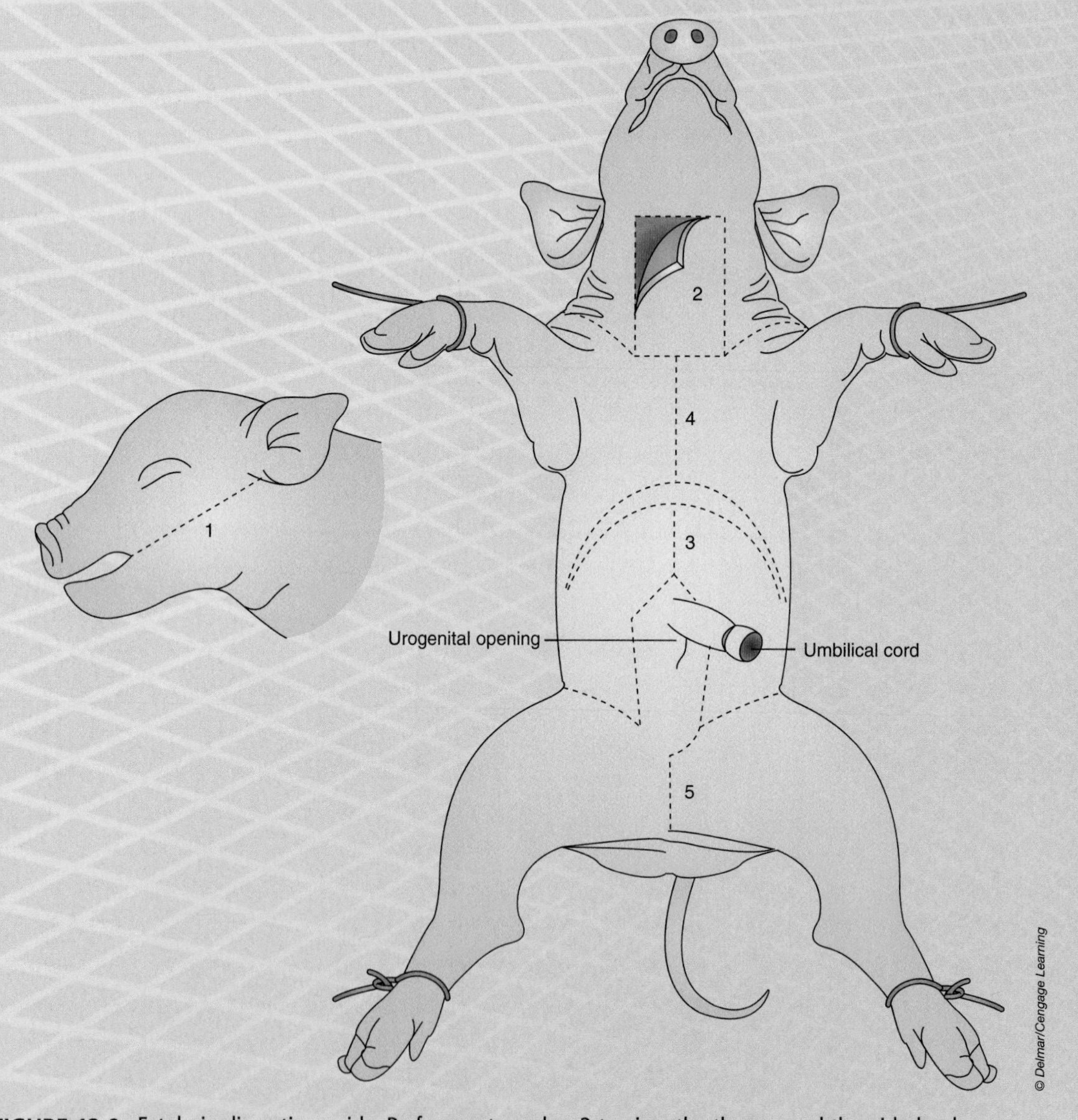

FIGURE 12-9. Fetal pig dissection guide. Perform cut number 2 to view the thymus and thyroid glands.

(continues)

THE ENDOCRINE SYSTEM (Continued)

will prevent your scalpel from damaging any tissues. Now remove the skin and muscle layer attached to the skin by gently pulling with your forceps and scraping with your scalpel.

2. When the skin is removed you will see exposed muscles and glands. Notice the large lengthwise muscles of the neck region. These can be removed to expose the thymus gland. Refer to Figure 12-10 of a fetal pig dissection for a view of this region. The thymus gland appears spongelike as opposed to the thick cords of muscle tissue.

3. In the pig, the thymus gland is two large lobes located just below the cartilaginous larynx or voice box. Now push the two lobes apart with a probe or forceps to expose the smaller, dark brown thyroid gland found on top of the trachea. While here, note the trachea with its cartilaginous rings.

4. Once you have found these glands of the endocrine system, return your fetal pig to its storage area. We will use it again in future dissections.

5. Your instructor may now show you a videotape or DVD on the endocrine system. There are a number to choose from. An excellent videotape is "Homeostasis: The Body in Balance" by Human Relations Media, 175 Tompkins Ave., Pleasantville, N.Y. 10570.

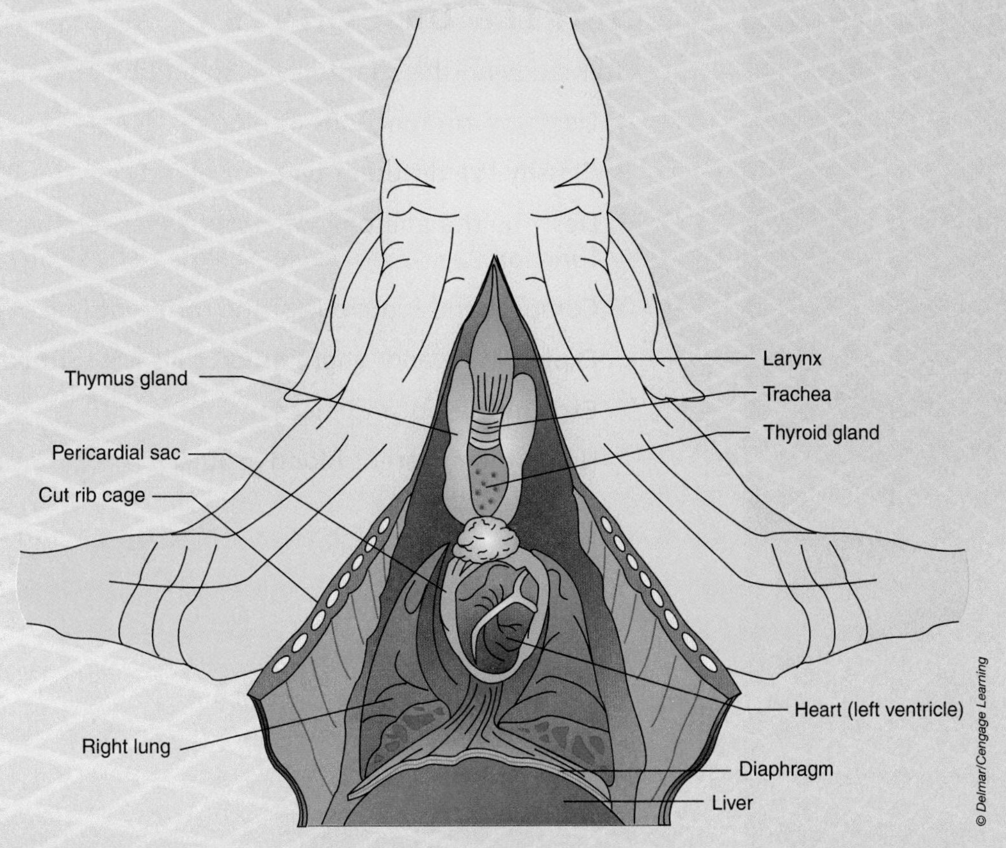

FIGURE 12-10. The throat and thoracic region of a dissected fetal pig.

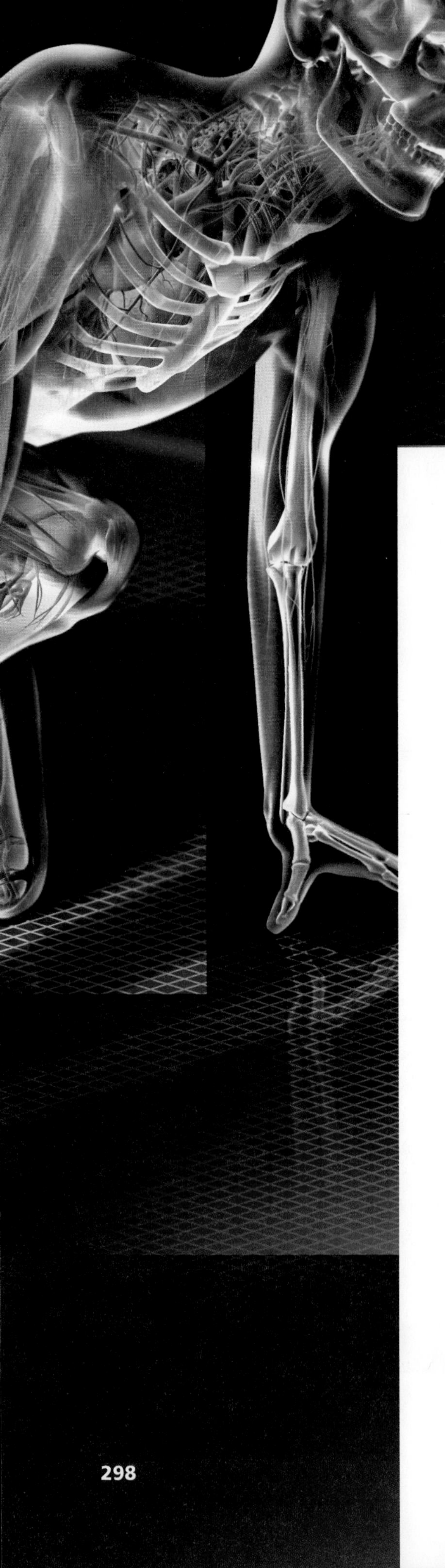

The Blood

CHAPTER OBJECTIVES

After studying this chapter, you should be able to:

1. Describe the functions of blood.
2. Classify the different types of blood cells.
3. Describe the anatomy of erythrocytes relative to their function.
4. Compare the functions of the different leukocytes.
5. Explain how and where blood cells are formed.
6. Explain the clotting mechanism.
7. Name the different blood groups.

13

KEY TERMS

INTRODUCTION

Blood is a uniquely specialized connective tissue in that it consists of two components: the formed elements of blood, or the blood cells, and the fluid part of blood or **plasma** (Figure 13-1). The formed elements of blood are the red blood cells (RBCs) or **erythrocytes** (eh-**RITH**-roh-sightz), the white blood cells (WBCs) or **leukocytes** (**LOO**-koh-sightz), and the **platelets** or **thrombocytes** (**THROM**-boh-sightz). Plasma, a viscous fluid, accounts for about 55% of blood; the formed elements make up about 45% of the total volume of blood. An average woman has about 5 liters of blood; an average man has approximately 6 liters of blood in the body. Blood accounts for about 8% of total body weight. See Concept Map 13-1: The Blood.

FUNCTIONS OF THE BLOOD

Pumped by the heart and carried by blood vessels throughout the body, blood is a complex liquid that performs a number of critical functions. These functions relate to maintaining homeostasis.

Blood transports oxygen from the lungs, where it enters the RBCs, to all cells of the body. Oxygen is needed by the cells for cellular metabolism. Blood also transports carbon dioxide from the cells, where it is produced as a waste product of cellular metabolism, to the lungs to be

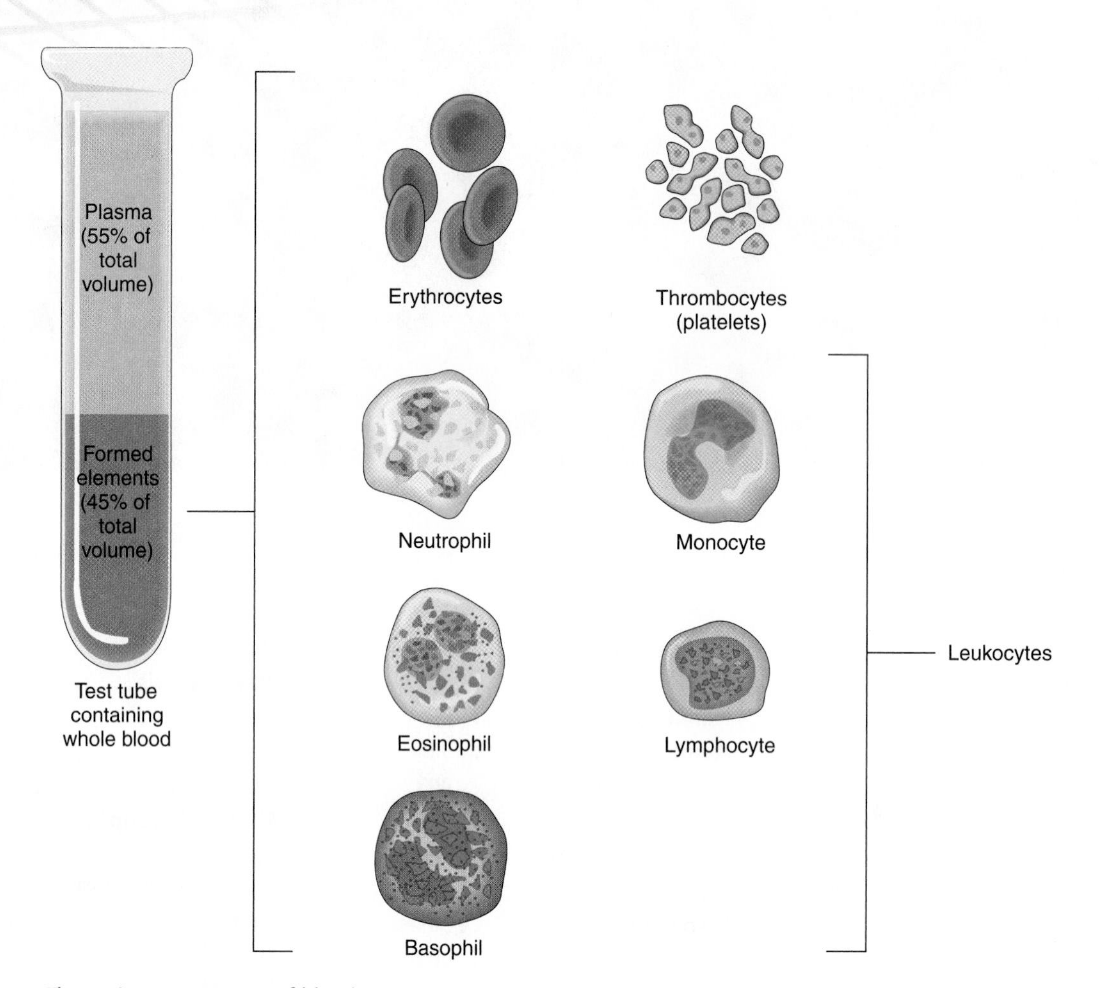

FIGURE 13-1. The major components of blood.

expelled from the body. Blood carries nutrients, ions, and water from the digestive tract to all cells of the body. It also transports waste products from the body's cells to the sweat glands and kidneys for excretion.

Blood transports hormones from endocrine glands to target organs in the body. It also transports enzymes to body cells to regulate chemical processes and chemical reactions. Blood helps regulate body pH through buffers and amino acids that it carries. The normal pH of blood is slightly basic (alkaline) at 7.35 to 7.45. Blood plays a role in the regulation of normal body temperature because it contains a large volume of water (the plasma), which is an excellent heat absorber and coolant.

Blood helps to regulate the water content of cells through its dissolved sodium ions; thus, it plays a role in the process of osmosis. It is through the clotting mechanism that blood helps prevent fluid loss when blood vessels and tissues are damaged. Finally, blood plays a vital role in protecting the body against foreign microorganisms and toxins through its special combat-unit cells, the leukocytes.

THE CLASSIFICATION OF BLOOD CELLS AND THE COMPOSITION OF PLASMA

The most common classification of the formed elements or cells of blood is:

A. Erythrocytes or red blood cells (RBCs), which make up about 95% of the volume of blood cells.

B. Leukocytes or white blood cells (WBCs) are divided into two subcategories: the granular leukocytes and the agranular or nongranular leukocytes.

   1. The granular leukocytes have granules in their cytoplasm when stained with Wright's stain. There are three types:

      a. **Neutrophils** (**NOO**-troh-fillz), which make up 60% to 70% of WBCs

      b. **Eosinophils** (**ee**-oh-**SIN**-oh-fillz), which make up 2% to 4% of WBCs

      c. **Basophils** (**BAY**-soh-fillz), which make up 0.5% to 1% of WBCs

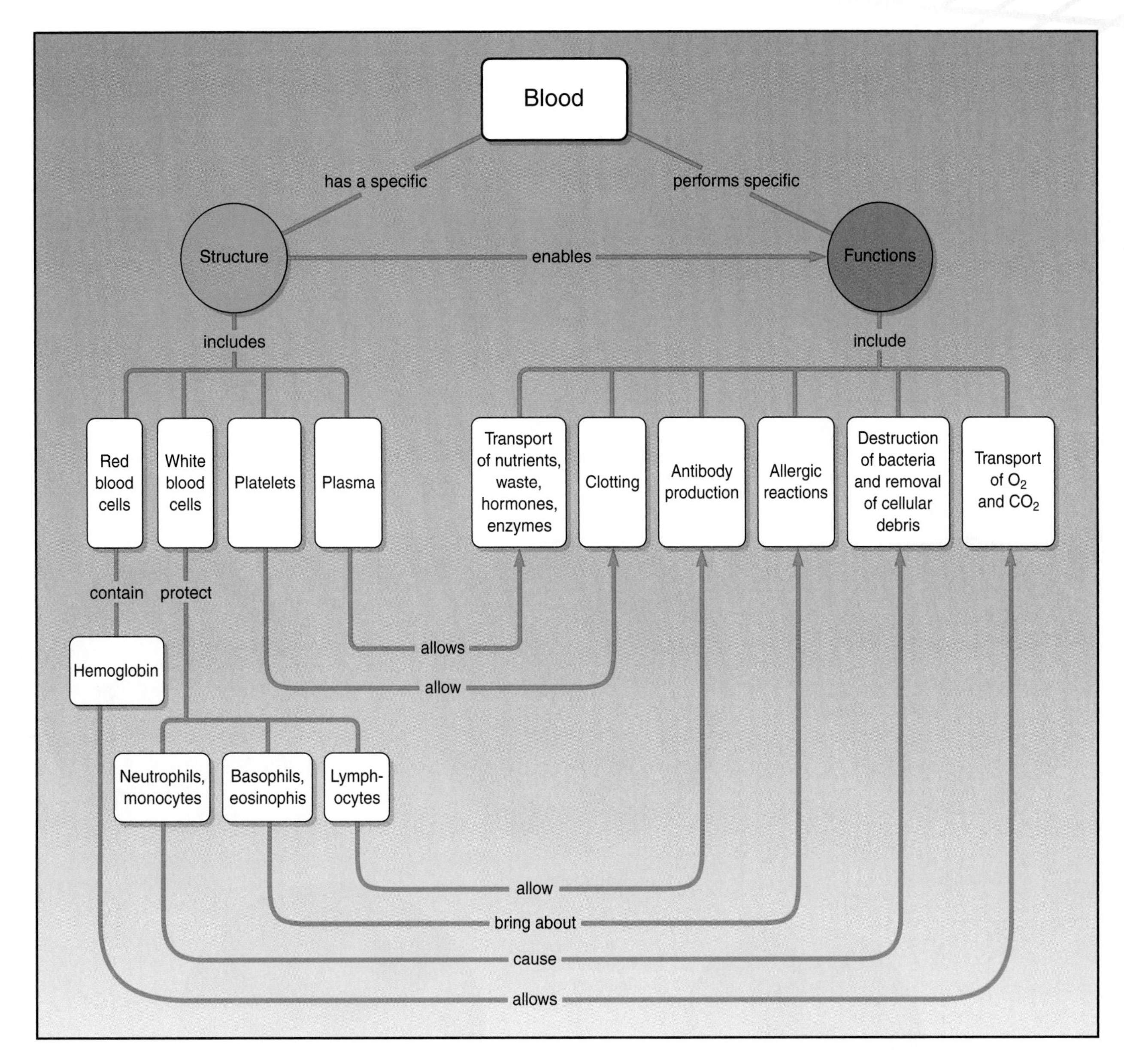

CONCEPT MAP 13-1. The blood.

2. The agranular or nongranular leukocytes do not show granules in their cytoplasm when stained with Wright's stain.

   There are two types:

   a. **Monocytes** (**MON**-oh-sightz), which make up about 3% to 8% of WBCs

   b. **Lymphocytes** (**LIM**-foh-sightz), which make up about 20% to 25% of WBCs

C. Thrombocytes or platelets (Figure 13-2).

There are 700 times more RBCs in blood than WBCs and at least 17 times more RBCs than platelets.

Plasma is the fluid component of blood; 91% of plasma is water. About 7% are the proteins **albumin** (al-**BYOO**-men), **globulins** (**GLOB**-yoo-linz), and **fibrinogen** (fih-**BRIN**-oh-jen). Albumin plays a role in maintaining osmotic pressure and water balance between blood and tissues. Some examples of globulins are antibodies and **complement**, which are important in the immune response of the body. Other globulins act as transport molecules for hormones and carry them to target organs. Fibrinogen plays a vital role in the clotting mechanism. The remaining 2% of plasma consist of solutes such as ions, nutrients, waste products, gases, enzymes, and hormones.

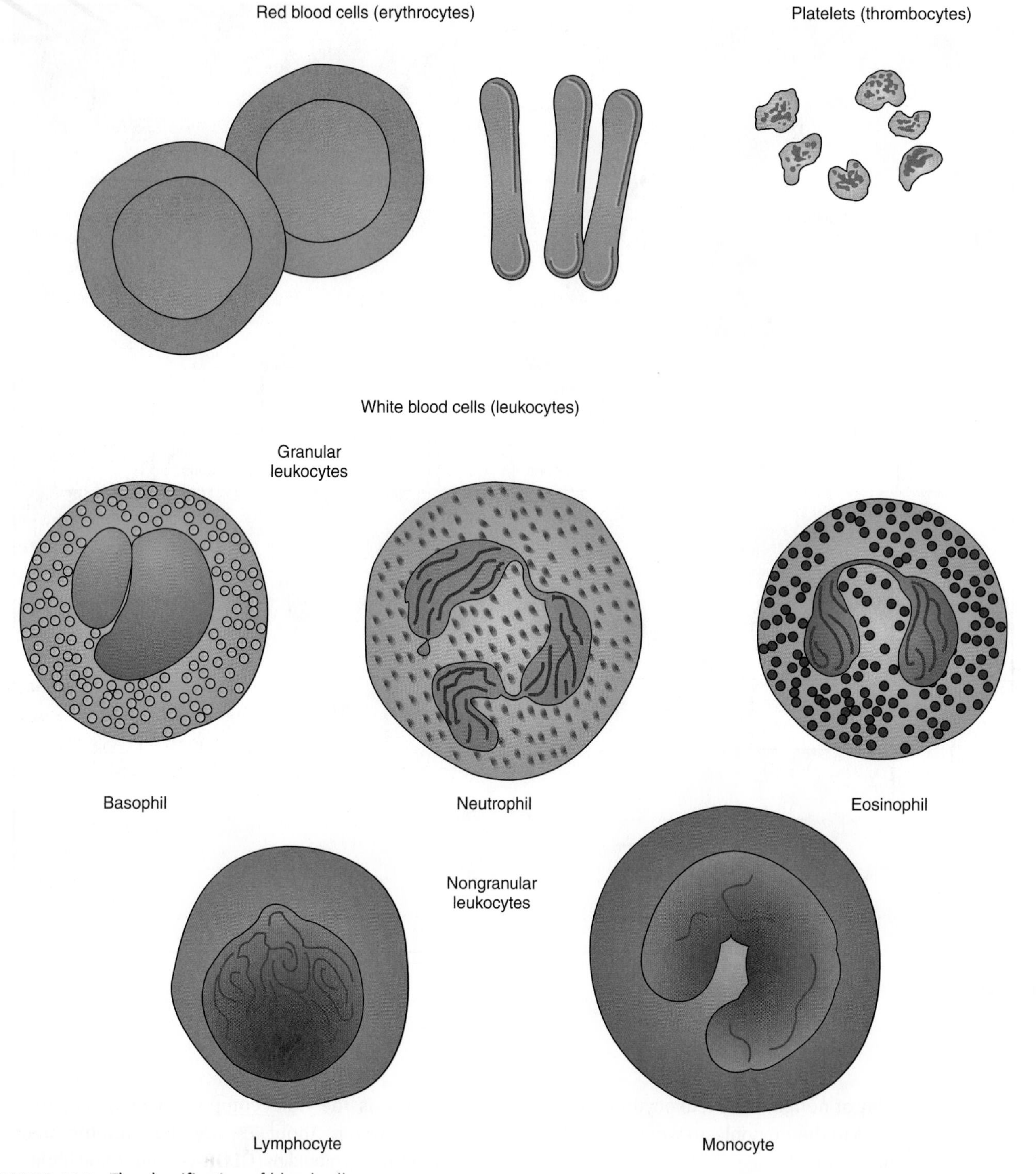

FIGURE 13-2. The classification of blood cells.

Blood storage techniques were developed by Dr. Charles Drew, an African American scientist. He is best known for his research on blood plasma. His blood preservation discoveries led to the formation of blood banks in the United States and Great Britain during World War II. He was director of the first American Red Cross Blood Bank.

FORMATION OF BLOOD CELLS: HEMATOPOIESIS

Blood cell formation is known as **hematopoiesis** (**hem**-ah-toh-poy-**EE**-sis). Hematopoiesis occurs in **red bone marrow,** which is also known as **myeloid tissue.** All blood cells are produced by red bone marrow.

However, certain lymphatic tissue such as the spleen, tonsils, and lymph nodes produce agranular leukocytes (lymphocytes and monocytes). They assist in the production of those blood cells. All blood cells develop from undifferentiated mesenchymal cells called **stem cells** or **hematocytoblasts** (hee-**MAT**-oh-**SIGHT**-oh-blastz) (Figure 13-3).

Some stem cells will differentiate into proerythroblasts, which will eventually lose their nuclei and become mature RBCs. Some stem cells will become myeloblasts. These cells will develop into progranulocytes. Some of these cells will develop into basophilic myelocytes and will mature into basophils, others will develop into eosinophilic myelocytes and will mature into eosinophils, while still others will develop into neutrophilic myelocytes and will mature into neutrophils. Other stem cells will become lymphoblasts and will mature into lymphocytes; some stem cells will become monoblasts and will mature into monocytes. Finally, some stem cells will become megakaryoblasts and will undergo multipolar mitosis of the nucleus to mature into blood platelets. All stages of blood cell development will be found in red bone marrow tissue.

BLOOD CELL ANATOMY AND FUNCTIONS

Erythrocytes appear as biconcave disks with edges that are thicker than the center of the cell, looking somewhat doughnut-shaped. They do not have a nucleus and are simple in structure. They are composed of a network of protein called stroma, cytoplasm, some lipid substances including cholesterol and a red pigment called **hemoglobin** (hee-moh-**GLOH**-bin). Hemoglobin constitutes 33% of the cell's volume. Erythrocytes contain about 280 million molecules of hemoglobin per erythrocyte. Because they have lost their nuclei, they do not divide. They live for approximately 120 days.

The primary function of erythrocytes is to combine with oxygen in the lungs and to transport it to the various tissues of the body. It then combines with carbon dioxide in tissues and transports it to the lungs for expulsion from the body. The pigment hemoglobin allows this to happen. Hemoglobin is made of a protein called **globin** (**GLOH**-bin) and a pigment called **heme** (**HEEM**). The pigment heme contains four iron atoms. The iron atoms of heme combine with the oxygen in the lungs. In the tissues of the body, the oxygen is released and the protein globin now combines with the carbon dioxide from the interstitial fluids and carries it to the lungs where it is released. Hemoglobin that is carrying oxygen is bright red in color, whereas hemoglobin not carrying oxygen is a darker red in color. A healthy man has 5.4 million RBCs/mm^3 of blood and a healthy woman has 4.8 million RBCs/mm^3 of blood. Due to menstruation and loss of blood, some women need more iron in their diet for the most efficient transport of oxygen by their blood.

Leukocytes have nuclei and no pigment. Their general function is to combat inflammation and infection. They are called white blood cells because they lack pigmentation. They are larger in size than RBCs and are carried by the blood to various tissues in the body. They have the ability to leave the blood and move into tissues by ameboid movement, sending out a cytoplasmic extension that attaches to an object while the rest of the cell's contents then flows into that extension. In this manner, the leukocytes attack invading microorganisms and clean up cellular debris by consuming this material by **phagocytosis** (**fag**-oh-sigh-**TOH**-sis), which means *eating cells.*

When stained with Wright's stain, the cytoplasm of leukocytes shows the presence or absence of granules. Therefore, leukocytes are divided into the granular leukocytes and the agranular (*a* means without) or nongranular leukocytes. The three types of granular leukocytes are neutrophils, basophils, and eosinophils. The two types of nongranular leukocytes are monocytes and lymphocytes.

Neutrophils are the most common of leukocytes. They are the most active WBCs in response to tissue destruction by bacteria. They stay in the blood for about 12 hours and then move into tissues where they phagocytize foreign substances and secrete the enzyme **lysozyme** (**LYE**-soh-zyme), which destroys certain bacteria. When pus accumulates at an infection area, it consists of cell debris, fluid, and dead neutrophils.

Monocytes are also phagocytotic. They phagocytize bacteria and any dead cells or cellular debris. They are the largest leukocytes. After they leave the blood and enter tissues, they increase in size and are called **macrophages** (**MACK**-roh-fay-jeez).

Eosinophils combat irritants, such as pollen or cat hair, that cause allergies. They produce antihistamines. Their chemical secretions also attack some worm parasites in the body.

Basophils are also involved in allergic reactions. They release heparin (an anticoagulant), histamine (an inflammatory substance), and serotonin (a vasoconstrictor) into tissues.

Lymphocytes are involved in the production of antibodies and play a crucial role in the body's immune response. They are the smallest of the leukocytes. There are several types of lymphocytes: the B lymphocytes and the T lymphocytes, which are discussed further in

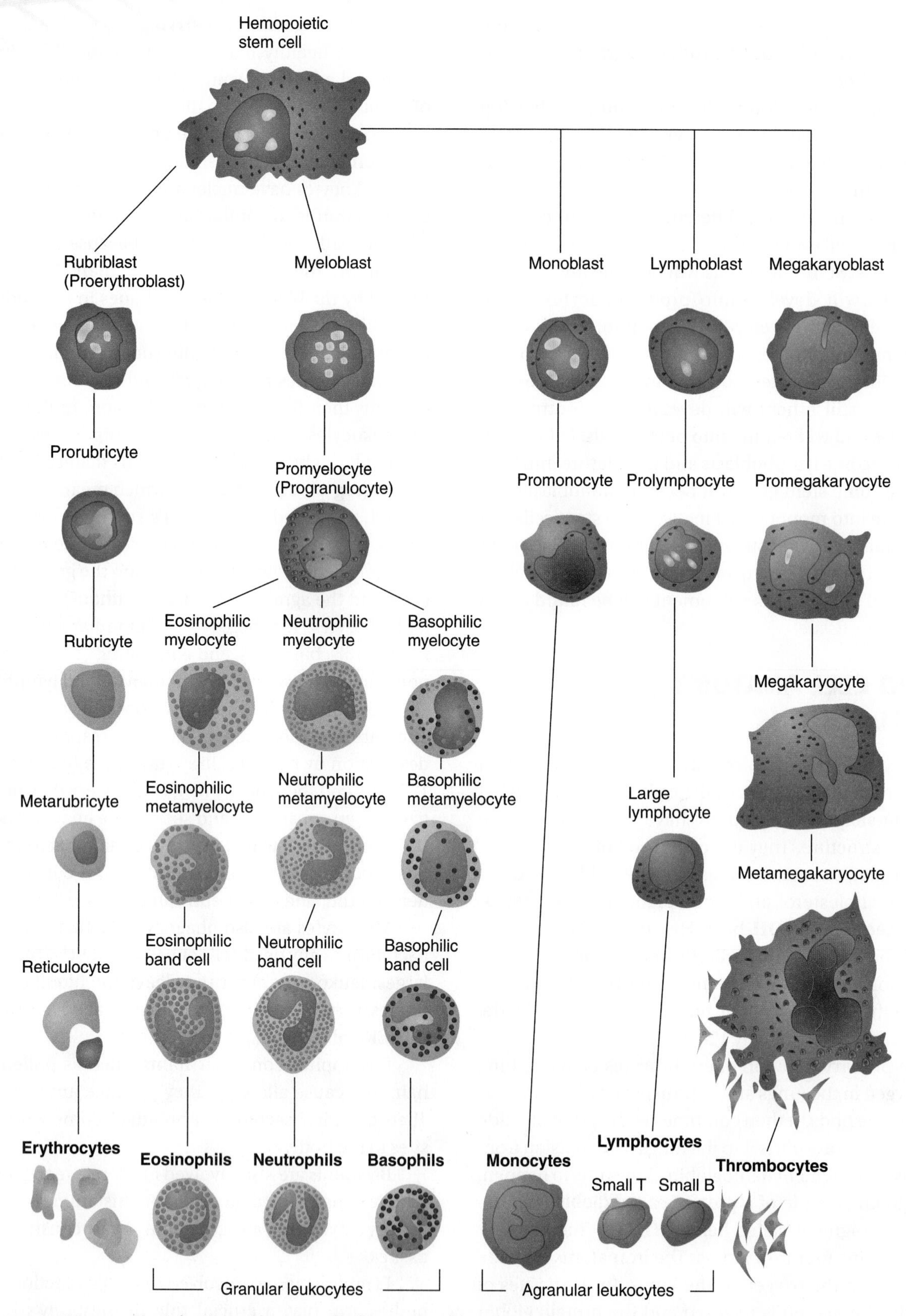

FIGURE 13-3. Hematopoiesis begins with stem cells (hematocytoblasts) in red bone marrow. There are numerous stages of development before the cell becomes a mature blood cell.

Chapter 15. They are involved in controlling cancer cells, destroying microorganisms and parasites, and rejecting foreign tissue implants.

Leukocytes are far less numerous than RBCs, averaging from 5000 to 9000 per mm^3 of blood. They can phagocytize only a certain number of substances before these materials interfere with the leukocyte's normal metabolic activity. Therefore, their life span is quite short. In a healthy body, some WBCs will live only a few days. During infections, they may live for only a few hours.

Thrombocytes or platelets are disk-shaped cellular fragments with a nucleus. They range in size from 2 to 4 micrometers in diameter. They prevent fluid loss when blood vessels are damaged by initiating a chain of reactions that result in blood clotting. They have a life span of about a week. They are produced in red bone marrow from large **megakaryocytes** (**meg**-ah-**KAIR**-ee-oh-sightz). Refer to Figure 13-4 for the life span and functions of blood cells.

*Study***WARE**™ Connection

Watch an animation about blood on your StudyWARE™ CD-ROM.

THE CLOTTING MECHANISM

When we injure ourselves through a fall or scrape, blood vessels are damaged and blood flows into tissues and can be lost from the body. Fortunately, the body has a mechanism to stop the loss of blood and repair the damaged blood vessels and tissues. The clotting mechanism is a process that the body uses to stop the loss of blood.

When small blood vessels are damaged, smooth muscles in the vessel's walls contract. This can stop blood loss. When larger vessels are damaged, the constriction of the smooth muscles in the vessel walls only slows down blood loss and the clotting mechanism takes over. A cut in a blood vessel causes the smooth walls of the vessel to become rough and irregular. Clotting or coagulation is a complex process that proceeds in three stages (Figure 13-5).

In the first stage, the roughened surface of the cut vessel causes the platelets or thrombocytes to aggregate, or clump together, at the site of injury. The damaged tissues release **thromboplastin** (**throm**-boh-**PLAST**-in). The thromboplastin causes a series of reactions that result in the production of prothrombin activator. These activities require the presence of calcium ions and certain proteins and phospholipids.

HEALTH ALERT

MAINTAINING CIRCULATION

Diet is important to maintaining healthy red blood cells. Since hemoglobin contains iron atoms, iron in our diet helps the RBCs carry more oxygen. Moderate amounts of red meat and liver are good sources of iron in our diet. At the same time, we need to maintain heart-healthy diets by eating the right kinds of meats. To maintain good blood circulation, we need to have low blood cholesterol levels. Low-fat meats, such as fish and poultry, and increased fiber in our diet from fresh fruits and vegetables help lower blood cholesterol levels. High cholesterol levels can lead to arterial plaque formation, which blocks blood flow by causing clots to form in unwanted places.

There are activities that can lead to preventing unwanted clotting. Regular exercise increases blood plasma volume, thus keeping the blood "thin." This results in less fibrinogen and RBCs per volume of blood, reducing the risk of blood clotting. People who exercise regularly are at low risk of heart attacks. Sedentary lifestyles produce the opposite effects. Based on this fact, patients are encouraged to walk very soon after operations to prevent the possibility of blood clot formation. If you smoke, try to quit. Smoking increases fibrinogen levels in the blood, thus increasing the possibility of unwanted clotting. Moderate alcohol intake is also associated with reduced heart disease possibilities. One or two drinks per day or one glass of red wine can help maintain a healthy heart and good blood circulation in many individuals. Remember, a combination of a healthy diet and exercise helps maintain good blood circulation. ■

Blood cell	Life span in blood	Function
Erythrocyte	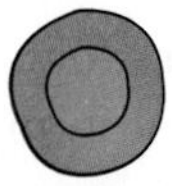 120 days	O_2 and CO_2 transport
Neutrophil	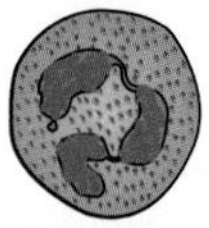 7–12 hours	Phagocytosis
Eosinophil	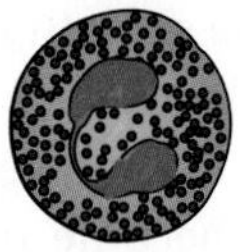 Unknown	Defense against allergens and worms
Basophil	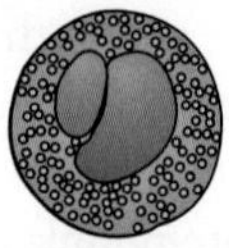 Unknown	Inflammatory response
Monocyte	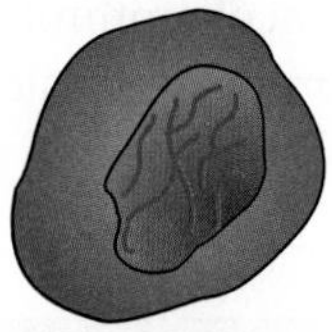 3 days	Immune surveillance (precursor of tissue macrophage) Phagocytosis
B Lymphocyte	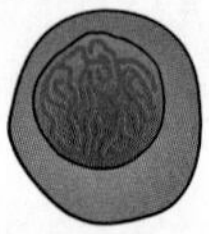 Unknown	Antibody production (precursor of plasma cells)
T Lymphocyte	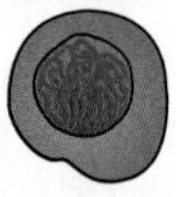 Unknown	Cellular immune response
Platelets	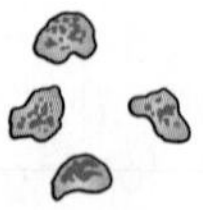 7–8 days	Blood clotting

FIGURE 13-4. The life span and functions of blood cells.

In the second stage, **prothrombin** (proh-**THROM**-bin), a plasma protein produced by the liver, is converted into **thrombin**. In the presence of the calcium ions, prothrombin activator converts the prothrombin into thrombin.

In the third stage, another plasma protein, soluble fibrinogen, is converted into insoluble **fibrin.** It is the thrombin that catalyzes the reaction that fragments fibrinogen into fibrin. Fibrin forms long threads that act like a fish net at the site of injury. The fibrin forms what

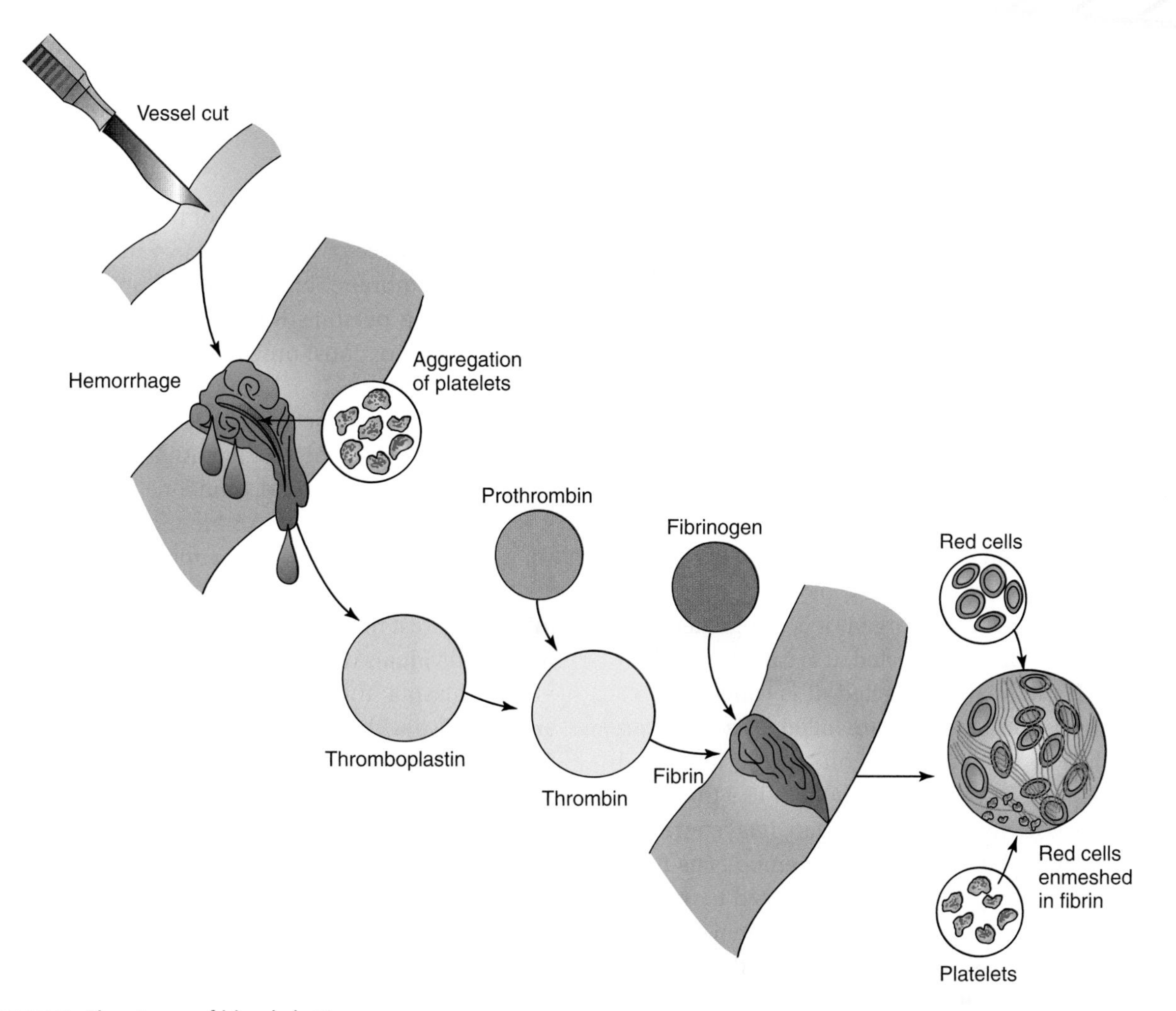

FIGURE 13-5. The stages of blood clotting.

Table 13-1 The Chemical Reactions in Blood Clotting

Reaction
1. Injured tissue $\xrightarrow{Ca^{+}}$ thromboplastin $\xrightarrow{Ca^{+}}$ prothrombin activator
2. Prothrombin $\xrightarrow{\text{Prothrombin activator} + Ca^{+}}$ thrombin
3. Soluble fibrinogen $\xrightarrow{\text{thrombin}}$ fibrin threads

we call the **clot**. As the clot forms, blood cells and platelets get enmeshed in the fibrin threads and the wound stops bleeding. Clot retraction or **syneresis** (sih-**NER**-eh-sis) is the tightening of the fibrin clot in such a way that the ruptured area of the blood vessel gets smaller and smaller, thus decreasing the hemorrhage. The clear yellowish liquid that is seen after the clot forms is called serum. Serum is blood plasma without the clotting factors. Now that the hemorrhage is stopped, blood vessel tissues repair themselves by mitotic cellular division. Once the tissue is repaired, **fibrinolysis** (**fi**-brih-**NOL**-ih-sis) or dissolution of the blood clot occurs. This is caused by a plasma protein that digests the fibrin threads and other proteins associated with the formation of the clot. Table 13-1 summarizes the chemical reactions in blood clotting.

Occasionally, unwanted clotting may occur in an undamaged blood vessel. This is brought about by a cholesterol-containing mass called **plaque** (**PLAK**) that adheres to the smooth walls of blood vessels. This results in a rough surface that is ideal for the adhesion of the platelets, thus starting the clotting mechanism. Too much cholesterol in the diet from eating too much fatty food contributes to the formation of these masses. Clotting in such an unbroken vessel is called **thrombosis** (throm-**BOH**-sis) and the clot itself is called a **thrombus** (**THROM**-bus). A thrombus may dissolve. However, if it remains intact, it can damage tissues beneath it by cutting off oxygen supplies. If a piece of a blood clot dislodges and gets transported by the bloodstream, it is called an **embolus** (**EM**-boh-lus). When an embolus becomes lodged in a vessel and cuts off circulation, it is called an **embolism** (**EM**-boh-lizm). When a blood clot forms in a vessel that supplies a vital organ, it is designated in a special way. If the brain is affected, it is called a cerebral thrombosis. If the heart is affected, it is called a coronary thrombosis. If the tissues are killed, it is called an **infarction** (in-**FARK**-shun) and is often fatal. If a blood clot dislodges and travels to a vital organ like the lungs and blocks a vessel supplying that organ, it is referred to as a pulmonary embolism. To prevent embolisms from occurring after surgery, patients are expected to walk or ambulate as soon as possible so that the normal destruction of cellular and tissue debris can occur through the activities of the WBCs and phagocytosis.

THE BLOOD GROUPS

Human blood is of different types and only certain combinations of these blood types are compatible. Procedures have been developed for typing blood. This ensures that donor and recipient blood transfusions are compatible. If blood groups are mismatched, **agglutination** (ah-**gloo**-tih-**NAY**-shun) or clumping of RBCs will occur. This is called a transfusion reaction and is caused by a reaction between protein antibodies in the blood plasma and RBC surface molecules called antigens. It is just a few of the many RBC antigens that can cause a serious transfusion reaction. These are the antigens of the ABO group and the Rh group.

Agglutination of RBCs is the result of a transfusion reaction caused by mismatched blood. The individual will experience headache and difficulty breathing; the face will appear flushed, and there will be accompanying pain in the neck, chest, and lower back. The RBCs will be destroyed, their hemoglobin converted to bilirubin, which accumulates, causing jaundice or yellowing of the skin. The kidneys may fail.

The ABO Blood Group

The **ABO blood group** consists of those individuals who have the presence or absence of two major antigens on the RBC membrane, antigen A and antigen B. Due to inheritance, a person's RBCs contain only one of four antigen combinations: only A, only B, both A and B, or neither A nor B.

Based on these facts, blood is typed. Someone with only antigen A has type A blood. An individual with only antigen B has type B blood. Someone with both antigen A and antigen B has type AB blood. If, however, a person has neither antigen A nor antigen B, then that individual has type O blood. Antibodies are formed during infancy against the ABO antigens *not* present in our own RBCs. Individuals with type A blood have antibody anti-B in their plasma; those with type B blood have antibody anti-A; those with type AB blood have *neither* antibody; finally, those with type O blood have both antibody anti-A and antibody anti-B. An antibody of one type will react with an antigen of the same type and cause agglutination. Therefore, a person with type A (anti-B) blood must not receive blood of type B or AB. Likewise, a person with type B (anti-A) must not receive type A or AB blood. Similarly, a person with type O (anti-A and anti-B) must not receive type A, B, or AB blood. However, a person with type AB blood, which lacks both anti-A and anti-B antibodies, can receive a transfusion of blood of any type and is therefore known as a universal recipient. A person with type O blood lacks antigens A and B and is known as a universal donor, because type O blood could be transfused into people with any of the blood groups.

The Rh Blood Group

The **Rh blood group** was named after the *Rhesus* monkeys, the animals in which one of the eight Rh antigens or factors was first identified and studied. This was antigen D or agglutinogen D, which was later discovered in humans. If antigen D and other Rh antigens are found on the RBC membrane, the blood is Rh positive. Most Americans are Rh positive. If the RBCs lack the antigens, the blood is Rh negative. The presence or absence of the antigens is an inherited trait.

Unlike the antibodies of the ABO system, anti-Rh antibodies do not develop spontaneously. Instead, they develop only in Rh-negative persons if an Rh-negative

COMMON DISEASE, DISORDER, OR CONDITION

DISORDERS OF THE BLOOD

HEMOPHILIA

Hemophilia is a genetically inherited clotting disorder associated with the expression of a recessive gene on the X chromosome inherited from the mother and passed down to male children. During the Middle Ages, an Arab, Albucacus, first described and recognized hemophilia as an inherited disease. Queen Victoria of England passed this gene to several of her offspring who passed this on to royal families in England, Germany, Spain, and Russia. Hemophilia A is the most common form of the disease, whose main symptom is a tendency to hemorrhage following minor injuries. Other symptoms include frequent nosebleeds, hematomas in muscles, and bloody urine.

LEUKEMIA

Leukemia is a type of cancer in which there is abnormal production of WBCs. These cells lack normal immunologic capabilities, so that individuals with the disease are susceptible to opportunistic infections. The excess production of leukocytes interferes with normal RBC and platelet formation, which results in anemia and excessive bleeding from minor injuries.

ANEMIA

Anemia can be caused by four factors: decrease in the normal number of erythrocytes; a decrease of normal amounts of hemoglobin in the RBCs; a deficiency of normal hemoglobin; or production of abnormal hemoglobin. Anemia reduces the amount of oxygen that RBCs can transport resulting in a lack of energy, shortness of breath on minor exertions, listlessness, pale skin, and a general feeling of fatigue. There are a number of types of anemia.

SICKLE-CELL ANEMIA

Sickle-cell anemia is a hereditary disease found mostly in African Americans and individuals of southern European ancestry. The erythrocytes have an abnormal sickle shape with abnormal hemoglobin and cannot carry sufficient oxygen. Variations include severity resulting in death before age 30, to mild cases with no symptoms.

IRON-DEFICIENCY ANEMIA

Iron-deficiency anemia results from vitamin B_{12} deficiency, nutritional deficiencies, or excessive iron loss from the body, resulting in lower than normal erythrocyte production.

HEMOLYTIC ANEMIA

Hemolytic anemia is an inherited condition in which erythrocytes rupture or are destroyed at a faster rate than normal. It can also be caused by drugs, autoimmune diseases, or snake venom.

THALASSEMIA

Thalassemia (thal-ah-**SEE**-mee-ah) is a hereditary disease found in people of African, Mediterranean, and Asian background. Hemoglobin production is suppressed and death can occur by the age of 20. Mild cases produce a mild anemia.

SEPTICEMIA

Septicemia, also known as blood poisoning, is caused by an infection of microorganisms and their toxins in the blood. These toxims cause a decrease in blood pressure, referred to as septic shock, and can lead to death.

(continues)

COMMON DISEASE, DISORDER, OR CONDITION

DISORDERS OF THE BLOOD (continued)

MALARIA

Malaria is caused by the injection of a protozoan, *Plasmodium*, by a female *Anopheles* mosquito. The microorganism spends part of its life cycle in the erythrocytes, eventually destroying them. Chills and fever are produced by toxins released when the RBCs rupture.

INFECTIOUS MONONUCLEOSIS

Infectious mononucleosis (in-**FEK**-shus **MON**-oh-**NOO**-klee-**OH**-sis), also known as mono, is caused by the Epstein-Barr virus, which infects B lymphocytes and the salivary glands. The virus alters the lymphocytes, causing the immune system to destroy them. Symptoms include sore throat, swollen lymph nodes, and fever.

THROMBOCYTOPENIA

Thrombocytopenia (**THROM**-bo-**SIGH**-to-**PEA**-knee-ah) is a serious decrease in the number of thrombocytes or blood platelets. Platelets can be destroyed by an autoimmune disease, a genetic disorder that produces less platelets, radiation and drug therapies, or infections. The condition results in long-term bleeding of capillaries, and other small vessels like arterioles and venules. It's the most common cause of bleeding disorders.

ERYTHROCYTOSIS

Erythrocytosis (ee-**RITH**-ro-sigh-**TOE**-sis) is caused by an excessive amount of red blood cells. It is also called polycythemia. It can be caused by a defect in stem cell production, decreased plasma volume due to dehydration, or by chronic exposure to high altitudes. It can result in reduced blood flow, clogging of the capillaries, and increased thickness of the blood. These conditions can lead to hypertension or high blood pressure.

CARBON MONOXIDE POISONING

Carbon monoxide poisoning can develop from working in places that accumulate large amounts of exhaust from combustion engines such as automobile engines. Thus, employment in parking garages, tunnels, and tollbooths can lead to this condition. In addition, a defective furnace in a home can produce excessive amounts of carbon monoxide gas (CO). This is why we utilize carbon monoxide detectors in our homes. The CO binds to the iron atoms in the hemoglobin of a red blood cell producing carboxyhemoglobin. This prevents the red blood cell from binding to oxygen gas and thus it does not transport oxygen to tissue cells and death can result.

person receives a blood transfusion of Rh-positive blood. Shortly after receiving the mismatched Rh-positive blood, the Rh-negative person begins to produce anti-Rh antibodies against the foreign blood. This initial mismatch has no immediate serious consequences because it takes the body time to react and produce antibodies. However, if a second mismatched transfusion occurs, the patient's antibodies will now attack and rupture the Rh-positive blood donor's RBCs and they will agglutinate.

A similar problem occurs when an Rh-negative mother carries an Rh-positive baby. This is the case in which the mother is Rh-negative and the father is Rh-positive. The first pregnancy is usually normal. Because the mother may now be sensitized to the Rh-positive antigens, she will produce anti-Rh positive antibodies in the future. These antibodies will cross through the placenta and destroy the child's RBCs, causing a condition known as **erythroblastosis fetalis** (eh-**RITH**-roh-blass-**TOH**-sis fee-**TAL**-is) or hemolytic disease of the newborn. The baby will be anemic and suffer brain damage due to lack of oxygen to nerve cells. Death may result. However, today this condition is rare. An Rh-negative woman can be given a drug called RhoGAM. This is, in actuality, anti-Rh antibodies that will bind to any Rh-positive fetal cells and shield them, thus protecting any of the child's RBCs that might contact the mother's cells. This sensitizes her immune system.

As the fetus develops in the womb, blood cell formation occurs in the spleen, liver, and yolk sac, but, by the third trimester, hematopoiesis is occurring in the red marrow and myeloid tissue and continues there throughout life. In addition, lymphoid tissue assists red bone marrow in providing lymphocytes and monocytes during one's lifetime.

Vitamin K plays a major role in many of the factors involved in blood clotting. We receive about half of our supply of vitamin K in our diet, while the rest is produced from friendly bacteria that inhabit our large intestine. Patients on prolonged antibacterial therapy may have these helpful bacteria destroyed and may develop bleeding problems. These individuals may need to supplement their nutrient intake with vitamin K pills.

Individuals who smoke cigarettes increase their intake of carbon monoxide gas, found in cigarette smoke. The carbon monoxide binds to the iron in the hemoglobin molecules of red blood cells to form carboxyhemoglobin. This interferes with the blood's ability to transport oxygen. Smokers can have 5% to 15% carboxyhemoglobin in their blood. This may cause some mental impairment.

During aging, a number of blood diseases may also develop, such as clotting disorders and anemias. The development of chronic leukemias in old age is a result of the decreased efficiency of the immune system. Many of the age-related disorders of the blood are intimately associated with problems of the blood vessels, heart, and immune system. ■

Career FOCUS

These are careers that are available to individuals interested in the blood.

- **Blood bank technologists** are allied health professionals who are responsible for the blood typing of blood donors, newborns, and expectant mothers to detect the presence of ABO antigens and the Rh factors. These individuals can also draw blood, support nurses and physicians in blood transfusions, and investigate blood abnormalities such as hemolytic anemias.
- **Hematologists** are medical specialists whose training is in the field of blood and the blood-forming tissues. Such individuals usually work in a laboratory, which is part of a medical center.
- **Infectious disease specialists** are physicians with advanced training in communicable infectious diseases, some of which are transmitted by the blood-sucking bite of an arthropod like a mosquito. Mosquitoes transmit pathogens that cause malaria and yellow fever into the bloodstream of the insect's victim. Many other infectious diseases such as AIDS and certain sexually transmitted diseases are transmitted by blood and bodily fluid contact.

SUMMARY OUTLINE

INTRODUCTION

1. Blood is specialized connective tissue consisting of a fluid part, called plasma, and the formed blood cells.
2. The formed cells of blood include the red blood cells (RBCs) or erythrocytes, the white blood cells (WBCs) or leukocytes, and the platelets or thrombocytes.

THE FUNCTIONS OF BLOOD

1. Transports oxygen from the lungs to the cells of the body.
2. Transports carbon dioxide from the cells to the lungs for excretion.
3. Transports nutrients, ions, and water from the digestive tract to cells.
4. Transports waste products from cells to kidneys and sweat glands.
5. Transports hormones to target organs and enzymes to body cells.
6. Regulates body pH through its buffers and amino acids in its plasma.
7. Helps regulate normal body temperature and the water content of cells.
8. Helps prevent fluid loss through the clotting mechanism.
9. Protects against foreign microbes and toxins through its combat cells or leukocytes.

THE CLASSIFICATION OF BLOOD CELLS AND THE COMPOSITION OF PLASMA

1. Blood is composed of the following elements:
   A. Erythrocytes or red blood cells
   B. Leukocytes or white blood cells, which are subdivided into
      1. Granular leukocytes (three types)
         a. Neutrophils
         b. Eosinophils
         c. Basophils
      2. Agranular or nongranular leukocytes (two types)
         a. Monocytes
         b. Lymphocytes
   C. Thrombocytes or platelets
2. Plasma is the fluid component of blood; 91% is water.
3. 7 percent of plasma are the proteins: albumin, globulin, and fibrinogen.
4. 2 percent of plasma are solutes: ions, nutrients, waste products, gases, enzymes, and hormones.

THE FORMATION OF BLOOD CELLS: HEMATOPOIESIS

1. Hematopoiesis occurs in red bone marrow or myeloid tissue where all blood cells are produced.
2. Lymphocytes and monocytes are also produced by lymph nodes, the spleen, and the tonsils.
3. Blood cells develop from undifferentiated mesenchymal cells called stem cells or hematocytoblasts.

THE BLOOD CELL ANATOMY AND FUNCTIONS

1. Erythrocytes appear as biconcave disks without a nucleus. They consist of a protein network or stroma and the red pigment hemoglobin.
2. Hemoglobin is made of the pigment heme, which has four iron atoms that combine with oxygen gas in the lungs, and the protein globin, which combines with carbon dioxide in tissues.
3. Leukocytes have nuclei and do not have hemoglobin. The two categories of leukocytes are the granular leukocytes and the agranular or nongranular leukocytes. They leave the blood and move into body tissues where they combat infection and inflammation.
4. Neutrophils are the most common granular leukocytes. They respond to tissue destruction from bacteria by phagocytizing foreign substances and destroying bacteria via their enzyme lysozyme.
5. Monocytes are nongranular leukocytes that phagocytize bacteria and cellular debris. In tissues, they are called macrophages because they are fairly large, about 18 μm wide.
6. Eosinophils are granular leukocytes that combat irritants that cause allergies and parasitic worms. They produce antihistamines.

7. Basophils are granular leukocytes that are also involved in allergic reactions. They produce heparin, histamine, and serotonin.
8. Lymphocytes are nongranular leukocytes that produce antibodies and are involved in the immune response. Two common lymphocytes are the T lymphocytes and the B lymphocytes.
9. Thrombocytes or platelets are very small disk-shaped, cellular fragments with a nucleus. They cause the clotting mechanism.

THE CLOTTING MECHANISM

1. A ruptured blood vessel attracts thrombocytes to the site of injury.
2. The damaged tissues release thromboplastin.
3. Thromboplastin, with the assistance of calcium ions, proteins, and phospholipids, causes the production of prothrombin activator.
4. Prothrombin activator with the assistance of calcium ions causes prothrombin, a plasma protein, to be converted into thrombin.
5. Thrombin causes soluble fibrinogen, another plasma protein, to be converted into insoluble fibrin.
6. Fibrin forms the threads of the clot, which enmesh the blood cells and platelets seeping from the wound.
7. Tightening of the clot (clot retraction) or syneresis occurs and hemorrhaging ceases.
8. After tissues are repaired, dissolution of the clot or fibrinolysis occurs.
9. Unwanted clotting, caused by masses of cholesterol known as plaque, in an unbroken blood vessel is known as a thrombosis. The clot is called a thrombus.
10. A piece of a blood clot, transported by the bloodstream, can get lodged in a vessel and block off circulation. It is called an embolus and the condition is called an embolism.

THE BLOOD GROUPS

1. The different types of human blood groups must be matched in a blood transfusion to prevent agglutination of RBCs.
2. Agglutination is caused by a reaction between protein antibodies in the blood plasma and surface antigens on the red blood cell membrane.
3. Agglutination of RBCs will cause headache, breathing difficulties, pain, and jaundice. Kidneys may fail.

The ABO Blood Group

1. Type A blood individuals have antibody anti-B in their blood plasma. Type B blood individuals have antibody anti-A. Type AB blood individuals have no antibodies. Type O blood individuals have antibody anti-A and antibody anti-B.
2. Type AB individuals are known as universal recipients because they can receive any blood type in a transfusion.
3. Type O individuals are known as universal donors because they have no antigens and their blood can be transfused into any blood group.

The Rh Blood Group

1. This blood group was named after the *Rhesus* monkey in which one of the eight Rh antigens was discovered.
2. The most important Rh antigen is antigen D. People with this antigen are Rh positive; those without it are Rh negative.
3. Most Americans are Rh positive.
4. Anti-Rh antibodies develop only after exposure to Rh-positive blood in an Rh-negative individual. Therefore, second transfusions can cause agglutination.
5. Rh-negative mothers carrying an Rh-positive baby can be treated with a drug called RhoGAM to protect the developing fetus.

REVIEW QUESTIONS

1. Classify blood cells into three major categories with subdivisions when appropriate.
2. List the functions of blood.
*3. How does the anatomy of the red blood cell allow it to carry oxygen and carbon dioxide gas?
4. Name the four categories of ABO blood groups.
*5. Why must blood groups be matched during a blood transfusion?
6. Explain the clotting mechanism.

***Critical Thinking Questions**

MATCHING

Place the most appropriate number in the blank provided.

_____ Hematopoiesis	1.	Macrophage
_____ Heme	2.	Destroys bacteria
_____ Globin	3.	Antibodies
_____ Syneresis	4.	Combat irritants/allergies
_____ Monocyte	5.	Red bone marrow
_____ Lysozyme	6.	Clotting
_____ Eosinophils	7.	Carries carbon dioxide
_____ Thrombocytes	8.	Universal donor
_____ Type O blood	9.	Iron atoms bond with oxygen
_____ Lymphocytes	10.	Heparin
	11.	Clot retraction
	12.	*Rhesus* monkey

Search the Internet with key words from the chapter to discover additional information and interactive exercises. Key words might include blood groups, blood clotting process, blood cell anatomy, or disorders of the blood.

CASE STUDY

Anna, a 19-year-old college student, volunteers to donate blood to a blood bank to help a friend injured in a car accident. Following blood tests, Anna learns that she cannot donate blood because her erythrocyte count and hemoglobin levels are too low. The blood bank recommends that Anna schedule an appointment with her primary health care provider for a follow-up examination and further blood tests.

Questions

1. What condition might Anna have that is characterized by a low erythrocyte count and hemoglobin level?
2. What are the major functions of erythrocytes, and what vital pigment do erythrocytes contain?
3. Why do women tend to have a lower erythrocyte count and hemoglobin level than men?
4. What general symptoms do individuals with this condition tend to develop?
5. How does diet play a role in preventing and treating this problem?

*Study*WARE™ Connection

Take a practice quiz or play interactive games that reinforce the content in this chapter on your StudyWARE™ CD-ROM

Study Guide Practice

Go to your **Study Guide** for more practice questions, labeling and coloring exercises, and crossword puzzles to help you learn the content in this chapter.

LABORATORY EXERCISE: THE BLOOD

Materials needed: A blood-typing kit, commercially prepared microscope slides of human blood stained with Wright's stain

1. Your instructor will supply you with a blood-typing kit to type your own blood. Carolina Biological Supply Company, Burlington, NC, has an Eldoncard System for blood typing that has a number of choices using or not using students' own blood.
2. Examine under your microscope a commercially prepared microscope slide stained with Wright's stain. Find and draw the seven types of blood cells: erythrocytes, monocytes, lymphocytes, eosinophils, basophils, neutrophils, and thrombocytes.

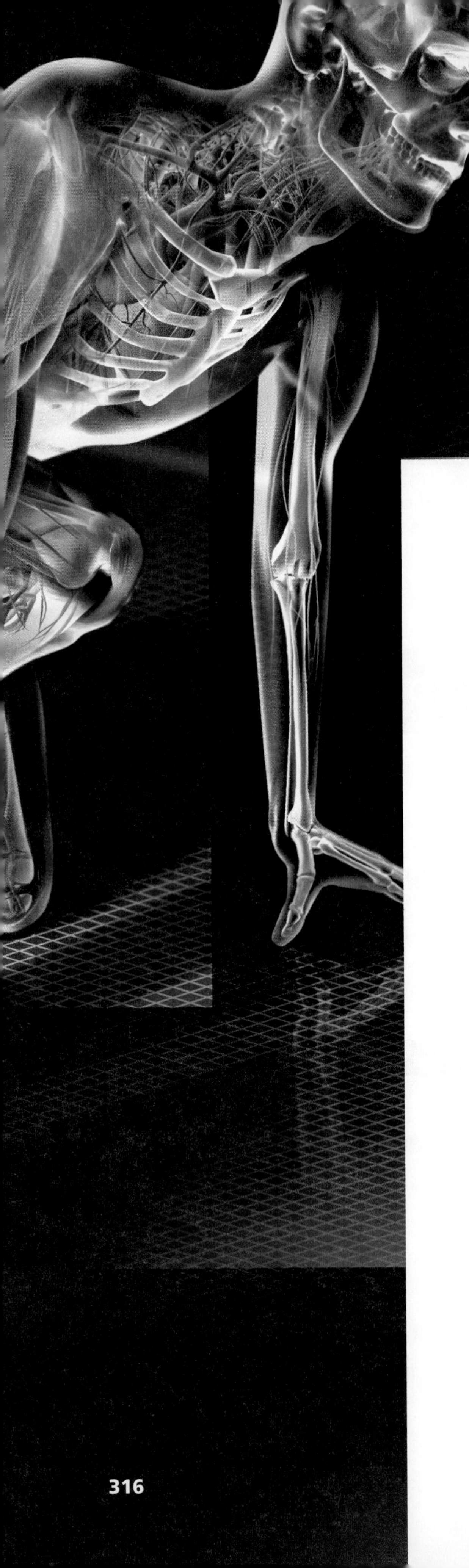

The Cardiovascular System

CHAPTER OBJECTIVES

After studying this chapter, you should be able to:

1. Describe how the heart is positioned in the thoracic cavity.
2. List and describe the layers of the heart wall.
3. Name the chambers of the heart and their valves.
4. Name the major vessels that enter and exit the heart.
5. Describe blood flow through the heart.
6. Explain how the conduction system of the heart controls proper blood flow.
7. Describe the stages of a cardiac cycle.
8. Compare the anatomy of a vein, artery, and capillary.
9. Name the major blood circulatory routes.

14

KEY TERMS

INTRODUCTION

The cardiovascular system consists of the heart and thousands of miles of blood vessels. Refer to Concept Maps 14-1 and 14-2: The Heart. The heart is the muscular pump that forces the blood through a system of vessels made of arteries, veins, and capillaries. These vessels transport the blood, which carries oxygen, nutrients, hormones, enzymes, and cellular waste to and from the trillions of cells that make up our bodies. These cells need oxygen and nutrients from digested food to make the chemical energy (ATP) that allows the cells to function properly. Enzymes assist in the chemical reactions inside the cells, and waste products from these reactions must be transported by the cardiovascular system to sites like the lungs and kidneys for excretion from the body. The force to transport the blood is provided by the cardiac muscle that makes up the bulk of the heart.

The function of transportation of blood by the cardiovascular system occurs 24 hours a day 7 days per week, nonstop for 70, 80, or 90 years or more. This is possible because the heart beats about 72 times a minute. It is a

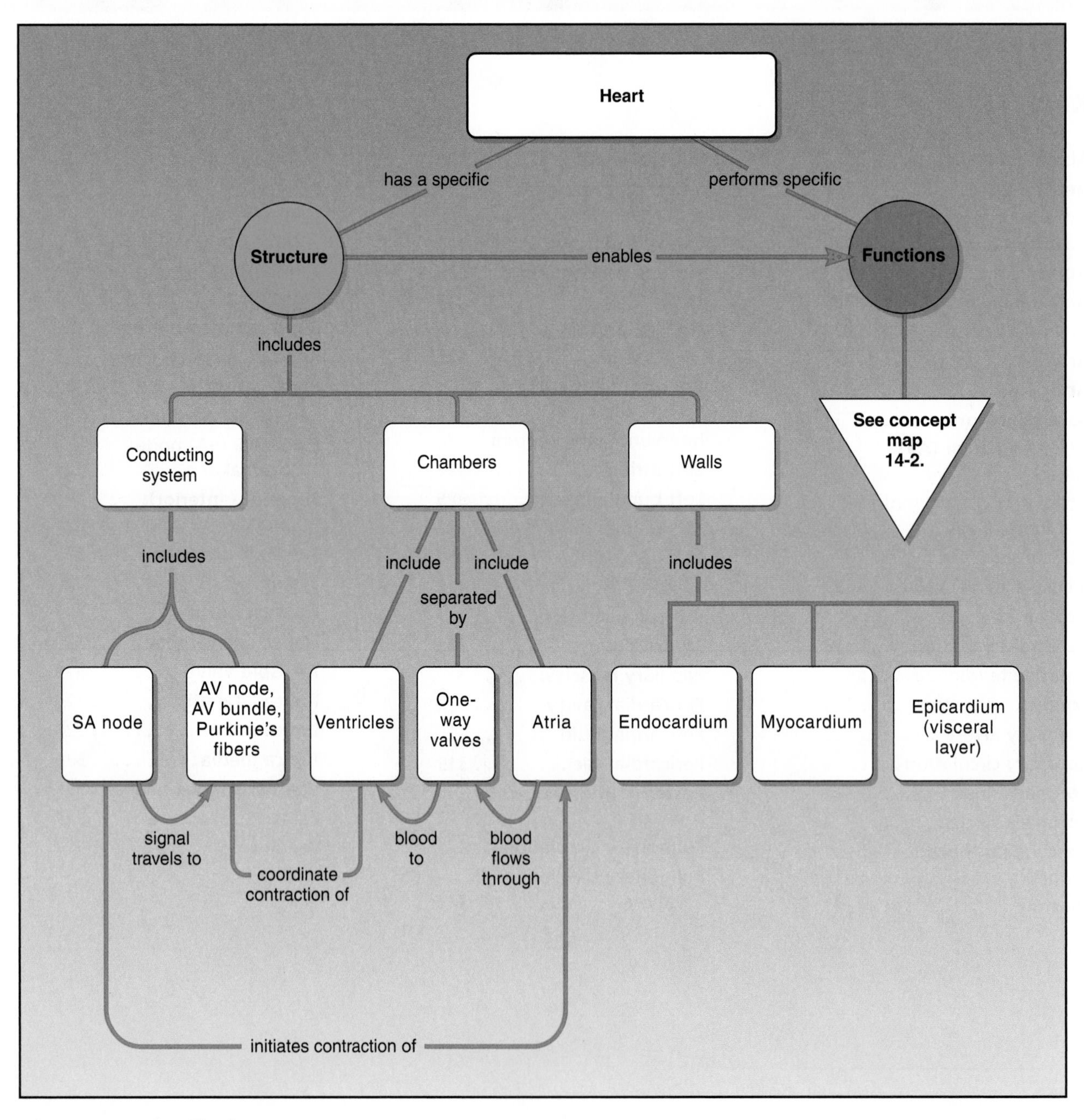

CONCEPT MAP 14-1. The heart.

unique organ in that it can contract, rest, and immediately contract again during our entire lifetime. The system has a series of valves that prevent blood from backflowing through the blood vessels. It is our pumping heart that helps keep us alive and healthy.

THE ANATOMY OF THE HEART

The **heart** is positioned obliquely between the lungs in the mediastinum (Figure 14-1). About two-thirds of its bulk lies to the left side of the midline of the body. It is shaped like a blunt cone. It is about the size of a closed fist. It is approximately 5 inches long (12 cm), 3.5 inches wide at its broadest point (9 cm), and 2.5 inches thick (6 cm).

It is enclosed in a loose fitting serous membrane known as the **pericardial** (**pair**-ih-**CAR**-dee-al) **sac**, which can also be referred to as the parietal pericardium. The pericardial sac is made up of two layers (Figure 14-2). The outermost layer is the fibrous layer or **fibrous pericardium** (**FYE**-bruss **pair**-ih-**CAR**-dee-um). It is made of tough fibrous connective tissue and connects to the large blood vessels that enter and leave the heart (the venae cavae, aorta, pulmonary arteries, and veins), to the diaphragm muscle, and to the inside of the sternal wall of the thorax. It prevents overdistention of the pumping heart by acting as a tough protective membrane surrounding the heart. It also anchors the heart in the mediastinum. The innermost layer of the pericardial sac is the serous layer or **serous pericardium**. This layer is thin and delicate. It is continuous with the outermost layer of the wall of the heart, called the epicardium, at the base of the heart. It is also continuous with the large blood vessels of the heart and is also known as the parietal layer of the pericardial sac.

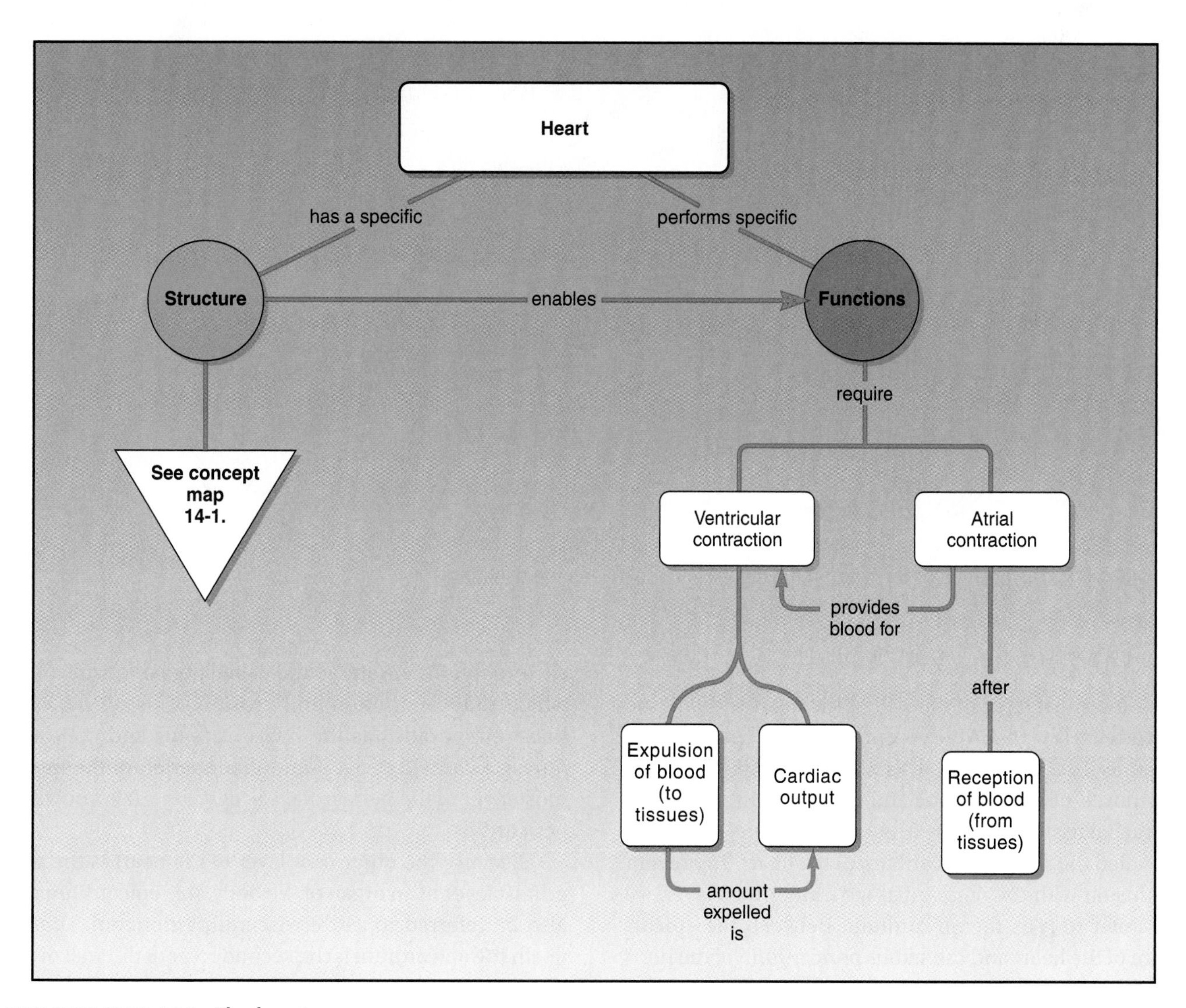

CONCEPT MAP 14-2. The heart.

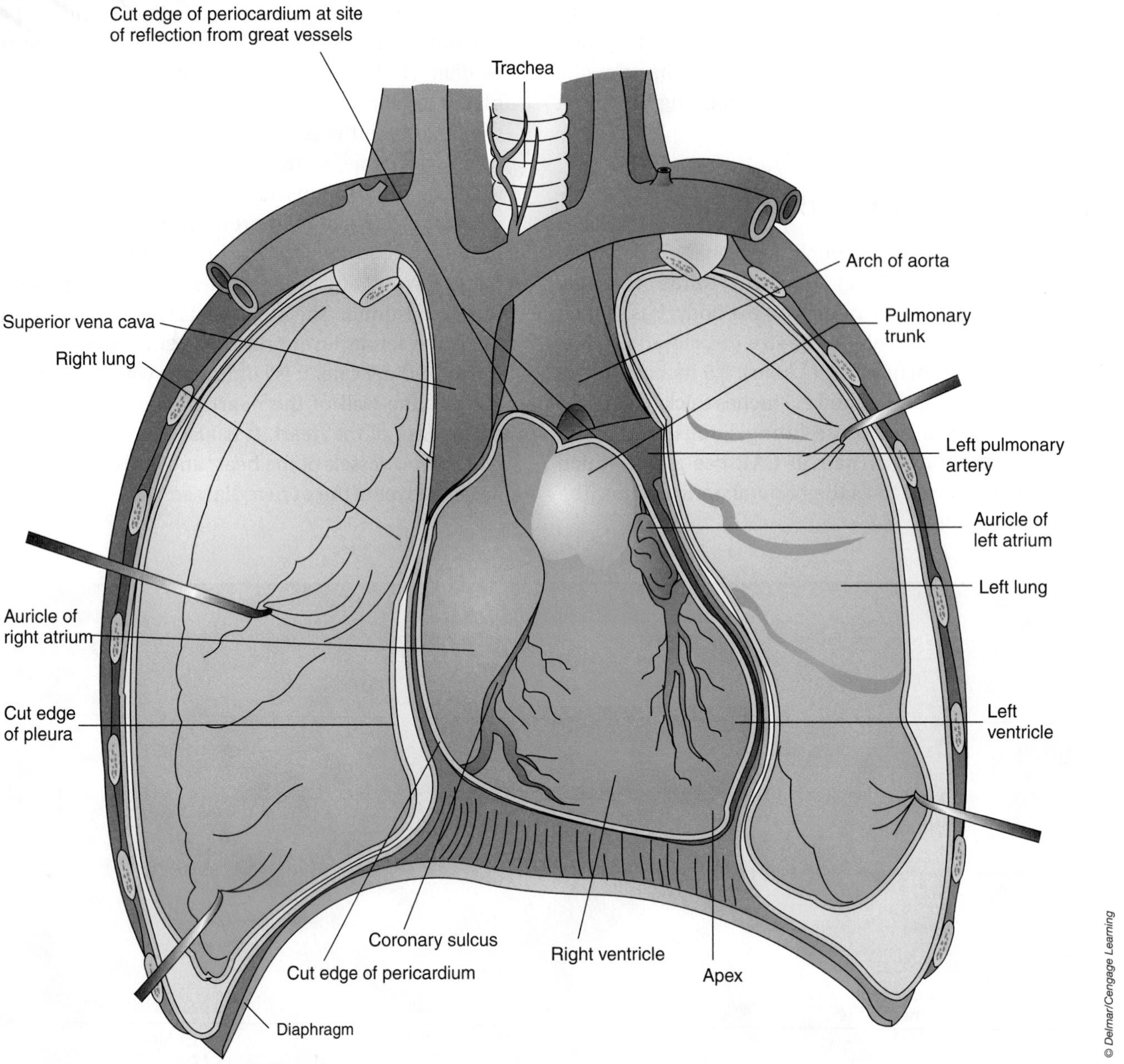

FIGURE 14-1. The position of the heart in the mediastinum between the two lungs.

The Layers of the Heart Wall

The outermost layer of the wall of the heart is called the **epicardium** (**ep**-ih-**CAR**-dee-um) or **visceral pericardium** (Figures 14-2 and 14-3). It is a thin, transparent layer composed of serous tissue and mesothelium (a type of epithelial tissue). Because of its serous nature, it can also be called the serous pericardium of the heart. To prevent confusion with the pericardial sac's innermost layer, we will refer to it as the epicardium. Between the epicardium of the heart and the serous pericardium of the pericardial sac is a space called the **pericardial cavity**. This cavity contains a watery fluid called the **pericardial fluid**, which reduces friction and erosion of tissue between these membranes as the heart expands and contracts during a cardiac cycle. If an inflammation of the innermost layer of the pericardial sac develops, it is known as pericarditis.

Because the outermost layer of the heart is the outermost layer of an organ of the body, this epicardium can also be referred to as the visceral peritoneum. Underneath the epicardium is the second layer of the wall of the heart. This makes up the bulk of the heart and is called

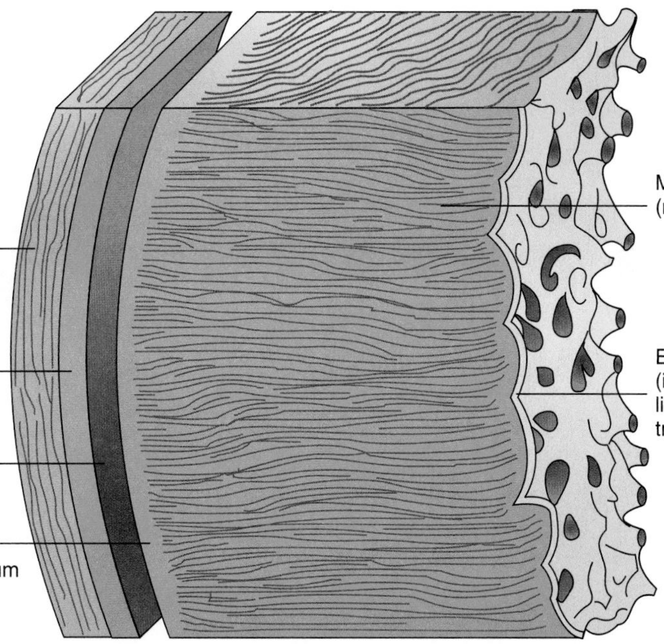

PERICARDIUM
Fibrous pericardium
Serous pericardium (parietal layer)
Space
Epicardium (visceral layer) (Serous pericardium of the heart)
MYOCARDIUM (muscle layer)
ENDOCARDIUM (inner endothelial lining covering trabeculae)

FIGURE 14-2. The layers of the pericardial sac (left) and the layers of the walls of the heart (right).

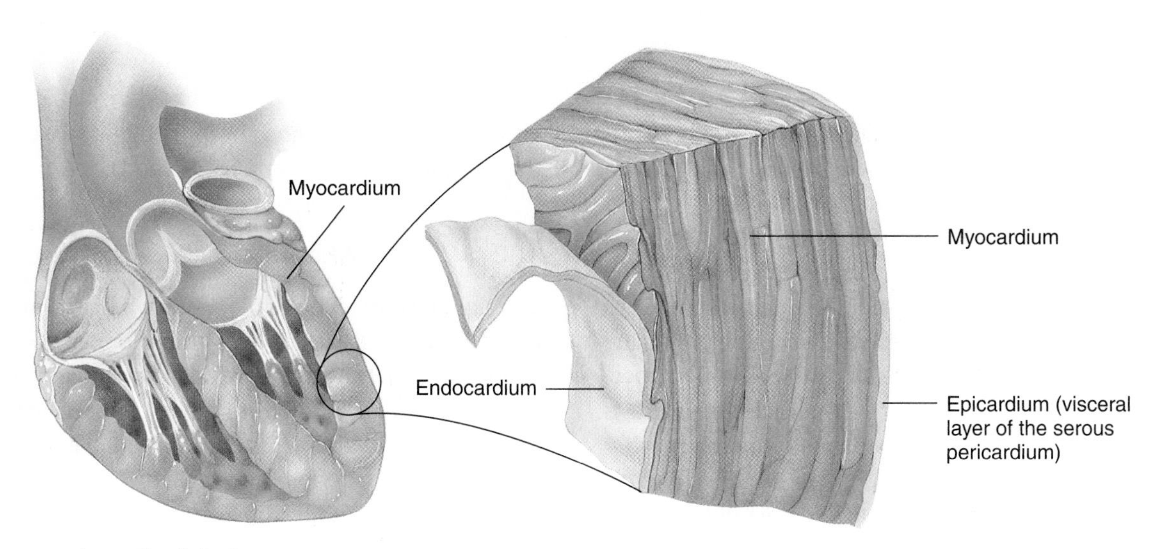

FIGURE 14-3. The walls of the heart.

the **myocardium** (my-oh-**CAR**-dee-um). This is the layer of cardiac muscle tissue. Its cells or fibers are involuntary, striated, and branched. Refer to Chapter 9 for a review of cardiac tissue anatomy. The tissue of this layer is arranged in interlacing bundles and is the layer responsible for contraction of the heart.

The third or innermost layer of the wall of the heart is called the **endocardium** (**en**-doh-**CAR**-dee-um). It is a thin layer of endothelium (a type of epithelial tissue) that overlies a thin layer of connective tissue penetrated by tiny blood vessels and bundles of smooth muscle. It acts as a lining for the myocardium. It covers the valves of the heart and the chordae tendineae of the valves.

The Chambers of the Heart

The inside of the heart is divided into four chambers that receive blood from various parts of the body (Figure 14-4).

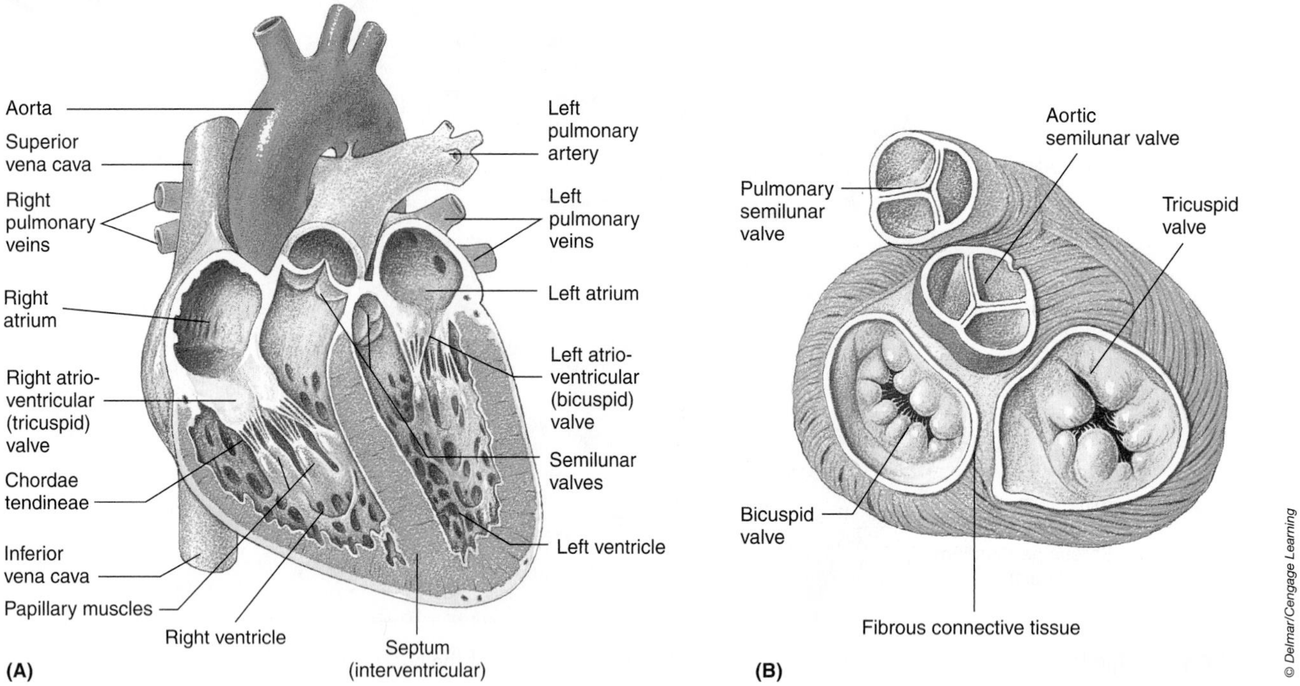

FIGURE 14-4. The interior of the heart. (A) Cross section of the heart showing the chambers, valves, septum, chordae tendineae, and vessels. (B) Superior view of valves.

The two upper chambers are called the **right atrium** (**RITE AY**-tree-um) and the **left atrium**. Each atrium has an external appendage called an **auricle** (**AW**-rih-kl), named because of its similarity to the ear of a dog (see Figure 14-1). The auricle increases the volume of the atrium. The lining of each atrium is smooth, except for the anterior atrial walls and the lining of the two auricles, which contain projecting muscle bundles called the **musculi** (**MUSS**-kyoo-lye) **pectinati** (pek-tin-**NAY**-tye) that give the auricles their rough appearance. The two atria are separated from each other by an internal interatrial septum.

The lower two chambers are called the **right ventricle** (**RITE VEN**-trih-kl) and the **left ventricle**. The two ventricles are separated from one another by an internal **interventricular septum**. The irregular ridges and folds of the myocardium of the ventricles are called the **trabeculae** (trah-**BEK**-yoo-lee) **carneae** (**KAR**-neh-ee). The muscle tissue of the atria and ventricles is separated by connective tissue that also forms the valves. This connective tissue divides the myocardium into two separate muscle masses. Externally, a groove called the **coronary sulcus** (**KOR**-ah-nair-ee **SULL**-kus) separates the atria from the ventricles (see Figure 14-1). Two other sulci (plural) are seen externally. The **anterior interventricular sulcus** and the **posterior interventricular sulcus** separate the right and left ventricles from one another. The sulci contain a varying amount of fat and coronary blood vessels (vessels that supply the heart tissue with blood).

The Great Vessels of the Heart

The right atrium receives blood from all parts of the body except the lungs. It receives this blood through three veins. The **superior vena cava**, also known as the **anterior vena cava**, brings blood from the upper parts of the body, the head, neck, and arms. The **inferior vena cava**, also known as the **posterior vena cava**, brings blood from the lower parts of the body, the legs, and abdomen. The **coronary sinus** drains the blood from most of the vessels that supply the walls of the heart with blood. This blood in the right atrium is then squeezed into the right ventricle.

The right ventricle pumps the blood into the next major vessel, the **pulmonary trunk**, which splits into the **right pulmonary artery** and the **left pulmonary artery**. These arteries each carry the blood to a lung. In the lungs, the blood releases the carbon dioxide it has been carrying and picks up oxygen. The oxygenated blood returns to the heart via four **pulmonary veins** that empty into the left atrium. The blood is then squeezed into the left ventricle.

The left ventricle pumps the blood into the next great vessel, the **ascending aorta**. From here the aortic blood goes to the **coronary arteries** (which supply the walls of the heart with oxygenated blood), the **arch of the aorta** (which sends arteries to upper parts of the body), and the **descending thoracic aorta**, which becomes the **abdominal aorta**. These arteries transport oxygenated blood to all parts of the body.

The size of the chambers and the thickness of the chamber walls vary, due to the amount of blood received and the distance this blood must be pumped. The right atrium, which collects blood coming from all parts of the body except the lungs, is slightly larger than the left atrium, which receives only the blood coming from the lungs. The thickness of the chamber walls also varies. Ventricles have thick walls, whereas the atria are thin walled. They are assisted with pumping blood by the reduced pressure caused by the expanding ventricles as they receive the blood. The thickness of the two ventricle walls varies also. The left ventricle has walls thicker than the right ventricle (see Figure 14-4) since it must pump the oxygenated blood at high pressure through thousands of miles of blood vessels in the head, trunk, and extremities.

The Valves of the Heart

The valves of the heart are designed in such a way as to prevent blood from flowing back into the pumping chamber. There are two atrioventricular valves between the atria and their ventricles. The valve between the right atrium and the right ventricle is called the **tricuspid** (try-**KUSS**-pid) **valve** because it consists of three flaps or cusps (see Figure 14-4B). These flaps are made of fibrous connective tissue that grows out of the walls of the heart and is covered with endocardium. The pointed ends of the cusps project down into the ventricle. Cords called **chordae** (**KOR**-dee) **tendineae** (**TIN**-din-ee) connect the pointed ends of the flaps or cusps to small conical projections called the **papillary** (**PAP**-ih-layr-ee) **muscles** located on the inner surface of the ventricle (see Figure 14-4A and B).

The atrioventricular valve between the left atrium and the left ventricle is known as the **bicuspid** (bye-**KUSS**-pid) or **mitral** (**MYE**-tral) **valve**. As the name indicates, it has two cusps or flaps, whose pointed ends project down into the ventricle with the same structures as the tricuspid valve. It is the only valve in the heart with only two cusps; all others have three cusps. The two arteries that leave the heart (the ascending aorta and the pulmonary trunk) also have valves that prevent blood from flowing back into the pumping chamber. These are called the semilunar valves. The **pulmonary semilunar valve** is found in the opening where the pulmonary trunk exits the right ventricle. The **aortic semilunar valve** is found in the opening where the ascending aorta leaves the left ventricle. Both of the valves are made of three semilunar cusps that allow blood to flow only in one direction.

> **StudyWARE™ Connection**
>
> Play an interactive game labeling the structures of the heart on your StudyWARE™ CD-ROM.

BLOOD FLOW THROUGH THE HEART

As we discuss blood flow through the heart, it is easier to study its path by beginning at one point, going in a one-way direction, and ending at the same point we began. We will do this, but remember, the heart actually pumps the blood in a slightly different manner. That is, the two atria contract simultaneously, while the two ventricles relax. Then the two ventricles contract simultaneously, while the two atria relax. Then all chambers rest before the cycle begins again.

Refer to Figures 14-4A and 14-5 as we discuss blood flow. Deoxygenated blood (blood high in carbon dioxide gas) returns from the upper portions of the body through the superior or anterior vena cava and from the lower portions of the body through the inferior or posterior vena cava to the right atrium of the heart. The blood is then squeezed by contraction of the right atrium through the tricuspid valve into the right ventricle. As the right ventricle contracts, it pumps the blood through the pulmonary semilunar valve into the pulmonary trunk, which branches into the right pulmonary artery that goes to the right lung and the left pulmonary artery that goes to the left lung. In the alveoli of the lungs surrounded by capillaries, the blood loses the carbon dioxide gas and picks up oxygen. Deoxygenated blood looks dark; hence, veins are usually depicted in textbooks as blue. Oxygenated blood looks bright red; hence, arteries are usually shown in red in textbooks. The oxygenated blood returns to the left atrium of the heart through four pulmonary veins. When the left atrium contracts, it squeezes the blood through the bicuspid or mitral valve into the left ventricle. As this ventricle with its thick muscular walls contracts, it pushes the blood through the aortic semilunar valve into the ascending aorta. The ascending aorta distributes the oxygenated blood to all parts of the body.

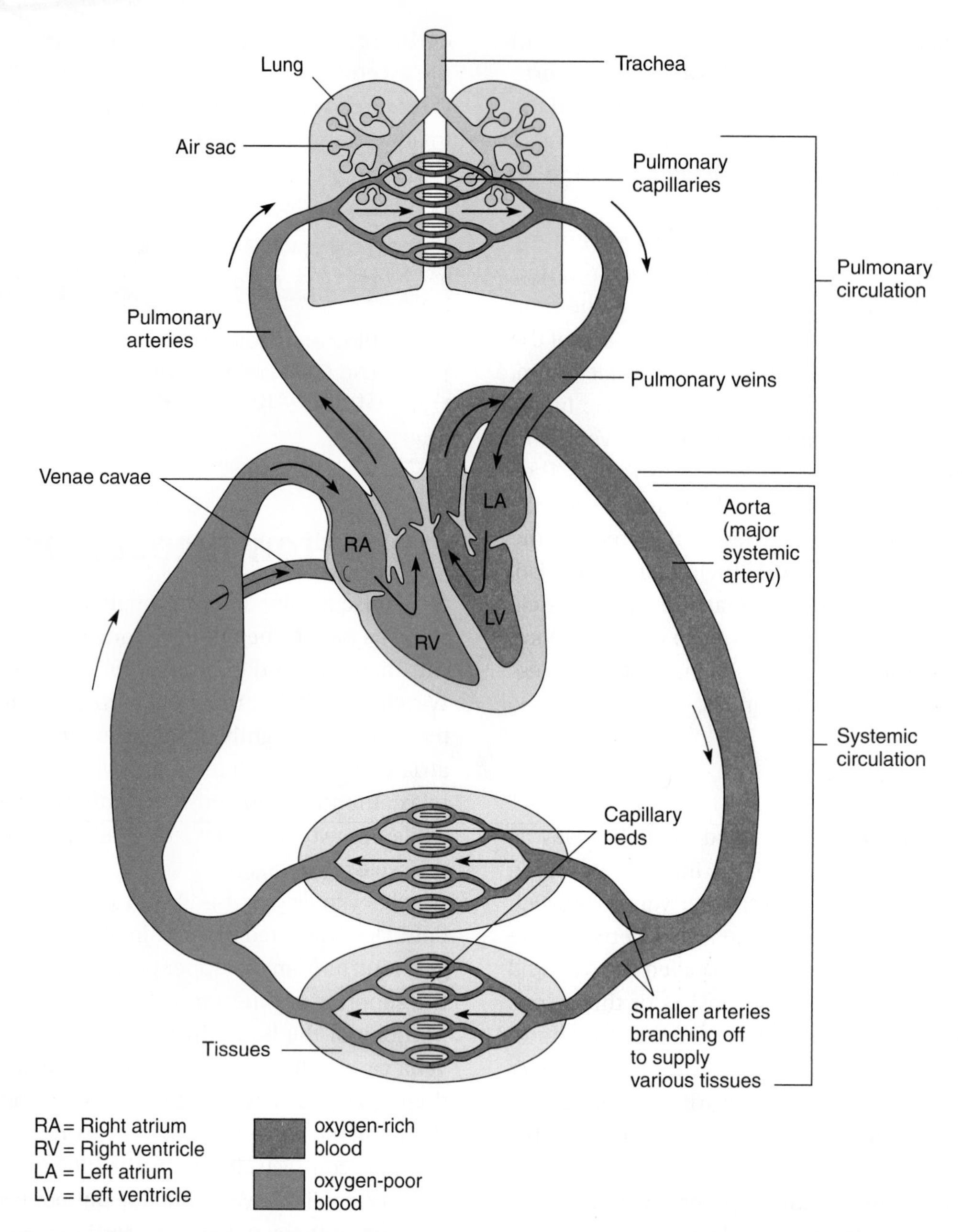

FIGURE 14-5. Schematic drawing of blood flow through the heart showing pulmonary and systemic circulation.

THE CONDUCTION SYSTEM OF THE HEART

The heart is innervated by the autonomic nervous system. However, it does not initiate a contraction but only increases or decreases the time it takes to complete a cardiac cycle. This is made possible because the heart has its own intrinsic regulating system called the **conduction system** (Figure 14-6) that generates and distributes electrical impulses over the heart to stimulate cardiac muscle fibers or cells to contract.

The system begins at the **sinoatrial** (**sigh**-no-**AY**-tree-al) **node**, known as the **SA node** or **pacemaker**, which initiates each cardiac cycle and sets the pace for the heart rate. It is located in the superior wall of the right atrium. It can be modified by nerve impulses from the autonomic nervous system; sympathetic impulses will speed it up, and parasympathetic impulses will restore or slow it down. Thyroid hormone and epinephrine carried by the blood will also affect the pacemaker. Once an impulse is initiated by the SA node, the impulse spreads out over both atria, causing them to contract simultaneously.

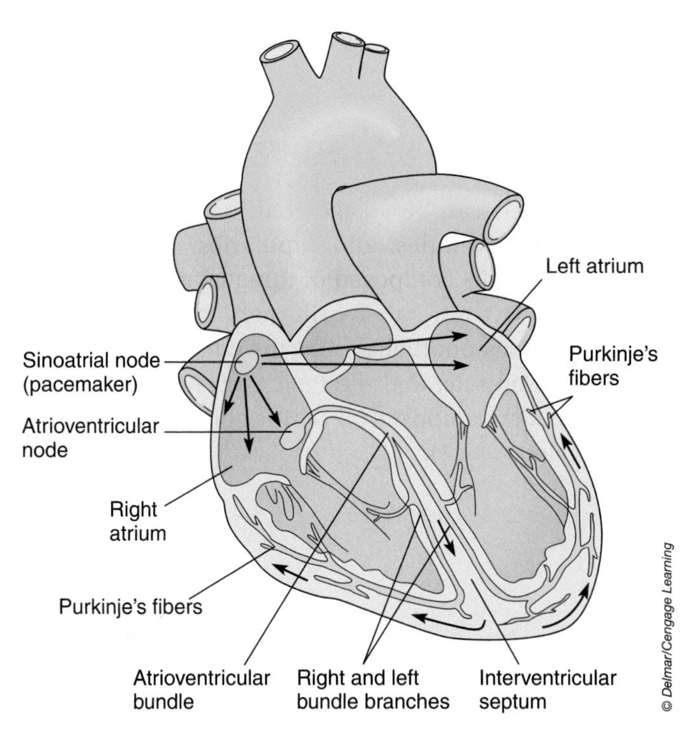

FIGURE 14-6. The conduction system of the heart.

At the same time, it depolarizes the **atrioventricular** (**AY**-tree-oh-vin-**TRIK**-yoo-lar) **(AV) node**. It is located in the lower portion of the right atrium.

From the AV node, a tract of conducting fibers called the **atrioventricular bundle** or **bundle of His** (**BUN**-dull of **HIZ**) runs through the cardiac mass to the top of the interventricular septum. It then branches and continues down both sides of the septum as the **right** and **left bundle branches**. Thus, the bundle of His distributes the electrical charge over the medial surfaces of the ventricles. The actual contraction of the ventricles is stimulated by the **Purkinje's** (pur-**KIN**-jeez) **fibers** (also known as the **conduction myofibers**) that emerge from the bundle branches and pass into the cells of the myocardium of the ventricles.

A CARDIAC CYCLE

In a normal heartbeat, the two atria contract simultaneously while the two ventricles relax. Then, when the two ventricles contract, the two atria relax. **Systole** (**SIS**-toh-lee) is the term used to refer to a phase of contraction and **diastole** (dye-**ASS**-toh-lee) is the term for a phase of relaxation. A cardiac cycle or complete heartbeat, therefore, consists of the systole and diastole of both atria and the systole and diastole of both ventricles. The pressure developed in a heart chamber is related to the chamber size and the volume of blood it contains. The greater the volume of blood, the higher the pressure.

The average heart beats approximately 72 times per minute. Therefore, we will assume that each cardiac cycle requires about 0.8 second. During the first 0.1 second, the atria contract and the ventricles relax. The atrioventricular valves (the bicuspid and tricuspid) are open and the semilunar valves (aortic and pulmonary) are closed. During the next 0.3 second, the atria are relaxing and the ventricles are contracting. During the first part of this period, all valves are closed to help build pressure. During the second part of this period, the semilunars are open. The last 0.4 second of the cycle is the relaxation or quiescent period. All chambers are in diastole. So for a full one-half period of a cycle, the heart muscle is resting. For the first part of this period all valves are closed. During the latter half, the tricuspid and bicuspid valves open to allow blood to start draining into the ventricles.

StudyWARE™ Connection

Watch an animation about the conduction system of the heart on your StudyWARE™ CD-ROM.

SOME MAJOR BLOOD CIRCULATORY ROUTES

The **systemic circulation** (see Figure 14-5) route includes all of the oxygenated blood that leaves the left ventricle of the heart through the aortic semilunar valve and goes to the aorta and the deoxygenated blood that returns to the right atrium of the heart via the superior and inferior venae cavae after traveling to all the organs of the body, including the nutrient arteries to the lungs. There are many subdivisions of this route. Systemic refers to the fact that it is carrying blood to all organs of all the systems of the body. Two of its significant subdivisions are the **coronary circulation** and the **hepatic portal circulation** routes. The coronary circulation route supplies the myocardium of the heart. The hepatic portal circulation route travels back and forth from the intestine of the digestive tract to the liver. This route is used to store excess sugars from digestion into the liver after a meal and to release sugars stored in the liver as glycogen between meals to maintain blood glucose levels.

The **pulmonary circulation** (see Figure 14-5) is the route that goes from the right ventricle of the heart through the pulmonary semilunar valve to the pulmonary trunk that branches into the right and left pulmonary arteries, which go to the lungs. Here the deoxygenated blood loses its carbon dioxide and picks up the oxygen and returns to the left atrium of the heart via the four pulmonary veins.

The **cerebral circulation** is the blood circulatory route that supplies the brain with oxygen and nutrients and disposes of waste.

The **fetal circulation**, a temporary circulation, is the circulation route that exists only between the developing fetus and its mother. It contains special structures that allow the fetus to exchange oxygen and nutrients with its mother and to get rid of fetal waste products through this connection.

ANATOMY OF BLOOD VESSELS

Blood vessels can be categorized into arteries, arterioles, veins, venules, and capillaries. Arteries and veins have walls composed of three layers: the **tunica intima** (**TYOO**-nih-kah **IN**-tih-mah) composed of a single layer of endothelial cells; the **tunica media** made of smooth muscle; and the **tunica adventitia** (**ad**-vin-**TISH**-ee-ah) composed of white fibrous connective tissue (Figure 14-7).

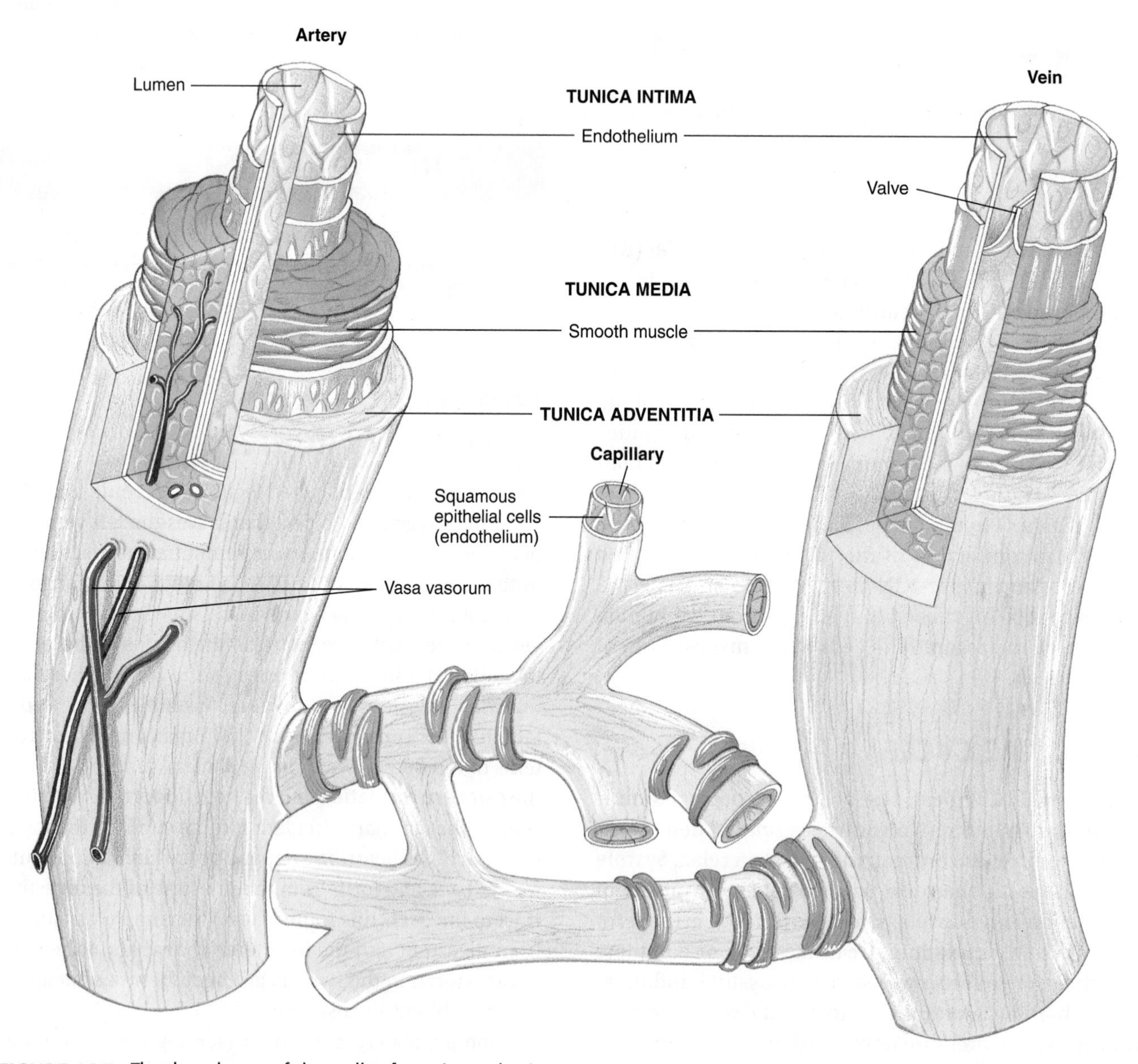

FIGURE 14-7. The three layers of the walls of arteries and veins.

HEALTH ALERT

HEALTHY HEART

In order to maintain a healthy heart, we need to concentrate on both consistent, moderate exercise and our diet. Our diets should contain high levels of fresh fruit and vegetables, which have a high fiber content. Cereal grains and nuts are also good sources of fiber. Our diet needs to be low in saturated fats; moderate amounts of fish and skinless poultry and small amounts of lean red meat will contribute to keeping cholesterol levels low. We need to avoid large intakes of both sugar and salt. A guiding rule for diets should be "moderation in all things." A healthy diet leads to lessening the risk of arterial disease, which leads to a healthy heart. Good side effects of a healthy diet are better blood pressure rates, lower blood cholesterol levels, and less excessive weight.

Exercise is also important to maintain a healthy heart. However, in older adults, strenuous exercise can cause a heart attack, usually by causing an unstable plaque to dislodge or rupture, initiating the clotting mechanism in an artery of the heart. Such risk can be minimized by exercising regularly at a low-to-moderate rate as we reach our older years (50 and above). ■

Arteries have walls made of these three coats or tunics surrounding a hollow core known as a **lumen** (**LOO**-men) through which blood flows. Arteries are thicker and stronger than veins with two major properties: elasticity and contractility. This is necessary because when the two ventricles of the heart contract, they inject a large amount of blood into the large aorta and pulmonary trunk. These arteries must be able to expand to accommodate the extra blood. Then while the ventricles relax, the elastic recoil of the arteries pushes the blood forward. Most parts of our body receive branches from more than one artery. In these areas, the distal ends of these branches unite to form one artery going into the organ. The junction of two or more blood vessels is called an **anastomosis** (ah-**nas**-toh-**MOH**-sis).

Arterioles (ar-**TEE**-ree-olz) are small arteries that deliver blood to capillaries.

Capillaries (**CAP**-ih-**lair**-eez) are microscopic vessels made of simple squamous epithelial cells, one cell layer thick, called endothelium. They are found in close proximity to nearly every cell of the body. They connect arterioles with venules. Their primary critical function is to permit the exchange of nutrients and oxygen and waste and carbon dioxide between the blood and the tissue cells of the body. Their unique wall structure of a single cell layer allows this to occur by diffusion. A substance in the blood must pass through the plasma membrane of just one cell to reach the tissue cell and vice versa. This vital exchange occurs only through capillary walls.

Venules (**VIN**-yoolz) are small vessels that connect capillaries to veins. They collect blood from capillaries and drain it into veins.

Veins are made of the same three coats or tunics as arteries but have less elastic tissue and smooth muscle but more white fibrous connective tissue in the outer layer or adventitia (see Figure 14-7). They also are capable of distention to adapt to variations of blood volume and blood pressure. Veins also contain valves that ensure blood flow in one direction, toward the heart.

There is another term that will be encountered when reading about blood vessels.That term is **vascular** or **venous sinuses**. These should not be confused with cavities in bones, which are also called sinuses. Venous sinuses are veins with thin walls.

MAJOR ARTERIES AND VEINS OF THE BODY

The **aorta** is the largest artery in the body. It begins as it exits from the left ventricle of the heart as the ascending aorta. It then arches to the left as the arch of the aorta (or aortic arch) and heads down along the spine through the thorax as the **thoracic aorta**.When it passes through the diaphragm muscle and enters the abdominal cavity, it is called the **abdominal aorta**. It is as thick as a finger and its branches go to various regions of the body. Its arterial branches are usually named according to the region of the body, the organ it goes to, or the bone it follows (Figure 14-8).

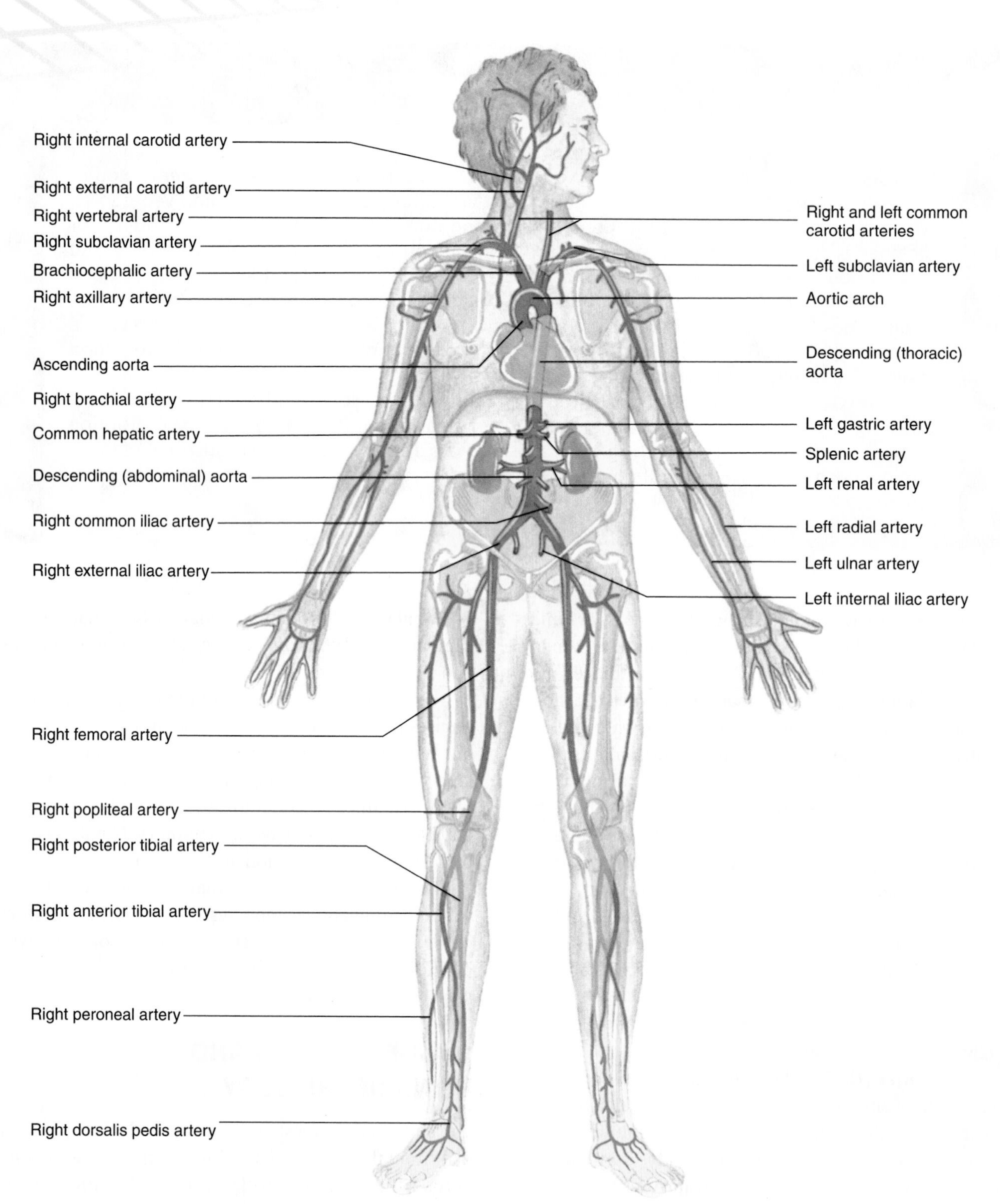

FIGURE 14-8. The major arteries of the systemic circulation.

Ascending Aorta Branches

The right and left coronary arteries branch off the ascending aorta and supply the heart.

Aortic Arch Branches

The first branch of the aortic arch is the *brachiocephalic artery*, which divides into the *right common carotid artery*, which transports blood to the right side of the head and neck, and the *right subclavian artery*, which transports blood to the upper right limb.

The second branch of the aortic arch is the *left common carotid artery*, which divides into the *left internal carotid artery* that supplies the brain and the *left external carotid artery* that supplies muscles and skin of the neck and head.

The third branch of the aortic arch is the *left subclavian artery*, which branches into the *vertebral artery* that supplies part of the brain. In the axillary area of the body, the subclavian artery is now known as the *axillary artery*, which continues down the arm as the *brachial artery*. Near the elbow joint it divides into the *radial* and *ulnar arteries*, which supply the forearm.

COMMON DISEASE, DISORDER, OR CONDITION

DISORDERS OF THE CARDIOVASCULAR SYSTEM

RHEUMATIC HEART DISEASE

Rheumatic (roo-**MAT**-ik) **heart disease** usually occurs in young children rather than adults. It results from untreated infections with the bacterium *Streptococcus* (**strep**-toh-**KOK**-us). The bacteria produce a toxin, which causes an immune reaction called rheumatic fever anywhere from two to four weeks following the infection. This causes an inflammation of the endocardium of the heart. Especially affected is the bicuspid valve, which can become narrowed, resulting in incompetence of the valve. Antibiotic treatments have reduced the frequency of this disease.

ENDOCARDITIS

Endocarditis (en-doh-car-**DYE**-tis) is an inflammation of the endocardium.

MYOCARDITIS

Myocarditis (my-oh-car-**DYE**-tis) is an inflammation of the myocardium, which can cause a heart attack.

PERICARDITIS

Pericarditis (pair-ih-car-**DYE**-tis) is an inflammation of the pericardium caused by viral or bacterial infection and is very painful.

ATHEROSCLEROSIS

Atherosclerosis (**ath**-er-**oh**-skleh-**ROH**-sis) is a disease of the arteries in which cholesterol-containing masses called plaque accumulate on the inside of arterial walls. They interfere with blood flow by blocking part of the lumen of the blood vessel. They form a roughened surface that can initiate the clotting mechanism, which can result in an embolus or thrombus. This can result in the death of tissues beneath the plaque. The walls of the artery affected can become hardened or sclerotic, lose their elasticity, and degenerate. Such vessels may rupture. A number of procedures are used to clear clogged arteries, such as angioplasty and bypass graft surgery. However, the best way to avoid atherosclerosis is by modifying one's behavior: avoid fatty diets, do not smoke, control your weight, and exercise. Elevated blood pressure may also be a risk for developing atherosclerosis.

CORONARY HEART DISEASE

Coronary heart disease results from reduced blood flow in the coronary arteries that supply the myocardium of the heart. As we age, the walls of these arteries thicken and harden, reducing blood volume. This reduced blood flow causes a sensation of pain in the chest, left arm, and shoulder called

(continues)

COMMON DISEASE, DISORDER, OR CONDITION

DISORDERS OF THE CARDIOVASCULAR SYSTEM (continued)

angina pectoris (an-**JYE**-nah **PEK**-toh-ris). Inadequate blood flow can cause an **infarct** (in-**FARKT**), an area of damaged cardiac tissue. A heart attack is commonly referred to as a **myocardial infarction**. The degenerative changes in the coronary arteries cause the walls to be roughened with platelet aggregation, resulting in a blood clot in the vessel called a **coronary thrombosis**. If the infarct is large, death can result. However, if it is small, although weakened, the heart can function even though the walls of the heart may be weakened due to the development of scar tissue in the infarct area. Myocardial infarctions can be prevented by moderate exercise, rest, good diet, and lowering stress. Small doses of aspirins and drug treatments can also help.

HEART FAILURE

Several heart diseases develop as people age. **Heart failure** is caused by progressive weakening of the myocardium and failure of the heart to pump adequate amounts of blood.

HYPERTENSION

Hypertension or high blood pressure can cause enlargement of the heart leading to heart failure. Besides hereditary factors, malnutrition, hyperthyroidism, chronic infections, anemia, and advanced age can lead to heart failure.

CONGENITAL HEART DISEASE

Heart disease present at birth is known as **congenital heart disease**. The heart does not develop properly. Two common congenital heart defects are septal defect and stenosis of the heart valves. **Septal defect** is a hole in the interatrial or interventricular septum between the left and right sides of the heart. This reduces the pumping effect of the heart. **Stenosis of the heart valves** is a narrowed opening through the valves. In pulmonary or aortic semilunar valve stenosis, the ventricles must work harder to pump blood. Stenosis of the mitral valve causes blood to backflow into the left atrium and the lungs, causing lung congestion. Stenosis of the tricuspid valve causes blood to back up in the right atrium and superior and inferior venae cavae.

Having regular checkups to monitor arterial pulse rates and blood pressure is a good way to monitor the efficiency of our cardiovascular system.

ARRHYTHMIA

An **arrhythmia** (ah-**RITH**-me-ah) is any deviation from a normal heartbeat rhythm. There are several kinds; two examples are: bradycardia is a slow heartbeat rate of less than 60 beats per minute; and tachycardia is a fast heartbeat rate of more than 100 beats per minute and can be brought on by excessive sympathetic innervation by the autonomic nervous system, elevated body temperature, or drug interactions.

HEART MURMUR

A **heart murmur** is an abnormal heart sound akin to a fluttering or a humming sound like a gentle blowing. A heart murmur is indicative of a heart abnormality such as a stenosis of a heart valve or an incompetent heart valve.

STENOSED HEART VALVE

A **stenosed heart valve** (stenosis of a heart valve) is one that has an abnormal opening that is narrowed. This produces a rushing sound just before the valve closes. This is due to the fact that blood flows in a very turbulent way producing an abnormal sound. Stenosed tricuspid or bicuspid valves produce the rushing sound just *before* the first thump in a heartbeat while a stenosed aortic or pulmonary semilunar valve produces the sound just *before* the second thump in a heartbeat.

COMMON DISEASE, DISORDER, OR CONDITION

DISORDERS OF THE CARDIOVASCULAR SYSTEM (continued)

INCOMPETENT HEART VALVE

An **incompetent heart valve** produces serious leakage of blood. Once the valve closes, blood still flows through it but in the reverse direction. This produces a swishing sound after the valve closes. An incompetent atrioventricular valve produces the sound *after* the first thump of a heartbeat while an incompetent semilunar valve produces the sound *after* the second thump of a heartbeat. An incompetent tricuspid valve causes the blood to flow back into the right atrium and superior and inferior venae cavae producing a swelling in the periphery of the body. An incompetent biscupid valve causes the blood to flow back into the left atrium and the pulmonary veins and lungs causing lung congestion and swelling.

ANGINA PECTORIS

Angina pectoris is also called cardiac pain. Minor symptoms are occasionally misdiagnosed as indigestion by the layman while major symptoms precede an actual heart attack. It results in episodes of pain in the thoracic region caused by a lack of oxygen to the muscle cells of the heart. The pain can progress along the jaw, neck, shoulder, and down the left arm. It causes an individual to experience chest pressure and feelings of suffocation and possible death. These attacks are usually caused by buildup of plaque in the coronary arteries (atherosclerosis) or by spasms of these arteries. Symptoms can be relieved by nitroglycerin tablets, which cause the coronary arteries to dilate allowing more oxygen-carrying blood to reach the heart. Symptoms seem to be related to overexertion, emotional stress, and eating certain foods. Moderate exercising, not smoking, weight control, avoidance of fatty foods, and low blood pressure are preventative measures that can inhibit the development of this condition.

MITRAL VALVE PROLAPSE (MVP) SYNDROME

Mitral valve prolapse syndrome develops when one or both of the bicuspid/mitral cusps project back into the left atrium during contraction of the ventricles. This condition can be acquired due to bacterial endocarditis or rheumatic fever, or it can be congential. Surgery may be performed to replace the defective valve with an artificial valve.

MYOCARDIAL INFARCTION

Myocardial infarction is also known as a heart attack. It results from the death of a portion of the heart muscle cells caused by an obstruction or blockage of a coronary artery. This occurs due to plaque buildup (atherosclerosis), a coronary spasm, or a thrombus (unwanted clotting in the artery). The first symptom is a great crushing pressure in the chest (occasionally referred to by the patients as an elephant sitting on their chest) with accompanying pain radiating down the left arm. This is followed by nausea, clamminess, shortness of breath, turning ashen in color, and feeling faint. Other symptoms include: a heightened heartbeat rate, low blood pressure, low pulse rate, elevated temperature, and an irregular heartbeat rate. If oxygen-carrying blood is supplied to the heart within 20 minutes, permanent damage to the heart is prevented. However, if lack of oxygen occurs for a longer period of time, cellular death will occur. Since aspirin inhibits platelet activation, people who are at risk for heart attacks take daily reduced doses of aspirin to reduce the likelihood of plaque formation in the coronary arteries.

ANGIOPLASTY AND STENTS

An **angioplasty** is the reconstruction of a diseased coronary artery by inserting a small balloon through the aorta and into a coronary artery. Once the balloon is positioned in the blocked artery, it is inflated, thus flattening the plaque mass and dilating or stretching the artery. This allows more blood to flow through the artery. After an angioplasty the dilation of the blood vessel can be reversed and again become blocked. Therefore, in some cases, a metal-mesh tube called a **stent**, which is stronger and permanent, is inserted into the vessel to prevent any future blockage.

StudyWARE™ Connection

Watch an animation on ventricular fibrillation on your StudyWARE™ CD-ROM.

Thoracic Aorta Branches

There are 10 pairs of *intercostal arteries* that supply muscles of the thorax. *Bronchial arteries* supply the two lungs, *esophageal arteries* go to the esophagus, and *phrenic arteries* supply the diaphragm muscle.

Abdominal Aorta Branches

The first branch is the *celiac trunk,* which has three branches: the *left gastric artery,* which goes to the stomach; the *splenic artery,* which supplies the spleen; and the *common hepatic artery,* which goes to the liver.

The *superior mesenteric artery* supplies the small intestine and the colon.

The *right* and *left renal arteries* go to the kidneys.

The *right* and *left gonadal* (ovarian in female and testicular in males) *arteries* serve the gonads.

The *lumbar arteries* are several pairs that go to the muscles of the abdomen and walls of the trunk of the body.

The *inferior mesenteric artery* is quite small and serves the rest of the large intestine.

The *right* and *left common iliac arteries* is the final branch of the abdominal aorta. Each divides into an *internal iliac artery,* which goes to the thigh. Here it is called the *femoral artery* and its branch is called the *deep femoral artery,* which supplies the thigh with blood. At the knee, the femoral artery is now called the *popliteal artery,* which divides into the *anterior* and *posterior tibial arteries,* which supply the leg and foot. The anterior tibial artery then terminates as the *dorsalis pedis artery,* which supplies the dorsal part of the foot.

HEALTH ALERT — BYPASS SURGERY

Bypass surgery or a coronary bypass is performed on those individuals who have obstructions in the coronary arteries that supply the heart. These obstructions can be caused by a buildup of deposits of cholesterol called plaque. Segments of healthy blood vessels from other parts of the patient's body are transplanted to the heart to bypass the obstructions located in various parts of the coronary arteries. ■

HEALTH ALERT — CONGESTIVE HEART FAILURE (CHF)

Congestive heart failure is caused by an improperly pumping heart. As a result, blood tends to accumulate in the ventricles at the end of each cardiac cycle. This buildup of blood results in an increase in blood volume in the chambers, causing chamber dilation and an increase in intracardiac pressure over time. Usually one ventricle fails before the other. When the right ventricle fails first, blood backs up in vessels leading to the heart, causing peripheral edema and symptoms such as swelling of the feet and ankles. When the left ventricle fails first, blood will back up and accumulate in the lungs, causing pulmonary edema. This can lead to suffocation and death. Some common causes of CHF are myocardial infarction (heart attack), disorders of the valves of the heart, high blood pressure over a long period of time, and cardiomyopathy. ■

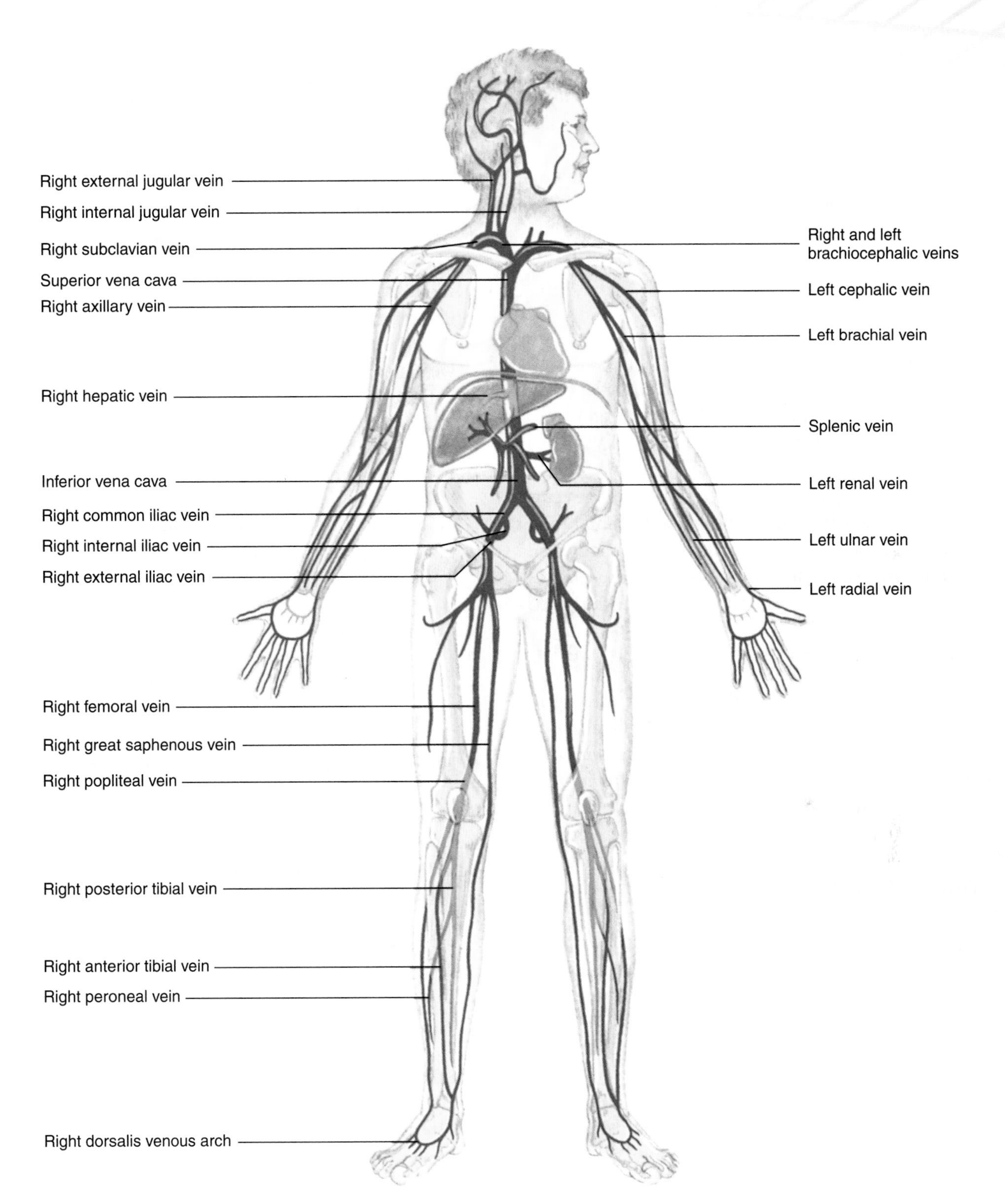

FIGURE 14-9. The major veins of the body.

Most of the arteries of the body are in deep and protected areas of the body. Veins, however, tend to be closer to the body surface and are easily seen through the skin. Deeper veins follow the courses of the major arteries and their names are identical to the arteries. Veins converge on either the superior or inferior vena cava. Veins draining the head and arms merge into the *superior vena cava*; those draining the lower parts of the body merge into the *inferior vena cava* (Figure 14-9).

Veins Merging into the Superior Vena Cava

The *radial* and *ulnar veins*, which drain the forearm, unite to form the *brachial vein*, which drains the arm and empties into the *axillary vein* in the armpit area.

The *cephalic vein* drains the lateral part of the arm and connects into the axillary vein.

The *basilic vein* drains the medial part of the arm and joins the brachial vein. The basilic and brachial veins are joined near the elbow by the *median cubital vein*. It is this vein that is usually used for drawing blood.

The *subclavian vein* drains blood from the arm via the *axillary vein* and drains blood from the muscles and skin of the head region via the *external jugular vein*.

The *vertebral vein* drains the back of the head.

The *internal jugular vein* also drains the dural sinus of the brain in the head.

The *right* and *left brachiocephalic* veins are large veins that receive blood from the subclavian, vertebral, and internal jugular veins. It then joins the superior vena cava.

The *azygos vein* drains the thorax and also merges with the superior vena cava just before it enters the heart.

Veins Merging into the Inferior Vena Cava

The *anterior* and *posterior tibial veins* and the *peroneal vein* drain the calf and the foot. The posterior tibial vein is called the *popliteal vein* at the knee and the *femoral vein* in the thigh. The femoral vein is called the *external iliac vein* as it goes into the pelvis.

The *great saphenous* veins are the longest veins of the body. They drain the superficial aspects of the leg and begin as the *dorsal venous arch* in the foot and eventually merge with the femoral vein in the thigh.

The *right* and *left common iliac veins* are formed by the union of the *external* and *internal iliac veins*. These drain the pelvis. The common iliac veins unite to form the inferior vena cava in the abdominal cavity.

The *right* and *left gonadal veins* drain the gonads and eventually join the left renal vein.

The *right* and *left renal veins* drain the kidneys.

The *hepatic portal vein* drains the organs of the digestive tract and goes to the liver.

The *right* and *left hepatic veins* drain the liver.

StudyWARE™ Connection

Watch an animation on congestive heart failure (CHF) on your StudyWARE™ CD-ROM.

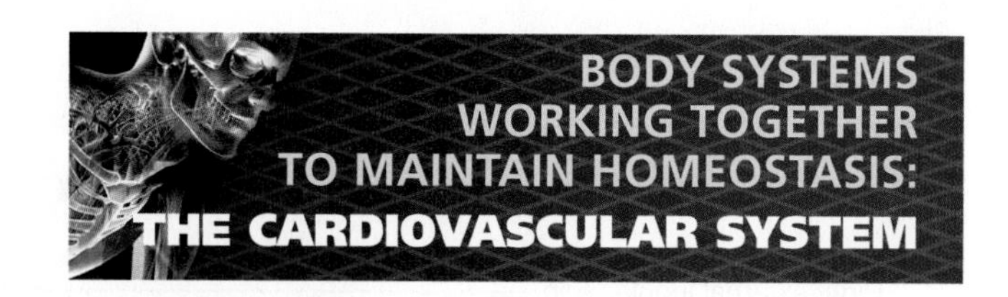

BODY SYSTEMS WORKING TOGETHER TO MAINTAIN HOMEOSTASIS: THE CARDIOVASCULAR SYSTEM

Integumentary System

- Blood flow to the skin aids in temperature control for the body.
- Blood flow to the skin brings oxygen and nutrients to and removes waste from skin tissue and glands.
- Dilation of blood vessels in the dermis occurs when we are embarrassed, resulting in blushing seen in light-skinned individuals.

Skeletal System

- Bones store and release calcium to maintain plasma levels of calcium.
- Bones are the sites of hematopoiesis.
- Bones (sternum and ribs) protect the cardiovascular organs.

Muscular System

- Exercising muscles receive increased blood flow delivering oxygen and nutrients and removing waste.
- Cardiac and smooth muscle contractions maintain blood flow and blood pressure.
- Exercise helps prevent cardiovascular disease.

Nervous System

- The brain and spinal cord depend on blood flow for survival and maximum function.
- The autonomic nervous system regulates heartbeat rate and blood pressure.

Endocrine System

- The bloodstream transports hormones to their target organs.
- Epinephrine, thyroxine, and antiduretic hormone affect blood pressure rates.

Lymphatic System

- The lymphatic system drains and returns interstitial fluids back to the bloodstream.
- Blood transports lymphocytes and antibodies.
- The immune system protects the heart and blood vessels from foreign microbes.

Digestive System

- The digestive system breaks down food and nutrients into forms that can be absorbed and transported by the bloodstream.

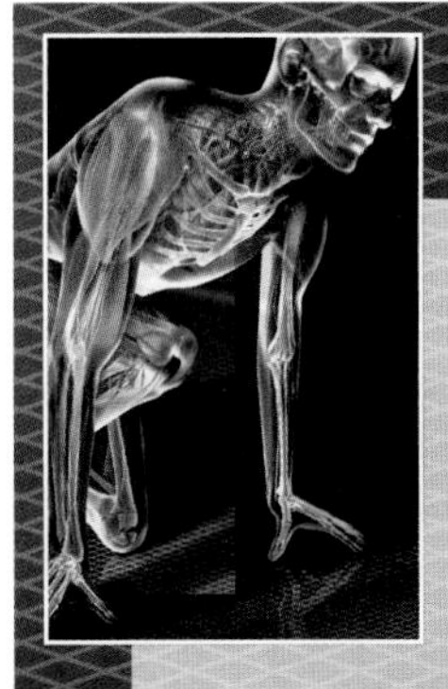

AS THE BODY AGES

As we age, our maximum heart rate gradually decreases. These conditions in cardiac muscle lead to this situation: a decrease in the rate of calcium transport and a corresponding increase in the rate of the breakdown of ATP molecules, a decrease in the rate of aerobic metabolism, and a decrease in the effects of adrenalin and noradrenaline on heartbeat rates. These conditions cause longer relaxation and contraction rates of the heart, decreasing in maximum heartbeat rates. By 70 years of age, there can be as much as a 75% decrease in cardiac output.

The left ventricle can enlarge as we age. There is increased pressure in the aorta caused by a decrease in its elasticity and that of other large arteries, causing the left ventricle to work harder thus building up mass. During the aging process, the heart cells also accumulate collagen fibers and lipid molecules, producing stiffness and less elasticity. The resulting increases in pressure due to increased volume of the left ventricle lead to individuals feeling out of breath when they exercise vigorously.

The heart valves are also affected by age. Calcium deposits increase and the connective tissue becomes stiff, resulting in abnormal valve function. A narrowing of the aortic semilunar valve and the bicuspid valve is especially common.

The development of coronary artery disease occurs as we age, as does congestive heart failure. Coronary artery disease is the major cause of death and heart disease in older Americans. This especially occurs in people in their 80s, resulting in an inability to deal with stress, blood loss, and infections. Changes in the frequency of the electrical conduction system of the heart leads to a higher rate of arrhythmias or irregular heartbeats in older adults.

The best way to maintain a healthy heart, at any age, is to exercise regularly. Assuming no heart disease conditions, maintaining a consistent level of aerobic exercising improves the working capability of the heart muscle. In older adults, a daily walking regimen is one of the best exercises to maintain good heart performance. ■

Career FOCUS

These are careers that are available to individuals who are interested in the cardiovascular system.

- **Cardiologists** are physicians whose specialty training is in diagnosing and treating the disorders of the heart. Further training can result in a career as a cardiac surgeon.
- **Cardiovascular technologists** are allied health professionals who, under the direction of a physician, perform diagnostic examinations on patients for peripheral vascular studies and invasive and noninvasive cardiology.
- **Electrocardiographic technicians** are allied health specialists who operate and maintain electrocardiographic equipment, which records the electrical impulse of the heart. These technicians provide recorded data on heart performance for review by a physician.
- **Cardiac sonographers** are allied health professionals with training in cardiovascular technology who take pictures of patients' hearts utilizing ultrasound technology. Those pictures are then reviewed by a physician for diagnosis.

- Iron and the B vitamins are provided by the digestive system for red blood cell formation.

Respiratory System
- The respiratory system provides lungs for the exchange of oxygen and carbon dioxide with the red blood cells.
- Respiratory movements aid in venous blood return to the heart.

Urinary System
- The kidneys filter the blood of wastes, excess electrolytes, and water.
- The kidneys help control blood volume and blood pressure.
- Blood pressure helps maintain kidney function.

Reproductive System
- Increases in blood volume to the penis maintains an erection.
- Estrogen maintains vascular health in females.

SUMMARY OUTLINE

INTRODUCTION

1. The heart, blood, and the blood vessels of the body constitute the cardiovascular system.
2. The cardiac muscle of the heart is the pumping organ of the system.
3. The function of the system is the transportation of blood.

THE ANATOMY OF THE HEART

1. The heart is situated in the mediastinum surrounded by the pericardial sac.
2. The pericardial sac is composed of two layers. The outer layer is the fibrous pericardium, made of tough fibrous connective tissue. It anchors the heart in the mediastinum and prevents overdistention of the heart.
3. The inner layer of the pericardial sac is the serous pericardium. It is thin and delicate and is also called the parietal layer.

The Layers of the Heart Wall

1. The outermost layer is called the epicardium or visceral pericardium. It consists of serous tissue and mesothelium.
2. A space called the pericardial cavity separates the epicardium of the heart from the serous pericardium of the pericardial sac.
3. The second layer of the heart is the myocardium. This makes up the bulk of the heart and consists of cardiac muscle tissue.
4. The innermost third layer of the heart is the endocardium. It is the endothelial lining of the heart.

The Chambers of the Heart

1. The heart is divided into four chambers: two upper and two lower.
2. The upper chambers are the right atrium and the left atrium separated internally by an interatrial septum.
3. Each atrium has an external appendage called an auricle, whose rough appearance is caused by the musculi pectinati.
4. The lower chambers are the right ventricle and the left ventricle, which are separated internally from one another by the interventricular septum.
5. Externally, the coronary sulcus, a groove, separates the atria from the ventricles.
6. The anterior and posterior interventricular sulci separate the ventricles from one another externally.

The Great Vessels of the Heart

1. The superior or anterior vena cava receives blood from the upper parts of the body.
2. The inferior or posterior vena cava receives blood from the lower parts of the body.
3. The coronary sinus drains blood from the heart.
4. The pulmonary trunk splits into the right and left pulmonary arteries, which carry deoxygenated blood to the lungs.
5. The four pulmonary veins return oxygenated blood to the heart.
6. The ascending aorta carries oxygenated blood away from the heart to all parts of the body. It is divided into the arch of the aorta, the descending thoracic aorta, and the abdominal aorta.
7. Atrial walls are thin, while ventricle walls are thick. Of the two ventricles, the left ventricle has the thickest walls of cardiac muscle because of the great distance it must transport blood.

The Valves of the Heart

1. Valves prevent blood from backflowing. The heart has two atrioventricular valves and two semilunar valves.
2. The tricuspid valve is found between the right atrium and the right ventricle. It is made of three cusps or flaps.
3. The bicuspid or mitral valve is found between the left atrium and the left ventricle. It is made of two cusps or flaps.
4. The cusps of the valves project into the ventricles by the chordae tendineae, which connect to papillary muscles in the ventricles, thus ensuring a one-way blood flow.
5. The pulmonary semilunar valve is found in the right ventricle where the pulmonary trunk exits the heart.
6. The aortic semilunar valve is found in the left ventricle where the ascending aorta leaves the heart.
7. The semilunar valves are made of three cusps or flaps.

BLOOD FLOW THROUGH THE HEART

1. Deoxygenated blood returns to the heart from all parts of the body via the superior (anterior) and inferior (posterior) venae cavae to the right atrium of the heart.
2. The blood is then squeezed through the tricuspid valve into the right ventricle.
3. From the right ventricle the blood is pumped through the pulmonary semilunar valve to the pulmonary trunk, which divides into the right and left pulmonary arteries.
4. The pulmonary arteries carry the blood to the lungs where it releases carbon dioxide gas and picks up oxygen.
5. The oxygenated blood returns to the left atrium of the heart through four pulmonary veins.
6. The blood is squeezed through the bicuspid or mitral valve into the left ventricle.
7. The left ventricle pumps the blood through the aortic semilunar valve to the ascending aorta, which distributes the blood to all organs of the body.

THE CONDUCTION SYSTEM OF THE HEART

1. The conduction system of the heart generates and distributes electrical impulses over the heart, which cause contraction of the heart.
2. The sinoatrial (SA) node, also known as the pacemaker, initiates each cardiac cycle, and is found in the superior wall of the right atrium. It spreads electrical impulses over both atria, causing them to contract simultaneously while depolarizing the AV node.
3. The atrioventricular (AV) node, in the lower part of the right atrium, sends electrical impulses through the atrioventricular bundle or bundle of His to the top of the interventricular septum.
4. The bundle of His branches down both sides of the septum as the right and left bundle branches, distributing the electrical impulses over the medial surface of the ventricles.
5. Purkinje's fibers emerge from the bundle branches and distribute the impulses to the cells of the myocardium of the ventricle causing actual contraction.

A CARDIAC CYCLE

1. In a cardiac cycle, the two atria contract simultaneously while the two ventricles relax, and the two ventricles contract simultaneously while the two atria relax.
2. The phase of contraction is called systole and the phase of relaxation is called diastole.
3. An average cardiac cycle takes 0.8 second: during the first 0.1 second, the atria contract and the ventricles relax and the atrioventricular valves are open and the semilunars are closed; for the next 0.3 second, the atria relax while the ventricles contract and all valves are closed at first and then the semilunars open; the last 0.4 second is the relaxation/quiescent period, during the first part of which all valves are closed and then the atrioventricular valves open to start blood draining into the ventricles.

SOME MAJOR BLOOD CIRCULATORY ROUTES

1. Systemic circulation includes all the oxygenated blood that leaves the left ventricle through the aortic semilunar valve to the aorta and all the deoxygenated blood that returns to the right atrium via the superior and inferior venae cavae.

2. Systemic circulation has many subdivisions. Two are the coronary circulation that goes to the heart and the hepatic portal circulation that travels between the intestine and the liver.
3. Pulmonary circulation includes the deoxygenated blood that leaves the right ventricle through the pulmonary semilunar valve to the pulmonary trunk, which branches and goes to the lungs. In the lungs, carbon dioxide gas is released and oxygen gas is picked up to return to the left atrium via the four pulmonary veins.
4. Cerebral circulation is the route in the brain.
5. Fetal circulation exists only between the developing fetus and its mother.

THE ANATOMY OF BLOOD VESSELS

1. Arteries and veins have walls made of three layers: the innermost, tunica intima, made of a single layer of endothelial cells; the middle, tunica media, made of smooth muscle; and the outer, tunica adventitia, made of white fibrous connective tissue.
2. The cavity of blood vessels is called the lumen.
3. Arteries are thicker and stronger than veins. They are elastic and can contract.
4. Arterioles are small arteries that deliver blood to capillaries.
5. Capillaries are microscopic vessels made of a single layer of endothelial cells with their basement membrane. They connect arterioles with venules and because of their structure, allow the exchange of gases, nutrients, and waste between blood and tissue cells.
6. Venules are small vessels that connect capillaries to veins.
7. Veins have less elastic and smooth muscle than arteries but have more fibrous connective tissue. They also have internal valves to ensure blood flow in one direction.
8. Venous sinuses are veins with thin walls.

MAJOR ARTERIES AND VEINS

1. The aorta is the largest artery of the body. It has numerous branches named either according to the region of the body or organ it goes to or according to the bone its branch may follow.
2. Most of the arteries of the body are in deep and protected areas of the body.
3. Veins are found closer to the surface of the body, and many can be seen superficially. Many of their names are identical to the arteries.
4. The veins of the body converge with either the superior or inferior vena cava, the two largest veins of the body, which empty into the right atrium of the heart.

REVIEW QUESTIONS

1. Explain blood flow through the heart, naming the major blood vessels entering and exiting the heart, the chambers, and valves.

*2. Compare the anatomy of the walls of arteries and veins and relate this to function.

3. Describe the systemic and pulmonary blood circulation routes.

*4. Explain how the conduction system of the heart functions and what factors may affect its rate.

5. Describe what occurs in the 0.8 second of a cardiac cycle.

***Critical Thinking Questions**

MATCHING

Place the most appropriate number in the blank provided.

_____ Parietal pericardium	1. Mitral valve
_____ Auricle	2. Pacemaker
_____ Trabeculae carneae	3. Visceral pericardium
_____ Bicuspid valve	4. Right atrioventricular valve
_____ Epicardium	5. Pericardial sac
_____ Sinoatrial node	6. Irregular folds of myocardium
_____ Tricuspid	7. Appendage of atrium
_____ Chordae tendineae	8. Ventricular contraction
_____ Purkinje's fibres	9. Bundle of His
_____ Systole	10. Relaxation phase fibers
	11. Project cusps into ventricle
	12. Contraction phase

Search and Explore

- Search the newspaper, a scientific magazine, or online for an article related to the anatomy of the heart. Bring to class to share.
- Visit the Centers for Disease Control and Prevention website at http://www.cdc.gov/HeartDisease to learn more about heart disease, including the signs and symptoms of a heart attack. Write a two to three paragraph entry in your notebook summarizing what you learned.
- Create a list of things you can do to maintain a healthy heart. Pick one item from the list and write an action plan including what steps you will take to achieve this goal.

*Study*WARE™ Connection

Take a quiz or play interactive games that reinforce the content in this chapter on your StudyWARE™ CD-ROM.

Study Guide Practice

Go to your **Study Guide** for more practice questions, labeling and coloring exercises, and crossword puzzles to help you learn the content in this chapter.

CASE STUDY

Enrico Shavez, an 80-year-old man, is admitted to the emergency room with severe pain in his chest and shoulder, which is radiating down his left arm. Enrico appears extremely apprehensive. He tells his health care provider that he has experienced similar episodes of chest pain in the past, which were apparently triggered by stress. Based on his symptoms, the care provider tells Enrico that he is experiencing an attack of angina pectoris. She treats his condition, and then instructs Enrico to see his primary health care provider as soon as possible. Meanwhile, Enrico is advised to avoid stressful situations and take a small dose of aspirin daily.

Questions

1. What underlying heart condition causes the chest pain associated with angina pectoris?
2. What life-threatening complication could Enrico develop?
3. What diet should Enrico follow to reduce his risk of further attacks?
4. What type of exercise program is safest for older adults with a heart condition?
5. What surgery might Enrico need to undergo to treat his heart condition?

LABORATORY EXERCISE: THE CARDIOVASCULAR SYSTEM

Materials needed: A dissecting kit, a preserved sheep heart, and a dissecting pan

1. Your instructor will supply you with a sheep heart. It is very close in size to a human heart. Examine the external anatomy of the heart. If the pericardial sac is still present, remove it. Otherwise notice the auricles, the appendages of the atria. Notice how irregular and gray they are. As you hold the heart in your hand, notice that most of the heart is ventricles; in fact, about three-quarters of what you are holding are the ventricles.
2. Find the coronary sulcus. This is the external horizontal groove that goes around the heart and separates the atria from the ventricles externally. Locate the anterior and posterior interventricular sulci that separate the right and left ventricles from one another externally. Note the sulci have fat deposits in them and blood vessels.
3. You will now cut the heart into two mirror images to view the internal anatomy. First, locate the two auricles. Now with your scalpel make a horizontal cut medially between the two auricles across the top of the heart. Continue cutting in a slicing motion down to the apex of the ventricles. Now separate the two halves. You will have cut the heart into two half sections, each a mirror image of the other. Refer to Figure 14-4 in your text.

THE CARDIOVASCULAR SYSTEM (Continued)

4. The easiest way to identify the correct chambers you are viewing is to note which ventricle has the thickest outermost wall. Remember, the thick interventricular septum is shared by both ventricles. Locate the outermost walls. The thickest outermost wall will belong to the left ventricle. Once you have identified this, chamber recognition should proceed easily. Above the left ventricle you will see the left atrium. The valve between the two is the bicuspid or mitral valve. It does not matter which half of your cut you are viewing because they are mirror images of one another. The ventricle with the thin *outermost* wall will be the right ventricle. Above the right ventricle is the right atrium and the valve between them will be the tricuspid valve.
5. Examine valve anatomy. Note how the chordae tendineae pull the cusps down into the ventricle and attach to the papillary muscles in the ventricles. Note the irregular folds in the ventricles, which are the trabeculae carneae. Force open the ventricular chamber. Note the space in the left ventricle is smaller than the space in the right ventricle because there is so much more myocardium in the walls of the left ventricle. Less space equals more muscle, equals more pressure to pump the blood through the thousands of miles of blood vessels. Note the thin walls of the atria.
6. If possible try to identify some of the great vessels entering and leaving the heart. This may be difficult and will depend on how you cut your heart. Your instructor may have a preserved pig's heart on demonstration with all parts labeled. These are available from biologic supply houses.

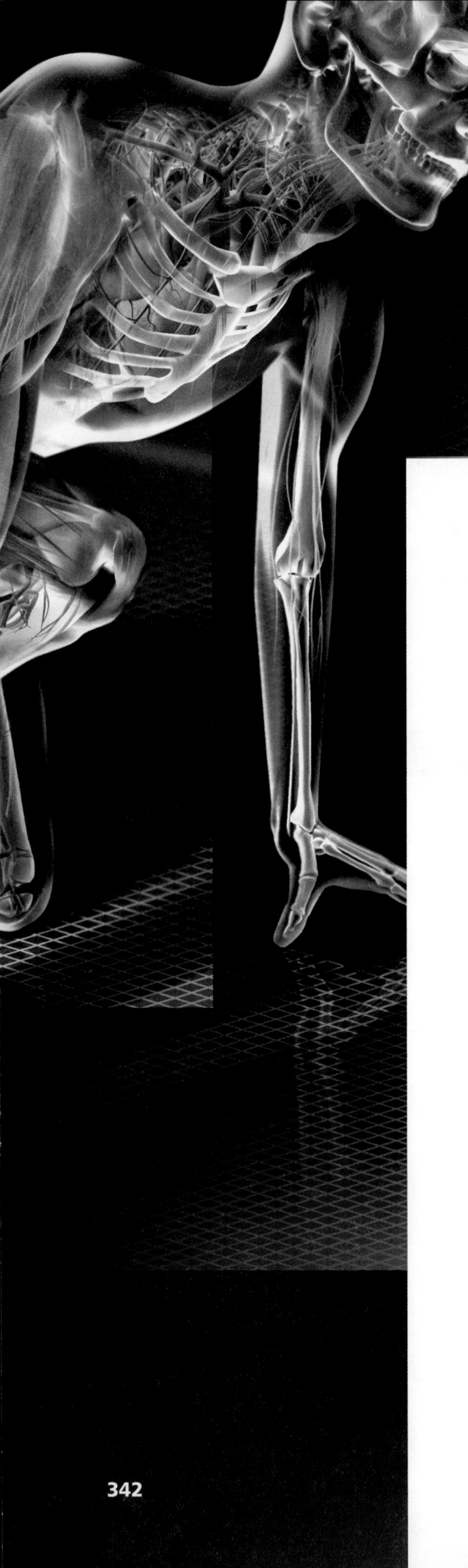

The Lymphatic System

CHAPTER OBJECTIVES

After studying this chapter, you should be able to:

1. Name the functions of the lymphatic system.
2. Explain what lymph is and how it forms.
3. Describe lymph flow through the body.
4. Name the principal lymphatic trunks.
5. Describe the functions of the tonsils and spleen.
6. Explain the unique role the thymus gland plays as part of the lymphatic system.
7. Describe the different types of immunity.
8. Explain the difference between blood and lymphatic capillaries.
9. Explain the difference between active immunity and passive immunity.
10. Define an antigen and an antibody.

15

KEY TERMS

INTRODUCTION

The lymphatic system is intimately associated with the blood and the cardiovascular system. Both systems transport vital fluids throughout the body, and both have a system of vessels that transport these fluids. The lymphatic system transports a fluid called lymph through special vessels called lymphatic capillaries and lymphatics. This lymph eventually gets returned to the blood from where it originated. In addition to fluid control, our lymphatic system is essential to helping us control and destroy a large number of microorganisms that can invade our bodies and cause disease and even death. The lymphatic system consists of lymph, lymph vessels, lymph nodes, and four organs. The organs are the tonsils, the spleen, the thymus gland, and Peyer's patches. See Concept Map 15-1: The Lymphatic System. Figure 15-1 shows the vessels and organs of the lymphatic system.

THE FUNCTIONS OF THE SYSTEM AND THE STRUCTURE AND FUNCTIONS OF THE LYMPHATIC VESSELS

The primary function of this system is to drain from tissue spaces protein-containing fluid that escapes from the

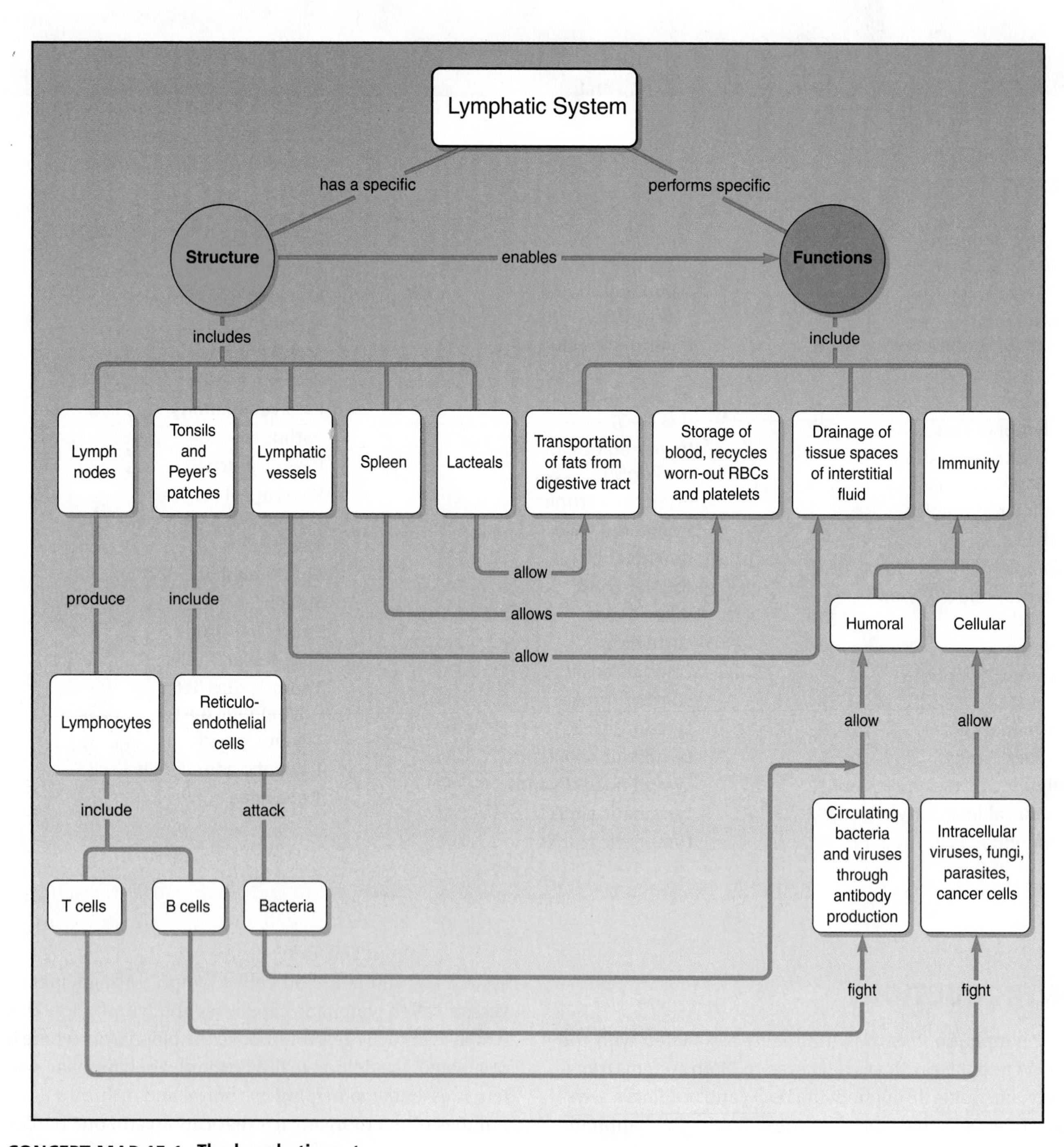

CONCEPT MAP 15-1. The lymphatic system.

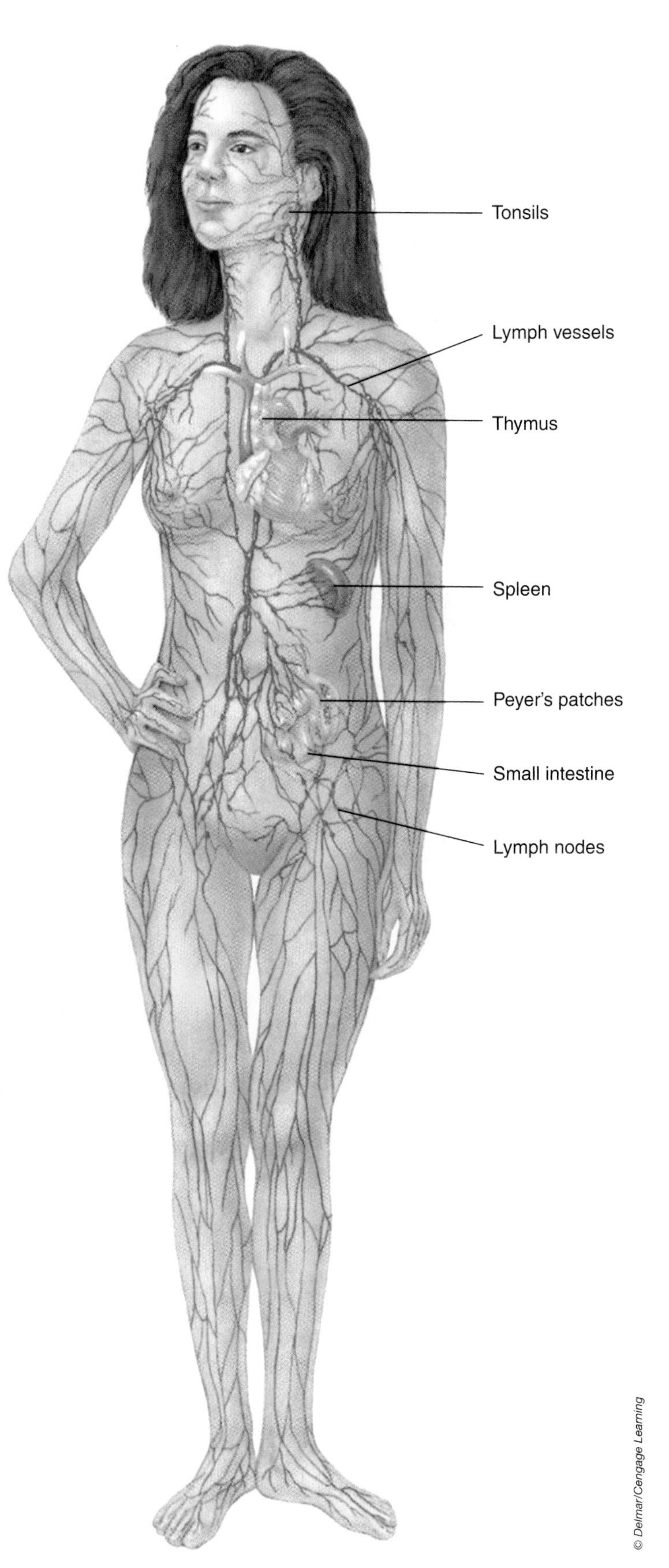

FIGURE 15-1. The vessels and organs of the lymphatic system.

blood capillaries. Other functions are to transport fats from the digestive tract to the blood, to produce lymphocytes and to develop immunities.

In our bodies where blood capillaries are close to the cells of tissues, the blood pressure in the cardiovascular system forces some of the plasma of blood through the single-celled capillary walls. When this plasma moves out of the capillaries and into the spaces between tissue cells, it gets another name and is called **interstitial** (in-ter-**STISH**-al) **fluid**. Most, but not all, of this fluid gets reabsorbed into the capillary by differences in osmotic pressure. However, some does not, and this interstitial fluid must be drained from the tissue spaces to prevent swelling or **edema** (eh-**DEE**-mah) from occurring. It is the role of the lymphatic capillaries to drain this fluid. Once the interstitial fluid enters a lymphatic capillary, it gets a third name and is now called **lymph** (**LIMF**).

In the villi of the small intestine, there are special lymphatic vessels called **lacteals** (**LACK**-teelz) whose role is to absorb fats and transport them from the digestive tract to the blood. Fats from the intestine travel through the lymphatic system, which delivers them to the blood, when the lymph rejoins the blood at the right and left subclavian veins. Lymph in the lacteals looks milky because of the fat content and is called **chyle** (**KYLE**).

Lymphatic Vessels

Lymphatic vessels originate as blind-end tubes that begin in spaces between cells in most parts of the body. The tubes, which are closed at one end, occur singly or in extensive plexuses and are called **lymphatic capillaries** (Figures 15-2 and 15-3). These vessels are not found in the central nervous system, red bone marrow, vascular tissue, or portions of the spleen. Lymphatic capillaries are much larger and more permeable than blood capillaries. Lymphatic capillaries will eventually unite to form larger and larger lymph vessels called **lymphatics** (**LIM**-fat-iks). Lymphatics resemble veins in structure but have thinner walls and more valves. The large number of valves helps to ensure that the lymph will not backflow but go in one direction only. Along lymphatics there are lymph nodes found at various intervals.

Lymphatics of the skin travel in loose subcutaneous connective tissue and generally follow the routes of veins. Lymphatics of the viscera generally follow the routes of arteries and form plexuses around the arteries. Eventually, all the lymphatics of the body converge into one of two main channels: either the **thoracic duct** (the main collecting channel), also known as the left lymphatic duct, or the **right lymphatic duct**.

Lymphatic capillaries
Pulmonary capillary network
Lymph node
Lymphatic vessels
Blood flow
Lymph node
Systemic capillary network
Lymph flow
Lymphatic capillaries
(A)

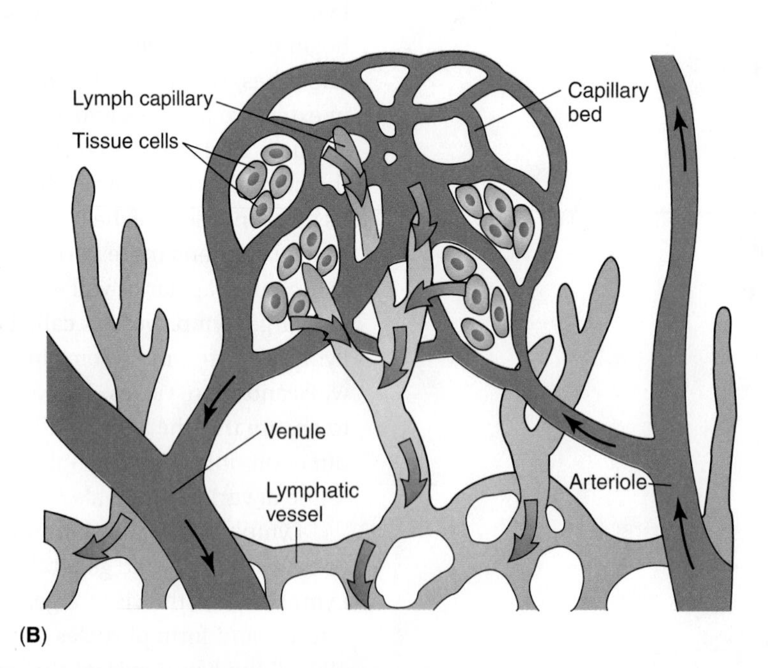

(B)

FIGURE 15-2. (A) Diagrammatic view of lymphatics transporting fluid from interstitial spaces to the bloodstream. (B) Lymphatic capillaries begin as blind-end tubes next to tissue cells and blood capillaries.

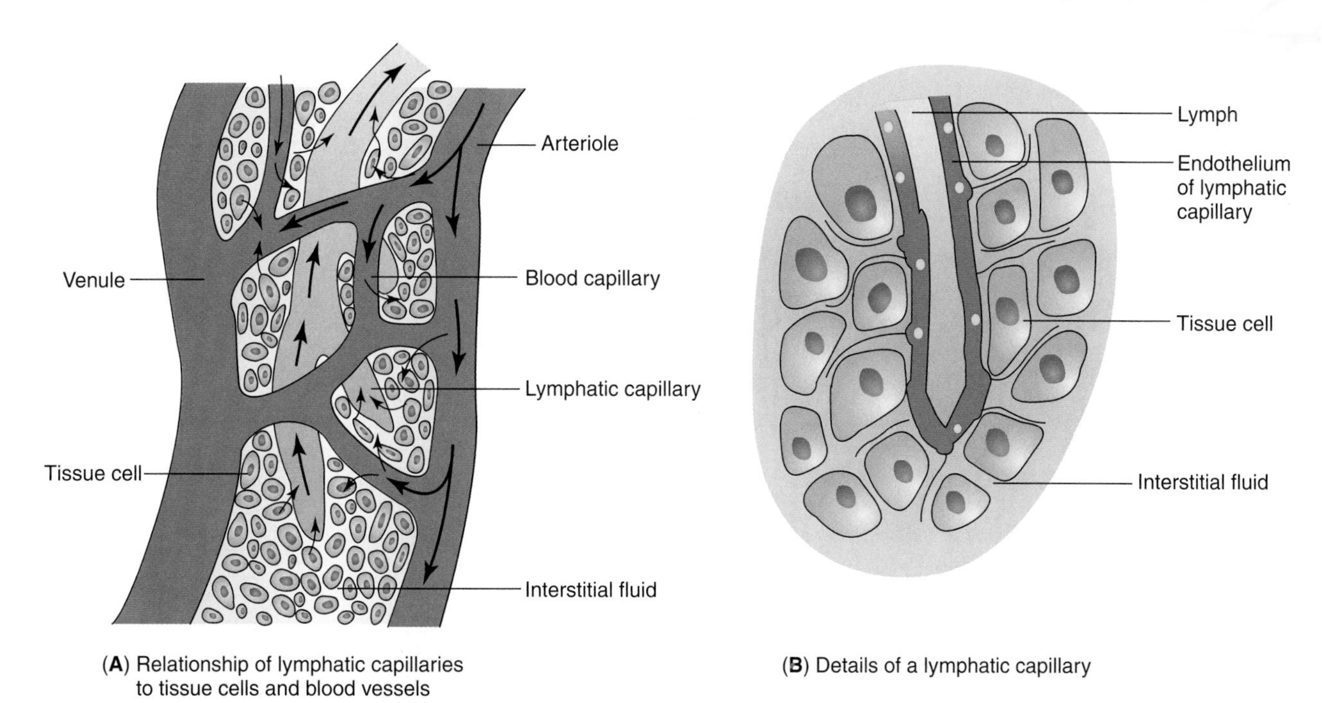

FIGURE 15-3. (A) The relationship between lymphatic capillaries and tissue cells and blood vessels. (B) Detail of a lymphatic capillary and tissue cells.

Lymph Nodes

Lymph nodes are oval to bean-shaped structures found along the length of lymphatics (Figure 15-4). They are also known as **lymph glands**. They range in size from 1 to 25 mm in length (about 0.04 to 1 inch), looking like small seeds or almonds. The three regions of aggregations of nodes in the body are the groin, armpits, and neck. A lymph node contains a slight depression on one side called the **hilum** (**HIGH**-lum) where **efferent** (**EE**-fair-ent) **lymphatic vessels** leave and a nodal artery enters and a nodal vein leaves the node. Each lymph node or gland is covered by a capsule of fibrous connective tissue that extends into the node. These capsular extensions are called **trabeculae** (trah-**BEK**-yoo-lee). The capsular extensions divide the lymph node internally into a series of compartments that contain lymphatic sinuses and lymphatic tissue. Lymphatic vessels that enter the lymph node at various sites are called **afferent** (**AFF**-er-ent) **lymphatic vessels**.

The lymphatic tissue of the node consists of different kinds of lymphocytes and other cells that make up dense aggregations of tissue called **cortical** or **lymphatic nodules** (Figure 15-5). The lymph nodule surrounds a **germinal center** that produces lymphocytes. **Lymphatic sinuses** are spaces between these groups of lymphatic tissue. They contain a network of fibers and the macrophage cells (see Figure 15-4). The capsule, trabeculae, and hilum make up the stroma or framework of the lymph node.

As lymph enters the node through the afferent lymphatics, the immune response is activated. Any microorganisms or foreign substances in the lymph stimulate the germinal centers to produce lymphocytes, which are then released into the lymph. Eventually, they reach the blood and produce antibodies against the microorganisms. The macrophages will remove the dead microorganisms and foreign substances by phagocytosis.

LYMPH CIRCULATION

As the plasma of blood is filtered by the blood capillaries, it passes into the interstitial spaces between tissue cells and is now known as interstitial fluid. When this fluid passes from the interstitial spaces into the lymphatic capillaries, it is called lymph. Lymph is primarily water but it also contains plasma solutes such as ions, gases, nutrients, and some proteins and substances from tissue cells such as hormones, enzymes, and waste products.

The lymph, drained by the lymphatic capillaries and the lymphatic plexuses, is then passed to the lymphatic vessels that have a beaded appearance due to the one-way valves that prevent backflow movement. The lymphatics head toward lymph nodes. At the lymph nodes, afferent vessels penetrate the capsules at various positions on the node and the lymph passes through the sinuses of the nodes. In the node, antigenic microorganisms, foreign substances, or cancer cells stimulate lymphocytes to divide, and the immune response is

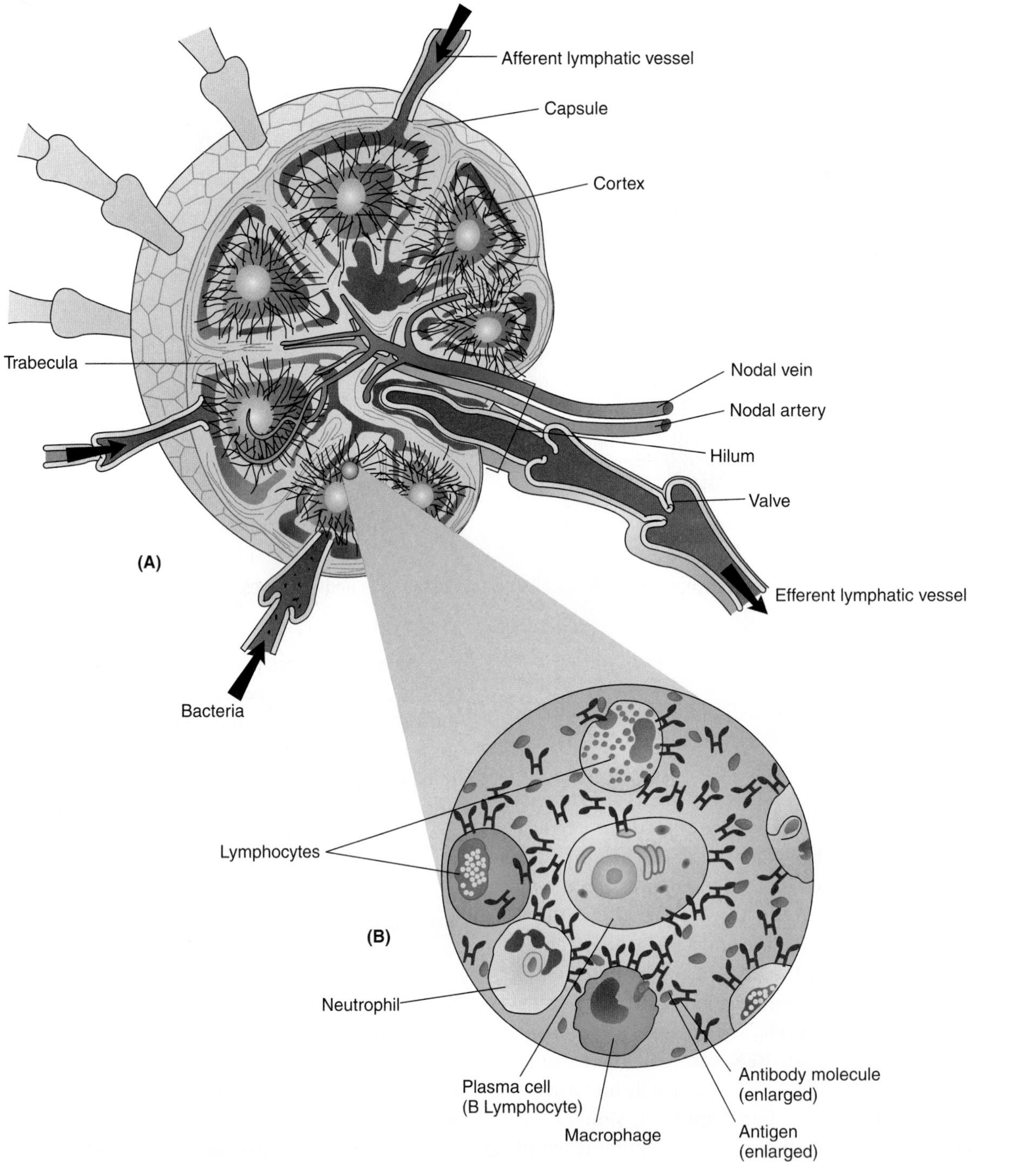

Afferent lymphatic vessel
Capsule
Cortex
Trabecula
Nodal vein
Nodal artery
Hilum
Valve
(A)
Efferent lymphatic vessel
Bacteria
Lymphocytes
(B)
Neutrophil
Plasma cell (B Lymphocyte)
Macrophage
Antibody molecule (enlarged)
Antigen (enlarged)

FIGURE 15-4. Lymph node. (A) Section through a lymph node showing the flow of lymph. (B) Microscopic detail of bacteria being destroyed within the lymph node.

activated. Macrophages phagocytize the attacked foreign substances. Efferent vessels leave nodes and pass on to other lymph nodes by either going with other afferent vessels into another node of the same group or passing on to another group of nodes. The efferent vessels will eventually unite to form **lymphatic trunks**.

This circulation of lymph through the various lymphatic vessels is maintained by normal skeletal muscle contractions. This action compresses the lymphatic vessels and because those vessels have one-way valves, the compression forces the lymph in one direction toward the subclavian veins. Normal movement helps circulate

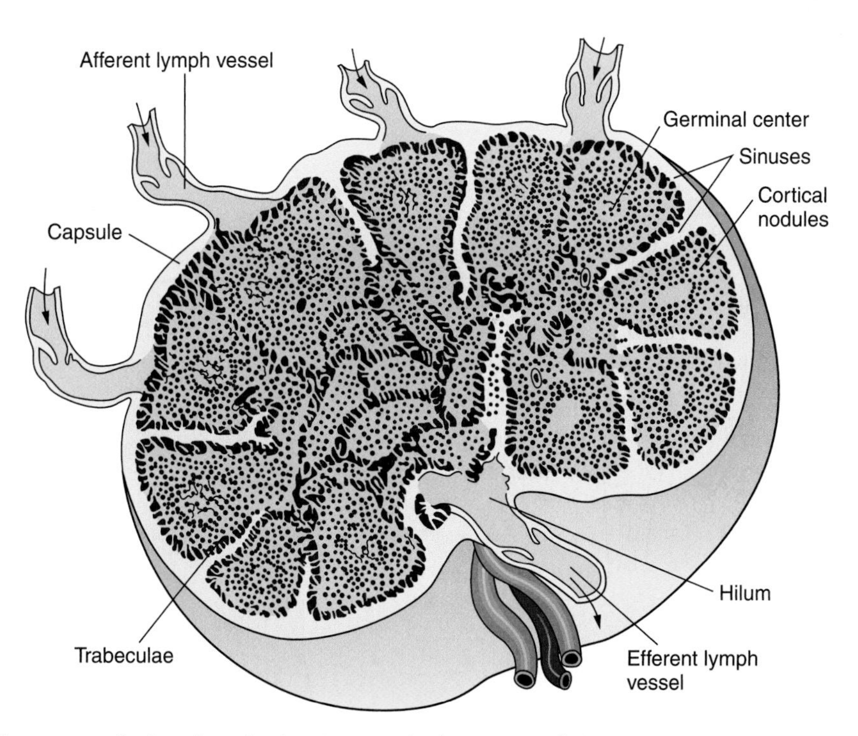

FIGURE 15-5. Internal anatomy of a lymph node showing germinal centers and sinuses.

COMMON DISEASE, DISORDER, OR CONDITION

ELEPHANTIASIS

In Africa, the bite of a mosquito can transmit a microscopic filarial nematode worm known as *Wuchereria bancrofti*. This nematode worm lives in lymphatics and lymph nodes where it reproduces and multiplies. After many years of continuous reinfections, the lymphatic system gets blocked, resulting in tremendous swelling of the arms or legs. This swelling or edema results in a condition known as elephantiasis. In addition to the limbs, the scrotum can also be affected, resulting in an almost basketball-like swelling of the scrotum or testicles. The disease also occurs in Malaysia but is caused by another nematode worm known as *Brugia malayi.* In that part of the world, only the limbs show the characteristic swelling of the disease.

the lymph. Another factor in lymph circulation is respiratory or breathing movements, which cause pressure changes in the thorax. Finally, smooth muscle contraction in the lymphatic vessel also pushes lymph along. However, if lymphatics become obstructed by blockage, then an excessive amount of interstitial fluid will develop in tissue spaces and result in swelling or edema.

Eventually, the efferent lymphatic vessels unite to form lymphatic trunks (Figure 15-6). The principal lymphatic trunks of the body are: the **lumbar trunk**, the **intestinal trunk**, the **bronchomediastinal** (**brong**-koh-mee-dee-ass-**TYE**-nal) **trunk**, the **intercostal trunk**, the **subclavian** (sub-**KLAY**-vee-an) **trunk**, and the **jugular trunk**.

The lumbar trunk drains lymph from the lower extremities, the walls and viscera of the pelvis, the kidneys and adrenal glands, and most of the abdominal wall. The intestinal trunk drains lymph from the stomach, intestines, pancreas, spleen, and the surface of the liver. The bronchomediastinal trunk drains the thorax, lungs, heart, diaphragm, and the rest of the liver. The intercostal trunk also helps drain lymph from portions of the thorax. The subclavian trunk drains the upper extremities, that is, arms, hands, and fingers. Finally, the jugular trunk drains the head, and neck.

These principal trunks now pass their lymph into two main channels: the **thoracic duct** which is the main

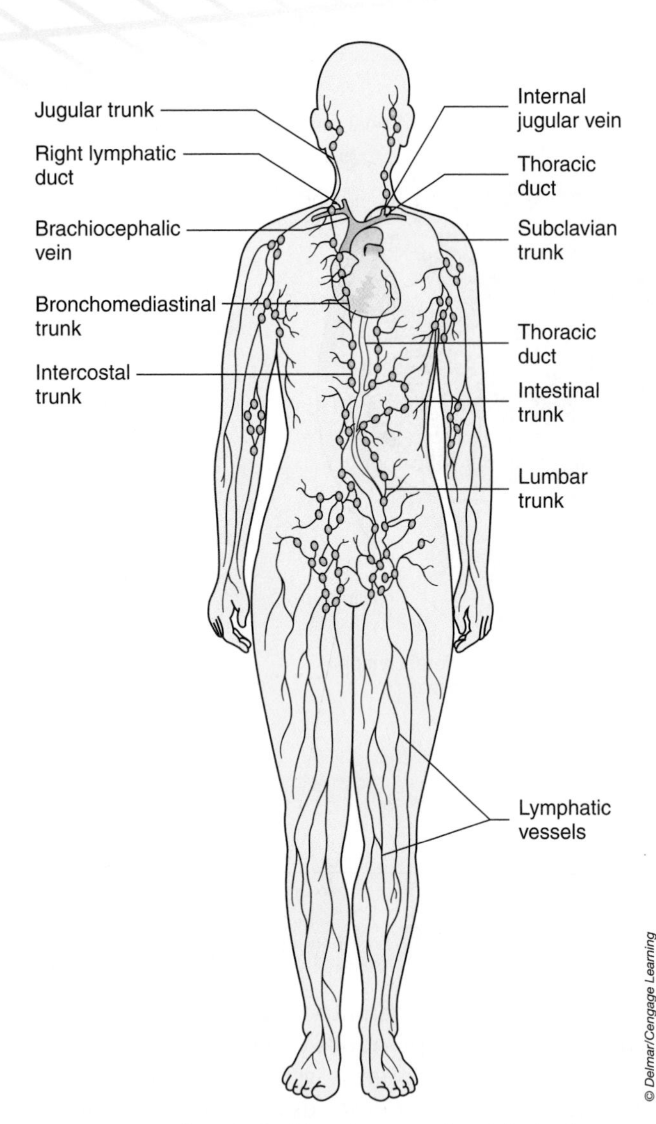

FIGURE 15-6. The principal lymphatic trunks of the body.

collecting duct of the system and is also known as the left lymphatic duct, and the right lymphatic duct (Figure 15-7). Ultimately, the thoracic duct empties all of its lymph into the left subclavian vein and the right lymphatic duct empties all of its lymph into the right subclavian vein, so the journey of the lymph is now completed. The lymph is drained back into the blood where it originally came from, and the cycle completes itself. This circulation repeats itself continuously, thus maintaining the proper levels of lymph, plasma, and interstitial fluids in the body.

THE ORGANS OF THE LYMPHATIC SYSTEM

The lymphatic system has four organs: the tonsils, spleen, thymus gland, and Peyer's patches. Tonsils are masses of lymphoid tissue embedded in mucous membrane. There

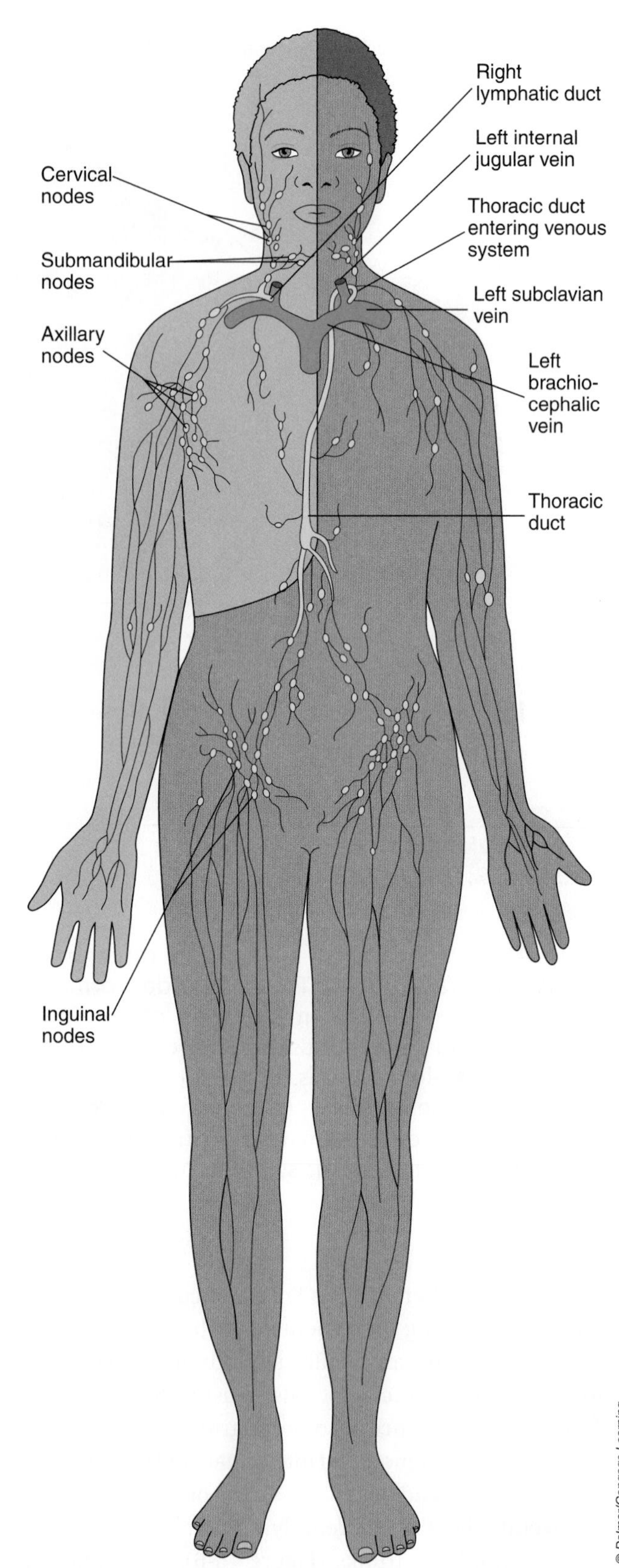

FIGURE 15-7. Lymphatic trunks pass their lymph into two main collecting ducts, the thoracic duct and the right lymphatic duct. These ducts empty into the left and right subclavian veins, respectively.

are three groups of tonsils. The **palatine** (**PAL**-ah-tyne) **tonsils** are the ones commonly removed in a tonsillectomy. They are located in the tonsillar fossae between the pharyngopalatine and glossopalatine arches on each side of the posterior opening of the oral cavity. The **pharyngeal** (fair-in-**JEE**-al) **tonsils** are also known as the **adenoids** (**ADD**-eh-noydz). They are located close to the internal opening of the nasal cavity. When they become swollen, they can interfere with breathing. The **lingual** (**LING**-gwall) **tonsils** are located on the back surface of the tongue at its base.

In these positions, the tonsils form a protective ring of reticuloendothelial cells against harmful microorganisms that might enter the nose or oral cavity. Occasionally, they become chronically infected and need to be removed. However, this operation is not as common as it once was because of the understanding of how important these organs are in protecting the body and as being part of the immune system. Tonsils are more functional in children. As we age, the tonsils decrease in size and may even disappear in some individuals.

The **spleen** (**SPLEEN**) is oval in shape and is the single largest mass of lymphatic tissue in the body (Figure 15-8). It measures about 12 cm, or 5 inches, in length. It is found in the left upper corner of the abdominal cavity. It filters blood via the splenic artery and splenic vein, which enter the spleen at a slightly concave border called the hilum. The spleen phagocytizes bacteria and worn-out platelets and red blood cells. This action releases hemoglobin to be recycled. It also produces lymphocytes and plasma cells. The spleen stores blood and functions as a blood reservoir. During a hemorrhage, the spleen releases blood into the blood circulation route. Serious injury to the spleen may require its removal.

The **thymus gland** is a bilobed mass of tissue located in the mediastinum along the trachea behind the sternum. Its role in the endocrine system was discussed in Chapter 12. It reaches maximum size during puberty and

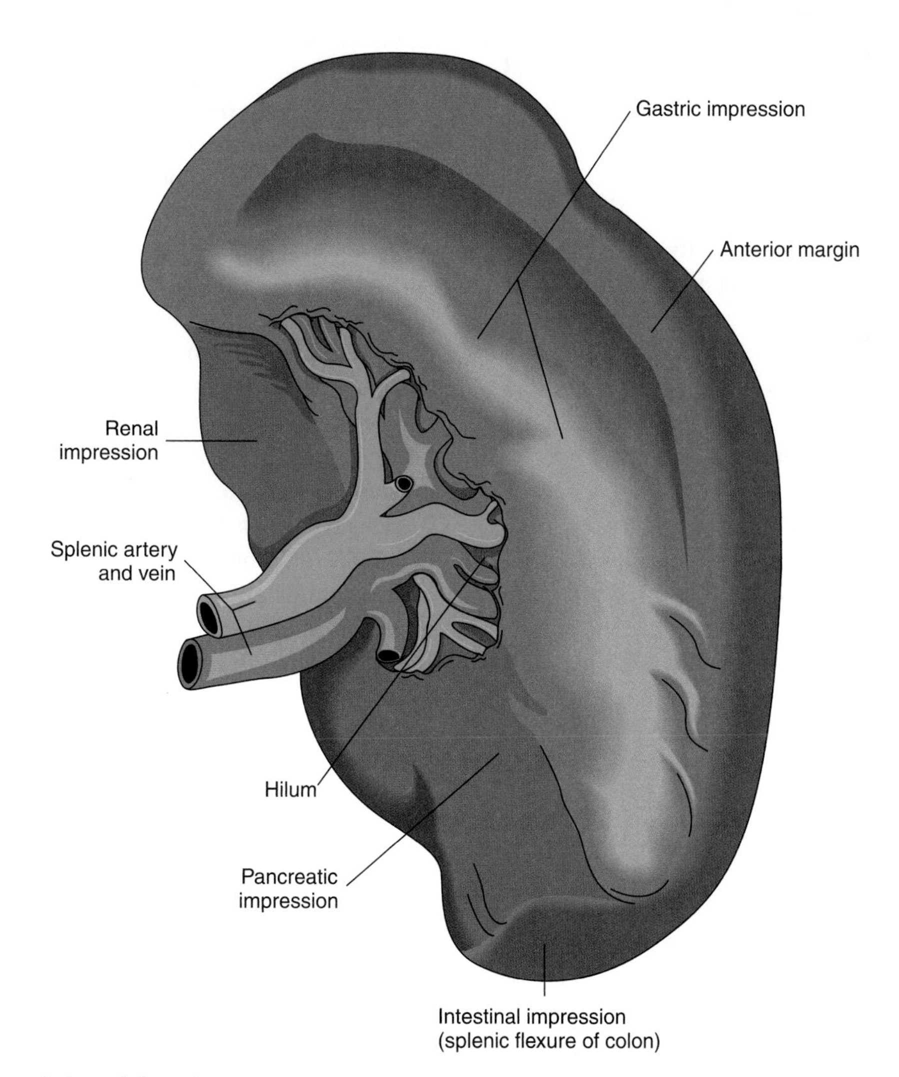

FIGURE 15-8. External view of the spleen.

then decreases. In older individuals, the thymus becomes small and is difficult to detect because it is replaced with fat and connective tissue. It is involved in immunity. The thymus is a site for lymphocyte production and maturation. The thymus helps develop T lymphocytes in the fetus and in infants for a few months after birth. A number of its lymphocytes degenerate, but those that mature leave the thymus and enter the blood to travel to other lymphatic tissues where they and their descendents protect against foreign substances and harmful microorganisms.

Peyer's (**PIE**-erz) **patches** (also known as aggregated lymphatic follicles) are found in the wall of the small intestine. They resemble tonsils. Their macrophages destroy bacteria. Bacteria are always present in large numbers in the intestine, and the macrophages prevent the bacteria from infecting and penetrating the walls of the intestine.

StudyWARE™ Connection

Watch an animation about lymph on your StudyWARE™ CD-ROM.

IMMUNITY

Immunity (im-**YOO**-nih-tee) is the ability of the body to resist infection from disease-causing microorganisms or **pathogens** (**PATH**-oh-jenz), damage from foreign substances, and harmful chemicals. **Humoral immunity** and **cellular immunity** are the results of the body's lymphoid tissue. The bulk of our lymphoid tissue is located in the lymph nodes. However, as mentioned, it is also found in the spleen, the tonsils, in the small intestine in Peyer's patches, and, to a lesser extent, in red bone marrow. This lymphoid tissue consists primarily of lymphocytes that can be categorized into two broad groups of cells: the B lymphocytes and the T lymphocytes.

The **B lymphocytes** are the cells that produce antibodies, and they provide humoral immunity. This type of immunity is particularly effective against circulating bacterial and viral infections. The B cells produce the circulating antibodies that attack the invading agent. B lymphocytes that enter tissues become specialized cells called plasma cells.

The **T lymphocytes** are responsible for providing cellular immunity. These cells come from the thymus gland, where immunologic competence is conferred on the T lymphocytes around birth. This type of immunity is particularly effective against fungi, parasites, intracellular (inside the cell) viral infections, cancer cells, and foreign tissue implants.

ANTIGENS AND ANTIBODIES

Antigens (**AN**-tih-jenz) are foreign proteins that gain access to our bodies via cuts and scrapes, through the respiratory system, through the digestive or circulatory systems, or through the urinary and reproductive systems. They cause the immune system to produce high molecular weight proteins, called **antibodies** or **immunoglobulins** (**im**-yoo-noh-**GLOB**-yoo-linz), to destroy the foreign invader.

These foreign proteins can be the flagella or cell membranes of protozoans, the flagella or cell membrane of bacteria, the protein coat of a virus, or the surface of a fungal spore. The B lymphocyte and the plasma cell recognize these antigens and produce antibodies that bind with the specific antigen. This binding causes the foreign cells to agglutinate (stick together) and precipitate within the circulatory system or tissues. Then the phagocytic white blood cells like neutrophils and macrophages come along and eat them up by phagocytosis, eliminating them from the body. Thus, we have an internal defense system to protect us from foreign microbes. Antibodies are formed in response to an enormous number of antigens. Antibodies have a basic structure consisting of four amino acid chains linked together by disulfide bonds. Two of the chains are identical, with about 400 amino acids, and are called the heavy chains; the other two chains are half as long, identical, and are called the light chains. When united, the antibody molecule is made of two identical halves, each with a heavy and light chain. The molecule has a Y shape and the tips of the Y are the antigen-binding sites. The binding site varies, thus allowing the antibody to bind with the enormous number of antigens. The stem of the Y is always constant. (See Figure 15-9.)

Five types of antibodies make up the gamma globulins of plasma proteins. **Immunoglobulin G (IgG)** is found in tissue fluids and plasma. It attacks viruses, bacteria, and toxins. It also activates **complement**, a set of enzymes that attack foreign antigens. **Immunoglobulin A (IgA)** is found in exocrine gland secretions, nasal fluid, tears, gastric and intestinal juice, bile, breast milk, and urine. **Immunoglobulin M (IgM)** develops in blood plasma as a response to bacteria or antigens in food. **Immunoglobulin D (IgD)** is found on the surface of B lymphocytes and is important in B cell activation. **Immunoglobulin E (IgE)** is also found in exocrine gland secretions and is associated with allergic reactions, attacking allergy-causing antigens. The most abundant antibodies are IgG, IgA, and IgM.

When B lymphocytes come in contact with antigens and produce antibodies against them, this is called **active immunity**. It can be acquired naturally, as when we

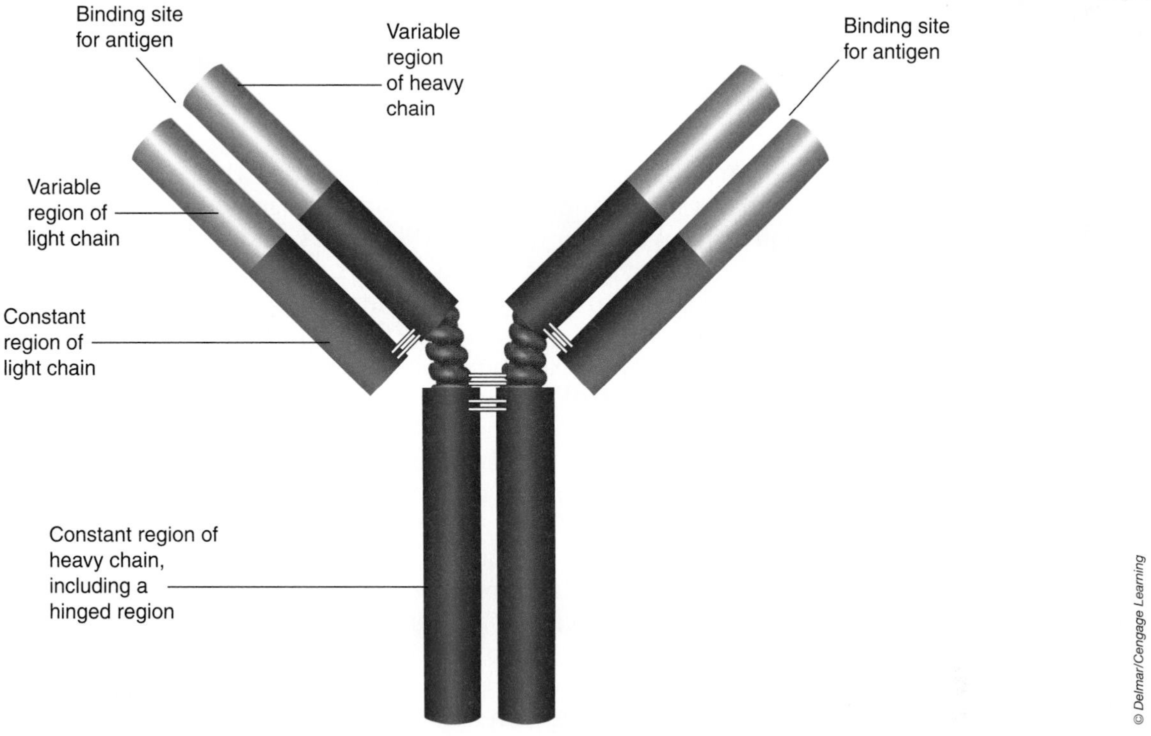

FIGURE 15-9 The structure of an antibody.

are exposed to a bacterial or viral infection, or it can be acquired artificially, as when we receive a vaccine. A vaccine contains either killed pathogens or live, but very weak pathogens. It does not matter whether the antigen is introduced to the body on its own or through a vaccine, the immune response is the same. The advantage of vaccines is that we do not experience the major symptoms of the disease, which would occur in the primary response to the pathogen, and the weakened antigen stimulates antibody production and immunologic memory. Future exposure keeps us immune to the pathogen. Vaccines are currently available against measles, smallpox, polio, tetanus, chickenpox, pneumonia, diphtheria, and various strains of flu.

Passive immunity can be conferred naturally when a fetus receives its mother's antibodies through the placenta and they become part of the fetal circulatory route. This immunity lasts for several months after birth. Passive immunity can be conferred artificially by receiving gamma globulin, breast milk, or immune serum. This is used after exposure to hepatitis. These donated antibodies provide immediate protection, but it only lasts 2 to 3 weeks. Other immune serums include antivenom for snakebites or botulism and rabies serum.

Like the B lymphocytes, T lymphocytes are activated to form clones by binding with an antigen. But T cells are not able to bind with free antigens. The antigens must first be engulfed by macrophages, processed internally, and then displayed on their surface to the T cells. Thus, *antigen presentation* is a major role for macrophages and is absolutely necessary for activation and clonal response of the T cells.

CELLS OF THE IMMUNE RESPONSE AND OTHER DEFENSES

The lymphocytes of the body are the precursors of a whole range of cells that are involved in the immune response. The following is a list of those cells and their functions.

B cells are lymphocytes found in the lymph nodes, spleen, and other lymphoid tissue where they replicate, induced by antigen-binding activities. Their clones or progeny form plasma cells and memory cells.

Plasma cells are formed by replicating B cells that enter tissue, and produce huge numbers of the same antibody or immunoglobulin.

Helper T cells are T cells that bind with specific antigens presented by macrophages. They stimulate the production of killer T cells and more B cells to fight the invading pathogen. They release lymphokines.

Killer T cells kill virus-invaded body cells and cancerous body cells. They are also involved in graft rejections.

Suppressor T cells slow down the activities of B and T cells once the infection is controlled.

HEALTH ALERT

HAPPINESS AND HEALTH

How often have you heard that laughter is the best medicine? This really does have some truth. Scientific research has shown that stress debilitates the immune system. People under long-term stress frequently get minor illnesses, like colds, more often than individuals who are leading stress-free lifestyles. Our moods, beliefs, and feelings can affect our level of health. For example, people with strong positive attitudes about life are often able to overcome illnesses much quicker than individuals who constantly harbor negative feelings. Stress is the great open door for the development of disease. During stressful situations, the adrenal cortex secretes cortisol. One of this hormone's effects is to suppress immune system activities. Feelings of despair, chronic melancholy, hopelessness, and inadequate abilities for self-help can all contribute to increased susceptibility to disease.

"A smile a day can help keep the doctor away." Individuals with feelings of strong social acceptance who enjoy their work and have a strong commitment to achievement and who believe they control their own destiny are much more likely to be happy. These individuals are resistant to any negative effects that stress may exert over their lives. In addition, scientists recommend that we eat balanced meals, get adequate sleep, and relax periodically. Why not take a "real" vacation to maintain a healthy immune system? The amount of sleep needed to relieve or control stress varies, with individuals averaging between 6 and 8 hours nightly, but everyone can smile and laugh on a daily basis. ■

Memory cells are descendents of activated T and B cells produced during an initial immune response. They will exist in the body for years, enabling it to respond quickly to any future infections by the same pathogen.

Macrophages engulf and digest antigens. They then present parts of these antigens in their cell membranes for recognition by T cells. This antigen-presentation function is crucial for normal T-cell responses.

In addition to these cells, certain chemicals are produced in the immune response that also help keep us healthy. The **lymphokines** (**LIM**-foh-kynz) are chemicals released by the sensitized T lymphocytes. There are a number of these chemicals. Chemotactic factors attract neutrophils, basophils, and eosinophils to the infected area. Macrophage migration-inhibiting factor (MIF) keeps macrophages in the local area of infection and inflammation. Helper factors stimulate plasma cells to produce antibodies. Interleukin-2 stimulates proliferation of T and B cells. Gamma interferon helps make tissue cells resistant to viruses, activates macrophages, and causes killer T cells to mature. Perforin causes cells to break down. Suppressor factors suppress antibody formation by T cells.

Activated macrophages also release chemicals called **monokines** (**MON**-oh-kynz). One is interleukin-1, which stimulates T-cell proliferation and causes fever. The body produces fever or elevated temperatures as a response to attempt to kill the invading pathogen by changing its environment. The other is tumor necrosis factor (TNF), which kills tumor cells and attracts the granular leukocytes to the area. Blood-borne proteins, called complement, cause the breakdown or lysis of microorganisms and enhance the inflammatory response.

In addition to these cellular and chemical barriers inside the body, the body has an external covering and other protective mechanisms. The skin's epidermis is a mechanical barrier to pathogens and toxins. It also has a so-called acid mantle (acidic pH) that inhibits bacterial growth. Sebum from the sebaceous glands has antifungal and antibacterial qualities. Tears from the lacrimal glands and saliva contain lysozyme, which destroys bacteria. Mucous membranes lining the digestive, respiratory, urinary, and reproductive tracts trap microorganisms and dust and prevent them from entering the circulatory system. In the nose and throat, the mucus-dust package is brought up to the throat to be swallowed by the action of cilia on the free edge of the epithelial tissue. The hydrochloric acid in the stomach then destroys most pathogens. Even the hairs in our nose have a role to trap large particles and filter them out before they enter the respiratory system. Refer to Figure 15-10 for an overview of the body's defense mechanisms.

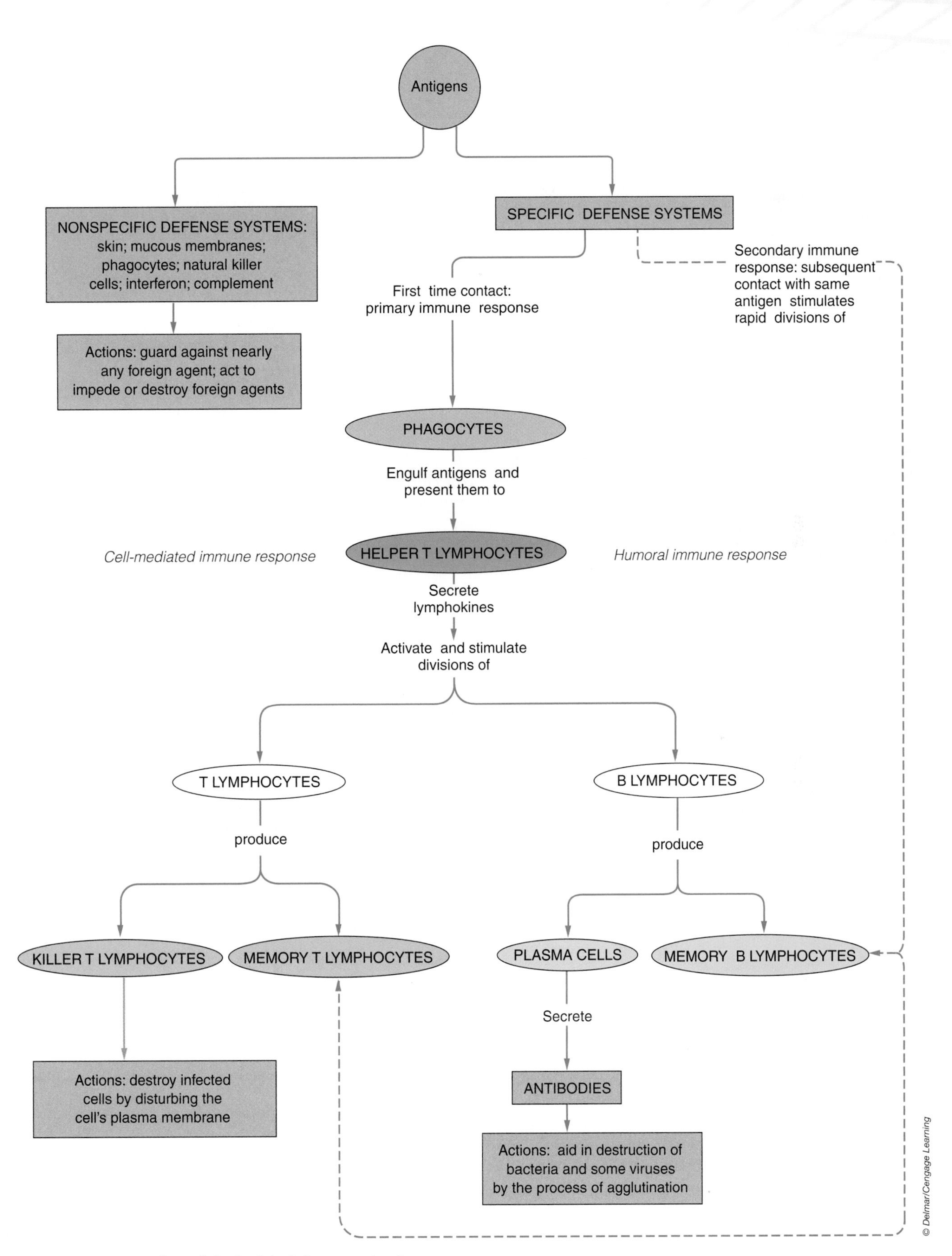

FIGURE 15-10. Overview of the body's defense mechanisms.

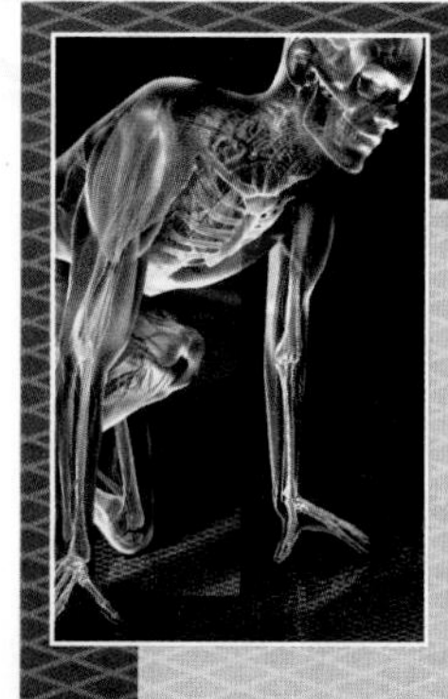

AS THE BODY AGES

At birth, the thymus gland is fairly large but decreases in size as we age. It eventually is replaced with adipose tissue, thus losing its ability to produce new mature T cells. The number of T cells in the body remains fairly stable, however, due to their constant replication in secondary lymphatic tissues. With advancing age, we become more susceptible to infections because our immune response decreases. Older adults also tend to produce more antibodies against their own body cells, resulting in autoimmune deficiencies.

As the aging process progresses, T cells become less responsive to antigens and fewer of them respond to infections. Since the T-cell population decreases as we age, B cells also become less responsive, so antibody levels do not increase as rapidly in response to antigens. It is therefore recommended that vaccinations be administered to older adults before the age of 60. Older adults also should be encouraged to receive yearly influenza shots because they are more susceptible to the flu.

With decreased immunity as we age, latent pathogens can be reactivated; for example, the chickenpox virus that erupts in children can remain in the body in a latent state within nerve cells. As we age, the virus can leave the nerve cells and enter the skin cells, resulting in painful lesions known as shingles.

The increased susceptibility of older adults to various types of cancers is also assumed to be related to the immune system's decreased response to cancer cells.

Career FOCUS

Individuals with an interest in the lymphatic system can pursue careers related to the study of the immune response.

- **Immunologists** are physician specialists who study reactions of tissues of the body's immune system to stimulation from antigens.
- **Oncologists** are physicians who specialize in the treatment and study of any abnormal growth of new tissue, benign or malignant. Such a physician is an authority on cancerous diseases.
- **Cytotechnologists** are individuals who work in a clinical laboratory utilizing microscopy in examining cell samples for signs of cancer.
- **Lymphedema therapists** are individuals who utilize massage therapy, exercises, and bandaging to relieve the swellings caused by blockage of the lymphatic vessels. They also train patients with lymphedema on how to care for themselves, utilizing water massage or hydrotherapy and exercise.

COMMON DISEASE, DISORDER, OR CONDITION

DISORDERS OF THE LYMPHATIC SYSTEM

ALLERGIES

Allergies (**AL**-er-jeez) are hypersensitive reactions to common, normal, usually harmless environmental substances referred to as allergens. Some examples are house dust, pollen, and cigarette smoke. The reactions to the allergens can damage body tissues. Over 20 million Americans have allergic reactions to inhaled allergens. When the nose is affected, we refer to it as hay fever. In the lungs, it is referred to as asthma. If the eyes are affected, it is called allergic conjunctivitis. If the material is swallowed and affects the digestive tract, it can result in diarrhea, vomiting, or cramps. Contact on the skin can result in contact dermatitis. Hives can develop when certain foods cause allergic reactions or when drugs, animal hairs, or insect stings are encountered.

When exposed to a certain allergen, antibody IgE is produced and attaches to basophils and mast cells. These cells release histamine and prostaglandins. Histamine causes the secretion of mucus from mucous membranes and causes capillaries to become more permeable. Prostaglandins cause smooth muscle to constrict, as in the bronchioles of the lungs. Such an inflammatory response produces a runny nose, sneezing, difficult breathing, and congestion as with asthma and hay fever.

Severe allergic reactions can result in anaphylactic shock, culminating in death. Some individuals are overly sensitive to bee stings and certain drugs, resulting in severe bronchial constriction, mucus production, and breathing difficulties. This lowers blood pressure and can lead to death. This severe allergic reaction is called anaphylaxis.

Treatment can use desensitization, which is exposure to minute amounts of the allergen over time, or the use of antihistamines, bronchial dilators, or steroids.

LYMPHOMA

Lymphoma (lim-**FOH**-mah) is a tumor of lymphatic tissue, which usually is malignant. It begins as an enlarged mass of lymph nodes, usually with no accompanying pain. The enlarged nodes will compress surrounding structures, causing further complications. The immune response becomes depressed, and the individual becomes susceptible to opportunistic infections. Lymphomas are classified into two groups: Hodgkin's disease and non-Hodgkin's lymphomas. The name Hodgkin's disease was named after an English physician in 1832 who first described a group of patients with lymph node swellings in the neck. The disease usually manifests itself in the 20s or 30s, more commonly in men than in women. The disease tends to involve the reticulum cells of the lymph node rather than the lymphocytes. Treatment with drugs and radiation is effective for most people with lymphomas.

LYMPHADENITIS

Lymphadenitis (lim-**fad**-en-**EYE**-tis) is an inflammation of lymph nodes or glands. They become enlarged and tender. When microorganisms are being trapped and attacked in the lymph nodes, they enlarge. Hence, a swollen lymph gland is indicative of an infection. Very often when you visit a physician when you are ill, the doctor will feel for swollen lymph nodes in the neck region.

LYMPHANGITIS

Lymphangitis (lim-**fan**-JYE-**tis**) is an inflammation of the lymphatic vessels with accompanying red streaks visible in the skin.

BUBONIC PLAGUE

Bubonic plague (boo-**BON**-ik **PLAYGH**) is a disease of the lymphatic system with historical implications. It is caused by the bacterium *Klebsiella pestis* transmitted to humans by the bite of the Asiatic rat flea *Xenopsylla cheopis*. The bacteria grow in the lymph nodes, causing them to enlarge, forming dark swellings called buboes. The bacteria also get into the bloodstream, causing septicemia. Without treatment, death follows quickly in about 80% of cases. In the Middle Ages, bubonic plague

(continues)

COMMON DISEASE, DISORDER, OR CONDITION

DISORDERS OF THE LYMPHATIC SYSTEM (continued)

epidemics wiped out a third of the population of Europe. Before effective treatment with antibiotics, like penicillin, bubonic plague outbreaks occurred throughout the world. Fortunately, few cases ever occur today.

AIDS

AIDS, or acquired immune deficiency syndrome, is caused by infection with the human immunodeficiency virus (HIV). This virus is transmitted by contact with body fluids containing the virus through sexual contact, including anal intercourse, through contaminated needles, during birth from an infected mother, or by receiving contaminated blood in a transfusion. The infection has three stages: initial symptoms, a latency period, and full-blown AIDS. Initial symptoms include weakness, fever, night sweats, weight loss, and swollen lymph glands in the neck region. These symptoms mimic the flu and last only a few days. The latency period may last 5 to 10 years with no symptoms. Full-blown AIDS occurs with the onset of opportunistic infections that can be fatal. Some such infections are pneumonia, skin cancer, diarrhea, tuberculosis, toxoplasma affecting the nervous system, and fungal infections in the lungs and throat.

The virus attacks the T cells, compromising the immune response. It invades the T cells, reproducing more viruses in the T cells, and eventually destroys the T cells. Hence, a person with the virus has the T-cell count constantly monitored by a physician. The virus also invades macrophages but does not destroy them utilizing them to produce more viral cells. Some common AIDS-related symptoms include weight loss due to diarrhea, persistent swollen lymph glands, chronic low-grade fever, fatigue, night sweats, and, occasionally, memory loss and confusion.

Research is constantly going on to try to understand and combat this dreaded disease. Some individuals have the virus but never develop any symptoms; others get exposed repeatedly but never become infected. By studying the genetics and backgrounds of these individuals, scientists may find a cure in the future. Meanwhile, drugs are used to combat the virus. Older drugs like AZT blocked viral replication but had serious side effects. New drugs, called protease inhibitors, inhibit the HIV from becoming functional, thus crippling the virus. Combinations of drugs, called the AIDS cocktail, have extended the life of many infected individuals by stemming the growth and activity of the virus, thus giving new hope for controlling and eventually curing AIDS.

BONE MARROW TRANSPLANT

Bone marrow transplants are done to treat various disorders of the human body including disorders of the lymphatic system such as leukemia, lymphomas, and immunological deficiencies. Red bone marrow contains blood cells in all stages of development, since its role is hematopoiesis. Red bone marrow from a healthy donor is transplanted intravenously into an ill recipient after chemotherapy and body radiation treatment. The bone marrow from the donor will produce whatever blood cells are needed by the recipient dependent upon the specific disease experienced by the patient.

CANCER AND LYMPH NODES

Cancer cells can be spread by growing in lymph nodes and being transported throughout the body by the lymphatic circulatory system. The cancer cells originally came from a tumor from which they metastasized and entered lymphatic vessels. The vessels carry them to lymph nodes where they become trapped and reproduce. Some escape from the lymph nodes and are carried by the lymphatics to the blood circulatory system, which then carries them to other parts of the body where they develop into more tumors. For example, when a woman develops advanced breast cancer, the axillary lymph nodes are also removed in addition to the cancerous breast in a radical mastectomy to prevent the possible spread of the cancer from the lymph nodes.

COMMON DISEASE, DISORDER, OR CONDITION

DISORDERS OF THE LYMPHATIC SYSTEM (continued)

SYSTEMIC LUPUS ERYTHEMATOSUS (SLE)

Systemic lupus erythematosus (SLE) is a chronic inflammatory disease in which cells and tissues are damaged by the immune system. One of the symptoms is a red butterfly shaped rash over the nose and cheeks. The exact cause of the disease is not known but a viral infection may disrupt the normal functioning of the immune system. Eight times as many women than men develop SLE. There also appears to be a genetic connection. In addition to severe inflammation of blood vessels, there can be kidney involvement leading to renal disease. Other systems of the body that can be affected are the respiratory and the nervous system. Symptoms first appear between the ages of 15 and 25. These symptoms may go into remission but periodic, unpredictable flare-ups will occur. If a patient survives this condition for 10 years or more, the survival rate can be as high as 90%. Death results from kidney failure, heart disease, central nervous system involvement with severe neurological abnormalities, and infections. Treatment involves the careful and monitored use of steroids, and antimalarial drugs to treat skin rashes and joint pain. Patients are also recommended to protect themselves from direct exposure to the rays of the sun by using sunscreen, and to avoid stressful situations. Avoiding fatigue and getting lots of rest also help prevent periodic flare-ups.

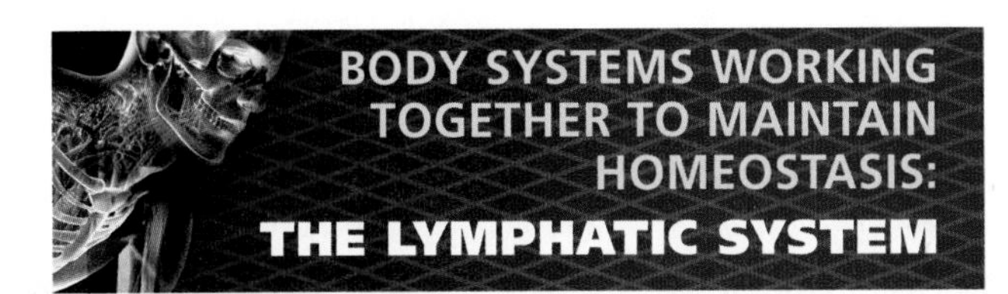

Integumentary System
- The skin's epidermis is a mechanical barrier to microorganisms.
- The acid pH mantle of the skin inhibits the growth of most bacteria.
- Sebum, produced by the sebaceous glands of the skin, has antifungal properties.
- Lymphatic vessels drain interstitial fluid from the dermis of the skin preventing edema.

Skeletal System
- Lymphocytes are produced in red bone marrow.

Muscular System
- Contraction of muscles compresses lymphatics and helps push the flow of lymph toward the right and left lymphatic ducts.

Nervous System
- Undue stress may suppress the immune response.
- The nervous system innervates large lymphatic vessels and helps regulate the immune response.

Endocrine System
- The thymus gland confers immunologic competence on the T lymphocytes.
- Hormones stimulate the production of lymphocytes.

Cardiovascular System
- Blood plasma is the source of interstitial fluid, which becomes lymph when drained by lymphatic capillaries.
- The lymphatic system returns this fluid to the bloodstream via the right and left subclavian veins, connecting with the right and left lymphatic ducts.

Digestive System
- Lacteals in the villi of the small intestine absorb fats.
- The hydrochloric acid in the gastric juice destroys most pathogens.
- The digestive system digests and absorbs nutrients for lymphatic tissues.
- Peyer's patches in the wall of the small intestine destroy bacteria.

Respiratory System
- The tonsils are located in the pharynx.
- Breathing and contraction of the respiratory muscles maintain lymph flow through lymphatics.
- Immune system cells receive their oxygen and get rid of carbon dioxide waste via the respiratory system.

Urinary System

- The kidneys maintain homeostasis by regulating the amounts of extracellular fluid.
- Electrolyte and acid-base levels of the blood are maintained by the urinary system for lymphoid tissue function.
- Urine can flush out certain microorganisms from the body.

Reproductive System

- The acid environment of the female vagina and male urethra prevent bacterial growth.
- In the female reproductive tract, the immune system does not attack the male sperm as a foreign antigen, ensuring the possibility of fertilization.

HEALTH ALERT

SPLENOMEGALY

Splenomegaly is an abnormal enlargement of the spleen that can develop from infections with diseases such as scarlet fever, syphilis, typhoid fever, and typhus fever. Infections with a microscopic blood fluke worm of the genus *Schistosoma* in Japan and in Africa can also cause splenomegaly. Eggs of the parasite get carried by the bloodstream and lodge in the spleen, causing irritation. This results in a swollen abdomen caused by the enlarged spleen.

*Study*WARE™ Connection

Watch an animation on cancer metastasizing on your StudyWARE™ CD-ROM.

HEALTH ALERT

SPLENIC ANEMIA

Splenic anemia is also characterized by an abnormal enlargement of the spleen caused by hemorrhages that develop from the stomach and fluid accumulating in the abdomen. This condition usually requires the surgical removal of the spleen.

SUMMARY OUTLINE

INTRODUCTION

1. The lymphatic system transports a fluid called lymph through lymphatic capillaries and vessels called lymphatics.
2. The system controls body fluids and destroys harmful microorganisms.
3. The system consists of lymph, lymph vessels, lymph nodes, the tonsils, the spleen, the thymus gland, and Peyer's patches in the intestine.

THE FUNCTIONS OF THE SYSTEM AND THE STRUCTURE AND FUNCTIONS OF THE LYMPHATIC VESSELS

1. The functions of the system are to drain interstitial fluid from tissue spaces, to transport fats from the digestive tract to the blood, to develop immunities, and to produce lymphocytes.
2. Some blood plasma gets forced through the blood capillary walls into spaces between tissue cells. This is called interstitial fluid.
3. Lymphatic capillaries drain interstitial fluid, which is now called lymph, and pass this fluid on to lymph vessels called lymphatics.
4. In the villi of the small intestine, special lymphatics called lacteals pick up fats and transport them to the blood.
5. Lymph in lacteals looks milky due to the fats, and is called chyle.

Lymphatic Vessels

1. Lymphatic vessels originate singly or in plexuses as blind-end tubes called lymphatic capillaries between cells in most parts of the body.
2. Lymphatic capillaries unite to form larger vessels called lymphatics, which resemble veins but are thinner and have more valves. Valves ensure a one-way flow of lymph.
3. Eventually, all lymphatics converge into two main channels: the thoracic duct and the right lymphatic duct.

Lymph Nodes

1. Lymph nodes, also called lymph glands, are found along the lengths of lymphatics. Groupings of lymph nodes are present in the groin, armpit, and neck region.
2. Efferent lymphatics exit the lymph node at the hilum, a slight depression on one side. Blood vessels and nerves also exit and enter at the hilum.
3. Afferent lymphatics enter the lymph node at various locations on the node.
4. Capsular extensions of the node, called trabeculae, divide the node internally into a series of compartments with germinal centers.
5. The germinal centers produce lymphocytes.

LYMPH CIRCULATION

1. Plasma, filtered by the blood capillaries, passes into interstitial tissue spaces and is now called interstitial fluid.
2. When this fluid passes into lymphatic capillaries, it is called lymph.
3. The lymph now passes into larger lymphatic vessels called lymphatics, which have many valves to prevent backflow of lymph and have lymph nodes along their lengths.
4. Afferent lymphatics enter the lymph nodes, and efferent lymphatics leave the nodes.
5. Circulation of lymph is maintained by muscular contractions, which compress lymphatics and push the lymph along.
6. Eventually, efferent lymphatics unite to form six lymphatic trunks.
7. The lumbar trunk drains the lower extremities and pelvis.
8. The intestinal trunk drains the abdominal region.
9. The bronchomediastinal trunk and the intercostal trunk drain the thorax.
10. The subclavian trunk drains the upper extremities.
11. The jugular trunk drains the head and neck.
12. These trunks drain their lymph into two main collecting ducts: the main duct, called the thoracic duct, and the right lymphatic duct.
13. The thoracic duct empties its lymph into the left subclavian vein. The right lymphatic duct drains into the right subclavian vein. This process returns lymph to the blood vessels from whence it originated.

THE ORGANS OF THE LYMPHATIC SYSTEM

1. The three groups of tonsils are the palatine tonsils (commonly removed in a tonsillectomy); the

pharyngeal tonsils, or adenoids; and the lingual tonsils.

2. The tonsils are composed of reticuloendothelial cells that protect the nose and oral cavity from pathogens.
3. The spleen is the single largest mass of lymphatic tissue in the body. It phagocytizes worn-out red blood cells and platelets. It destroys bacteria. It produces lymphocytes and plasma cells and functions as a blood storage organ.
4. The thymus gland is the site for T lymphocyte production and maturation.
5. Peyer's patches resemble tonsils but are found in the walls of the small intestine where their macrophages destroy bacteria.

IMMUNITY

1. Immunity is the ability to resist infection from microorganisms, damage from foreign substances, and harmful chemicals.
2. Humoral immunity and cellular immunity are produced by the body's lymphoid tissues.
3. Lymphoid tissue produces two main groups of lymphocytes: the B lymphocyte cells and the T lymphocyte cells.
4. The B cells produce antibodies and provide humoral immunity, which is effective against circulating bacterial and viral infections.
5. The T cells are responsible for providing cellular immunity, which is effective against intracellular viruses, fungi, parasites, cancer cells, and foreign tissue implants.
6. B cells that enter tissues and become specialized cells are known as plasma cells.

ANTIGENS AND ANTIBODIES

1. An antigen is a foreign protein that gains access to our bodies. Some examples are the cell membrane or flagella of protozoans, the protein coat of a virus, the surface of a fungal spore, and the flagella or cell membranes of bacteria.
2. The B lymphocytes recognize antigens and produce antibodies, which bind to specific antigens, causing the foreign antigens to agglutinate and precipitate.
3. Phagocytic white blood cells then eat up the invading microorganism.
4. The antibody molecule has a Y shape. The binding sites are the tips of the Y.
5. Antibodies are also called immunoglobulins (Ig). There are five types that make up the gamma globulins of blood plasma: IgG found in tissue fluids and plasma; IgA found in exocrine gland secretions, nasal fluid, tears, gastric and intestinal juice, bile, breast milk, and urine; IgM found in plasma as a response to bacteria in food; IgD found on the surface of B cells; and IgE associated with allergic reactions found in exocrine gland secretions.
6. Active immunity occurs when B cells contact antigens and produce antibodies against them. It is acquired naturally when we are exposed to a viral or bacterial infection. It is acquired artificially when we receive a vaccine.
7. Passive immunity occurs naturally when a fetus receives antibodies from its mother through the placenta. Passive immunity is conferred artificially by receiving gamma globulin or immune serum via injection. Passive immunity is short-lived.
8. T cells cannot bind with free antigens like B cells. They must go through antigen presentation via macrophages. They engulf the antigen, process it internally, and then display the antigen on the surface of the macrophage.

CELLS OF THE IMMUNE RESPONSE AND OTHER DEFENSES

1. B cells, found in lymphoid tissue, induce antigen-antibody-binding activities. Their replication produces clones that form plasma cells and memory cells.
2. Plasma cells produce huge amounts of antibodies.
3. Helper T cells bind with specific antigens presented by macrophages. They release lymphokines and stimulate the production of killer T cells and more B cells.
4. Killer T cells attack virus-invaded body cells and cancer cells. They also reject body grafts.
5. Suppressor T cells slow down the activity of B and T cells once the infection is controlled.
6. Memory cells are the descendants of activated B and T cells that remain in the body

for years, allowing the body to respond to future infections.

7. Macrophages engulf and digest antigens and present them to T cells for recognition.
8. Lymphokines are chemicals released by T cells: chemotactic factors attract neutrophils, eosinophils, and basophils; MIF keeps macrophages in the inflamed and infected area; helper factors stimulate plasma cells to produce antibodies; interleukin-2 stimulates proliferation of T and B cells; gamma interferon makes tissue cells resistant to viruses, activates macrophages, and matures killer T cells; suppressor factors stop antibody production by T cells.
9. Monokines are chemicals produced by macrophages: interleukin-1 stimulates T-cell production and fever; TNF kills tumor cells and attracts granular leukocytes; complement causes the lysis of microorganisms and enhances the inflammatory response.
10. Skin is a mechanical barrier, and its acid mantle inhibits bacterial growth. Sebum has antibacterial and antifungal properties.
11. Lysozyme, in tears and saliva, attacks bacteria.
12. Mucous membranes trap microorganisms and debris.
13. Hydrochloric acid in the stomach destroys most microorganisms.

REVIEW QUESTIONS

1. Name the functions of the lymphatic system.
*2. Compare the anatomy of a vein to a lymphatic vessel.
*3. Discuss what factors keep lymph flowing in a one-way direction without the aid of a pumping organ?
4. Name the major lymphatic trunks of the body and what they drain.
5. Beginning at a lymph capillary, describe the flow of lymph, ending at the two main collecting ducts, and name the vessels of the system.
6. Name the five types of antibodies that make up the gamma globulins of plasma proteins.
7. Give examples of active and passive immunity.

***Critical Thinking Questions**

FILL IN THE BLANK

Fill in the blank with the most appropriate term.

1. At a lymph node, ________________ vessels penetrate the capsule while ________________ vessels pass on to another node and unite to form lymph trunks.
2. The ________________ tonsils are the ones commonly removed by a tonsillectomy.
3. The ________________ is the largest single mass of lymphatic tissue in the body.
4. The ________________ confers immunologic competency on the T cells. As we age, it disintegrates and may disappear in adults.
5. If interstitial fluid builds up between tissue cells, ________________ or swelling will develop.

MATCHING

Place the most appropriate number in the blank provided.

_____ B cells	1. Immunoglobulin
_____ T cells	2. Vaccine
_____ Adenoid	3. Macrophage
_____ Antibody	4. Gamma globulin
_____ Blood storage	5. Lysis of microorganisms
_____ Active immunity	6. Humoral immunity
_____ Passive immunity	7. Lymphokines
_____ Antigen presentation	8. Pharyngeal tonsils
_____ Helper T cells	9. Lingual tonsils
_____ Complement	10. Spleen
	11. Cellular immunity
	12. Palatine tonsils

Search and Explore

- Write about a family member or someone you know that has one of the common diseases, disorders, or conditions introduced in this chapter, and tell about the disease.
- Visit the National Cancer Institute at http://www.cancer.gov. Choose a cancer related to this chapter from the *A to Z List of Cancers.* Give an oral presentation about your findings.

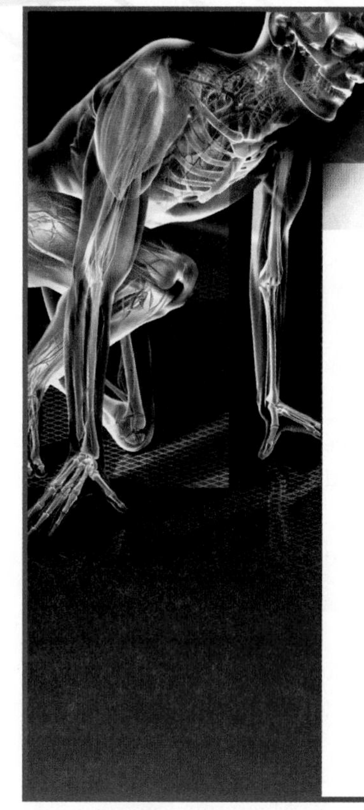

CASE STUDY

Quinn, a 20-year-old man, visits an outpatient clinic. He appears to be weak and lethargic. Quinn tells the health care provider that he feels like he has a *bad case* of the flu. He states he has night sweats that drench the bed, and that he has lost around 10 pounds since becoming ill. He seems anxious and admits that he has had unprotected sex with two male partners over the last few months. The care provider examines Quinn and documents that he has a fever and swollen lymph nodes in his neck.

Questions

1. Based on this information, what disorder might Quinn have?
2. How is this disorder transmitted from one person to another?
3. Why must the care provider continually monitor the T-cell count?
4. What complications can individuals with this disorder eventually develop?
5. What is the prognosis for individuals with this condition?

*Study*WARE™ Connection

Take a quiz or play interactive games on your StudyWARE™ CD-ROM.

Study Guide Practice

Go to your **Study Guide** for more practice questions, labeling and coloring exercises, and crossword puzzles to help you learn the content in this chapter.

LABORATORY EXERCISE: THE LYMPHATIC SYSTEM

Materials needed: A microscope slide of a lymph node, an anatomic model of the head or throat showing the location of the tonsils, a fetal pig for dissection, a dissecting pan, and a dissection kit

1. Examine a microscope slide of a lymph node. Notice that the lymph node is enclosed by a sheath of tissue called the *capsule.* The outer part of the node is called the *cortex* and the inner part is called the *medulla.* Within the medulla, note the darker *nodules* where the germinal centers that produce the lymphocytes are found. Remember that *afferent vessels* bring lymph into the node, and *efferent lymphatic vessels* take lymph out of the node. Compare your slide to Figure 15-5 in your text.
2. Your instructor will supply you with an anatomic model. Locate the three tonsils. The *palatine tonsils* are on the sides of the oral cavity. The *lingual tonsils* are at the back of the tongue, and the *pharyngeal tonsils* or *adenoids* are in the nasopharynx.
3. You will dissect a fetal pig in Chapter 16. You will then find and identify the spleen located near the stomach in the abdominal cavity. Remember, we located the thymus gland when we dissected the throat region in the lab on the endocrine system. Review this also when you perform your dissection on the digestive system.

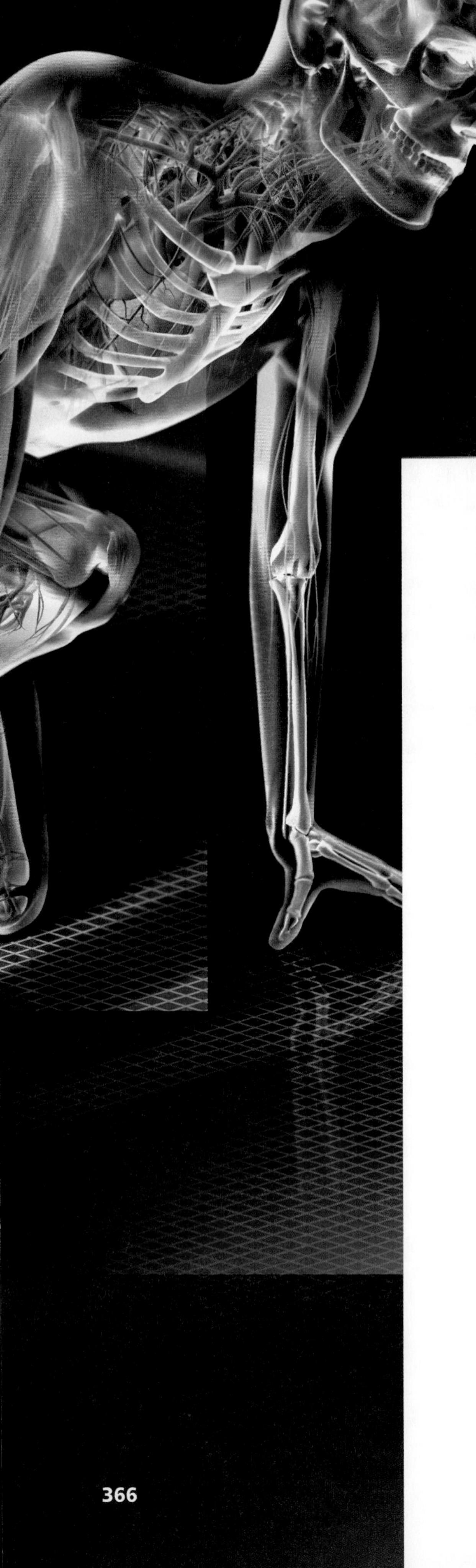

Nutrition and the Digestive System

CHAPTER OBJECTIVES

After studying this chapter, you should be able to:

1. List and describe the five basic activities of the digestive process.
2. List the four layers or tunics of the walls of the digestive tract.
3. Name the major and accessory organs of the digestive tract and their component anatomic parts.
4. Explain the major digestive enzymes and how they function.
5. Explain the functions of the liver.
6. Explain how absorption of nutrients occurs in the small intestine and how feces form in the large intestine.
7. Name and describe the functions of the organs of the digestive tract.

16

KEY TERMS

Absorption. 368
Acini 380
Adventitia 369
Alimentary canal. 368
Ampulla of Vater/ hepatopancreatic ampulla 380
Amylase 375
Anal canal 385
Anal columns. 386
Anus 386
Apical foramen 376
Ascending colon 385
Bicuspids 375
Bile duct. 384
Bowel. 385
Brunner's glands/ duodenal glands 384
Buccal glands. 375
Canine teeth 375
Cardia. 380
Cecum 385
Cementum 376
Cervix/neck 376
Chyme 384
Circumvallate papillae . . 374
Colon 385
Crown 376
Crypts of Lieberkuhn . . . 384
Cuspids. 375
Defecation 368
Deglutition 378
Dentes 375
Dentin 376
Descending colon 386
Diaphragm. 379
Digestion 368
Duct of Wirsung/ pancreatic duct 380
Duodenum. 380
Enamel. 376
Entamoeba histolytica . . 389
Escherichia coli 392
Esophageal hiatus. 379
Esophagus 379
Falciform ligament 383
Feces. 392
Filiform papillae 374
Food bolus. 378
Fundus 380
Fungiform papillae 374
Gallbladder 384
Gastroesophageal sphincter 380
Gastrointestinal tract . . . 368
Gingivae. 375
Glucagon 380
Glycogen 383
Hard palate 373
Haustrae 385
Helicobacter pylori 382
Heparin 383
Ileocecal valve. 384
Ileum 384
Incisors. 375
Ingestion 368
Insulin 380
Intestinal glands 384
Islets of Langerhans/ pancreatic islets. 380
Jejunum 384
Kupffer's cells 383
Lamina propria 371
Large intestine/bowel . . 385
Left colic (splenic) flexure 386
Lingual frenulum. 374
Lips. 373
Liver 383
Lower esophageal sphincter 380
Mastication 375
Mediastinum 379
Mesentery 373
Mesocolon 385
Microvilli 385
Molar teeth 375
Mucous cells 380
Mumps. 376
Muscularis mucosa 371
Nasopharynx. 378
Oropharynx. 378
Pancreas. 380
Pancreatic juice 380
Papillae 374
Parietal cells 380
Parotid gland. 375
Pepsin. 380
Pepsinogen 380

(*continues*)

INTRODUCTION

The function of the digestive system is to break down food (complex carbohydrates, proteins, and fats) via hydrolysis into simpler substances or molecules that can be used by the body's cells (Table 16-1). This process is called digestion. Digestion allows the body's cells to convert food energy into the high-energy adenosine triphosphate (ATP) molecules that run the cell's machinery. The major organs and accessory structures that perform this function are collectively referred to as the digestive system. The digestive system prepares food for use by cells through five basic activities:

1. **Ingestion** or the taking of food into the body
2. **Peristalsis** or the physical movement or pushing of food along the digestive tract
3. **Digestion** or the breakdown of food by both mechanical and chemical mechanisms
4. **Absorption** or the passage of digested food from the digestive tract into the cardiovascular and lymphatic systems for distribution to the body's cells
5. **Defecation** or the elimination from the body of those substances that are indigestible and cannot be absorbed.

See Concept Maps 16-1 and 16-2: The Digestive System.

GENERAL ORGANIZATION

The organs of digestion are part of two main groups. The first is the **gastrointestinal tract**, or **alimentary canal**, which is a long continuous tube that runs through the ventral cavity of the body and extends from the mouth to the anus (Figure 16-1). The length of this tube is approximately 30 feet, or 9 meters. Its organs include the mouth or oral cavity, oropharynx, esophagus, stomach, and the small and large intestine. Muscular contractions in the tube break down food physically by churning it; enzymes from cells in the tube's wall break down food chemically. The second group of organs consists of accessory structures. They include the teeth, tongue, salivary glands, liver and gallbladder, and pancreas.

Table 16-1 Digestion: An Example of Hydrolysis

Digestion is an example of a hydrolytic chemical reaction that uses water and enzymes to cause the breakdown of ingested food into smaller, simpler compounds.

Food Categories + Enzyme + Water ⟶	Simpler Compounds
1. Complex carbohydrates + Amylase + Water →	Simple sugars (from the complex carbohydrates: starch, glycogen)
2. Proteins + Proteases + Water →	Amino acids
3. Fats + Lipases + Water →	Fatty acids and glycerol

HISTOLOGY

The walls of the alimentary canal from the esophagus to the anal canal have the same arrangement of tissue layers. These layers are referred to as coats or tunics (Figure 16-2). There are four tunics of the canal. From the inside out they are called: the tunica mucosa (**TYOO**-nih-kah myoo-**KOH**-sah), the tunica submucosa, the tunica muscularis, and the **adventitia** (**ad**-vin-**TISH**-ee-ah), or **tunica serosa** (see-**ROH**-sah).

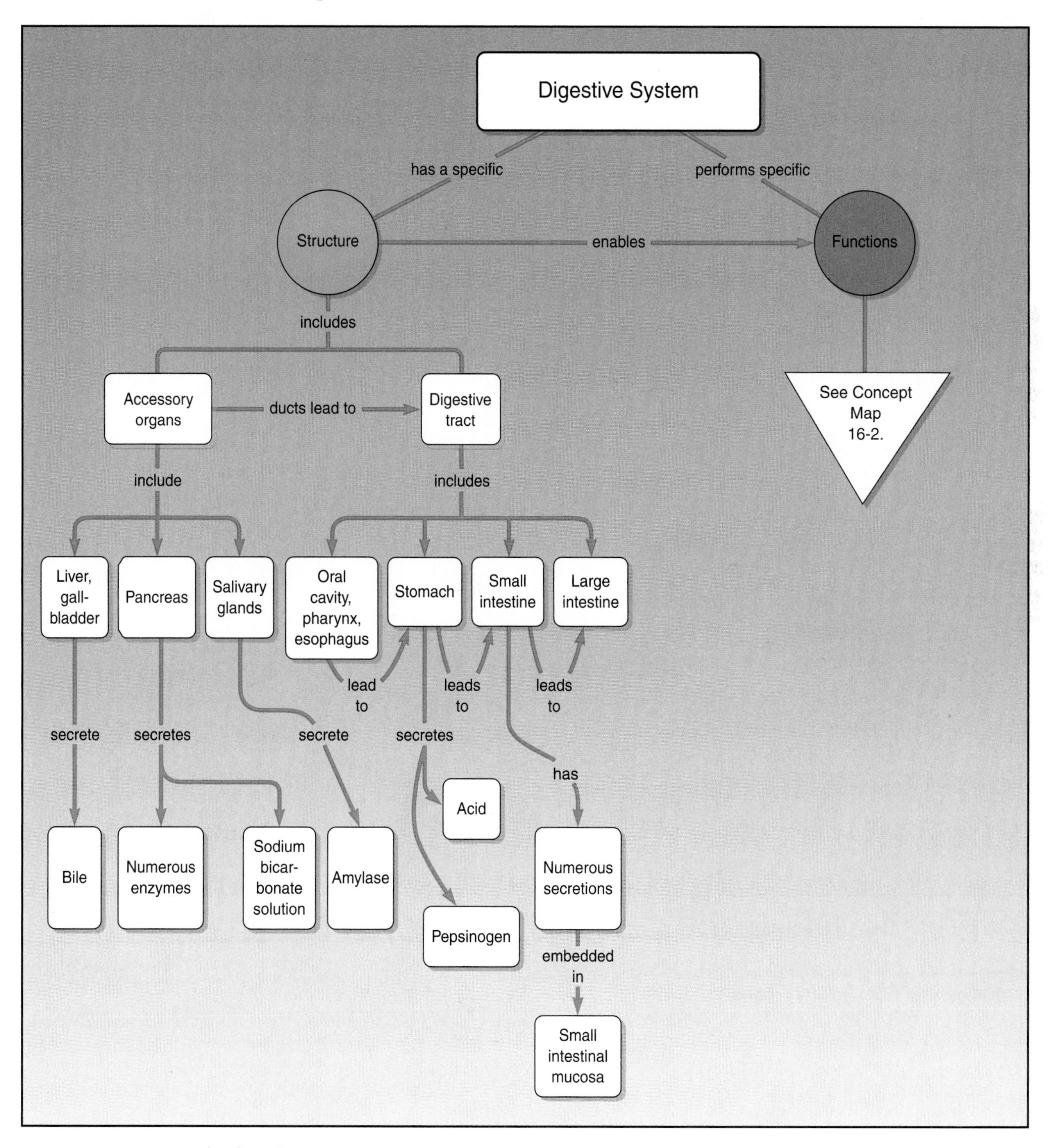

CONCEPT MAP 16-1. The digestive system.

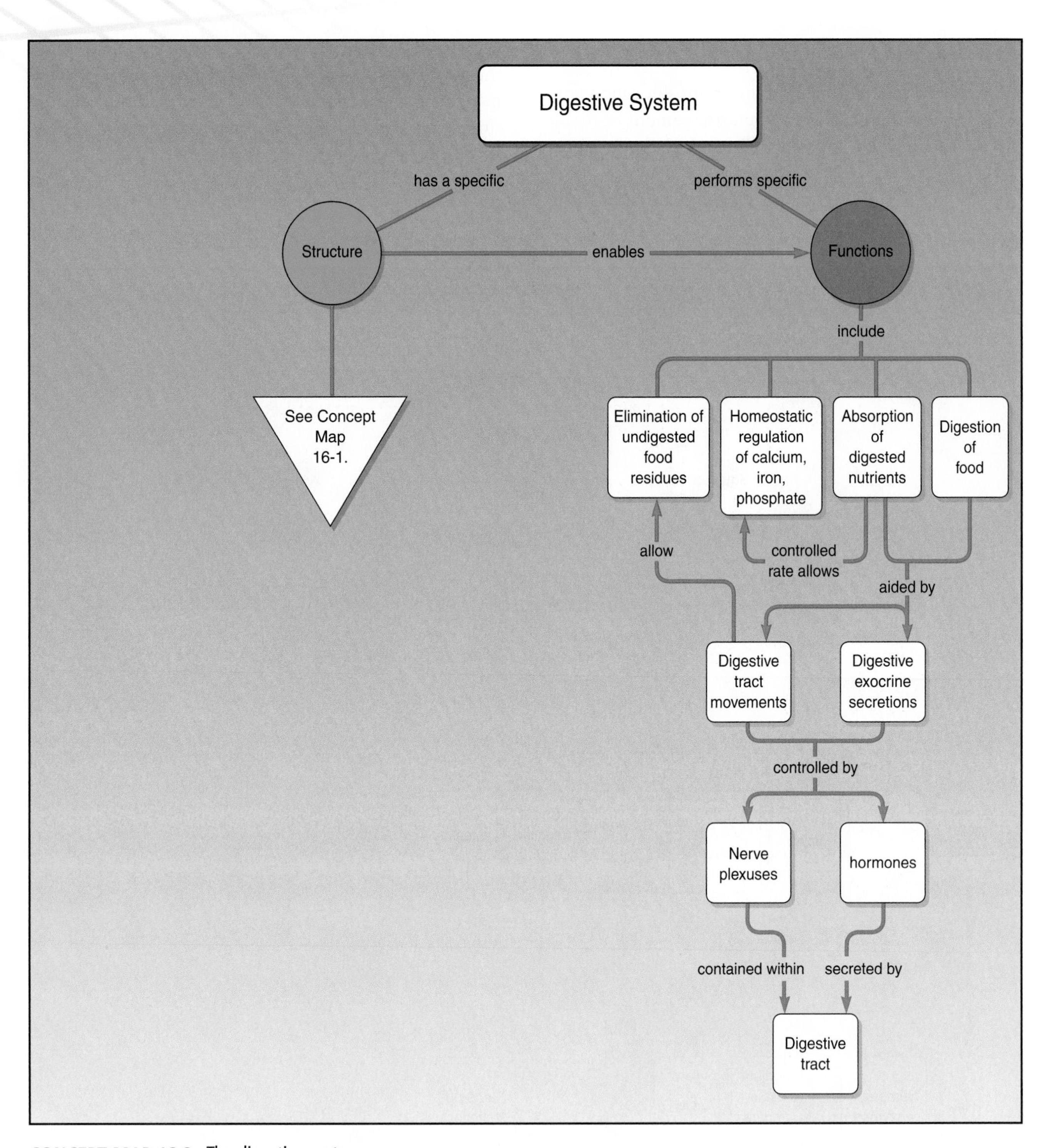

CONCEPT MAP 16-2. The digestive system.

HEALTH ALERT

ESSENTIAL NUTRIENTS

There are six classes of nutrients: carbohydrates, lipids (fats), proteins, vitamins, minerals, and water. These nutrients in combination contain the essential elements carbon, hydrogen, nitrogen, oxygen, phosphorus, and sulfur (CHNOPS). These are essential nutrients that a person must obtain from food because the body cannot manufacture them in sufficient amounts to meet its physiological needs.

It is essential that we choose good dietary guidelines to maintain a healthy body. These guidelines have been established by the U.S. Department of Agriculture and the U.S. Department of Health and Human Services:

- Eat a variety of foods.
- Balance the food you eat with physical activity and maintain or improve your weight.
- Pick a diet with lots of fresh fruit and vegetables and grain products.
- Choose a diet low in fats, saturated fat, and cholesterol.
- Pick a diet that has moderate salt, sodium, and sugars.
- If you drink alcoholic beverages, do so in moderation.

To assist you in choosing a healthy diet, read food labels. Food labels help consumers select foods with less fat, saturated fats, cholesterol, and sodium, and foods with more dietary fiber and complex carbohydrates. Nutrition fact labels can include information on serving size and number of servings per container; kilocalorie (kcal) information and quantities of nutrients per serving, in actual amounts; quantities of nutrients as "% Daily Values" based on a 2000 kcal energy intake; the daily values for selected nutrients for a 2000 and a 2500 kcal diet; kilocalorie per gram reminder; and the ingredients in descending order of predominance by weight. Packages with less than 12 square inches of surface area do not have to carry any nutrient information but will have a telephone number or address to contact to obtain that information.

All the food groups provide valuable nutrients. MyPyramid, developed by the Department of Agriculture, provides personalized eating plans and tools to help you make food choices that are right for you. Visit www.mypyramid.gov to learn more about the various pyramids based on different lifestyles and nutritional needs, and to take advantage of the many tools offered on the website.

1. The **tunica mucosa** is the innermost lining of the canal and consists of a mucous membrane attached to a thin layer of visceral muscle. Three layers make up the mucous membrane. The first is an epithelial layer, which is in direct contact with the contents of the canal; the second is an underlying layer of loose connective tissue called the **lamina propria** (**LAM**-ih-nah **PROH**-pree-ah); and the third is the **muscularis mucosa**. The epithelial layer functions in protection, secretion of enzymes and mucus, and absorption of nutrients. The lamina propria supports the epithelium, binds it to the muscularis mucosa, and provides it with its lymph and blood supply. The tunica mucosa of the small intestine has another special layer, the muscularis mucosa, made of muscle fibers that produce folds to tremendously increase the digestive and absorptive area of the small intestine.
2. The **tunica submucosa** consists of loose connective tissue that binds the tunica mucosa to the next layer, the tunica muscularis.
3. The **tunica muscularis** of the mouth, pharynx, and the first part of the esophagus consists of skeletal muscle that allows the voluntary act of swallowing. The rest of the tract consists of smooth muscle: an inner circular layer and an outer longitudinal layer of fibers. Involuntary contractions of these smooth muscle fibers break down food physically, mix it with the digestive secretions that break down food chemically, and propel the food through the canal.

GRAINS Make half your grains whole	VEGETABLES Vary your veggies	FRUITS Focus on fruits	MILK Get your calcium-rich foods	MEAT & BEANS Go lean with protein
Eat at least 3 oz. of whole-grain cereals, breads, crackers, rice, or pasta every day 1 oz. is about 1 slice of bread, about 1 cup of breakfast cereal, or 1/2 cup of cooked rice, cereal, or pasta	Eat more dark-green veggies like broccoli, spinach, and other dark leafy greens Eat more orange vegetables like carrots and sweetpotatoes Eat more dry beans and peas like pinto beans, kidney beans, and lentils	Eat a variety of fruit Choose fresh, frozen, canned, or dried fruit Go easy on fruit juices	Go low-fat or fat-free when you choose milk, yogurt, and other milk products If you don't or can't consume milk, choose lactose-free products or other calcium sources such as fortified foods and beverages	Choose low-fat or lean meats and poultry Bake it, broil it, or grill it Vary your protein routine – choose more fish, beans, peas, nuts, and seeds
For a 2,000-calorie diet, you need the amounts below from each food group. To find the amounts that are right for you, go to MyPyramid.gov.				
Eat 6 oz. every day	Eat 2 1/2 cups every day	Eat 2 cups every day	Get 3 cups every day; for kids aged 2 to 8, it's 2	Eat 5 1/2 oz. every day

Find your balance between food and physical activity

- Be sure to stay within your daily calorie needs.
- Be physically active for at least 30 minutes most days of the week.
- About 60 minutes a day of physical activity may be needed to prevent weight gain.
- For sustaining weight loss, at least 60 to 90 minutes a day of physical activity may be required.
- Children and teenagers should be physically active for 60 minutes every day, or most days.

Know the limits on fats, sugars, and salt (sodium)

- Make most of your fat sources from fish, nuts, and vegetable oils.
- Limit solid fats like butter, margarine, shortening, and lard, as well as foods that contain these.
- Check the Nutrition Facts label to keep saturated fats, *trans* fats, and sodium low.
- Choose food and beverages low in added sugars. Added sugars contribute calories with few, if any, nutrients.

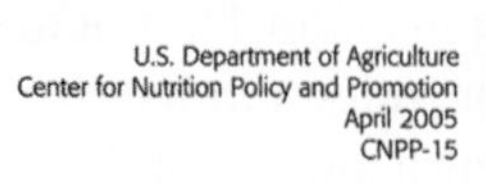

USDA is an equal opportunity provider and employer.

MyPyramid/(U.S. Department of Agriculture).

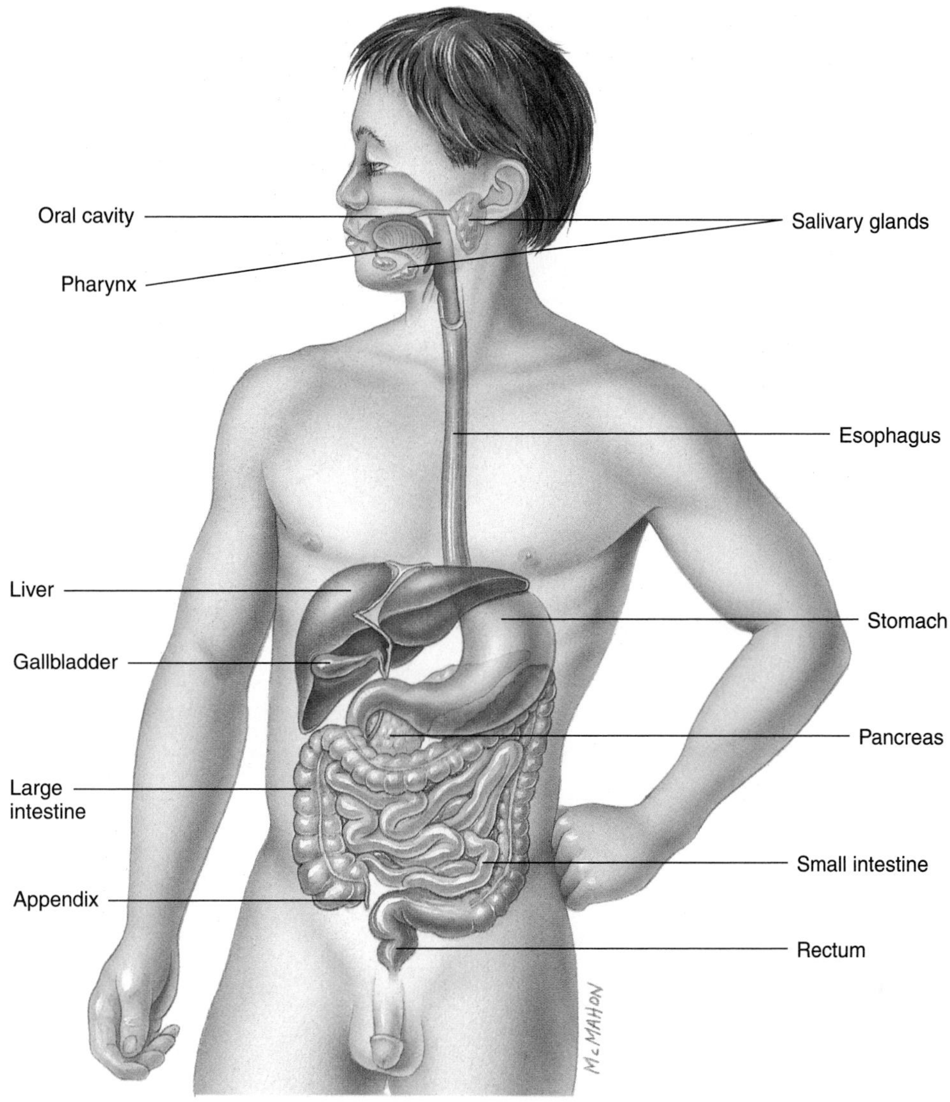

FIGURE 16-1. The gastrointestinal tract, or alimentary canal, and its accessory organs.

The tunica muscularis also contains the major nerve supply to the alimentary canal, the plexus of Auerbach.

4. The **tunica serosa** is the outermost layer. It consists of serous membrane made up of connective and epithelial tissue. It is also known as the **visceral peritoneum** (**VISS**-er-al **pair**-ih-**TOH**-nee-um). This layer covers organs and has large folds that weave in and between the organs, thus binding the organs to each other and to the walls of the cavity. This layer also contains blood vessels, lymph vessels, and nerves that supply the organs. One extension of the visceral peritoneum forms the **mesentery** (**MEZ**-in-**tehr**-ee).

THE MOUTH OR ORAL CAVITY

The mouth or oral cavity can also be called the buccal (**BUCK**-ull) cavity. Its sides are formed by the cheeks. The roof consists of the hard and soft palates, and its floor is formed by the tongue. The **lips** are fleshy folds that surround the opening or orifice of the mouth. On the outside the oral cavity is covered by skin and on the inside by mucous membrane. During the chewing of food, the lips and cheeks help keep food between the upper and lower teeth. They also assist in speech. The **hard** (bony) **palate** forms the anterior part of the roof of the mouth. The **soft** (muscular) **palate** forms the posterior portion of the roof of the mouth. Hanging from its posterior border is a

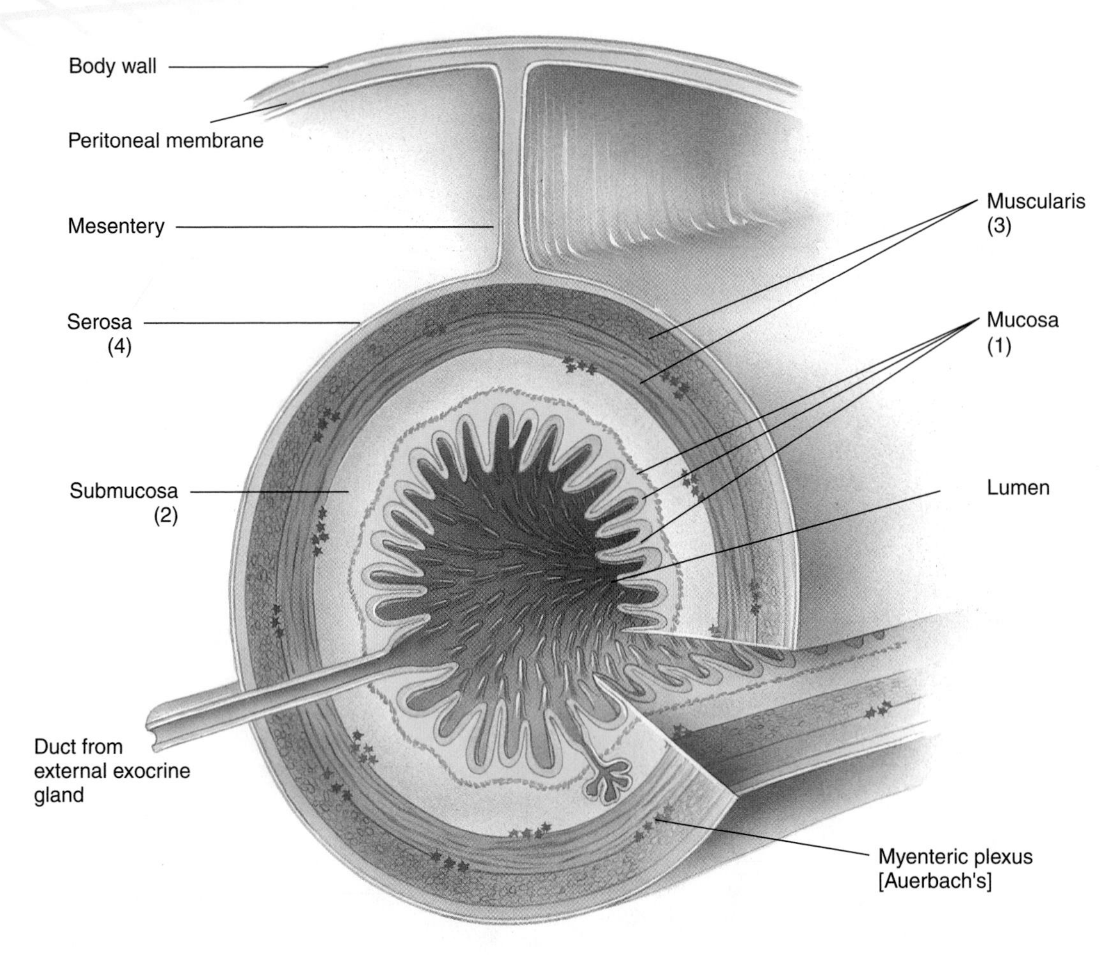

FIGURE 16-2. The tunics, or coats, of the alimentary tract.

cone-shaped muscular structure called the **uvula** (**YOO**-vyoo-lah), which functions in the swallowing process and prevents food from backing up into the nasal area.

The **tongue** and its associated muscles form the floor of the oral cavity (Figure 16-3). It consists of skeletal muscle covered with mucous membrane. It is divided into symmetrical halves by a septum called the **lingual frenulum** (**LING**-gwall **FRIN**-yoo-lum). The tongue is attached to and supported by the hyoid bone. There are two types of skeletal muscle found in the tongue: extrinsic and intrinsic. Extrinsic muscles originate outside the tongue and insert into it, moving the tongue from side to side and in and out to manipulate food. Intrinsic muscles originate and insert within the tongue, altering the size and shape of the tongue for speech and swallowing. The upper surface and sides of the tongue are covered with **papillae** (pah-**PILL**-ee), which are projections of the lamina propria covered with epithelium (Figure 16-4). They produce the rough surface of the tongue. The anterior two-thirds contain taste buds and are most numerous at the tip of the tongue and on the posterior surface of the tongue. The **filiform** (**FILL**-ih-form) **papillae**, found at the front of the tongue, are rough and are important in licking. The **fungiform** (**FUN**-jih-form) **papillae** and the **circumvallate** (**sir**-kum-**VAL**-ate) **papillae**, found toward the back of the tongue, all contain taste buds. There are five tastes: sweet, sour, salt, umami, and bitter. The taste of umami was identified by Japanese researchers. It detects

*Study*WARE™ Connection

Play an interactive game labeling the tongue and oral cavity on your StudyWARE™ CD-ROM.

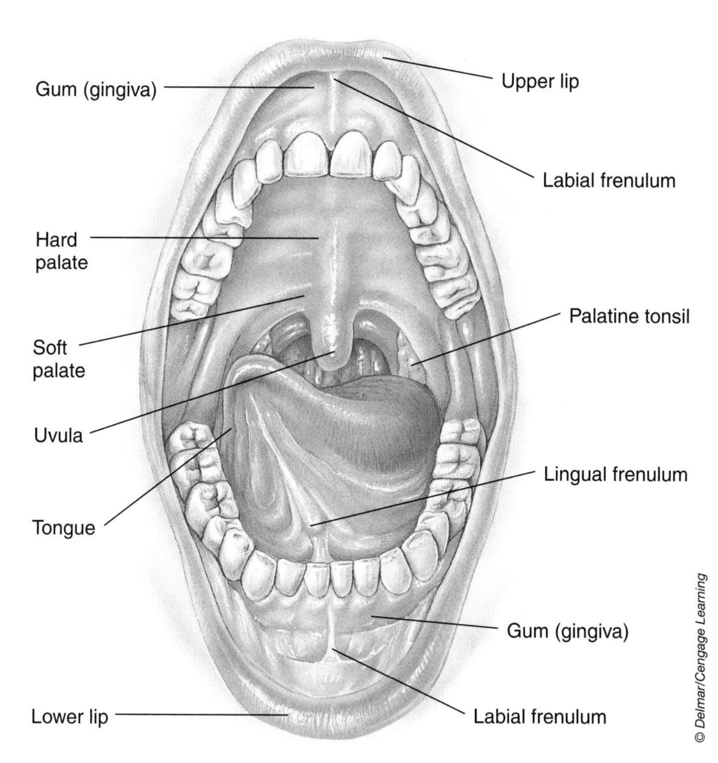

FIGURE 16-3. The anatomy of the tongue and oral cavity.

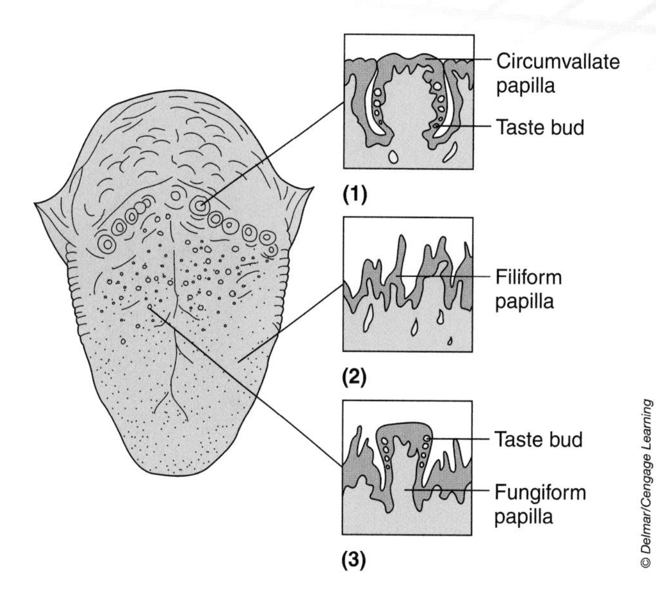

FIGURE 16-4. The three types of papillae found on the tongue.

MSG (monosodium glutamate), a distinct flavor popular in Asian foods.

THE SALIVARY GLANDS

The major portion of saliva is secreted by the large salivary glands (Figure 16-5). The mucous membrane lining of the mouth contains many small glands, called the **buccal** (**BUCK**-ull) **glands**, which secrete small amounts of saliva. The large salivary glands are found outside of the oral cavity and pour their secretions into ducts that empty into the mouth. The three pairs of salivary glands are the **parotid** (pah-**ROT**-id) **gland**, the **submandibular** (**sub**-man-**DIB**-yoo-lar) or **submaxillary** (**sub**-**MACK**-sih-**lair**-ee) **gland**, and the **sublingual** (**sub**-**LING**-gwall) **gland**.

Saliva is 99.5% water, which provides a medium for dissolving foods. The remaining 0.5% consists of the following solutes:

- Chlorides activate the salivary enzyme **amylase** (**AM**-ih-lays).
- Amylase initiates the breakdown of complex carbohydrates like starch and glycogen into simple sugars.
- Bicarbonates and phosphates, which are buffer chemicals, keep the saliva at a slightly acidic pH of 6.35 to 6.85.
- Urea and uric acid are waste products.
- Mucin forms mucus to lubricate food.
- The enzyme lysozyme destroys bacteria, thus protecting the mucous membrane from infection and the teeth from possible decay.

TEETH

The teeth, also known as the **dentes** (**DEN**-teez), are located in the sockets of the alveolar processes of the mandible and maxillae bones. The teeth break up food by chewing. Chewing is called **mastication** (mass-tih-**KAY**-shun). There are 20 temporary or deciduous teeth that form in infants between the ages of 6 months to 2 years. By the age of 13, there will develop 32 permanent teeth to replace the deciduous ones (Figure 16-6). The eight front teeth are called **incisors** (in-**SIGH**-zors) and are used to cut food. The four **canine teeth** are used to tear food. Because they have one cusp, they are also called **cuspids** (**KUSS**-pids). The **molar teeth** grind food. There are two kinds of molar teeth. The eight **premolars** have two cusps or projections and are also called **bicuspids**; some of the 12 molars have three cusps and are called **tricuspids** but many have four cusps and the maxillary first molar actually has five cusps.

The alveolar processes are covered by the gums or **gingivae** (**JIN**-jih-vee) that extend slightly into each socket. The sockets are lined by the periodontal ligament that anchors the teeth in position and acts as a shock

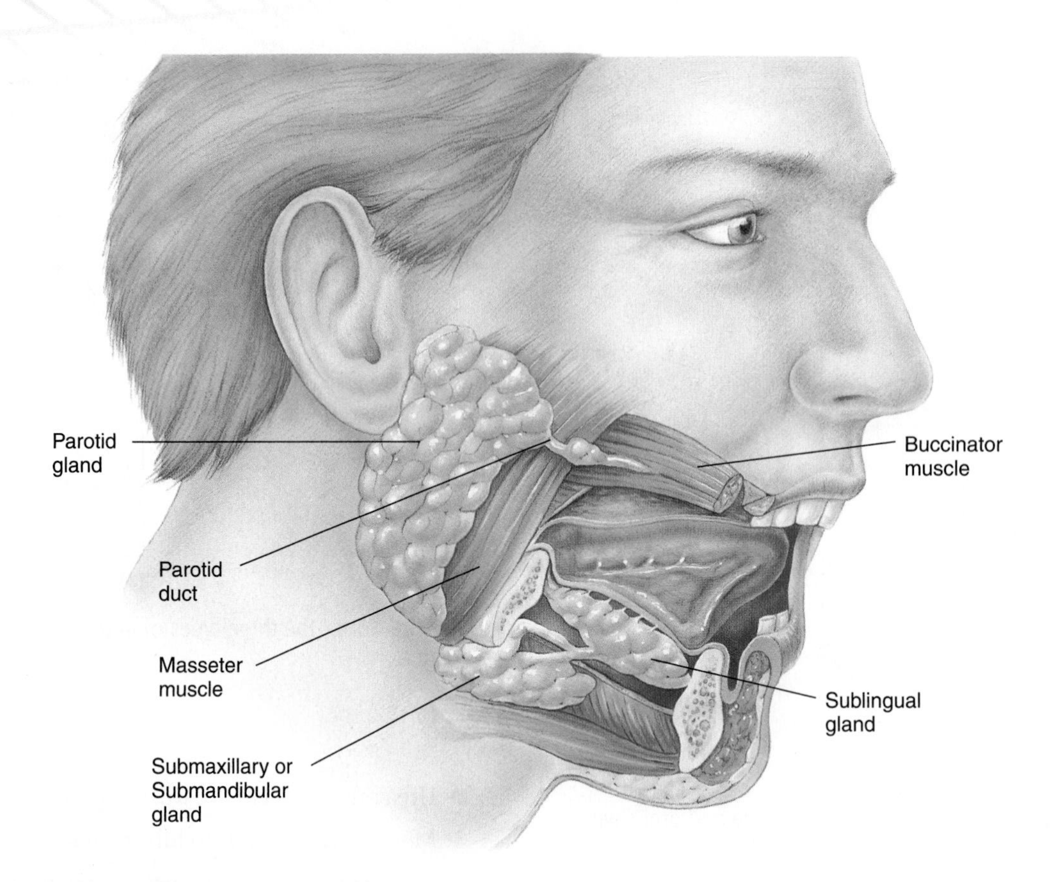

FIGURE 16-5. The salivary glands.

COMMON DISEASE, DISORDER, OR CONDITION

MUMPS

Mumps, caused by a virus that infects the salivary glands and, in particular, the parotid salivary glands, affects children between the ages of 5 and 9. Initial exposure stimulates antibody production and results in permanent immunity. In older males, the virus may infect the testes, causing sterility. If the virus infects the pancreas, diabetes can result. Today, an effective vaccine (part of an early childhood series) is readily available. Thus, the incidence of mumps in the United States has been greatly reduced.

absorber to soften the forces created during chewing. A tooth can be divided into three principal portions:

- The **crown** is the portion above the level of the gums and is covered with **enamel**, the hardest substance in the body that protects the tooth from wear and acids.
- The **cervix** or **neck** is the constricted junction between the crown and the root.
- The **root** can consist of one, two, or three projections embedded in the socket. Larger teeth, like molars, will have more than one root.

Teeth are made of **dentin**, a bonelike substance that encloses the **pulp cavity** in the crown. The exposed surface of the crown is covered with enamel. Narrow extensions of the pulp cavity project into the root, called **root canals**. At the base of each root canal is an opening, the **apical foramen**, through which blood vessels and nerves enter the tooth and become part of the pulp. The dentin of the root is covered with another substance called **cementum**, which attaches the root to the **periodontal ligament**.

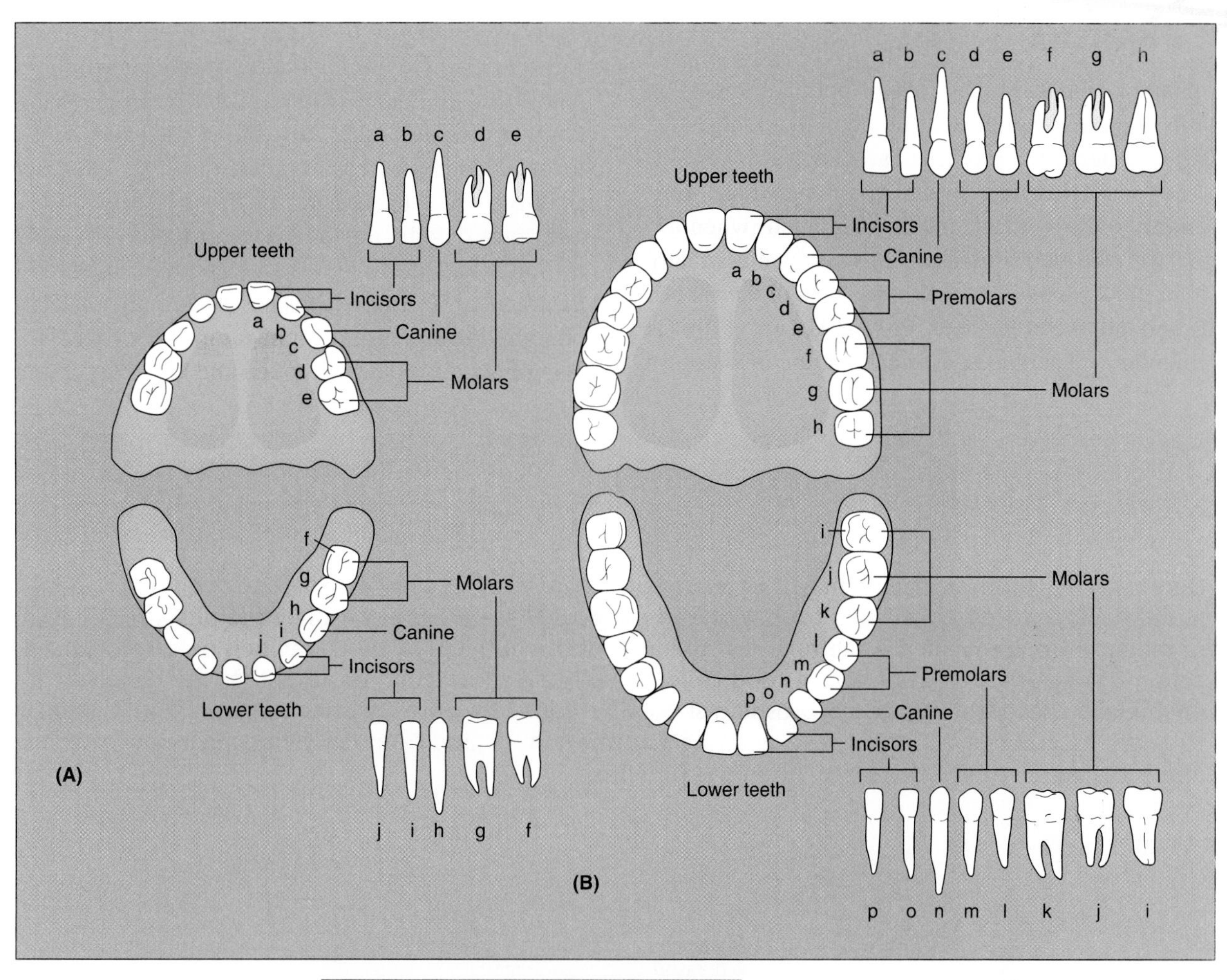

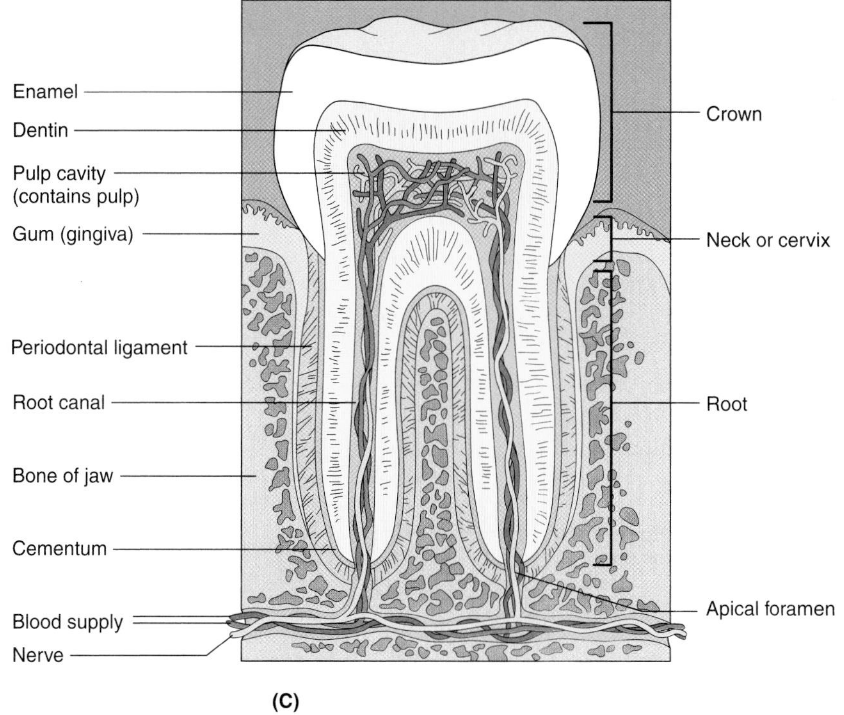

FIGURE 16-6. (A) Deciduous teeth. (B) Permanent teeth. (C) The anatomy of a typical tooth.

THE PHARYNX

The **pharynx** (**FAIR**-inks) is part of both the digestive and the respiratory systems. Its function in the digestive system is to begin the process of swallowing or **deglutition** (**deg**-loo-**TISH**-un). Swallowing moves food from the mouth to the stomach. Swallowing begins when the tongue, with the teeth and saliva, forms a soft mass called the **food bolus**. Food is forced to the back of the mouth cavity and into the **oropharynx** (or-oh-**FAIR**-inks). This is the voluntary stage of swallowing. Next, the involuntary stage begins (Figure 16-7). First, the respiratory passageways close and breathing is temporarily interrupted. The food bolus stimulates oropharyngeal receptors that send impulses to the brain. This causes the soft palate and the uvula to move upward and close off the **nasopharynx** (**nay**-zoh-**FAIR**-inks). Now the larynx is pulled forward and upward under the tongue where it meets the epiglottis and seals off the glottis (the common opening into the trachea). The food bolus passes through the laryngopharynx and enters the esophagus in about 1 second. The respiratory passageways reopen and breathing resumes.

COMMON DISEASE, DISORDER, OR CONDITION

CAVITIES AND TOOTH DECAY

Dental caries (tooth decay) form at the tooth surface in areas where bacteria and food debris accumulate and remain undisturbed for prolonged periods. These areas are usually the pits and grooves of molar teeth, between the teeth, and in the area of the gumline. A tooth cavity is produced by the action of bacteria on carbohydrate food residues with the production of acids that can dissolve the enamel. To prevent decay, tooth brushing should be done immediately after eating, when possible, to exert its greatest effort in destroying acid formation. In addition, daily flossing removes food deposits from between the teeth along the gumline.

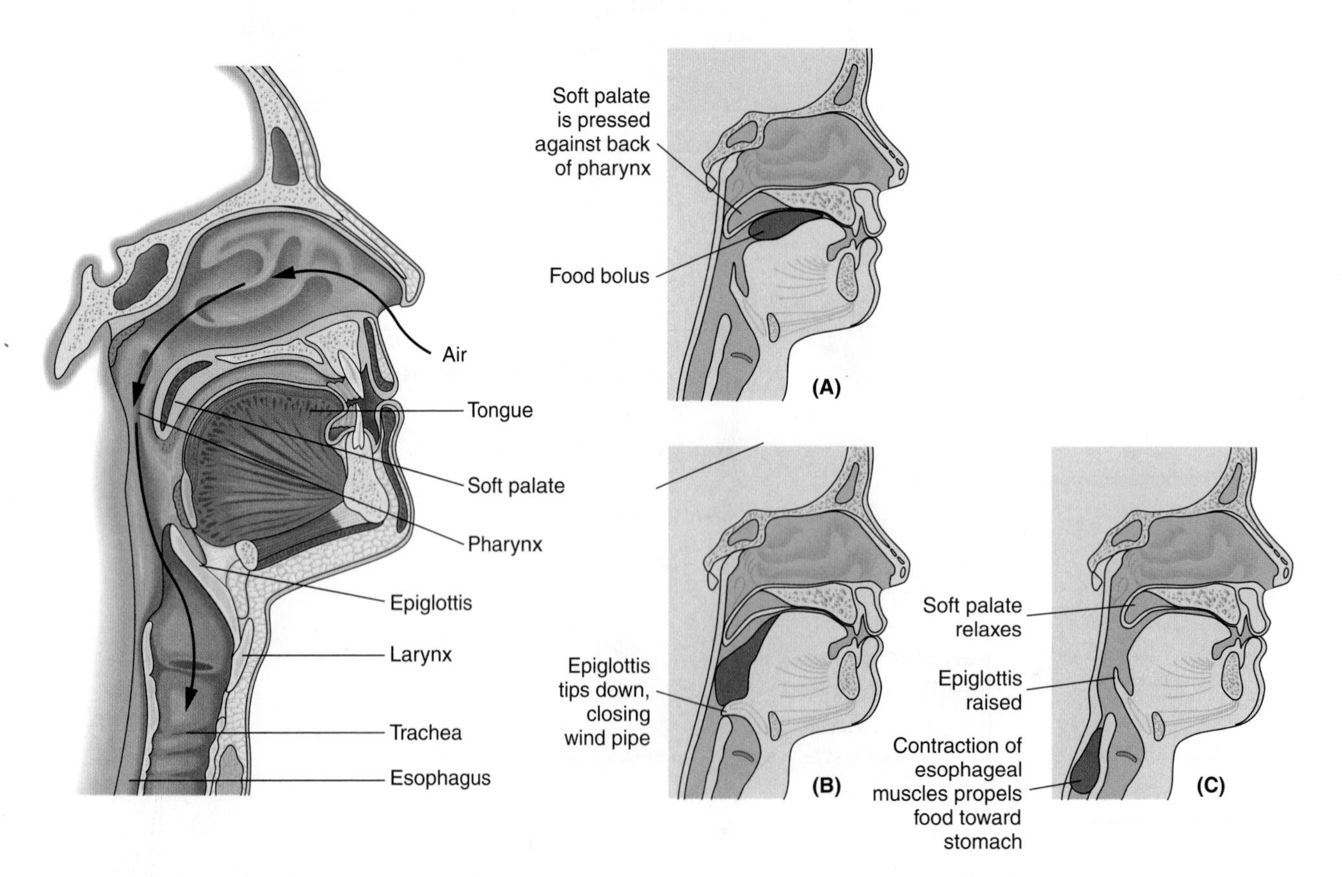

FIGURE 16-7. The swallowing sequence into the esophagus.

COMMON DISEASE, DISORDER, OR CONDITION

HIATAL HERNIA

A hiatal hernia is caused by a widening of the esophageal hiatus (the opening where the esophagus penetrates through the diaphragm muscle to connect with the stomach). This results in a portion of the stomach protruding upward through the diaphragm. This condition occurs most often in adults, with 40% of the population affected. The major symptom of this condition is gastroesophageal reflux or acid reflux, the backflow of the acidic contents of the stomach into the esophagus. This condition is also called heartburn and is quite uncomfortable. A hiatal hernia can also compress blood vessels of the mucosa of the stomach lining resulting in ulcers or gastritis, which can be quite painful.

THE ESOPHAGUS

The **esophagus** (eh-**SOFF**-ah-gus) is a collapsible, muscular tube that is situated behind the trachea or windpipe. It is about 10 inches (23–25 cm) long and begins at the end of the laryngopharynx. It passes through the **mediastinum** (mee-dee-ass-**TYE**-num) (the space between the lungs), pierces the **diaphragm** (**DYE**-ah-fram) through an opening called the **esophageal hiatus,** and ends at the superior portion of the stomach. The function of the esophagus is to secrete mucus and transport food to the stomach. It does not produce any digestive enzymes and it does not absorb food. Food is pushed through the esophagus by smooth muscle contractions, called peristalsis, repeated in wavelike motions that push the food toward the stomach (Figure 16-8). Movement of solid or semisolid foods

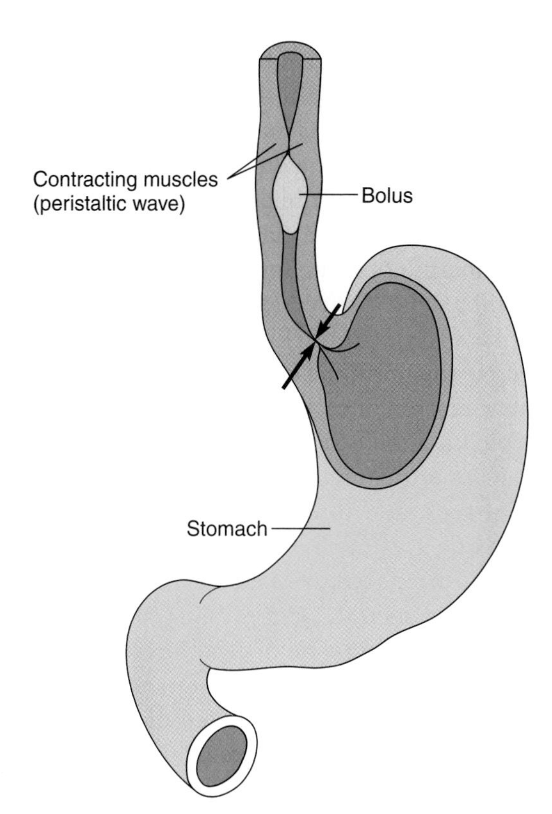

(A) Bolus is swept along esophagus by peristaltic contractions; lower esophageal sphincter (arrows) is closed

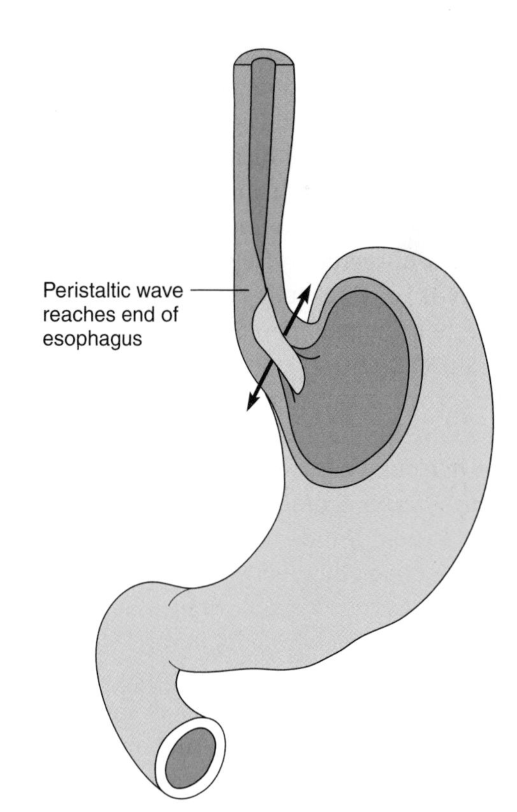

(B) Lower esophageal sphincter (arrows) opens; bolus moves into stomach

FIGURE 16-8. Swallowing. (A) Bolus is pushed through the esophagus by peristaltic waves; lower esophageal sphincter is closed. (B) Lower esophageal sphincter is open, and the bolus enters the stomach.

from the mouth to the stomach takes 4 to 8 seconds; liquids pass in about 1 second. Just above the diaphragm muscle, the esophagus is slightly narrowed by the **lower esophageal** or **gastroesophageal sphincter** (**gas**-troh-eh-**soff**-ah-**JEE**-al **SFINGK**-ter). This sphincter connects the esophagus with the stomach and controls the passage of food into the stomach.

THE STOMACH

The stomach is an enlargement of the gastrointestinal tract. It lies in the upper part of the abdominal cavity just under the diaphragm muscle. It has the shape of the letter J. When it is empty, it is about the size of a large sausage. However, it can be stretched to accommodate large amounts of food.

The stomach is divided into four parts (Figure 16-9A): (1) the **cardia** surrounds the gastroesophageal sphincter; (2) the **fundus** is the rounded portion above and to the left of the cardia; (3) below the fundus is the large central portion of the stomach known as the body; and (4) the **pylorus** or **antrum** is the narrow inferior region that connects with the duodenum of the small intestine via the pyloric sphincter.

When there is no food in the stomach, the mucosa lies in large folds called **rugae** (**ROO**-ghee), which are visible with the unaided eye. As the stomach fills, the rugae smooth out and disappear, like an accordion when it is extended with air. The mucosa of the stomach contains many pits or gastric glands that have three kinds of secreting cells: (1) the **zymogenic** (**zye**-moh-**JEN**-ik) or **chief cells** secrete the principal gastric enzyme **pepsinogen** (pep-**SIN**-oh-gen); (2) the **parietal** (pah-**RYE**-eh-tal) **cells** secrete hydrochloric acid, which activates the pepsinogen to become **pepsin**, the enzyme that begins to break down proteins; and (3) the **mucous cells**, which secrete mucus that protects the stomach from being digested. The secretions of these gastric glands collectively are referred to as gastric juice (see Figure 16-9B). The muscularis coat of the stomach has uniquely three, not just two, layers of smooth muscle: an inner oblique, a middle circular, and an outer longitudinal. These three layers allow the stomach to contract in a variety of ways to break up food into small pieces, churn it, and mix it with the gastric juice. When the stomach is empty and this activity occurs, we experience the stomach *growling*.

The main chemical activity of the stomach is to begin the digestion of proteins by the enzyme pepsin. The protein components of the stomach cells themselves are protected from being digested by the mucus secreted by the mucous cells. The stomach then empties all its contents into the **duodenum** (doo-oh-**DEE**-num) of the small intestine approximately 2 to 6 hours after ingestion. Foods high in carbohydrates pass through the stomach first because their digestion begins in the mouth via the salivary enzyme amylase. Protein foods pass through somewhat more slowly because their digestion begins in the stomach. Foods containing large amounts of fats take the longest to pass into the duodenum. The stomach participates in the absorption of some water and salts. Certain drugs, such as aspirin and alcohol, can also be absorbed in the stomach.

The next step is chemical digestion in the small intestine. This process depends on secretions from intestinal glands and on secretions of the two large accessory glands of the system, the pancreas, and the liver and its gallbladder.

THE PANCREAS

The **pancreas** is a soft, oblong gland about 6 inches long and 1 inch thick (Figure 16-10). It is found beneath the great curvature of the stomach and is connected by a duct to the duodenum of the small intestine. The pancreas is divided into a head (the part closest to the duodenum), the body (the main part), and the tail. Internally, the pancreas is made up of clusters of glandular epithelial cells. One group of these clusters, the **islets of Langerhans**, or the **pancreatic islets**, form the endocrine portions of the gland and are therefore part of the endocrine system. Some of these clusters consist of alpha cells that secrete the hormone **glucagon** (**GLOO**-kah-gon). Other clusters consist of beta cells that secrete the hormone **insulin**. (Review Chapter 12.) The other masses of cells are called the **acini** (**AS**-in-eye), which are the exocrine glands of the organ. The acini release a mixture of digestive enzymes (lipases, carbohydrases, and proteases) called the **pancreatic juice**, which leaves the pancreas through a large main tube called the **pancreatic duct**, or **duct of Wirsung**. The duct cells secrete sodium bicarbonate. In most individuals the pancreatic duct unites with the common bile duct of the liver and enters the duodenum in a common duct, originally called the **ampulla** (am-**PULL**-lah) **of Vater** but now is called the **hepatopancreatic ampulla**.

The functions of the pancreas are, therefore, twofold. The acini secrete enzymes that continue the digestion of food in the small intestine, and the alpha and beta cells secrete the hormones glucagon and insulin that regulate and control blood sugar levels.

Esophagus
Fundus
Lower esophageal sphincter
Serosa
Cardia
Antrum (pylorus)
Lesser curvature
Muscularis externa: outer longitudinal
middle circular
inner oblique
Duodenum of small intestine
Pyloric sphincter
Greater curvature
Body
Rugae
(A)

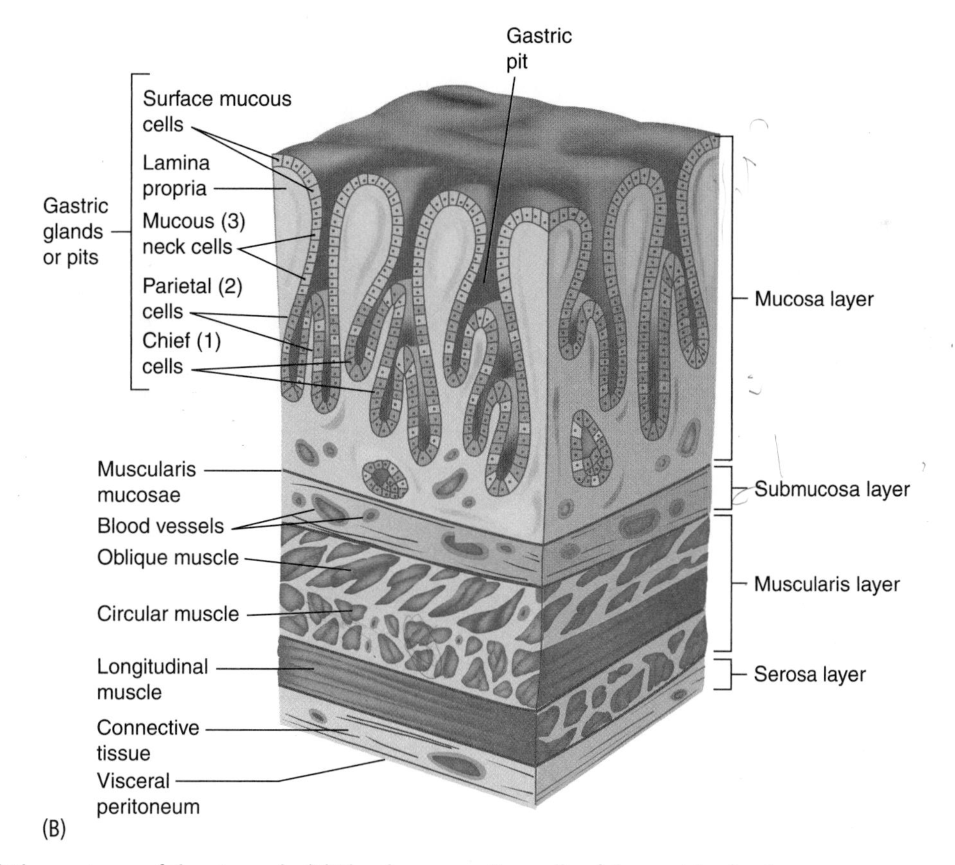

FIGURE 16-9. (A) The anatomy of the stomach. (B) The three secreting cells of the gastric glands.

HEALTH ALERT

ABSORPTION IN THE STOMACH

Some aspirins are marketed as "coated" to delay their absorption until they reach the intestine for individuals with sensitive stomachs. Alcohol is also absorbed through the stomach. Hence, alcohol should not be consumed when a person has not eaten. Eating food when drinking alcohol delays the speed at which the alcohol is absorbed. ■

COMMON DISEASE, DISORDER, OR CONDITION

ULCERS

Ulcers occur when the hydrochloric acid and digestive enzymes erode the layers of the stomach or duodenum. This is caused by either the excessive production of acid (which can be caused by stress) or the inadequate production of the alkaline mucus that protects the epithelial lining of the tract. Current research indicates that a bacterium ***Helicobacter pylori*** is associated with the development of stomach or peptic ulcers in about 65% of ulcer cases. Antibiotics are administered to treat these ulcers. In the past, antacids have been used to inhibit acid production by the parietal cells of the stomach.

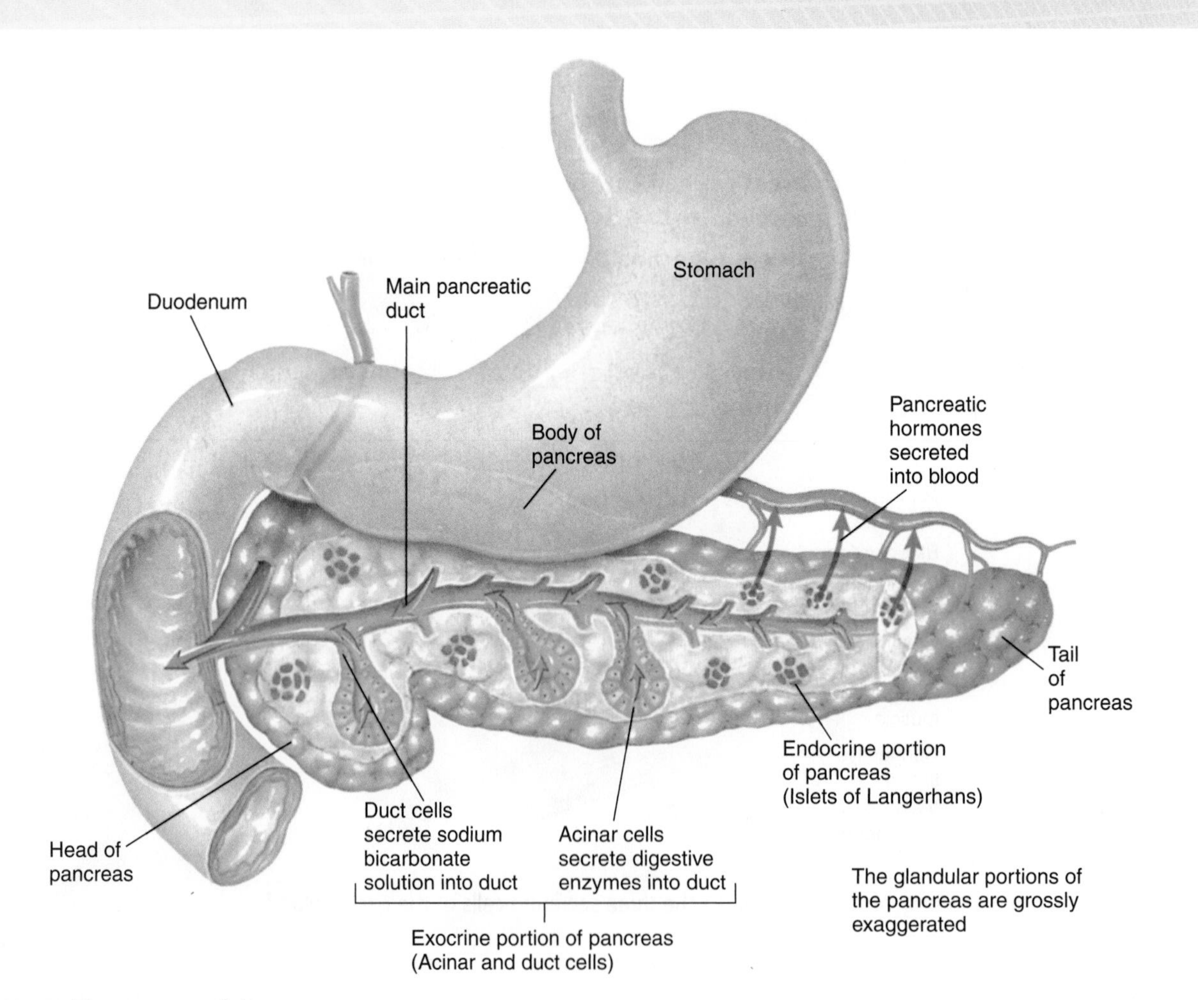

FIGURE 16-10. The anatomy of the pancreas.

THE LIVER

The **liver** is one of the largest organs of the digestive system (Figure 16-11). The liver weighs approximately 4 pounds and is divided into two principal lobes: the right lobe and the left lobe, each separated from one another by the **falciform ligament**. The lobes of the liver are made up of numerous functional units called lobules.

The functions of the liver are so numerous and important that we cannot survive without it. The liver has six major functions:

1. The liver manufactures the anticoagulant **heparin** and most of the other plasma proteins, such as **prothrombin** and **thrombin**, that are involved in the blood clotting mechanism.
2. **Kupffer cells** of the liver phagocytose (eat) certain bacteria and old, worn-out white and red blood cells.
3. Liver cells contain various enzymes that either break down poisons or transform them into less harmful substances. If the body cannot break down and excrete certain poisons, it stores those poisons. When we digest proteins into amino acids, the amino acids go to the mitochondria to be converted into ATP. This process produces ammonia as a waste product, which is toxic to cells. The liver cells convert ammonia to urea (harmless) that is then excreted by the kidneys or the sweat glands.
4. Excessive newly absorbed nutrients are collected in the liver. Excess glucose and other monosaccharides can be stored as **glycogen** (animal starch) or converted to fat. When needed, the liver can then transform glycogen and fat into glucose.
5. The liver stores glycogen, copper, and iron, as well as vitamins A, D, E, and K.
6. The liver produces bile salts that break down fats. These bile salts are sent to the duodenum of the small intestine for the emulsification (breakup) and absorption of fats.

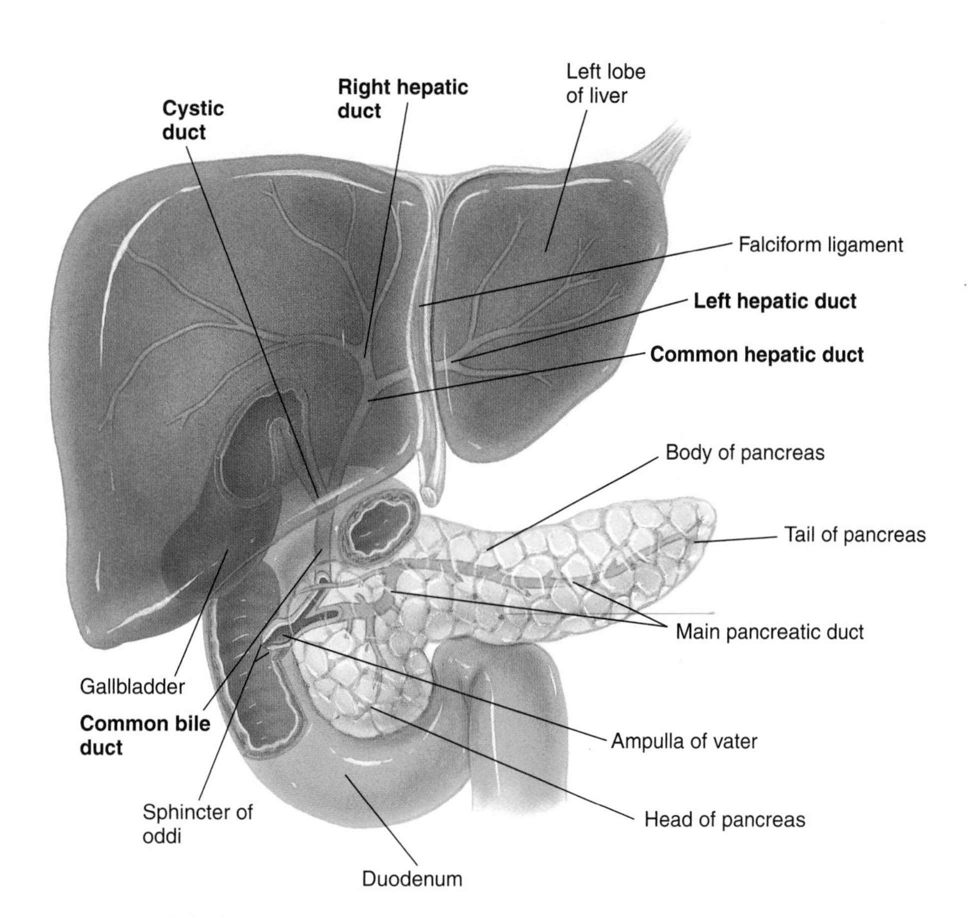

FIGURE 16-11. The anatomy of the liver.

The Gallbladder

The **gallbladder** is a pear-shaped sac about 3 to 4 inches long located in a depression of the surface of the liver. Its lining, like the stomach, has rugae that allow it to expand and fill with stored bile. The gallbladder's function is to store and concentrate the bile produced by the liver lobules until it is needed in the small intestine. The bile enters the duodenum through the common **bile duct**.

THE SMALL INTESTINE

The major portion of absorption and digestion occurs in the **small intestine**. It is approximately 21 feet in length and averages 1 inch in diameter (Figure 16-12). The small intestine is divided into three portions. First is the **duodenum**, which is the shortest part, and is about 10 inches long. The duodenum originates at the **pyloric sphincter** and joins the second portion, the **jejunum** (jee-**JOO**-num). The jejunum is about 8 feet long and extends to the third part, the **ileum** (**ILL**-ee-um), which measures 12 feet and joins the large intestine at the **ileocecal** (**ill**-ee-oh-**SEE**-kal) **valve** (sphincter).

The mucosa of the small intestine contains many pits lined with glandular epithelium. These pits are known as the **intestinal glands** or **crypts of Lieberkuhn** (**KRIPTZ** of **LEE**-ber-koon). They secrete the intestinal digestive enzymes that supplement the bulk of the digestive enzymes secreted by the liver and the pancreas. The **submucosa** of the duodenum contains numerous **Brunner's glands**, now called **duodenal glands**, which secrete an alkaline mucus. Additional mucus is secreted by goblet cells. This mucus protects the walls of the small intestine from being digested by enzymes and neutralizes the acid found in the **chyme** (**KIGHM**). Chyme is the term used to describe the digested, viscous, semifluid contents of the intestine.

Approximately 80% of all absorption of nutrients (simple sugars, amino acids, fatty acids, water, vitamins,

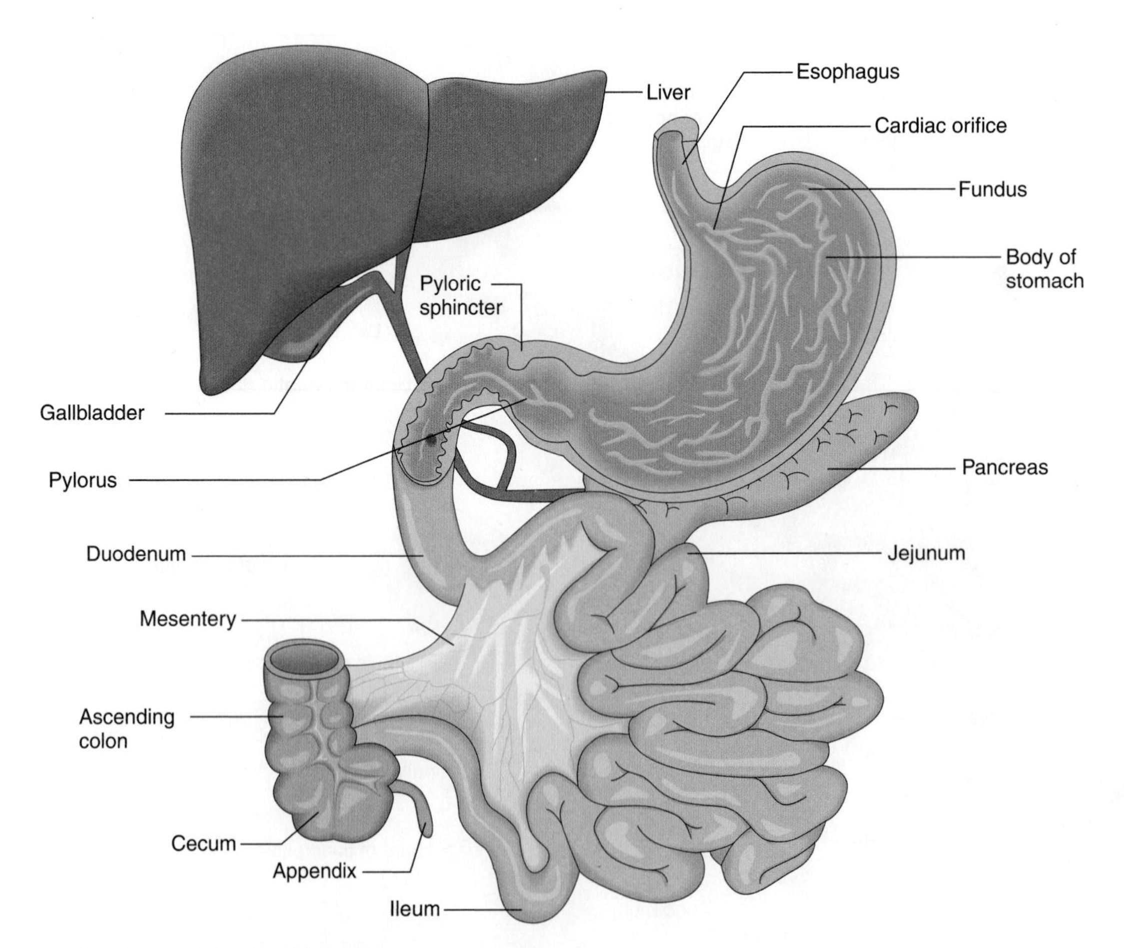

FIGURE 16-12. Structures of the small intestine.

and minerals) occurs in the small intestine. The anatomic structure of the small intestine is highly specialized for this function (Figure 16-13). First of all, the tract is 21 feet long. Second, an even larger surface for absorption of nutrients is provided by the structure of the walls of the tract, which are thrown into a series of folds called **plicae** (**PLYE**-kee). Third, the mucosal coat is transformed into projections called **villi**, which look like microscopic eye dropper bulbs approximately 0.5 to 1 mm long. A tremendous number of villi line the intestine, about 4 to 5 million. These villi vastly increase the surface area of the epithelium for absorption of nutrients. The structure of each villus contains a capillary network where blood picks up nutrients, a venule or small vein to transport the nutrients, an arteriole or small artery, and a lacteal of the lymphatic system to pick up fats. In addition, the individual epithelial cells that cover the surface of a villus have a brush border of **microvilli** to further increase the absorptive capability of the small intestine. Nutrients that pass through the epithelial cells covering the villus are able to pass through the endothelial cells of the capillary walls and through the lacteals to enter the blood and lymphatic circulatory systems. From there they are transported to the billions of cells in the body.

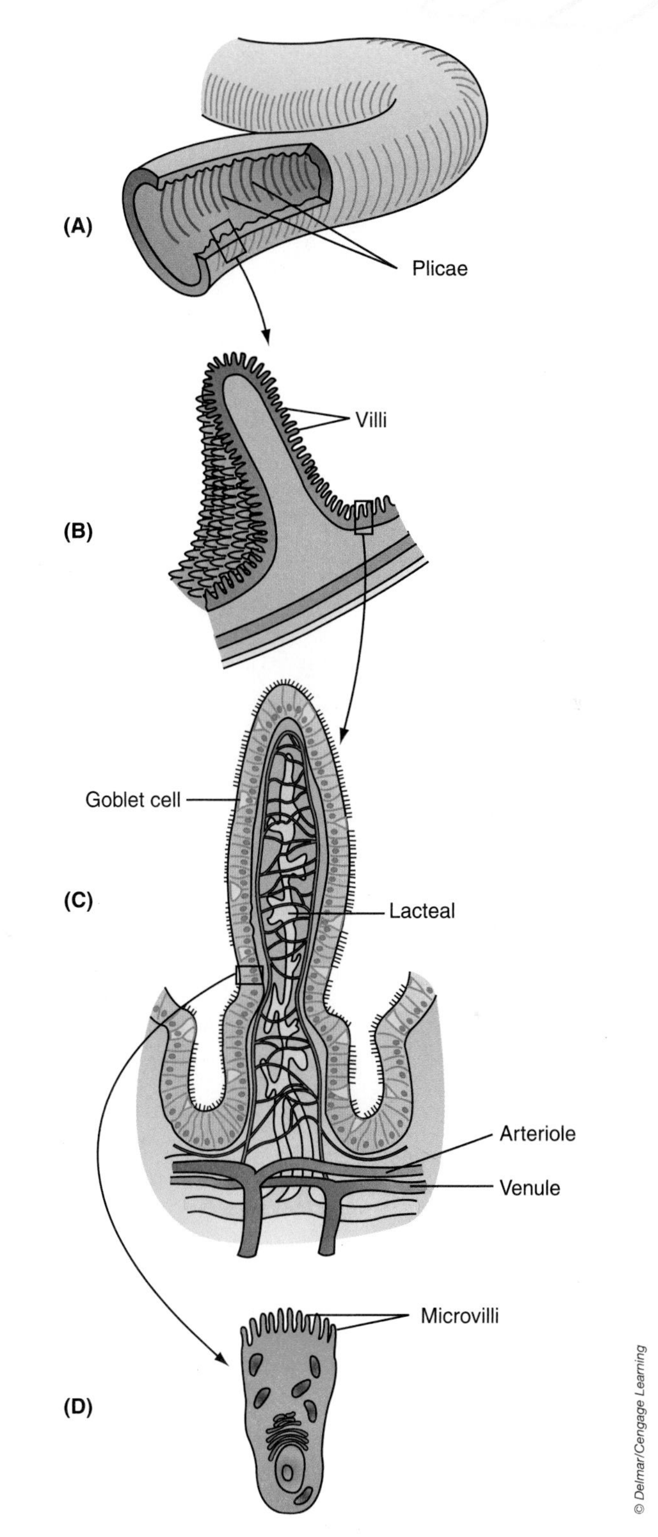

FIGURE 16-13. Absorptive features of the small intestine. (A) Plicae circulares; (B) villi; (C) microscopic anatomy of a villus; (D) microvilli.

THE LARGE INTESTINE

The functions of the **large intestine** are the absorption of water, the manufacturing and absorption of certain vitamins, and the formation and expulsion of the feces (**FEE**-seez). The large intestine is about 5 feet in length and averages 2.5 inches in diameter (Figure 16-14). It is also referred to as the **bowel**. It is attached to the posterior wall of the abdomen by extensions of its visceral peritoneum known as the **mesocolon**. It is divided into four principal regions: (1) the **cecum** (**SEE**-kum), the pouchlike first part of the large intestine; (2) the **colon** (**KOH**-lon), the largest part; (3) the **rectum**; and (4) the **anal canal**.

The opening from the ileum of the small intestine into the cecum of the large intestine is a fold of mucous membrane known as the ileocecal valve. This valve allows material to pass from the small intestine into the large intestine. The cecum, a blind pouch (one end is closed), is 2 to 3 inches long and hangs below the ileocecal valve. Attached to the closed end of the cecum is the twisted tube known as the **vermiform** (**VER**-mih-form) **appendix**, about 3 inches in length. The open end of the cecum merges with the long tube called the colon.

The colon looks like a tube of consecutive pouches. The pouches are called **haustrae** (**HAW**-stree). The first part of the colon is known as the **ascending colon**. It rises

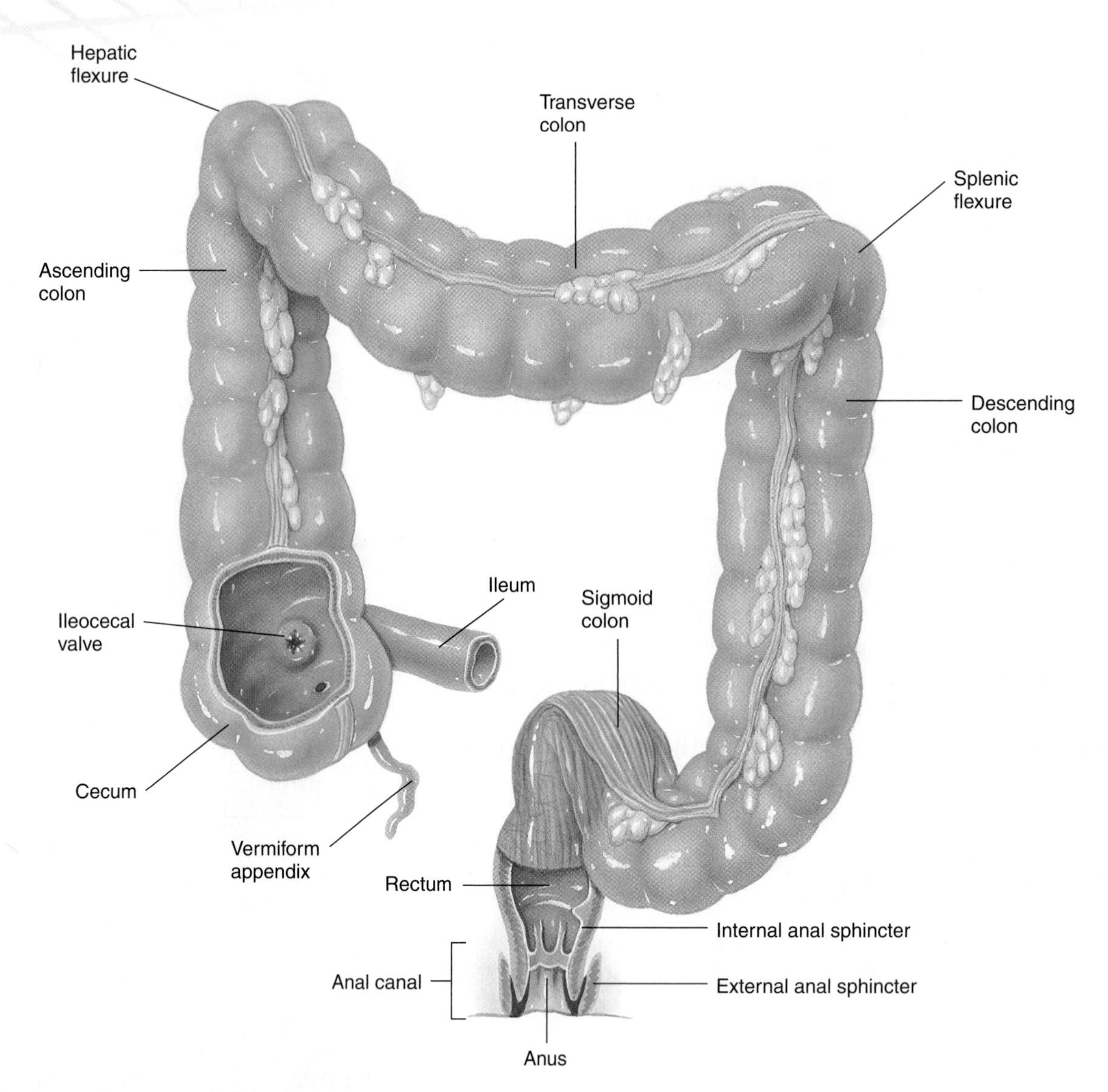

FIGURE 16-14. The anatomy of the large intestine.

on the right side of the abdomen, reaches the undersurface of the liver, and turns to the left at the **right colic (hepatic) flexure**. The right colic flexure continues across the abdomen to the left side as the **transverse colon**. It then curves beneath the lower end of the spleen to the left side as the **left colic (splenic) flexure**. Next it passes downward as the **descending colon**. The last part of the colon is called the **sigmoid** (**SIG**-moyd) **colon**, where the colon joins the rectum.

Three mechanical movements occur in the large intestine:

- Haustral churning
- Peristalsis at the rate of 3 to 12 contractions per minute
- Mass peristalsis

Food in the stomach initiates mass peristalsis, which is a strong peristaltic wave that begins in the middle of the transverse colon and drives its contents into the rectum.

The rectum is the last 7–8 inches of the gastrointestinal tract. It is situated anterior to the sacrum and the coccyx. (Review Chapter 7.) The terminal 1 inch of the rectum is called the **anal canal**. The mucous membrane of the anal canal is arranged in a series of longitudinal folds called the **anal columns** that contain a network of arteries and veins.

The opening of the anal canal to the exterior is called the **anus**. It is guarded by an internal sphincter of smooth muscle and an external sphincter of skeletal muscle.

The absorption of water is an important function of the large intestine. In addition, bacteria in the colon manufacture three important vitamins that are also absorbed in the colon: vitamin K needed for clotting, biotin needed for glucose metabolism, and vitamin B5 needed to make certain hormones and neurotransmitters. Mucus is also produced by glands in the intestine. Intestinal water absorption is greatest in the cecum and ascending colon.

StudyWARE™ Connection

- Watch an animation about digestion.
- Play an interactive game labeling the gastrointestinal tract and its accessory organs on your StudyWARE™ CD-ROM.

COMMON DISEASE, DISORDER, OR CONDITION

DISORDERS OF THE DIGESTIVE SYSTEM

HEPATITIS

Hepatitis (**hep**-ah-**TYE**-tis) is an inflammation of the liver caused by excessive alcohol consumption or by a virus infection. If not treated, liver cells die and are replaced by scar tissue. This results in liver failure and death can result. Hepatitis A is caused by the hepatitis A virus (HAV) and is known as infectious hepatitis. It is contracted through contact with fecally contaminated water or food like raw oysters or clams that are filter feeders. Hepatitis B is caused by the hepatitis B virus (HBV). It is known as serum hepatitis and is transmitted by contaminated serum in blood transfusions, by contaminated needles, or by sexual contact with an infected individual. Vaccines are available for immunization. Infected individuals experience nausea, fever, loss of appetite, abdominal pain, and a yellowing of the skin and sclera of the eyes called jaundice. Jaundice is due to the accumulation of bile pigment in the skin. Viral hepatitis is the second most frequently contracted infectious disease in the United States.

CIRRHOSIS

Cirrhosis (sih-**ROH**-sis) is a long-term degenerative disease of the liver in which the lobes are covered with fibrous connective tissue. The parenchyma of the liver degenerates and the lobules are filled with fat. Blood flow through the liver is obstructed. Cirrhosis is most commonly the result of chronic alcohol abuse. Recovery is very slow but the liver can regenerate and treatment depends on the cause.

GALLSTONES

Gallstones affect about 20% of the population over 40 years of age and is more prevalent in women than men. Cholesterol is secreted by the liver into the bile. When the cholesterol precipitates in the gallbladder, it produces gallstones. If a stone leaves the gallbladder and gets caught in the bile duct, it will block the release of bile and interfere with normal digestion (Figure 16-15). Under these conditions, the patient will complain of unlocalized abdominal discomfort, intolerance to certain foods, and frequent belching. Usually, the gallbladder has to be removed surgically to remove the buildup of gallstones.

APPENDICITIS

Appendicitis (ap-**pin**-dih-**SIGH**-tis) is an inflammation of the vermiform appendix caused by an obstruction. Obstructions can be a hard mass of feces, a foreign body in the lumen of the appendix, an adhesion, or a parasitic infection. Secretions from the appendix cannot get past the obstruction. They accumulate and cause enlargement and pain. Bacteria in the area cause infection. If the appendix

(*continues*)

COMMON DISEASE, DISORDER, OR CONDITION

DISORDERS OF THE DIGESTIVE SYSTEM (continued)

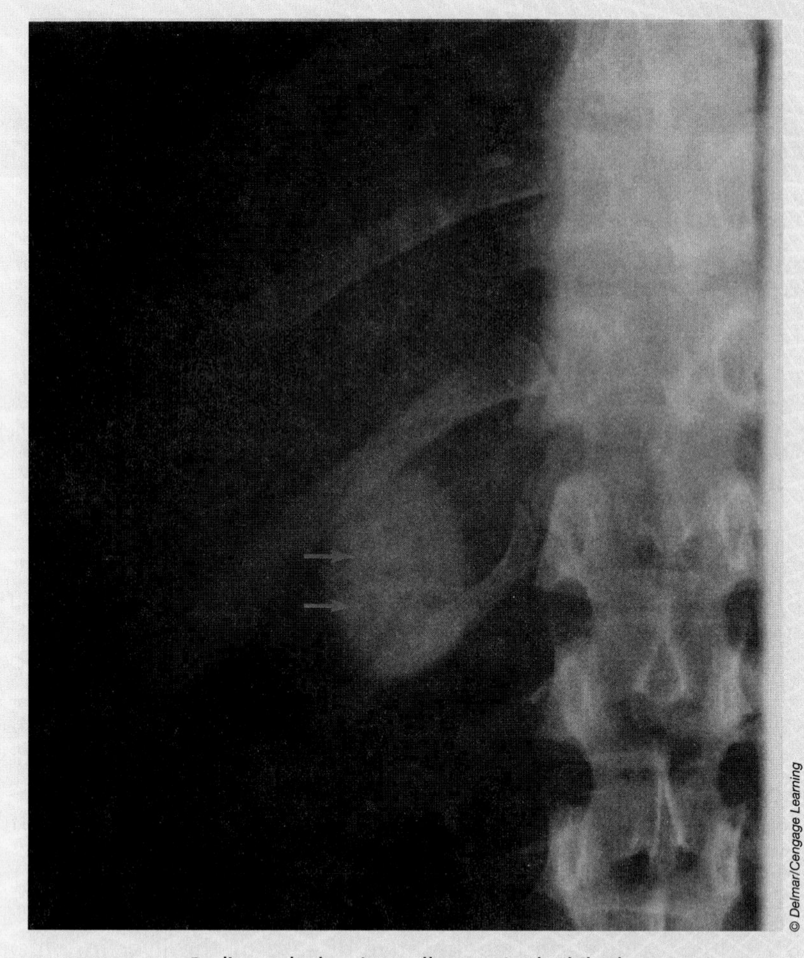

FIGURE 16-15. Radiograph showing gallstones in the bile duct.

bursts, the bacterial infection will spread throughout the peritoneal cavity, resulting in peritonitis and possible death. Appendicitis is more common in teenagers and young adults and is more frequent in males.

CROHN'S DISEASE

Crohn's disease (**KROHNZ** dih-**ZEEZ**) is a chronic, inflammatory bowel disease of unknown origin. It usually affects the ileum, the colon, or another part of the gastrointestinal tract. Diseased segments are usually separated from normal bowel segments, giving the characteristic appearance of regional enteritis or skip lesions. Symptoms include frequent bouts of diarrhea, severe abdominal pain, fever, chills, nausea, weakness, anorexia, and weight loss. Treatment is aimed at the symptoms. Surgical removal of a diseased bowel segment provides temporary relief. However, recurrence after surgery usually occurs.

DIVERTICULOSIS

Diverticulosis is the presence of pouchlike herniations through the muscular layer of the colon, particularly the sigmoid colon. It occurs most frequently in people over 50 years of age and may result from low-fiber diets. The major symptom is bleeding from the rectum. An increase in fiber in the diet

COMMON DISEASE, DISORDER, OR CONDITION

DISORDERS OF THE DIGESTIVE SYSTEM (continued)

will aid in propelling the fecal material through the colon. Avoiding nuts and foods with seeds will decrease the risk of fecal material becoming lodged in the diverticula. Hemorrhage from bleeding diverticulae can be severe and may require surgery. This condition can develop into diverticulitis, an inflammation of one or more of the diverticulae.

COLORECTAL CANCER

Colorectal cancer is cancer of the large intestine and rectum. It is the second most prevalent cancer in the United States with 60,000 deaths annually. A variety of symptoms can indicate this condition, including blood in the feces, abdominal pain, and a change in the consistency and frequency of bowel movements. A hemoccult test should become a normal part of yearly physicals to detect any blood in the feces. Some colon cancers take years to develop. Intestinal cells begin to divide more frequently than normal. They may form benign polyps at first and then lead to malignant growths. In severe cases, sections of the intestine may be removed surgically with a new opening created to exit the feces. This procedure is called a colostomy and the patient must wear a colostomy bag to collect and dispose of the fecal material.

HEMORRHOIDS

Hemorrhoids, or **piles**, are caused by the inflammation and enlargement of rectal veins. Those that are initially contained within the anus are known as first-degree hemorrhoids. If these enlarge so that they extend outward on defecation (prolapse), they are then called second-degree hemorrhoids. When they finally are so enlarged that they remain prolapsed through the anal opening, they are called third-degree hemorrhoids and usually require surgical removal. Strain on defecation causes inflammation of these veins. A diet rich in fiber helps produce softer stools, which results in less strain on defecation.

DIARRHEA

Diarrhea (dye-ah-**REE**-ah) literally means a flowing through. It occurs when the mucosa of the colon is unable to maintain its usual levels of water absorption and secretes larger than normal amounts of water, ions, and mucus. Food poisoning with bacteria of the salmonella group, certain viral infections like those that cause intestinal flu, and protozoan infections with the amoeba ***Entamoeba histolytica*** caused by drinking untreated water causes bouts of severe diarrhea called dysentery. These organisms invade the intestinal lining and destroy the normal cells that carry on the absorption of water. The excessive loss of water caused by bouts of diarrhea can lead to serious dehydration. Without proper treatment, a victim could die. Diarrhea is a common cause of death in young children in poor, undeveloped countries.

THRUSH

Thrush is a fungal infection of the tissues of the mouth caused by *Candida albicans* (Figure 16-16). It can also infect the skin, intestines, and the vagina of women. In the mouth it is called oral candidiasis. It is characterized by creamy white plaques on the mucous membrane lining of the mouth. It is more common in children and is usually benign and can be treated with medications. In adults who have human immunodeficiency virus (HIV), it is a common opportunistic infection that can spread to the esophagus.

TONSILLITIS

Tonsillitis (**ton**-sih-**LYE**-tis) is a bacterial infection of the tonsils frequently caused by strains of *Streptococcus*. It is mainly characterized by a condition of severe throat pain with difficulty in the

(*continues*)

COMMON DISEASE, DISORDER, OR CONDITION

DISORDERS OF THE DIGESTIVE SYSTEM (continued)

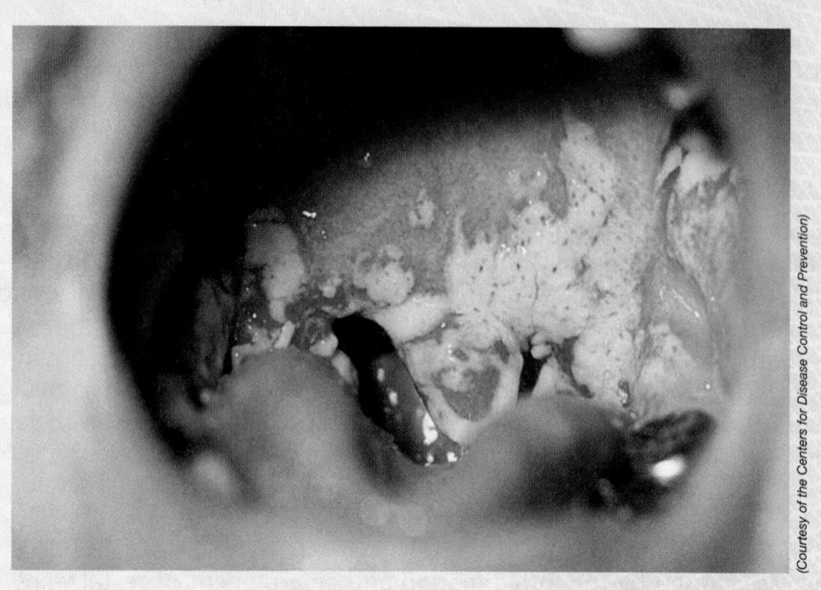

(Courtesy of the Centers for Disease Control and Prevention)

FIGURE 16-16. Oral thrush. *Candida albicans.*

ability to swallow. Other symptoms include fever and chills, enlarged lymph nodes of the neck, ear-ache, and general malaise. This condition is treated with antibiotics. If the condition becomes recurrent, a tonsillectomy may be performed.

FOOD POISONING

Food poisoning is caused by the digestion of food contaminated with the bacteria *Salmonella,* resulting in salmonellosis (**SAL**-mon-el-**OH**-sis). There is an incubation period of 6 to 48 hours, which is followed by bouts of abdominal pain, diarrhea, fever, and nausea. There is no specific treatment other than prevention by adequately cooking food, keeping cooked food refrigerated, and thorough washing of one's hands before eating. Symptoms may persist anywhere from 4 to 7 days. Frequent diarrhea may lead to dehydration.

TAPEWORM INFECTIONS

Tapeworm infections are caused by ingesting the larval forms of various tapeworms in improperly cooked pork, beef, or freshwater fish. Human tapeworms can range in size from three feet (pork and beef tapeworm) to 60 feet (broad fish tapeworm). They live in the small intestine attached by a head structure called a scolex with hooks and suckers. Symptoms are very mild or absent. Presence of a tapeworm can be detected by noticing the white, small, cubelike cast-off proglotlids or segments of the tapeworm's body, which are full of eggs in the fecal material in the toilet. Tapeworm infections can be treated with various drugs like albendazole.

PANCREATITIS

Pancreatitis (**pan**-cree-ah-**TY**-tis) is an inflammation of the pancreas caused by damage to the organ due to alcohol abuse, infectious disease, or drugs. The condition can be acute or chronic. Symptoms include severe abdominal pain, nausea with accompanying vomiting, and fever.

(*continues*)

COMMON DISEASE, DISORDER, OR CONDITION

DISORDERS OF THE DIGESTIVE SYSTEM (continued)

Occasionally jaundice may develop if the common bile duct gets obstructed. A serious complication is the development of abscesses in the pancreas. Treatment includes diet modification, vitamin supplements, certain drugs, and analgesics for pain.

GASTRITIS

Gastritis (gas-**TRY**-tis) is an inflammation of the stomach lining that occurs after eating certain foods producing discomfort, nausea, and vomiting. It can be caused by food allergens, certain drugs, or bacterial or viral infections, or chemical toxins. Chronic gastritis is indicative of a disease condition like stomach ulcers or cancer.

GASTROESOPHAGEAL REFLUX DISEASE (GERD)

Gastroesophageal reflux disease (**gas**-troh-eh-**sof**-ah-**JEE**-al **REE**-flucks dih-**ZEEZ**) is also known as GERD. It is caused by stomach contents going back into the esophagus due to a weakening or incompetency of the lower esophageal sphincter. Since the gastric juice is high in hydrochloric acid, this causes a burning sensation in the esophagus, commonly referred to as "heartburn." Consistent episodes of GERD may lead to severe damage to the esophageal lining. Preventative measures include sleeping with two pillows to elevate the head, drinking a glass of water before retiring, avoiding highly acidic foods at evening meals, and the use of antacids like TUMS.

GASTRIC CANCER

Gastric or stomach cancer produces gastric tumors called adenocarcinomas. The exact cause is unknown. However, dietary factors that have been implicated include nitrates as preservatives in foods, smoked or salted meats and fish, and the ingestion of moldy foods. Infection with the bacterium *Helicobacter pylori* has also been implicated. Symptoms include chronic discomfort, a loss of appetite with developing anorexia, difficulty eating, weight loss, and developing anemia. Yearly medical examinations usually include a test for occult blood in the stool to check for early signs of cancer in the digestive system. Other diagnoses include an upper GI series and an endoscopy. Individuals exposed to excessive chemicals like asbestos or living in a highly polluted industrialized area can also develop stomach or gastric cancer. Surgery is usually recommended for removal of the tumors.

PANCREATIC CANCER

Pancreatic cancer is an uncommon but deadly cancer. Individuals diagnosed with the disease rarely live for more than one year. It occurs more often in men than in women (three to four times more often in men). It occurs more often in industrialized areas, among smokers, people with diabetes, and individuals exposed to polychlorinated biphenyl compounds. Symptoms include weakness, weight loss, jaundice, gas, epigastric pain, and an onset of diabetes. Occasionally localized tumors may be removed surgically by a partial pancreatectomy and removal of part of the stomach, the duodenum, and common bile duct.

ORAL CANCER

Oral cancer is a malignancy found either in the mouth or on the lips that can occur after age 60. It occurs more frequently in men than women. Contributing factors include smoking, the use of chewing tobacco, poor oral hygiene, alcoholism, and even poor-fitting dentures. Pipe smoking contributes to the development of lip cancer as does overexposure to the wind and sun. Usually a painless, nonhealing ulcer is the first sign of the cancer. Frequent dental checkups can detect the development of oral cancer of which only 5% are fatal.

THE FORMATION OF THE FECES

By the time the chyme has remained in the large intestine 3–10 hours, it is a semisolid mass of material as a result of the absorption of water and is now known as the **feces**. The feces consist of water, inorganic salts, and epithelial cells from the mucosa of the tract that were scraped away as the chyme moved through the tract. In addition the feces have bacteria, in particular ***Escherichia coli***, a normal inhabitant of our intestine that feeds on undigested materials. The products of bacterial decomposition, such as gas and odor (hydrogen sulfide gas, H_2S, which produces a "rotten egg" odor), and undigested parts of food not attacked by bacteria are also found in the feces. The more fiber (the cellulose of plant cell walls from eating fruits and vegetables) in the diet, the more undigestible materials in the feces and the softer the stool.

When mass peristalsis pushes the fecal material into the rectum, it causes distention of the rectal walls. This triggers pressure-sensitive receptors in the walls of the rectum, sending an impulse to the nervous system, which initiates the reflex for defecation. Defecation is the act of emptying the rectum and is the final activity of the digestive system.

BODY SYSTEMS WORKING TOGETHER TO MAINTAIN HOMEOSTASIS: THE DIGESTIVE SYSTEM

Integumentary System

- The skin protects the organs of the system and provides vitamin D needed for calcium absorption in the intestine.
- The digestive system provides nutrients for growth and repair of the integument.
- Fat for insulation is deposited beneath the skin in the subcutaneous tissue.

Skeletal System

- The bones protect some digestive organs and the hyoid bone provides support for the tongue.
- Yellow bone marrow stores fat.
- The digestive system supplies calcium and nutrients for bone growth and repair.

Muscular System

- Smooth muscle pushes food and nutrients along the digestive tract via peristalsis.
- Skeletal muscles protect and support the abdominal organs.

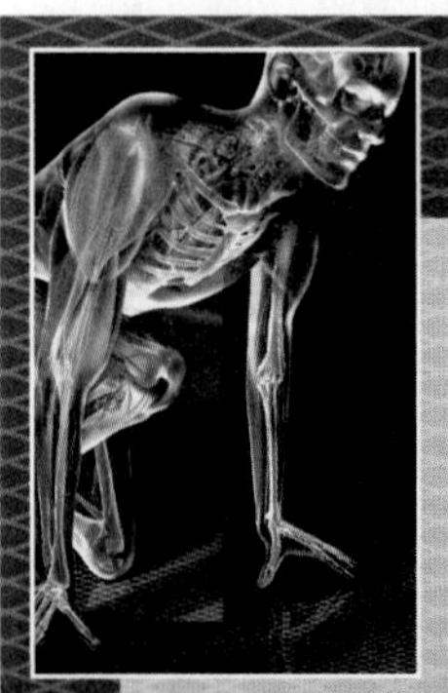

AS THE BODY AGES

As we age, the entire lining of the gastrointestinal tract changes. The smooth muscle layer decreases, causing less mobility and thus a decrease in time for peristalsis to occur. Less mucus is secreted, resulting in less lubrication and some difficulty pushing food easily along the tract. Enzyme secretions diminish from the liver, pancreas, and the gastric pits of the stomach.

Older adults experience a diminished sense of taste. Daily oral hygiene is critical to avoid the inevitable reduced natural defenses against microorganisms that can cause gum disease and mouth sores. Older adults also experience diminished internal sensations and response to pain along the tract. They tend to develop difficulty with swallowing.

Since the digestive tract is exposed to environmental contaminants over the years, with reduced mucous membrane lining and less protective connective tissue, the alimentary canal becomes less protected from toxic contaminants. Older adults are more susceptible to infections and are more likely to develop cancers and ulcerations of the tract, especially in the colon and rectum.

Older adults' liver declines in its ability to detoxify chemicals and contaminants and store excess sugar in the form of glucose. They experience more frequent constipation and develop hemorrhoids. In addition, a higher incidence of diverticulosis develops as we age.

Career FOCUS

There are numerous career opportunities available to individuals who are interested in the digestive system and nutrition.

- **Dentists** are individuals who have a college degree and have trained for 4 years at a dental college. They are qualified and licensed by the state to diagnose and treat abnormalities of the teeth, gums, and underlying bones.
- **Dental hygienists** are health care professionals who work under the supervision of a dentist. They have special training to provide teeth cleaning, taking X-rays, and administering medications.
- **Dental laboratory technicians** are individuals who supply dentists with orthodontic appliances and dental prostheses.
- **Dental assistants** are individuals who assist dentists in performing such tasks as dental office laboratory work, chairside assistance, some assistance with radiography, or clerical and reception responsibilities.
- **Nutritionists** are professionals who have completed at least a bachelor's degree or more advanced work in foods and nutrition. They are employed in nursing homes, schools, hospitals, and other public service agencies to monitor and prepare nutritious menus.
- **Registered dieticians** are professionals with a bachelor's degree in the management of foods and diets, providing nutritional care services and food service supervision.
- **Gastroenterologists** are physicians whose specialty is the diagnosis and treatment of diseases affecting the gastrointestinal tract.
- **Enterostomal therapists** are registered nurses with special training to provide care for patients who have an artificial anus or a fistula in the intestine through surgery into the abdominal wall.
- **Proctologists** are physicians with special training to diagnose and treat disorders of the colon, rectum, and anus.
- **Hepatologists** are physicians who specialize in treating diseases and disorders of the liver.

- The digestive system provides nutrients such as glucose for muscle contraction, growth, and repair.
- The liver metabolizes lactic acid after anaerobic muscle contraction.

Nervous System

- The nervous system sends impulses for muscular contractions in the walls of the gastrointestinal tract for peristalsis to occur.
- Nerve impulses coordinate swallowing and defecation.
- The digestive system provides nutrients for the growth, maintenance, and functioning of neurons and neuroglia cells.

Endocrine System

- Hormones help regulate the metabolism of nutrients for growth and development.
- Insulin and glucagon control sugar metabolism.
- The digestive system provides nutrients to maintain the endocrine glands.

Cardiovascular System

- The cardiovascular system distributes, via the blood, nutrients absorbed in the small intestine to all tissues of the body.
- The digestive system provides the nutrients for maintaining the organs of the circulatory system and

absorbs iron for hemoglobin production and water for blood plasma formation.

Lymphatic System

- The lacteals of the villi of the small intestine absorb fats and transport them to the blood.
- Lymphoid tissues protect the digestive organs from infection.
- The digestive system provides nutrients for the lymphoid organs for growth and repair.
- The hydrochloric acid of the stomach destroys most pathogens that may enter the body with food.

Respiratory System

- The respiratory system provides the cells of the digestive tract with oxygen needed for metabolism and takes away the waste product carbon dioxide gas.
- Breathing can occur through the mouth due to the pharynx, which is shared by both systems.
- The digestive system provides nutrients for the respiratory organs.

Urinary System

- The kidneys convert vitamin D to a form needed for calcium absorption and reabsorb water lost in the digestive tract.
- The digestive system provides nutrients to the organs of the urinary system.
- The liver converts harmful ammonia from the digestion of proteins to harmless urea and provides bile to emulsify fats.

Reproductive System

- When a woman is pregnant, the fetus crowds the abdominal organs and the mother may experience constipation.
- The digestive system provides nutrients for maintenance, growth, and repair of the reproductive organs and supplies the developing fetus with nutrients.

SUMMARY OUTLINE

INTRODUCTION

1. Digestion is the process by which food is broken down mechanically and chemically into simpler substances that can be used by the body's cells and converted into high-energy ATP molecules.
2. The five basic activities of the digestive system are ingestion, peristalsis, digestion, absorption, and defecation.

GENERAL ORGANIZATION

1. The digestive tract includes the mouth, pharynx, esophagus, stomach, small intestine, large intestine, and anus.
2. The accessory organs of the tract include the teeth, tongue, salivary glands, liver, gallbladder, and pancreas.
3. The four coats, or tunics, of the tract, from the inside out, are the mucosa, submucosa, muscularis, and adventitia or serosa.

THE ORAL CAVITY

1. The functions of the oral cavity are taste, mechanical breakdown of food using the teeth, and chemical digestion of carbohydrates using the salivary enzyme amylase.
2. The oral cavity is lined with mucous membrane. The floor of the cavity is formed by the tongue, the roof by the hard and soft palate, and the sides by the cheek. The opening is guarded by the lips.
3. The functions of the tongue are manipulation of the food, taste through some of its papillae, and assistance in speech.
4. The three pairs of salivary glands are the parotid, the submandibular or submaxillary, and the sublingual. Saliva lubricates the food, begins the digestion of complex carbohydrates, and controls certain bacteria.
5. A tooth is composed of the crown, the neck or cervix, and the root. The crown of the tooth is covered with enamel. A tooth is made up of dentin. The periodontal ligament anchors the tooth into the alveolar socket.

THE PHARYNX

1. The pharynx is a common passageway for food and air. It is divided into the nasopharynx, the oropharynx, and the laryngopharynx.
2. Its function is to begin the process of swallowing or deglutition.

THE ESOPHAGUS

1. The function of the esophagus is to secrete mucus and transport food to the stomach through an opening in the diaphragm called the esophageal hiatus.
2. Peristalsis, caused by smooth muscle contractions, pushes the food bolus into the stomach through the lower esophageal sphincter.

THE STOMACH

1. The main function of the stomach is to begin the chemical breakdown of proteins through the enzyme pepsin. It also breaks up food mechanically by churning its contents. It absorbs some water, salts, alcohol, and certain drugs like aspirin.
2. The four parts of the stomach are the cardia, fundus, body, and pylorus. The pyloric sphincter guards the connection into the small intestine. The empty stomach lining has many folds or rugae that allow the stomach to expand and hold large amounts of food.
3. The gastric glands of the stomach mucosa contain three kinds of secretory cells: (1) the zymogenic or chief cells that secrete pepsinogen; (2) the parietal cells that secrete hydrochloric acid; and (3) the mucous cells that secrete mucus.

THE PANCREAS

1. The pancreas has a dual function. Its acini produce digestive enzymes that get carried by the pancreatic duct to the duodenum of the small intestine. Its pancreatic islets secrete the hormones insulin and glucagon into the blood to control blood sugar levels.
2. The pancreas is divided into the head, body, and tail.

THE LIVER

1. The liver is the largest organ of the body. It is divided into a right and a left lobe. The functional units of the liver are called lobules.
2. The functions of the liver are so numerous and important that we cannot live without it. It produces heparin, prothrombin, and thrombin. Its Kupffer cells phagocytose bacteria and worn-out blood cells. It stores excess carbohydrates as glycogen. It stores copper, iron, and vitamins A, D, E, and K. It stores or transforms poisons into less harmful substances. It produces bile salts that emulsify or break down fats.
3. The gallbladder stores and concentrates bile produced by the liver lobules. The bile enters the duodenum of the small intestine through the common bile duct.

THE SMALL INTESTINE

1. The main function of the small intestine is the completion of absorption of the digested food. It is divided into three portions: the duodenum (10 inches), the jejunum (8 feet), and the ileum (12 feet). The ileocecal valve connects the small intestine with the large intestine.
2. In addition to its length, the walls of the small intestine are thrown into folds called plicae that are covered with millions of villi. Nutrients are absorbed through the villi. Each villus has an epithelial cell covering whose free edge is covered with microscopic folds called microvilli to further increase the absorptive capabilities of the villus. Each villus contains an arteriole, venule, capillary network, and a lacteal that picks up fats.
3. The intestinal glands or crypts of Lieberkuhn secrete the intestinal digestive enzymes. Brunner's glands secrete an alkaline mucus.
4. Chyme is the name of the digested contents of the small intestine.

THE LARGE INTESTINE

1. The functions of the large intestine are the reabsorption of water, the manufacture and absorption of certain vitamins, and the formation and expulsion of the feces.
2. The four regions of the large intestine are the cecum, colon, rectum, and anus.
3. The colon is divided into the ascending colon, transverse colon, and descending colon. It has pouches or haustrae.
4. The rectum terminates at the anus. The anus is controlled by an internal sphincter of smooth muscle and an external sphincter of skeletal muscle.
5. The three mechanical movements that occur in the large intestine are haustral churning, peristalsis at the rate of 3 to 12 contractions per minute, and mass peristalsis triggered by distention of the stomach.
6. Distention of the rectal walls initiates the defecation reflex.

REVIEW QUESTIONS

1. List the five basic activities of the digestive system.
2. Name the major organs and the five accessory organs of the digestive tract.
3. List the four walls or tunics of the digestive tract.
4. Name the three pairs of salivary glands and discuss why they are important for digestion.

5. What are the three portions of a tooth and what substance forms a tooth?
*6. Why is the pharynx considered essential to the process of swallowing?
7. Name the four parts of the stomach.
8. List the three types of secretory cells of the mucosa of the stomach and what they secrete. Explain the functions of those secretions.
*9. Name the two functions of the pancreas and explain what its secretions accomplish.
*10. Discuss why we cannot live without the liver.
11. Name the three parts of the small intestine.
*12. List the two types of glands of the small intestine and compare their function in the process of digestion.
13. Describe the anatomy and physiology of a typical villus, including its function in the absorption of nutrients.
*14. How does the large intestine contribute to the digestive process?
15. Name the four principal regions of the large intestine.
16. List the major digestive enzymes and how they function.

***Critical Thinking Questions**

FILL IN THE BLANK

Fill in the blank with the most appropriate term.

1. The movement of food along the digestive tract is called ________________.
2. The four tunics or coats of the gastrointestinal tract are the ________________, ________________, ________________, and the ________________ or adventitia.
3. The three kinds of papillae found on the tongue are the ________________, the ________________, and the circumvallate papillae.
4. The three kinds of salivary glands are the ________________, the submandibular or ________________, and the ________________ glands.
5. The small intestine is divided into three portions: the ________________, the ________________, and the ________________.
6. The ________________ stores and concentrates bile until it is needed in the small intestine.
7. A typical tooth consists of three principal portions: the ________________, the portion above the level of the gums; the ________________ consisting of one to three projections embedded in the socket; and the ________________.
8. The dentin of a tooth is covered by ________________, the hardest substance in the body.
9. Deglutition is the term used for ________________.
10. The name given to the digested material moving through the intestine is ________________.

Search and Explore

- Go to the United States Department of Agriculture website at http://www.mypyramid.gov and explore the *Inside the Pyramid* section to learn more about the food categories and recommended daily servings.
- Create a list of all the foods you eat in an ordinary day. Compare the foods on the list you eat regularly with the recommendations you learned about at the MyPyramid website.

MATCHING

Place the most appropriate number in the blank provided.

____ Zymogenic cells	1. Hydrochloric acid
____ Pancreatic islets	2. Pancreatic juice
____ Heparin	3. Regulates blood sugar level
____ Crypts of Lieberkuhn	4. Phagocytosis
	5. Pepsinogen
____ Parietal cells	6. Absorb fats
____ Brunner's glands	7. Chyme
____ Lacteals	8. Neutralizes acid in chyme
____ Kupffer cells	9. Intestinal glands
____ Acini	10. Anticoagulant
____ Insulin	11. Secrete hormones
	12. Amylase

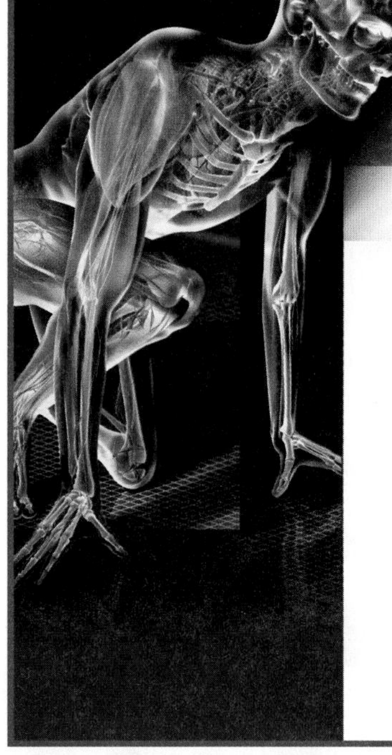

CASE STUDY

Judith Goldberg, a 17-year-old woman, is admitted to the emergency room for the third time this year for the same symptoms, which have now increased in severity. Judith states that she is experiencing abdominal pain, frequent bouts of diarrhea, fever, chills, nausea, and weakness. She reports that she has no appetite, and has been steadily losing weight. The health care provider instructs Judith to see her gastroenterologist immediately to discuss surgery as an option to control the symptoms.

Questions

1. What do you think might be causing Judith's recurring symptoms?
2. What areas of the digestive system does this disorder affect?
3. What is the prognosis for individuals with this condition?

*Study***WARE**™ Connection

Play interactive games or take a practice quiz on your StudyWARE™ CD-ROM.

Study Guide Practice

Go to your **Study Guide** for more practice questions, labeling and coloring exercises, and crossword puzzles to help you learn the content in this chapter.

LABORATORY EXERCISE: THE DIGESTIVE SYSTEM

Materials needed: a dissecting kit, a 9 to 11-inch fetal pig from a biologic supply company and a dissecting pan.

A. EXTERNAL FEATURES

1. **The head:** locate the nostrils or snout, the closed eyes, the mouth, and the external ears.
2. **The trunk:** examine the four legs. Notice that there are four toes on each foot and that the pig walks on only two. Hence, the pig is referred to as an eventoed ungulate mammal, which means that it is hoofed and that it walks on its toenails. Observe the umbilical cord, which is seen on the ventral surface of the abdomen. This allows the fetal pig to exchange nutrients, oxygen, and waste materials with its mother through the fetal blood circulatory system. There are also pairs of nipples on the ventral surface of the abdomen. They are the openings from the mammary (milk) glands. All sexes have nipples. Male mammary glands are functionless. Note the presence of hair at various places on the pig's body. Hair can usually be seen under the chin and around the closed eyes.
3. To sex your pig, look first at the anus, which is the posterior opening of the digestive tract. If you have a female, you will see a small projecting papilla coming out of what looks like the anus. Actually, the papilla is part of the urogenital opening that is just ventral to the anus. Later in development, the urogenital opening will separate from the digestive opening and the single opening will become two. If you have a male, the urogenital opening will be located posterior to the umbilical cord. A muscular tube, the penis, can be felt just under the skin running forward from between the back legs forward to the urogenital opening.

B. INTERNAL FEATURES

1. The oral cavity: open the jaws as much as you can without cutting. Identify the oral cavity with the tongue forming the ventral floor. The hard palate forms the dorsal area or roof of the mouth. The hard palate separates the nasal cavities from the oral cavity. Use scissors to make a cut from one corner of the mouth to the bottom of the ear. Use Figure 16-17 as your guide and make cut number 1. You must cut through soft tissue and then through bone (the ramus of the mandible). Repeat this cut on the other side of the mouth, being careful not to rip the tissues of the soft palate in the posterior region of the mouth. Once you have cut through the pig's jawbone, you will be able to open the mouth very widely and you will be able to examine the internal features of the oral cavity.
2. Examine the tongue and feel the papillae scattered over the surface of the tongue. Many of these contain taste buds. Feel the roof of the mouth. It is divided into the anterior hard palate with its wavelike hard tissues and the posterior soft and delicate soft palate. In humans, a soft extension of the soft palate called the uvula hangs down into the throat. Note that the pig does not have a uvula. The small opening at the back of the soft palate in the roof of the mouth is the nasopharynx. This is the opening from the nasal cavities, which allows air to enter the pharynx on its way to the larynx and trachea of the respiratory system.
3. Now observe the gums and the teeth that have erupted through the gum tissue. They are probably canines and incisors. You may cut into the gums and remove some of the embryonic teeth. If you have cut back far enough, you will see at the base of the tongue a small median flap, the epiglottis. It will pop out as you pull the jaw down gently. It partially covers the glottis, which is the entrance into the trachea, which leads to the lungs. It prevents food from entering the respiratory system when swallowing. Posterior and dorsal to the glottis is the esophagus, which leads to the stomach. Try to push your probe down into the esophagus. We will observe

(*continues*)

THE DIGESTIVE SYSTEM (Continued)

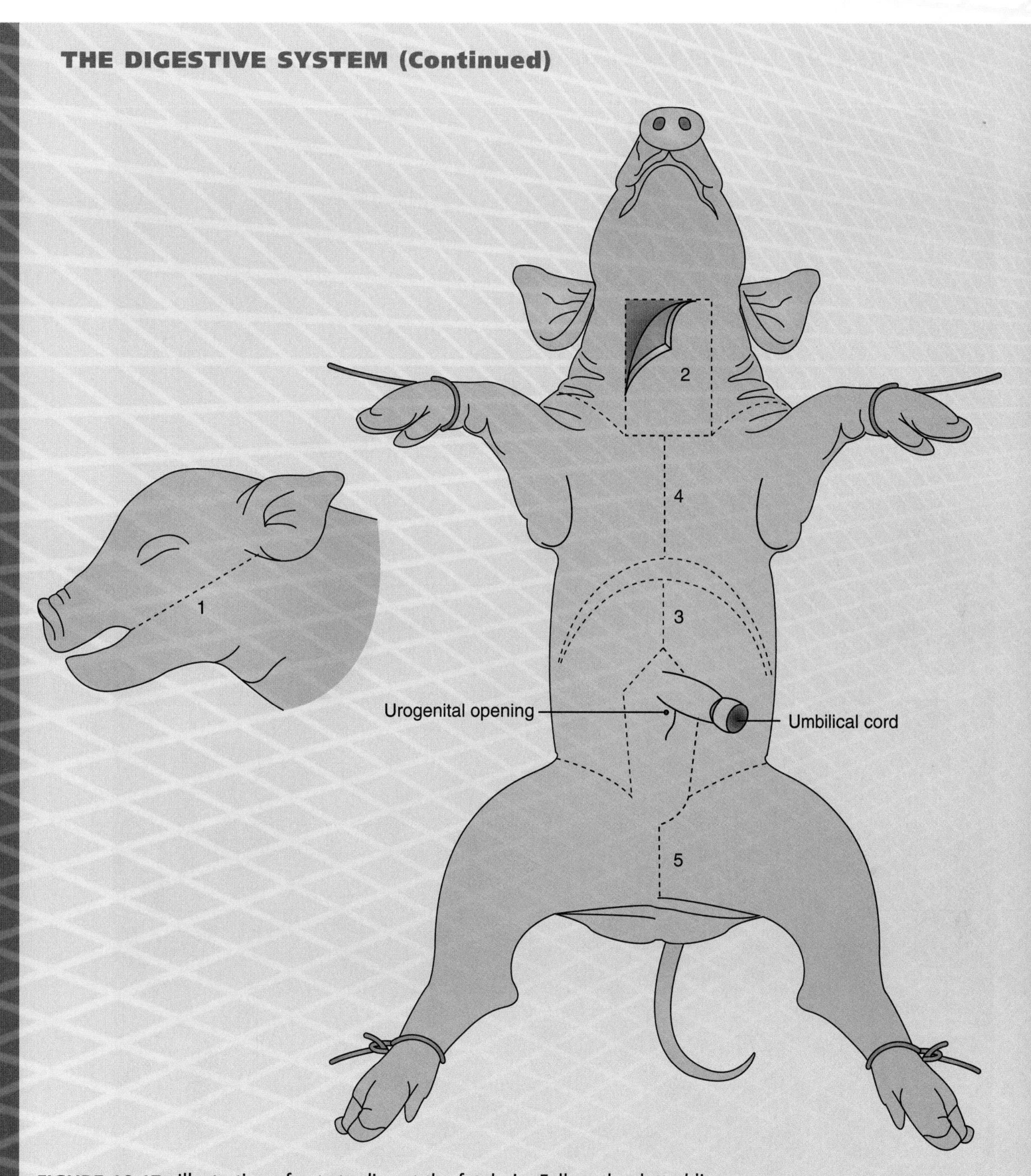

FIGURE 16-17. Illustration of cuts to dissect the fetal pig. Follow the dotted lines.

the esophagus when we dissect the thoracic and neck region of the pig in Chapter 17. Refer to Figure 16-18 of the fetal pig dissection for the following.

4. The abdominal cavity: the abdominal cavity is posterior to the ribs where the ventral body wall is very soft due to the absence of any bony support. The body wall encloses the large peritoneal cavity, which contains most of the digestive, excretory, and reproductive organs. In opening the peritoneal cavity, use the umbilical cord as a landmark and follow the dotted lines numbers 3 and 5 of Figure 16-17. Cut through the soft, thin body wall very carefully with your scalpel. Once you have cut through, continue with your scissors. The quarter moon cut just below the ribs is done to protect

(continues)

THE DIGESTIVE SYSTEM (Continued)

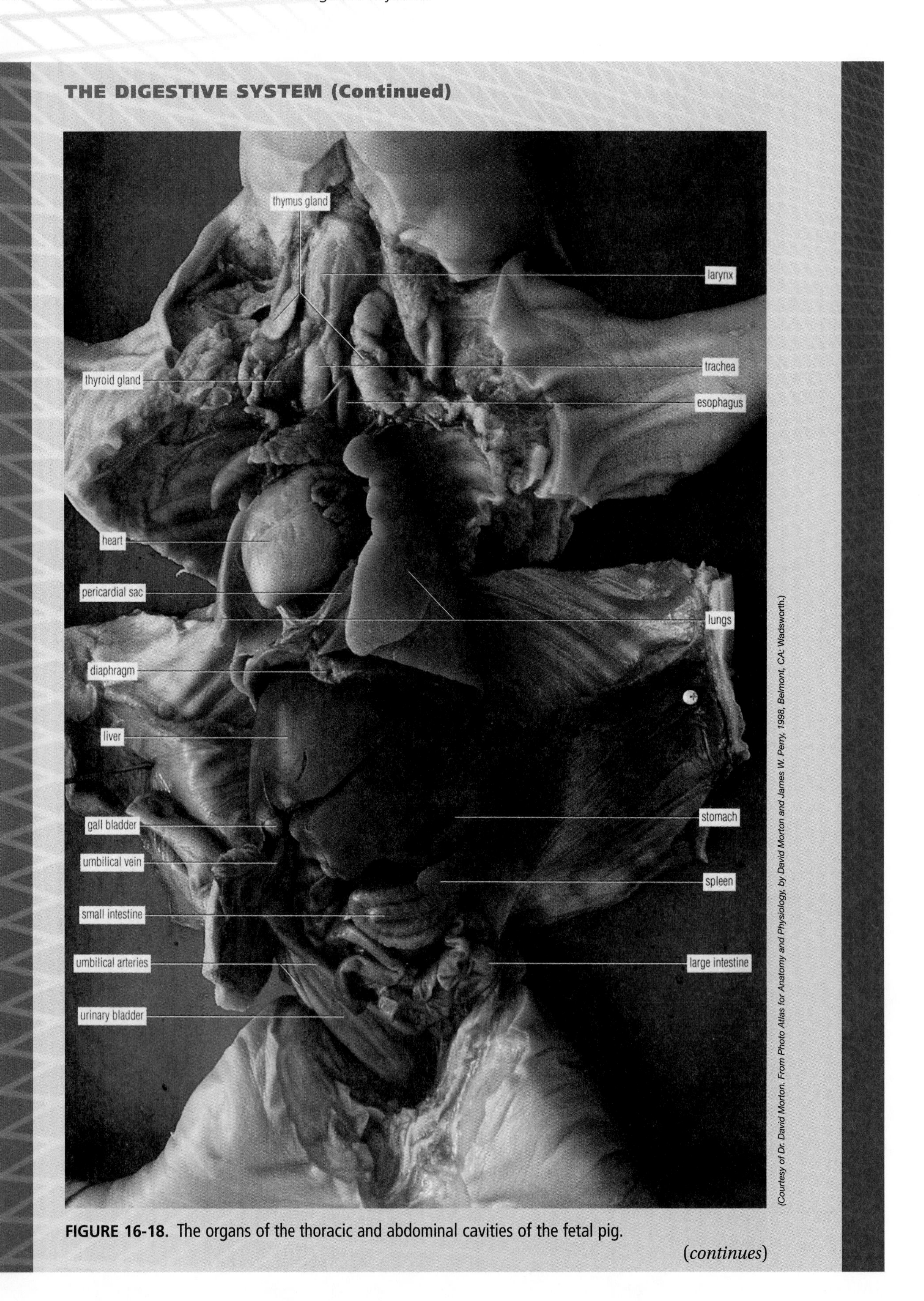

FIGURE 16-18. The organs of the thoracic and abdominal cavities of the fetal pig.

(continues)

THE DIGESTIVE SYSTEM (Continued)

the diaphragm muscle that separates the thoracic cavity (to be observed in Chapter 17) from the abdominal cavity. Once you have completed the cut, pin back the lateral flaps of skin of the abdominal area. If your cavity is filled with dark brown clotted blood, take your specimen to the sink and flush it out. It does not harm or alter your specimen. It should just be cleaned out to make your observations clearer.

5. Examine one of the flaps of skin. It is composed of three layers of tissue. The outermost is the skin, followed by a middle layer of abdominal muscle. The very thin transparent innermost layer is the parietal peritoneum, which lines the inside of the wall of the peritoneal cavity. A similar tissue, the visceral peritoneum, covers the organs of the cavity. The parietal and visceral peritoneum are connected by thin sheets of tissue called the mesentery. Several mesenteries suspend and support visceral organs and allow the passage of blood vessels and nerves to the organs.
6. Now observe the major organs of the abdominal cavity. Note the diaphragm muscle that lies directly above the large brown liver. The liver is the largest organ in the abdomen. Lift up the right lobe of the liver to observe the green pear-shaped gallbladder, which stores bile. Note the bile duct, which leads from the gallbladder to the duodenum of the small intestine. Push the liver aside and identify the stomach, which lies on the left side of the liver. Note its J shape. Note the point at the anterior end of the stomach where the esophagus penetrates the diaphragm to join the stomach. The posterior end of the stomach narrows to join the anterior end of the small intestine, the duodenum. The constriction at the junction between the stomach and the duodenum is the pylorus. While in this area observe the long flaplike structure called the spleen, which is part of the lymphatic system and whose function is blood storage. Our spleen is more kidney bean shaped as opposed to the long, flat spleen of the pig.
7. Examine the duodenum of the small intestine more closely. Observe how the liver and the duodenum are connected by the bile duct. Now lift the stomach and locate the pancreas, a light-colored spongy gland. The pancreas also has a duct that empties into the duodenum. Note the parts of the pancreas: the head, body, and tail. Observe that a large amount of intestine in the pig is the small intestine. At the distal end of the small intestine, a blind sac, the cecum, can be seen on the left side of the abdominal cavity of the pig. It is found at the point of juncture where the small intestine joins the large intestine.
8. From the cecum, follow the large intestine (about double the diameter of the small intestine), now known as the colon, as it tightly coils to join the large dorsal rectum, which ends at the anus. Before we complete our dissection of the digestive tract of the fetal pig, cut open a piece of the small intestine to observe the plicae. These are the folds of the small intestine that increase the absorptive capabilities of the intestine.

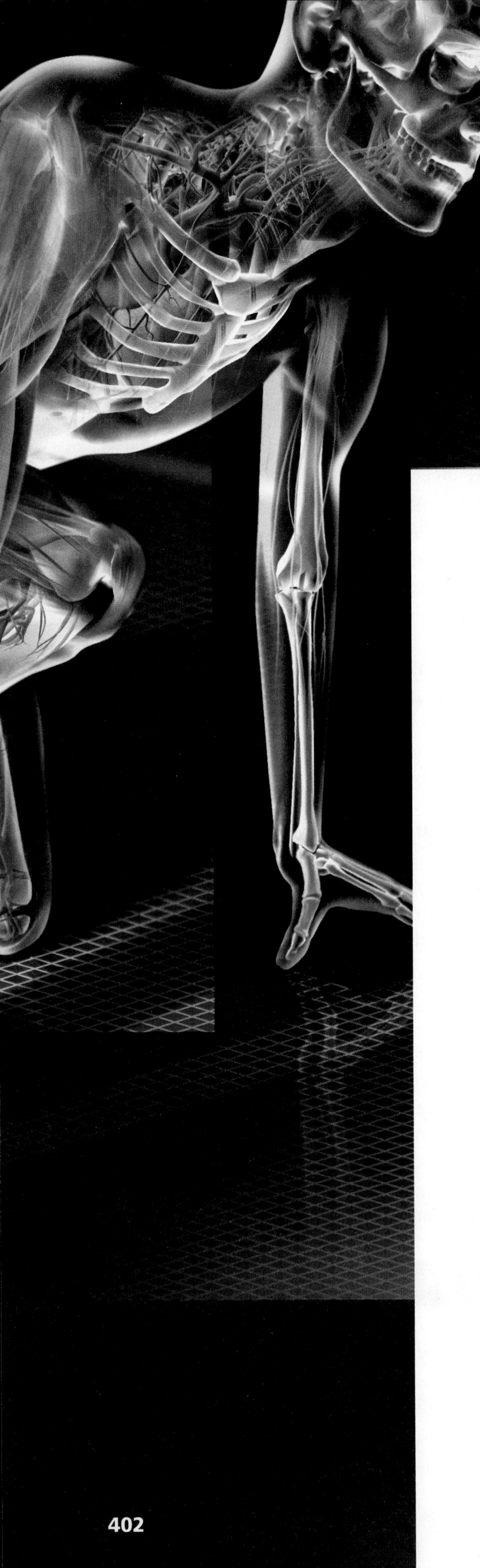

The Respiratory System

CHAPTER OBJECTIVES

After studying this chapter, you should be able to:

1. Explain the function of the respiratory system.
2. Name the organs of the system.
3. Define the parts of the internal nose and their functions.
4. Name the three areas of the pharynx and explain their anatomy.
5. Name the cartilages and membranes of the larynx and how they function.
6. Explain how the anatomy of the trachea prevents collapse during breathing and allows for esophageal expansion during swallowing.
7. Explain what is meant by the term *bronchial tree*.
8. Describe the structure and function of the lungs and pleura.
9. Describe the overall process of gas exchange in the lungs and tissues.
10. Define *ventilation*, *external respiration*, and *internal respiration*.

17

KEY TERMS

Alveolar-capillary membrane/respiratory membrane 413
Alveolar ducts/atria 412
Alveolar sacs 412
Alveoli 412
Arytenoid cartilages 408
Auditory/eustachian tube 407
Bronchial tree 410
Bronchioles 410
Bronchopulmonary segment 412
Corniculate cartilage.... 408
Cricoid cartilage 407
Cuneiform cartilage 408
Epiglottis 407
Exhalation/expiration... 413
External respiration 413
Fauces 407
Glottis 407
Inferior meatus 406
Inhalation/inspiration... 413
Internal nares 405
Internal respiration..... 413
Laryngopharynx 407
Larynx 407
Left primary bronchus .. 410
Lobules.............. 412
Middle meatus 406
Nasal cavities.......... 406
Nasal septum.......... 406
Nasopharynx.......... 407
Nostrils/external nares.............. 404
Olfactory stimuli 406
Oropharynx........... 407
Parietal pleura......... 411
Partial pressure 417
Pharynx 405
Pleural cavity.......... 411
Pleural membrane...... 411
Pleurisy/pleuritis....... 411
Respiration 404
Respiratory bronchioles . 412
Right primary bronchus . 410
Secondary/lobar bronchi............. 410
Superior meatus 406
Surfactant 413
Terminal bronchioles.... 410
Tertiary/segmental bronchi............. 410
Thyroid cartilage/ Adam's apple........ 407
Trachea............... 408
Ventilation/breathing... 413
Vestibular folds/ false vocal cords 408
Vestibules 406
Visceral pleura......... 411
Vocal folds/true vocal cords............... 408

INTRODUCTION

The trillions of cells of our body need a continuous supply of oxygen to carry out the various and vital processes that are necessary for their survival. Cellular respiration, which converts food into the chemical energy of adenosine triphosphate (ATP), produces large quantities of carbon dioxide gas. An excess accumulation of this gas in tissue fluids produces acidic conditions in the form of carbonic acid that can be poisonous to cells. Thus, this gas must be quickly eliminated.

The two systems of the body that share the responsibility of supplying oxygen and eliminating carbon dioxide gas are the cardiovascular system and the respiratory system. The respiratory system consists of the organs that exchange these gases between the

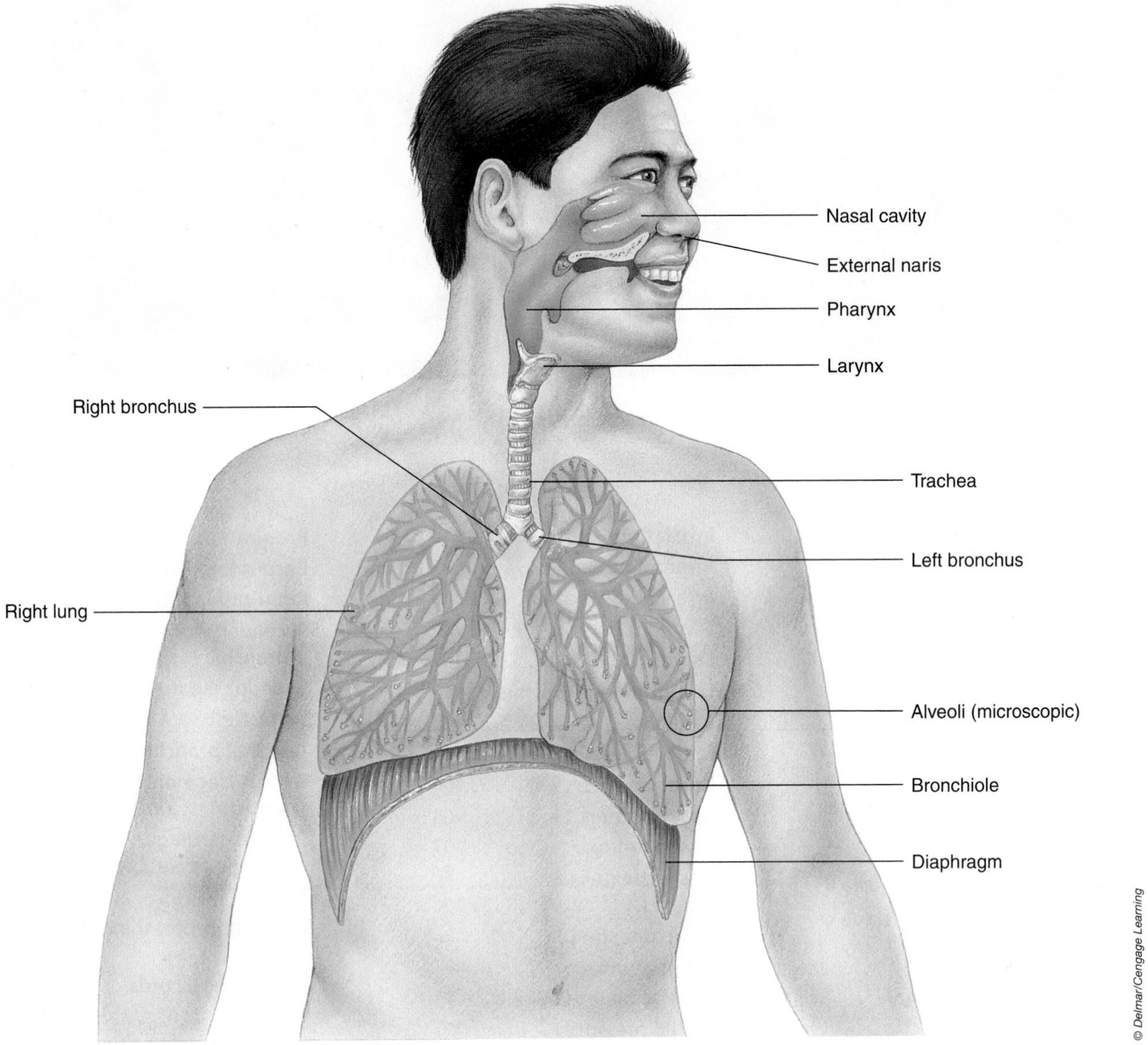

FIGURE 17-1. The organs of the respiratory system.

atmosphere and the blood. Those organs are the nose, pharynx, larynx, trachea, bronchi, and lungs (Figure 17-1). In turn, the blood in the cardiovascular system transports these gases between the lungs and the cells. The overall exchange of gases between the atmosphere, the blood, and the cells is called **respiration**. This term is to be distinguished from the biochemical meaning of respiration discussed in Chapter 4. The respiratory and cardiovascular systems participate equally in respiration. If either system malfunctions, the body cells will die from oxygen deprivation and accumulation of carbon dioxide gas and death will be inevitable. See Concept Map 17-1: The Respiratory System.

THE ANATOMY AND FUNCTIONS OF THE NOSE

The nose has an external part and an internal part that is inside the skull (Figure 17-2). Externally, the nose is formed by a framework of cartilage and bone covered with skin and lined internally with mucous membrane. The bridge of the nose is formed by the nasal bones that help support the external nose and hold it in a fixed position. On the undersurface of the external nose are two openings called the **nostrils** or **external nares** (ex-**TER**-nal **NAIREZ**). The hard palate of the mouth forms the floor of the nasal cavity, separating the nasal cavity from the oral cavity.

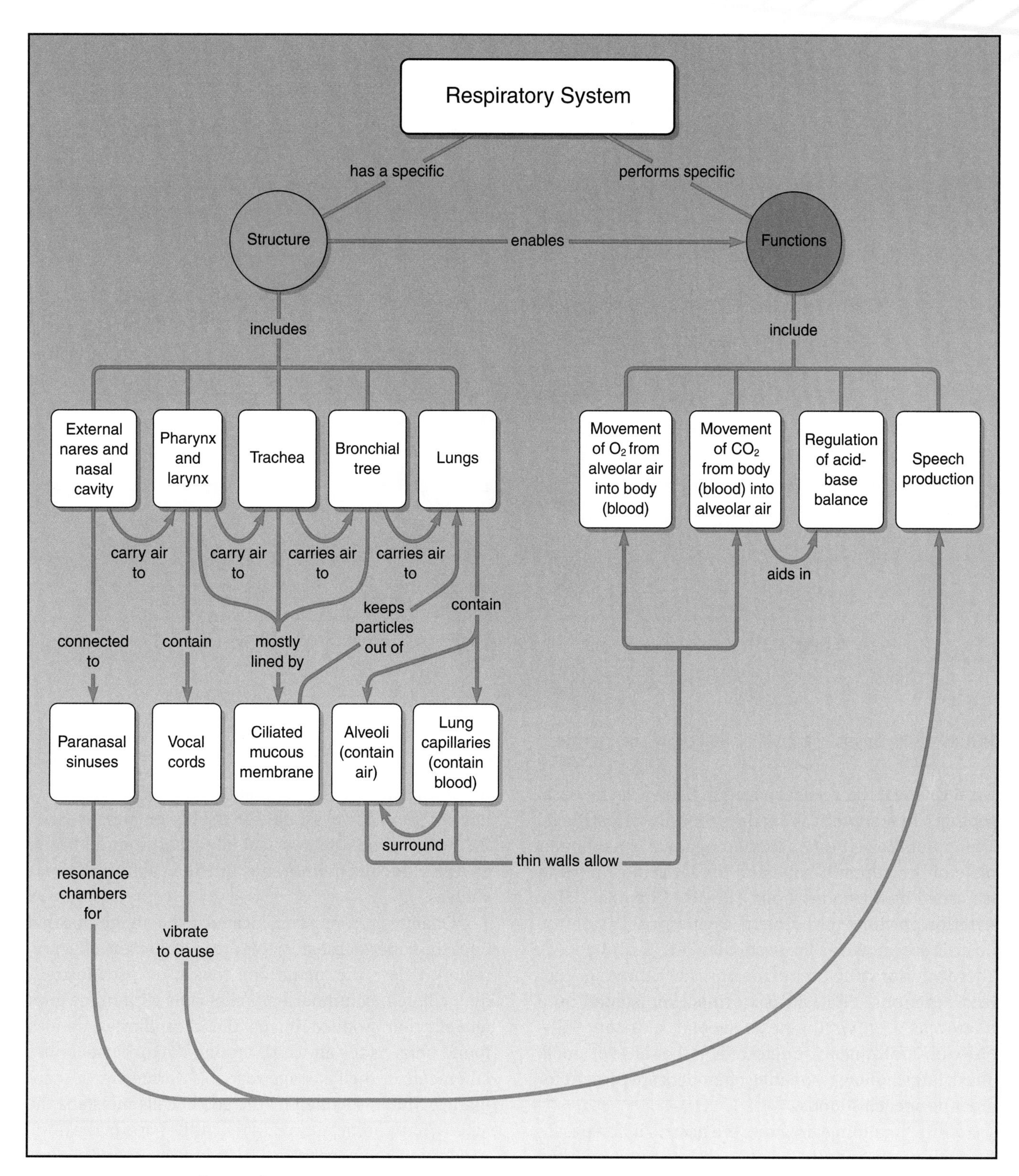

CONCEPT MAP 17-1. The respiratory system.

Anteriorly, the internal nose merges with the external nose. Posteriorly, it connects with the **pharynx** (**FAIR**-inks) or throat via two openings called the **internal nares**. The nasolacrimal ducts from the lacrimal or tear sacs empty into the nose, as well as four paranasal sinuses (air-filled spaces inside bone): sphenoidal, frontal, ethmoidal, and maxillary. The inside of both the internal and external nose is divided into right and left

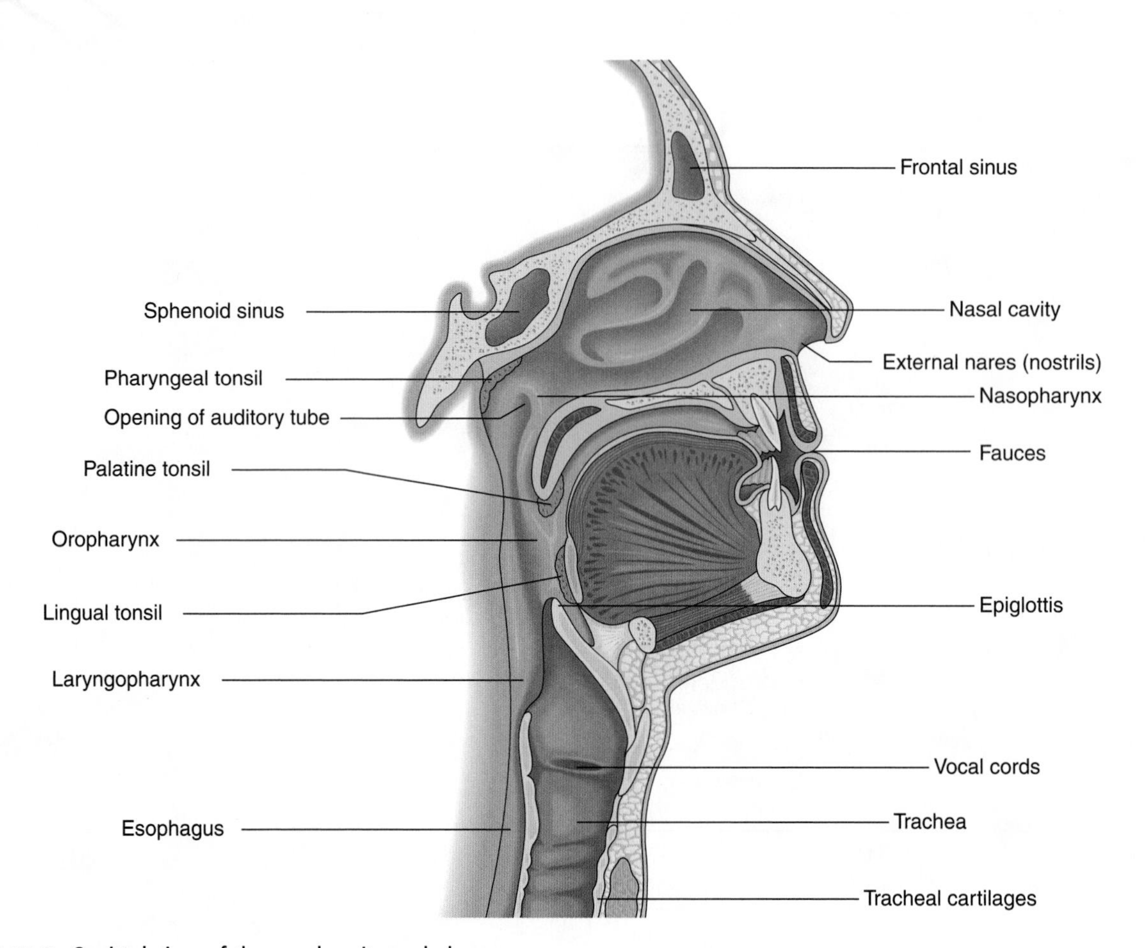

FIGURE 17-2. Sagittal view of the nasal cavity and pharynx.

nasal cavities by a vertical partition known as the **nasal septum**. This septum is made primarily of cartilage. The top of the septum is formed by the perpendicular plate of the ethmoid bone and the lowermost portion is formed by the vomer bone. (Review Chapter 7.) The anterior portions of the nasal cavities just inside the nostrils are known as the **vestibules** (**VESS**-tih-byoolz). These interior structures of the nose have three specialized functions. First, air is warmed, moistened, and filtered as it enters the nose. Second, **olfactory** (**olh-FAK**-toh-ree) **stimuli** are detected for the sense of smell. Third, large hollow resonating chambers are present for creating speech sounds.

As the incoming air enters the nostril, it first passes through the vestibule. Because the vestibule is lined with coarse hairs, it filters out large dust particles. This is the body's first line of defense to prevent foreign objects from entering the respiratory system. The air then moves into the rest of the cavity. Three shelves are formed by the projections of the superior, middle, and inferior conchae or turbinate bones. These extend out from the lateral wall of the cavity and almost reach the nasal septum. The cavity is subdivided into a series of narrow passageways called the **superior meatus** (soo-**PEER**-ih-or mee-**AY**-tus), **middle meatus**, and **inferior meatus**. Mucous membranes line the cavity and those shelves.

Olfactory receptors are located in the membrane that lines the superior meatus; this area is called the olfactory region. Below, the membrane consists of pseudostratified, ciliated, columnar epithelial cells with many goblet cells that produce mucus. Blood capillaries are also found here. As the air whirls around the turbinate bones and meati, or shelf passageways, it is warmed by the capillaries. Mucus secreted by the goblet cells moistens the air and traps particles not filtered by the hairs in the nose. In addition, drainage from the lacrimal ducts and sinuses help moisten the air. The cilia on the free edge of the epithelial cells move this mucus-dust package back toward the throat so it can be swallowed and eliminated from the body through the digestive system. Its enzymes and acidic environment will destroy most microorganisms that may have entered with the air. The cold virus and the flu virus are not destroyed.

THE STRUCTURE AND FUNCTIONS OF THE PHARYNX

The pharynx is also called the throat. It is a tube approximately 5 inches (13 cm) long that begins at the internal nares and extends part way down the neck. Its position in the body is noted just posterior to the nasal and oral cavities and just anterior to the cervical vertebrae. Its walls are made of skeletal muscle lined with mucous membrane. The pharynx is a passageway for both air and food and forms a resonating chamber for speech sounds. It is divided into three portions (see Figure 17-2).

The uppermost portion is called the **nasopharynx** (**nay**-zoh-**FAIR**-inks). It has four openings in its walls: the two internal nares and, just behind those, the two openings that lead into the **auditory** or **eustachian** (you-**STAY**-shen) **tubes.** In its posterior wall the pharyngeal or adenoid tonsils are located.

The second portion is called the **oropharynx** (**or**-oh-**FAIR**-inks). It has only one opening, the **fauces** (**FOH**-sez), which connects with the mouth. Hence, the oropharynx is a common passageway for both food and air. The palatine and lingual tonsils are found in the oropharynx.

The lowermost portion is called the **laryngopharynx** (lah-**ring**-go-**FAIR**-inks). It connects with the esophagus posteriorly and with the larynx anteriorly. The pharynx or throat serves as both a connection between the mouth and the digestive tract and as a connection between the nose and the respiratory system.

THE LARYNX OR VOICE BOX

The **larynx** (**LAIR**-inks) is also called the voice box (Figure 17-3). It is a very short passageway that connects the pharynx with the trachea. Its walls are supported by nine pieces of cartilage. Three of the pieces are single and three are paired. The three single pieces are the **thyroid** (**THIGH**-royd) **cartilage**, the **epiglottis** (**ep**-ih-**GLOT**-iss), and the **cricoid** (**KRYE**-koyd) **cartilage.**

The thyroid cartilage is the largest piece of cartilage and is also known as the **Adam's apple.** It is larger in males than in females and can be easily seen externally, moving up and down when a person is speaking or swallowing. The epiglottis is a large, leaf-shaped piece of cartilage. It lies on the tip of the larynx. It can be viewed in its entirety from a posterior view, but, anteriorly, one can only see its tip. The *stem* part is attached to the thyroid cartilage, but the *leaf* part is unattached and is free to move up and down like a trap door. When we swallow, this free edge or leaflike part pulls down and forms a lid over the **glottis** (**GLOT**-iss). The glottis is the space between the vocal cords in the larynx. The larynx is closed off when we swallow, so that foods and liquids get routed posteriorly into the esophagus and are kept out of the trachea anteriorly. If anything other than air passes into the larynx, a cough reflex should dislodge the foreign material.

When we try to talk and swallow at the same time, we choke and the cough reflex functions. Sensory receptors in the larynx detect the foreign substance and send a signal to the medulla oblongata, which triggers the cough reflex. Air is taken in and the vestibular folds and vocal cords tightly close trapping the air in the lungs. Muscular contractions increase the pressure in the lungs and the cords open,

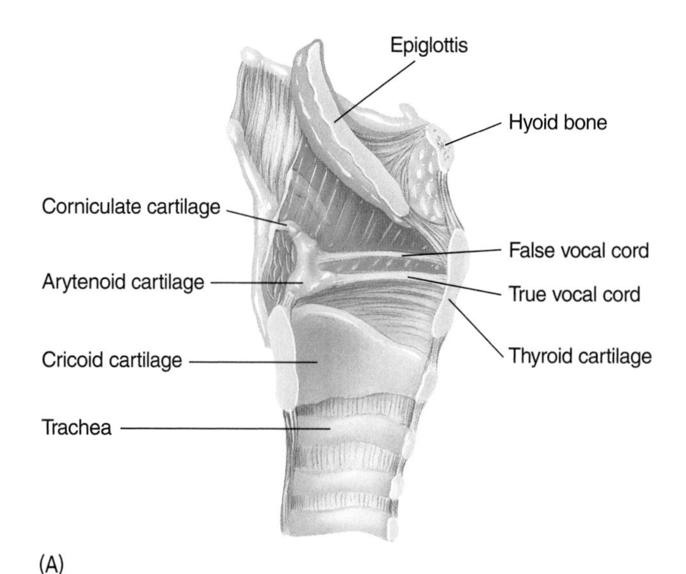

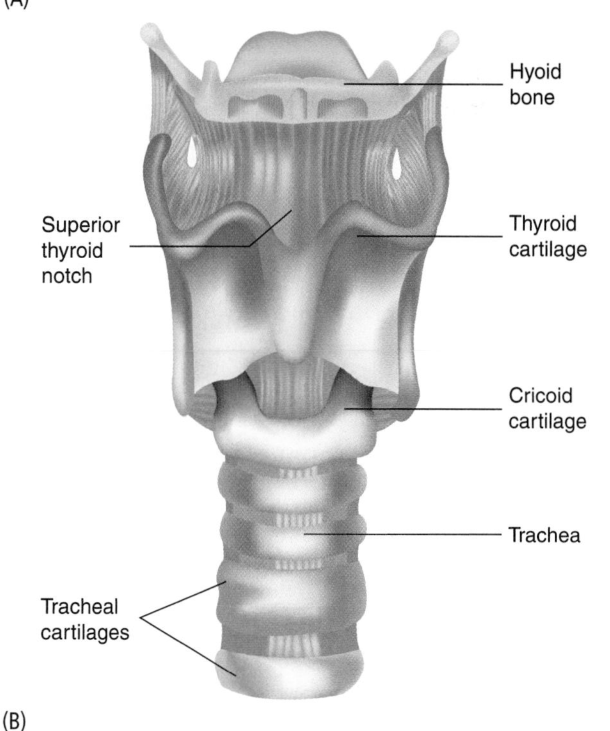

FIGURE 17-3. The larynx. (A) Lateral view, (B) anterior view, (C) posterior view.

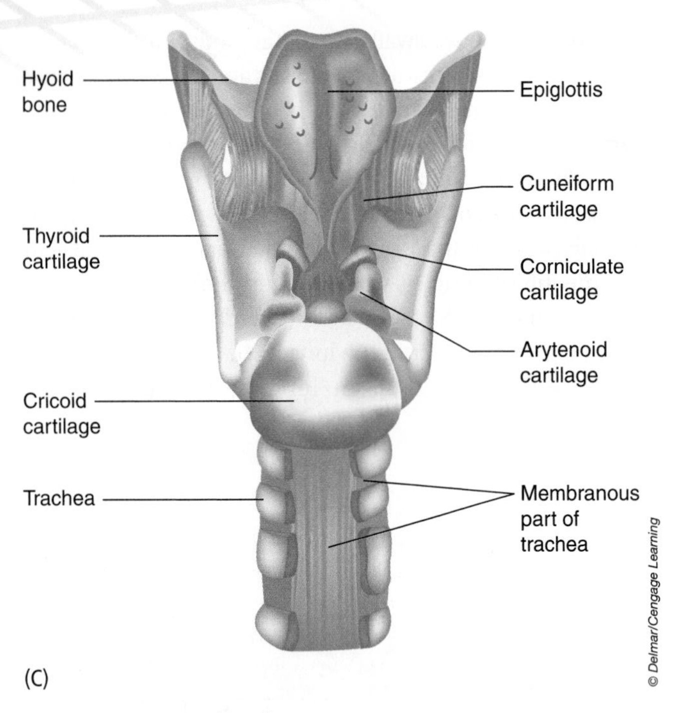

FIGURE 17-3. continued

forcing air from the lungs at a very high velocity and carrying any foreign substance with it.

The cricoid cartilage is a ring of cartilage that forms the lowermost or inferior walls of the larynx. It attaches to the first ring of cartilage of the trachea. This is the last of the three unpaired cartilages. The six paired cartilages consist of three cartilages on either side of the posterior part of the larynx. The paired **arytenoid** (**ahr**-ih-**TEE**-noyd) **cartilages** are ladle-shaped and attach to the vocal cords and laryngeal muscles and by their action they move the vocal cords. The **corniculate** (kor-**NIK**-yoo-late) **cartilages** are cone-shaped; the paired **cuneiform** (kyoo-**NEE**-ih-form) **cartilages** are rod-shaped. The cuneiforms are located in the mucous membrane fold that connects the epiglottis to the arytenoid cartilages.

The mucous membrane of the larynx is arranged into two pairs of folds: an upper pair called the **vestibular** (vess-**TIB**-yoo-lar) **folds** or **false vocal cords**; and a lower pair called the **vocal folds** or **true vocal cords** (Figure 17-4). When the vestibular folds come together, they prevent air from exiting the lungs as when you hold your breath. Along with the epiglottis, the vestibular folds can prevent food or liquids from getting into the larynx. Under the mucous membrane of the true vocal cords lie bands of elastic ligament, stretched between pieces of rigid cartilage like the strings of a guitar. Skeletal muscles of the larynx are attached internally to the pieces of rigid cartilage and to the vocal folds. When the muscles contract, the glottis or opening is narrowed.

As air exits the lungs and is directed against the vocal cords, they vibrate and set up sound waves in the column of air in the pharynx, nose, and mouth. The greater the pressure of air, the louder the sound. Take a full breath of air in and force it out all at once. You will create a very loud sound.

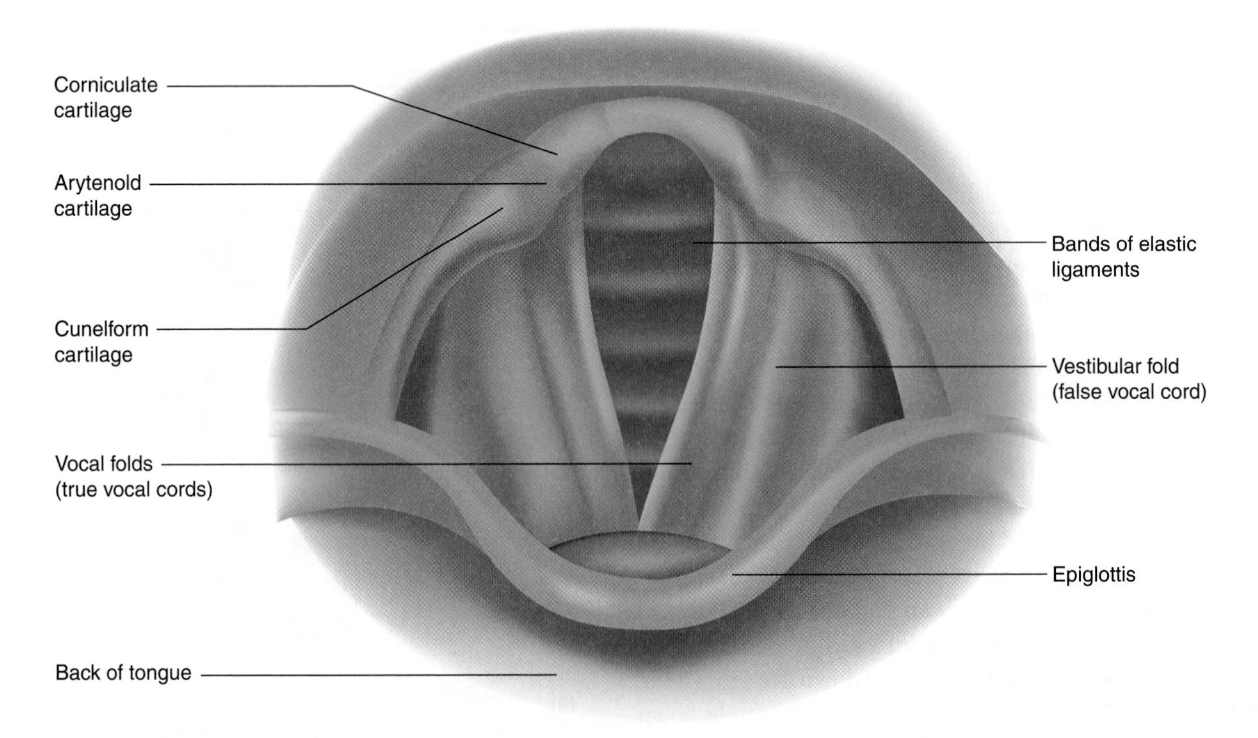

FIGURE 17-4. The position of the vocal cords in the larynx.

However, if you take in a full breath of air and let it out slowly with less pressure, the sound you create will be much softer sounding. Pitch is controlled by tension on the true vocal cords. When the cords are pulled taut by the muscles, they vibrate more rapidly and a higher pitch results. Decreasing the muscular tension produces lower pitch sounds. Try it. Because the true vocal cords are usually thicker and longer in men than in women, they vibrate more slowly so men have a lower range of pitch than women.

Sound originates from the vibrations of the true vocal cords. In humans, this sound is converted into speech. The pharynx, mouth, nasal cavities, and the paranasal sinuses all function as resonating chambers. The movement of the tongue and cheeks also contributes to creating the individual quality of human speech.

THE TRACHEA OR WINDPIPE

The **trachea** (**TRAY**-kee-ah) is also referred to as the windpipe (Figure 17-5). It is a tubular passageway for air approximately 4.5 inches in length and about 1 inch in diameter. It is found anterior to the esophagus and extends from the cricoid cartilage of the larynx to the fifth thoracic vertebra, where it divides into the right and left primary bronchi.

The tracheal epithelium is pseudostratified, ciliated columnar cells with goblet cells and basal cells. The goblet cells produce mucus, and the ciliated cells provide the same protection against dust particles as does the membrane in the larynx and pharynx. The cilia beat upward and move the mucus-dust package to the throat for elimination from the body. The smooth muscle and elastic connective tissue of the trachea are encircled by a series of 16 to 20 horizontal incomplete rings of hyaline cartilage that resemble a stack of Cs. The open part of the Cs face the esophagus and allow it to expand into the trachea during swallowing. When we swallow, we stop breathing to permit the large food bolus to expand into the trachea on its way to the stomach. The solid part of the Cs provides a strong rigid support for the tracheal walls so that they do not collapse inward and obstruct the air passageway. Varying pressure, as air moves in and out of the trachea, would collapse the tube if the cartilaginous rings were not present.

If a foreign object becomes caught in the trachea and cannot be expelled by the cough reflex, a tracheostomy

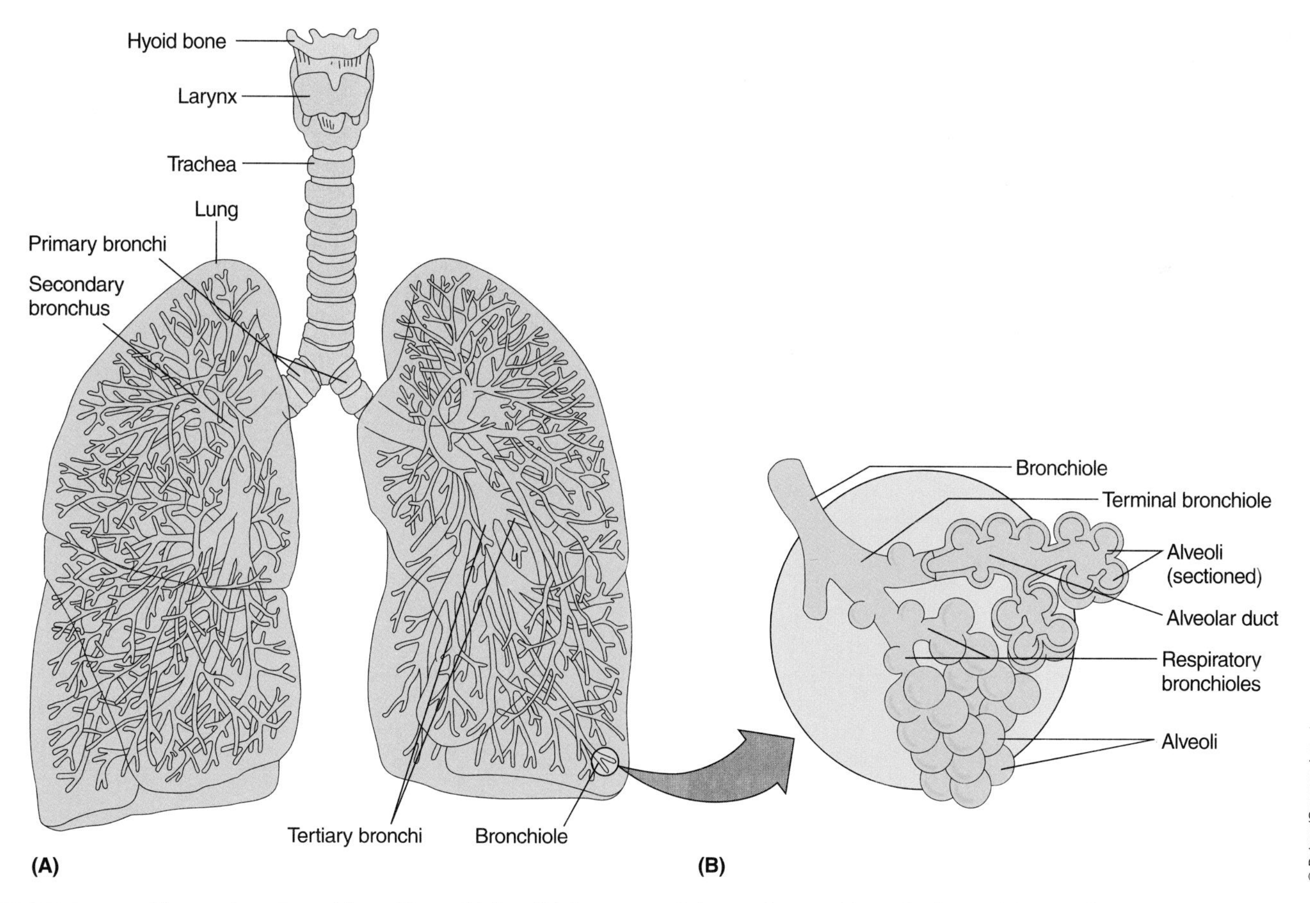

FIGURE 17-5. The trachea, bronchi, and bronchioles. (A) Anatomy of the trachea and bronchial tree. (B) End of the bronchial tree showing the terminal bronchioles, alveolar duct, and alveoli.

may be necessary to save the person's life. A tracheostomy is an incision into the trachea creating a new opening for air to enter. It is usually done between the second and third tracheal cartilages. This temporary opening can be closed, once the blocking object has been removed.

THE BRONCHI AND THE BRONCHIAL TREE

The trachea terminates in the chest by dividing into a **right primary bronchus** (**RITE PRYE**-mary **BRONG**-kus) that goes to the right lung and a **left primary bronchus** that goes to the left lung (see Figure 17-5 [A]). The right primary bronchus is more vertical, shorter, and wider than the left. Consequently, if a foreign object gets past the throat into the trachea, it will frequently get caught and lodge in the right primary bronchus. The bronchi, like the trachea, also contain the incomplete rings of hyaline cartilage and are lined with the same pseudostratified, ciliated columnar epithelium.

On entering the lungs, the primary bronchi divide to form smaller bronchi called the **secondary** or **lobar bronchi**, one for each lobe of the lung. The right lung has three lobes and the left lung has two lobes (Figure 17-6). The secondary bronchi continue to branch forming even smaller bronchi called **tertiary** or **segmental bronchi**. These branch into the segments of each lobe of the lung. Tertiary or segmental bronchi divide into smaller branches called **bronchioles** (**BRONG**-kee-olz). Bronchioles finally branch into even smaller tubes called **terminal** (end) **bronchioles** (see Figure 17-5 [B]). This continuous branching of the trachea resembles a tree trunk with branches. For this reason this branching is commonly referred to as a **bronchial tree** (see Figure 17-5 [A]).

As the branching becomes more and more extensive, the rings of cartilage get replaced with plates of cartilage. These finally disappear completely in the bronchioles. As the cartilage decreases, the amount of smooth muscle in the branches increases. In addition the pseudostratified, ciliated columnar epithelium changes to a simple, cuboidal epithelium.

StudyWARE™ Connection

Play an interactive game labeling structures of the bronchi and lobes of the lungs on your StudyWARE™ CD-ROM.

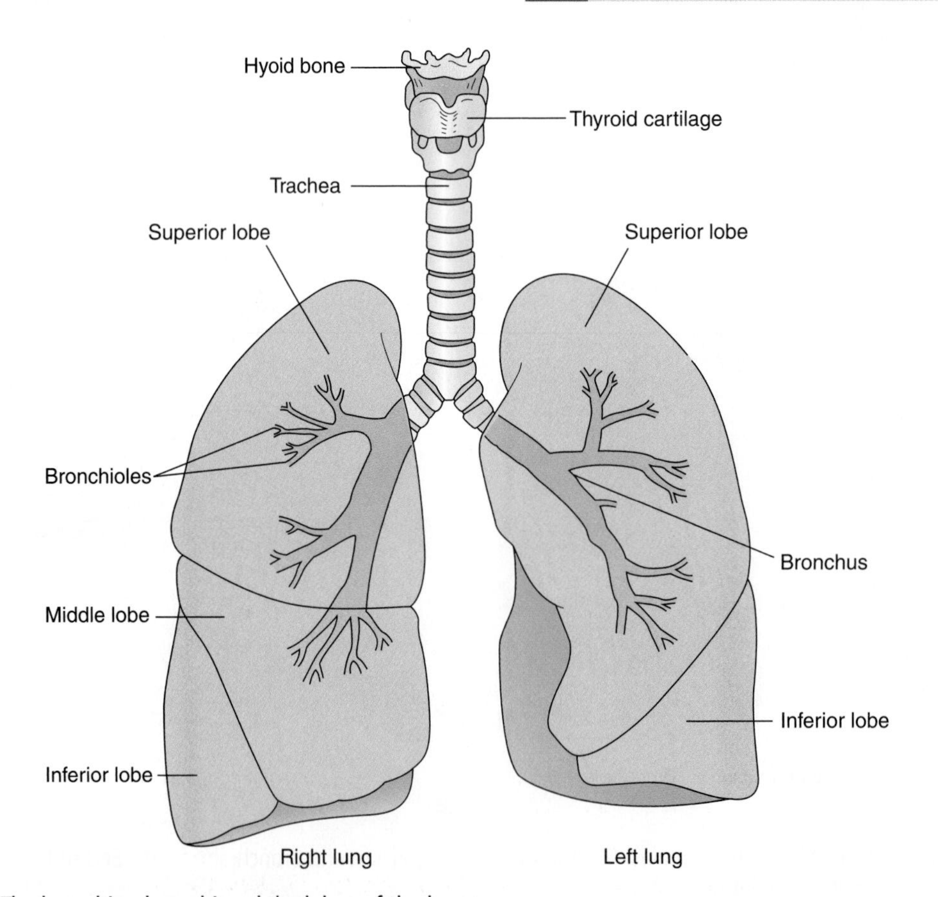

FIGURE 17-6. The branching bronchi and the lobes of the lungs.

HEALTH ALERT

SMOKING AND THE TRACHEA

Individuals who smoke or are constantly exposed to secondhand smoke create a constant irritation to the trachea. Over time, this irritation from smoke can cause the epithelium of the trachea to change from a pseudostratified, ciliated columnar epithelium to a stratified, squamous epithelium. Without cilia, the epithelium cannot clear the passageway of mucus and debris. This provides an environment ideal for the growth of microorganisms, leading to respiratory infections. This constant irritation and respiratory inflammation triggers the cough reflex, resulting in what we call a smoker's cough.

THE ANATOMY AND FUNCTION OF THE LUNGS

The lungs are paired, cone-shaped organs located in and filling the pleural divisions of the thoracic cavity. Two layers of serous membrane, known as the **pleural** (**PLOO**-rah) **membrane**, enclose and protect each lung (Figure 17-7). The outer layer attaches the lung to the wall of the thoracic cavity and is called the **parietal** (pah-**RYE**-eh-tal) **pleura**. The inner layer is called the **visceral** (**VISS**-er-al) **pleura** and covers the lungs. Between these two layers is a small space called the **pleural cavity**, which contains a lubricating fluid that is secreted by the membranes. This pleural fluid prevents friction between the two membranes and allows them to slide past each other during breathing, as the lungs and thorax change shape. It also assists in holding the pleural membranes together. **Pleurisy** (**PLOOR**-ih-see), or **pleuritis**, is an inflammation of this area and is very painful. The right lung with its three lobes is thicker and broader than the left lung with its two lobes. The right lung is also a bit shorter than the left because the diaphragm muscle is higher on the right side, as it must

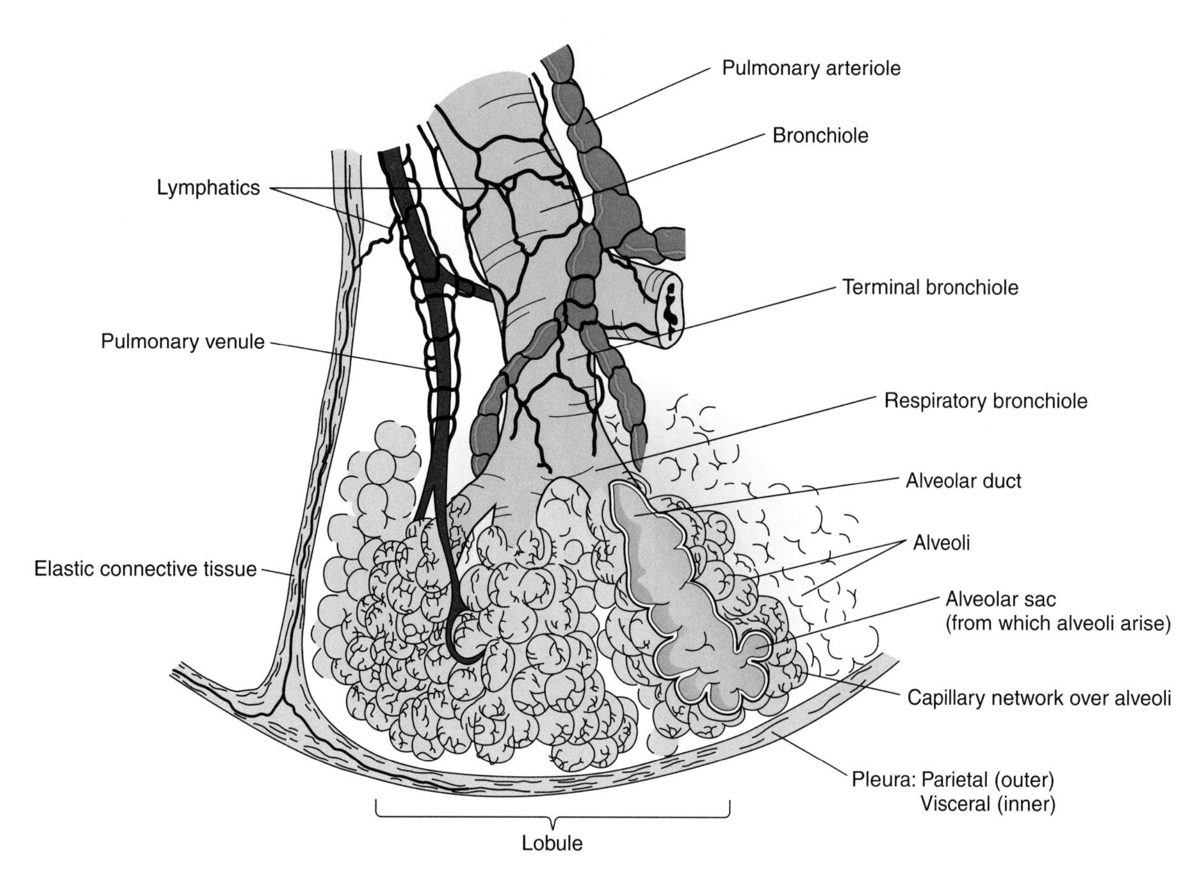

FIGURE 17-7. Anatomy of a lobule of a lung.

make room for the liver that is found below it. The left lung is thinner, longer, and narrower than the right.

The segment of lung tissue that each tertiary or segmental bronchi supplies is called a **bronchopulmonary** (**brong**-koh-**PULL**-mon-air-ree) **segment**. Each of these segments is divided into many small compartments called **lobules** (**LOB**-yoolz) (see Figure 17-7). Every lobule is wrapped in elastic connective tissue and contains a lymphatic vessel, an arteriole, a venule, and bronchioles from a terminal bronchiole.

Terminal bronchioles subdivide into microscopic branches called **respiratory bronchioles**. These respiratory bronchioles further subdivide into 2 to 11 **alveolar** (al-**VEE**-oh-lar) **ducts** or **atria**. Around the circumference of the alveolar ducts are numerous **alveoli** (al-**VEE**-oh-lye) and **alveolar sacs**. An alveolus (singular) is a cup-shaped or grapelike out-pouching lined with epithelium and supported by a thin, elastic basement membrane. Alveolar sacs are two or more alveoli that share a common opening. Refer to Figure 17-8 for the anatomy of an alveolus.

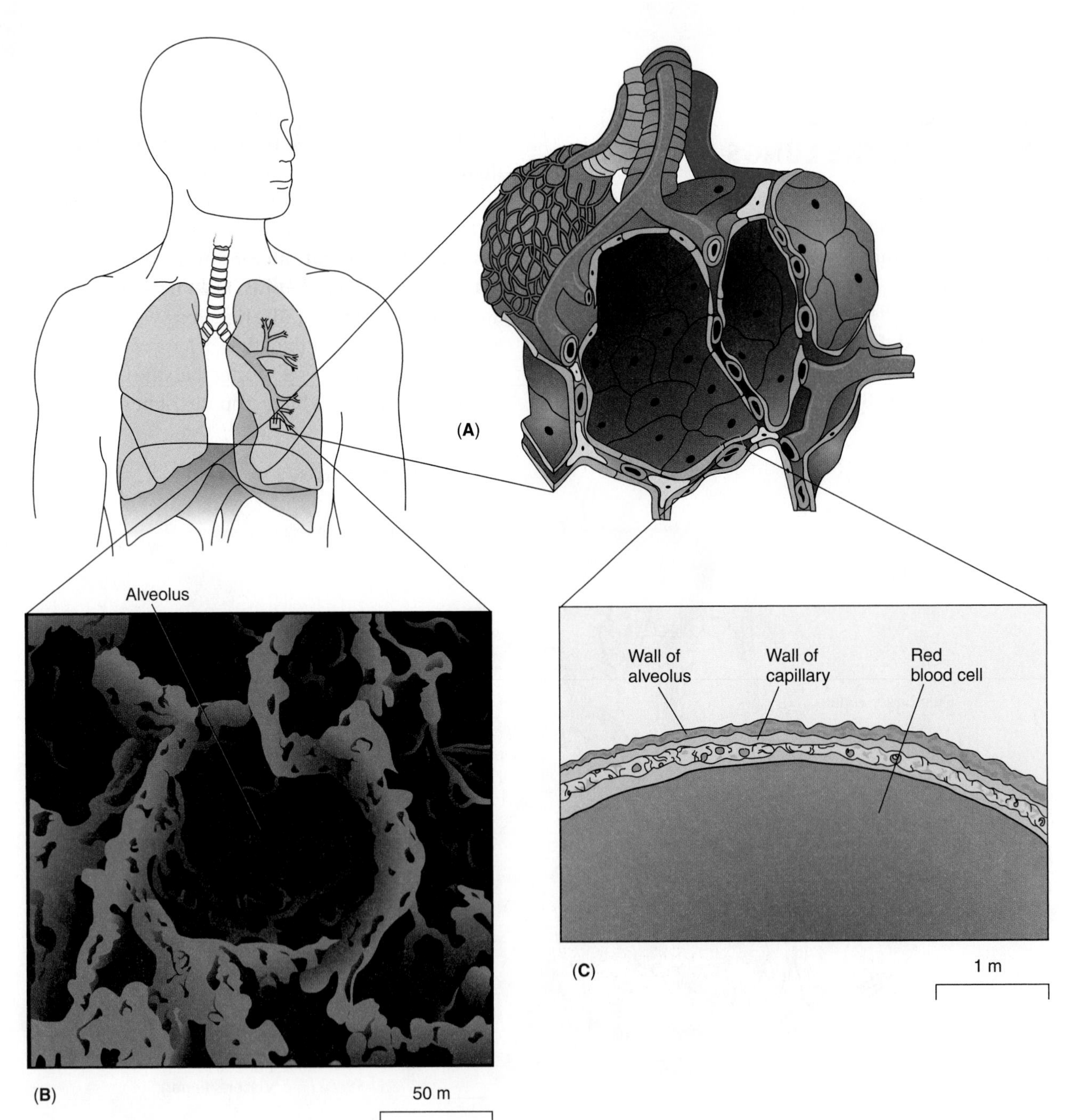

FIGURE 17-8. The anatomy of an alveolus.

The actual exchange of the respiratory gases between the lungs and blood occurs by diffusion across the alveoli and the walls of the capillary network that surrounds the alveoli. This membrane, through which the respiratory gases move, is referred to as the **alveolar-capillary** or **respiratory membrane**. The surface of the respiratory membrane inside each alveolus is coated with a fluid, consisting of a mixture of lipoproteins called **surfactant** (sir-**FAK**-tant). This material is secreted by certain alveolar cells (alveolar type II cells). Surfactant helps reduce surface tension (the force of attraction between water molecules) of the fluid. Therefore, surfactant helps prevent the alveoli from collapsing or sticking shut as air moves in and out during breathing. The gases need only diffuse through a single squamous epithelial cell of an alveolus and the single endothelial cell of the capillary to reach the red blood cell inside the capillary. It has been estimated that the lungs contain over 300 million alveoli. This is an immense surface area of 70 square meters (753 square feet) for the exchange of oxygen and carbon dioxide. This is about the square footage of a small house or cottage.

THE RESPIRATION PROCESS

The principal purpose of respiration is to supply the trillions of cells of the body with oxygen and to remove the carbon dioxide gas produced by cellular activities. There are three basic processes of respiration. The first process is **ventilation** or **breathing**, which is the movement of air between the atmosphere and the lungs. Ventilation has two phases: **inhalation** or **inspiration** to move air into the lungs and **exhalation** or **expiration** to move air out of the lungs (Figure 17-9).

The second and third processes of respiration involve the exchange of the gases within the body. **External respiration** is the exchange of gases between the lungs and the blood, the second process. The third process is called **internal respiration**, which is the exchange of gases between the blood and the body cells.

When the diaphragm and external intercostal muscles contract, we breathe in. This occurs because as the dome-shaped diaphragm contracts, it moves downward and flattens and the height of the thoracic cavity increases.

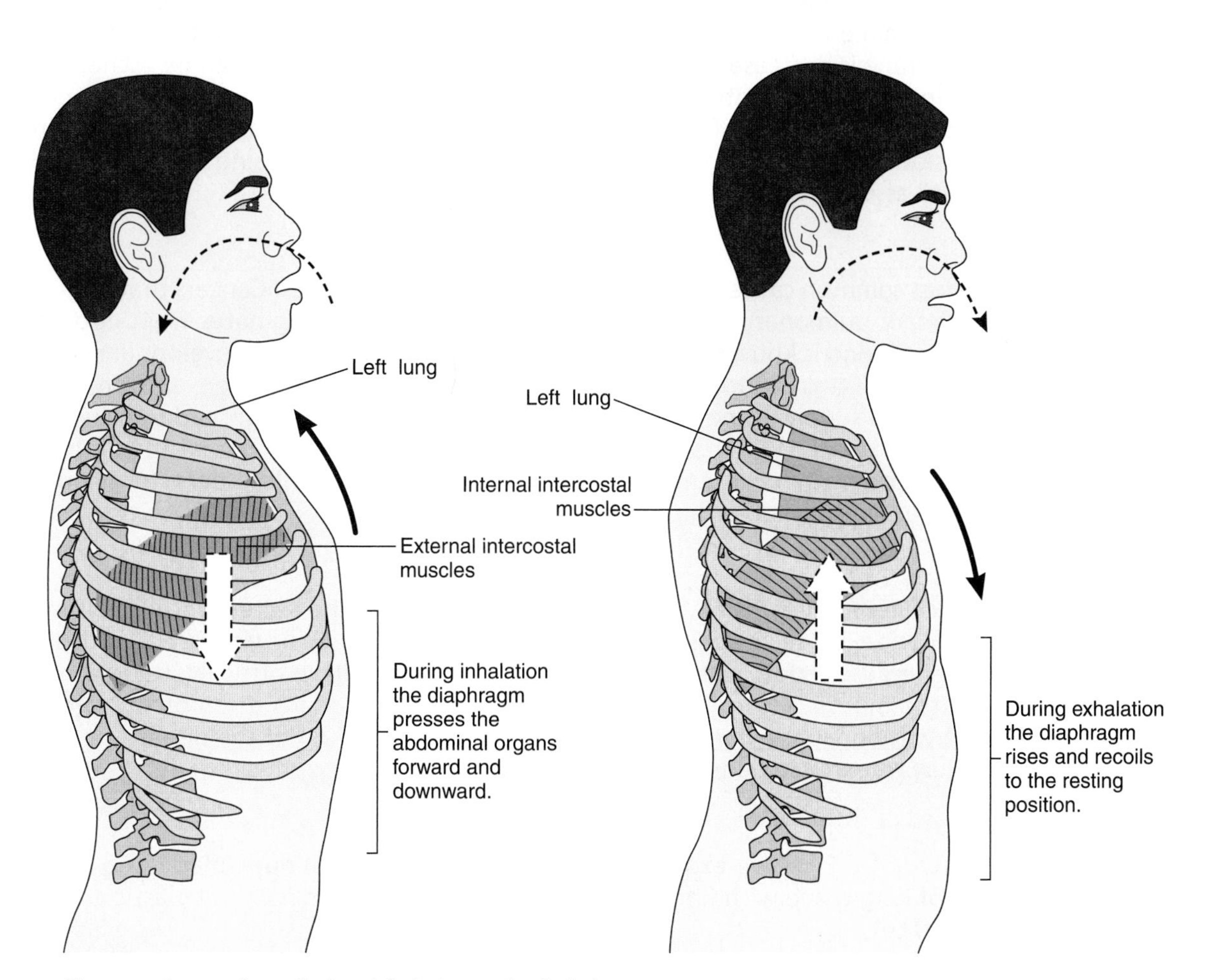

FIGURE 17-9. The two phases of ventilation, inhalation and exhalation.

Simultaneous contraction of the external intercostals lifts the rib cage and pushes the sternum forward. The lungs get stretched to the larger size of the thorax. Gases within the lungs spread out to fill the larger space, resulting in a decrease in gas pressure, causing a vacuum that sucks air into the lungs. This is inspiration.

As the diaphragm and external intercostals relax, the rib cage descends, the space decreases, and the gases inside the lungs come closer together. Pressure increases, causing the gases to flow out of the lungs. This is expiration and we breathe out. This is mainly a passive activity. When we force air out, the internal intercostal muscles contract to help further decrease the size of the rib cage.

COMMON DISEASE, DISORDER, OR CONDITION

DISORDERS OF THE RESPIRATORY SYSTEM

BRONCHITIS

Bronchitis (brong-**KIGH**-tis) is an inflammation of the bronchi. The inflamed tissue causes a swelling of the mucous membrane. This produces increased mucous production with a decrease in the ability of the cilia to move the mucus up to the throat. This results in a decrease in the diameter of the bronchial tubes, impairing breathing. Bronchitis can be caused by an infection with bacteria or viruses. It can also develop from increased exposure to irritants like air pollutants or cigarette smoke.

EMPHYSEMA

Emphysema (em-fih-**SEE**-mah) is characterized by the destruction of the walls of the alveoli. It is a progressively degenerative disease with no cure. It develops from prolonged exposure to respiratory irritants like tobacco smoke and air pollutants. As alveolar walls are destroyed, the surface area of the respiratory membrane is decreased. This decreases the amount of gas that can be exchanged. The alveolar walls also lose some elasticity, which decreases the ability of the lungs to recoil and expel air. Symptoms include enlargement of the thoracic cavity and shortness of breath. The progress of the disease can be slowed by removing the source of irritants, such as stopping smoking, or the use of bronchodilators to assist in breathing.

LUNG CANCER

Lung cancer is the most common cause of cancer deaths in the United States. Cancers that begin in the lungs are called primary pulmonary cancers. The most common form originates from uncontrolled growth of epithelial cells and is known as bronchogenic carcinoma. This form develops in response to prolonged exposure to irritants like tobacco, smoke, CO_2, dust, asbestos, radiation, and vinyl chloride. Because of the rich blood supply in the lungs, cancer in the lung can readily spread to other parts of the body. Lung cancer is treated with surgery, radiation, and chemotherapy but is difficult to curtail. Survival rates for patients with lung cancer remain low. Symptoms include a persistent cough, difficulty breathing, excessive sputum, or blood containing sputum.

CYSTIC FIBROSIS

Cystic fibrosis (**SIS**-tik fye-**BROH**-sis) is an inherited disease. It affects the secretory cells of the lungs. Due to abnormal chloride ion secretions, the mucus becomes very thick or viscous. It tends to accumulate in the lungs because it does not get moved by cilia. This results in difficulty in breathing due to obstructions by the mucus of the airways and severe coughing, which attempts to remove the mucus. The disease was once fatal in childhood, but today individuals with the disease can live into early adulthood. New research in genetic engineering may one day cure this disease.

PNEUMOCONIOSIS

Pneumoconiosis is caused by excessive exposure to asbestos, silica, or coal dust (black lung disease). It is the replacement of lung tissue with fibrous connective tissue. The lungs are not elastic and breathing becomes very difficult.

COMMON DISEASE, DISORDER, OR CONDITION

DISORDERS OF THE RESPIRATORY SYSTEM (continued)

RESPIRATORY DISTRESS SYNDROME

Respiratory distress syndrome, or **hyaline membrane disease**, is common in premature infants. This is caused by too little surfactant being produced and the lungs tend to collapse (surfactant is not produced in adequate amounts until after 7 months of development). Without special treatment, most premature babies will die shortly after birth as a result of inadequate ventilation and tiring of respiratory muscles.

PNEUMONIA

Pneumonia (new-**MOH**-nee-ah), or **pneumonitis**, refers to any infection in the lungs. Most pneumonias are caused by bacterial infections. However, some are viral, fungal, or protozoan in origin. Individuals with AIDS commonly are infected with a protozoan, *Pneumocystis carinii*, which causes pneumocystsis pneumonia. Symptoms of pneumonia include fever, chest pain, fluid in the lungs, and difficulty in breathing.

WHOOPING COUGH

Whooping cough, or **pertussis**, is caused by an infection with the bacterium *Bordetella pertussis*. It results in a loss of the cilia of the epithelium that lines the respiratory tract. Mucus accumulates and severe coughing attempts to expel the material. A childhood vaccine is now available to prevent this disease.

LARYNGITIS

Laryngitis (**lar**-in-**JIGH**-tis) is an inflammation of the mucosal membrane lining of the larynx. Symptoms include a swelling of the vocal cords with either a complete loss of voice or a very hoarse and rasping voice. This condition can develop as an accompanying symptom of a cold, by excessive use of the voice (as the condition occasionally develops in professional singers and orators), or by bacterial or viral infections of the respiratory tract. It can also be caused by excessive exposure to smoke or irritating fumes. The acute condition can be accompanied by a painful and scratchy feeling in the throat along with a cough. A common treatment is to inhale aromatic vaporized steam with menthol, pine oil, or tincture of benzoin. Children under five years of age who develop this condition can progress to serious respiratory distress. Chronic laryngitis is treated by using an astringent antiseptic throat spray, avoiding smoking and secondhand smoke, avoiding other irritants, resting the vocal cords periodically, and occasionally using a vaporizer.

PLEURISY

Pleurisy (**PLOOR**-ih-see) is an inflammation of the parietal pleura of the lungs. Symptoms include difficulty in breathing and a stabbing pain when the lungs are inflated. Simple pleurisy, also called dry or fibrinous pleurisy, does not produce excessive fluid in the pleural cavity between the parietal (outer layer) pleura and the visceral (inner layer) pleura, whereas pleural effusion pleurisy produces considerable amounts of fluid in the pleural cavity with extensive inflammation. Causes of pleurisy include pneumonia, tuberculosis, abscesses of the lung or chest wall, and bronchial tumors. Treatment consists of pain relievers and therapy for the disease that caused the condition.

ATELECTASIS

Atelectasis (**at**-ee-**LEK**-tah-sis) is a condition of a collapsed lung or reduction in the volume of a part of a lung. This results from an accumulation of either air or fluid in the pleural cavity. It can also result from a loss of pressure in the lung or reduced elastic recoil of a lung.

(continues)

COMMON DISEASE, DISORDER, OR CONDITION

DISORDERS OF THE RESPIRATORY SYSTEM (continued)

LEGIONNAIRE'S DISEASE

Legionnaire's disease, also known as legionellosis, is caused by exposure to the gram-negative bacterium *Legionella pneumophila,* which produces an acute pneumonia. Symptoms include a flu-like illness followed by chills, high fever, headache, and muscle aches within a week. The disease may progress to pleurisy, a dry cough, and occasionally diarrhea. Contaminated whirlpool spas, air-conditioning towers, and stagnant warm water can be sources of the bacteria. The disease got its name from the first episode of the disease that occurred in a hotel during an American Legion convention in 1976. Treatment includes the use of antibiotics like tetracycline and erythromycin.

SUDDEN INFANT DEATH SYNDROME (SIDS)

Sudden infant death syndrome or **SIDS** is also known as crib death. It is the unexpected death of a healthy infant that happens during sleep when the child stops breathing. It is the most frequent cause of death in infants between two weeks and one year old. It occurs in one out of every 300 to 350 births. The cause remains unknown and controversial. Abnormal development of the respiratory centers in the brain may be a factor. Other proposed causes are prolonged apnea, a defect in the respiratory mucosa, and immunoglobulin deficiencies. There are no preventative treatments. As a preventative measure, infants should be placed for sleep on their backs or on their sides. Children at risk are those between 10 and 14 weeks of age, premature babies, infants with respiratory infections, those whose mothers are less than 20 years of age, and mothers who smoke or use drugs. It occurs more often in female babies than in males.

COMMON COLD

The **common cold** is a contagious infection of the upper respiratory tract caused by a form of the rhinovirus. Its symptoms include nasal dripping, sneezing, excessive tearing, and malaise. It may be accompanied by a low-grade fever and may affect the lower respiratory tract resulting in occasional coughing. The common cold usually lasts about a week. It is treated by rest, increased fluid intake, and decongestants.

INFLUEZENA (FLU)

Influenza (**in**-flew-**EN**-zah) or **flu** is a highly contagious viral infection of the respiratory tract. It is caused by a myxovirus. Three main strains have been identified: Type A, B, and C. New strains of the virus continually evolve hence why yearly vaccinations with the current prevalent virus is recommended, especially for the very young, the elderly, and debilitated persons at risk. The strains are usually named for either the area (Asian flu) or the organism from which the strain evolved (bird flu). Symptoms include fever and chills, sore throat, cough, headache, muscle aches, and fatigue. The virus is transmitted in airborne droplets.

TUBERCULOSIS (TB)

Tuberculosis (too-**ber**-kew-**LOH**-sis) or **TB** is a chronic bacterial infection that usually affects the lungs. It is caused by the bacillus bacterium *Mycobacterium tuberculosis.* It can be transmitted by the ingestion or inhalation of infected droplets. The bacterium can also infect other organs of the body such as the spleen, liver, lymph nodes, bone marrow, as well as the meninges of the central nervous system. Early symptoms of pulmonary tuberculosis include chest pain, fever, loss of appetite with accompanying weight loss, and pleurisy. The tissue of the lungs react to the presence of the bacterium by producing cells that phagocytize the organism forming tubercles, hence, the name tuberculosis. If untreated the tubercles can enlarge and merge forming clumps of dead tissue filling the lung cavity. This results in the patient coughing up blood.

HEALTH ALERT

CHRONIC OBSTRUCTIVE PULMONARY DISEASE (COPD)

Chronic obstructive pulmonary disease is a progressive disorder characterized by long-term obstruction of airflow, which results in diminished inspiration and expiration capabilities of the lungs. The disease includes emphysema, asthma, and chronic bronchitis. In most cases, COPD is preventable since smoking and breathing in secondhand smoke are its most common cause. Other causes of COPD include exposure to dusts and gases at the workplace, chronic air pollution, and pulmonary infections. Pulmonary infections can be treated with antibiotics to slow down the progress of the disease. However, no cure exists if emphysema has set in. Patients with COPD experience difficulty with breathing during physical exertion. They cannot inhale or exhale deeply and usually have a chronic cough. The disease is also referred to as chronic obstructive lung disease.

HEALTH ALERT

ASTHMA

Asthma is characterized by recurring spasms of difficulty with breathing. An individual with asthma has symptoms such as wheezing while exhaling and inhaling, shortness of breath, and coughing. These are caused by a narrowing of the bronchial passageway and a buildup of a viscous mucoid secretion in the bronchial tubes. The exact causes of asthma are unknown although allergies are commonly associated with asthma attacks. Inhalation of pollen, dust mites, and animal dander as well as vigorous exercise, emotional stress, and inhalation of very cold air can trigger an asthma attack. Approximately 3–6% of the population experiences asthma. Symptoms may reverse spontaneously or with therapy. Children who experience asthma can become symptom-free after the onset of adolescence (this occurs in about 25–50% of cases). Treatment may include elimination of the causative agent, hyposensitization, use of an aerosol bronchodilator, or short-term use of corticosteroids.

The pressure of a gas will determine the rate at which it diffuses from one area to another. Review the discussion of diffusion in Chapter 2. Molecules move from an area of high concentration to an area of low concentration. In a mixture of gases, like the air, each gas contributes a portion of the total pressure of the mixture. The **partial pressure** of a gas is the amount of pressure that gas contributes to the total pressure and is directly proportional to the concentration of that gas in the mixture. Air is 78% nitrogen, 21% oxygen, and 0.04% carbon dioxide, and the rest a mixture of other gases. Because air is 21% oxygen, it makes up 21% of atmospheric pressure (21% of 760 mm Hg). We can abbreviate the partial pressure of oxygen as $PO_2 = 160$ mm Hg and carbon dioxide as $PCO_2 = 0.3$ mm Hg in air.

When a mixture of gases dissolves in blood, the resulting concentration of each gas is proportional to its partial pressure. Each gas diffuses between the blood and its surrounding tissues from areas of high partial pressure to areas of low partial pressure, until the partial pressure in the areas reaches equilibrium. The PCO_2 in capillary blood is 45 mm Hg. The PCO_2 in alveolar air is 40 mm Hg. Because of these differences in partial pressures, carbon dioxide diffuses from blood, where its partial pressure is higher at 45 mm Hg, across the respiratory membrane into alveolar air, where its partial pressure is lower at 40 mm Hg. Similarly, the PO_2 of capillary blood is 40 mm Hg, while that of alveolar air is 104 mm Hg. Therefore, oxygen diffuses from alveolar air, where the partial pressure is higher at 104 mm Hg, into the blood, where the partial pressure is lower at 40 mm Hg. The blood then leaves the lungs with a PO_2 of 104 mm Hg (Figure 17-10).

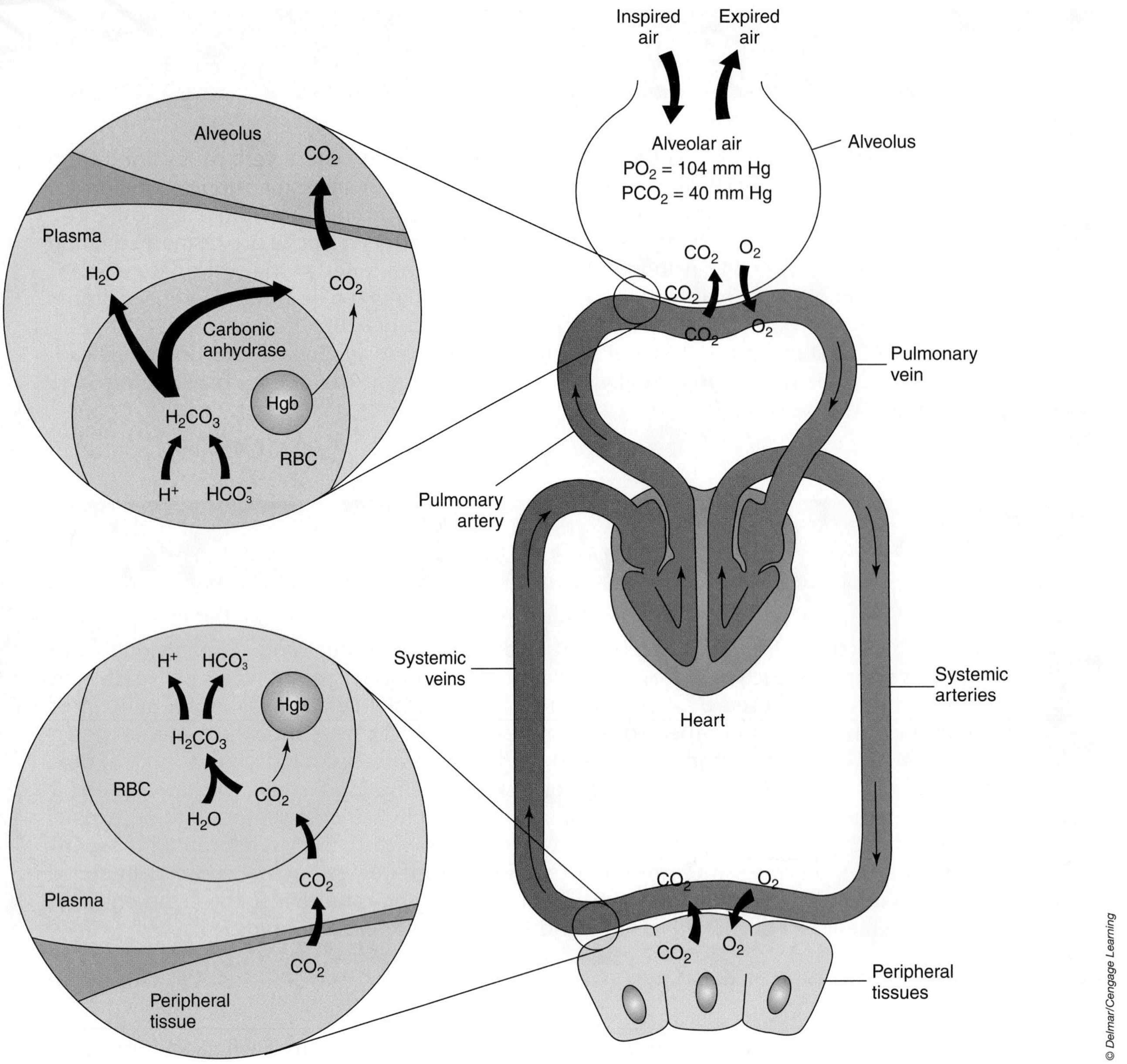

FIGURE 17-10. The respiratory pathway of oxygen and carbon dioxide.

The blood then transports oxygen to tissue cells and picks up carbon dioxide waste from the tissue cells. The tissue cells are high in carbon dioxide from cellular metabolic activities and low in oxygen, because it is used up in those activities. The pressure of CO_2 is higher in tissue cells than in blood cells and diffuses from tissues to blood cells. The blood cell is higher in O_2 levels than the tissue cells; thus, the pressure of O_2 in blood is higher and diffuses into tissue cells, where it is lower. Recall that it is the iron atoms in heme that carry the oxygen and the protein globin that carries the carbon dioxide. The hemoglobin molecule in the red blood cell transports these gases.

StudyWARE™ Connection

Watch an animation about respiration on your **StudyWARE™ CD-ROM**.

LUNG CAPACITY

Lung capacity is the lung volume that is the sum of two or more of the four primary, nonoverlapping lung volumes. There are four lung capacities. The first is

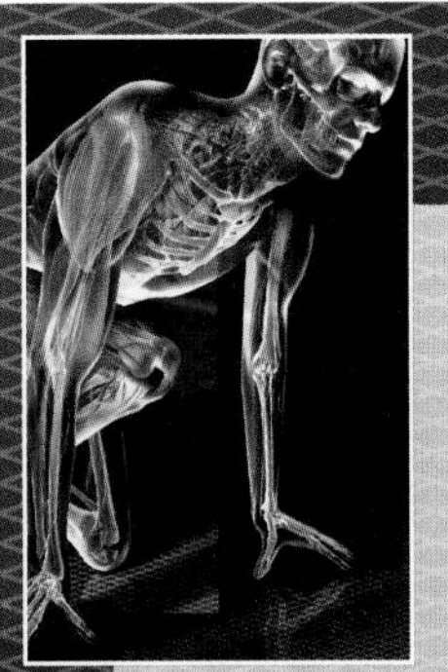

AS THE BODY AGES

As we advance in age, the respiratory muscles weaken and the chest wall becomes more rigid due to a stiffening of the costal cartilages and ribs. The tissues of the respiratory tract become less elastic and more rigid. This includes the alveolar sacs, resulting in a decrease in the lung capacity. This decrease can amount to almost 35% when individuals reach their 70s.

The levels of oxygen gas being carried by the blood also decrease as we age, and gas exchange across the respiratory membranes of the alveoli decreases. In spite of these changes, older adults are capable of light exercise regimens and are encouraged to do so in order to maintain their muscle tone, strength, and endurance.

The ciliary action of the epithelium lining the respiratory tract decreases with age, resulting in a buildup of mucus inside the respiratory passageways. This is why older adults become much more susceptible to bronchitis, pneumonia, emphysema, and other respiratory infections.

functional residual capacity (FRC) and is the volume of gas in the lungs at the end of a normal tidal volume exhalation. The FRC is equal to the residual volume plus the expiratory reserve volume. The second is inspiratory capacity (IC) and is the maximum volume of gas that can be inhaled from the end of a resting exhalation. It is measured with a spirometer and is equal to the sum of the tidal volume and the inspiratory reserve volume. The third capacity is total lung capacity (TLC). It is the volume of gas in the lungs at the end of a maximum inspiration. It equals the vital capacity and the residual capacity. The fourth capacity is the vital capacity (VC). It is the maximum volume of air that can be expelled at the normal rate of exhalation after a maximum inspiration. This represents the greatest possible breathing or lung capacity. It equals the inspiratory reserve volume plus the tidal volume plus the expiratory reserve volume.

*Study*WARE™ Connection

Watch animations on asthma on your **StudyWARE™ CD-ROM.**

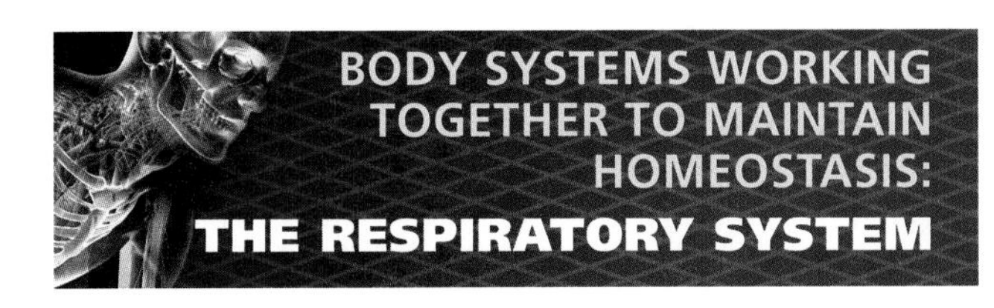

Integumentary System
- The skin is the first line of defense because it forms a barrier to protect respiratory organs and tissues from microorganisms.
- Stimulation of receptors in the skin can alter respiratory rates.

Skeletal System
- Bones provide attachments for the muscles involved in breathing, for example, the intercostals.
- The ribs and sternum enclose and protect the lungs and bronchi in the thoracic cavity.

Muscular System
- The diaphragm and intercostal muscles produce changes in the volume of the thorax and lungs, resulting in the ability to inhale and exhale.
- The respiratory system eliminates the carbon dioxide produced by contracting muscle cells.

Nervous System
- The brainstem has control centers that regulate the respiratory rate.
- The respiratory system supplies nerve cells with needed oxygen for maximum efficiency.

There are a number of career options available to individuals who are interested in the respiratory system.

- **Otorhinolaryngologists** are physicians who specialize in the diagnosis and treatment of diseases and disorders of the ears, nose, and throat.
- **Respiratory therapists** are allied health specialists with knowledge of the theory of clinical problems of respiratory care.
- **Pulmonologists** are physicians who specialize in the diagnosis and treatment of diseases of the lungs.
- **Perfusionists** are allied health professionals who assist in performing procedures that are concerned with extracorporeal circulation necessary during the treatment of hypothermia and during open heart surgery.
- **Thoracic surgeons** are physician specialists who are concerned with diseases and disorders of the thoracic area, utilizing operative and manipulative treatments.

Endocrine System

- Hormones stimulate red blood cell production, and the blood cells carry the oxygen and carbon dioxide for the respiratory system.
- Epinephrine dilates bronchioles, increasing breathing abilities.
- Testosterone causes the enlargement of the thyroid cartilage, producing the prominent Adam's apple in men.

Cardiovascular System

- The heart pumps the oxygen carrying red blood cells from the lungs through its system of arteries and veins to tissue cells where oxygen is exchanged with carbon dioxide.

Lymphatic System

- The immune system protects respiratory organs from infection and cancers.
- The tonsils in the pharynx produce immune cells.

Digestive System

- The pharynx is used by both the digestive and the respiratory systems.
- The digestive system provides nutrients to respiratory organs and tissues.

Urinary System

- The kidneys and the respiratory system help maintain blood pH.
- The kidneys reabsorb the water lost through breathing by filtering water from the blood.

Reproductive System

- Breathing rates increase during sexual activities.
- Fetal respiration occurs through the placenta with the mother.

SUMMARY OUTLINE

INTRODUCTION

1. The organs of the respiratory system are the nose, pharynx, larynx, trachea, bronchi, and lungs.
2. Respiration is the overall exchange of the gases oxygen and carbon dioxide between the atmosphere, the blood and the cells.
3. The cardiovascular and respiratory systems equally share the responsibility of supplying oxygen to and eliminating carbon dioxide gas from cells.

THE ANATOMY AND FUNCTIONS OF THE NOSE

1. The openings into the external nose are called the nostrils or external nares.
2. The internal nose connects with the throat or pharynx via the two internal nares.

3. The nose is separated into a right and left nasal cavity by the nasal septum.
4. Coarse hairs line the vestibules of the nostrils to filter out large dust particles in the air.
5. The internal nose has three shelves formed by the turbinate bones: the superior, middle, and inferior meatus lined with mucous membranes.
6. The olfactory receptors are found in the superior meatus.
7. The internal nose has three functions: air is warmed, moistened, and filtered; olfactory stimuli are detected; and large hollow resonating chambers are provided for speech sounds.

THE STRUCTURE AND FUNCTIONS OF THE PHARYNX

1. The pharynx or throat has two functions. It is a passageway for both food and air, and it forms a resonating chamber for speech sounds.
2. The pharynx is divided into the nasopharynx, the oropharynx, and the laryngopharynx.
3. The nasopharynx has four openings in its walls: the two internal nares and the openings to the two eustachian tubes. It also houses the pharyngeal tonsils.
4. The oropharynx has one opening, the fauces or connection to the mouth. It houses the palatine and lingual tonsils.
5. The laryngopharynx connects with the esophagus posteriorly and the larynx anteriorly.

THE LARYNX OR VOICE BOX

1. The walls of the larynx are supported by nine pieces of cartilage; three are single and three are paired.
2. The thyroid cartilage is the largest single piece. It is also called the Adam's apple and is usually larger in men.
3. The epiglottis is a large, single leaf-shaped piece of cartilage. It pulls down over the glottis when we swallow to keep food or liquids from getting into the trachea.
4. The cricoid cartilage is a single ring of cartilage that connects with the first tracheal ring.
5. The paired arytenoid cartilages are ladle-shaped and are attached to the vocal cords and laryngeal muscles.
6. The paired corniculate cartilages are cone-shaped, and the paired cuneiforms are rod-shaped.
7. The mucous membrane of the larynx is arranged in two pairs of folds. The upper pair is the vestibular folds or false vocal cords, and the lower pair is the vocal folds or true vocal cords.
8. The glottis is the opening over the true vocal cords.
9. Air coming from the lungs causes the vocal cords to vibrate and produce sound. The greater the volume of air, the louder the sound.
10. Pitch is controlled by tension on the true vocal cords. The stronger the tension, the higher the pitch. True vocal cords are thicker in men; they vibrate more slowly and produce a lower pitch than that in women.

THE TRACHEA OR WINDPIPE

1. The trachea is a 4.5-inch tubular passageway for air and is located anterior to the esophagus.
2. Its epithelium is pseudostratified, ciliated columnar cells with goblet cells that produce mucus, and basal cells.
3. Its smooth muscle and connective tissue are encircled by incomplete rings of hyaline cartilage shaped like a stack of Cs.
4. The open part of the Cs faces the esophagus and allows it to expand into the trachea during swallowing.
5. The closed part of the Cs forms a solid support to prevent collapse of the tracheal wall.
6. If a foreign object gets caught in the trachea, a cough reflex expels it.

THE BRONCHI AND THE BRONCHIAL TREE

1. The right primary bronchus branches from the trachea and goes to the right lung; the left primary bronchus branches and goes to the left lung.
2. The primary bronchi branch into secondary or lobar bronchi that go into the lobes of the lungs. The right lung has three lobes and the left lung has two.
3. The secondary bronchi branch into tertiary or segmental bronchi, which branch into the segments of the lobes of the lungs.
4. Tertiary or segmental bronchi branch into smaller branches called bronchioles.
5. Bronchioles finally branch into the smallest branches called terminal bronchioles.
6. Because this continuous branching of the bronchi resembles a tree and its branches, it is referred to as a bronchial tree.

THE ANATOMY AND FUNCTION OF THE LUNGS

1. The pleural membrane encloses and protects each lung. It is composed of two layers of serous membranes: the outer is the parietal pleura and the inner is the visceral pleura.
2. Between these two layers is the pleural cavity, which contains a lubricating fluid to prevent friction as the lungs expand and contract during breathing.
3. The segment of lung tissue that each tertiary or segmental bronchi supplies is called a bronchopulmonary segment.
4. Each of these segments is divided into a number of lobules wrapped in elastic connective tissue with a lymphatic, an arteriole, a venule, and bronchioles from a terminal bronchiole.
5. Terminal bronchioles subdivide into microscopic respiratory bronchioles, which further divide into 2 to 11 alveolar ducts or atria.
6. Around the circumference of the alveolar ducts are alveoli and alveolar sacs.
7. Alveoli are grapelike outpouchings of epithelium and elastic basement membrane surrounded externally by a capillary network.
8. An alveolar sac is two or more alveoli that share a common opening.
9. The microscopic membrane through which the respiratory gases move is this alveolar-capillary (respiratory) membrane.

THE RESPIRATION PROCESS

1. There are three basic processes of respiration.
2. The first process is called ventilation or breathing, which is the movement of air between the atmosphere and the lungs.
3. The two phases of ventilation are inhalation or inspiration, which moves air into the lungs, and exhalation or expiration, which moves air out of the lungs.
4. The second process of respiration is external respiration, which is the exchange of gases between the lungs and the blood.
5. The third process is internal respiration, which is the exchange of gases between the blood and body cells.
6. Breathing in occurs when the diaphragm and external intercostal muscle contract, causing decreased pressure and a vacuum in the lungs.
7. When the diaphragm and external intercostal muscles relax, we breathe out due to increased pressure in the lungs forcing the air out. This is mainly a passive activity.
8. The partial pressure of a gas is the amount of pressure that gas contributes to the total pressure and is directly proportional to the concentration of that gas in the mixture.
9. The partial pressure of oxygen is $PO_2 = 160$ mm Hg and of carbon dioxide $PCO_2 = 0.3$ mm Hg in air.
10. Each gas diffuses between blood and its surrounding tissues from an area of high partial pressure to an area of low partial pressure until equilibrium is reached.
11. The PCO_2 in capillary blood is 45 mm Hg, but it is 40 mm Hg in the alveolar blood of the lungs. Therefore, carbon dioxide diffuses from blood into the lungs.
12. The PO_2 in capillary blood is 40 mm Hg, but it is 104 mm Hg in the alveolar sacs of the lungs. Therefore, oxygen diffuses from the lungs into the blood cells.
13. As the blood cells transport their high levels of oxygen to tissue cells, the tissue cells are low in oxygen but high in carbon dioxide; therefore, carbon dioxide diffuses into the blood cell and oxygen diffuses from the blood cell into the tissue cell.

LUNG CAPACITY

1. Lung capacity is the lung volume that is the sum of two or more of the four primary, nonoverlapping lung volumes.
2. There are four lung capacities: functional residual capacity (FRC); inspiratory capacity (IC); total lung capacity (TLC); and vital capacity (VC).

REVIEW QUESTIONS

1. Name the three functions performed by the internal structures of the nose.
2. Name the three parts of the pharynx and their functions.
3. Name and describe the three processes in respiration.
4. Name the cartilages that support the trachea.

***5.** Explain how breathing depends on muscular contraction, relaxation, and changes in lung pressure.

***6.** Explain how the anatomy of the tracheal walls accommodates both breathing and swallowing.

***7.** What does the partial pressure of a gas mean?

***Critical Thinking Questions**

MATCHING

Place the most appropriate number in the blank provided.

_____ Eustachian tubes	**1.** Surface for respiration
_____ Fauces	**2.** Thyroid cartilage
_____ Vestibular folds	**3.** Segmental bronchi
_____ Alveoli	**4.** Blocks glottis during swallowing
_____ Trachea	**5.** Auditory/ nasopharynx
_____ Epiglottis	**6.** Breathing
_____ Adam's apple	**7.** Lobar bronchi
_____ Tertiary bronchi	**8.** Windpipe
_____ Secondary bronchi	**9.** Opening from mouth
_____ Ventilation	**10.** True vocal cords
	11. False vocal cords
	12. Cricoid cartilage

Search and Explore

- Choose one of the Health Alerts in this chapter. Perform an Internet search to learn more about the condition. Imagine that you have just been diagnosed with this condition. Write about how your life would be affected and how you feel about that. As a health care provider, do you think it is important to consider the whole individual rather than treating just the disease? Explain your answer in your notebook.
- Visit the American Cancer Society website at http://www.cancer.org to learn more about lung cancer. Based on your research and in your own words, write a description of lung cancer based on your research and list some possible causes. Discuss two ways lung cancer can be prevented.
- Choose one of the structures of the respiratory system. In your own words, briefly describe the structure and explain its importance to this system.

*Study*WARE™ Connection

- Choose one of the Health Alerts or Disorders in this chapter. Perform an Internet search to learn more about the condition. Imagine that you have just been diagnosed with this condition. Write about how your life would be affected and how you feel about that. As a health care provider, do you think it is important to consider the whole individual rather than treating just the disease? Explain your answer in your notebook.
- Visit the American Cancer Society website at http://www.cancer.org to learn more about lung cancer. Based on your research and in your own words, write a description of lung cancer based on your research and list some possible causes. Discuss two ways lung cancer can be prevented.
- Choose one of the structures of the respiratory system. In your own words, briefly describe the structure and explain its importance to this system.

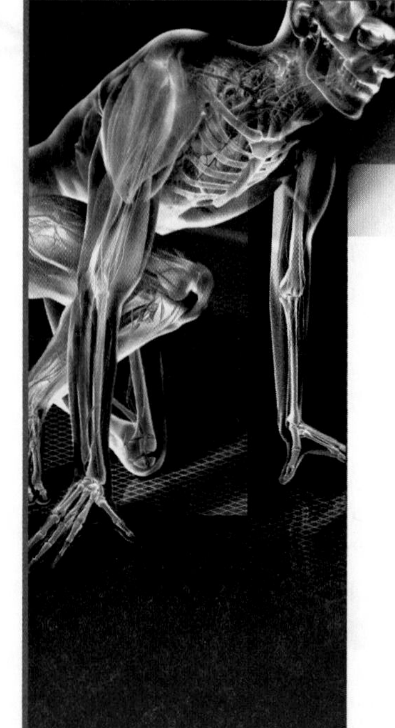

CASE STUDY

Cadence, a 6-year-old girl, is admitted into the emergency room with shortness of breath, wheezing when exhaling and inhaling, and coughing. Cadence's mother states that she has had recurring attacks of difficulty with breathing since she was around 8 months old. Just prior to the attack, Cadence had been helping the little boy who lives next door build a snowman.

Questions

1. What type of attack do you think Cadence might be experiencing?
2. What are some of the factors that trigger these attacks?
3. What caused the respiratory symptoms associated with Cadence's attack?
4. How is this respiratory problem generally treated?
5. What is the likely long-term outcome for children with this condition?

*Study*WARE™ Connection

Take a practice quiz or play interactive games that reinforce the content in this chapter on your **StudyWARE™ CD-ROM**.

Study Guide Practice

Go to your **Study Guide** for more practice questions, labeling and coloring exercises, and crossword puzzles to help you learn the content in this chapter.

LABORATORY EXERCISE: THE RESPIRATORY SYSTEM

Materials needed: A dissecting kit, a fetal pig, and a dissecting pan.

1. Take your fetal pig out of its storage area. You have already made the cut in the neck region when you did your lab exercise on the endocrine system. We will begin in this area. Find the larynx or voice box that you exposed in a previous lab and now note its connection to the trachea. Refer to your fetal pig dissection Figure 16-16 in Chapter 16. Take your scissors and make a cut from the top of the sternum down to the diaphragm muscle; cut number 4 in Figure 17-11. This will open the thoracic cavity and expose the heart and the lungs. Carefully remove the heart with your scissors and forceps; be careful not to cut into the trachea. You want to expose the trachea beneath the heart.

THE RESPIRATORY SYSTEM (Continued)

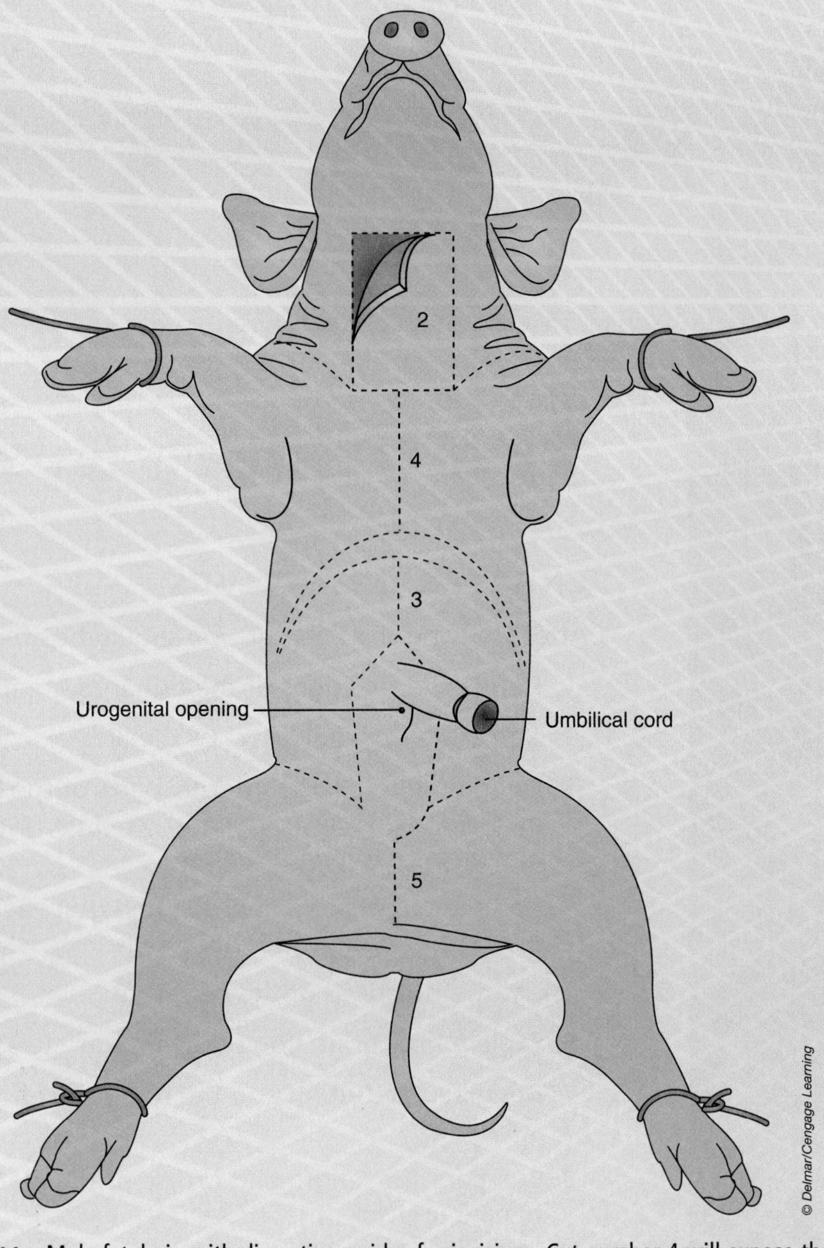

FIGURE 17-11. Male fetal pig with dissecting guides for incisions. Cut number 4 will expose the thoracic cavity.

2. Once the heart and its corresponding blood vessels are removed, follow the trachea down until it branches into the right and left primary bronchi. Clear tissue away from the trachea, being careful not to cut into any lung tissue until you see the bronchial branches that go into the lungs. Review all of the parts of the respiratory system that you can see in the thoracic cavity. We observed the nasopharynx, oropharynx, and epiglottis when we did the dissection on the digestive system. Observe these structures one more time.

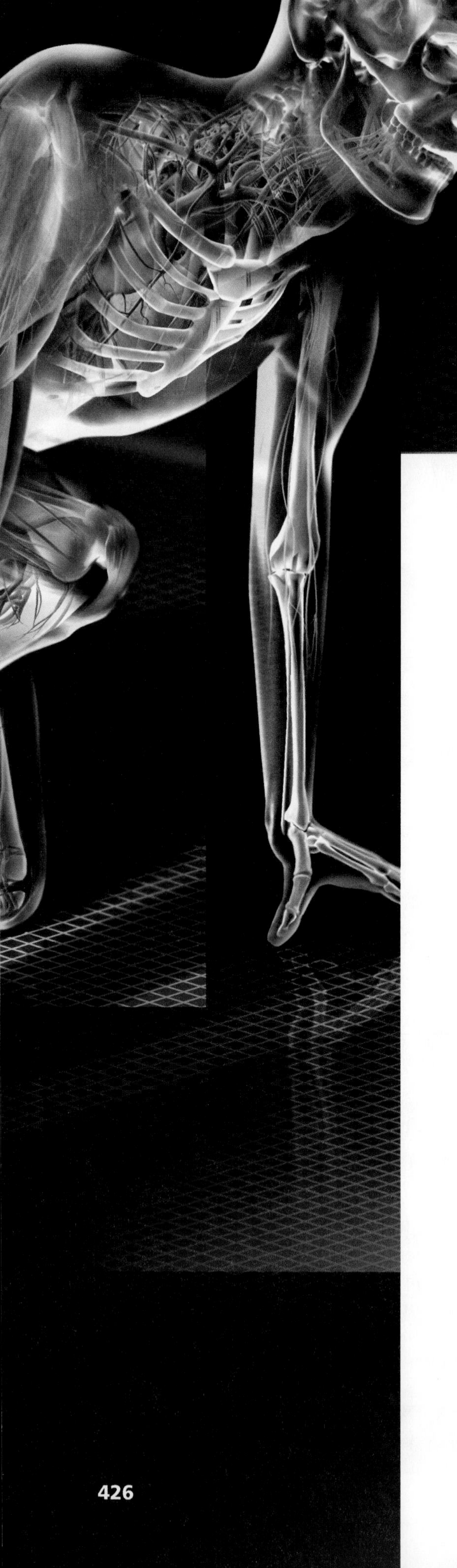

The Urinary System

CHAPTER OBJECTIVES

After studying this chapter, you should be able to:

1. Define the functions of the urinary system.
2. Name the external layers of the kidney.
3. Define the following internal parts of the kidneys: *cortex*, *medulla*, *medullary pyramids*, *renal papillae*, *renal columns*, and *major* and *minor calyces*.
4. Name the parts of a nephron and describe the flow of urine through this renal tubule.
5. List the functions of the nephrons.
6. Explain how urine flows down the ureters.
7. Describe micturition and the role of stretch receptors in the bladder.
8. Compare the length and course of the male urethra to the female urethra.
9. Name the normal constituents of urine.

18

KEY TERMS

Adipose capsule 431
Afferent arteriole 434
Arcuate arteries. 434
Arcuate veins. 434
Ascending limb of Henle 432
Bowman's glomerular capsule 432
Collecting duct 432
Cortex 431
Descending limb of Henle 432
Detrusor muscle 438
Distal convoluted tubule. 432
Efferent arteriole 434
Endothelial-capsular membrane 432
Erythropoietin. 431
External urinary sphincter 438
Glomerulus 432
Hilum 431
Interlobar arteries. 434
Interlobar veins. 434
Interlobular arteries 434
Interlobular veins 434
Internal urinary sphincter 438
Kidneys 431
Left renal artery 434
Left renal vein. 434
Loop of Henle 432
Major calyces. 431
Medulla 431
Micturition. 438
Micturition reflex 438
Minor calyx 431
Nephrons 431
Papillary ducts. 432
Parenchyma. 431
Peritubular capillaries. . . 434
Podocytes 432
Proximal convoluted tubule. 432
Renal capsule 431
Renal columns. 431
Renal corpuscle 432
Renal fascia 431
Renal papillae 431
Renal pelvis. 431
Renal plexus 434
Renal pyramids 431
Renal sinus 431
Renal tubule 432
Renin 431
Right renal artery 434
Right renal vein. 434
Trigone. 438
Ureters 437
Urethra. 438
Urinary bladder. 438
Urinary system 428
Urine 435

INTRODUCTION

As the body metabolizes the various foods and nutrients taken in through the digestive tract, body cells produce metabolic wastes in the form of carbon dioxide gas, heat, and water. The breakdown of proteins into amino acids and the subsequent metabolism of the amino acids produces nitrogenous wastes like ammonia. The harmful ammonia is converted by liver enzymes into less harmful urea. In addition, the body accumulates excess ions of sodium, chloride, potassium, hydrogen, sulfate, and phosphate.

It is the role of the urinary system to maintain a balance of these products and to remove excesses from the blood. This system helps to keep the body in homeostasis by both removing and restoring selected amounts of solutes and water from the blood. See Concept Maps 18-1 and 18-2: The Urinary System. The **urinary** (**YOO**-rih-nair-ee) **system** consists of two kidneys, two ureters, the urinary bladder, and the urethra (Figure 18-1). The kidneys regulate the composition and volume of the blood and remove wastes from the blood in the form of urine. The urine consists of the metabolic waste urea, excess water, excess ions, and toxic wastes that may have been consumed with food. Urine is excreted from each kidney through the kidney ureter. It is then stored in the urinary bladder, until it is expelled from the body through the urethra.

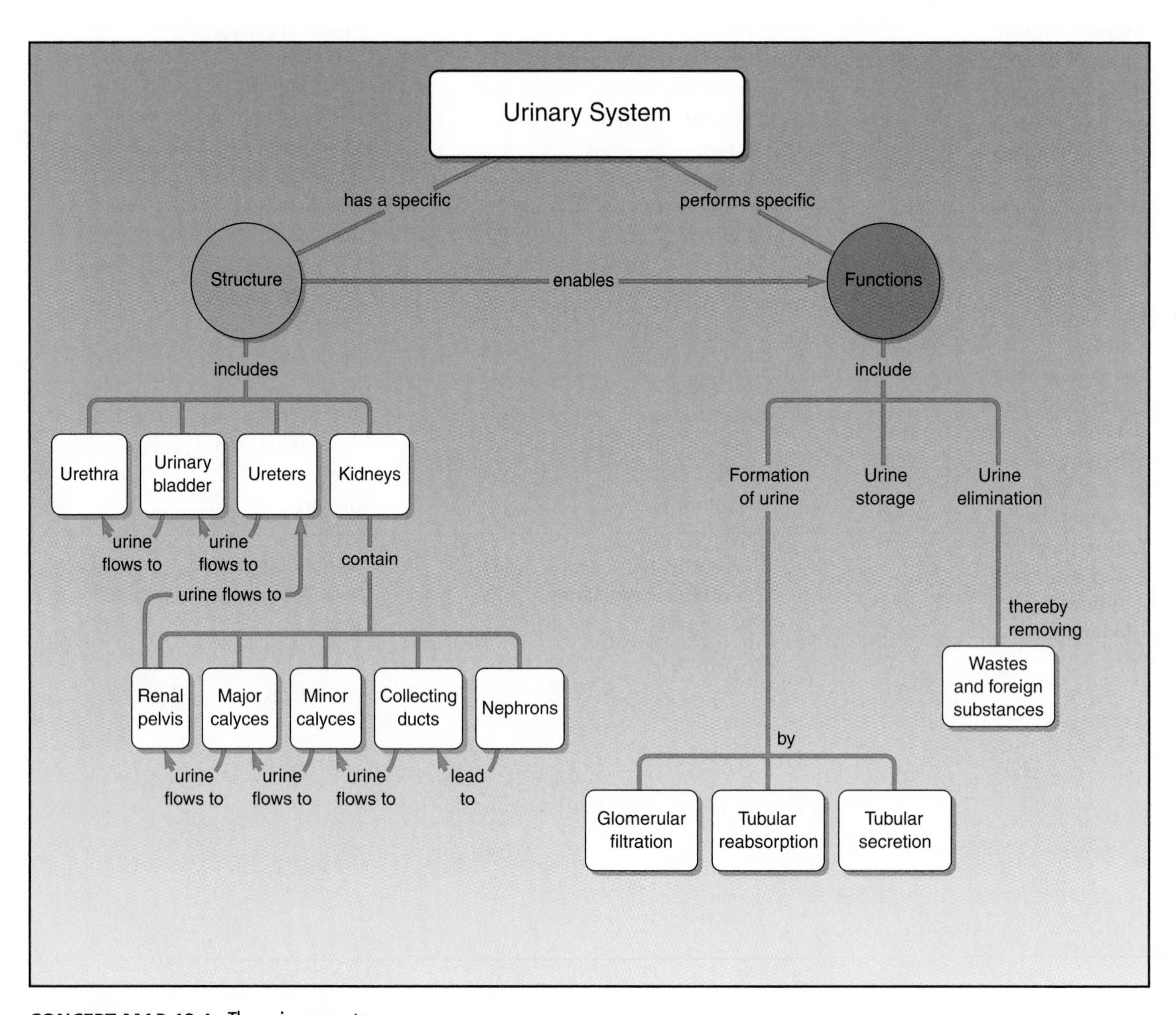

CONCEPT MAP 18-1. The urinary system.

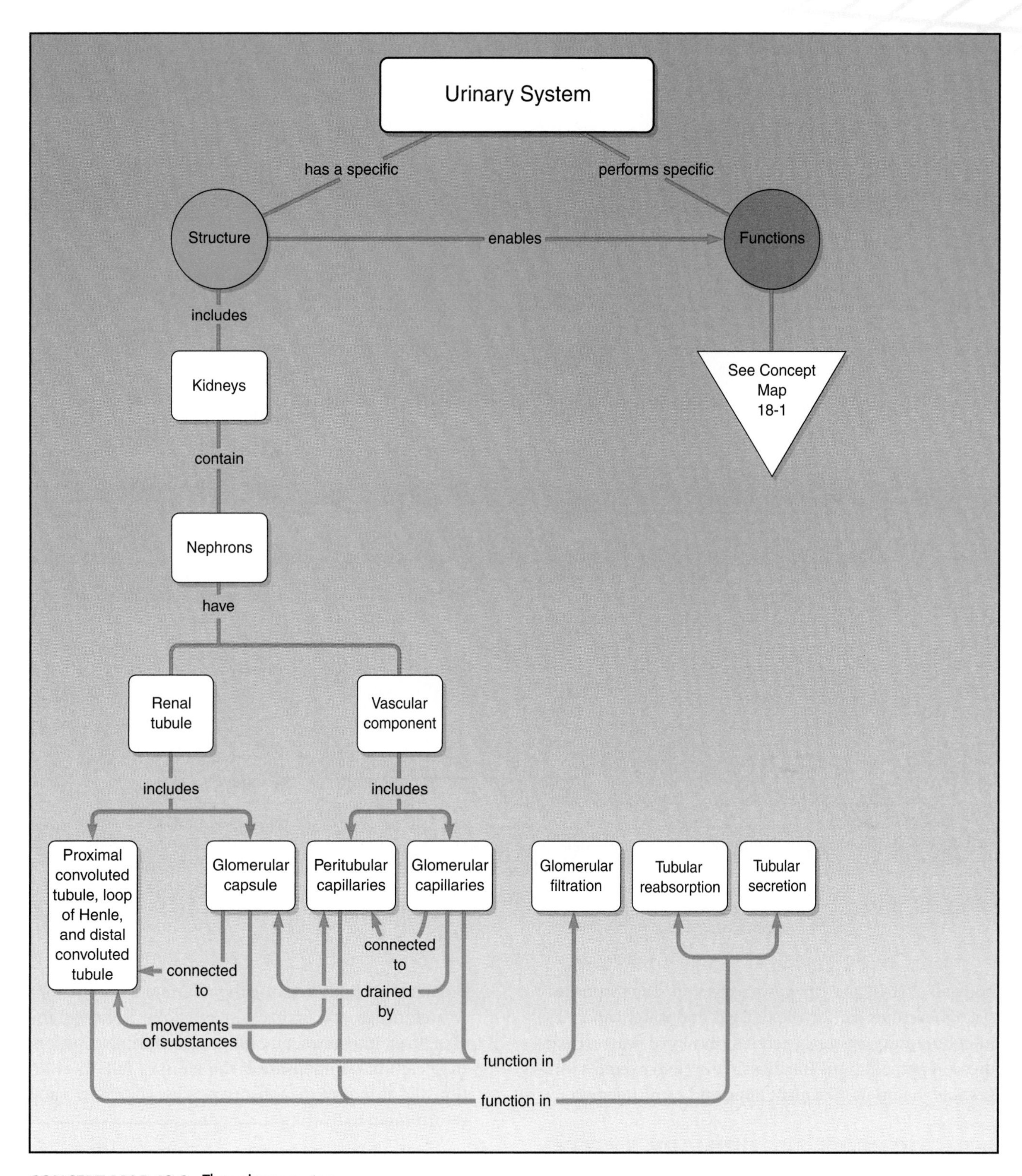

CONCEPT MAP 18-2. The urinary system.

The kidneys are extremely efficient organs and are crucial in maintaining homeostasis in the body. A person can function very well with only one kidney, as we know from hearing about kidney donations among family members. In fact, as long as at least one-third of the kidney is functional, a person can survive. However, if kidney failure occurs, death is inevitable without medical treatment through kidney dialysis. Other systems of the

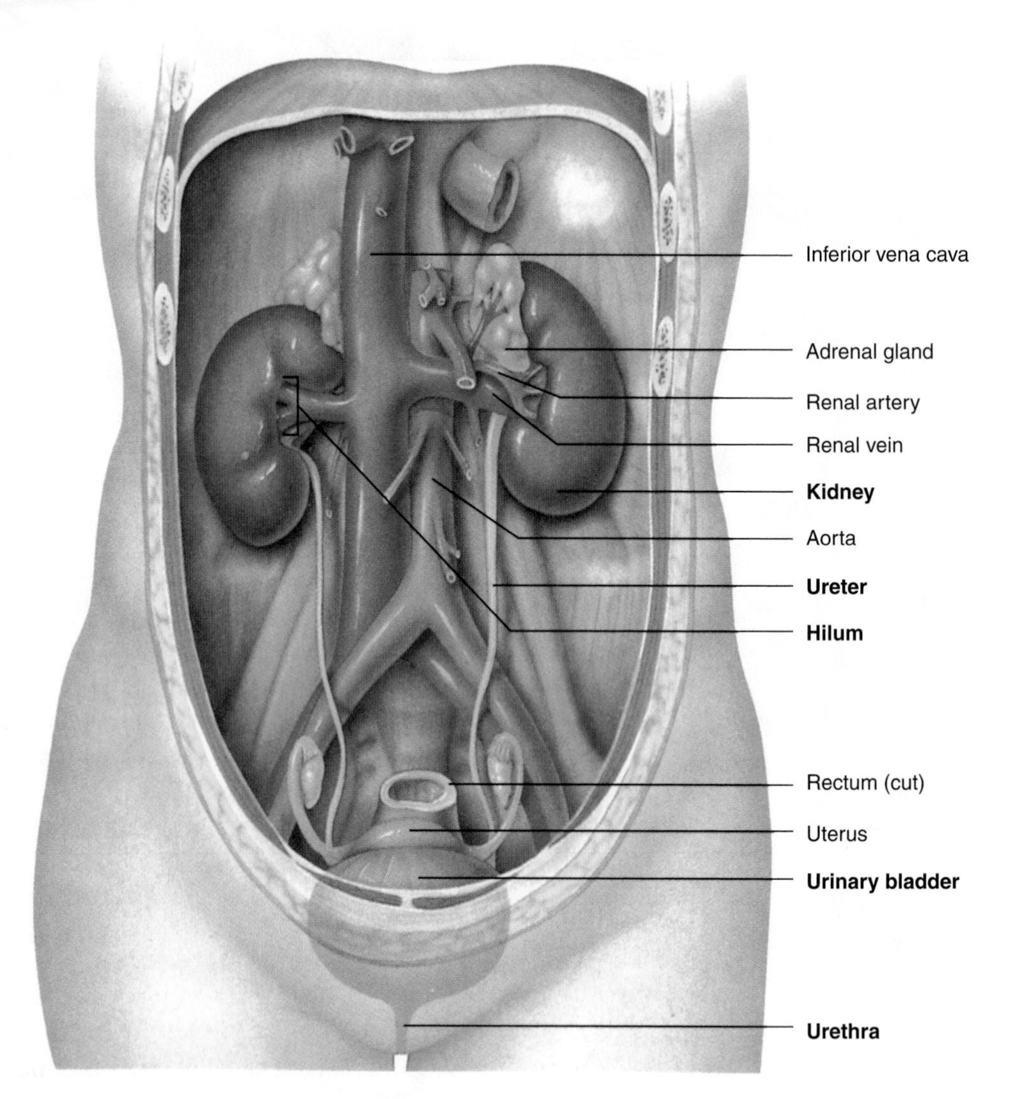

FIGURE 18-1. The organs of the urinary system of a female.

body also participate in waste excretion. The respiratory system excretes carbon dioxide gas and water vapor. The integumentary system excretes dissolved wastes (e.g., urea) in perspiration. The digestive system excretes indigestible materials, like plant fiber, and some bacteria.

FUNCTIONS OF THE URINARY SYSTEM

The major role of maintaining homeostasis with respect to the composition and volume of blood and body fluids is controlled by the kidneys, which perform various functions:

- **Excretion:** The kidneys filter large amounts of fluid from the bloodstream. They are the major excretory organs of the body because they eliminate nitrogenous wastes, drugs, and toxins from the body. Although the skin, liver, intestines, and lungs also eliminate wastes, they cannot compensate if the kidneys fail. In addition, the kidneys can reabsorb needed substances and return them to the blood.
- **Maintain blood volume and concentration:** The kidneys control blood volume by regulating the proper balance in the blood between salts and water. They regulate the volume of urine produced. They also regulate the concentration of ions in body fluids and blood, so the proper balance of sodium, chloride, potassium, calcium, and phosphate ions is maintained.

- **pH regulation:** The kidneys control the proper balance of hydrogen ions in the blood, thus helping to regulate the proper pH levels in the body along with buffers in the blood and the respiratory system.
- **Blood pressure:** The kidneys produce the enzyme **renin** (**REN**-in), which helps adjust filtration pressure.
- **Erythrocyte concentration:** The kidneys produce **erythropoietin** (eh-**rith**-roh-**POY**-eh-tin), a hormone that stimulates red blood cell production in red bone marrow. They help regulate the concentration of erythrocytes in the blood in cases of chronic hypoxia (inadequate oxygen in tissue cells).
- **Vitamin D production:** The kidneys convert vitamin D to its active form (calciferol). Vitamin D is important for normal bone and teeth development. It also helps control calcium and phosphorus metabolism. The kidneys participate, along with the liver and the skin, in vitamin D synthesis.

THE EXTERNAL ANATOMY OF THE KIDNEYS

The **kidneys** are paired organs that are reddish in color and resemble kidney beans in shape. They are about the size of a closed fist. They are located just above the waist between the parietal peritoneum and the posterior wall of the abdomen. This placement of the kidneys is also referred to as retroperitoneal. The right kidney is slightly lower than the left because of the large area occupied by the liver (see Figure 18-1).

The average adult kidney measures about 11.25 cm (4 inches) long, 5.0 to 7.5 cm (2–3 inches) wide and 2.5 cm (1 inch) thick. Near the center of the concave border of the kidney is a notch called the **hilum** (**HIGH**-lum) through which the ureter leaves the kidney. Blood vessels, nerves, and lymph vessels also enter and exit the kidney through this hilum. The hilum is the entrance to a cavity in the kidney called the **renal sinus**, which consists of connective tissue and fat.

Three layers of tissue surround each kidney. The innermost layer is the **renal capsule**. It is a smooth, transparent, fibrous connective tissue membrane that connects with the outermost covering of the ureter at the hilum. It functions as a barrier against infection and trauma to the kidney. The second layer, on top of the renal capsule, is the **adipose capsule**. It is a mass of fatty tissue that protects the kidney from blows. It also firmly holds the kidney in place in the abdominal cavity. The outermost layer is the **renal fascia** (**REE**-nal **FASH**-ee-ah), which consists of a thin layer of fibrous connective tissue that also anchors the kidneys to their surrounding structures and to the abdominal wall.

THE INTERNAL ANATOMY OF THE KIDNEYS

A frontal section through a kidney will reveal an outer area called the **cortex** and an inner area known as the **medulla** (Figure 18-2). In a freshly dissected kidney, the cortex would be reddish in color and the medulla reddish-brown. Within the medulla are 8 to 18 striated, triangular structures called the **renal pyramids**. The striated appearance is caused by an aggregation of straight tubules and blood vessels. The bases of the pyramids face the cortex and their tips, called the **renal papillae** (**REE**-nal pah-**PILL**-ee), point toward the center of the kidney.

The cortex is the smooth textured area that extends from the renal capsule to the bases of the renal pyramids. It also extends into the spaces between the pyramids. This cortical substance in between the renal pyramids is called the **renal columns**. Together, the cortex and the renal pyramids make up the **parenchyma** (par-**EN**-kih-mah) of the kidney. Structurally, this parenchyma consists of millions of microscopic collecting tubules called nephrons (**NEFF**-ronz). The nephrons are the functional units of the kidney. They regulate the composition and volume of blood and form the urine.

A funnel-shaped structure called the **minor calyx** (**MYE**-nohr **KAY**-liks) surrounds the tip of each renal pyramid. There can be 8 to 18 minor calyces. Each minor calyx collects urine from the ducts of the pyramids. Minor calyces join to form **major calyces**. There are two or three major calyces in the kidney. The major calyces join together to form the large collecting funnel called the **renal pelvis**, which is found in the renal sinus. It is the renal pelvis that eventually narrows to form the ureter (**YOO**-reh-ter). Urine drains from the tips of the renal pyramids into the calyces. It then collects in the renal pelvis and leaves the kidney through the ureter.

THE ANATOMY OF THE NEPHRONS

The functional units of the kidney are the **nephrons**. There are two types of nephrons: Juxtamedullary nephrons have loops of Henle that extend deep into

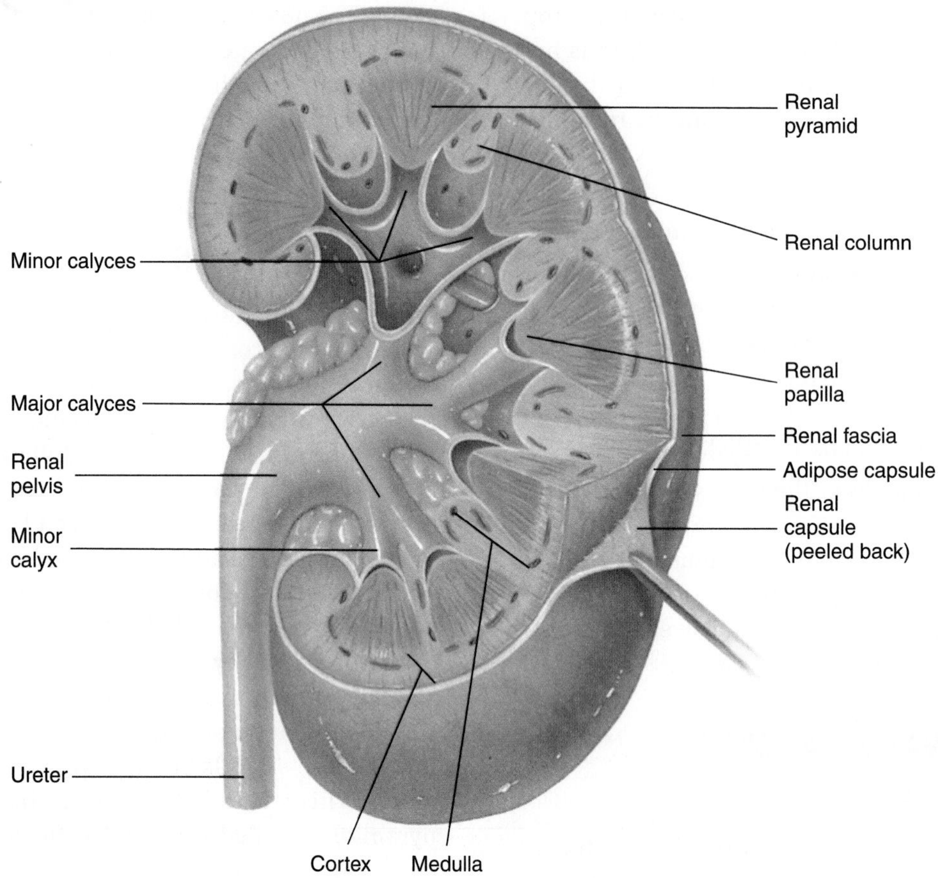

FIGURE 18-2. The internal anatomy of a kidney.

the medulla; cortical nephrons have loops of Henle that do not extend deep into the medulla. Basically, a nephron is a microscopic renal tubule, which functions as a filter, and its vascular (surrounding blood vessels) component (Figure 18-3). The nephron begins as a double-walled globe known as **Bowman's glomerular capsule**. This is located in the cortex of the kidney. The innermost layer of the capsule is known as the visceral layer and consists of epithelial cells called **podocytes** (**POH**-doh-sightz). This visceral layer of podocytes surrounds a capillary network known as the **glomerulus** (glom-**AIR**-youlus). The outer wall of Bowman's glomerular capsule is known as the parietal layer. A collecting space separates the inner visceral layer from the outer parietal layer of the capsule. Together, Bowman's glomerular capsule and the enclosed glomerulus make up what is called a **renal corpuscle**.

The visceral layer of Bowman's capsule and the endothelial capillary network of the glomerulus form an **endothelial-capsular membrane**, which is the site of filtration of water and solutes from the blood. This filtered fluid now moves into the **renal tubule**.

Bowman's capsule opens into the first part of the renal tubule, called the **proximal convoluted tubule**, located in the cortex. The next section of the tubule is called the **descending limb of Henle**, which narrows in diameter as it dips into the medulla of the kidney. The tubule then bends into a U-shaped structure known as the **loop of Henle**. As the tubule straightens, it increases in diameter and ascends toward the cortex of the kidney. Here it is called the **ascending limb of Henle**. In the cortex, the tubule again becomes convoluted and is now called the **distal convoluted tubule**. The distal convoluted tubule ends by merging with a large straight **collecting duct**. In the medulla, collecting ducts connect with the distal tubules of other nephrons. The collecting ducts now pass through the renal pyramids and open into the calyces of the pelvis through a number of larger **papillary ducts**, which empty urine into the renal pelvis.

To facilitate filtration, most of the descending limb has thin walls of simple squamous epithelium,

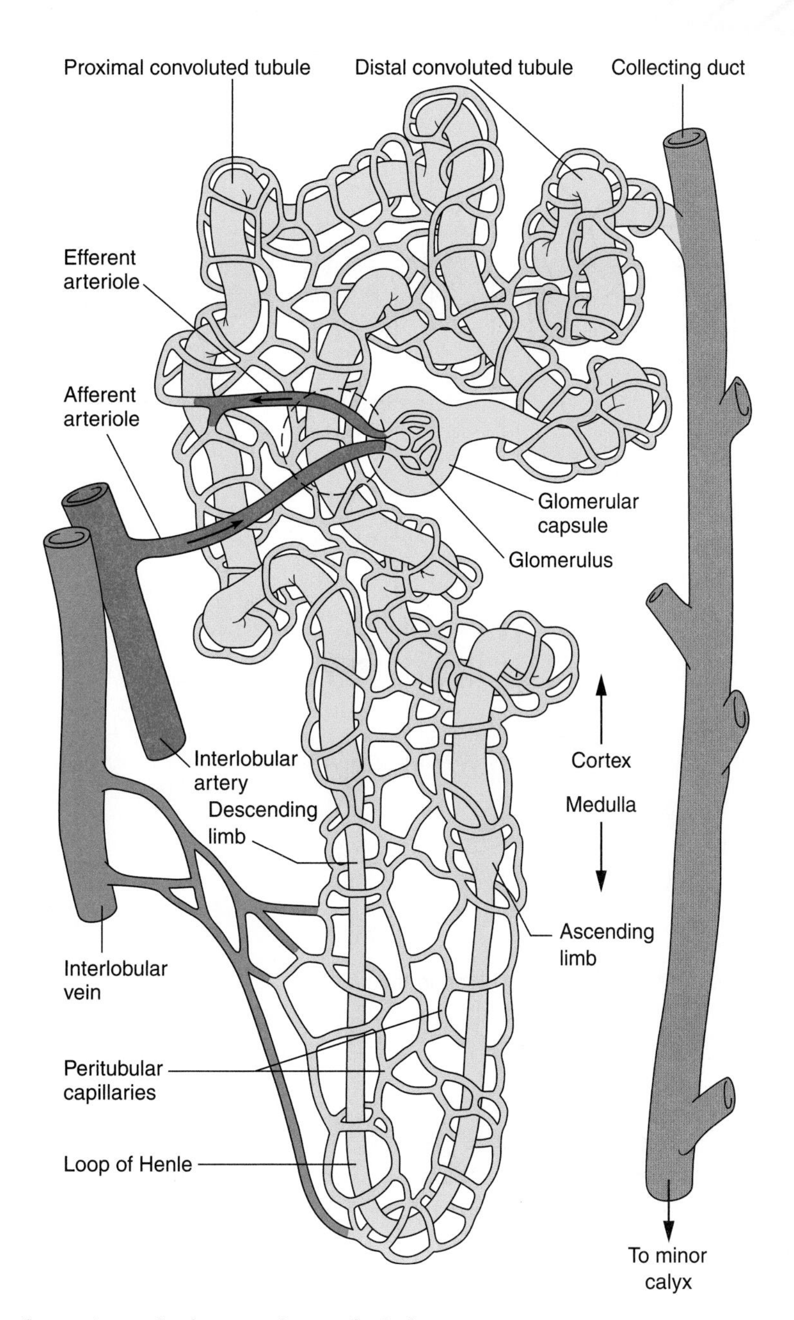

FIGURE 18-3. The anatomy of a nephron, the functional unit of a kidney.

and the rest of the nephron and collecting duct is composed of simple cuboidal epithelium. The proximal tubule, ascending limb of Henle, and the collecting duct transport molecules and ions across the wall of the nephron. The descending limb of Henle is highly permeable to water and solutes (Figure 18-4).

BLOOD AND NERVE SUPPLY TO THE NEPHRONS

Because the nephrons are mainly responsible for removing wastes from the blood and regulating its electrolytes (which are responsible for the acid or alkaline components of blood) and fluid content, they are richly supplied

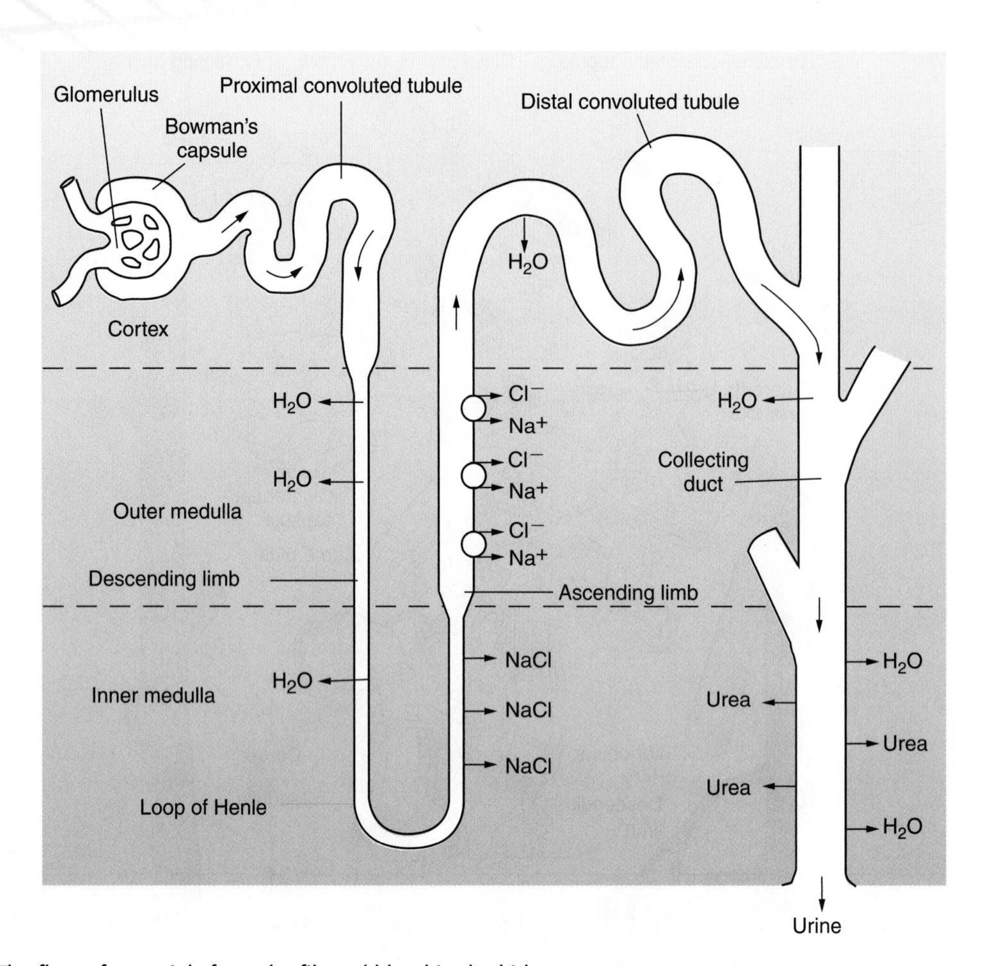

FIGURE 18-4. The flow of materials from the filtered blood in the kidney.

with blood vessels. The **right** and **left renal arteries** (see Figure 18-1) transport one-quarter of the total cardiac output directly to the kidneys. About 1200 mL of blood passes through the kidneys every minute. This amounts to blood being filtered of wastes approximately 60 times a day.

Just before or immediately after entering the hilum, the renal artery divides into several branches that enter the parenchyma of the kidney in between the renal pyramids. In the renal columns, these branches are called the **interlobar arteries** (Figure 18-5). At the base of the pyramids, the interlobar arteries arch between the cortex and the medulla. Here they are called the **arcuate arteries.** Branches of the arcuate arteries produce a series of interlobular arteries (see Figure 18-3) that enter the cortex and divide into afferent arterioles. Each **afferent arteriole** takes blood from the renal artery to Bowman's glomerular capsule, where the arteriole divides into the tangled capillary network known as the glomerulus.

The glomerular capillaries then reunite to form an **efferent arteriole** that carries blood away from the glomerular capsule. Each efferent arteriole further divides to form a network of capillaries called the **peritubular capillaries**, which surround the convoluted tubules of the nephron. Eventually, the peritubular capillaries reunite to form an **interlobular vein**. The filtered blood then drains into the **arcuate vein** at the base of the pyramid. From the arcuate veins, the blood travels through the **interlobar veins** that run between the pyramids in the renal columns. The interlobar veins unite at the single **right** and **left renal vein** that leave the right and left kidney at the hilum.

The nerve supply to the kidney comes from the **renal plexus** of the autonomic nervous system. Sympathetic neurons, using norepinephrine, innervate the blood vessels of the kidneys. This stimulation causes constriction of the arteries, resulting in a decrease in blood flow and a decrease in filtrate formation. Thus, there is a decrease in urine formation. Urine volume production increases

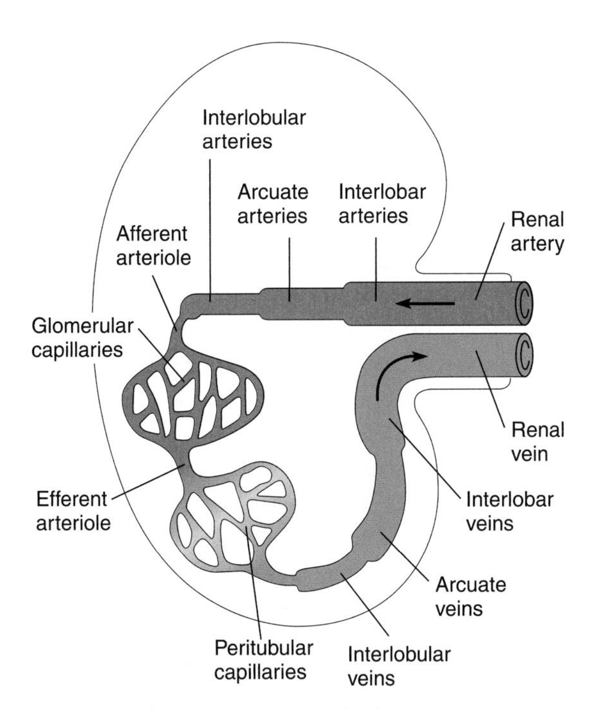

FIGURE 18-5. Blood flow through the kidney.

in response to a decrease in sympathetic innervation to the renal arteries.

Trauma or increased physical activity causes an increase in sympathetic stimulation, resulting in low levels of urine production.

PHYSIOLOGY OF THE NEPHRONS

The nephrons carry out a number of important functions. They control blood concentration and volume by removing selected amounts of water and solutes, they help regulate blood pH, they remove toxic wastes from the blood, and they stimulate red blood cell production in red bone marrow by producing a hormone called erythropoietin. The eliminated materials are collectively called **urine**. Urine is formed by three processes in the nephrons: glomerular filtration, tubular reabsorption, and tubular secretion (see Figure 18-6).

In *glomerular filtration*, the glomerulus filters water and certain dissolved substances from the plasma of blood. This process of glomerular filtration results in increased blood pressure. This increased pressure forces the fluid to filter from the blood. The dissolved substances include: positively charged ions of sodium, potassium, calcium, and magnesium; negatively charged ions of chloride, bicarbonate, sulfate, and phosphate; and glucose, urea, and uric acid. This filtrate is mainly water with some of the same components as the blood plasma. No large proteins are filtered. Both kidneys filter about 45 gallons of blood plasma per day. Yet only a small portion of the glomerular filtrate leaves the kidneys as urine. Most fluid gets reabsorbed in the renal tubules and reenters the plasma.

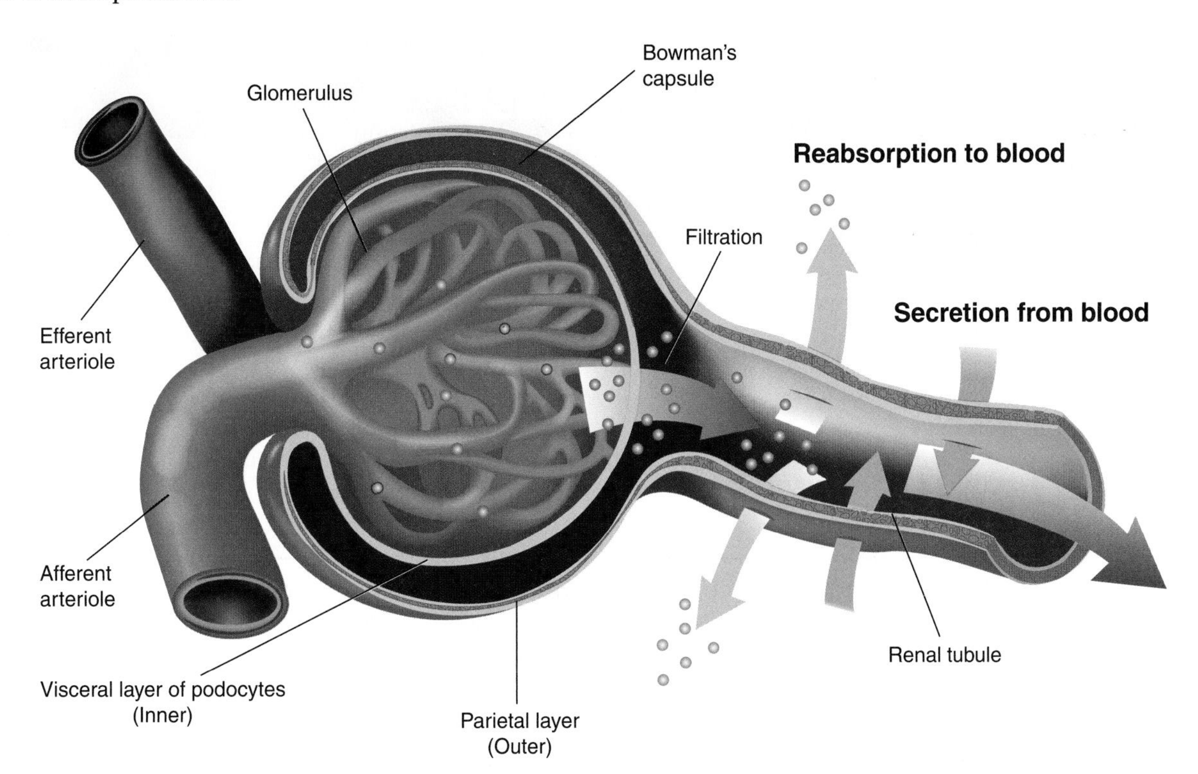

FIGURE 18-6. The main functions of the nephrons: filtration, reabsorption, and secretion.

The *tubular reabsorption* process transports substances out of the tubular fluid and back into the blood of the peritubular capillary. This reabsorption occurs throughout the renal tubule, but the majority of reabsorption occurs in the proximal convoluted tubule. Active transport reabsorbs glucose while osmosis rapidly reabsorbs water. Active transport reabsorbs amino acids, creatine, lactic acid, uric acid, citric acid, and ascorbic acid. Active transport also reabsorbs phosphate, calcium, sulfate, sodium, and potassium ions. Chloride ions and other negatively charged ions are reabsorbed by electrochemical attraction. The descending limb of Henle reabsorbs water by osmosis. The ascending limb reabsorbs sodium, potassium, and chloride ions by active transport. The distal convoluted tubule reabsorbs sodium ions by active transport and water by osmosis. The collecting ducts of the nephrons also will reabsorb water by osmosis. About 95% of water is reabsorbed back into the bloodstream. Hormones, such as vasopressin and aldosterone, are essential to help control this process.

In *tubular secretion*, substances will move from the plasma in the peritubular capillary into the fluid of the renal tubule. The amount of a certain substance excreted into the urine may eventually exceed the amount originally filtered from the blood plasma in the glomerulus. The proximal convoluted tubule actively secretes penicillin, creatinine, and histamine into the tubular fluid. The entire renal tubule actively secretes hydrogen ions (H+), thus helping to regulate the pH of the body fluids. The distal convoluted tubule and the collecting duct secrete potassium ions (K+).

Urine consists of water and solutes that the kidneys either eliminate or retain in the body to maintain homeostasis. Urine is about 95% water with urea, uric acid, some amino acids, and electrolytes. The daily production of urine is between 0.6 and 2.5 liters per day. This depends on a person's fluid intake, environmental temperature and humidity, respiratory rates, body temperature, and emotional conditions. Urine production of 56 mL an hour is considered normal; 30 mL an hour indicates possible kidney failure.

StudyWARE™ Connection

Watch an animation on the formation of urine on your StudyWARE™ CD-ROM.

HEALTH ALERT

PREVENTING URINARY TRACT INFECTIONS

Urinary tract infections, referred to as UTIs, are more common in women than in men. The bacterium *Escherichia coli* is a commensalistic bacterium that is part of our normal intestinal contents and is a necessary component of the digestive system. *E. coli* is harmless long as it remains in the intestine. However, if it is transferred from the anal area to the urethra, it leads to infection of the urinary system. Since the anus and urethral opening are closer in females than in males, and since the urethra is much shorter in females than in males, bacteria can easily enter the urinary bladder, grow and reproduce, and cause much pain and annoyance.

There are a number of ways to prevent UTIs. Drinking 3 to 4 quarts of fluid daily, besides helping to prevent kidney stones, will cause one to urinate every 2 to 3 hours. This will flush out any invading bacteria and buildup of urine that is needed for the growth of bacteria. In addition, drinking cranberry juice and eating blueberries helps decrease any bacterial growth in the bladder. Good personal hygiene is also essential to preventing bacterial contamination of the urinary system from the anal area. Women should be taught to wipe from front to back and to wash from front to back. Thoroughly washing hands with hot water and soap after toilet use is also very important. Frequent urination flushes out the system and helps retard the growth of any bacteria. Thus, both men and women should be conscious of the necessity for fluid intake and hygiene to prevent UTIs. ■

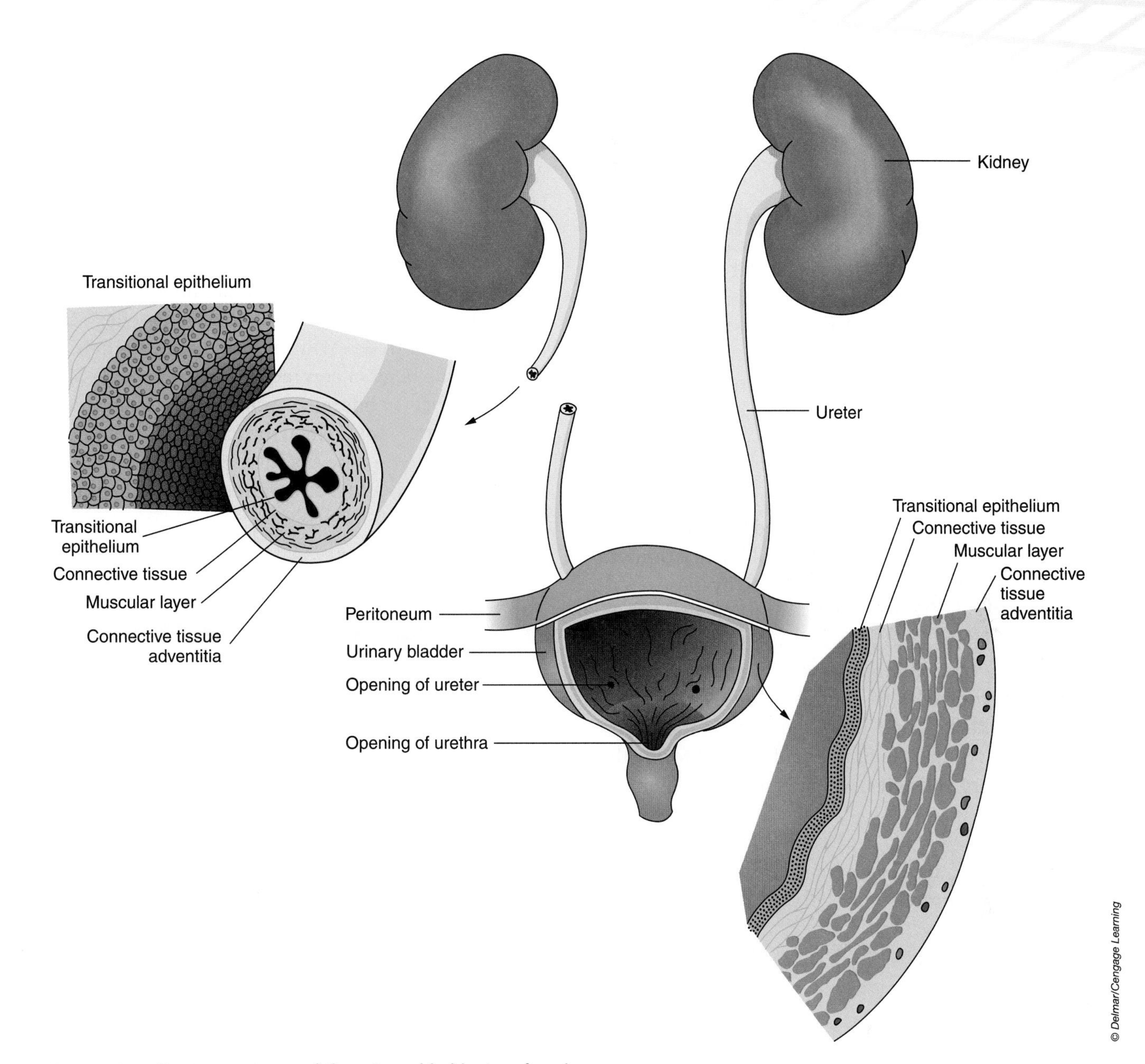

FIGURE 18-7. The two ureters and the urinary bladder in a female.

THE URETERS: ANATOMY AND FUNCTION

The body has two **ureters** (**YOO**-reh-terz) each one descending from a kidney. Each ureter is basically an extension of the pelvis of a kidney and extends about 25 to 30 cm (10 to 12 inches) down to the urinary bladder (Figure 18-7). Each begins as the funnel-shaped renal pelvis and descends parallel on each side of the vertebral column to the bladder. They connect to the urinary bladder posteriorly.

The principal function of the ureters is to transport urine from the renal pelvis into the urinary bladder. The ureters are lined with a mucous coat of transitional epithelium that can stretch. Connective tissue binds the epithelium to a layer of smooth muscle. The urine is carried through the ureters primarily by peristaltic contractions of the smooth muscular walls of the ureters, but gravity and hydrostatic pressure also contribute. The outermost layer of the ureter is composed of connective tissue called the adventitia. Peristaltic waves pass from the kidney to the urinary bladder varying from one to five waves per minute, depending on the amount of urine formation. Consuming excess liquids will cause more urine formation per unit of time.

THE URINARY BLADDER AND THE MICTURITION REFLEX

The **urinary bladder** is a hollow muscular organ located in the pelvic cavity posterior to the pubic symphysis. It consists of the same tissue layers as the ureters. It is a movable organ held in position by folds of peritoneum (see Figure 18-7). When empty, it resembles a deflated balloon. It assumes a spherical shape when slightly full of urine. As urine volume increases, it becomes pear-shaped and ascends into the abdominal cavity.

The interior of the bladder has three openings, the two openings from the two ureters and the single opening to the urethra that will drain the bladder. A smooth triangular region of the bladder outlined by these openings is called the **trigone** (**TRY**-gohn) (Figure 18-8). Bladder infections tend to develop in this region. The bladder wall contains three layers of smooth muscle collectively known as the **detrusor** (**dee-TRUE**-sohr) **muscle**. At the junction of the urinary bladder and urethra, smooth muscle of the bladder wall forms the **internal urinary sphincter**, which is under involuntary control.

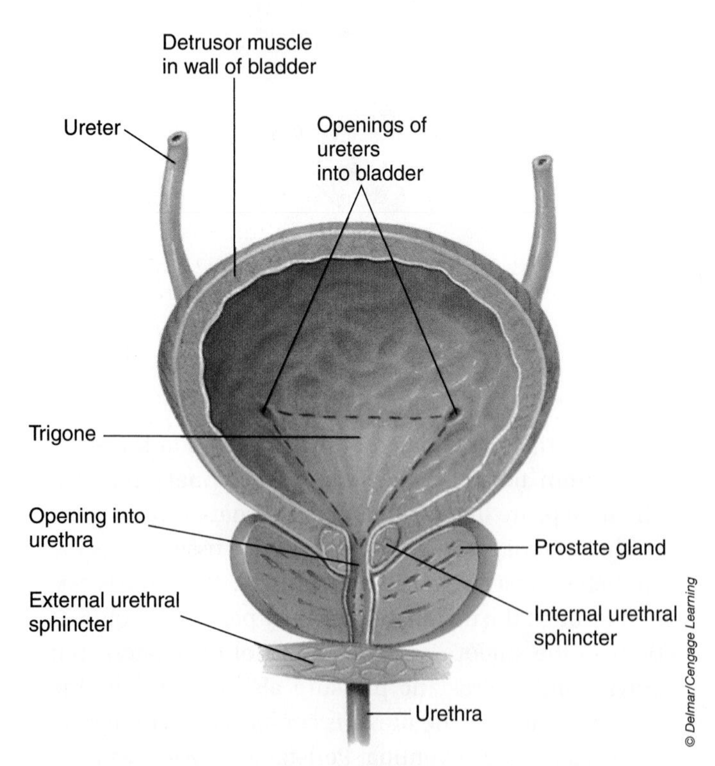

FIGURE 18-8. The anatomy of the urinary bladder in a male.

Urine is expelled from the bladder by an act known as **micturition** (mik-tyoo-**RIH**-shun), commonly referred to as urination or voiding. This response is caused by a combination of involuntary and voluntary nerve impulses. The average capacity of the bladder is approximately 500 mL. When the amount of urine reaches 200 to 400 mL, stretch receptors in the bladder wall transmit nerve impulses to the lower portion of the spinal cord. It is these impulses that initiate a conscious desire to expel urine and an unconscious reflex called the **micturition reflex**. During urination, the detrusor muscle of the bladder contracts as do the muscles of the pelvic floor and abdominal wall. The **external urinary sphincter**, formed of skeletal muscle that surrounds the urethra as it leaves the bladder, must relax and urine then leaves the bladder and moves through the urethra to the outside.

> **StudyWARE™ Connection**
>
> Play an interactive game labeling structures of the urinary system on your StudyWARE™ CD-ROM.

THE URETHRA: MALE AND FEMALE POSITIONS

The **urethra** (yoo-**REE**-thrah) is a small thin-walled tube leading from the floor of the urinary bladder to the outside of the body. It transports urine by peristalsis. Its position in the two sexes differs slightly as does its function.

In females, it lies directly posterior to the pubic symphysis and is located in the wall of the vagina in an anterior position just above the vaginal opening. Its length is about 3.8 cm (1.5 inches). Its opening to the outside is called the urethral orifice and is located between the clitoris and the vaginal opening.

In males, the urethra is 20 cm (8 inches) long. Directly below the bladder, it passes vertically through the prostate gland. It then passes through the urogenital diaphragm and enters the penis. It opens at the tip of the penis at the urethral orifice. In the male, the urethra has a dual function as part of both the urinary and reproductive systems. It carries urine out of the body and functions as a passageway for semen to be discharged from the body.

COMMON DISEASE, DISORDER, OR CONDITION

DISORDERS OF THE URINARY SYSTEM

KIDNEY STONES

Kidney stones, also known as renal calculi, are composed of the precipitates of uric acid, magnesium or calcium phosphate, or calcium oxalate. They can form in the renal pelvis or in the collecting ducts. When a stone passes through the ureter, it can be very painful with accompanying nausea. The pain radiates from the area of the kidney, abdomen, and pelvis. Most kidney stones will pass on their own. Today, larger ones are pulverized by a process called ultrasound lithotripsy. The patient is placed in water where ultrasound waves are focused on the kidney stones, which are then crushed and easily passed in the urine.

CYSTITIS

Cystitis is an inflammation of the urinary bladder, usually caused by a bacterial infection. The infection causes a frequent urge to urinate, with an accompanying burning sensation during urination. This infection can be treated with antibiotics. Early treatment will prevent the bacteria from ascending into the ureters and causing ureteritis (inflammation of the ureters) and possible kidney infection.

GOUT

Gout is a condition caused by high concentrations of uric acid in the plasma. This condition was once believed to be caused by excess food intake, but it may be inherited. The crystals of uric acid get deposited in joints of the hands and feet, causing inflammation and great pain. Gout is treated with drugs that inhibit uric acid reabsorption. Uric acid forms from the metabolism of certain nitrogen bases in nucleic acids.

GLOMERULONEPHRITIS

Glomerulonephritis (glom-**air**-yoo-loh-neh-**FRYE**-tis) is an inflammation of the kidneys, where the filtration membrane within the renal capsule is infected with bacteria. This can be acute following streptococcal sore throat or scarlet fever infection, or it can be a chronic condition resulting in kidney failure.

RENAL FAILURE

Renal failure can result from almost any condition that interferes with kidney function. As urea and other metabolites accumulate in the blood, acidosis develops and death can occur within 1 to 2 weeks. This type of acute renal failure can result from acute glomerulonephritis or blockage of the renal tubules. Chronic renal failure is caused by damage to so many nephrons that the remaining ones cannot accommodate normal kidney function. It can be caused by chronic glomerulonephritis, tumors, obstructions of the urinary tract, or lack of blood supply to the kidneys caused by arteriosclerosis. The toxic effects of the accumulated metabolic waste products result in coma and eventually death. Renal failure can be treated by a procedure called **hemodialysis**. A dialysis machine filters blood taken from an artery and then returns it to a vein. In this procedure, a machine substitutes for the excretory functions of the kidneys. In peritoneal dialysis, the peritoneum is used as a diffusable membrane to correct an imbalance of electrolytes or fluid in the blood or to remove wastes, toxins, or drugs normally filtered by the kidneys. A catheter is sutured into the peritoneum and connected to inflow and outflow tubing containing dialysate.

HEMATURIA

Hematuria (hee-mah-**TOO**-ree-ah) is blood in the urine, specifically an abundance of red blood cells. This can develop from kidney stones in the system or from bacterial infections of the urinary tract. Inflammation of the bladder, urethra, or the prostate gland can also cause hematuria.

(continues)

COMMON DISEASE, DISORDER, OR CONDITION	DISORDERS OF THE URINARY SYSTEM (continued)

OLIGURIA

Oliguria (ol-ig-**YOO**-ree-ah) is a condition in which only a small amount of urine is being produced, less than 500 mL per day. This results in an inability to effectively excrete waste products from the blood. It can be caused by urinary tract obstructions, lesions in the kidney, or imbalances in body electrolytes and fluids.

POLYURIA

Polyuria (pall-ee-**YOO**-ree-ah) is the production of an excessive amount of urine. This is a result of both diabetes mellitus and diabetes insipidus. It can also occur due to the intake of an excessive amount of fluids and the use of diuretics (drugs that promote the formation and excretion of urine).

PYURIA

Pyuria (pye-**YOO**-ree-ah) is a condition in which there is an excessive number of white blood cells in the urine (pus). It results from a bacterial infection of the urinary tract.

UREMIA

Uremia (yoo-**REE**-mee-ah) is a condition in which there is an excessive amount of urine (specifically urea and nitrogenous waste) in the blood. It is also known as azotemia, a toxic condition produced by renal failure when the kidneys cannot remove the urea from the blood.

POLYCYSTIC KIDNEY DISEASE (PKD)

Polycystic (pol-ee-**SIS**-tic) **kidney disease** or **PKD** is a condition in which the kidneys are abnormally enlarged and contain numerous cysts. There are three forms of this disease. The first is congenital polycystic disease (PKD), which is a rare congenital defect in which all or part of one or both kidneys fail to develop properly. Severe defective development results in death shortly after birth. The second is childhood polycystic disease (PKD), which is a rare condition resulting in death after a few years of age. It develops from kidney and liver failure as well as portal hypertension (increased venous pressure in the hepatic portal circulatory route). The third is adult polycystic disease (PKD), which is characterized by lower back pain and high blood pressure. The kidneys eventually fail resulting in uremia and death. This condition can be acquired or congenital and can involve either one or both kidneys.

URINARY INCONTINENCE

Urinary incontinence (**YOO**-rih-nair-ee in-**CON**-tin-ens) is a condition in which an individual experiences an uncontrollable and continued flow of urine. It can be caused by neurological dysfunctions resulting in not realizing that the bladder is full, independent contractions of the detrusor muscle of the bladder as a result of surgery (like removal of a cancerous prostate gland), or a disease that affects the nerves of the spinal cord that go to the bladder.

HEALTH ALERT

URINARY TRACT INFECTION (UTI)

Urinary tract infections affect one or more structures of the urinary tract. They are more common in females than in males due to the short length of the urethra in females. Most infections are caused by gram-negative bacteria, most commonly *Escherichia coli* or species of *Klebsiella*, *Pseudomonas*, *Proteus*, or *Enterobacter*. Bacteria can reproduce in the bladder causing UTI in women much easier than in men. By the time a woman urinates and flushes bacteria from the urethra, those bacteria may have already invaded the bladder. The long urethra in males makes it more difficult for the bacteria to get to the bladder. Symptoms of UTI include frequent urination and a burning sensation on urination, accompanied by pain. In severe infections, pus and blood may be visible in the urine. UTI in men may not show typical symptoms. Prevention can be achieved by increasing fluid intake, having good perineal hygiene, and urinating frequently. UTI is treated with antibacterial and antiseptic drugs. ■

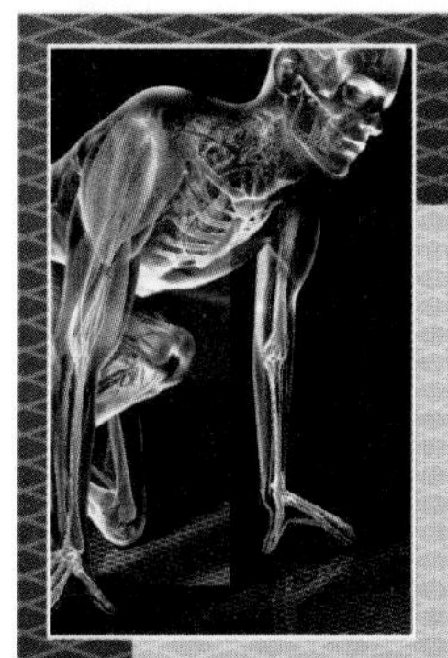

AS THE BODY AGES

As we age, the kidneys decrease in size, in fact, beginning as early as age 20. At this age, our kidneys weigh about 260 grams. By age 80, they weigh about 200 grams. This is related to decreased blood flow to the kidneys. After 20 years of age, there is about a 10% decrease in blood flow to the kidneys, occcurring approximately every 10 years. During this time, functional glomeruli decrease, and efferent and afferent arterioles to the glomeruli become twisted and irregular, thus inhibiting normal blood filtration at Bowman's capsule. The kidneys' capacity for absorption decreases as does their ability to secrete various substances. By age 80, almost half of the kidneys' glomeruli have ceased to function.

Kidney diseases are also more common in older adults and include kidney stones, urinary tract infections, and inflammations of the kidney. As the ability to concentrate urine declines with age, and as the sense of thirst diminishes, a more common risk of dehydration occurs with individuals in their golden years.

With aging, the kidneys have a reduced ability to eliminate urea, uric acid, toxins, and creatinine from the blood. Older adults also tend to experience frequent urination at night, excessive daily urination, occasional blood in the urine, and even painful urination. These age-related changes in the kidneys cause the later-in-life acquired conditions of diabetes and high blood pressure to have a greater effect on the functioning of kidneys in older adults. ■

There are a number of careers available to individuals with a special interest in the urinary system.

- **Urologists** are physicians who specialize in the study and treatment of disorders of the urinary tract of both women and men and who also specialize in the study of the male genital tract.
- **Dialysis technicians** are allied health professionals who maintain and operate dialysis equipment to treat patients who have various kidney disorders.
- **Nephrologists** are physicians who specialize in the study of the kidney and the treatment of its disorders and pathology.
- **Medical laboratory technicians** are individuals who perform bacteriological tests and microscopic examinations of blood, fluids, and tissues of the body under the supervision of a physician or medical technologist.

Integumentary System

- The skin and the kidneys are involved in vitamin D production.
- The skin is a protective barrier and a site for water loss via perspiration.
- The urinary system compensates for water loss caused by perspiration.

Skeletal System

- The lower ribs provide some protection to the kidneys.
- Both the kidneys and bones help maintain calcium levels in the blood.

Muscular System

- Muscles control elimination of urine from the bladder in the voluntary action of micturition.
- Muscle cells produce creatinine as a nitrogenous waste product of metabolism that the kidneys excrete.

Nervous System

- The nervous system controls urine production and micturition.

Endocrine System

- Antidiuretic hormone (ADH) and aldosterone help regulate urine production by influencing renal reabsorption of electrolytes and water.

Cardiovascular System

- Blood volume is controlled by the urinary system.
- Blood pressure controls glomerular filtration.
- Blood carries nutrients and oxygen to and eliminates waste from the urinary tissues.

Lymphatic System

- The kidney helps maintain extracellular fluid composition and volume.
- Lymphatic vessels help maintain blood pressure by returning lymph to the plasma of blood.
- The lymphocytes help protect the urinary structures from infection and cancer.

Digestive System

- The liver transforms toxic ammonia (a by-product of amino acid metabolism) into less harmful urea that is then excreted by the kidneys.
- The kidneys restore fluids lost by the digestive process.

Respiratory System

- The lungs and the kidneys help maintain the proper pH of the body.
- The respiratory system provides the oxygen needed by the cells of the kidneys to function and eliminates the carbon dioxide waste product.

Reproductive System

- The urethra of the male functions as both an organ to eliminate urine from the bladder and as a tube to transfer sperm to the outside.
- The kidneys replace fluid lost from the normal activities of the reproductive system.

SUMMARY OUTLINE

INTRODUCTION

1. The urinary system helps keep the body in homeostasis by removing and restoring selected amounts of solutes and water from the blood.
2. The system consists of two kidneys, two ureters, the bladder, and the urethra.
3. The kidneys are the main filtering organs of the system, producing the urine.
4. Urine consists of urea, excess water, excess ions, and toxic wastes that may have been consumed with food.

FUNCTIONS OF THE URINARY SYSTEM

The kidneys perform six functions as they daily filter the blood:

1. **Excretion.** They are the major excretory organs of the body, filtering large amounts of fluids from the bloodstream, including nitrogenous wastes, drugs, and toxins.
2. **Maintain blood volume and concentration.** They regulate the proper balance of water and dissolved salts by maintaining proper ion concentrations.
3. **pH regulation.** They control the proper hydrogen ion concentration of the blood.
4. **Blood pressure.** They produce the enzyme renin, which helps maintain blood pressure.
5. **Erythrocyte concentration.** They produce the protein hormone erythropoietin, which stimulates red blood cell production.
6. **Vitamin D production.** They convert vitamin D to its active form (calciferol).

THE EXTERNAL ANATOMY OF THE KIDNEYS

1. The kidneys are located just above the waist between the parietal peritoneum and the posterior wall of the abdomen.
2. The hilum is a notch in the concave center of each kidney through which a ureter leaves the kidney and blood vessels, nerves, and lymph vessels enter and exit the kidney.
3. Three layers of tissue surround each kidney.
4. The innermost layer is the renal capsule that acts as a barrier against infection and trauma.
5. The second layer is the adipose capsule, a mass of fatty tissue that protects the kidney from blows.
6. The outermost layer is the renal fascia that anchors the kidney to the abdominal wall.

THE INTERNAL ANATOMY OF THE KIDNEYS

1. The outer area of a kidney is called the cortex.
2. The inner area of a kidney is called the medulla.
3. Within the medulla are striated, triangular structures called the renal pyramids whose bases face the cortex and whose tips are called renal papillae that point to the center of the kidney.
4. The cortical material that extends between the pyramids is called the renal columns.
5. The cortex and renal pyramids make up the parenchyma of the kidney, which is composed of millions of microscopic units called nephrons.
6. The nephrons are the functional units of the kidneys.
7. The tip of each renal pyramid is surrounded by a funnel-shaped structure called a minor calyx, which collects urine from the ducts of the pyramids.
8. Minor calyces join to form a few major calyces. The major calyces join to form the large collecting funnel called the renal pelvis, which narrows to form the ureter.

THE ANATOMY OF THE NEPHRONS

1. A nephron is a microscopic renal tubule and its vascular component.
2. The nephron begins as a double-walled globe known as Bowman's glomerular capsule.
3. The innermost layer of the capsule is the visceral layer made of podocytes. These epithelial podocytes surround a capillary network called the glomerulus.
4. The outermost layer of the capsule is called the parietal layer.
5. A renal corpuscle is made up of Bowman's glomerular capsule and the enclosed capillary glomerulus.
6. The visceral layer of Bowman's capsule and the capillary network of the glomerulus form the endothelial-capsular membrane, which filters water and solutes from the blood and moves it into the renal tubule.
7. The first part of the renal tubule is called the proximal convoluted tubule, located in the cortex.
8. The following part is the descending limb of Henle, which narrows as it dips into the medulla. The tubule then bends into the U-shaped loop of Henle.
9. As the loop straightens, it increases in diameter and ascends toward the cortex as the ascending limb of Henle.

10. In the cortex, the renal tubule again becomes convoluted and is known as the distal convoluted tubule, which ends by merging with a large, straight collecting duct.
11. In the medulla, collecting ducts connect with the distal convoluted tubules of a number of nephrons.
12. Collecting ducts now pass through the renal pyramids and open into the calyces of the pelvis through a number of larger papillary ducts. They empty urine into the renal pelvis.

BLOOD AND NERVE SUPPLY TO THE NEPHRONS

1. The right and left renal arteries transport 1200 mL of blood to the kidneys every minute.
2. The arteries branch and pass between the renal pyramids in the renal columns as the interlobar arteries. At the base of the pyramids, they arch as the arcuate arteries found between the cortex and the medulla.
3. Branches of the arcuate arteries become the interlobular arteries, which branch into afferent arterioles in the cortex. Afferent arterioles divide into the capillary network called the glomerulus.
4. Glomerular capillaries reunite to form the efferent arteriole, which exits the capsule of the glomerulus.
5. Efferent arterioles divide to form peritubular capillaries, which surround the convoluted tubules of the nephron.
6. Peritubular capillaries reunite to form an interlobular vein, which connects with the arcuate vein at the base of a pyramid.
7. Arcuate veins connect to interlobar veins found between the pyramids in the renal columns.
8. Interlobar veins unite at the right and left renal veins that exit the right and the left kidney at the hilum.
9. The nerve supply to the kidney is the renal plexus of the autonomic nervous system.

PHYSIOLOGY OF THE NEPHRONS

1. The three major functions of nephrons are to control blood concentration and volume by removing and restoring selected amounts of water and solutes, help regulate blood pH, and remove toxic waste from the blood.
2. Urine forms by glomerular filtration, tubular reabsorption, and tubular secretion, all of which occur in the nephrons.
3. Glomerular filtration removes water and these dissolved substances from the plasma of blood: sodium, potassium, calcium, and magnesium positive ions; negative ions of chloride, bicarbonate, sulfate, and phosphate; and glucose, urea, and uric acid. Ninety-nine percent of the fluid gets reabsorbed in the renal tubules.
4. Tubular reabsorption transports substances from the tubular fluid into the blood of the peritubular capillaries. Active transport reabsorbs glucose, and osmosis reabsorbs water. Active transport reabsorbs positively charged ions, amino acids, creatinine, and lactic, uric, citric, and ascorbic acids. Negatively charged ions are reabsorbed by electrochemical attraction.
5. Tubular secretion moves these substances from the plasma in the peritubular capillary into the fluid of the renal tubule: penicillin and other drugs, creatinine, histamine, hydrogen ions, and potassium ions.
6. Urine consists of 95% water with urea, uric acid, some amino acids, and electrolytes.

THE URETERS: ANATOMY AND FUNCTION

1. Each of the two ureters begins as an extension of the renal pelvis of a kidney and connects to the urinary bladder.
2. The function of the ureters is to transport urine from the renal pelvis to the urinary bladder.
3. Urine moves mainly by peristaltic contractions of the smooth muscle walls, but gravity and hydrostatic pressure also contribute.

THE URINARY BLADDER AND THE MICTURITION REFLEX

1. The urinary bladder is held in position by folds of peritoneum in the pelvic cavity.
2. The two openings from the ureters and the single opening into the urethra outline a smooth triangular region called the trigone.
3. The bladder wall is composed of three layers of smooth muscle called the detrusor muscle.
4. At the junction of the urinary bladder and the urethra is the internal urinary sphincter under involuntary control.
5. Urine is expelled from the bladder by an act known as micturition.
6. The external urinary sphincter, formed by skeletal muscle surrounding the urethra as it leaves the bladder, relaxes and urine leaves the bladder.

7. The bladder can hold 700 to 800 mL of urine. When it reaches 200 to 400 mL, stretch receptors in the bladder wall transmit impulses to the lower spinal cord, which initiate a conscious desire to urinate and an unconscious reflex called the micturition reflex.

THE URETHRA: MALE AND FEMALE POSITIONS

1. The urethra is a small thin-walled tube connecting to the floor of the urinary bladder that leads to the outside.
2. In females, it is located in the wall of the vagina just above the vaginal opening. It is 3.8 cm long and its opening, called the urethral orifice, is located between the clitoris and the vaginal opening.
3. In males, the urethra is 20 cm long, and just below the bladder it passes through the prostate gland and enters the penis, opening at the tip of the male penis as the urethral orifice. In the male, the urethra not only transports urine but also transfers semen to the outside.

REVIEW QUESTIONS

1. Name the organs of the urinary system.
2. Explain six functions of the urinary system.

*3. What other systems of the body perform excretion and what do they excrete?

*4. Explain the role of glomerular filtration, tubular reabsorption, and tubular secretion in the nephron in maintaining homeostasis.

5. Name three functions of the nephrons.

*6. Explain the micturition reflex in terms of changes in the urinary bladder.

7. Compare the length and position of the urethra in the male and in the female.
8. Name the parts of a nephron's renal tubule.
9. Name three constituents of urine.

***Critical Thinking Questions**

FILL IN THE BLANK

Fill in the blank with the most appropriate term.

1. Within the medulla of the kidney are 8 to 18 striated triangular structures called ___________.
2. The ___________ are functional microscopic units of the kidney.
3. Urine is expelled by an act called ___________, commonly known as urination or voiding.
4. The innermost layer of the kidney is called the ___________, a fibrous connective tissue membrane that is a barrier against infection and trauma to the kidney.
5. A nephron begins as a double-walled globe called ___________ capsule.
6. In the medulla, the renal tubule bends into a U-shape known as the loop of ___________.
7. The kidneys produce a protein hormone called ___________, which stimulates hematopoiesis in red bone marrow.
8. The active form of vitamin D is called ___________.
9. Kidneys produce an enzyme called ___________, which helps regulate blood pressure.
10. When crystals of uric acid get deposited in the joints of the hands and feet, the condition is called ___________.

MATCHING

Place the most appropriate number in the blank provided.

_____ Podocytes
_____ Cortex
_____ Medulla
_____ Capillary network
_____ Hilum
_____ Trigone
_____ Renal column
_____ Detrusor muscle
_____ Renal plexus
_____ Calyx

1. Inner region of kidney
2. Area in bladder
3. Collect urine
4. Bladder wall
5. Cortical area in medulla
6. Epithelial cells in inner wall of Bowman's capsule
7. Sympathetic neurons
8. Renal pelvis
9. Notch through which the ureter leaves the kidney
10. Outer layer of kidney
11. Glomerulus
12. Sphincter muscle

Search and Explore

- Search the Internet with key words from the chapter to discover additional information. Key words might include kidneys, nephrons, urinary system anatomy, or one of the disorders of the urinary system introduced in this chapter.

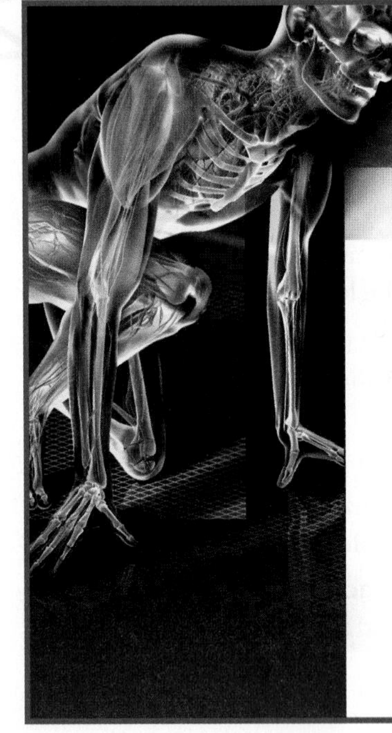

CASE STUDY

Philip Gomez, a 50-year-old man, is doing some weekend gardening when he suddenly experiences severe pain that radiates from his left side to his abdomen. In addition, Mr. Gomez begins to feel extremely nauseated. His alarmed wife rushes him to the emergency room. Following a period of careful evaluation, Philip's health care provider decides on a course of treatment for his condition.

Questions

1. What problem do you think is causing Philip's severe pain and nausea?
2. Where in the urinary system does this condition develop?
3. How is this problem treated?
4. Describe one activity that will help prevent this disorder from recurring.

*Study*WARE™ Connection

Play interactive games or take a practice quiz on your StudyWARE™ CD-ROM.

Study Guide Practice

Go to your **Study Guide** for more practice questions, labeling and coloring exercises, and crossword puzzles to help you learn the content in this chapter.

LABORATORY EXERCISE: THE URINARY SYSTEM

Materials needed: A dissecting kit, your fetal pig, and a dissecting pan.

1. Remove your fetal pig from its storage area. You have already opened the abdominal cavity when you performed your dissection for the digestive system. Locate the large paired kidneys, attached dorsally to the abdominal wall behind the intestines. Midpoint on the medial side of each kidney is a depression called the hilum. Note the renal artery and vein entering and leaving the kidney. Find the ureter as it leaves each kidney through this hilum. Follow a ureter as it leaves the kidney. It goes posteriorly under the parietal peritoneum. Follow its reconnection to the urinary bladder, which normally would be found in the posterior ventral part of the abdomen. In your fetal pig, it will be located on the inner surface of the flap of tissue to which the umbilical cord was attached. The urethra arises from the bladder medially between the two ureters and runs posteriorly, parallel to the rectum.
2. Although the kidneys of the fetal pig are small, remove one and cut it in a frontal section. You will be able to observe the cortex, the medulla, and the renal pyramids with their striations.
3. Examine the models of the urinary system provided by your instructor and identify the major organs of the system.

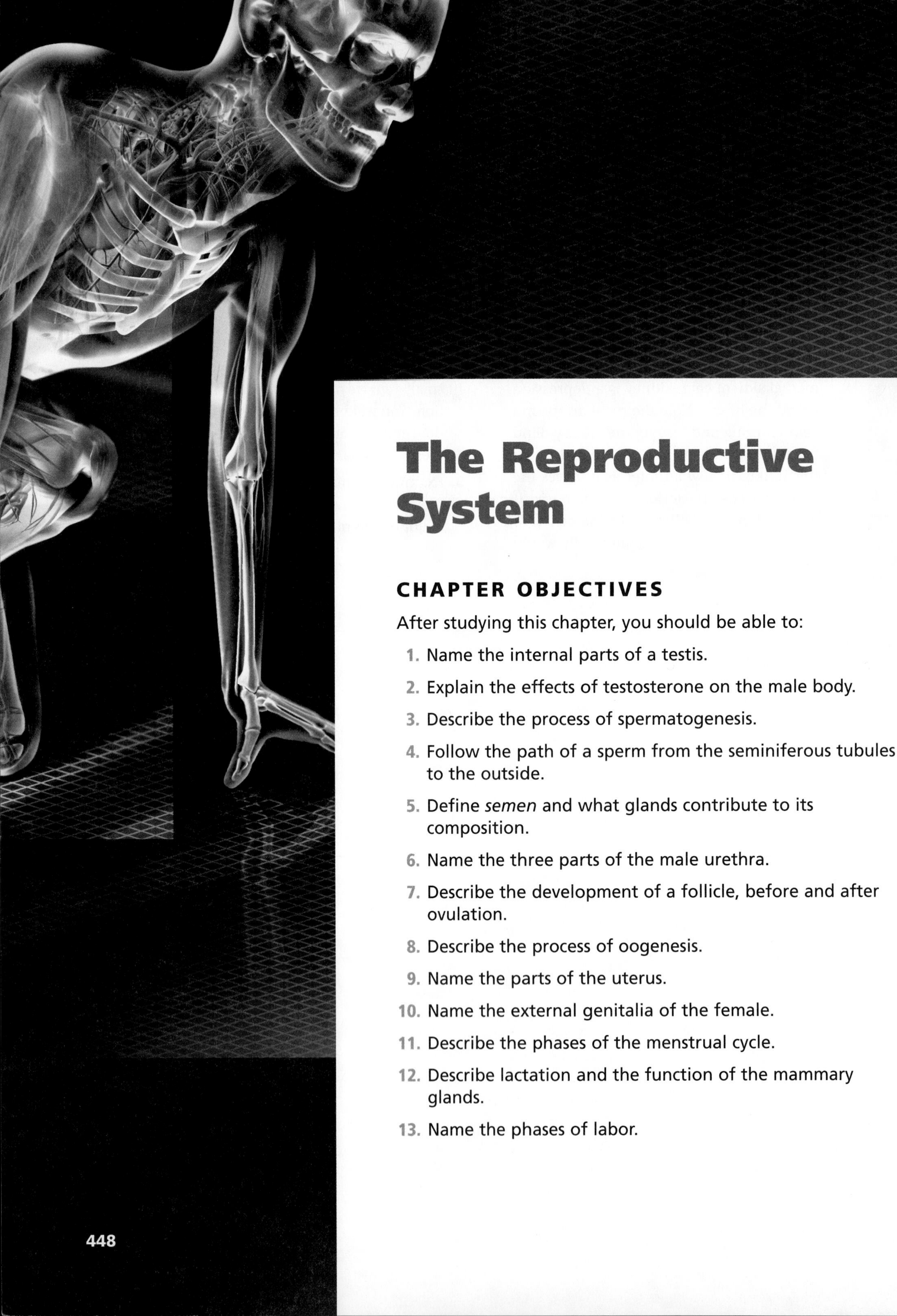

The Reproductive System

CHAPTER OBJECTIVES

After studying this chapter, you should be able to:

1. Name the internal parts of a testis.
2. Explain the effects of testosterone on the male body.
3. Describe the process of spermatogenesis.
4. Follow the path of a sperm from the seminiferous tubules to the outside.
5. Define *semen* and what glands contribute to its composition.
6. Name the three parts of the male urethra.
7. Describe the development of a follicle, before and after ovulation.
8. Describe the process of oogenesis.
9. Name the parts of the uterus.
10. Name the external genitalia of the female.
11. Describe the phases of the menstrual cycle.
12. Describe lactation and the function of the mammary glands.
13. Name the phases of labor.

19

KEY TERMS

Term	Page
Acrosome	453
Alveoli	466
Amnion	467
Ampullae/lactiferous sinuses	466
Areola	466
Blastula/blastocyst	467
Body of the uterus	462
Bulbourethral/Cowper's glands	456
Cervical canal	462
Cervix	462
Chorionic vesicle	467
Chorionic villi	467
Clitoris	465
Coitus	464
Corpus albicans	457
Corpus hemorrhagicum	464
Corpus luteum	457
Cremaster muscle	450
Ductus (vas) deferens	455
Ductus epididymis	455
Ectoderm	467
Efferent ducts	455
Ejaculatory duct	455
Endoderm	467
Endometrium	462
Erection	457
Estrogen	457
External os	462
Fetus	467
Fimbriae	461
Fornix	464
Fundus	462
Germinal epithelium	457
Glans	465
Glans penis	456
Graafian follicle	457
Greater vestibular/ Bartholin's glands	465
Hymen	465
Infundibulum	460
Internal os	462
Interstitial cells of Leydig	452
Isthmus	462
Labia majora	465
Labia minora	465
Labor	467
Lactation	466
Lactiferous ducts	466
Lesser vestibular/ Skene's glands	465
Mammary glands	466
Membranous urethra	455
Menarche	464
Menopause	464
Menses	462
Menstrual cycle	462
Menstruation	462
Mesoderm	467
Mons pubis/veneris	465
Myometrium	462
Nipple	466
Oocyte	457
Oogenesis	460
Oogonia	460
Ootid	460
Ova	457
Ovarian cycle	460
Ovarian follicles	457
Ovaries	457
Ovulation	457
Parturition	467
Penis	456
Perimetrium	462
Perineum	466
Placenta	467
Polar body	460
Prepuce/foreskin	457
Primary oocytes	460
Primary spermatocytes	452
Progesterone	457
Prostate gland	456
Prostatic urethra	455
Raphe	450
Rete testis	454
Scrotum	450
Secondary oocyte	460
Secondary spermatocytes	452
Semen/seminal fluid	456
Seminal vesicles	455
Seminalplasmin	456
Seminiferous tubules	450
Sertoli cells	452
Shaft	456
Spermatic cord	455
Spermatids	452

(continues)

INTRODUCTION

At its most basic level, reproduction is the process by which a single cell duplicates its genetic material. This is the process of mitosis, discussed in Chapter 4. Mitosis allows us to grow and repair damaged or old tissues. In this sense, cellular reproduction enables us to maintain ourselves. However, reproduction is also the process by which our genetic material is passed on from one generation to the next. This process requires a special kind of cellular reproduction that produces special cells, the sperm from the male and the egg from the female. These join in the process of fertilization to produce a fertilized egg, or zygote. The special type of cellular division that produces the sex cells is called meiosis. Meiosis, you will recall from Chapter 4, is a reduction division of the genetic material. This results in an egg carrying 23 chromosomes and the sperm carrying 23. When fertilization occurs, the resulting zygote will possess the full complement of 46 chromosomes.

This chapter discusses the organs of the reproductive system that produce the sex cells, transport and nurture their development. Once an egg is fertilized by a sperm, the resulting zygote will develop into an embryo in the uterus of the female and grow by the process of mitosis into a fetus. The fetus will continue to develop until birth. The purpose of the reproductive system is to produce offspring and ensure the perpetuation of the human species.

THE MALE REPRODUCTIVE SYSTEM

The primary sex organs of the male reproductive system are the **testes** (**TES**-teez) or male gonads. These organs produce sperm and the male sex hormones. There are also accessory organs, like the scrotum, that support the testes. Other accessory structures nurture the developing sperm cells and various ducts store or transport the sperm to the exterior or into the female reproductive tract. Accessory glands add secretions that make up the semen. A transporting and supporting structure is the penis. See Concept Map 19-1: The Male Reproductive System.

The Scrotum

The **scrotum** (**SKROH**-tum) is an outpouching of the abdominal wall. It consists of loose skin and superficial fascia. It is the supporting structure of the testes. Externally, it appears as a single pouch of skin separated into lateral portions by a median ridge known as the **raphe** (**RAY**-fee) (Figure 19-1). Internally, it is separated into two sacs by a septum. Each sac contains a single **testis** (**TES**-tis). The testes (plural) produce the sperm and the male sex hormones. Because sperm and hormone production and survival require a temperature lower than normal body temperature, the scrotum lies outside the body cavity. Its environment is about 3°F below body temperature.

Exposure to cold, as in winter, causes contraction of the smooth muscle fibers, moving the testes closer to the pelvic cavity so they can absorb more body heat. The whole scrotal sac contracts and a muscle, the **cremaster** (kree-**MASS**-ter) **muscle**, located in the spermatic cord, elevates the testes. Exposure to heat, as on a hot summer day, reverses the process, and the scrotal sac hangs well below the pelvic cavity to avoid body warmth.

The Testes

The testes are paired oval glands measuring about 5 cm (2 inches) in length and 2.5 cm (1 inch) in diameter (Figure 19-2). They are covered by a dense layer of white fibrous connective tissue called the **tunica** (**TYOO**-nih-kah) **albuginea** (al-byoo-**JEN**-ee-ah) that extends inward and divides each testis into a number of smaller, internal compartments known as lobules. Each lobule contains one to three tightly coiled tubules called the convoluted **seminiferous** (sem-in-**IF**-er-us) **tubules**. These seminiferous tubules actually produce the sperm by a process called **spermatogenesis** (**spur**-mat-oh-**JEN**-eh-sis) (Figure 19-3).

Spermatogenesis begins in the seminiferous tubules, as the most immature sperm cells called **spermatogonia**

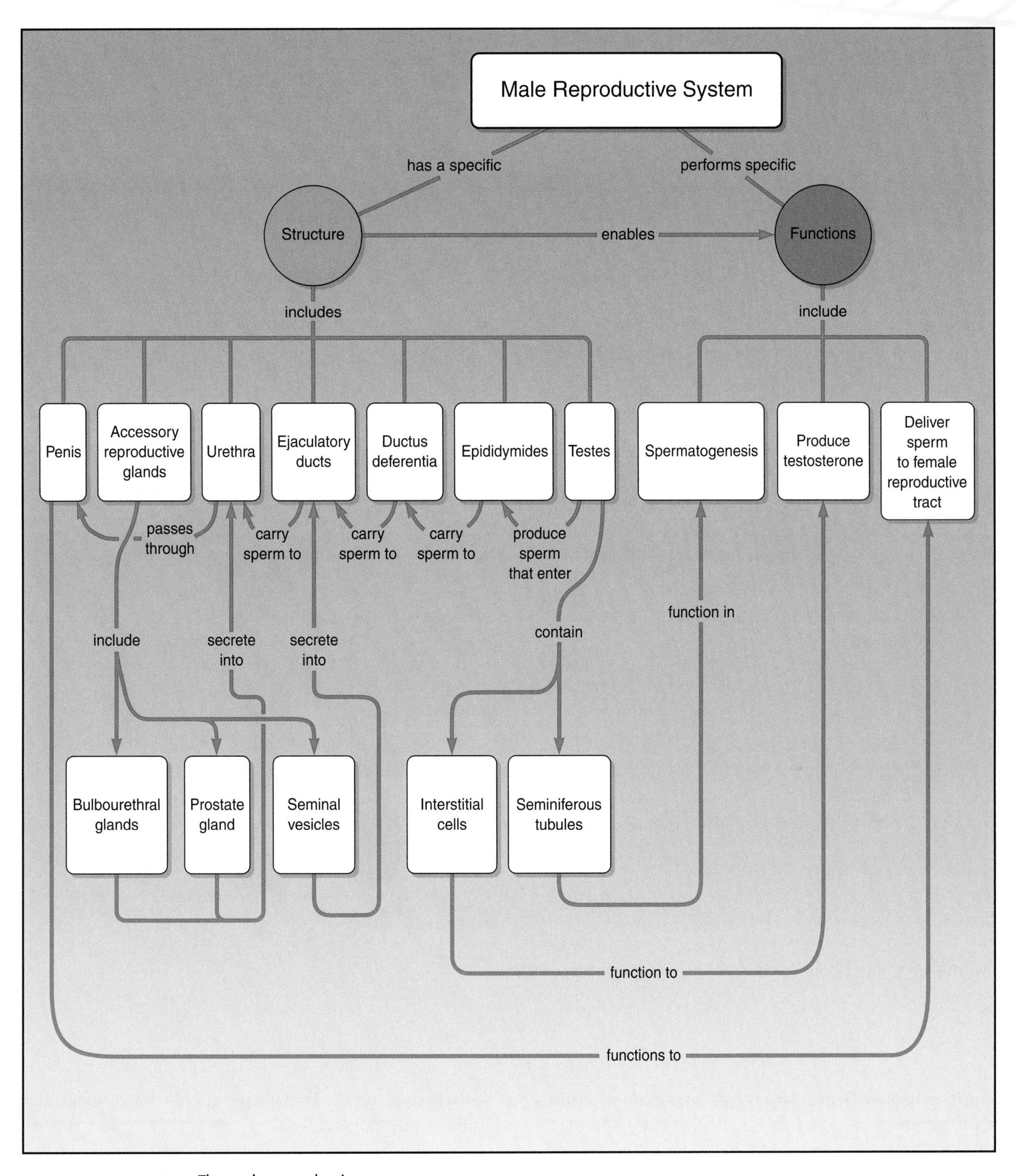

CONCEPT MAP 19-1. The male reproductive system.

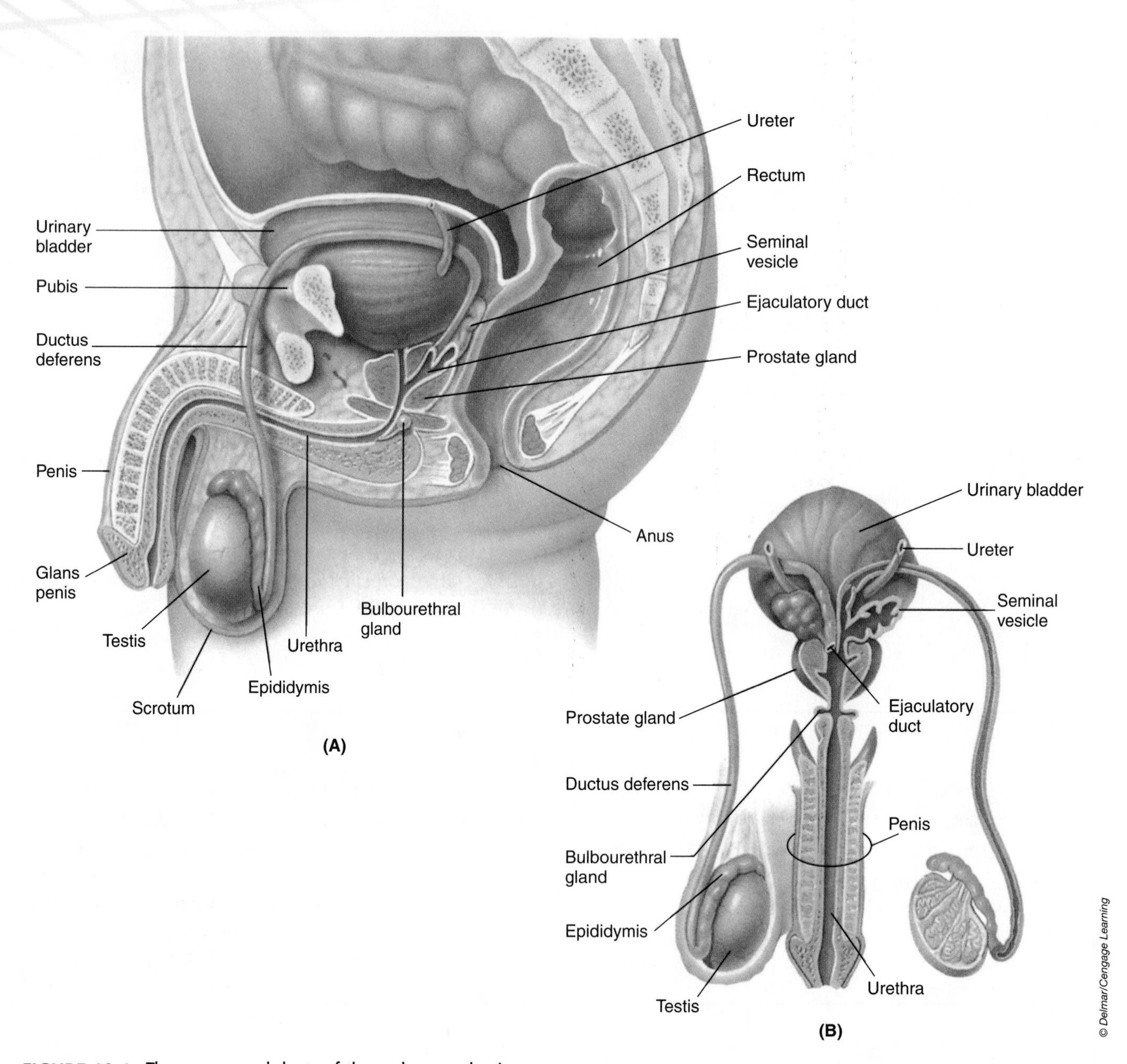

FIGURE 19-1. The organs and ducts of the male reproductive system.

(**spur**-mat-oh-**GO**-nee-ah) begin to divide by mitosis to produce daughter cells called **primary spermatocytes** (**PRYE**-mary spur-**MAT**-oh-sightz). The primary spermatocytes undergo the cellular reduction division called meiosis. After the first meiotic division (review Chapter 4), the primary spermatocytes become **secondary spermatocytes** and their genetic information has been reduced in half (chromosomes are reduced from 46 to 23). These secondary spermatocytes now undergo the second meiotic division and become **spermatids** (**SPUR**-mah-tidz). These spermatids will eventually mature into sperm cells.

Found among the developing sperm cells in the seminiferous tubules are the **Sertoli** (Sir-**TOH**-lee) **cells**. They produce secretions that supply nutrients for the developing sperm cells, or **spermatozoa** (**spur**-mat-oh-**ZOH**-ah). In the lobules of the testes, between the seminiferous tubules in the soft connective tissue are clusters of **interstitial** (in-ter-**STISH**-al) **cells of Leydig** (**SELZ** of **LYE**-dig). The interstitial cells of Leydig produce the male

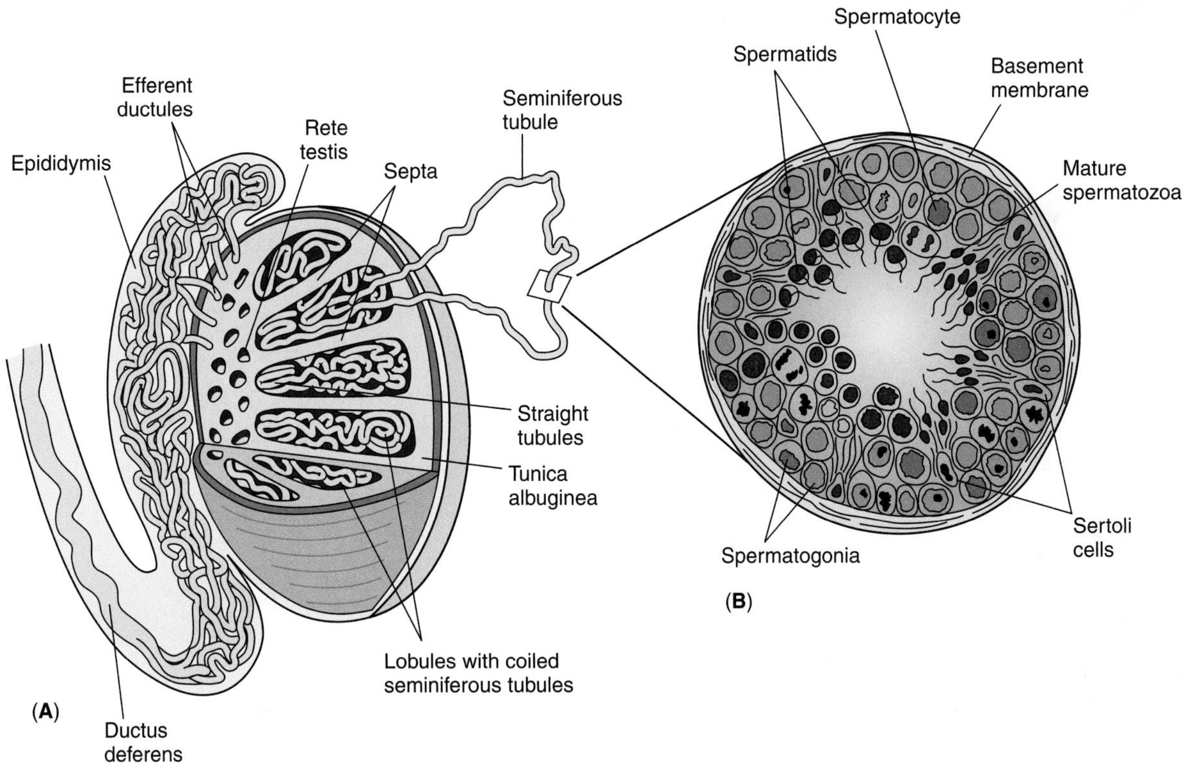

FIGURE 19-2. The anatomy of a testis.

sex hormone testosterone. Thus, we see that two different areas of the testes produce two different products through two different groups of cells. The primary spermatocytes in the seminiferous tubules produce sperm, and the interstitial cells of Leydig in the tissue of the lobules around the tubules produce testosterone. Because the testes produce both sperm and testosterone, they are both exocrine (glands with ducts) and endocrine (without ducts) glands.

The Anatomy of the Spermatozoa

The spermatozoa, or mature sperm cells, are produced at a rate of about 300 million per day. Once ejaculated, they have a life expectancy of about 48 hours in the female reproductive tract. They will not survive very long at all outside the female reproductive tract in the external environment. They are highly adapted for reaching and penetrating a female egg or ovum. Each sperm is composed of a head, a middle piece, and a tail (Figure 19-4).

The head, which developed from the nucleus of a spermatid cell, contains the genetic material and an **acrosome** (ak-roh-**SOHM**). The acrosome contains enzymes that aid the sperm cell in penetrating the covering of the female egg cell or ovum. The rest of the sperm cell develops from the cytoplasm of the spermatid cell. The middle piece or collar contains numerous mitochondria, which produce the high-energy molecule adenosine triphosphate (ATP) that provides the energy for locomotion. The tail of the sperm cell is a typical flagellum. The flagellum beats, from the energy of the ATP molecule, and propels the sperm as it swims its way up the female reproductive tract in search of an ovum.

The Functions of Testosterone

Testosterone (tess-**TOSS**-ter-ohn) has a number of effects on the male body. It controls the development, growth, and maintenance of the male sex organs. Just before birth, it causes the descent of the testes from the abdominal cavity into the scrotal sac. During puberty, it stimulates bone growth, resulting in the development of broad shoulders and narrow hips. It stimulates protein buildup in muscles, producing muscular development, typical of males with more bulk and firmness in their muscular physique. Testosterone stimulates maturation of sperm cells. It causes enlargement of the thyroid cartilage, resulting in the visible Adam's apple

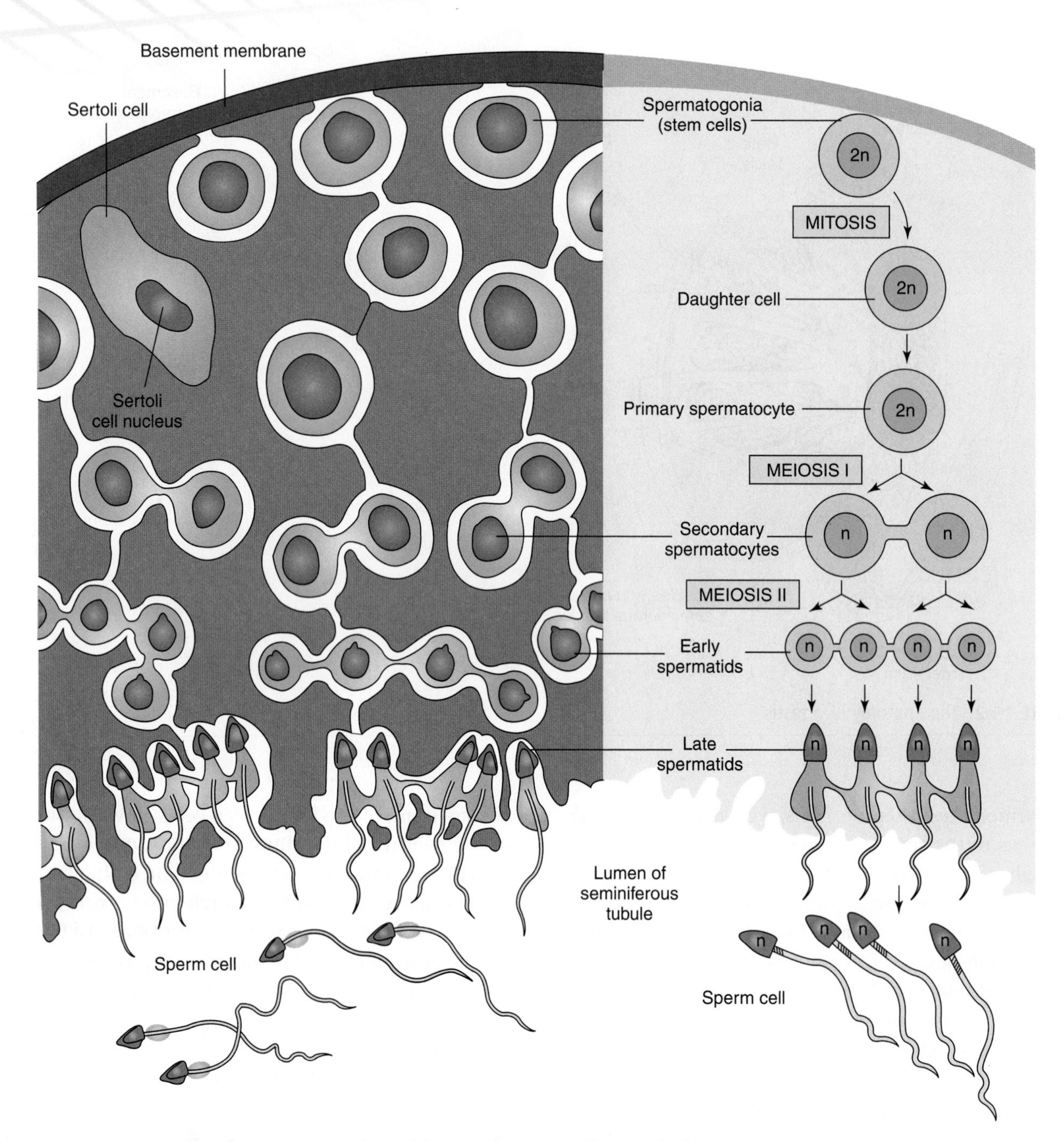

FIGURE 19-3. The process of spermatogenesis, which occurs in a seminiferous tubule.

in males, and the deepening of the voice with its low range of pitch due to thick vocal cords. Other secondary male sex characteristics are also influenced by the production of testosterone. These include aggressive behavior and body hair patterns. Body hair patterns include the development of chest hair and axillary hair within hereditary limits, facial hair, and temporal hairline recession.

The Ducts of the System

As the sperm cells are formed (see Figure 19-2), they are moved from the convoluted seminiferous tubules in the testis to the **straight tubules** at the tip of each lobule. Here the convoluted seminiferous tubules become linear and lose their convolutions. These lead to a network of ducts in the testis called the **rete** (**REE**-tee) **testis**.

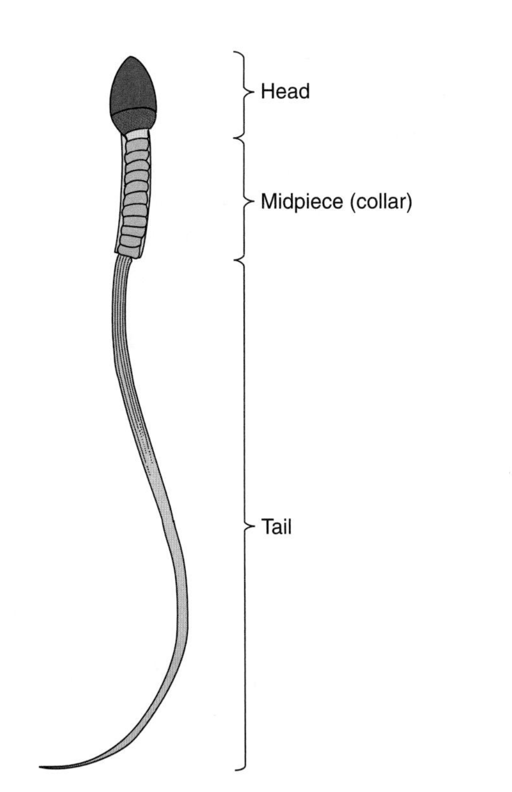

FIGURE 19-4. The anatomy of a sperm cell.

The sperm are now transferred out of the testis through a series of coiled **efferent ducts** that empty into a single tube known as the **ductus epididymis**.

The epididymis is a comma-shaped structure located along the posterior border of a testis, consisting principally of a tightly coiled tube called the ductus epididymis. The epididymis is the site where the sperm cells continue to mature. The tube is about 20 feet long and it takes the sperm about 20 days to move through this tube. The tube stores the maturing sperm cells while they develop their swimming capability via their flagella. The tube propels the sperm toward the next part of the system during ejaculation, when the smooth muscle of the wall of the tube contracts by peristalsis.

The next part of the duct system is the **ductus deferens** (**DUCK**-tus **DEF**-er-enz), or the **vas deferens**. Within the tail of the epididymis, the ductus epididymis becomes less and less convoluted. At this point it is now referred to as the ductus or vas deferens. It is also commonly called the seminal duct. It is approximately 18 inches long and ascends the posterior border of the testis, penetrates the inguinal canal, and enters the pelvic cavity where it loops over the side and down the posterior surface of the urinary bladder (see Figure 19-1). The tube is enclosed in a connective tissue sheath along with nerves and blood vessels. This sheath is called the **spermatic cord**. The end of the ductus or vas deferens has a dilated terminal portion known as the ampulla. Each ductus deferens empties into its ejaculatory duct, the next part of the system. When a **vasectomy** (vas-**EK**-toh-mee) is performed for birth control, the physician performs this minor operation by making an incision into the scrotal sac to cut or cauterize the ductus or vas deferens. The testes still produce sperm but they are now unable to make their way to the outside of the system. They eventually die and are reabsorbed by the body. A man becomes sterile by this procedure, but fluids are still produced and are ejaculated during intercourse. The male libido is not affected, erections and ejaculations still occur, but there are no sperm in the semen. In addition, because testosterone is still being produced, all the secondary male sex characteristics are maintained.

The next part of the tube is the **ejaculatory duct**. Posterior to the urinary bladder, each ductus deferens joins its ejaculatory duct (see Figure 19-1). Each duct is about 1 inch long. The ducts eject spermatozoa into the urethra. The urethra is the terminal duct of the system. It serves as a common passageway for both spermatozoa coming from the testes and urine coming from the bladder.

In the male, the **urethra** (yoo-**REE**-thrah) passes through the prostate gland, the urogenital diaphragm, and the penis. It is about 8 inches long and is subdivided into three parts: the **prostatic urethra**, which is surrounded by the prostate gland and is about 1 inch long; the **membranous urethra** about 1/2 inch long, which runs from the prostatic urethra to the penis; and the **spongy** or **cavernous urethra**, found within the penis and about 6 inches long, but varies according to the size of the penis. The spongy urethra enters the head or bulb of the penis and terminates at the male urethral orifice.

The Accessory Glands

The accessory glands include the two seminal vesicles, the prostate gland, and the paired bulbourethral glands. These glands secrete the liquid portion of the semen, the sperm-containing fluid that is produced during ejaculation.

The paired **seminal vesicles** (**SEM**-ih-nal **VESS**-ih-kulz) are convoluted pouchlike structures approximately 2 inches in length. They are located posterior to and at the base of the urinary bladder in front of the rectum (see Figure 19-1). They produce the alkaline, viscous component of semen that is rich in the sugar fructose and other nutrients for the sperm cells and pass it into the ejaculatory duct. They produce about 60% of the volume of semen. Because the duct of each seminal vesicle joins the ductus deferens on each side to form the ejaculatory duct, sperm and seminal fluid together enter the urethra during ejaculation.

The **prostate gland** is a single, doughnut-shaped gland about the size and shape of a chestnut. It surrounds the superior portion (the prostatic urethra) of the urethra just below the bladder. It also secretes an alkaline fluid that makes up about 13% to 33% of semen. Its fluid plays a role in activating the sperm cells to swim. The fluid enters the prostatic urethra through several small ducts. The prostate gland is located anterior to the rectum and a physician can feel its size and texture by digital examination through the anterior wall of the rectum (to check for an enlarged prostate, which could indicate cancer).

The paired **bulbourethral** (**BUL**-boh-yoo-**REE**-thral) glands, also known as **Cowper's glands**, are about the size of peas. They are located beneath the prostate gland on either side of the membranous urethra. They secrete thick, viscous, alkaline mucus. Their ducts connect with the spongy urethra. This secretion is the first to move down the urethra when a man becomes sexually aroused and develops an erection. It functions as both a lubricant for sexual intercourse and as an agent to clean the urethra of any traces of acidic urine.

Semen

Semen, also known as **seminal fluid**, is a mixture of sperm cells and the secretions of the seminal vesicles, the prostate, and the bulbourethral glands. The fluid is milky in color and sticky, due to the fructose sugar that provides the energy for the beating flagellum of each sperm cell. The semen is alkaline, with a pH of 7.2 to 7.6. This neutralizes the acidity of the female vagina and the male urethra and helps protect the sperm cell. The semen provides a transport medium for the swimming sperm cells.

The average volume of semen per ejaculation is 2.5 to 6 mL, and the average range of spermatozoa ejaculated is 50 to 100 million/mL. If the number of spermatozoa falls below 20 million/mL the man is considered to be sterile. Semen contains enzymes that activate sperm after ejaculation. The semen also contains an antibiotic called **seminalplasmin** (**SEM**-ih-null-**PLAZ**-min), which has the capability of destroying certain bacteria. Because the female reproductive tract and the semen contain bacteria, the seminalplasmin helps keep these bacteria under control and thus helps protect the sperm and ensure fertilization.

*Study*WARE™ Connection

Watch an animation on sperm formation on your StudyWARE™ CD-ROM.

The Penis

The **penis** (**PEE**-nis) is used to introduce or deliver spermatozoa into the female reproductive tract by being inserted into the vagina. The penis consists of a **shaft** whose distal end is a slightly enlarged region called the **glans penis** or head, which means "shaped like an acorn." Covering the glans penis is a section of loose skin called

COMMON DISEASE, DISORDER, OR CONDITION

CONDITIONS OF THE PROSTATE GLAND

ENLARGED PROSTATE

The prostate gland enlarges in almost every older adult man. If this occurs, it constricts the urethra, making urination difficult. This condition can lead to kidney damage and bladder infections. A physician can palpate the prostate gland through the rectal wall to detect any changes in the size of the gland. It should be checked regularly as a man ages. Enlarged prostate glands can be treated surgically. The prostate gland can also hypertrophy or enlarge due to infections or tumors.

PROSTATE CANCER

Cancer of the prostate gland is the third most prevalent type of cancer in men. It is a slow-growing cancer that is not easily detected. However, if undetected, it can result in death. Normal and regular checkups for the detection of enlargement or any abnormal changes in the gland by a physician are an important procedure used to detect prostate cancer.

the **prepuce** (**PRE**-pyoose) or **foreskin**. Occasionally, the foreskin is removed at birth by a surgical procedure called circumcision. This will be done if the foreskin does not pull back completely over the glans penis, resulting in future hygiene problems. If the skin is loose and pulls back, circumcision is not necessary. It is also done as a rite in certain religions.

Internally, the penis is composed of three cylindrical masses of spongy tissue containing blood sinuses (see Figure 19-1). During sexual stimulation, the arteries that supply the penis dilate, and large quantities of blood enter these blood sinuses. Expansion of these sinuses compresses the veins that would normally drain the penis so that most of this entry blood is retained. These changes in blood vessels produce an **erection**, which helps the penis penetrate the female vagina. Once sexual stimulation ceases, the arteries supplying the blood constrict and the veins drain the blood, resulting in the penis going limp and the end of the erection.

During ejaculation, the smooth muscle sphincter at the base of the urinary bladder is closed. This ensures that urine is not expelled during ejaculation and that semen does not enter the urinary bladder.

*Study***WARE**™ Connection

- Watch an animation about the male reproductive system.
- Play an interactive game labeling the structures of the male reproductive system on your StudyWARE™ CD-ROM.

THE FEMALE REPRODUCTIVE SYSTEM

The primary sex organs of the female reproductive system are the **ovaries** (**OH**-vah-reez), or female gonads.These organs produce eggs, or **ova**, as exocrine glands and as endocrine glands produce the female sex hormones **estrogen** (**ESS**-troh-jen) and **progesterone** (proh-**JESS**-ter-ohn). The accessory organs of the system are the uterine or fallopian tubes, the uterus, the vagina, and the external genitalia. Some accessory glands also produce mucus for lubrication during sexual intercourse. See Concept Map 19-2: The Female Reproductive System, and Figure 19-5. The female system is more complex hormonally than the male system because it must also nurture the developing fetus during pregnancy.

The Ovaries

The ovaries or female gonads are paired glands about the size of unshelled almonds. They are found in the upper pelvic cavity, one on each side of the uterus (Figure 19-6). They are held in position by a series of ligaments. Suspensory ligaments secure the ovaries to the lateral walls of the pelvis. Ovarian ligaments anchor the ovaries medially. In between, they are held in place and enclosed by the broad ligament, which is a fold of peritoneum. A microscopic view of an ovary reveals that each one consists of a number of parts (Figure 19-7).

The surface of an ovary is covered with **germinal epithelium**. The capsule of an ovary consists of collagenous connective tissue known as the tunica albuginea. This is divided into an outer area called the cortex of the stroma and contains ovarian follicles in various stages of development and an inner area called the stroma of the medulla. **Ovarian follicles** are eggs or ova and their surrounding tissues in various stages of development. Each follicle contains an immature egg or **oocyte** (**OH**-oh-sight) and at this stage is referred to as a primary follicle. As the developing egg begins to mature, the follicle increases in size and develops a fluid-filled central region called the antrum. At this stage the follicle is called a secondary follicle as it begins to develop the fluid. A mature follicle with a mature egg is called a **graafian** (**GRAF**-ee-an) **follicle**. This is basically an endocrine gland that begins to secrete estrogen and is ready to eject the mature egg, an event known as **ovulation**. After the egg ruptures from the mature graafian follicle, the follicle changes into the **corpus luteum** (**KOR**-pus **LOO**-tee-um), or yellow body, which produces estrogen and progesterone and eventually degenerates to become the **corpus albicans** (**KOR**-pus **AL**-bih-konz), or white body.

The function of the ovaries is to produce eggs or ova, discharge the ova in ovulation, and secrete the female sex hormones estrogen and progesterone. The rest of the female system consists of ducts that transport and nurture the egg and, if fertilization occurs, delivers the fetus

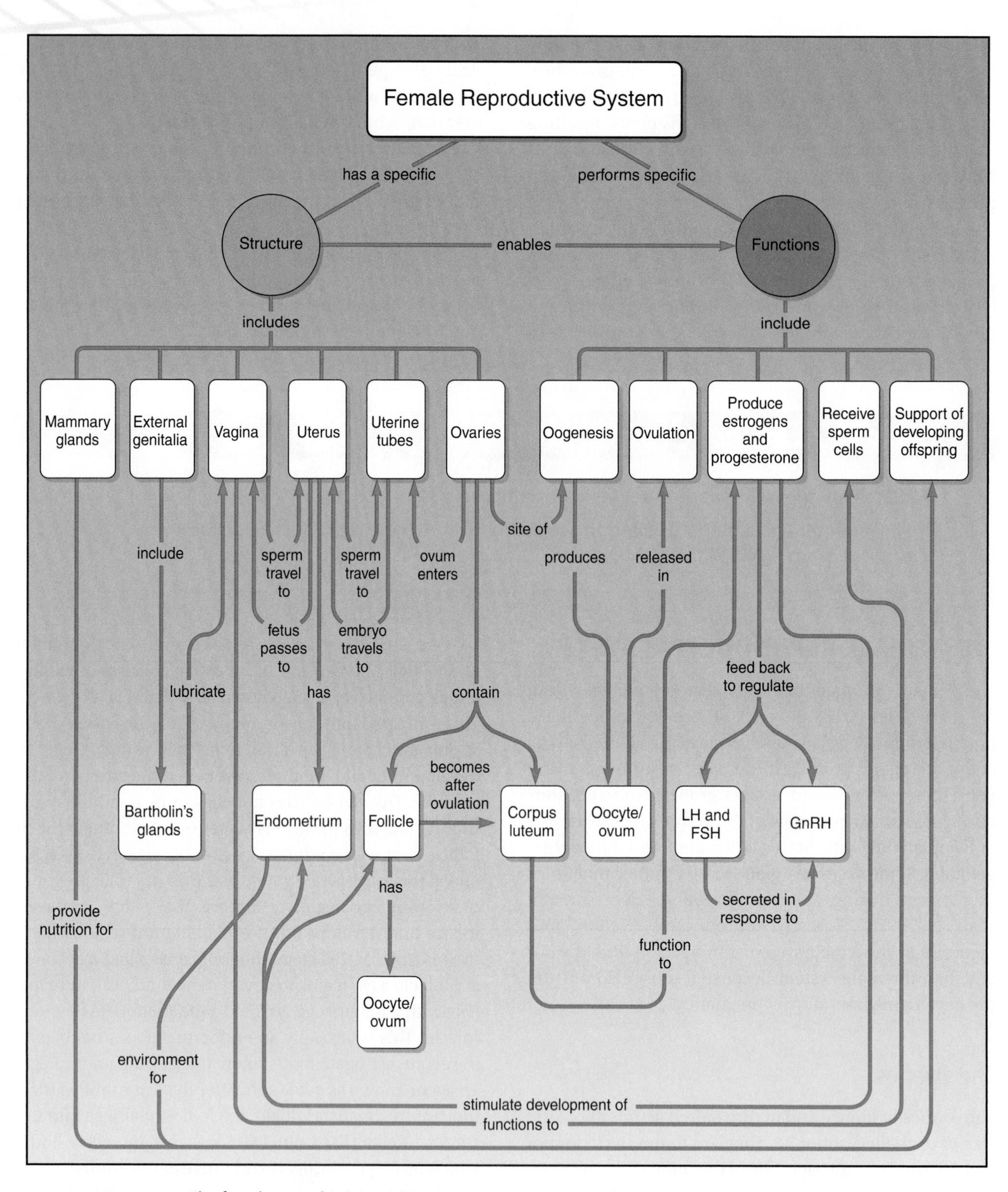

CONCEPT MAP 19-2. The female reproductive system.

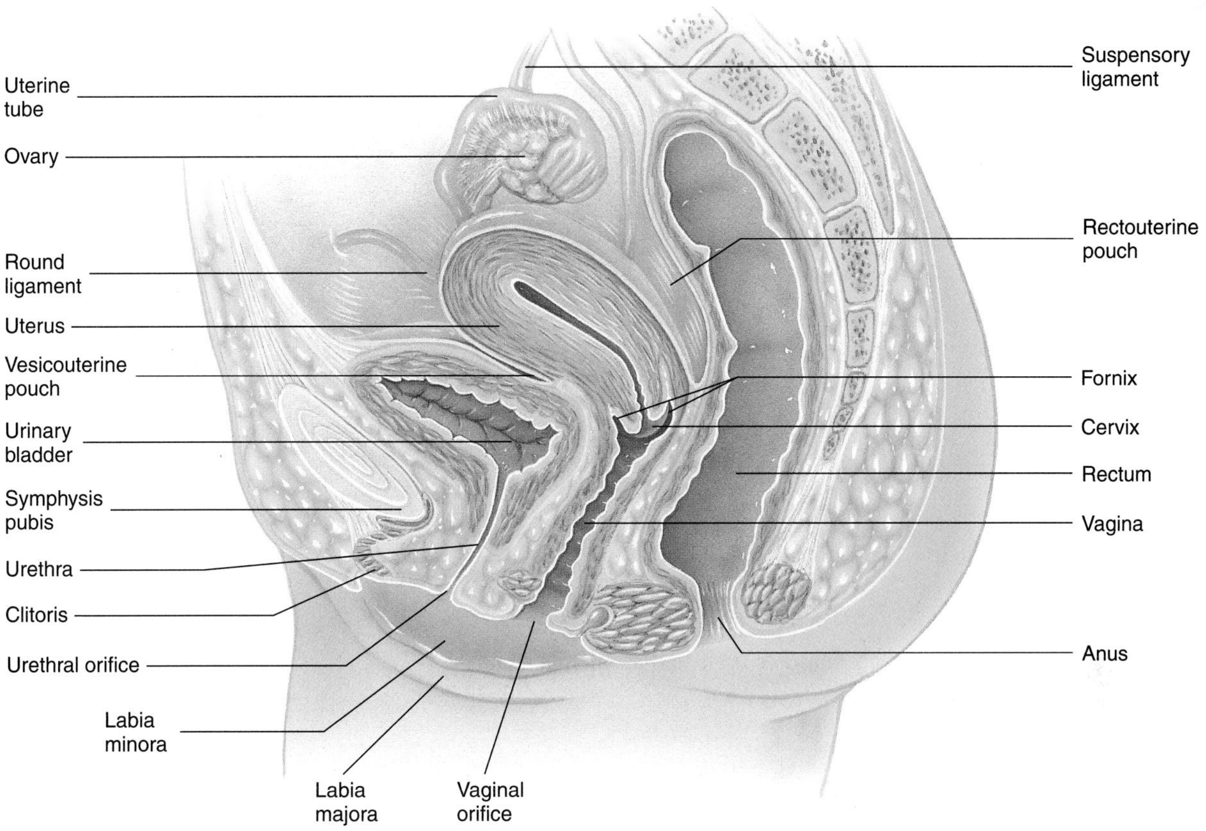

FIGURE 19-5. The organs of the female reproductive system.

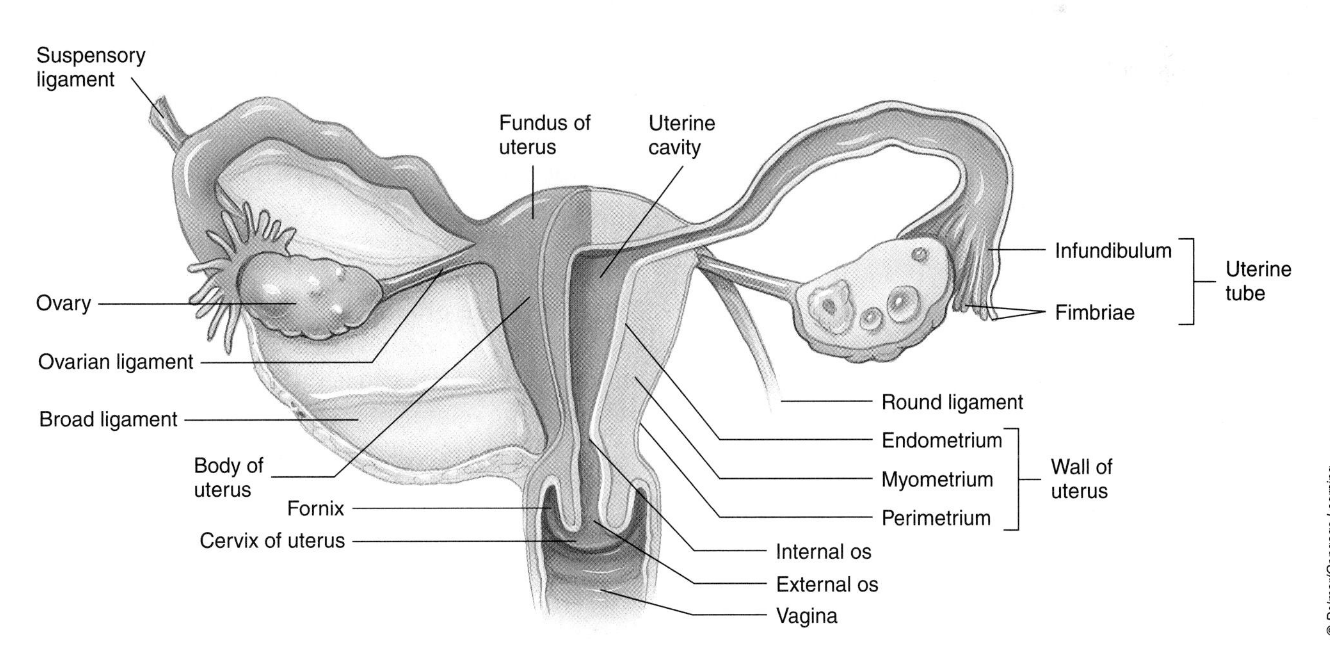

FIGURE 19-6. The position of the ovaries, uterine tubes, uterus, and vagina of the female reproductive system.

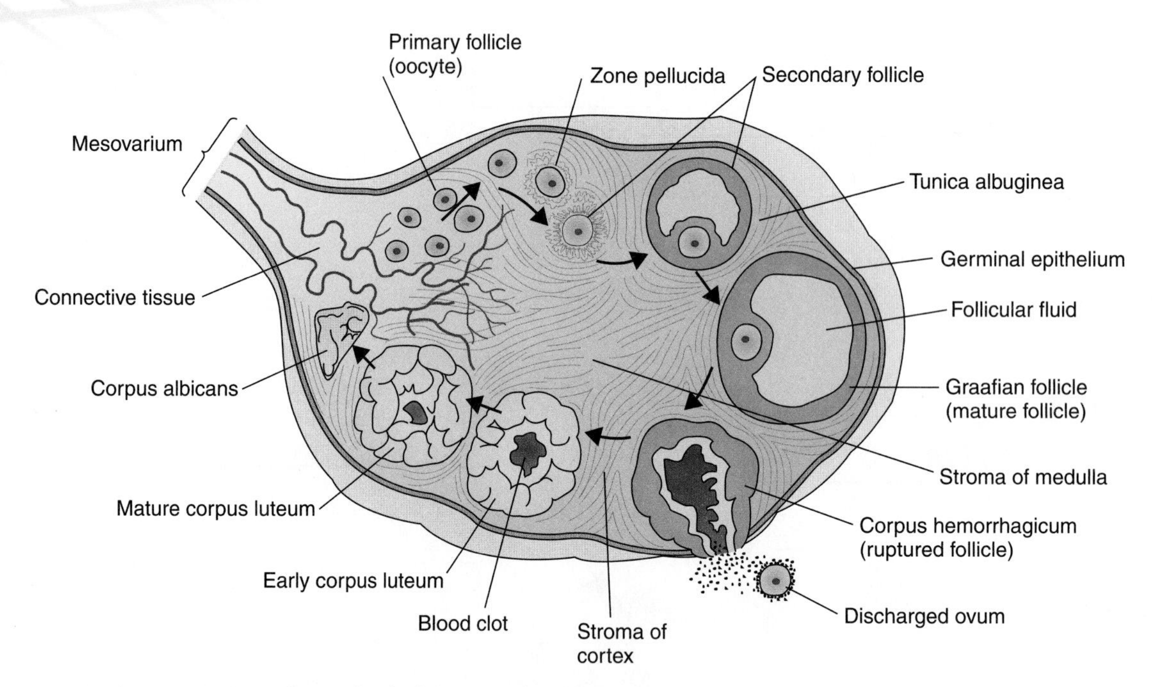

FIGURE 19-7. A microscopic view of an individual ovary and its internal anatomy.

to the outside world during birth. The duct system consists of the uterine or fallopian tubes, the uterus, and the vagina.

Oogenesis (oh-oh-**JEN**-eh-sis) or formation of the female sex cells or eggs occurs in the ovaries (Figure 19-8). In males the process is called spermatogenesis and occurs in the seminiferous convoluted tubules of the testes. In males, the process begins in the fetus, continues at puberty, and men can produce sperm throughout their lives. In women, the process begins in the fetus, continues at puberty, but ends at menopause around the age of 50. In addition, the total number of eggs that a woman can produce and release is determined at birth.

In a developing female fetus, the female stem cells called **oogonia** (oh-oh-**GO**-nee-ah) divide by mitosis to produce **primary oocytes** (**PRYE**-mary **oh**-oh-sightz). They become surrounded by follicular cells in the ovary and are now part of the primary follicles. Around 700,000 are produced at this time and represent the total number of eggs that a female could produce during her reproductive years. They now await further development until puberty.

At puberty, the **ovarian cycle** is stimulated when the anterior pituitary gland secretes follicle-stimulating hormone (FSH). Only a small number of primary follicles will grow and develop, with only one egg being released each month. Only about 450 eggs will be produced from the store of 700,000 primary oocytes through the process of meiosis.

After the first meiotic division the primary oocyte becomes two cells. The **secondary oocyte** is the larger of the two cells; a very small cell called the **polar body** is nonfunctional. During the second meiotic division, which occurs only after fertilization, the secondary oocyte and the polar body divide again and the secondary oocyte becomes an **ootid** (**OH**-oh-tid), or mature egg cell, and due to unequal division of the cytoplasm another polar body is formed. The first polar body from the first meiotic division again divides in the second meiotic division to become two nonfunctional polar bodies. After the second meiotic division, one functional egg cell is produced and three nonfunctional polar bodies. This is very different from meiosis in men, in whom four functional sperm cells are produced. However, the sperm cells are very tiny with few stored nutrients and will not survive very long because they must get their nutrients from the seminal fluid. The egg cell, however, has lots of stored food because it is a large cell and can supply the developing embryo until it embeds itself in the endometrial lining of the uterus.

The Uterine or Fallopian Tubes

The female body contains two **uterine** (**YOO**-ter-in) or **fallopian** (fah-**LOH**-pee-an) **tubes** that transport the ova from the ovaries to the uterus (see Figure 19-6). There is a funnel-shaped open end to each tube called the **infundibulum**

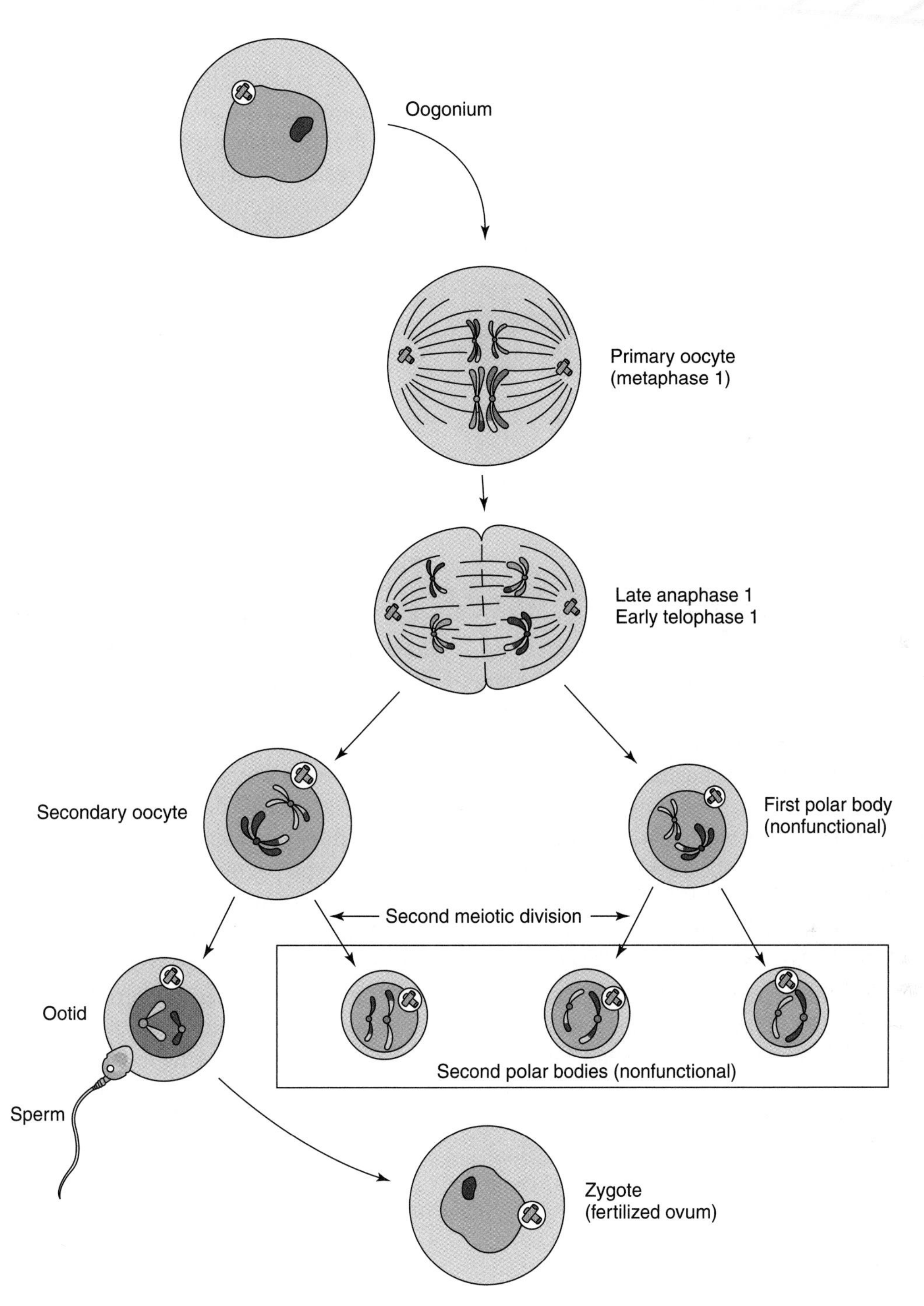

FIGURE 19-8. Oogenesis in the ovary.

(**in**-fun-**DIB**-yoo-lum). This lies close to an ovary but does not attach to it. The infundibulum is surrounded by a fringe of finger-like projections called the **fimbriae** (**FIM**-bree-ee) that partially surround an ovary. Approximately once a month an ovum ruptures from the surface of an ovary near the infundibulum of the uterine tube, a process called ovulation.

The ovum is swept by ciliary action of the epithelium of the infundibulum and by the waving fimbriae, which create a current that carries the egg into the uterine tube. The ovum is then moved along the uterine tube by the action of the cilia and by wavelike peristaltic contractions of the smooth muscle of the uterine tube. If the egg is fertilized, it usually occurs in the upper third of the uterine tube. This means that the sperm must swim up through the vagina, into the uterus, and then up two-thirds of the uterine tube. Fertilization may occur at any time up to about 24 hours following ovulation. If fertilized, the ovum will make its journey down the uterine tube and enter the uterus within 7 days.

StudyWARE™ Connection

Watch an animation about ovulation on your StudyWARE™ CD-ROM.

The Uterus

The **uterus** (**YOO**-ter-us) or womb is located in the pelvic cavity between the rectum and the urinary bladder (see Figure 19-5). It is held in position by a series of ligaments, and it is the site of menstruation. The uterus is where the fertilized egg is implanted, where the fetus develops during pregnancy, and where labor begins during delivery. It is shaped like an inverted pear and can greatly increase in size during pregnancy to accommodate the developing fetus. It will extend well above the navel or umbilicus in the late stages of pregnancy.

Its anatomic divisions include the dome-shaped portion above the uterine tubes called the **fundus** (**FUN**-dus) (see Figure 19-6). Its major portion is the central tapering region known as the **body of the uterus**. The narrow inferior portion that opens into the vagina is called the **cervix**. Between the body and the cervix is a small constricted region called the **isthmus** (**ISS**-mus). The interior of the body of the uterus is known as the **uterine cavity**; the interior of the narrow cervix is known as the **cervical canal**. The junction of the uterine cavity with the cervical canal is called the **internal os** and the opening of the cervix into the vagina is called the **external os**.

The wall of the uterus is made of three layers of tissue. The innermost layer is the **endometrium** (endoh-**MEE**-tree-um). This mucosal layer is where the fertilized egg burrows into the uterus, a process called implantation. The middle layer of the uterus is called the **myometrium** (my-oh-**MEE**-tree-um), which consists of smooth muscle important during delivery to move the child out of the womb. The outermost layer is the **perimetrium** (pair-ih-**MEE**-tree-um) made of serous membrane and also known as the visceral peritoneum.When a woman is not pregnant, the endometrial lining of the uterus is shed approximately every 28 days in the process called menstruation.

The Menstrual Cycle

The **menstrual** (**MEN**-stroo-al) **cycle**, also known as the **menses** (**MEN**-seez) or **menstruation** (men-stroo-**AY**-shun), is the cyclical shedding of the lining of the uterus in response to changes in hormonal levels. The cycle varies from woman to woman within a range of 24 to 35 days. To discuss the events occurring during the cycle, we will assume an average duration of 28 days. Events occurring during the cycle can be divided into three phases: the menstrual phase, the preovulatory or proliferative phase, and the postovulatory or secretory phase (Figure 19-9).

The *menstrual phase*. This phase is also known as menstruation or menses. It lasts from day 1 to 5. During this time, the thick endometrial lining of the uterus is shed along with tissue fluid, blood, mucus, and epithelial cells. Bleeding during this period can last from 3 to 5 days. The detached tissues and blood exit through the vagina as the menstrual flow.

During this phase, the ovarian cycle is also in operation. The ovarian follicles, known as primary follicles, begin their development. During the early phase of each

COMMON DISEASE, DISORDER, OR CONDITION

CERVICAL CANCER

Cancer of the cervix is a slow-growing cancer that is common in women who are between the ages of 30 and 50. Conditions that can lead to the development of this type of cancer include frequent sexual intercourse with multiple sex partners, sexually transmitted diseases (e.g., gonorrhea and syphilis), frequent cervical inflammation, and multiple pregnancies. Yearly Pap smear tests are significant factors in early detection and treatment of this type of cancer.

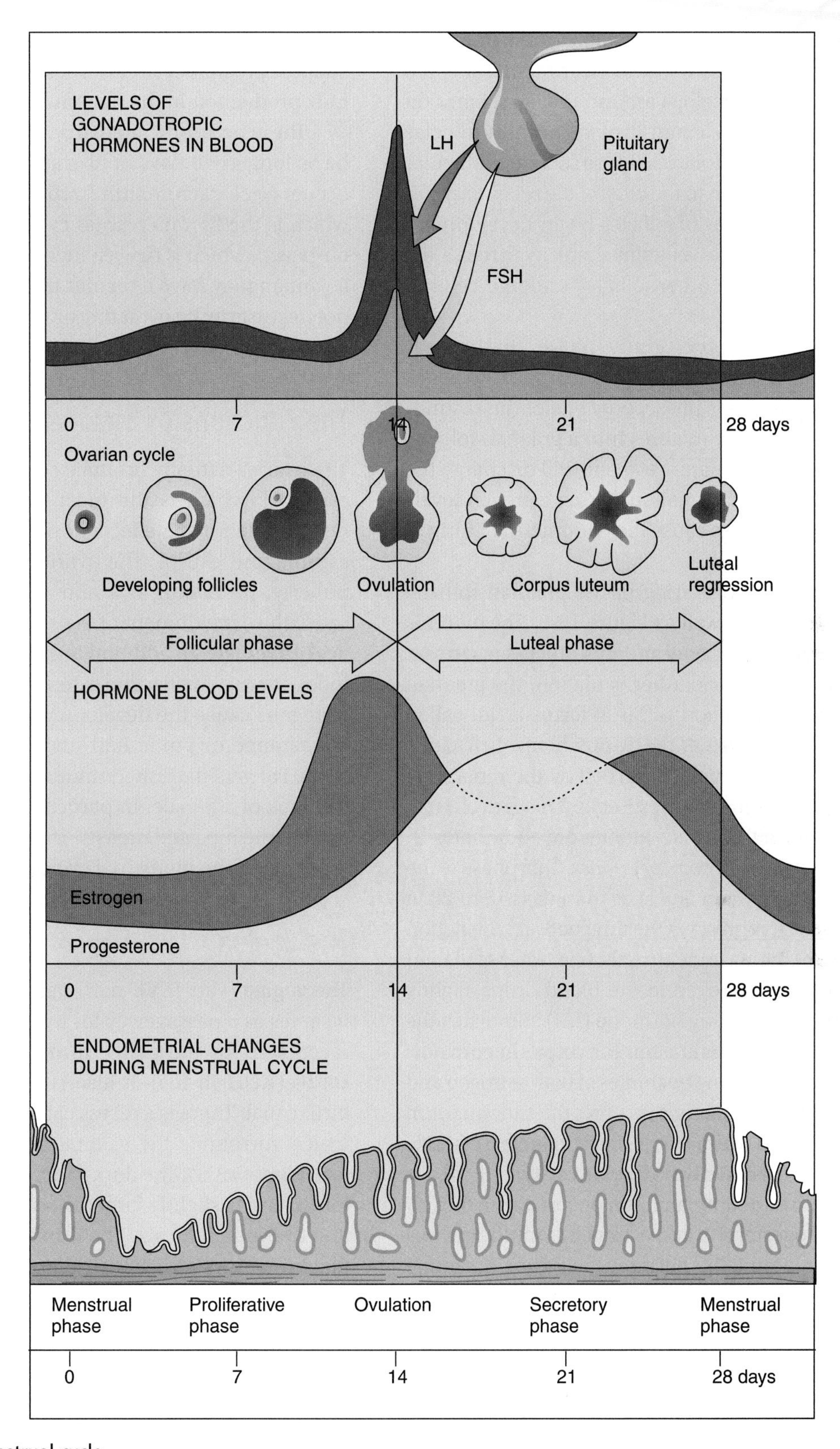

FIGURE 19-9. The menstrual cycle.

menstrual cycle, 20 to 25 primary follicles begin to produce very low levels of estrogen. A clear membrane, the zona pellucida, also develops around the eggs. Later on in the phase at day 4 to 5, about 20 of the primary follicles develop into secondary follicles. These secrete a follicular fluid that forces the ovum to the edge of the secondary follicle. Although a number of follicles begin development during each cycle, only one attains maturity through the process of meiosis. The other follicles undergo cellular death or atresia.

The *preovulatory* or *proliferative* phase. This phase is more variable in length. It will last from day 6 to 14 in our 28-day cycle. During this phase, only one of the secondary follicles in the ovary matures into a graafian follicle. This follicle contains a mature egg and will discharge the egg in a process called ovulation. Rising estrogen levels produced by the follicles cause the endometrial lining to thicken during this phase.

Ovulation is the rupturing of the graafian follicle. Refer to Figure 19-7 as well as Figure 19-9. The ovum is released into the pelvic cavity and this process occurs on day 14 in our 28-day cycle. After ovulation, the graafian follicle collapses and blood within it forms a clot called the **corpus hemorrhagicum** (**KOR**-pus hem-oh-**RAJ**-ih-kum). This clot is eventually absorbed by the remaining follicular cells. Eventually these cells enlarge, change structure, and form the corpus luteum or yellow body.

The *postovulatory* or *secretory phase.* This phase is the most constant in duration and lasts from days 15 to 28 in our 28-day cycle. It represents the time between ovulation and the onset of the next menstrual cycle. After ovulation occurs, the level of estrogen in the blood drops slightly and secretion of luteinizing hormone (LH) stimulates the development of the corpus luteum. The corpus luteum now begins to secrete increasing quantities of both estrogen and progesterone. The progesterone prepares the endometrium to receive a fertilized ovum by causing it to increase in size and to secrete nutrients into the uterine cavity.

If fertilization and implantation do not occur, the rising levels of progesterone and estrogen from the corpus luteum inhibit luteinizing hormone-releasing hormone (LHRH) from the hypothalamus and LH from the anterior pituitary gland. As a result, the corpus luteum degenerates and becomes the corpus albicans. This will initiate another menstrual cycle.

If fertilization and implantation do occur, the corpus luteum will be maintained for about 4 months. During this time, it continues to secrete estrogen and progesterone. The corpus luteum is maintained by human chorionic gonadotropin, a hormone produced by the developing placenta. Once the placenta is developed, it will secrete estrogen to support pregnancy and progesterone to both support pregnancy and to cause breast development for milk production in the mammary glands.

The length of a menstrual cycle is variable. It can be as long as 40 days or as short as 21 days. It normally occurs once each month from **menarche** (men**AR**-kee), which is the first menstrual cycle, to **menopause** (**MEN**-oh-pawz), which is the last menstrual cycle. Even though a woman may have a regular menustrual cycle, she may not necessarily be ovulating or releasing an egg. This may lead to problems with fertility and conception.

The Functions of Estrogen

The ovaries actually produce several types of estrogens: estradiol, which is the most abundant and is mainly responsible for the effects of estrogen on the body, and estrone and estriol. The ovaries become active during puberty, producing ova and estrogens. The estrogens cause the development of the secondary sex characteristics of a female. In addition to enlargement of the uterine tubes, uterus, vagina, and external genitalia of the female, estrogens cause the development of the breasts and the appearance of pubic hair and axillary hair under the arms. Fat gets deposited under the skin, resulting in the soft look of a female. In particular, more fat is deposited around the hips and breasts. The pelvic bone widens and the onset of the menstrual cycle begins.

The Vagina

The **vagina** (vah-**JEYE**-nah) has a number of functions. It serves as a passageway for the menstrual flow. It is the receptacle for the penis during sexual intercourse, or **coitus** (**KOH**-ih-tus). It also is the lower portion of the birth canal. There is a recess called the **fornix** (**FOR**-niks), which surrounds the vaginal attachment to the cervix (see Figure 19-6). The dorsal recess is called the posterior fornix and is slightly larger than the two lateral fornices and the ventral fornix. The fornix can accommodate the placing of a contraceptive diaphragm, which prevents sperm from entering the uterus.

The External Genitalia of the Female

The reproductive structures located external to the vagina are referred to as the external genitalia of the female. The term **vulva** (**VULL**-vah), or **pudendum** (pyoo-**DEN**-dum), is a collective term for these structures. They include the mons pubis, labia majora, labia minora, clitoris, urethral and vaginal openings, and the vestibular glands (Figure 19-10).

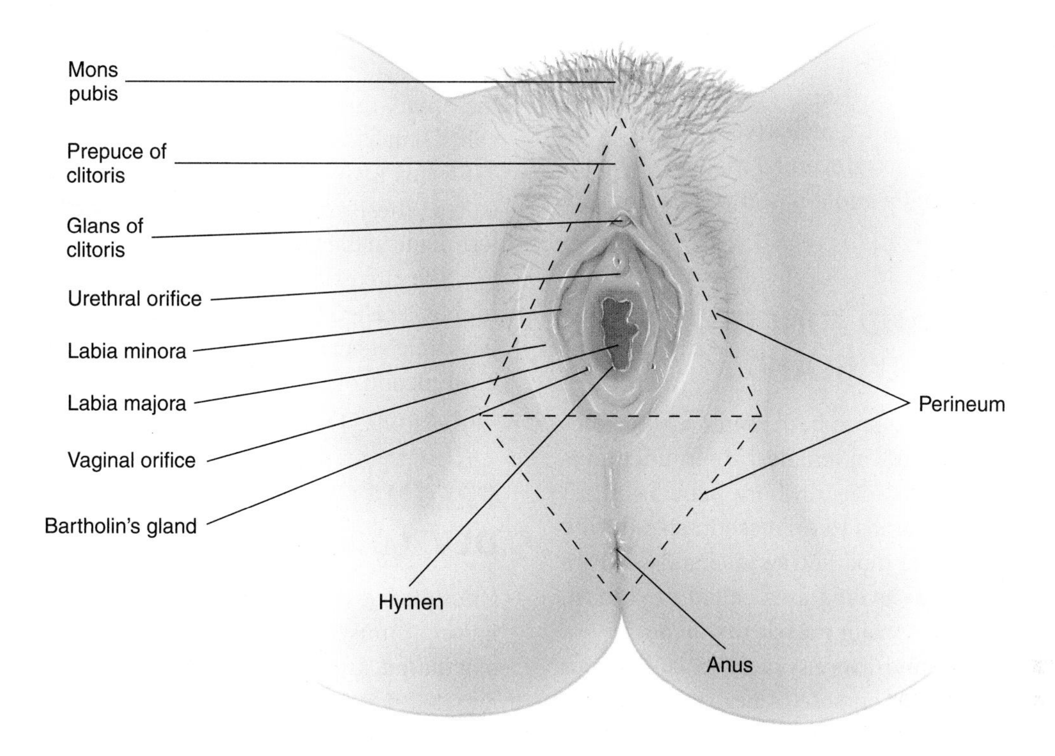

FIGURE 19-10. The female external genitalia or vulva.

The **mons pubis**, also called the **veneris** (veh-**NEER**-is), is a mound of elevated adipose tissue that becomes covered with pubic hair at puberty. It is situated directly over the pubic symphysis. From the mons pubis, extending posteriorly and inferiorly are two longitudinal folds of hair-covered skin called the **labia majora** (**LAY**-bee-ah mah-**JOR**-ah). This is the female homologue to the male scrotum. These two folds of skin contain an abundance of adipose tissue and sweat glands. Medial to the labia majora are two other delicate folds of skin called the **labia minora** (**LAY**-bee-ah mih-**NOR**-ah). The labia minora do not have hair and have just a few sweat glands, but possess numerous sebaceous glands.

The **clitoris** (**KLIT**-oh-ris) is a small, cylindrical mass of erectile tissue with nerves found at the anterior junction of the labia minora. There is a layer of skin called the prepuce or foreskin formed at the point where the two labia minora join and cover the body of the clitoris. The exposed portion of the clitoris is the **glans**. The clitoris is the female homologue to the male penis. It, too, is capable of enlargement by becoming swollen with blood during sexual stimulation and excites the female. Unlike the penis, it does not have an internal duct.

The opening or region between the two labia minora is called the **vestibule**. Within the vestibule is a thin fold of tissue called the **hymen** (**HIGH**-men), which partially closes the distal end of the vagina. This fold of mucosa is highly vascularized and bleeds when ruptured during the first sexual intercourse. It occasionally is torn during a sports activity or on insertion of a tampon. Also located in the vestibule are the **vaginal orifice** and the **urethral orifice** plus the openings of several ducts coming from the vestibular glands.

Posterior to and on either side of the urethral orifice are the two openings of the ducts of the **lesser vestibular** or **Skene's glands**. These glands are homologous to the male prostate gland and they secrete mucus. On both sides of the vaginal orifice are the openings of two small glands called the **greater vestibular** or **Bartholin's glands**. These glands are homologous to the male's Cowper's glands and also secrete mucus. The mucus secreted by these vestibular glands lubricates the distal end of the vagina during sexual intercourse.

*Study*WARE™ Connection

Play an interactive game labeling structures of the female reproductive system on your StudyWARE™ CD-ROM.

THE PERINEUM

The **perineum** (pair-ih-**NEE**-um) is a diamond-shaped area at the inferior end of the trunk between the buttocks and thighs of both males and females (see Figure 19-10). It is divided into an anterior urogenital triangle that contains the external genitalia and a posterior anal triangle that contains the anus.

THE ANATOMY AND FUNCTION OF THE MAMMARY GLANDS

Mammary glands are present in both males and females but normally function only in females. Their function is to produce milk to nourish the newborn baby. Estrogen causes the mammary glands to increase in size during puberty. These glands are modified sweat glands and are located in a round skin-covered area called the breast, anterior to the pectoralis major muscle of the thorax.

Each mammary gland consists of 15 to 20 lobes or compartments separated by adipose tissue. It is the amount of adipose tissue present in the breast that determines the size of the breast (Figure 19-11). In each lobe are several smaller compartments known as lobules, which contain the milk-secreting cells called **alveoli** (al-**VEE**-oh-lye). These alveolar glands are arranged in grapelike clusters. They convey the milk into a series of secondary tubules. From here the milk passes into the mammary ducts. As the mammary ducts approach the nipple, expanded sinuses called **ampullae** (am-**PULL**-ee) or **lactiferous sinuses** (lak-**TIF**-er-us **SIGH**-nuh-sez) are found where milk may be stored. These ampullae continue as **lactiferous ducts** that terminate in the **nipple** (**NIP**-l).

The circular pigmented area of skin surrounding the nipple is called the **areola** (ah-**REE**-oh-lah). It looks and feels rough because it contains modified sebaceous glands. The function of the mammary glands is to secrete and eject milk, a process known as **lactation** (lak-**TAY**-shun).

PREGNANCY AND EMBRYONIC DEVELOPMENT

Once the egg cell or ovum ruptures from the ovary in ovulation, it must be fertilized within 12 to 24 hours. Once ejaculated, the sperm cell remains viable within the female reproductive tract for 12 to 48 hours. Some sperm can remain viable for up to 72 hours. For fertilization to occur, sexual intercourse must occur no more than 72 hours before ovulation or no later than 24 hours after ovulation. It takes the ovum 24 hours to go approximately

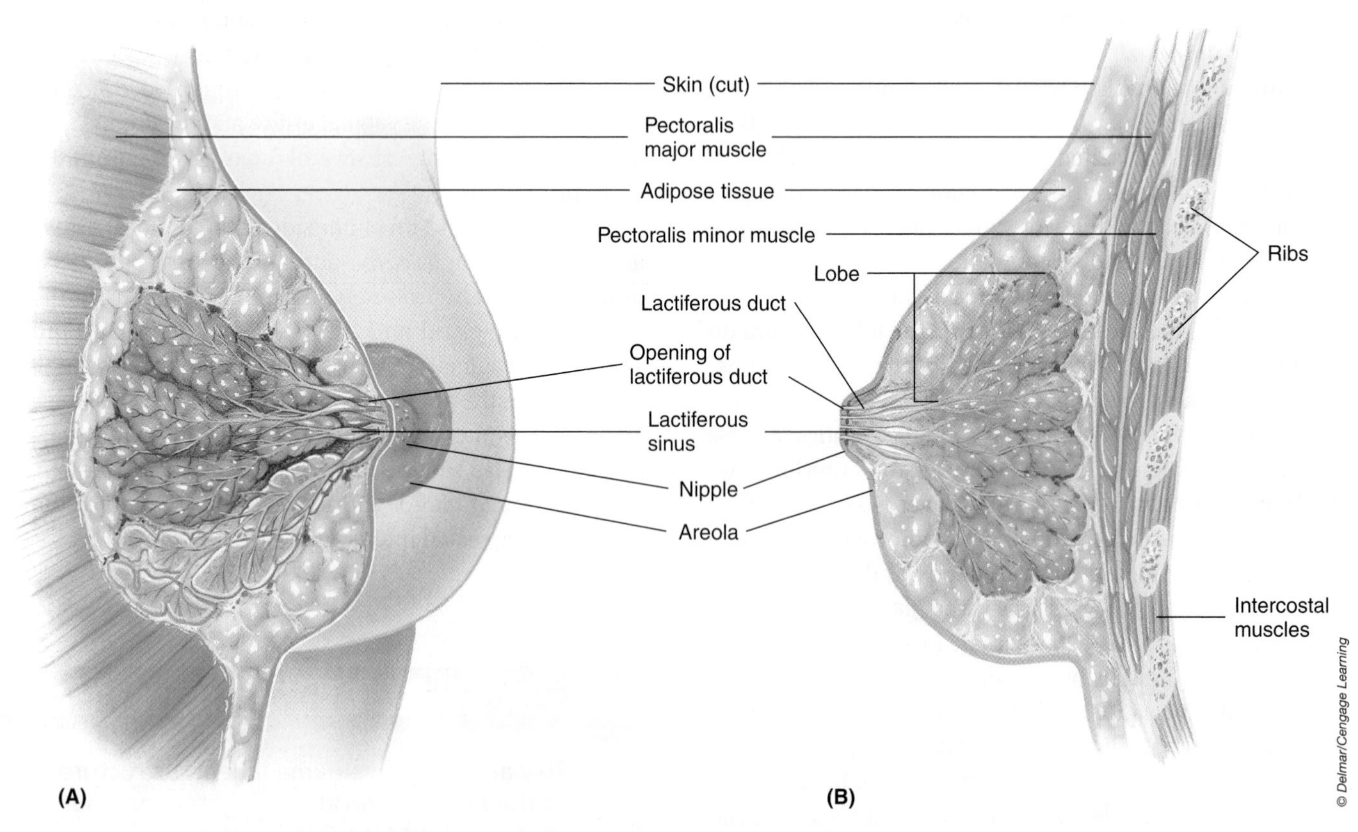

FIGURE 19-11. The mammary glands of a lactating breast. (A) Anterior view and (B) sagittal view.

HEALTH ALERT

PREVENTION OF SEXUALLY TRANSMITTED DISEASES

The one absolute way to prevent the transmission of sexually transmitted diseases (referred to as STDs), like syphilis and gonorrhea, is through abstinence. HIV, which causes AIDS, can also be transmitted sexually. However, sexually active individuals can help prevent the occurrence of STDs by practicing what is called "safe sex." Safe sex practices include: always using a latex condom (HIV can pass through other types of condoms, such as animal skin condoms); avoiding the exchange of bodily fluids like blood, semen, or vaginal secretions; and avoiding contact with genital or oral sores (which may contain the genital herpes virus). Good personal hygiene and yearly medical checkups, along with self-awareness and examination of the genital areas, will help prevent the spread of disease-causing microorganisms. Any abnormalities, like sores or discharges from the reproductive tract, should be immediately brought to the attention of a physician.

one-third down the uterine or fallopian tube. If fertilization is going to take place, it will occur in the upper two-thirds of the uterine tube.

The sperm cells swim by means of their beating flagella up the vagina and uterus and into the uterine tube. They are attracted to the egg cell by chemicals secreted by the ovum. Although hundreds of sperm cells are surrounding the ovum and rupturing their acrosomes to release enzymes to penetrate the egg cell, only one sperm will penetrate the egg and join its genetic material with the genetic material of the egg to produce a fertilized egg or **zygote** (**ZYE**-gote).

As the zygote moves down the uterine tube, it undergoes a series of rapid mitotic divisions, resulting in a hollow ball-like mass of cells called a **blastula** (**BLASS**-tyoo-lah) or **blastocyst**. By the time it reaches the uterine cavity, it consists of about 100 cells. At this stage, part of the blastocyst develops into the **chorionic vesicle** (**KOH**-ree-on-ik), and it secretes human chorionic gonadotropin, a hormone that causes the corpus luteum of the ovary to continue producing its hormones to maintain the lining of the uterus.

By the seventh day following ovulation, the developing embryo has embedded itself in the endometrial lining of the uterus. Meanwhile, the three primary germ layers are being formed by mitotic divisions. The **ectoderm** will form skin and the nervous system, the **endoderm** will form the linings of internal organs and glands and the **mesoderm** will form muscles, bone, and the rest of the body tissues. The blastocyst's inner cell mass forms these primary germ layers and its trophoblast, the large fluid-filled sphere, now begins to form projections called **chorionic villi**, which will interact with the uterine tissues to form the **placenta** (plah-**SEN**-tah) (Figure 19-12).

Once the placenta is formed, the embryo, which looks like a three-layered plate of cells, becomes surrounded by a fluid-filled sac called the **amnion** (**AM**-nee-on). The embryo is attached by a connecting stalk of tissue called the **umbilical cord**. The placenta delivers nutrients and oxygen to and removes wastes and carbon dioxide from the embryonic blood. All exchanges with the mother are made through the placenta. By the ninth week of development, the embryo is now called a **fetus**.

By the ninth week of development the embryo looks definitely human, the placenta has become an endocrine organ secreting estrogen and progesterone to maintain pregnancy, and the corpus luteum is now inactive. Later on in development, the umbilical cord will become the structure that will allow the exchange of nutrients and wastes between the mother and the fetus.

As pregnancy progresses, the uterus enlarges to accommodate the developing fetus. It eventually pushes up into the abdominal cavity and occupies most of this area. The abdominal organs push against the diaphragm muscle, causing the ribs to expand and the thorax to widen. During this time the center of gravity of the mother moves, resulting in an accentuated curvature of the lumbar vertebrae called lordosis, which may cause backaches. At this time, it is essential for the mother to practice good nutrition, eating high-quality food, not just more food. The mother should also avoid any harmful substances that could pass through the placenta into the fetal blood such as alcohol, drugs, and nicotine.

Childbirth is called **parturition** (par-tyoo-**RISH**-un). The fetus is expelled from the uterus through a process called **labor**. The hormone oxytocin causes contraction of the smooth muscles of the uterus. At this time the placenta releases prostaglandins. The combination of these hormones produces more powerful and more frequent

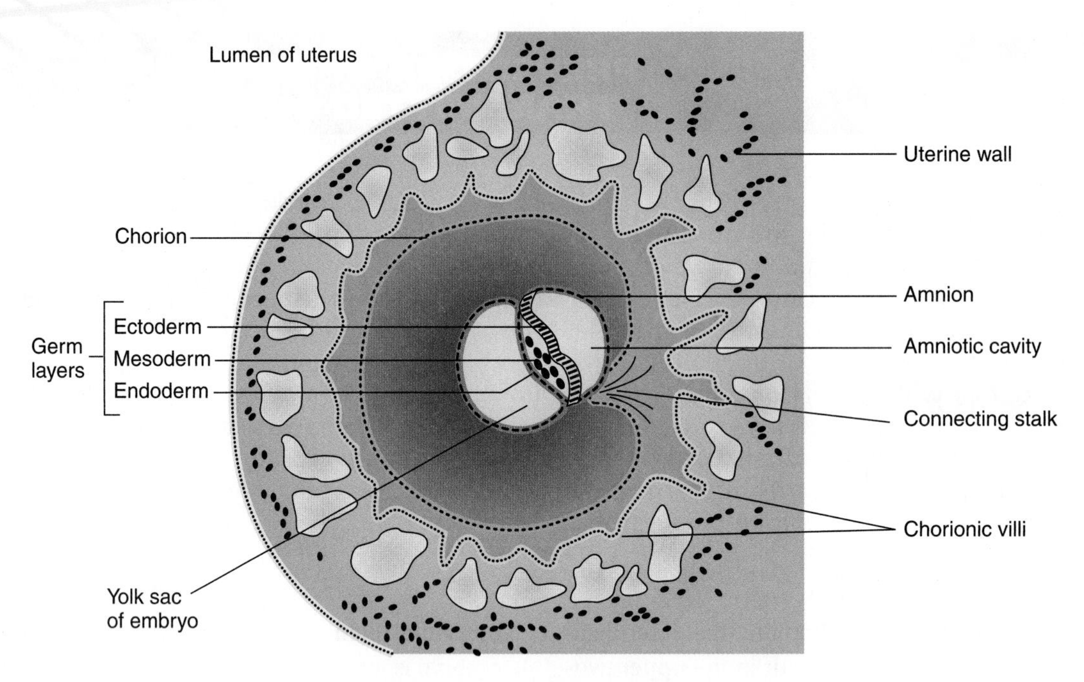

FIGURE 19-12. The early developmental stages of the embryo showing germ layers, chorionic villi, and other structures.

COMMON DISEASE, DISORDER, OR CONDITION

BREAST CANCER

Breast cancer, which occurs most often in females, is a leading cause of death in women. However, men can also get breast cancer. One in every eight women will develop this serious, often fatal disease. Symptoms of this disease include lumps, leakage, or puckering of the nipple and changes in skin texture of the breast. Early detection is crucial to survive this cancer. Women should perform self-examinations on a regular basis. The American Cancer Society recommends that women have mammograms done every 2 years between 40 to 49 years old and once a year thereafter.

Mammography uses low intensity x-rays to detect tumors in the soft tissues of the breast. This procedure detects tumors less than 1 cm that are too small to be detected by self-examination. If a tumor is detected, a biopsy is performed to determine whether it is malignant or benign. Fortunately, most tumors of the mammary glands are benign. However, those that are malignant can spread to other parts of the body and eventually lead to death. Yet early detection of breast cancer can lead to effective treatment and survival.

contractions of the uterus, forcing the fetus out of the uterus. The stages of labor include the dilation stage, the expulsion stage, and the placental stage (Figure 19-13).

During the dilation stage, the cervix of the uterus is fully dilated by the head of the fetus. The amnion ruptures releasing the amniotic fluid. This is commonly referred to as the water breaking. The dilation stage is the longest stage of labor, lasting up to 12 hours. During the expulsion stage, the child moves through the cervix and vagina to the outside world. This stage usually lasts about 50 minutes in the first birth to about 20 minutes in future births. Usually, the head of the child emerges first, and the nose and mouth are cleared of mucus so the child can breathe. The umbilical cord is cut and clamped after the rest of the body of the child emerges. A breech birth is one in which the buttocks emerge first and delivery is more difficult. During the placental stage, the placenta detaches from the uterus within 15 minutes after birth. This placenta and its attached fetal membranes are called the afterbirth. The removal of all placental material will prevent prolonged bleeding after delivery.

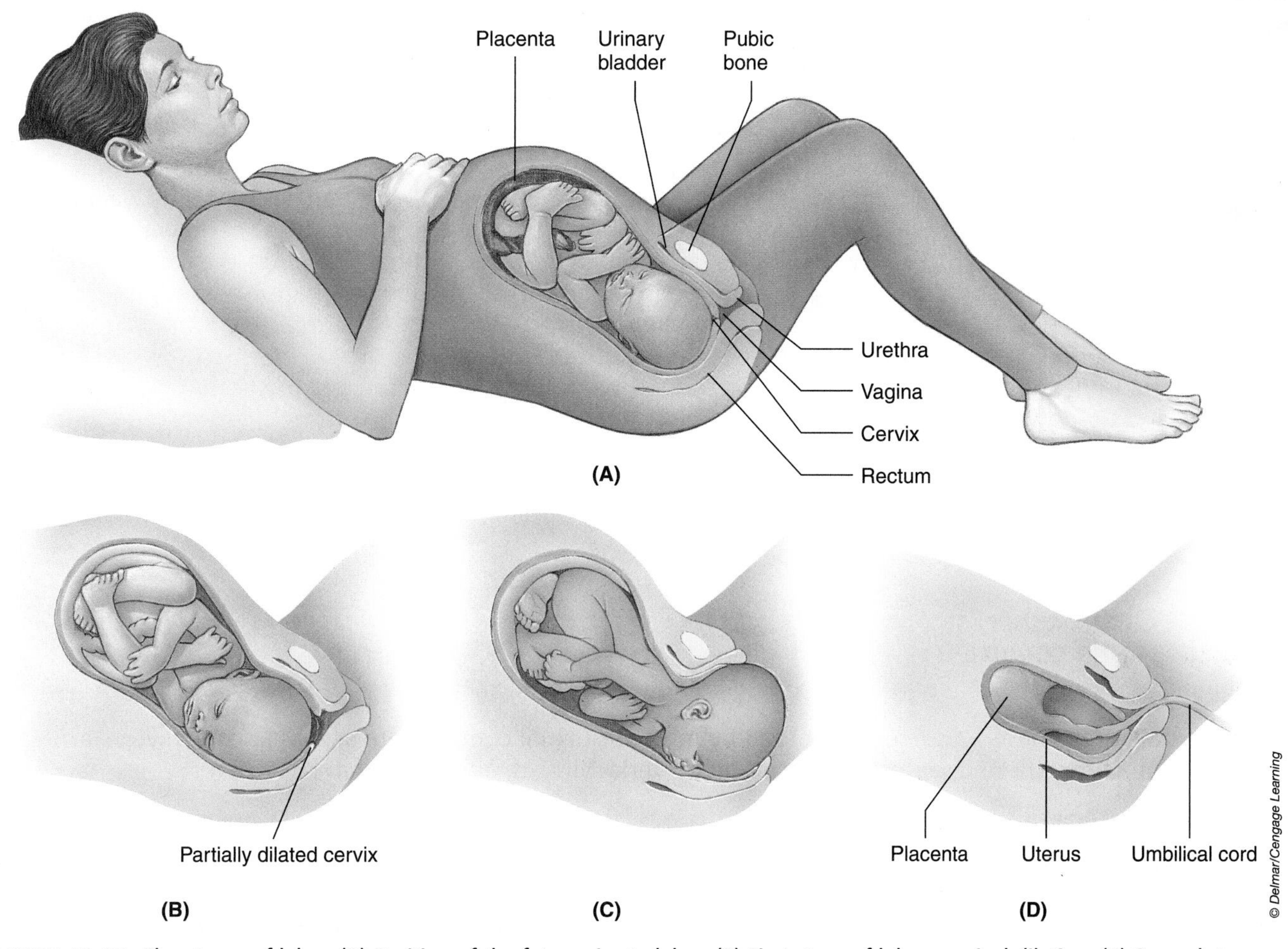

FIGURE 19-13. The stages of labor. (A) Position of the fetus prior to labor. (B) First stage of labor, cervical dilation. (C) Second stage of labor, fetal delivery. (D) Third stage of labor, placental delivery.

HEALTH ALERT

GENETIC INHERITANCE

Sperm and egg cells are produced by the process of meiosis. Review Chapter 4 where this is discussed. Humans have 23 pairs of chromosomes, one member of each pair inherited from the father through the sperm cell, and one member of each pair from the mother through the egg cell. Meiosis reduces the number of chromosomes in half, so the sperm cell has 23 chromosomes and the egg cell has 23 chromosomes. When fertilization of the egg by the sperm occurs, the number of chromosomes is restored to 46 (23 pairs). Genes are located on these chromosomes.

After Watson and Crick published the structure of the DNA molecule, the human genome project was started to identify all the genes on the 46 human chromosomes. It took almost 40 years to complete this study. We now know that there are about 30,000 genes on those 46 chromosomes. We inherit our genes from our parents at the moment of conception, when the sperm from our father fertilizes our mother's egg. These genes determine all of our characteristics, such as eye and hair color, personality, and immunities.

(continues)

HEALTH ALERT

GENETIC INHERITANCE (continued)

Of the 23 pairs of chromosomes, 1 pair are the sex chromosomes that, among other things, determines our sex. Females have two X chromosomes (XX) and males have one X and one Y (XY). Thus, a female's eggs all carry an X chromosome, while the males sperm will have 50% with an X and 50% with a Y. Therefore, it is the father who determines the sex of the child. If a Y-bearing sperm fertilizes an X egg, the child will be a male. If an X-bearing sperm fertilizes an X egg, the child will be a female. There are many other genes, such as the recessive genes for hemophilia and color blindness, that are found on the X chromosome. Male children express the gene and have the trait, while females are carriers of these conditions and do not express the trait. ■

COMMON DISEASE, DISORDER, OR CONDITION

DISORDERS OF THE REPRODUCTIVE SYSTEM

Sexually transmitted diseases are spread by intimate sexual contact with sexual partners who harbor the infection. These diseases are caused by bacteria, viruses, and protozoa.

TRICHOMONAS

Trichomonas is a flagellate protozoan that is more commonly found in women, where it erodes the tissues of the vagina (Figure 19-14). It can be transmitted to men, where it infects the urethra. The acidity of the vaginal tract can control this organism. If, however, it grows and reproduces, it results in inflammation of vaginal tissue with an odorous yellow-green discharge.

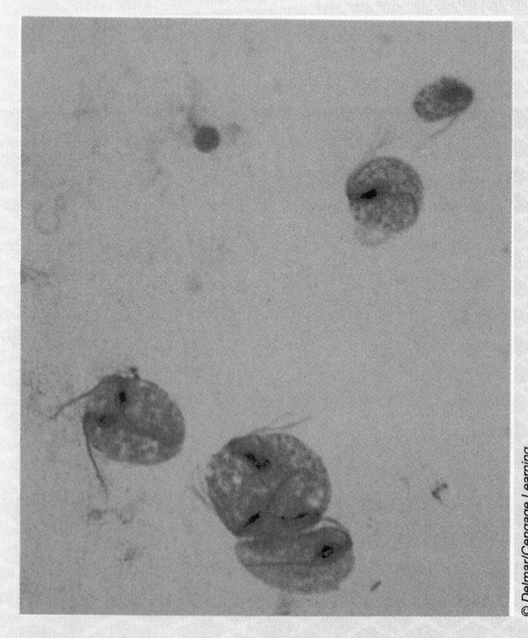

FIGURE 19-14. *Trichomonas vaginalis.*

(continues)

COMMON DISEASE, DISORDER, OR CONDITION

DISORDERS OF THE REPRODUCTIVE SYSTEM (continued)

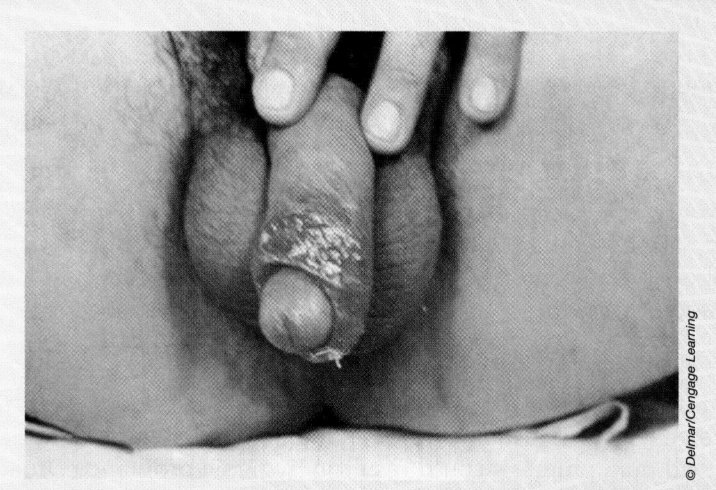

FIGURE 19-15. Syphilis chancre on the foreskin of the penis.

GONORRHEA

Gonorrhea is caused by the bacterium *Neisseria gonorrhoeae*. The bacterium invades the epithelial lining of the vagina and the male urethra, resulting in a discharge of pus. Men experience burning sensations during urination and notice a periodic discharge of pus from the urethral orifice. Women may not notice any symptoms in the early stages of infection, but if untreated it can lead to pelvic inflammatory disease. Treatment is very successful with antibiotics. Symptoms occur within a week after exposure.

SYPHILIS

Syphilis is caused by the bacterium *Treponema pallidum* (Figure 19-15). It can have an incubation period anywhere from a few weeks to several months. There are several stages to this dangerous disease. The primary stage results in a sore called a **chancre** (**SHANG**-ker), which develops at the site of infection. The sore may eventually disappear if untreated. Several weeks later during the secondary stage, fever and skin rashes will appear and last for several weeks. The disease then enters a latent period with no symptoms. The tertiary stage develops many years later. This stage produces neural lesions, resulting in extensive damage to nervous tissue, causing paralysis, insanity, and eventually death. King Henry VIII of England died from a syphilis infection. Antibiotics are used today to treat syphilis, although a number of very resistant strains of the bacterium have evolved.

GENITAL HERPES

Genital herpes is an infection with simplex herpes II virus, which causes lesions and blister-like eruptions of the skin. Contact with the infected areas can be very painful and produces painful urination. It takes about 2 weeks for the blisters to heal but they can recur. The virus becomes dormant in the infected tissues but can become active and produce lesions during periods of stress, such as illness or during female menstruation. There is no cure for this condition. Infection occurs through contact with infected individuals.

GENITAL WARTS

Genital warts are caused by a virus and are highly contagious. The warts can vary from small growths to large clusters. They are usually not painful but can result in painful intercourse and can bleed. There are treatments for genital warts, including surgery or the use of antiviral topical medications.

(*continues*)

COMMON DISEASE, DISORDER, OR CONDITION

DISORDERS OF THE REPRODUCTIVE SYSTEM (continued)

AIDS

AIDS or acquired immune deficiency syndrome can be a sexually transmitted disease. It is discussed in Chapter 15.

PELVIC INFLAMMATORY DISEASE

Pelvic inflammatory disease is a bacterial infection of the uterus, uterine tubes, and/or ovaries. It can result from infection with a number of bacteria but mainly infections with *Chlamydia* or gonorrhea will produce vaginal discharges and pelvic pain. Antibiotics can treat the disease. If not treated, it can lead to sterility and even death.

PHIMOSIS

Phimosis is the condition where the foreskin of the penis fits too tightly over the head of the penis and cannot be retracted. This is usually treated at birth by circumcision. Severe phimosis could obstruct urine flow, and mild phimosis could result in hygienic problems. Organic matter and debris can accumulate under the foreskin, creating an environment suitable for the growth of bacteria or other microorganisms resulting in infections. In a circumcision, the foreskin is cut along the base of the glans and removed.

PREMENSTRUAL SYNDROME (PMS)

Premenstrual syndrome (PMS) is a series of symptoms that develop in many women during the premenstrual phase of the menstrual cycle. Women become irritable, tire easily, become highly nervous, and feel depressed. The cause of PMS is not known so treatment is aimed at relieving the symptoms. The symptoms can become so severe that they can affect relationships and disrupt family life.

ENDOMETRIOSIS

Endometriosis (en-doh-**MEE**-tree-**OH**-**SIS**) occurs when endometrial tissue is found growing outside the lining of the uterus in such places as the abdominal wall, on the surface of the ovaries, the urinary bladder, the kidneys, the outer surface of the uterus, or the sigmoid colon. It usually is caused by endometrial cells passing through the fallopian tubes from the uterus and entering the pelvic cavity where they successfully establish themselves on these organs. Since endometrial tissue is sensitive to the hormones estrogen and progesterone, the tissue first proliferates and grows, but then it degenerates with resulting bleeding. Such a cycle is a cause of abdominal pain, inflammation, and tissue scarring. Individuals with this condition experience unusually severe premenstrual and menstrual pain. This condition is also often a source of infertility.

ERECTILE DYSFUNCTION (ED) OR IMPOTENCE

Erectile dysfunction or **impotence** (**IM**-poh-tens) is the inability of a male to maintain an erection. It also can be classified as the inability to ejaculate after having achieved an erection. It can be caused by possessing physically defective genitalia, certain diseases like Type II diabetes, neuromuscular dysfunctions, or various psychological issues. In addition, impotence can develop from stress, fatigue, advanced age, and the use of certain drugs. Currently the use of Viagra can enhance the maintenance of an erection by causing dilation of blood vessels. However, a side effect can be the overworking of the heart and lowering of blood pressure. Therefore, individuals with certain heart conditions are advised not to use this product.

OVARIAN CANCER

Ovarian cancer is a malignant growth in the ovaries that is usually not detected until it has well advanced. It usually develops in women in their 50s. Some symptoms include abdominal pain and

(*continues*)

COMMON DISEASE, DISORDER, OR CONDITION

DISORDERS OF THE REPRODUCTIVE SYSTEM (continued)

swelling, abnormal bleeding from the vagina, weight loss, and frequent urination. Some risk factors are infertility, not having children, delaying child birth to later on in life, and endometriosis. Yearly pelvic examination after age 40 can lead to early diagnosis and treatment.

MENSTRUAL CRAMPS

Menstrual cramps are experienced by many women with the onset of the menses, resulting from strong contractions of the myometrial layer of the uterus. These contractions produce a low abdominal pain that can extend to the legs and lower back. The pain can range from a dull ache to a sharp visceral pain that peaks in 24 hours but can last for a number of days. Treatment includes the taking of non-steroid drugs such as aspirin and ibuprofen, just before or after the start of the menstrual cycle.

ECTOPIC PREGNANCY

An **ectopic pregnancy** occurs when the fertilized egg implants in tissues other than the lining of the uterus, usually outside of the uterine cavity. The most common ectopic pregnancies occur in the fallopian tubes. This results in the death of the fetus and the possibility of hemorrhage from a rupturing of the tube. If the egg implants in tissues of the abdominal cavity, the fetus may be able to develop but must be delivered by caesarian section. This is known as an abdominal pregnancy.

FEMALE INFERTILITY

Female infertility is the inability of the female to produce a child. The cause could be immature or defective sexual organs, decreased hormonal secretions from the ovaries or from the anterior pituitary gland, or from an inability of the fertilized egg to implant in the endometrial lining of the uterus. Blockages of the fallopian tubes due to adhesions from various infections are the most common cause of female infertility.

HEALTH ALERT

INFERTILITY

Male infertility is most commonly caused by a low sperm cell count. The average range of spermatozoa ejaculated is 50–100 million per milliliter. If that number falls below 20 million per milliliter, the male is considered to be infertile. Decreased sperm count can result from a number of causes resulting in damage to the testes: radiation, trauma, undescended testes still in the abdominal cavity, higher than normal scrotal temperatures, and an infection with mumps. In addition, inadequate secretions of follicle stimulating hormone (which causes sperm cells to mature) and leutenizing hormone, and low testosterone levels also reduce sperm cell counts. Fertility can also be reduced due to abnormal sperm cell development.

Female infertility occurs in 10% of women. This can be caused by a number of factors such as obstruction of the uterine or fallopian tubes, a diseased ovary or conditions that prevent implantation of the fertilized egg into the uterine wall lining, and reduced hormone secretions from the ovaries and pituitary gland.

Today, there are numerous techniques utilized to assist infertile couples to have a child. One includes collecting and concentrating the male's sperm cells from several ejaculations and introducing them into the female reproductive tract via artificial insemination. ■

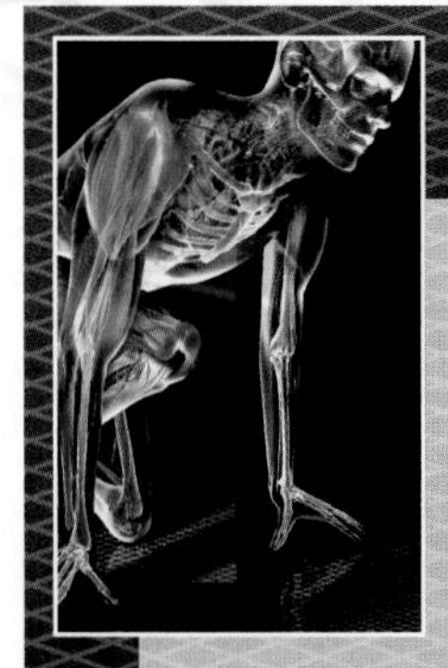

AS THE BODY AGES

The reproductive systems of both sexes mature at puberty, which begins earlier in females at about age 10 and a year or so later in males. Prior to that time, the system is in its juvenile or immature state.

For females, the reproductive cycle occurs once a month from menarche (the first menstrual cycle) to menopause (the cessation of the cycle). Menopause in women varies, occurring somewhere between the mid 40s and 50. The number of follicles decreases dramatically until very few remain. This results in a decrease in the female sex hormones estrogen and progesterone. The lining of the uterus becomes quite thin, and the uterus decreases in size over time. The wall of the vagina thins with less production of mucus, the connective tissue becomes more elastic, and the vaginal opening narrows. Sexual excitement during intercourse takes longer to develop, but sexual desire does not decline in any major way.

During menopause, many women experience hot flashes with profuse sweating. In addition, women may experience depression, mood swings, weight gain, inability to sleep, and headaches. About 10% of all women will experience some form of breast cancer. Risk is greater for those women with a history of breast cancer in their family. This type of cancer risk increases between the ages of 45 and 65. The risk for uterine cancer peaks at age 65 but cervical cancer is more common in younger women. Ovarian cancer is the second most common cancer in older women.

For males, the change in the reproductive system is not as dramatic as it is in females. Healthy males can still produce sperm well into their 80s and have been known to father children well into their 60s. Some men do experience a small decrease in the size of their testes, but sperm cell production does not cease.

After the age of 60, approximately one-third of males experience an enlargement of the prostate gland. This is known as benign prostatic hyperplasia (BPH). This enlargement compresses the prostatic urethra, resulting in difficulty in urination. These males experience frequent urinations with small amounts of urine, a decrease in the strength of the urinary stream, occasional bed-wetting, and a sense of incomplete voiding. Approximately 15% of men with BPH require medical treatment. Prostatic cancer is the third leading cause of cancerous death in males 55 years of age or older. Prostatic cancer also causes enlargement of the prostate gland, producing difficulty in urination.

Impotence is the inability of a male to achieve penile erection and, less frequently, the inability to ejaculate after achieving an erection. Impotence increases in men between 60 and 80 years of age. Since the fibrous connective tissue of the penis increases with advanced age, this causes a decrease in the speed of an erection, occasionally causing impotence in some men. However, there is great variation in the sexual activity of older men. Much of this variation can be related to genetic factors as well as psychological changes that develop over time. Some men remain very active sexually, while others experience a decrease in their sexual activity and sexual interests. ■

Integumentary System

- Pressure receptors in the skin get stimulated during sexual activity, resulting in sexual pleasure.
- Male sex hormones activate the skin's sebaceous glands to produce oil for dermal lubrication.
- Sex hormones cause the development of pubic and axillary hair during puberty.
- Female sex hormones cause fat to be deposited on the hips and breasts during puberty.
- The skin protects the reproductive organs by being the first line of defense against microorganisms.

Skeletal System

- The bones are a source of calcium needed during lactation, or breastfeeding, of the newborn infant.
- The pelvis encloses and protects the reproductive organs.
- The sex hormones cause the development of broad hips in women and narrow hips and broad shoulders in men.

Muscular System

- The heart pumps blood to maintain an erection.
- Smooth muscle contractions of the uterus result in delivery of the newborn.
- Skeletal muscle contractions result in an erection.
- Male sex hormones cause the development of more muscle mass in men.

There are many careers available to individuals who are interested in the reproductive system.

- **Gynecologists** are physicians who specialize in caring for women by diagnosing and treating disorders of the reproductive organs and breasts. This specialty requires surgical and nonsurgical expertise.
- **Obstetricians** are physicians who specialize in obstetrics, the branch of medicine concerned with caring for women during pregnancy, childbirth, and puerperium (the time after childbirth, about 6 weeks).
- **Pediatricians** are physicians who specialize in pediatrics, the branch of medicine that treats and diagnoses disease and prevents disorders of infants and children.
- **Pediatric nurse practitioners** are nurse practitioners who specialize in the nursing care of infants and children.
- **Neonatalists** are physicians who specialize in the care of the neonate and are trained to diagnose and treat disorders of the newborn.
- **Genetic counselors** are professionals who have special training in counseling and genetics, the branch of biology that deals with the principles and mechanics of heredity, in particular the process by which genes, which control traits, are passed from parents to offspring. These individuals can determine the occurrence of a genetic disorder within a family and provide information and advice about possible courses of action to potential parents.

Nervous System

- The nervous system's sensory and motor neurons play a major role in sexual pleasure and activity.
- The hypothalamus triggers the onset of puberty.
- Sex hormones influence the development of the brain in the fetus.

Endocrine System

- Estrogens and progesterone control the production and development of the ova in females and secondary female sexual characteristics.
- Testosterone controls the development of sperm and the secondary sexual characteristics of males.
- Placental hormones maintain pregnancy.

Cardiovascular System

- Blood pressure maintains erections in both men and women.
- Blood transports sex hormones to target organs.
- Pregnancy results in the heart working harder to maintain circulation between the mother and the developing fetus.

Lymphatic System

- The female immune system does not destroy the male sperm cell, thus ensuring fertilization.
- The immune system does not reject the developing fetus.
- The immune system protects the reproductive organs from disease.

Digestive System

- Proper nutrients are made available to the developing fetus through the mother's digestive system via the placenta and umbilical cord.

Respiratory System

- The interaction between the respiratory system and the placenta provides the fetus with oxygen and removes carbon dioxide.

Urinary System

- The male urethra functions in both the urinary system to transport urine and the reproductive system to transport sperm.
- The kidneys compensate for fluid loss in the reproductive system.
- Pregnancy can result in fluid retention, so the kidneys compensate by working harder to eliminate the excess fluid.
- The developing fetus causes compression of the bladder, resulting in messages to the brain initiating frequent and urgent urination.

SUMMARY OUTLINE

THE MALE REPRODUCTIVE SYSTEM

1. The primary sex organs of the male reproductive system are the testes or male gonads, which produce sperm and the male sex hormones.
2. Accessory glands produce secretions that make up the semen.
3. Accessory organs, like the scrotum, support the testes and ducts transport the sperm.
4. The penis is a transporting and supporting structure.

The Scrotum

1. The scrotum is an outpouching of the abdominal wall that supports the testes.
2. Internally, it is divided by a septum into two lateral pouches, each containing a single testis.
3. The testes produce both sperm as exocrine glands and the male sex hormones as endocrine glands.
4. The scrotal sac will elevate or descend on exposure to changes in temperature to ensure sperm survival.

The Testes

1. Each testis is covered by a capsule of connective tissue called the tunica albuginea, which extends inward to form a series of compartments called lobules.
2. Each lobule contains one to three convoluted seminiferous tubules in which spermatogenesis occurs.
3. Spermatogenesis begins as immature sperm cells, called spermatogonia, divide by mitosis to form primary spermatocytes.
4. Primary spermatocytes divide by meiosis to form secondary spermatocytes that develop into spermatids.
5. Spermatids develop into mature sperm cells or spermatozoa.
6. Sertoli cells supply nutrients to the developing sperm cells.
7. In the lobules, interstitial cells of Leydig produce the male sex hormone testosterone.

The Anatomy of the Spermatoza

1. Three hundred million spermatozoa are produced daily and can live up to 48 hours in the female reproductive tract.

2. Each spermatozoa consists of a head, which contains the nuclear genetic material and an acrosome containing enzymes, a middle piece or collar containing mitochondria and a tail that is a flagellum to propel the sperm cell.

The Functions of Testosterone

1. It controls the development, growth, and maintenance of the male sex organs.
2. It stimulates muscle buildup and bone development.
3. It causes sperm maturation.
4. It causes enlargement of the thyroid cartilage, or Adam's apple, and thickening of the vocal cords, resulting in a deep voice.
5. It produces body hair patterns, like facial and chest hair, and a receding hairline.
6. It stimulates aggressive behavior.

The Ducts of the System

1. The formed sperm cells move from the convoluted seminiferous tubules of the testis to the straight tubules at the top of the lobule.
2. They then move to a network of ducts in the testis called the rete testis.
3. They move out of the testis through coiled efferent ducts that connect to a single tube called the ductus epididymis.
4. The tightly coiled epididymis is located on the posterior border of a testis.
5. As the epididymis straightens, it is called the ductus deferens or vas deferens. It is 18 inches long.
6. The vas deferens is enclosed in a sheath called the spermatic cord. It empties into its ejaculatory duct.
7. Each ejaculatory duct ejects the spermatozoa into the single urethra.
8. The urethra is the terminal duct of the system. It is about 8 inches long and is a common passageway for sperm and urine.
9. The urethra passes through the prostate gland, the urogenital diaphragm, and the penis. It is divided into three parts.
10. The prostatic urethra is surrounded by the prostate and is about 1 inch long; the membranous urethra is about 1/2 inch long and connects to the penis; the spongy or cavernous urethra is within the penis and ends at the male urethral orifice at the head of the penis. Its size varies according to the size of the penis but is about 6 inches long.

The Accessory Glands

1. The paired seminal vesicles produce an alkaline viscous part of semen rich in fructose and nutrients and pass it into each ejaculatory duct.
2. The single prostate gland produces an alkaline fluid of semen that activates the sperm cells to swim.
3. The small paired bulbourethral glands or Cowper's glands secrete a thick, viscous mucus that enters the spongy urethra and is a lubricant for sexual intercourse.

Semen

1. Semen or seminal fluid is a mixture of sperm cells and the secretions of the accessory glands.
2. It is milky in color and rich in the sugar fructose, which provides energy for the beating flagellum of each sperm.
3. Its alkaline pH neutralizes the acidity of the male urethra and female vagina.
4. It provides a transport medium for the swimming sperm.
5. Semen contains enzymes that activate sperm after ejaculation and an antibiotic called seminalplasmin to control bacterial growth in the male and female reproductive tract.

The Penis

1. The penis delivers spermatozoa to the female reproductive tract.
2. It consists of a shaft whose end is called the glans penis or head covered with loose skin called the prepuce or foreskin.
3. Internally, it is composed of three cylindrical masses of spongy tissue containing blood sinuses.
4. Swelling of the blood sinuses during sexual stimulation results in an erection.

THE FEMALE REPRODUCTIVE SYSTEM

1. The primary sex organs of the female reproductive system are the ovaries, or female gonads. They produce eggs and the female sex hormones.
2. Accessory organs of the system are the uterine or fallopian tubes, uterus, vagina, and the external genitalia.
3. Accessory glands produce mucus for lubrication during sexual intercourse.

The Ovaries

1. The ovaries are paired glands located in the upper pelvic cavity on each side of the uterus. They are held in position by a series of suspensory ligaments.

2. The surface of an ovary is covered with germinal epithelium.
3. The capsule of an ovary consists of connective tissue called the tunica albuginea, whose outer area is called the cortex and contains ovarian follicles.
4. Ovarian follicles are eggs in various stages of development.
5. Each follicle contains an immature egg or oocyte and is called a primary follicle.
6. As the egg matures through meiosis, the follicle develops a fluid-filled central area called the antrum and is now called a secondary follicle.
7. A mature follicle with a mature egg is called a graafian follicle, ready for ovulation.
8. When the egg ruptures from the graafian follicle in ovulation, the follicle changes into the corpus luteum or yellow body, which secretes estrogen and progesterone.
9. The corpus luteum eventually degenerates, if fertilization does not occur, into the corpus albicans, or white body.
10. The ovaries produce and discharge eggs in ovulation. They also secrete the female sex hormones estrogen and progesterone.
11. Oogenesis, or formation of the female sex cells, begins in the developing female fetus where female stem cells called oogonia divide by mitosis to produce primary oocytes.
12. About 700,000 primary oocytes are produced at this time and represent the total number of eggs a female will produce. They lie dormant until puberty.
13. At puberty, the ovarian cycle begins and approximately 450 of the 700,000 primary oocytes will develop into eggs by meiosis during the female's reproductive years.
14. After the first meiotic division a primary oocyte will develop into two cells: the secondary oocyte is the larger of the two with a smaller polar body cell.
15. After the second meiotic division, which occurs only after fertilization, the secondary oocyte becomes an ootid or mature egg with another nonfunctional polar body. The polar body from the first meiotic division divides into two nonfunctional polar bodies. Thus, one mature egg and three polar bodies are produced.
16. The one mature egg cell has a large supply of stored food to supply the developing embryo, if fertilization occurs.

The Uterine or Fallopian Tubes

1. The two uterine or fallopian tubes transport the ova from the ovaries to the uterus.
2. The funnel-shaped open end is called the infundibulum and is surrounded by a fringe of finger-like projections called the fimbriae.
3. Cilia on the epithelium of the infundibulum and the waving fimbriae sweep an ovum into the uterine tube after ovulation.
4. The egg is moved by peristalsis and the action of cilia toward the uterus. Fertilization usually occurs in the upper one-third of the tube within 24 hours after ovulation.

The Uterus

1. The uterus is the site of menstruation, it is where the fertilized egg is implanted and where the fetus develops, and it is where labor begins during delivery.
2. It is shaped like an inverted pear: the dome-shaped portion above the uterine tubes is the fundus, the major tapering portion the body, and the narrow inferior portion the cervix. Between the body and the cervix is a narrow region called the isthmus.
3. The interior of the body is the uterine cavity; the interior of the cervix is the cervical canal.
4. The opening between the uterine cavity and the cervical canal is called the internal os and the opening between the cervical canal and the vagina is the external os.
5. The wall of the uterus is composed of three layers: the innermost is the endometrium where the fertilized egg implants, the second is the myometrium of smooth muscle, and the outermost is the perimetrium or visceral peritoneum.

The Menstrual Cycle

1. The menstrual cycle, also called the menses or menstruation, is the cyclical shedding of the endometrial lining of the uterus.
2. The three phases are the menstrual phase, the proliferative phase, and the secretory phase.
3. During the menstrual phase, the endometrial lining of the uterus, tissue fluid, blood, and mucus are shed. Twenty to 25 primary follicles also begin their development and produce low levels of estrogen. The zona pellucida develops around each egg and about 20 primary follicles become secondary follicles, but only one attains maturity while the others die.

4. During the proliferative phase, one of the secondary follicles matures into a graafian follicle with a single mature egg. The egg ruptures from the follicle in a process called ovulation, and rising estrogen levels cause the endometrial lining of the uterus to thicken. After ovulation, the graafian follicle collapses with a clot inside called the corpus hemorrhagicum, which is eventually absorbed. The follicle eventually changes character and becomes the corpus luteum.
5. During the secretory phase, the corpus luteum begins to secrete estrogen and progesterone. If fertilization and implantation do not occur, the corpus luteum degenerates and becomes the corpus albicans. If fertilization and implantation do occur, the corpus luteum is maintained for 4 months by human chorionic gonadotropin produced by the developing placenta.
6. Once the placenta is developed, it will secrete estrogen to support pregnancy and progesterone to support pregnancy and breast development for milk production in the mammary glands.

The Functions of Estrogen

1. The ovaries become active during puberty producing ova and estrogen.
2. Estrogen causes the development of the female secondary sex characteristics: the development of breasts; the appearance of pubic and axillary hair; fat deposits on the hips, breasts, and under the skin; and widening of the pelvic bone producing a broad hip.
3. Estrogen causes enlargement of the uterine tubes, uterus, vagina, and the external genitalia.

The Vagina

1. The vagina is a passageway for the menstrual flow, is a receptacle for the penis during sexual intercourse, and is the lower portion of the birth canal.
2. A recess called the fornix surrounds the vaginal attachment to the cervix.

The External Genitalia of the Female

1. The vulva or pudendum is the collective term for the external genitalia of the female.
2. The mons pubis or veneris is a mound of adipose tissue covered with pubic hair at puberty.
3. Two longitudinal folds of hair-covered skin extend posteriorly and inferiorly from the mons pubis called the labia majora. They contain adipose tissue and sweat glands.
4. Medial to the labia majora are two other delicate folds of skin called the labia minora. They do not have hair but have numerous sebaceous glands.
5. The clitoris is a small mass of erectile tissue located at the anterior junction of the labia minora covered with a layer of skin called the prepuce. The exposed portion of the clitoris is called the glans.
6. The opening between the two labia minora is called the vestibule containing a thin fold of tissue called the hymen, which is ruptured during the first sexual intercourse.
7. Also in the vestibule are two openings, the vaginal orifice and the urethral orifice.
8. On each side of the urethral orifice are the two openings of the ducts of the lesser vestibular or Skene's glands, which secrete mucus.
9. On each side of the vaginal orifice are the two openings of the greater vestibular or Bartholin's glands that also secrete mucus for lubrication during sexual intercourse.

The Perineum

1. The perineum is a diamond-shaped area between the buttocks and thighs of males and females.
2. It is divided into an anterior urogenital triangle that contains the external genitalia and a posterior anal triangle that contains the anus.

The Anatomy and Function of the Mammary Glands

1. Mammary glands, found in both males and females, are functional to produce milk only in the female. They increase in size during puberty due to estrogen.
2. Each gland consists of 15 to 20 lobes separated by adipose tissue.
3. Each lobe contains smaller compartments called lobules, which contain the milk-secreting cells or alveoli arranged like a cluster of grapes.
4. The alveoli convey the milk into secondary tubules, which join into mammary ducts.
5. As the ducts approach the nipple, they expand into milk storage sinuses called ampullae.
6. Ampullae continue as lactiferous ducts that terminate in the nipple.
7. The circular pigmented area around each nipple is called the areola and contains modified sebaceous glands.

8. The function of the mammary glands is to produce milk and to eject it out the nipple, a process called lactation.

PREGNANCY AND EMBRYONIC DEVELOPMENT

1. An egg cell must be fertilized within 12 to 24 hours after ovulation; sperm remains viable for 12 to 48 hours in the female reproductive tract.
2. It takes the egg 24 hours to move down one-third of the uterine tube. Fertilization will occur in the upper two-thirds of the tube.
3. A fertilized egg is called a zygote. As it moves down the uterine tube, it divides by mitosis to form a hollow sphere of cells called the blastocyst or blastula. By the time it reaches the uterine cavity, it is called a chorionic vesicle.
4. It secretes chorionic gonadotropin, which stimulates the corpus luteum to maintain the uterine lining via its hormones. It embeds in the endometrial lining by the seventh day.
5. The three primary germ layers are now being developed. The ectoderm will develop into skin and the nervous system; the endoderm will form the linings of internal organs; and the mesoderm will form muscles, bones, and other tissues. These tissues come from the blastocyst's inner cell mass.
6. The blastocyst's fluid-filled sphere, the trophoblast, forms projections called chorionic villi, which will interact with uterine tissue to form the placenta.
7. Once the placenta is formed, the three-layered embryo becomes surrounded by a fluid-filled sac called the amnion.
8. The embryo becomes attached by a connecting stalk called the umbilical cord.
9. The placenta exchanges nutrients, oxygen, and wastes between the embryo and the mother. By the ninth week of development, the embryo is called a fetus. Later in development the umbilical cord will become the major exchange structure between fetus and mother.
10. As pregnancy continues, the uterus expands into the abdominal cavity to accommodate the growing fetus.
11. Childbirth is called parturition and begins with contractions of the smooth muscles of the uterus, called labor.
12. Labor is divided into three stages: the dilation stage, the expulsion stage, and the placental stage.
13. During the dilation stage, the cervix is fully dilated by the head of the fetus and the amnion ruptures releasing amniotic fluid.
14. During the expulsion stage, the child moves through the cervix and vagina, usually head first into the outside world.
15. During the placental stage, the placenta detaches from the uterus within 15 minutes following birth, called the afterbirth.

REVIEW QUESTIONS

1. Name the organs, accessory glands, and ducts of the male reproductive system.
*2. Discuss the significance of meiosis and spermatogenesis.
3. Beginning at the seminiferous tubule of a testis, describe the path of sperm to the outside.
4. Name the three parts of the male urethra and their location.
*5. What physiologic processes maintain an erection?
6. Name the organs of the female reproductive system.
*7. Discuss the significance of meiosis and oogenesis.
*8. Describe and discuss the significance of the events in an ovarian cycle.
9. Name the phases of the menstrual cycle and what occurs during each phase.
10. Describe the three stages of labor.

***Critical Thinking Questions**

FILL IN THE BLANK

Fill in the blank with the most appropriate term.

1. The ______________ muscle in the spermatic cord elevates the testes on exposure to cold.
2. Each lobule of a testis contains one to three tightly coiled tubules called the convoluted tubules that produce sperm by a process called ______________.
3. The ______________ is the site of sperm maturation and storage.
4. The funnel-shaped open end of each uterine tube is called the ______________ and is surrounded by a fringe of finger-like projections called the ______________.

5. A follicle ready for ovulation is known as a _____________ follicle.
6. The term _____________, or pudendum, is a collective term for the external genitalia of the female.
7. The circular pigmented area of skin surrounding the nipple is called the _____________; it appears rough because it contains modified sebaceous glands.
8. The _____________ is a diamond-shaped area between the thighs and buttocks of both males and females.
9. The milk-secreting cells of the mammary glands are grapelike clusters of cells called _____________.
10. Childbirth is called _____________.

MATCHING

Place the most appropriate number in the blank provided.

_____ Sertoli cells	1. Surrounds part of urethra
_____ Interstitial cells of Leydig	2. Antibiotic
_____ Seminal vesicles	3. First mensis
_____ Prostate gland	4. Sebaceous glands
_____ Cowper's gland	5. Adipose tissue and sweat glands
_____ Seminalplasmin	6. Progesterone and estrogen
_____ Corpus luteum	7. Nutrients for spermatozoa
_____ Menarche	8. Alkaline component of semen rich in fructose
_____ Labia majora	9. Mucus secretion
_____ Labia minora	10. Testosterone
	11. Last menses
	12. Lysozyme

Search and Explore

- Visit the Centers for Disease Control and Prevention website at http://www.cdc.gov/std and research one of the sexually transmitted diseases introduced in this chapter. Summarize what you learn by writing a two to three page entry in your notebook.
- Visit the Prostate Cancer Foundation website at http://www.prostatecancerfoundation.org to learn more about prostate cancer. Based on your research and in your own words, write a description of prostate cancer and list some possible symptoms and the two tests used for screening for prostate cancer.
- Visit the Human Anatomy Online website at http://www.innerbody.com and explore the female or male reproductive system or both.

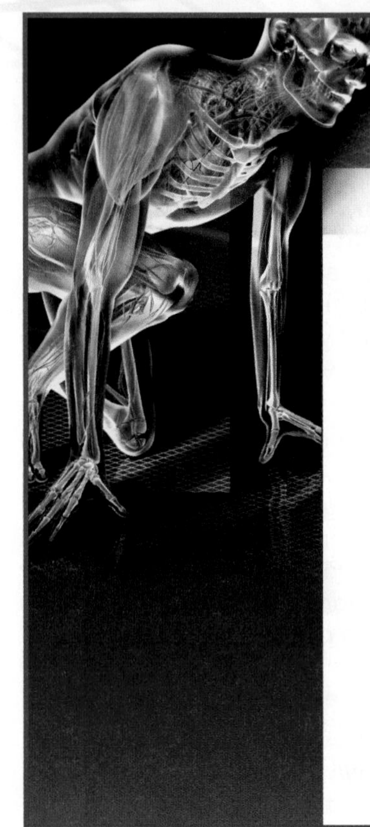

CASE STUDY

Monica, a 53-year-old woman, is scheduled to undergo a breast biopsy for a small tumor that was discovered in her right breast during a recent mammogram. Monica tells her primary health care provider that she performs a breast self-examination every month, and has undergone a mammogram every year since she turned 50 years old. She states she is very concerned about her breast lump, and is anxious to learn the outcome of the biopsy.

Questions

1. What is the purpose of a breast biopsy, and why is Monica so concerned about the outcome?
2. What are the mammary glands, where are they located, and what is their purpose?
3. How does a breast self-examination compare to a mammogram as a method for detecting breast lumps?
4. What are a woman's chances of developing breast cancer?

*Study*WARE™ Connection

Take a practice quiz or play interactive games that reinforce the content in this chapter on your StudyWARE™ CD-ROM.

Study Guide Practice

Go to your **Study Guide** for more practice questions, labeling and coloring exercises, and crossword puzzles to help you learn the content in this chapter.

LABORATORY EXERCISE: THE REPRODUCTIVE SYSTEM

Materials needed: A dissecting kit, your fetal pig, and a dissecting pan.

The Female Fetal Pig

If your animal is a female, locate the reproductive structures described. Refer to Figure 19-16. If your specimen is a male, examine another learner's female specimen. Each learner should be responsible for locating the reproductive structures of both sexes.

Locate the paired female gonads, or the *ovaries*. They are small bodies suspended from the peritoneal wall in mesenteries, posterior to the kidneys. Examine one ovary closely, and notice the small short coiled *oviduct*, also called the fallopian tube. The oviduct does not attach directly to the ovary, but ends in an open, funnel-shaped structure that partially encloses the ovary. This is visible only with the dissecting microscope. When ovulation occurs, the eggs are carried by ciliary currents into the mouth of the funnel and pass down the oviduct to the uterus.

In your pig, most of the *uterus*, or womb, is divided into two sections called *uterine horns*. At

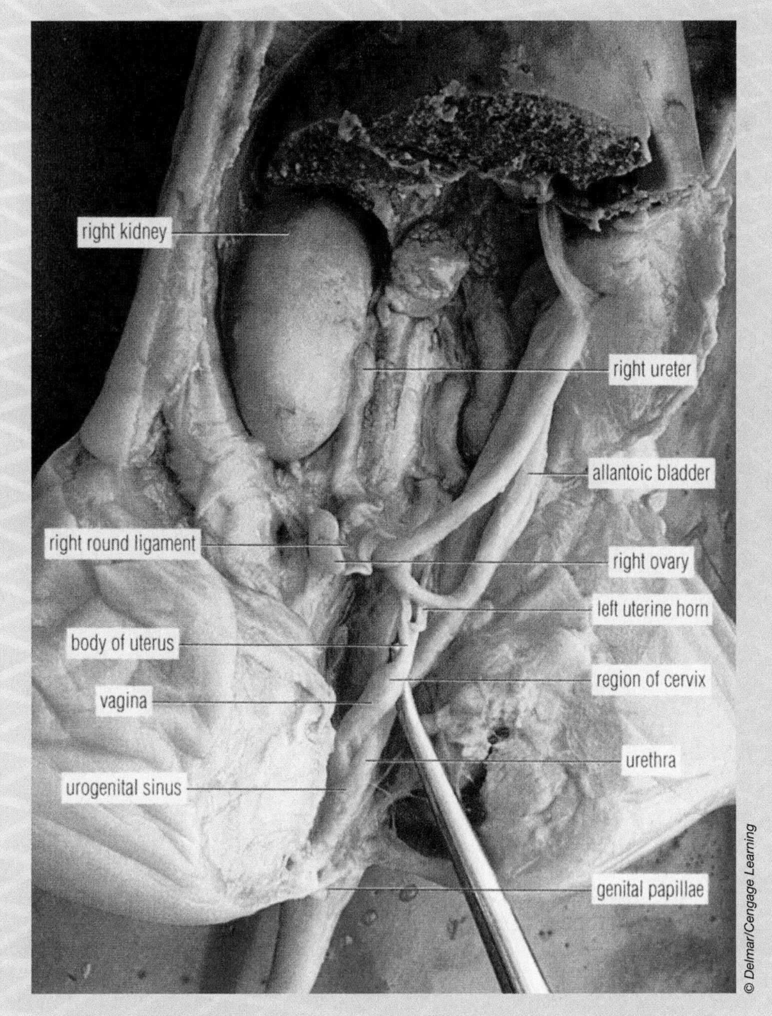

FIGURE 19-16. The reproductive organs of a female fetal pig. (Courtesy of Dr. David Morton. From *Photo Atlas for Anatomy and Physiology*, by David Morton and James W. Perry, 1998, Belmont, CA: Wadsworth.)

(*continues*)

THE REPRODUCTIVE SYSTEM (Continued)

its anterior end, each of the uterine horns connects with an oviduct; at their posterior ends the uterine horns join to form the smaller, median body of the uterus. (In the human female, the uterus is a single, median structure.)

The body of the uterus continues posteriorly with the *vagina*, slightly larger in diameter than the uterus, and marked from it by a slight constriction, the *cervix*. The vagina lies between the rectum and the urethra and disappears from view into the ring of the pelvic girdle.

The Male Fetal Pig

The structures described here refer to the male; if your specimen is a female, examine another student's male specimen. Refer to Figure 19-17.

The paired male gonads, or *testes*, are located by finding the ducts that carry sperm from the testes to the urethra. In the space bounded ventrally by the urinary bladder, laterally by the ureters, and dorsally by the rectum, you will see two thin ducts that join medially and then attach to the dorsal side of the urethra. Follow these ducts, each a *vas deferens*, from their medial junction, and note that they diverge and loop over the ureters. Follow one vas deferens posteriorly until it enters a small opening in the posterior peritoneal wall. The passageway into which this opening leads is the *inguinal canal*.

Each testis first develops in the peritoneal cavity near the kidney. As the fetus matures, the testes move posteriorly, or *descend*, passing

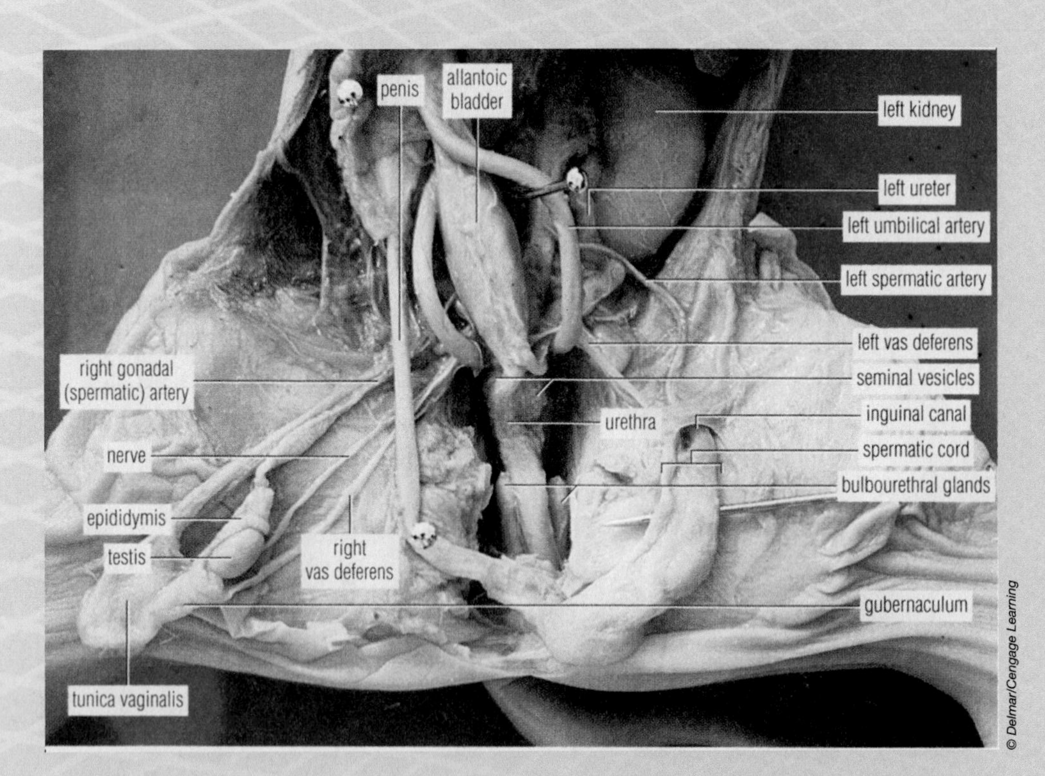

FIGURE 19-17. The reproductive organs of a male fetal pig. (Courtesy of Dr. David Morton. From *Photo Atlas for Anatomy and Physiology*, by David Morton and James W. Perry, 1998, Belmont, CA: Wadsworth.)

(*continues*)

THE REPRODUCTIVE SYSTEM (Continued)

through the inguinal canal and into a pocket at the base of the hind leg. This pocket is the *scrotum*. Next carefully remove the skin at the base of one hind leg. Just beneath the skin, you will notice a white bulbous structure, covered with a transparent membrane. The testis of the adult will lie in this *scrotal sac*. Carefully clean away all the surrounding tissue.

Located on or around the testis, find the *epididymis*, a very tightly coiled tubule, usually lighter in color than the testis. The coils of the epididymis serve for temporary storage of sperm cells. Trace the epididymis to its junction with the vas deferens.

The following is a list of companies that can supply materials needed for the laboratory exercises covered in the text and for audiovisual materials to supplement lectures.

1. An excellent, inexpensive color chart for student use during the fetal pig and the brain/heart dissections can be purchased from:

   BIOCAM Communications Inc.

   PMB 12, 250 H St.

   Blain, WA 98230 1-800-667-3316

   http://www.biocam.com

   #CC 122 The Concise Fetal Pig

   #CC 129 The Concise Brain/Heart

2. Chemical models illustrating bonding, and pH meters can be purchased from a chemical supply house such as:

   Fisher Scientific Company

   2000 Park Lane Dr.

   Pittsburgh, PA 15275-9952 1-800-766-7000 Fax:1-800-926-1166

   http://www.fisherscientific.com/college

3. Microscope slides of tissues and cells; osmosis kits; fetal pigs; sheep eyes, hearts, and brains; dissecting kits; and latex gloves can be purchased from a biological supply company such as:

   Carolina Biological Supply Company

   2700 York Road

   Burlington, NC 27215 1-800-334-5551 Fax: 1-800-222-7112

   http://www.carolina.com

4. Anatomical charts, bones, skeletons or anatomical models can be ordered from anatomical suppliers such as:

   Anatomical Chart Company

   P.O. Box 1600

   Hagerstown, MD 21741-1600 1-800-621-7500

   http://www.anatomical.com

5. Videos and CD-ROMs can be purchased from video suppliers such as:

   HRM Video

   41 Kensico Drive

   Mt. Kisco, NY 10549 1-800-431-2050 Fax: 914-244-0485

   http://www.hrmvideo.com and

   Films for the Humanities and Sciences

   200 American Metro Blvd. Suite 124

   Hamilton, N.J. 08619 1-800-257-5126 Fax: 1-800-329-6687 or www.films.com

A

A bands the dark, thick bands of the protein myosin in muscle cells

Abdominal aorta the part of the aorta located in the abdomen

Abdominopelvic cavity second subdivision of the ventral cavity that contains the kidneys, stomach, liver and gallbladder, small and large intestines, spleen, pancreas, and the ovaries and uterus (in women)

Abducens nerve VI controls movement of the eyeball

Abduction movement of a bone or limb away from the midline of the body

Abductor digiti minimi muscle that abducts little toe

Abductor hallucis muscle that abducts, flexes great toe

Abductor pollicis muscle that abducts the thumb

ABO blood group one of the blood groups

Absorption the passage of digested food from the digestive tract into the cardiovascular and lymphatic systems for distribution to the body's cells

Accessory nerve XI helps control swallowing and movements of the head

Acetabulum deep socket on the lateral side of the hipbone just above the obturator foramen

Acetaldehyde an intermediate product of fermentation

Acetic acid an intermediate product of the citric acid cycle

Acetylcholine neurotransmitter substance

Acetylcholinesterase an enzyme in the postsynaptic neuron that breaks down acetylcholine

Acetyl-CoA an intermediate product of the citric acid cycle

Acid a substance that dissociates and forms an excess of H ions when dissolved in water

Acidosis a condition caused by lowering of blood pH

Acid rain rain with a high acid concentration resulting from pollutants emitted from cars and coal-fired power plants

Acini exocrine glands of the pancreas

Acromial process bony prominence on the scapula

Acromegaly overdevelopment of bones of hands, feet, face, and jaw

Acrosome contains enzymes that aid the sperm cell in penetrating the ovum

Actin thin filaments of protein in a muscle cell

Action potential when a muscle generates its own impulse to contract

Active immunity a type of immunity acquired naturally when exposed to a bacterium or virus or acquired artificially through a vaccine

Active transport mechanism requiring energy by which cells acquire materials against a concentration gradient

Adam's apple thyroid cartilage

Addison's disease condition in which the adrenal cortex fails to produce enough hormones

Adduction movement of a bone or limb toward the midline of the body

Adductor pollicis muscle that adducts the thumb

Adenine a purine nitrogen base

Adenoids pharyngeal tonsils

Adenosine triphosphate (ATP) high-energy fuel molecule the cell needs to function

Adipose loose connective tissue full of fat cells

Adipose capsule second layer of tissue that surround the kidney

Adrenal cortex outermost part of the adrenal gland

Adrenal glands small glands found on top of each kidney; suprarenal glands

Adrenal medulla inner part of the adrenal gland

Adrenaline a neurotransmitter also called epinephrine used by the autonomic nervous system; a hormone

Adrenocorticotropic hormone (ACTH) stimulates the adrenal cortex to secrete the hormone cortisol

Adrenogenital syndrome excessive secretion of androgens producing male characteristics in females and an early enlarged penis in young males

Adventitia the outermost layer of the wall of the alimentary canal

Aerobic requires oxygen

Afferent arteriole takes blood from the renal artery to Bowman's glomerular capsule

Afferent lymphatic vessels lymphatic vessels that enter the lymph nodes at various sites

Afferent peripheral system consists of afferent or sensory neurons that convey information from receptors in the periphery of the body to the brain and spinal cord

Agglutination clumping of red blood cells

Agonists muscles performing the actual movement

AIDS acquired immunodeficiency syndrome

Albinism the absence of skin color

Albumin a protein found in blood plasma that maintains osmotic pressure in blood and tissues

Aldosterone hormone that regulates sodium reabsorption and potassium excretion by the kidneys

Aldosteronism excessive aldosterone causing high blood pressure

Alimentary canal the name given to the digestive tube that runs from the mouth to the anus

Allergies hypersensitive reactions to common, normally harmless environmental substances

All-or-none law a contraction or nervous transmission either occurs or does not occur

Alpha cells secrete the hormone glucagon

Alpha-ketoglutaric acid an intermediate product of the citric acid cycle

Alveolar-capillary/respiratory membrane membrane through which respiratory gases move in the lungs

Alveolar ducts branches of respiratory bronchioles

Alveolar sacs two or more alveoli that share a common opening

Alveoli 1. milk-secreting cells; 2. cup-shaped outpouchings lined with epithelium in the lungs

Alveolus a socket for articulation with a tooth

Alzheimer's disease results in severe mental deterioration

Amine group NH_2 found in amino acids

Ammonia molecule that comes from the decomposition of proteins via the digestive process, and the conversion of amino acids in cellular respiration to ATP molecules

Amnion fluid-filled sac that surrounds the embryo

Amphiarthroses joints that allow only slight movement

Ampulla of Vater common duct of the pancreas and liver that enters the duodenum

Amylase the salivary enzyme that breaks down carbohydrates

Anabolism an energy-requiring process that builds larger molecules by combining smaller molecules

Anaerobic respiration respiration that does not require oxygen

Anal canal the terminal one inch of the rectum

Anal columns longitudinal folds of mucous membrane of the anal canal

Anaphase third and shortest stage of mitosis

Anaphase I stage of meiosis in which the centromere does not divide

Anaphase II stage of meiosis in which the centromeres of the chromosomes divide

Anastomosis junction of two or more blood vessels

Anatomy the study of the structure and organization of the body

Anconeus muscle that extends the forearm

Androgens male sex hormones

Anemia a decrease of hemoglobin in the blood

Aneurysm dilation of a blood vessel wall

Angina pectoris a sensation of pain in the chest

Angioplasty reconstruction of a coronary artery

Antagonists muscles that relax while the agonist contracts

Anterior toward the front

Anterior (ventral) gray horn part of the spinal cord

Anterior interventricular sulcus separates the right and left ventricles from each other

Anterior (ventral) root point of attachment of the spinal nerve to the cord, also known as the motor root

Anterior tibial arteries supply blood to the leg and foot

Anterior tibial veins drains the calf and foot

Antibodies immunoglobulins; destroy foreign proteins

Antidiuretic hormone (ADH) maintains the body's water balance; vasopressin

Antigens foreign proteins that gain access to our bodies through cuts and scrapes, digestive or circulatory systems, or the urinary and reproductive systems

Anus the opening of the anal canal to the exterior

Aorta largest artery in the body

Aortic arch the part of the aorta that arches to the left and heads down the spine through the thorax

Aortic semilunar valve found in the opening where the ascending aorta leaves the left ventricle

Apical foramen opening at the base of each root canal

Aponeurosis wide and flat tendon

Appendicitis inflammation of the vermiform appendix

Aqueous humor fluid in the anterior compartment of the eye located in front of the lens

Arachnoid mater the middle spinal or cranial meninx

Arch of the aorta sends arteries to upper parts of the body

Arcuate arteries the interlobar arteries that arch between the cortex and medulla

Arcuate veins drain filtered blood at the base of the pyramid

Areola circular pigmented area of skin surrounding the nipple

Areolar a type of loose connective tissue

Arrector pili muscle consists of smooth muscle fibers attached to a hair follicle that causes the goose flesh appearance on the skin when we get scared or get a chill

Arrhythmia an irregular heartbeat

Arteries blood vessels that carry blood away from the heart

Arterioles small arteries that deliver blood to capillaries

Arthritis inflammation of the whole joint

Articulation a place of junction between two or more bones

Arytenoid cartilages move the vocal cords

Ascending aorta leaves the left ventricle of the heart

Ascending colon first part of the colon

Ascending limb of Henle name given to the loop of Henle as it ascends toward the cortex

Aster a starburst cluster of microtubules of tubulin produced by the centriole

Astrocytes star-shaped cells that twine around nerve cells to form a supporting network in the brain and spinal cord

Atherosclerosis disease of the arteries in which plaque accumulates on the inside of arterial walls

Atlas first cervical vertebra that supports the head by articulation with the condyles of the occipital bone

Atelectasus a collapsed lung

Atom the smallest particle of an element that maintains all the characteristics of that element

Atomic number the number of protons or electrons in an atom

Atrioventricular bundle bundle of His; part of the conduction system of the heart running through the top of the interventricular septum

Atrioventricular (AV) node part of the conduction system of the heart located in the lower portion of the right atrium

Atrophy a decrease in muscle bulk due to a lack of exercise

Auditory ossicles another name for the ear bones

Auditory tubes the ear canals located in the middle ear that equalize air pressure preventing hearing distortion; eustachian tubes

Auricle external appendage of an atrium

Autolysis the process of self-destruction in old or weakened cells

Autonomic nervous system (ANS) conducts impulses from the brain and spinal cord to smooth muscle tissue, cardiac muscle tissue, and glands

Axillary artery part of the subclavian artery that runs down the arm

Axillary vein found in the armpit

Axis the second vertebra

Axon the long extension of a nerve cell body; a neuron has only one axon

Axon endings the terminal portions of axons

Axon terminals the endings of axons

Azygos vein drains the thorax

B

B cells lymphocytes found in lymph nodes, spleen, and other lymphoid tissue where they replicate

B lymphocytes cells that produce antibodies and provide humoral immunity; also known as B cells

Ball-and-socket joint a type of synovial or diarthrosis joint, also called a multiaxial joint, like the shoulder or hip joint

Basal cell carcinoma most common type of skin cancer

Base a substance that combines with H^+ ions when dissolved in water

Basement membrane anchors epithelial cells to each other and to underlying tissues

Basilic vein drains the medial part of the arm

Basophils WBCs that release heparin, histamine, and serotonin during an allergic reaction

Bell's palsy paralysis of the facial nerve on one side of the face

Beta cells secrete the hormone insulin, found in the pancreas

Biceps brachii muscle that flexes the arm and forearm and supinates the hand

Biceps femoris muscle that flexes the leg; rotates laterally after flexed

Bicuspids molar teeth with two cusps

Bicuspid valve mitral valve; valve between the left atrium and left ventricle

Bile duct part of the gallbladder that transports bile to the duodenum of the small intestine

Bipolar neurons sensory neurons that consist of one dendrite and one axon

Black eye pariorbital bruise from a blow to supraorbital ridge

Blastula a hollow ball of cells produced by mitotic divisions of the zygote; blastocyst

Blood specialized connective tissue

Body term for the large central portion of the stomach and the pancreas

Body of the uterus central tapering region of uterus

Bonds formed when atoms combine chemically with one another

Bone specialized connective tissue

Bone marrow transplant used to treat leukemia lymphomas, and immunological deficiencies

Bowel large intestine

Bowman's glomerular capsule double-walled globe located in the cortex of the kidney

Brachial artery extension of the axillary artery that runs down the arm

Brachial vein drains the arm and empties into the axillary vein

Brachialis muscle that flexes the forearm, assists the biceps muscle

Brachiocephalic artery first branch of the aortic arch

Brachioradialis muscle that flexes the forearm, assists the biceps muscle

Brainstem one of the four major parts of the brain; it connects the brain to the spinal cord

Breast cancer leading cause of death in women

Bronchial arteries supply blood to the lungs

Bronchial tree branching of the bronchi

Bronchioles smaller branches of segmental bronchi

Bronchitis inflammation of the bronchi

Bronchomediastinal trunk drains lymph from the thorax, lungs, heart, diaphragm, and portions of the liver

Bronchopulmonary segment segment of lung tissue that each of the tertiary bronchi supplies

Brownian movement the random collision of diffusing molecules

Brunner's glands secrete an alkaline mucus in the intestine

Bubonic plague disease of the lymphatic system

Buccal glands secrete small amounts of saliva; found in the mouth

Buccinator muscle that compresses the cheek

Buffer a substance that acts as a reservoir for hydrogen ions

Bulbourethral glands Cowper's glands; they produce an alkaline mucus

Bulla or vesicle a blister on the skin with serous fluid

Bursae closed sacs with a synovial membrane lining

Bursitis inflammation of the synovial bursa

C

Calcaneus the heel

Calcitonin hormone secreted by the thyroid that lowers the calcium and phosphate ion concentration of the blood

Callus a thickened area of skin developed from an excessive amount of friction

Calorie unit used to measure energy

Canaliculi small canals in compact bone that connect lacunae with one another

Cancellous bone forms the inner spongy tissue underneath compact bone

Cancer and lymph nodes lymph nodes can spread cancer cells to other parts of the body

Canine teeth teeth used to tear food

Capillaries microscopic blood vessels where exchange of nutrients and oxygen and waste and carbon dioxide gas occurs between blood and tissue cells

Capitate one of the bones of the wrist

Carbohydrate made of atoms of carbon, hydrogen, and oxygen in a 1:2:1 ratio

Carbon dioxide chemical produced as a waste product of cellular respiration

Carbon monoxide poisoning CO binds to hemoglobin in RBC's preventing oxygen transport

Carboxyl group the COOH group found in amino acids and fatty acids

Carcinogens cancer-causing agents

Carcinomas tumors developing from epithelial tissue

Cardia part of the stomach that surrounds the gastroesophageal sphincter

Cardiac muscle muscle found only in the heart

Cardiovascular system body system consisting of the heart and vessels that pump and distribute blood to and from all cells

Carotene a carotenoid pigment in plant cells that produces a red-orange color

Carpals bones of the wrist

Cartilage a type of specialized connective tissue

Catabolism an energy-releasing process that breaks down large molecules into smaller ones

Catalyst substance that increases the rate of a chemical reaction without being affected by that reaction

Cataracts cloudy film over eye lens due to protein buildup

Caudal synonymous with inferior; toward the tail

Cecum pouch-like first part of the large intestine

Celiac trunk the first branch of the abdominal artery

Cell body contains the nucleus of a neuron

Cell cycle process by which a cell divides in two and duplicates its genetic material

Cell plate first stage of a new cell wall forming at the equator of a dividing plant cell

Cellular immunity results of the body's lymphoid tissue

Cellular respiration/metabolism the energy changes that occur in cells

Cellulose carbohydrate material that makes the cell wall of plant cells, fiber in our diet

Cementum substance that covers the dentin of the root of a tooth

Central nervous system (CNS) consists of the brain and spinal cord

Centrioles two centrioles make up a centrosome; they produce spindle fibers during cell division

Centromere portion of a duplicated chromosome that holds the two daughter chromatids together

Centrosome area near the nucleus made of two centrioles

Cephalad toward the head

Cephalic vein drains the lateral part of the arm and connects with the axillary vein

Cerebellum second largest portion of the brain concerned with coordinating skeletal muscle movements and balance

Cerebral aqueduct connects the third and fourth ventricles of the brain; also called aqueduct of Sylvius

Cerebral circulation blood circulatory route that supplies the brain with oxygen and nutrients, and disposes of waste

Cerebral cortex surface of the cerebrum

Cerebral hemispheres the right and left halves of the cerebrum

Cerebral palsy condition caused by brain damage during brain development or the birth process

Cerebral peduncles convey impulses from the cerebral cortex to the pons and spinal cord

Cerebrovascular accident (CVA) caused by a thrombus or embolus that blocks circulation resulting in cellular death

Cerebrum the bulk of the brain consisting of two cerebral hemispheres

Cerumen earwax

Ceruminous glands glands that produce earwax

Cervical canal interior of the cervix

Cervical vertebrae the seven smallest vertebrae found in the neck

Cervix 1. narrow, inferior portion of uterus that opens into the vagina; 2. the constricted junction between the crown and the root of a tooth, also known as the neck of a tooth

Chancre a sore on the penis caused by syphilis

Chiasmata figures of chromosomes during crossing-over

Chickenpox a childhood disease of the skin caused by a virus

Chief cells 1. secreting cells of the parathyroid glands; 2. in the stomach secrete pepsinogen; also known as zymogenic cells

Chloroplasts organelles found only in plant cells where photosynthesis occurs; contain the pigment chlorophyll

Chondrocytes cells of cartilage

Chordae tendineae connect the pointed ends of the flaps or cusps to the papillary muscles in the ventricles of the heart

Chorionic vesicle a 100-cell blastocyst

Chorionic villi projections of the trophoblast

Choroid the second layer of the wall of the eye containing blood vessels and pigment cells

Chromatids duplicated copies of a chromosome

Chromatin term used to describe the genetic material inside a nucleus before duplication

Chromoplasts plastids in plant cells that contain the carotenoid pigments

Chyle lymph in the lacteals that has a high fat content and looks milky

Chyme the digested, viscous, semifluid contents of the intestine

Cilia small hairs found on cells that function in movement of materials across the cell's outer surface

Ciliary body consists of smooth muscles that hold the lens of the eye in place

Circumduction moving the bone in such a way so that the end of the bone or limb describes a circle in the air and the sides of the bone describe a cone in the air

Circumvallate papillae projections of the lamina propria that are covered with epithelium and contain taste buds; found toward the back of the tongue

Cirrhosis long-term degenerative disease of the liver in which the lobes are covered with fibrous connective tissue

Cisternae cavities of an endoplasmic reticulum that are sac or channel-like

Citric acid an intermediate product of the citric acid cycle

Clavicle collar bone

Cleavage furrow pinching in of an animal cell membrane during cell division

Clitoris small, cylindrical mass of erectile tissue with nerves found at the anterior junction of the labia minora

Clones exact duplicates

Clot formed by fibrin at the site of a cut in a blood vessel

Co enzyme A converts acetic acid to acetyl-CoA

Coccygeal vertebrae/coccyx the vertebrae of the tailbone

Coitus sexual intercourse

Cold sores small, fluid-filled blisters caused by the herpes simplex virus

Collagen a tough fiber found in the matrix of connective tissue

Collecting duct connects with the distal tubules of other nephrons

Colon largest part of the large intestine

Color blindness inability to perceive one or more colors

Colorectal cancer cancer of the large intestine and rectum

Columnar epithelium epithelial cells that are tall and rectangular; found lining the ducts of certain glands and in mucous-secreting tissues

Common hepatic artery supplies blood to the liver

Compact bone forms the outer layer of bone and is very dense

Complement a set of enzymes that attack foreign antigens

Compound formed when two or more elements combine via bonding

Compound exocrine glands glands made of several lobules with branching ducts

Concussion momentary loss of consciousness due to blow to the head

Conduction system generates and distributes electrical impulses over the heart to stimulate cardiac muscle fibers or cells to contract

Condyle rounded prominence found at the point of articulation with another bone

Condyloid joint a type of synovial joint, also called an ellipsoidal joint, like the wrist

Congenital heart disease heart disease present at birth

Conjunctivitis pinkeye, inflammation of the conjunctiva

Connective tissue a type of tissue that supports or binds

Contracture condition in which a muscle shortens its length in the resting state

Coracoid process bony projection on the scapula that functions as an attachment for muscles that move the arm

Corium true skin; another name for the dermis

Cornea transparent part of the outermost layer of the eye that permits light to enter the eye

Corniculate cartilage cone-shaped, paired cartilages of the larynx

Corns caused by abrasion on bony prominences on the foot

Coronal synonymous with frontal

Coronal plane dividing anterior and posterior portions of the body at right angles to the sagittal plane

Coronal suture found where the frontal bone joins the two parietal bones

Coronary arteries supply the walls of the heart with oxygenated blood

Coronary circulation supplies blood to the myocardium of the heart

Coronary heart disease results from reduced blood flow in the coronary arteries that supply the myocardium of the heart

Coronary sinus drains the blood from most of the vessels that supply the walls of the heart with blood

Coronary sulcus groove separating the atria from the ventricles externally

Coronary thrombosis blood clot in the vessel

Corpus albicans white body

Corpus callosum deep bridge of nerve fibers that connects the cerebral hemispheres

Corpus hemorrhagicum a ruptured graafian follicle

Corpus luteum yellow body

Cortex 1. smooth-textured area of the kidney extending from the renal capsule to the bases of the renal pyramids; 2. principal portion of the hair

Cortical nodule a dense aggregation of tissue in a lymph node

Cortisol hormone that stimulates the liver to synthesize glucose from circulating amino acids

Cortisone steroid closely related to cortisol given to reduce inflammation

Costae another name for ribs

Covalent bond a bond in which the atoms share electrons to fill their outermost energy levels

Cowper's glands bulbourethral glands

Cramp spastic and painful contraction of a muscle that occurs because of an irritation within the muscle

Cranial another name for cephalad

Cranial cavity cavity containing the brain

Cremaster muscle muscle in the spermatic cord that elevates the testes

Crest narrow ridge of bone

Cretinism a lack of or low level of thyroid hormones in children, resulting in mental and sexual retardation

Cricoid cartilage a ring of cartilage of the larynx that attaches to the first tracheal cartilage

Cristae the folds of the inner membrane of a mitochondrion

Crohn's disease chronic, inflammatory bowel disease of unknown origin

Crossing-over the exchange of genetic material during prophase I of meiosis

Crown the portion of a tooth above the level of the gums

Crust a hard dry layer on the skin's surface

Crypts of Lieberkuhn pits in the mucosa of the small intestine, also called intestinal glands

Cuboid a tarsal bone of the ankle

Cuboidal epithelium epithelial cells that look like small cubes; their function is secretion, protection, and absorption

Cuneiform cartilage rod-shaped, paired cartilages of the larynx

Cuneiforms the tarsal bones of the forefoot

Cushing's syndrome condition resulting from too much secretion from the adrenal cortex, resulting in obesity and puffiness in the skin

Cuspids another name for the canine teeth

Cuticle outermost portion of the hair

Cyanosis bluish discoloration of the skin caused by lack of oxygen in the blood

Cyst an encapsulated sac in the skin

Cystic fibrosis an inherited disease of the respiratory system, usually fatal by early adulthood

Cystitis inflammation of the urinary bladder

Cytochrome system an electron carrier complex

Cytokinesis the phase of cell division in which division and duplication of the cytoplasm occur

Cytoplasm the protoplasm outside the nucleus of a cell

Cytosine a pyrimidine nitrogen base

D

Decussation of pyramids crossing of the tracts in the brain stem

Deep femoral artery supplies blood to the thigh

Defecation the elimination from the body of those substances that are indigestible and cannot be absorbed

Deglutition the process of swallowing

Deltoid muscle that abducts the arm

Dendrites receptive areas of the neuron; extensions of the nerve cell body

Dentes teeth

Dentin bone-like substance found in teeth

Deoxyribonucleic acid (DNA) genetic material of cells located in the nucleus of the cell that determines all the functions and characteristics of the cell

Deoxyribose a five-carbon sugar found in DNA

Depolarization reversal of electrical charge

Depression 1. lowering a part of the body; 2. abnormal emotional state

Dermis second layer of skin; also called the corium

Descending colon part of the colon on the left side of the body

Descending limb of Henle name given to the proximal convoluted tubule as it dips into the medulla

Descending thoracic aorta part of the aorta located in the thorax

Desmosomes interlocking cellular bridges that hold skin cells together

Detrusor muscle three layers of smooth muscle in the bladder wall

Deviated septum abnormal left growth of nasal septum

Diabetes insipidus caused by insufficient ADH resulting in excessive urination and dehydration

Diabetes mellitus disease caused by a deficiency in insulin production

Diaphragm the muscle that separates the thoracic from the abdominal cavity used in breathing

Diaphysis shaft composed mainly of compact bone

Diarrhea the passing of loose, watery stools affecting the function of the colon

Diarthroses freely moving joints or articulations; also called synovial joints

Diastole phase of relaxation of the heart

Diencephalon one of the four major parts of the brain consisting of the thalamus and the hypothalamus

Diffusion the movement of molecules through a medium from an area of high concentration of those molecules to an area of low concentration of those molecules

Digestion the breakdown of food by both mechanical and chemical mechanisms

Digestive system consists of the alimentary canal with its associated glands

Diploid the full complement of chromosomes

Dislocated hip the head of the femur torn out of the acetabulum

Distal away from the point of attachment or origin

Distal convoluted tubule name given to the ascending limb of Henle as it enters the cortex and becomes convoluted

Diverticulosis the presence of pouch-like herniations through the muscular layer of the colon

Dopamine a neurotransmitter

Dorsal toward the back

Dorsal tectum reflex center that controls the movement of the eyeballs and head in response to visual stimuli

Dorsal venous arch drains blood in the foot

Dorsalis pedis artery supplies blood to the dorsal part of the foot

Dorsiflexion raising the foot up at the ankle joint

Duct of Wirsung large main duct of the pancreas; also called pancreatic duct

Ductus deferens vas deferens

Ductus epididymis a single tube in the testis into which the coiled efferent ducts empty

Duodenum shortest and first part of the small intestine

Dura mater the outermost spinal or cranial meninx

Dwarfism the result of inadequate ossification at the epiphyseal plates of long bones that causes an individual to be abnormally small

E

Ectoderm a primary germ layer that forms the skin and nervous system of a developing fetus

Ectopic pregnancy implantation of the fertilized egg outside the uterine cavity

Edema swelling

Efferent arteriole carries blood away from the glomerular capsule

Efferent ducts series of coiled tubes that transfer the sperm out of the testes

Efferent lymphatic vessels lymphatic vessels that leave a lymph node at the hilum

Efferent peripheral system consists of efferent or motor neurons that convey information from the brain and spinal cord to muscles and glands

Ejaculatory duct duct formed from the joining of the seminal vesicle and ductus deferens that ejects spermatozoa into the urethra

Elastic cartilage forms the external ear, ear canals, and epiglottis

Elastin flexible fibers found in the matrix of connective tissue

Electrical potential caused by a rapid influx of sodium ions into a muscle cell

Electron negatively charged particle that orbits the nucleus of an atom at some distance from its center

Electron acceptors molecules that gain electrons during a reaction

Electron carriers molecules that gain electrons only to lose them to some other molecule in a very short time

Electron donors molecules furnishing electrons during a reaction

Electron transfer/transport system the aerobic mechanism of respiration which produces most of the ATP molecules from the breakdown of glucose

Element a substance whose atoms all contain the same number of protons and electrons

Elevation raising a part of the body

Embolism embolus that becomes lodged in a vessel and cuts off circulation

Embolus piece of blood clot that dislodges and gets transported by the bloodstream

Emphysema a degenerative disease with no cure that results in the destruction of the walls of the alveoli of the lungs

Enamel protects teeth from wear and acids, found on the crown of a tooth

Encephalitis inflammation of brain tissue usually caused by a virus

Endocarditis inflammation of the endocardium

Endocardium innermost layer of the heart wall, including epithelial cells that line the heart

Endochondral ossification the formation of bone in a cartilagenous environment

Endocrine glands ductless glands that secrete hormones directly into the bloodstream

Endocrine system consists of the endocrine glands

Endoderm a primary germ layer that forms the lining of internal organs and glands of a developing fetus

Endometrium innermost layer of the uterine wall

Endomysium delicate connective tissue that surrounds the sarcolemma of a muscle cell

Endoplasmic reticulum a complex system of membranes that form a collection of membrane-bound cavities in a cell

Endorphins neurotransmitters

Endosteum a fibrovascular membrane that lines the medullary cavity of a long bone

Endothelial-capsular membrane formed by the visceral layer of Bowman's capsule and the endothelial capillary network of the glomerulus

Endothelium epithelial cells that line the circulatory system

Energy the ability to do work

Energy levels the levels in which electrons are grouped

Enlarged prostate causes constriction of the urethra making urination difficult, usually occurring in elderly males

Entamoeba histolytica amoeba in protozoan infections caused by drinking untreated water; causes severe diarrhea

Enzymes protein catalysts

Eosinophils WBCs that produce antihistamines

Ependymal cells line the fluid-filled ventricles of the brain; produce and move cerebrospinal fluid through the CNS.

Epicardium/visceral pericardium outermost layer of the heart wall

Epidermis top layer of skin

Epiglottis large, leaf-shaped piece of cartilage of the larynx that blocks food from entering the trachea when we swallow

Epilepsy a disorder of the brain resulting in seizures

Epimysium coarse, irregular connective tissue that surrounds the whole muscle

Epiphyseal line place where longitudinal growth of bone takes place

Epiphysis the extremity of a long bone

Epithelial tissue type of tissue that protects, absorbs, or secretes

Erectile dysfunction inability of a male to maintain an erection

Erection swelling and hardening of the penis due to retained entry blood

Erythroblastosis fetalis hemolytic disease of the newborn

Erythrocytes red blood cells (RBCs)

Erythrocytosis excessive red blood cells reducing blood flow

Erythropoietin hormone that stimulates red blood cell production in red bone marrow

Escherichia coli normal bacteria in the intestine

Esophageal arteries supply blood to the esophagus

Esophageal hiatus an opening in the diaphragm for passage of the esophagus to join the stomach

Esophagus collapsible, muscular tube located behind the trachea that transports food to the stomach

Estrogen female sex hormone

Ethmoid bone the principal supporting structure of the nasal cavities; forms part of the orbits

Ethyl alcohol a final product of fermentation

Eukaryotic refers to higher cells, like those of the human body, with membrane-bound organelles

Eversion moving the sole of the foot outward at the ankle

Exhalation expiration; movement of air out of the lungs

Exocrine glands glands that have ducts

Exophthalmia bulging of the eyeballs

Extension increasing the angle between bones

Extensor carpi muscles that extend the wrist

Extensor digitorum communis muscles involved in abducting and adducting the wrist; also extends toes and fingers

Extensor hallucis muscle that extends great toe; dorsiflexes ankle

Extensor pollicis muscle involved in extending the thumb

External auditory meatus ear canal

External iliac veins drain the pelvis

External intercostals muscles that draw adjacent ribs together

External jugular vein drains the muscle and skin of the head region

External oblique muscle that compresses abdominal contents

External occipital crest a projection of the occipital bone for muscle attachment

External occipital protuberance a projection of the occipital bone for muscle attachment

External os opening of the cervix into the vagina

External respiration the exchange of gases between the lungs and blood

External urinary sphincter surrounds the urethra as it leaves the bladder; made of skeletal muscle

F

Facial nerve VII controls the muscles of facial expression and conveys sensations related to taste

Falciform ligament separates the two lobes of the liver

Fascia layer of areolar tissue covering the whole muscle trunk

Fascicle individual bundle of muscle cells

Fasciculi skeletal muscle bundles

Fatty acids along with glycerol, a building block of fats

Fauces opening to the oropharynx

Feces semi-solid mass of indigestible material in the large intestine

Female infertility inability of the female to produce a child

Femoral artery supplies blood to the thigh

Femoral vein drains blood from the thigh

Femur thigh bone

Fermentation process in which yeast breaks down glucose anaerobically (in the absence of oxygen)

Fertilized egg zygote; cell produced by the union of two gametes

Fetal circulation circulation route that exists only between the developing fetus and its mother

Fetus embryo at nine weeks

Fibrillation rapid, uncontrolled contraction of individual cells in the heart

Fibrin long threads that form a clot

Fibrinogen plasma protein

Fibrinolysis dissolution of a blood clot

Fibroblasts small, flattened cells with large nuclei and reduced cytoplasm that produce fibrin in connective tissue

Fibrocartilage forms the intervertebral disks that surround the spinal cord

Fibromyaigia tendon and muscle pain and stiffness near a joint

Fibrous pericardium outermost layer of the pericardial sac

Fibula lateral calf bone

Filiform papillae found at the front of the tongue; important in licking

Fimbriae finger-like projections surrounding the infundibulum

First-degree burn burn involving just the epidermis and heals with no scarring

Flagella long fibers that push a cell, like the flagellum of a sperm cell

Flavin adenine dinucleotide (FAD) an electron carrier

Flexion bending or decreasing the angle between bones

Flexor carpi muscle that flexes the wrist

Flexor digitorum muscle that flexes toes and fingers

Flexor hallucis muscle that flexes great toe

Flexor pollicis muscle involved in flexing the thumb

Fluid mosaic pattern term used to describe the arrangement of protein and phospholipid molecules in a plasma or cell membrane

Follicle-stimulating hormone (FSH) stimulates development of the follicles in the ovaries of females, and the production of sperm cells in the seminiferous tubules of the testes

Fontanelle soft spot on top of a baby's head

Food bolus the soft mass of chewed food

Food poisoning a bacterial infection of the digestive tract

Foramen opening in a bone through which blood vessels, nerves, and/or ligaments pass

Foramen magnum inferior portion of the occipital bone through which the spinal cord connects with the brain

Foramen of Monroe connects each lateral ventricle with the third ventricle of the brain

Foreskin prepuce

Fornix recess in the lower portion of the birth canal

Fossa any depression or cavity in or on a bone

Fovea centralis a depression in the retina

Fracture a break in a bone

Fractured clavicle the most common broken bone

Frontal plane dividing anterior and posterior portions of the body at right angles to the sagittal plane

Frontal bone a single bone that forms the forehead and part of the roof of the nasal cavity

Frontal lobe forms the anterior portion of each cerebral hemisphere

Frontalis muscle that raises the eyebrows and wrinkles the skin of the forehead

Fructose a six-carbon sugar

Full-thickness burn burn in which the epidermis and dermis are completely destroyed; also called third-degree burn

Fundus 1. the rounded portion of the stomach above and to the left of the cardia; 2. dome-shaped portion of the uterus above the uterine tubes

Fungiform papillae found toward the back of the tongue; contain taste buds

Furuncle or Boli a pus-forming infection of a hair follicle or gland

G

Gallbladder pear-shaped sac located in a depression of the surface of the liver

Gallstones collection of precipitated cholesterol in the gallbladder

Gametogenesis the formation of the gametes

Ganglia nerve cell bodies grouped together outside the central nervous system

Gastric cancer stomach cancer

Gastritis inflammation of the stomach lining

Gastrocnemius calf muscle

Gastroesophageal reflux disease backflow of gastric juice into the esophagus

Gastroesophageal sphincter connects the esophagus with the stomach

Gastrointestinal tract the name given to the digestive tube that runs from the mouth to the anus; the alimentary canal

Gene a sequence of organic nitrogen base pairs that codes for a polypeptide or protein

Genital herpes infection with simplex herpes II virus that causes lesions and blister-like eruptions on the skin of the genitals

Genital warts highly contagious infection caused by a virus

Germinal center part of a lymphatic node that produces lymphocytes

Germinal epithelium surface of an ovary

Gigantism the result of abnormal endochondral ossification at the epiphyseal plates of long bones, giving the individual the appearance of a very tall giant

Gingivae the gums

Gingivitis an inflammation of the gingivae

Gladiolus part of the sternum bone resembling the blade of a sword

Glandular epithelium forms glands

Glans exposed portion of the clitoris

Glans penis head of the penis

Glaucoma condition causing destruction of the retina or optic nerve resulting in blindness

Glenoid fossa a depression in the scapula for articulation with the head of the humerus

Glial cells cells that perform support and protection

Gliding joint a type of synovial joint found in the spine

Globin protein in hemoglobin

Globulins blood plasma proteins like antibodies and complement

Glomerulonephritis inflammation of the kidneys

Glomerulus a capillary network surrounded by Bowman's capsule

Glossopharyngeal nerve IX controls swallowing and carries taste impulses

Glottis space between the vocal cords in the larynx

Glucagon a hormone produced by the pancreas that regulates blood glucose levels

Glucose a six-carbon sugar

Gluteus maximus muscle that extends and rotates the thigh laterally

Gluteus medius muscle that abducts and rotates the thigh medially

Gluteus minimus muscle that abducts and rotates the thigh medially

Glycerol a simple molecule similar to a sugar except that it has only a three-carbon chain, part of a fat

Glycogen animal starch

Glycolysis the first step in cellular respiration in which a glucose molecule gets broken down into two molecules of pyruvic acid; does not require oxygen; occurs in the cytoplasm

Glycosuria large amount of sugar in the urine

Goblet cells unicellular glands that secrete mucus

Goiter enlargement of the thyroid gland due to an inadequate amount of iodine in the diet

Golgi body/apparatus consists of an assembly of flat sac-like cisternae that look like a stack of saucers or pancakes; used as a storage area in the cell

Gomphosis a joint in which a conical process fits into a socket and is held in place by ligaments

Gonorrhea venereal disease caused by a bacterial infection

Gout an accumulation of uric acid crystals in the joint at the base of the large toe and other joints of the feet and legs

Graafian follicle a mature follicle with a mature egg

Gracilis muscle that adducts thigh, flexes leg

Granum stacks of membranes found in chloroplasts

Graves' disease a type of hyperthyroidism caused by overproduction of thyroid hormone

Gray matter gray areas of the nervous system

Great saphenous veins longest veins in the body

Greater vestibular glands secrete mucus; Bartholin's gland

Growth hormone (GH) stimulates cell metabolism in most tissues of the body

Guanine a purine nitrogen base

Gyri folds on the surface of each hemisphere of the cerebrum

H

H band slightly darker section in the middle of the dark A band; also called H zone

Hair one of the main characteristics of mammals

Hair follicle an epidermal tube surrounding an individual hair

Hamate one of the bones of the wrist

Haploid half the number of chromosomes

Hard palate anterior part of the roof of the mouth

Haustrae pouches in the colon

Haversian canals a feature of compact bone containing capillaries, also called central canals

Head 1. part of the pancreas closest to the duodenum; 2. terminal enlargement, like the head of the humerus

Headache cephalalgia; pain in the head

Heart major pumping organ of the cardiovascular system

Heart failure caused by progressive weakening of the myocardium and failure of the heart to pump adequate amounts of blood

Heart murmur an abnormal heart sound

Helicobacter pylori bacterium associated with the development of stomach or peptic ulcers

Helper T cells stimulate the production of killer T cells and more B cells to fight invading pathogens

Hematocytoblasts undifferentiated mesenchymal cells

Hematopoiesis blood cell formation

Hematopoietic tissue specialized connective tissue that produces blood cells

Hematuria blood in the urine

Heme pigment in hemoglobin

Hemodialysis procedure in which a dialysis machine filters blood taken from an artery and sends it back to a vein

Hemoglobin red pigment in erythrocytes

Hemolytic anemia inherited condition in which erythrocytes rupture or are destroyed at a faster rate than normal

Hemophilia genetically inherited clotting disorder

Hemorrhoids caused by inflammation and enlargement of rectal veins

Heparin anticoagulant manufactured by the liver and mast cells

Hepatic portal circulation route between the digestive tract and the liver

Hepatic portal vein drains the organs of the digestive tract

Hepatitis inflammation of the liver caused by excessive alcohol consumption or a virus infection

Herniated disk rupture of the fibrocartilage surrounding an intervertebral disk that cushions the vertebrae above and below

Hilum 1. notch in the center of the concave border of the kidney through which the ureter leaves the kidney; 2. depression on one side of a lymph node

Hinge joint a type of synovial joint, like the knee or elbow

Histamine an inflammatory substance produced in response to allergies

Histiocytes large, stationary phagocytic cells

Histology the study of tissue

Homeostasis maintaining the body's internal environment

Horizontal plane dividing the body into superior and inferior portions

Hormones chemical secretions from an endocrine gland

Horns the areas of gray matter in the spinal cord

Humerus largest and longest bone of the upper arm

Humoral immunity results of the body's lymphoid tissue

Hyaline cartilage a type of cartilage that forms the early skeleton of the embryo

Hydrogen bond a type of bond that helps hold water molecules together by forming a bridge between the negative oxygen atom of one water molecule and the positive hydrogen atoms of another water molecule

Hydroxyl group the OH group found in sugars

Hymen thin fold of tissue that partially closes the distal end of the vagina

Hyoid bone bone that supports the tongue

Hyperextension increases the joint angle beyond the anatomic position

Hyperglycemia chronic elevations of glucose in the blood

Hyperopia farsightedness

Hyperparathyroidism an abnormally high level of PTH secretion

Hypertension high blood pressure

Hyperthyroidism too much secretion of thyroid hormone

Hypertonic solution solution in which water molecules will move out of a cell and the cell will shrink, as in a 5% salt solution

Hypertrophy an increase in the bulk of a muscle caused by exercise

Hypodermis subcutaneous tissue

Hypoglossal nerve XII controls the muscles involved in speech and swallowing; its sensory fibers conduct impulses for muscle sense

Hypoparathyroidism an abnormally low level of PTH

Hypophysis another name for the pituitary gland

Hypothalamus part of the brain that controls secretions from the pituitary gland

Hypothyroidism a lack of or low level of thyroid hormone

Hypotonic solution solution in which water molecules will move into a cell and the cell will swell, as in pure distilled water

I

I bands light, thin bands of the protein actin in muscle cells

Ileocecal valve the opening from the ileum of the small intestine into the cecum of the large intestine

Ileum the third part of the small intestine measuring 12 feet in length

Iliacus muscle involved in flexing the thigh

Ilium the uppermost and largest portion of a hip bone

Immunity the ability of the body to resist infection from disease-causing microorganisms

Immunoglobulin A (IgA) type of antibody found in exocrine gland secretions, nasal fluid, tears, gastric and intestinal juice, bile, breast milk, and urine

Immunoglobulin D (IgD) type of antibody found on the surface of B lymphocytes

Immunoglobulin E (IgE) type of antibody found in exocrine gland secretions that is associated with allergic reactions

Immunoglobulin G (IgG) type of antibody found in tissue fluids and plasma

Immunoglobulin M (IgM) type of antibody that develops in blood plasma as a response to bacteria or antigens in food

Impetigo highly contagious skin disease of children caused by the bacterium *Staphylococcus aureus*

Incisors front teeth used to cut food

Incompetent heart valve a valve that leaks blood

Incus ear bone referred to as the anvil

Infarct an area of damaged cardiac tissue

Infarction death of tissues

Infectious mononucleosis caused by the Epstein-Barr virus; infects lymphocytes and the salivary glands

Inferior lowermost or below

Inferior meatus one of three narrow passageways in the nasal cavity formed by the turbinate bones

Inferior mesenteric artery supplies blood to the large intestine

Inferior oblique muscle that rotates the eyeball on axis

Inferior rectus muscle that rolls the eyeball downward

Inferior vena cava brings blood to the heart from the lower parts of the body; also called posterior vena cava

Influenza (Flu) viral infection of the respiratory tract

Infraspinatus muscle that rotates the humerus outward

Infundibulum 1. part of the hypothalamus that connects to the pituitary gland; 2. open end of fallopian tube

Ingestion the taking of food into the body

Inhalation inspiration; movement of air into the lungs

Insertion the movable attachment where the effects of contraction are seen

Insula lobe in the brain that separates the cerebrum into frontal, parietal, and temporal lobes

Insulin hormone produced by the pancreas that regulates blood glucose levels

Integumentary system consists of the epidermis and dermis and the appendages of the skin

Intercalated disks structures that connect the branches of cardiac muscle cells with one another

Intercostal arteries supply blood to the muscles of the thorax

Intercostal trunk helps drain lymph from portions of the thorax

Interlobar arteries branches of the renal arteries in the renal columns

Interlobar veins run between the pyramids in the renal columns

Interlobular arteries arteries that divide into afferent arterioles

Interlobular veins veins that are formed by reuniting peritubular capillaries

Internal iliac artery supplies blood to the thigh

Internal intercostals muscles that draw adjacent ribs together

Internal jugular vein drains the dural sinus of the brain

Internal nares two internal openings in the nose

Internal oblique muscle that compresses abdominal contents

Internal os junction of the uterine cavity with the cervical canal

Internal respiration exchange of gases between the blood and body cells

Internal urinary sphincter located at the junction of the urinary bladder and urethra; made of smooth muscle

Internuncial neurons transmit the sensory impulse to the appropriate part of the brain or spinal cord for interpretation and processing; also known as association neurons

Interossei muscles that cause abduction of the proximal phalanges of the fingers

Interphase a stage of the cell cycle

Interstitial cells of Leydig found in the testes that produce male sex hormone testosterone

Interstitial fluid blood plasma found in the spaces between tissue cells

Interventricular foramen another name for the foramen of Monroe

Interventricular septum separates the right and left ventricles of the heart

Intestinal glands pits in the mucosa of the small intestine; also known as crypts of Lieberkuhn that secrete intestinal digestive enzymes

Intestinal trunk drains lymph from the stomach, intestines, pancreas, spleen, and surface of the liver

Intramembranous ossification formation of bone by a process in which dense connective tissue membranes are replaced by deposits of inorganic calcium salts

Inversion moving the sole of the foot inward at the ankle

Ion charged atom

Ionic bond a bond that is formed when one atom gains electrons while the other atom loses electrons from its outermost energy level

Iris colored part of the eye

Iron-deficiency anemia results from nutritional deficiencies or excessive iron loss from the body

Ischium the strongest portion of a hip bone

Islets of Langerhans pancreatic islets; form the endocrine portions of the gland

Isometric activity contraction in which a muscle remains at a constant length while tension against the muscle increases

Isotonic contraction contraction in which tone or tension remains the same as the muscle becomes shorter and thicker

Isotonic solution solution in which water molecules diffuse into and out of a cell membrane at equal rates, as in normal saline solution

Isotopes different kinds of atoms of the same element

Isthmus small, constricted region between body of uterus and cervix

J

Jejunum the second part of the small intestine measuring 8 feet in length

Jugular trunk drains lymph from the head and neck

K

Keratin a protein material

Keratinization a process by which epidermal cells of the skin change shape, composition and lose water as they move to the upper layers and become mainly protein and die

Kidney stones stones made of precipitates of uric acid, magnesium, calcium phosphate, or calcium oxalate that can accumulate in the kidney

Kidneys paired organs that regulate the composition and volume of blood and remove wastes from the blood in the form of urine

Killer T cells types of lymphocytes that kill virus-invaded body cells and cancerous body cells

Kinetochore a disk of protein on the centromere

Krebs citric acid cycle the step after glycolysis that takes place in the mitochondria during which pyruvic acid gets broken down into carbon dioxide gas and water; requires oxygen

Kupffer cells eat bacteria and old white and red blood cells; found in the liver

Kyphosis condition commonly referred to as hunchback

L

Labia majora two longitudinal folds of hair-covered skin; part of the vulva

Labia minora two delicate folds of skin medial to the labia majora; part of the vulva

Labor process by which the fetus is expelled from the uterus

Lacrimal bones bones that make up part of the orbit at the inner angle of the eye; contain the tear sac

Lactation secretion of milk from the mammary glands

Lacteals lymphatic vessel that absorbs fats and transports them from the digestive tract to the blood

Lactic acid the final product of anaerobic production of ATP in muscle cells

Lactiferous ducts continuations of ampullae that terminate at the nipple

Lactiferous sinuses expanded sinuses that store milk

Lactogenic hormone (LTH) stimulates milk production in the mammary glands after delivery; also called prolactin

Lacunae tiny cavities between the lamellae or rings of compact bone that contain bone cells

Lambdoid suture a line where the two parietal bones connect with the occipital bone

Lamella 1. system of membranes that connect grana in a chloroplast; 2. layer of concentric rings surrounding the Haversian canals

Lamina propria the second layer of the tunica mucosa consisting of loose connective tissue

Large intestine the last part of the digestive tract measuring 5 feet in length; the bowel

Laryngitis inflammation of the mucosal lining of the larynx

Laryngopharynx lowermost portion of the pharynx

Larynx voice box

Lateral toward the side or away from the midline of the body

Lateral rectus muscle that rolls the eyeball laterally

Latissimus dorsi muscle that extends, adducts, and rotates the arm medially

Left atrium one of the upper chambers of the heart

Left bundle branch branch of the bundle of His; part of the conduction system of the heart

Left colic (splenic) flexure position where the transverse colon curves down beneath the spleen

Left common carotid artery second branch of the aortic arch

Left external carotid artery supplies blood to the muscles and skin of the neck and head

Left gastric artery supplies blood to the stomach

Left internal carotid artery supplies blood to the brain

Left primary bronchus the first left division of the trachea

Left pulmonary artery carries blood to the left lung

Left renal artery transports one-quarter of the total cardiac output directly to kidneys

Left renal vein carries filtered blood from the interlobar veins to the hilum

Left subclavian artery third branch of the aortic arch

Left ventricle one of the lower chambers of the heart

Legionnaires disease a bacterial induced acute pneumonia

Lens the crystalline part of the eye

Lesser vestibular glands Skene's glands; secrete mucus

Leucoplast plastid in plant cells that contains no pigment but stores sugar or starch

Leukemia type of cancer in which there is abnormal production of white blood cells

Leukocytes white blood cells (WBCs)

Levator labii superioris muscle that raises the upper lip and dilates the nostril

Levator scapulae muscle that elevates the scapula

Ligament connective tissue that attaches bone to bone

Line a less prominent ridge of bone than a crest

Lingual frenulum septum dividing tongue into symmetrical halves

Lingual tonsils located on the back surface of the tongue at its base

Lipids substances that are insoluble in water like fats

Lips fleshy folds that surround the opening of the mouth

Liver largest organ of the digestive system

Lobules divisions of a bronchopulmonary segment

Longitudinal fissure fissure separating the cerebrum into right and left halves

Loop of Henle the U-shaped structure of the limb of Henle

Lordosis an abnormal accentuated lumbar curvature

Lower esophageal sphincter gastroesophageal sphincter, connects the esophagus with the stomach, and controls the passage of food into the stomach

Lumbar arteries supply blood to the muscles of the abdomen and walls of the trunk of the body

Lumbar trunk drains lymph from the lower extremities, walls, and viscera of the pelvis, kidneys, and adrenal glands and most of the abdominal wall

Lumbar vertebrae the five vertebrae of the lower back

Lumen a hollow core

Lunate a bone of the wrist

Lung cancer a common type of fatal cancer mainly due to smoking

Lunula the white crescent at the proximal end of each nail

Luteinizing hormone (LH) stimulates ovulation in the ovary and production of the female sex hormone progesterone

Lymph the name given to interstitial fluid when it enters a lymphatic capillary

Lymphatic capillaries blind end tubes that are the origin of lymphatic vessels

Lymph glands lymph nodes

Lymph nodes lymph glands

Lymphatic sinus space between groups of lymphatic tissue

Lymphatic trunks the main draining vessels of the lymphatic system

Lymphadenitis inflammation of lymph nodes or glands

Lymphangitis inflammation of the lymphatic vessels

Lymphatic system consists of the lymph nodes, thymus gland, spleen, and the lymphatic vessels

Lymphatics lymphatic vessels that resemble veins but have more valves

Lymphocytes WBCs involved in the production of antibodies

Lymphoid tissue specialized connective tissue

Lymphokines chemicals released by the sensitized T lymphocytes

Lymphoma tumor of lymphatic tissue that is usually malignant

Lysosomes small bodies in the cytoplasm that contain powerful digestive enzymes that enhance the breakdown of cellular components

Lysozyme enzyme that destroys bacteria

M

Macrophages engulf and digest antigens; monocytes

Macule a small discoloration of the skin

Major calyces minor calyces joined together

Malaria disease caused by the injection of a protozoan by mosquitoes

Malic acid an intermediate product of the citric acid cycle

Malignant melanoma skin cancer associated with a mole on the skin

Malleus ear bone referred to as the hammer

Mammillary bodies part of the diencephalon involved in memory and emotional responses to odor

Mammary glands produce milk in females

Mammography procedure in which low intensity X-rays are used to detect tumors in the soft tissues of the breast

Mandible bone the strongest and longest bone of the face; forms the lower jaw

Manubrium part of the sternum resembling the handle of a sword

Masseter muscle that closes the jaw

Mast cells roundish-shaped cells found close to small blood vessels that produce heparin

Mastication chewing

Mastoid portion located behind and below the auditory meatus or opening of the ear; part of the temporal bone

Matrix intercellular material in connective tissue

Maxillary bones make up the upper jaw

Meatus/canal long tube-like passage

Medial nearest the midline of the body

Medial rectus muscle that rolls the eyeball medially

Median cubital vein vein used to draw blood from the arm

Mediastinum the space between the lungs

Medulla 1. middle or central portion of the hair; 2. the inner part of a kidney

Medulla oblongata part of the brainstem that contains all the ascending and descending tracts that connect between the spinal cord and various parts of the brain

Medullary cavity center of the shaft of long bone filled with yellow bone marrow

Megakaryocytes produce thrombocytes or platelets

Meiosis a reduction division that occurs in the gonads to produce egg and sperm cells

Melanin pigment responsible for variations in skin color

Melanocyte-stimulating hormone (MSH) increases the production of melanin in melanocytes in the skin, causing a darkening of the skin

Melanocytes cells responsible for producing melanin

Melatonin hormone produced by the pineal gland

Membrane potential the ionic and electrical charge around a nerve fiber that is not transmitting an impulse; also called resting potential

Membranous urethra connects the prostatic urethra to the penis

Memory cells descendants of activated T and B cells

Menarche first menstrual cycle

Meninges a series of connective tissue membranes that surround the brain and spinal cord

Meningitis inflammation of the meninges caused by bacterial or viral infection

Menopause last menstrual cycle

Menses cyclical shedding of the lining of the uterus

Menstrual cramps contractions of the myometrial layer of the uterus

Menstrual cycle cyclical shedding of the lining of the uterus

Menstruation cyclical shedding of the lining of the uterus

Mesentery extensions of the visceral peritoneum

Mesocolon an extension of the visceral peritoneum of the colon

Mesoderm a primary germ layer that forms the muscles, bone, and other tissues in a developing fetus

Mesothelium type of epithelial tissue based on function, also called serous tissue, that lines the cavities of the body that have no openings to the outside

Messenger RNA a type of RNA that transcribes the genetic code of a DNA molecule

Metabolism the total chemical changes that occur inside a cell

Metacarpal bones bones of the palm of the hand

Metaphase second stage of mitosis

Metaphase I stage in meiosis in which the spindle microtubules attach to the kinetochore only on the outside of each centromere

Metaphase II stage in meiosis in which the spindle fibers bind to both sides of the centromere

Metaphysis flared portion at each end of a long bone composed of cancellous or spongy bone

Metastasis movement beyond the place of origin

Metastasize defective cancerous cells spread to other parts of the body

Metatarsals bones of the sole of the foot; form the arch of the foot

Microglia phagocytic cell found in the central nervous system; also called neuroglia

Microglia cells small cells that protect the central nervous system by engulfing and destroying microbes like bacteria and cellular debris

Micrometer more common term used instead of microns

Micron a cellular measurement equal to one thousandth of a millimeter

Microtubules long, hollow cylinders made of tubulin

Microvilli found on the free edge of villi of intestinal epithelial cells to increase the absorptive surface area of the cell

Micturition urination

Micturition reflex an unconscious reflex and conscious desire to urinate

Midbrain mesencephalon; contains the ventral cerebral peduncles

Middle meatus one of three narrow passageways in the nasal cavity formed by the turbinate bones

Midsagittal plane vertically dividing the body into equal right and left portions

Mineral salts composed of small ions, they are essential for the survival and functioning of the body's cells

Minor calyx funnel-shaped structure that surrounds the tip of each renal pyramid

Mitochondrion small oblong-shaped structure composed of two membranes; the powerhouse of the cell where ATP is made

Mitosis process in which nuclear material is exactly replicated

Mitral value The atrioventricular valve between the left atrium and the left ventricle, also known as the bicuspid valve

Molars teeth that grind food; also known as tricuspids

Molecular oxygen necessary to convert food into chemical energy (ATP)

Molecule the smallest combination or particle retaining all the properties of a compound

Monocytes largest leukocytes; phagocytize bacteria and dead cells; histiocytes; macrophages

Monokines chemicals released by activated macrophages involved in the immune response

Mons pubis mound of elevated adipose tissue that becomes covered with pubic hair at puberty; also called mons veneris

Motion sickness caused by constant stimulation of the semicircular canals of the inner ear due to motion, resulting in nausea and weakness

Motor neuron neuron that connects with muscles or glands to bring about a reaction to a stimulus; also called efferent neuron

Motor unit all of the muscle cells or fibers innervated by one motor neuron

Mucous cells secrete mucus

Mucous membrane/epithelium lines the digestive, respiratory, urinary, and reproductive tracts; produces mucus

Multiple selerosis incurable disease of brain and spinal cord

Multipolar neurons neurons that have several dendrites and one axon

Mumps disease caused by a virus that infects the salivary glands, especially in children between ages 5 and 9

Muscle type of tissue that contracts and allows movement

Muscle fibers muscle cells

Muscle tissue tissue that can shorten and thicken or contract

Muscle twitch the analysis of a muscle contraction

Muscular dystrophy an inherited muscular disorder in which muscle tissue degenerates over time

Muscular system consists of muscles, fasciae, tendon sheaths, and bursae

Muscularis mucosa a third layer of the tunica mucosa of the small intestine

Musculi pectinati muscles that give the auricles their rough appearance

Mutation a mistake in the copying of genetic material

Myalgia muscle pain

Myasthenia gravis condition characterized by the easy tiring of muscles or muscle weakness

Myelin sheath a fatty sheath surrounding some axons

Myeloid tissue red bone marrow; produce blood cells by hematopoiesis

Myocardial infarction heart attack

Myocarditis inflammation of the myocardium that can cause a heart attack

Myocardium second layer of the wall of the heart

Myometrium middle layer of wall of the uterus

Myopia nearsightedness

Myosin thick filaments of protein in a muscle cell

Myositis inflammation of muscle tissue

Myxedema accumulation of fluid in subcutaneous tissues

N

Nail bed area from which the nail grows

Nail body visible part of the nail

Nail root the part of the nail body attached to the nail bed

Nasal bones thin, delicate bones that join to form the bridge of the nose

Nasal cavities cavities of the nose

Nasal septum divides the nose into right and left nasal cavities

Nasopharynx uppermost portion of the pharynx located in the nose

Navicular one of the bones of the wrist; also called the scaphoid bone

Neck the part of a bone that connects the head or terminal enlargement to the rest of a long bone

Negative feedback loop mechanism by which hormonal systems function

Nephrons functional units of the kidney

Nerve bundle of nerve cells or fibers

Nervous system consists of the brain, spinal cord, cranial nerves, peripheral nerves, and the sensory and motor structures of the body

Nervous tissue a type of tissue that transmits impulses

Neuroglia nerve cells that perform support and protection

Neuron nerve cell that transmits impulses

Neutron part of the central nucleus that makes up an atom; carries no charge

Neutrophils most common leukocytes; they secrete lysozyme

Nicotinamide adenine dinucleotide (NAD) an electron carrier

Nipple the terminal point of the mammary glands

Nissl bodies ribosomes attached to the rough ER in a neuron; also called chromatophilic substance

Nodes of Ranvier gaps in the myelin sheath; also called neurofibral nodes

Nonpolar compounds with unpolarized bonds

Norepinephrine hormone produced by the adrenal medulla; noradrenaline

Nostrils openings on the undersurface of the external nose; also called external nares

Nuclear membrane double-layered membrane that surrounds the nucleus

Nucleic acid the genetic material of a cell, either DNA or RNA

Nucleolus a spherical particle within the nucleoplasm that does not have a covering membrane around it

Nucleoplasm that protoplasm inside the nucleus of a cell

Nucleotides complex molecules made up of a sugar, a phosphate, and a nitrogen base; the building blocks of nucleic acids

Nucleus 1. a mass of nerve cell bodies and dendrites inside the central nervous system; 2. part of an atom

O

Obturator foramen a large opening in the hip bone for passage of nerves, blood vessels, and ligaments

Occipital bone a single bone that forms the back and base of the cranium

Occipital condyle a process for articulation with the first cervical vertebra

Occipital lobe part of the cerebrum that functions in receiving and interpreting visual input

Occipitalis muscle that draws the scalp backward

Oculomotor nerve III controls movements of the eyeball and upper eyelid and conveys impulses related to muscle sense

Olecranon process a projection of the ulna known as the funny bone

Olfactory nerve I conveys impulses related to smell

Olfactory sense sense of smell

Olfactory stimuli odors

Oligodendroglia provide support by forming semirigid connective-like tissue rows between neurons in the brain and spinal cord, also called oligodendrocytes

Oliguria a scant amount of urine produced daily

Onychocryptosis an ingrown tocnail

Onychomycosis a fungal infection of the naüs

Oocyte immature egg

Oogenesis formation of the female sex cells

Oogonia female stem cells in a developing female fetus

Ootid mature egg cell

Opponens pollicis muscle that flexes and opposes the thumb

Opposition movement that occurs only with the thumb

Optic chiasma part of the diencephalon where optic nerves cross each other

Optic disk where nerve fibers leave the eye as the optic nerve

Optic nerve II conveys impulses related to sight

Optic tracts part of the diencephalon involved with the sense of sight

Orbicularis oris muscle that closes the lips

Orbital margin a definite ridge above each orbit

Orbitals the paths that electrons travel in an energy level

Organelles structures within the protoplasm

Origin the more fixed attachment of a muscle that serves as a basis for the action

Oropharynx second portion of the pharynx located at the back of the mouth

Osmosis kind of diffusion that pertains only to the movement of water molecules through a selectively permeable membrane

Ossification formation of bone by osteoblasts

Osteoarthritis degenerative joint disease

Osteoblasts cells involved in the formation of bony tissue

Osteoclasts bone cells present in almost all cavities of bone responsible for reabsorbing bone during remodeling

Osteocytes mature bone cells

Osteomalacia softening of bone

Osteon Haversian canal

Osteoporosis disorder of the skeletal system characterized by a decrease in bone mass with accompanying susceptibility to fractures

Osteoprogenitor cell undifferentiated bone cell

Otitis media middle ear infection

Ova female eggs

Oval window one of the two openings in the middle ear

Ovarian cancer malignant growth in the ovaries

Ovarian cycle the cycle beginning at puberty that produces mature eggs

Ovarian follicles ova and their surrounding tissues in various stages of development

Ovaries primary sex organs of the female reproductive system

Ovulation ejection of a mature egg

Oxaloacetic acid an intermediate product of the citric acid cycle

Oxygen gaseous element required by all organisms that breathe air

Oxyphil cells secreting cells of the parathyroid glands; also called chief cells

Oxytocin (OT) hormone that stimulates contraction of smooth muscles in the wall of the uterus, also causes milk secretion

P

Paget's disease common disease of bone with symptoms that include irregular thickening and softening of the bones

Palatine bones form the posterior part of the roof of the mouth or part of the hard palate

Palatine tonsils tonsils commonly removed in a tonsillectomy

Pancreas a large digestive gland of the alimentary canal

Pancreatic islets islets of Langerhans; endocrine portion of the pancreas

Pancreatic cancer fatal cancer of the pancreas

Pancreatic juice a mixture of digestive enzymes in the pancreas

Papillae projections of the lamina propria covered with epithelium; produces the rough surface of the tongue

Papillary ducts ducts that empty urine into the renal pelvis

Papillary muscles small conical projections on the inner surface of the ventricles

Papillary portion the layer of the dermis that is adjacent to the epidermis

Papule a solid skin lesion

Parasympathetic division part of the autonomic nervous system that operates under normal nonstressful conditions

Parathyroid glands four glands embedded in the thyroid gland

Parathyroid hormone (PTH) parathormone; the hormone of the parathyroid glands

Parenchyma composed of the cortex and the renal pyramids of the kidney

Parietal refers to the walls of a cavity

Parietal bones form the upper sides and roof of the cranium

Parietal cells secrete hydrochloric acid in the stomach

Parietal lobe control center in the brain for evaluating sensory information of touch, pain, balance, taste, and temperature

Parietal pleura outer layer of the pleural membrane of the lung

Parkinson's disease a nervous disorder characterized by tremors of the hand and a shuffling walk

Paronychia infected fold of skin at edge of nail

Parotid gland one of the salivary glands

Partial pressure the amount of pressure that gas contributes to the total pressure

Partial-thickness burns first and second-degree burns

Parturition childbirth

Passive immunity occurs naturally when a fetus receives its mother's antibodies through the placenta

Patella kneecap

Pathogens disease-causing microorganisms

Pathology the study of diseases of the body

Pectoralis major muscle that flexes and adducts the arm

Pectoralis minor muscle that depresses the shoulder and rotates scapula downward

Pelvic girdle formed by the two hip bones

Pelvic inflammatory disease bacterial infection of the uterus, uterine tubes, or ovaries

Penis the male reproductive organ used to deliver spermatozoa into the female reproductive tract

Pepsin enzyme that begins to break down proteins

Pepsinogen principal gastric enzyme

Peptide bonds covalent bonds that form between different amino acids to form proteins

Pericardial cavity space between the epicardium of the heart and the inner layer of the pericardial sac

Pericardial fluid fluid in the pericardial cavity

Pericardial sac membrane covering the heart

Pericarditis inflammation of the pericardium

Pericardium membrane covering the heart

Perimetrium outermost layer of wall of the uterus

Perimysium layer of connective tissue surrounding the fascicle of a muscle

Perineum diamond-shaped area at the inferior end of the trunk between the buttocks and thighs of both males and females

Periodic table table that arranges the elements in such a way that similar properties repeat at periodic intervals

Periodontal ligament anchors the root of a tooth in its socket

Periosteum fibrovascular membrane that covers a bone

Peripheral nervous system (PNS) consists of all the nerves that connect the brain and spinal cord with sensory receptors, muscles, and glands

Peristalsis the physical movement or pushing of food along the digestive tract

Peritoneum membrane lining the abdominal cavity

Peritubular capillaries surround the convoluted tubules of a nephron

Peroneal vein drains the calf and foot

Peroneus longus muscle that everts, plantar-flexes the foot

Peroneus tertius muscle that dorsally flexes the foot

Petrous part part of the temporal bone found deep within the base of the skull where it protects and surrounds the inner ear

Peyer's patches aggregated lymphatic follicles found in the wall of the small intestine

pH the negative logarithm of the hydrogen ion concentration in a solution

Phagocytic describing the process by which a cell eats debris and microorganisms

Phagocytosis the process in which phagocytes eat cellular debris and other substances

Phalanges the bones of the fingers and toes

Phalanx a single bone of a finger or toe

Pharyngeal tonsils adenoids

Pharynx part of the digestive tract (throat) involved in swallowing

Phimosis condition where the foreskin of the penis fits too tightly over the head of the penis and cannot be retracted

Phosphocreatine found in muscle tissue; provides a rapid source of high-energy ATP for muscle contraction

Phosphoglyceraldehyde (PGAL) an intermediate product of glycolysis

Phosphoglyceric acid (PGA) an intermediate product of glycolysis

Phosphorylation process by which a phosphate is added to a molecule

Phrenic arteries supply blood to the diaphragm muscle

Physiology the study of the functions of the body parts

Pia mater the innermost spinal or cranial meninx

Pineal gland an endocrine gland located in the epithalamus of the diencephalon that produces the hormone melatonin

Pisiform a bone of the wrist

Pituitary gland hypophysis; the major gland of the endocrine system

Pivot joint a type of synovial joint, like the joint between the atlas and axis vertebrae

Placenta a structure in the uterus through which the fetus exchanges nutrients and wastes with the mother

Plantar fasceitis inflammation of fascea of arches

Plantar flexion pushing the foot down at the ankle joint

Plantaris muscle that plantar-flexes the foot

Plaque cholesterol-containing masses

Plasma the fluid part of blood

Plasma cells B lymphocyte cells that enter tissues and become specialized cells

Plasma membrane membrane surrounding cells

Plasmalemma membrane surrounding cells; also called the plasma membrane

Pleura membrane that lines the thoracic cavity

Pleural cavity small space between the pleural membranes

Pleural membrane membrane that encloses and protects the lung

Pleurisy inflammation of the pleural cavity; also called pleuritis

Plicae folds in the small intestine

Pneumonia pneumonitis; an infection in the lungs

Podocytes epithelial cells in the innermost layer of the Bowman's glomerular capsule

Poison ivy itching condition of the skin caused by leaves of *Rhus*

Polar a molecule with an unequal distribution of bonding electrons

Polar body nonfunctional cell produced in oogenesis

Polio virus infection causing muscle paralysis

Polydipsia excessive thirst

Polyphagia intense food cravings

Polcystic kidney disease (PKD) abnormally enlarged kidneys with numerous cysts

Polyuria increase in urine production

Pons varolii bridge that connects the spinal cord with the brain and parts of the brain with each other

Popliteal artery the name of the femoral artery at the knee

Popliteal vein the name of the posterior tibial vein at the knee

Popliteus muscle that flexes and rotates the leg

Posterior toward the back

Posterior gray horn part of the spinal cord; also known as dorsal gray horn

Posterior interventricular sulcus separates the left and right ventricles externally

Posterior root the sensory root containing only sensory nerve fibers; also known as dorsal root

Premenstrual syndrome (PMS) series of symptoms that develop in many women during the premenstrual phase of the menstrual cycle

Premolars teeth with two projections or cusps; bicuspid teeth

Prepuce foreskin; a section of loose skin covering the glans penis

Presbyopia a decrease in the ability of the eye to accommodate for near vision

Primary fibrositis inflammation of the fibrous connective tissue in a joint

Primary oocytes produced by mitotic division of female stem cells or oogonia

Primary spermatocytes produced by mitotic division of immature sperm cells or spermatogonia

Primary structure protein structure based on amino acid sequence

Process any obvious bone projection

Progesterone female sex hormone

Prokaryotic refers to cells that do not have membrane bound organelles, such as bacteria

Pronation moving the bones of the forearm so that the radius and ulna are not parallel

Pronator quadratus muscle involved in pronating the forearm

Pronator teres muscle involved in pronating the forearm

Prophase the first phase of mitosis

Prophase I the first stage of the first meiotic division

Prophase II in each of the two daughter cells produced in the first meiotic division, the nuclear membrane disappears but no duplication of DNA occurs

Prostate cancer cancer of the prostate gland that can be fatal if not detected

Prostate gland secretes an alkaline fluid that is part of the semen

Prostatic urethra surrounded by the prostate gland

Protein covalently bonded amino acids composed of carbon, hydrogen, oxygen, and nitrogen

Protein synthesis process in which cells produce proteins

Prothrombin plasma protein produced by the liver that is involved in blood clotting

Proton part of the central nucleus that makes up an atom; has a positive charge

Protoplasm the liquid part of a cell

Protraction moving a part of the body forward on a plane parallel to the ground

Proximal nearest the point of attachment or origin

Proximal convoluted tubule first part of the renal tubule

Pseudostratified epithelium cells that have a layered appearance but actually extend from the basement membrane to the outer free surface

Psoas muscle involved in flexing the thigh

Psoriasis common chronic skin disorder characterized by red patches covered with thick, dry, silvery scales

Pterygoid muscles involved in raising the mandible

Pubis a part of the hip bone found superior and slightly anterior to the ischium

Pulmonary circulation circulatory route that goes from the heart to the lungs and back to the heart

Pulmonary fibrosis black lung disease; caused by excessive exposure to asbestos, silica, or coal dust

Pulmonary semilunar valve found in the opening where the pulmonary trunk exits the right ventricle

Pulmonary trunk the artery that leaves the right ventricle

Pulmonary veins enter the left atrium of the heart

Pulp cavity a cavity in the crown of a tooth

Pupil a circular opening in the iris of the eye

Purine a nitrogen base consisting of a fused double ring of nine atoms of carbon and nitrogen

Purkinje's fibers/conduction myofibers cause actual contractions of the ventricles; they emerge from the bundle branches

Pustule a pus-filled elevation of the skin

Pyloric sphincter the connection between the stomach and the beginning of the duodenum

Pylorus narrow inferior region of the stomach; also called the antrum

Pyrimidine a nitrogen base consisting of a single ring of six atoms of carbon and nitrogen

Pyruvic acid the final product of glycolysis

Pyuria pus in the urine

Q

Quadriceps femoris muscle that extends the knee

Quaternary structure protein structure determined by spatial relationships between amino acids

Quinone an electron carrier

R

Rabies fatal viral disease transmitted by the bite of a rabid mammal

Radial arteries supply blood to the forearm

Radial veins drain blood from the forearm

Radius shorter, lateral bone of the forearm

Raphe external median ridge of the scrotum

Rectum the last 7 to 8 inches of the gastrointestinal tract

Rectus abdominis muscle that flexes the vertebral column and assists in compressing abdominal wall

Rectus femoris muscle that extends the leg and flexes the thigh

Red bone marrow found within cancellous bone; makes blood cells

Reflex an involuntary reaction to an external stimulus

Reflex arc the pathway that results in a reflex

Releasing hormones produced by the hypothalamus, they stimulate the release of hormones from the pituitary gland

Releasing inhibitory hormones produced by the hypothalamus, they inhibit the release of hormones from the pituitary gland

Renal capsule innermost layer of tissue that surrounds the kidney

Renal columns cortical substance between the renal pyramids

Renal corpuscle made up of Bowman's glomerular capsule and the enclosed glomerulus

Renal failure results from any condition that interferes with kidney function

Renal fascia outermost layer of tissue that surrounds the kidney

Renal papillae the tips of the renal pyramids

Renal pelvis a large collecting funnel formed where the major calyces join

Renal plexus the nerve supply to the kidneys

Renal pyramids triangular structures within the medulla

Renal sinus cavity in the kidney

Renal tubule the general name of the tubule of a nephron

Renin enzyme produced by the kidneys that helps regulate blood pressure

Repolarization restoration of electrical charge

Reposition occurs when the digits return to their normal positions

Reproductive system consists of the ovaries, uterine tubes, uterus, and vagina in the female; the testes, vas deferens, seminal vesicles, prostate gland, penis, and urethra in the male

Respiration the overall exchange of gases between the atmosphere, blood, and cells

Respiratory bronchioles microscopic divisions of terminal bronchioles

Respiratory distress syndrome condition in infants in which too little surfactant is produced, causing the lungs to collapse; also known as hyaline membrane disease

Respiratory system consists of nasal cavities, pharynx, larynx, trachea, bronchi, and lungs

Resting potential normal electrical distribution around a muscle cell when it is not contracting

Rete testis network of ducts in the testis

Reticular type of loose connective tissue that forms the framework of the liver, bone marrow, spleen, and lymph nodes

Reticular formation area of dispersed gray matter in the medulla of the brain

Reticular portion the layer of the dermis between the papillary portion and the subcutaneous tissue beneath

Reticuloendothelial (RE) system specialized connective tissue involved in phagocytosis

Retina innermost layer of the eye

Retraction moving a part of the body backward on a plane parallel to the ground

Reye's syndrome brain cell swelling leading to coma and respiratory failure

Rh blood group one of the blood groups

Rheumatic fever disease involving a mild bacterial infection

Rheumatic heart disease caused by infection with a bacterium in young children

Rheumatoid arthritis connective tissue disorder resulting in severe inflammation of small joints

Rhodopsin pigment found in the rod cells of the eye

Rhomboids muscles involved in moving the scapula

Ribonucleic acid (RNA) a type of nucleic acid

Ribose a five-carbon sugar found in RNA

Ribosomes tiny granules distributed throughout the cytoplasm where protein synthesis occurs

Rickets disease caused by deficiencies in calcium and phosphorus or by deficiencies in vitamin D

Right atrium one of the upper chambers of the heart

Right bundle branch branch of the bundle of His; part of the conduction system of the heart

Right colic (hepatic) flexure where the ascending colon reaches the undersurface of the liver and turns to the left

Right common carotid artery transports blood to the right side of the head and neck

Right lymphatic duct one of two collecting channels or ducts of the lymphatic system

Right primary bronchus the first right division of the trachea

Right pulmonary artery carries blood to the right lung

Right renal artery transports one-quarter of the total cardiac output directly to the kidneys

Right renal vein carries filtered blood from the interlobar veins to the hilum

Right subclavian artery transports blood to the upper right limb

Right ventricle lower chamber of the heart

Rigor mortis sustained muscle contraction for 24 hours after death

Ringworm caused by several species of fungus; its symptoms include itchy, patchy, scale-like lesions with raised edges

Root 1. a projection of a tooth embedded in a socket; 2. the lowermost portion of a hair found in the hair follicle

Root canals narrow extensions of the pulp cavity that project into the root

Rotation moving a bone around a central axis

Rough (granular) ER granular endoplasmic reticulum

Round window an opening on the medial side of the middle ear that connects the middle ear to the inner ear

Rugae large mucosal folds of the stomach

S

Sacral vertebrae lower part of the vertebral column that forms the sacrum

Saddle joint a type of synovial joint, the carpal metacarpal joint in the thumb

Sagittal any plane parallel to the midsagittal or median plane vertically dividing the body into unequal left and right portions

Sagittal suture line where the two parietal bones join superiorly

Sarcolemma the electrically polarized muscle cell membrane

Sarcomas tumors developing from connective tissue

Sarcomere the area between two adjacent Z lines in a muscle cell

Sarcoplasmic reticulum an irregular curtain around muscle fibrils

Sarcotubular system membranes of vesicles and tubules that surround muscle fibrils

Sartorius muscle that flexes the thigh, rotating it laterally

Saturated a fatty acid that contains only single covalent bonds

Scaphoid one of the wrist bones; also called the navicular bone

Scapula shoulder blade

Schwann cells/neurolemmocytes form myelin sheaths around nerve fibers in the peripheral nervous system

Sclera outermost layer of the wall of the eye

Scoliosis an abnormal lateral curvature of the spine

Scrotum outpouching of the abdominal wall containing the testes

Seasonal affective disorder excessive melatonin in winter causing depression

Sebaceous glands glands that secrete sebum

Sebum oily substance that lubricates the skin's surface

Second-degree burn a burn that involves the epidermis and dermis; may form scars

Secondary (lobar) bronchi divisions of the primary bronchi

Secondary oocyte produced by the first meiotic division of the primary oocyte

Secondary spermatocyte produced by the first meiotic division of the primary spermatocytes

Secondary structure protein structure determined by hydrogen bonds between amino acids, resulting in a helix or a pleated sheet

Selectively permeable membrane allows only certain materials to pass through, like water through a plasma membrane

Semen mixture of sperm cells and secretions of the seminal vesicles, prostate, and bulbourethral glands; also called seminal fluid

Semimembranosus muscle that flexes the leg, extends the thigh

Seminalplasmin an antibiotic in semen that can destroy certain bacteria

Seminal vesicles produce an alkaline, viscous component of semen rich in fructose

Seminiferous tubules tightly coiled tubules in each lobule of a testis

Semitendinosus muscle that flexes the leg, extends the thigh

Sensory neuron a neuron in contact with receptors, it detects changes in the external environment; also called afferent neuron

Septal defect a hole in the interatrial or interventricular septum between the left and right sides of the heart

Septicemia blood poisoning

Serosa the outermost layer of the wall of the alimentary canal

Serotonin hormone secreted by the pineal gland that acts as a neurotransmitter and vasoconstrictor

Serous pericardium innermost layer of the pericardial sac

Serous tissue lines the great cavities of the body that have no opening to the outside; also called mesothelium

Serratus anterior muscle that moves scapula forward

Sertoli cells produce secretions that supply nutrients for the developing sperm cells

Sesamoid bones enclosed in a tendon and fascial tissue, located adjacent to joints

Shaft 1. visible portion of the hair; 2. that part of the penis behind the head

Shingles painful, vesicular skin eruptions caused by the herpes zoster or chickenpox virus

Sickle-cell anemia hereditary disease found mostly in African Americans

Sigmoid colon last part of the colon

Simple epithelium one cell-layer thick

Simple exocrine glands glands with ducts that do not branch

Sinoatrial (SA) node initiates each cardiac cycle and sets the pace for the heart rate; also called pacemaker

Sinus/antrum a cavity within a bone

Sinusitus inflammation of the paranasal sinuses

Skeletal muscle muscle attached to bone through its tendon, under voluntary control

Skeletal system composed of bones, cartilage, and the membranous structures associated with bones

Small intestine place where absorption and digestion occur

Smooth (agranular) ER agranular endoplasmic reticulum

Smooth muscle found in hollow structures of the body like the intestines; cannot be influenced at will

Smooth muscle tissue made of spindle-shaped cells with a single nucleus and no striations

Snoring uvula and soft palate vibrations producing noise while sleeping

Soft palate posterior portion of the roof of the mouth

Soleus muscle that plantar-flexes foot

Solute substance that is dissolved in a solution

Solvent a medium allowing other reactions to occur in

Somatic nervous system conducts impulses from the brain and spinal cord to skeletal muscle, causing us to respond or react to changes in our external environment

Spermatic cord connective tissue sheath enclosing the vas deferens

Spermatids secondary spermatocytes that undergo the second meiotic division

Spermatogenesis production of sperm

Spermatogonia immature sperm cells

Spermatozoa mature sperm cells

Sphenoid bone forms anterior portion of the base of the cranium

Spina bifida congenital defect in the development of the posterior vertebral arch in which the laminae do not unite at the midline

Spinal cavity cavity containing the spinal cord

Spinal meninges a series of connective tissue membranes specifically associated with the spinal cord

Spindle fibers group of microtubules formed by the centrioles to guide the daughter chromatids to opposite poles

Spine any sharp, slender projection such as the spinous process of the vertebrae

Spleen the largest single mass of lymphatic tissue

Splenic artery supplies blood to the spleen

Spongy (cavernous) urethra located within the penis; about 6 inches long

Squamous cell carcinoma skin cancer in the epidermis

Squamous epithelium epithelial cells that are flat and slightly irregular in shape and serve as a protective layer

Squamous portion largest part of the temporal bone

Stapes ear bone referred to as the stirrup

Stem cells undifferentiated mesenchymal cells that develop into blood cells; also known as hematocytoblasts

Stenosis a narrowed opening through the heart valves

Sternocleidomastoid main muscle that moves the head

Sternum the breastbone

Straight tubules located at the tip of each lobule of a testis

Strata (stratum) layers

Stratified epithelium several layers of cells thick

Stratum basale the lowermost or basal layer of the stratum germinativum

Stratum corneum outermost layer of epidermis, consisting of dead cells

Stratum germinativum regenerative layer of epidermis, lowermost layer

Stratum granulosum a layer of epidermis made of flattened cells containing granules

Stratum lucidum clear layer of epidermis

Stratum spinosum spiny or prickly layer of epidermis

Stenosed heart valve valve with an abnormal narrow opening

Stress environmental influences resulting from excessive secretion of epinephrine and cortisol causing psychological and physiological problems

Stent a metal mesh tube inserted into a blood vessel

Striated muscle tissue that causes movement; multinucleated with striations; skeletal muscle

Subclavian trunk drains lymph from the upper extremities

Subclavian vein drains blood from the arm

Subcutaneous bursae found under the skin

Subfascial bursae located between muscles

Sublingual gland one of the three salivary glands

Submandibular gland one of the three salivary glands; submaxillary gland

Submaxillary gland one of the three salivary glands; submandibular gland

Submucosa layer of connective tissue beneath a mucous membrane

Subtendinous bursae found where one tendon overlies another tendon

Succinic acid an intermediate product of the citric acid cycle

Sudden infant death syndrome (SIDS) unexpected death of a healthy infant between 10 and 14 weeks of age due to respiratory failure

Sulcus a furrow or groove

Superior uppermost or above

Superior meatus one of three narrow passageways in the nasal cavity formed by the turbinate bones; known as the olfactory region of the nose

Superior mesenteric artery supplies blood to the small intestine and colon

Superior oblique muscle that rotates the eyeball on axis

Superior rectus muscle that rolls the eyeball upward

Superior vena cava brings blood from the upper parts of the body to the right atrium of the heart

Supination moving the bones of the forearm so that the radius and ulna are parallel

Supinator muscle that supinates the forearm

Suppressor T cells slow down the activities of B and T cells once infection is controlled

Supraorbital ridge overlies the frontal sinus and can be felt in the middle of the forehead

Supraspinatus muscle that abducts the arm

Surfactant fluid inside the respiratory membrane

Suture articulation in which the bones are united by a thin layer of fibrous tissue

Sweat glands simple tubular glands found in most parts of the body that secrete sweat

Sympathetic division part of the autonomic nervous system that prepares the body for stressful situations that require energy expenditure

Symphysis joints in which the bones are connected by a disc of fibrocartilage

Synapses areas where the terminal branches of an axon are anchored close to, but not touching, the ends of the dendrites of another neuron

Synapsis the lining up of homologous chromosomes in meiosis

Synarthroses joints between bones that do not allow movement

Synchondrosis joint in which two bony surfaces are connected by hyaline cartilage

Syndesmosis joint in which bones are connected by ligaments between the bones

Syneresis clot retraction

Synergists muscles that assist the prime movers

Synovial membranes line the cavities of freely moving joints; produce synovial fluid

Syphilis a venereal disease caused by a bacterial infection

Systemic circulation blood circulation to the body not including the lungs

Systemic lupus erythematosus (SLE) chronic inflammatory disease in which the immune system destroys cells and tissues of the body

Systole phase of contraction

T

T lymphocytes T cells; responsible for providing cellular immunity

T system tubules; part of the sarcotubular system

Tail terminal portion of the pancreas

Tapeworm an infection of the intestines with a parasitic worm

Talus ankle bone

Tarsal bones bones of the ankle

Taste buds sensory structures that detect taste stimuli

Taste cells interior of the taste bud

Tay-Sachs disease fatal genetically inherited disease of Eastern-European Jews

Telophase final stage of mitosis

Telophase I stage in meiosis in which the homologous chromosome pairs have separated with a member of each pair at opposite ends of the spindle

Telophase II stage in meiosis that produces four haploid daughter cells, each containing one-half of the genetic material of the original parent cell

Temporal bones form the lower sides and base of cranium

Temporal lobe part of the cerebral hemisphere that evaluates hearing input and smell

Temporalis muscle that raises the mandible and closes the mouth

Tendinitis inflammation of a tendon

Tendon dense connective tissue that attaches muscle to bone

Tensor fascia lata muscle that tenses fascia lata

Teres minor muscle that adducts and rotates the arm

Terminal bronchioles divisions or branches of bronchioles

Tertiary (segmental) bronchi divisions of secondary bronchi

Tertiary structure protein structure with a secondary folding

Testes primary sex organs of the male reproductive system that produce sperm and male sex hormones

Testis singular form of testes

Testosterone the principal male sex hormone

Tetanus infection caused by a bacterium that produces a neurotoxin affecting motor neurons

Tetrad the figure formed by the lining up of homologous chromosomes consisting of four chromatids in meiosis

Thalamus the second part of the diencephalon

Thalassemia hereditary disease found in African, Mediterranean, and Asian individuals that suppresses hemoglobin production

Third-degree burn a burn in which the epidermis and dermis are completely destroyed; also called full-thickness burns

Thoracic aorta the name given to the aorta in the thorax

Thoracic cavity first subdivision of the ventral cavity that is surrounded by the rib cage and contains the heart and lungs

Thoracic duct main collecting duct of the lymphatic system

Thoracic vertebrae 12 vertebrae that connect with the ribs

Thrombin an enzyme formed from prothrombin that is necessary for the clotting mechanism

Thrombocytes platelets; blood cells involved in blood clotting

Thrombocytopenia decrease in platelets producing chronic bleeding

Thromboplastin a substance released from blood platelets that is involved in the clotting reaction

Thrombosis clotting in an unbroken blood vessel

Thrombus a blood clot

Thrush a fungal infection of the mouth caused by a yeast

Thylakoid individual double membranes that make up a granum in a chloroplast

Thymine a pyrimidine nitrogen base

Thymosin hormone secreted by the thymus gland that causes the production of T lymphocytes

Thymus gland an endocrine gland located beneath the sternum that is also involved in immunity as a site for lymphocyte production and maturation

Thyroid cartilage Adam's apple

Thyroid gland an endocrine gland located along the trachea

Thyroid-stimulating hormone (TSH) stimulates the thyroid gland to produce its hormone

Thyroxine a hormone of the thyroid gland that regulates the metabolism of carbohydrates, fats, and proteins, also known as tetraiodothyroxine (T_4)

Tibia larger of the two bones forming the lower leg

Tibialis anterior muscle that dorsally flexes the foot

Tibialis posterior muscle that plantar-flexes the foot

Tissue groups of cells similar in size, shape, and function

Tone a property of muscle whereby a steady or constant state of partial contraction is maintained in a muscle

Tongue organ that manipulates food and forms the floor of the oral cavity

Tonsillitis a bacterial infection of the tonsils

Trabeculae fibrous connective tissue; extension of the capsule of a lymph node

Trabeculae carneae irregular ridges and folds of the myocardium of the ventricles

Trachea windpipe

Tract a bundle of fibers inside the central nervous system

Transcription process by which messenger RNA copies the genetic code in a DNA molecule

Transfer RNA a type of RNA that translates the code of a DNA molecule that was copied by messenger RNA

Transitional epithelium several layers of closely packed, flexible, easily stretched cells; appear flat when stretched and saw-toothed when relaxed

Translation process by which transfer RNA reads the code on messenger RNA and gets the amino acids to make a protein

Transverse plane dividing the body into superior and inferior portions

Transverse colon the second part of the colon found under the liver

Transversus abdominis muscle that compresses abdominal contents

Trapezium a bone of the wrist, also called the greater multiangular

Trapezius muscle that draws the head to one side and rotates the scapula

Trapezoid a bone of the wrist, also called the lesser multiangular

Triacylglycerols types of fats found in the human body

Triceps brachii muscle that extends and adducts the forearm

Trichomonas a flagellated protozoan that causes inflammation of vaginal tissue with an odorous yellow-green discharge

Tricuspids molar teeth with three cusps

Tricuspid valve valve between the right atrium and right ventricle

Trigeminal nerve V largest of the cranial nerves; controls chewing movements

Trigone a triangular region of the bladder formed by the two openings from the ureters and the single urethral opening

Triiodothyronine a hormone of the thyroid gland that regulates the metabolism of carbohydrates, fats, and proteins

Triquetral a bone of the wrist

Trochanter very large projection on a bone

Trochlea process shaped like a pulley on a bone

Trochlear nerve IV controls movement of the eyeball and conveys impulses related to muscle sense

Tropomyosin an inhibitor substance found in muscle cells

Troponin an inhibitor substance found in muscle cells

Tubercle small round process on a bone

Tuberculosis bacterial infection of the lungs

Tubulin protein fibers that make up the spindle during cell division

Tumor abnormal and uncontrolled growth of a cell

Tunica adventitia the outermost wall of an artery or vein

Tunica albuginea a layer of white, fibrous, connective tissue that covers each testis

Tunica intima the innermost layer of the wall of an artery or vein

Tunica media the middle layer of the wall of an artery or vein

Tunica mucosa innermost lining of the alimentary canal

Tunica muscularis the third layer of the wall of the alimentary canal

Tunica serosa the fourth or outermost layer of the wall of the alimentary canal

Tunica submucosa the second layer of the wall of the alimentary canal

Turbinates thin, fragile bones found on the lateral sides of the nostrils; nasal conchae

Tympanic membrane eardrum

Tympanic plate forms the floor and anterior wall of the external auditory meatus

U

Ulna longer, medial bone of the forearm

Ulnar arteries supply blood to the forearm

Ulnar veins drain blood from the forearm

Umbilical cord a connecting stalk of tissue that connects the developing fetus to the placenta

Unipolar neurons neurons that have only one process extending from the cell body; most sensory neurons are unipolar

Unsaturated a fatty acid that contains one or more double covalent bonds between the carbon atoms

Uremia excessive urine in the blood

Ureters transport urine from the renal pelvis into the urinary bladder

Urethra the tube that leads from the bladder to the outside

Urethral orifice the terminal opening of the urethra

Urinary bladder hollow muscular organ located in the pelvic cavity posterior to the pubic symphysis

Urinary incontinence an uncontrollable flow of urine

Urinary system consists of two kidneys, two ureters, urinary bladder, and urethra

Urine the eliminated materials from the filtered blood

Uterine cavity interior of the body of the uterus

Uterine tubes transport ova from the ovaries to the uterus; also called fallopian tubes

Uterus womb

Uvula a conical projection hanging from the posterior border of the soft palate; functions in the swallowing process and prevents food from backing up into the nasal area

V

Vacuole an area within the cytoplasm that is surrounded by a vacuolar membrane

Vagina the opening into the female reproductive system that leads to the uterus

Vaginal orifice the opening into the vagina

Vagus nerve X controls skeletal muscle movements in the pharynx, larynx, and palate

Vas deferens ductus deferens

Vascular (venous) sinuses veins with thin walls

Vasectomy an operation for birth control that severs the vas deferens, preventing sperm from reaching the exterior

Vastus intermedius muscle that extends the leg and flexes the thigh

Vastus lateralis muscle that extends the leg and flexes the thigh

Vastus medialis muscle that extends the leg and flexes the thigh

Veins blood vessels that carry blood to the heart

Ventilation breathing; movement of air between the atmosphere and lungs

Ventral the belly side

Ventral cerebral peduncles convey impulses from the cerebral cortex to the pons and spinal cord

Ventricles cavities within the brain that connect with each other

Venules small vessels that connect capillaries to veins

Vermiform appendix twisted tube attached to the closed end of the cecum

Vertebral artery supplies blood to part of the brain

Vertebral vein drains the back of the head

Vestibular folds upper folds in the mucous membrane of the larynx; also called false vocal cords

Vestibule 1. opening between the labia minora; 2. anterior portion of the nasal cavities just inside the nostrils

Vestibulocochlear nerve VIII transmits impulses related to equilibrium and hearing

Villi projections on the plicae of the mucosal coat of the small intestine that increase absorptive surface area

Viscera the organs of a cavity

Visceral refers to the covering of an organ

Visceral peritoneum another name for the tunica serosa

Visceral pleura covers the lungs

Vitiligo skin disease resulting in irregular patches of skin of various sizes completely lacking any pigmentation

Vitreous humor fluid that fills the posterior compartment of the eye behind the lens

Vocal folds lower folds in the mucous membrane of the larynx; also called true vocal cords

Volkmann's/perforating canals canals that run horizontally to the Haversian canals

Vomer bone flat bone that makes up the lower posterior portion of the nasal septum

Vulva external genitalia of the female; also called the pudendum

W

Warts uncontrolled growth of epidermal tissue caused by human papillomavirus

Water the most abundant substance in living cells

Wheal a pale or red swollen elevation on the skin

White matter groups of myelinated axons from many neurons supported by neuroglia

Whiplash violent shaking of the cervical vertebrae

Whooping cough respiratory disorder caused by a bacterial infection, resulting in severe coughing; also called pertussis

Wormian/sutural bones located within the sutures of the cranial bones

X

Xanthophyll a carotenoid pigment in plant cells that produces a yellow color

Xiphoid process the terminal portion of the sternum

Y

Yellow bone marrow connective tissue consisting of fat cells

Z

Z line narrow, dark-staining band found in the central region of the I band

Zygomatic or malar bones form the prominence of the cheek

Zygomaticus muscle that draws the lip upward and outward

Zygote fertilized egg

Zymogenic cells found in the stomach, they secrete the principal gastric enzyme

Note: The Letter f denotes a figure on that page.

A

B

C

D

E

H

M

N

O

P

T

U

V

W

X

Y

Z

6.0 FEDERAL GOVERNMENT CLIENTS

6.1 Except as expressly authorized by Cengage Learning, Federal Government clients obtain only the rights specified in this Agreement and no other rights. The Government acknowledges that (i) all software and related documentation incorporated in the Licensed Content is existing commercial computer software within the meaning of FAR 27.405(b)(2); and (2) all other data delivered in whatever form, is limited rights data within the meaning of FAR 27.401. The restrictions in this section are acceptable as consistent with the Government's need for software and other data under this Agreement.

7.0 DISCLAIMER OF WARRANTIES AND LIABILITIES

7.1 Although Cengage Learning believes the Licensed Content to be reliable, Cengage Learning does not guarantee or warrant (i) any information or materials contained in or produced by the Licensed Content, (ii) the accuracy, completeness or reliability of the Licensed Content, or (iii) that the Licensed Content is free from errors or other material defects. THE LICENSED PRODUCT IS PROVIDED "AS IS," WITHOUT ANY WARRANTY OF ANY KIND AND CENGAGE LEARNING DISCLAIMS ANY AND ALL WARRANTIES, EXPRESSED OR IMPLIED, INCLUDING, WITHOUT LIMITATION, WARRANTIES OF MERCHANTABILITY OR FITNESS FOR A PARTICULAR PURPOSE. IN NO EVENT SHALL CENGAGE LEARNING BE LIABLE FOR: INDIRECT, SPECIAL, PUNITIVE OR CONSEQUENTIAL DAMAGES INCLUDING FOR LOST PROFITS, LOST DATA, OR OTHERWISE. IN NO EVENT SHALL CENGAGE LEARNING'S AGGREGATE LIABILITY HEREUNDER, WHETHER ARISING IN CONTRACT, TORT, STRICT LIABILITY OR OTHERWISE, EXCEED THE AMOUNT OF FEES PAID BY THE END USER HEREUNDER FOR THE LICENSE OF THE LICENSED CONTENT.

8.0 GENERAL

8.1 Entire Agreement. This Agreement shall constitute the entire Agreement between the Parties and supercedes all prior Agreements and understandings oral or written relating to the subject matter hereof.

8.2 Enhancements/Modifications of Licensed Content. From time to time, and in Cengage Learning's sole discretion, Cengage Learning may advise the End User of updates, upgrades, enhancements and/or improvements to the Licensed Content, and may permit the End User to access and use, subject to the terms and conditions of this Agreement, such modifications, upon payment of prices as may be established by Cengage Learning.

8.3 No Export. The End User shall use the Licensed Content solely in the United States and shall not transfer or export, directly or indirectly, the Licensed Content outside the United States.

8.4 Severability. If any provision of this Agreement is invalid, illegal, or unenforceable under any applicable statute or rule of law, the provision shall be deemed omitted to the extent that it is invalid, illegal, or unenforceable. In such a case, the remainder of the Agreement shall be construed in a manner as to give greatest effect to the original intention of the parties hereto.

8.5 Waiver. The waiver of any right or failure of either party to exercise in any respect any right provided in this Agreement in any instance shall not be deemed to be a waiver of such right in the future or a waiver of any other right under this Agreement.

8.6 Choice of Law/Venue. This Agreement shall be interpreted, construed, and governed by and in accordance with the laws of the State of New York, applicable to contracts executed and to be wholly preformed therein, without regard to its principles governing conflicts of law. Each party agrees that any proceeding arising out of or relating to this Agreement or the breach or threatened breach of this Agreement may be commenced and prosecuted in a court in the State and County of New York. Each party consents and submits to the nonexclusive personal jurisdiction of any court in the State and County of New York in respect of any such proceeding.

8.7 Acknowledgment. By opening this package and/or by accessing the Licensed Content on this Web site, THE END USER ACKNOWLEDGES THAT IT HAS READ THIS AGREEMENT, UNDERSTANDS IT, AND AGREES TO BE BOUND BY ITS TERMS AND CONDITIONS. IF YOU DO NOT ACCEPT THESE TERMS AND CONDITIONS, YOU MUST NOT ACCESS THE LICENSED CONTENT AND RETURN THE LICENSED PRODUCT TO CENGAGE LEARNING (WITHIN 30 CALENDAR DAYS OF THE END USER'S PURCHASE) WITH PROOF OF PAYMENT ACCEPTABLE TO CENGAGE LEARNING, FOR A CREDIT OR A REFUND. Should the End User have any questions/comments regarding this Agreement, please contact Cengage Learning at Delmar.help@cengage.com.

StudyWare™ to Accompany
Fundamentals of Anatomy and Physiology, Third Edition

Minimum System Requirements

- Operating systems: Microsoft Windows XP w/SP 2, Windows Vista w/ SP 1
- Processor: Minimum required by Operating System
- Memory: Minimum required by Operating System
- Hard Drive Space: 225 MB
- Screen resolution: 800 × 600 pixels
- CD-ROM drive
- Sound card & listening device required for audio features
- Flash Player 9. The Adobe Flash Player is free, and can be downloaded from http://www.adobe.com/products/flashplayer/

Setup Instructions

1. Insert disc into CD-ROM drive. The StudyWare™ installation program should start automatically. If it does not, go to step 2.
2. From My Computer, double-click the icon for the CD drive.
3. Double-click the *setup.exe* file to start the program.

Technical Support

Telephone: 1-800-648-7450
8:30 A.M.-6:30 P.M. Eastern Time
E-mail: delmar.help@cengage.com